ADVANCES IN CHEMICAL PHYSICS

VOLUME LXIX

ADVANCES IN CHEMICAL PHYSICS—VOLUME LXIX

I. Prigogine and Stuart A. Rice—Editors

AB INITIO METHODS IN QUANTUM CHEMISTRY—II

Edited by

K. P. LAWLEY

Department of Chemistry
Edinburgh University

See also Volume LXVII, Ab Initio Methods in Quantum Chemistry—I

A WILEY–INTERSCIENCE PUBLICATION

JOHN WILEY & SONS

CHICHESTER · NEW YORK · BRISBANE · TORONTO · SINGAPORE

Library of Congress Cataloging-in-Publication Data:
(Revised for vol. 2)

Ab initio methods in quantum chemistry.
(Advances in chemical physics; v. 67,)
'A Wiley-Interscience publication.'
Includes index.
1. Quantum chemistry. I. Lawley, K. P.
II. Series: Advances in chemical physics; v. 67, etc.
QD453.A27 [QD462.5] 541.2'8 86-9168

ISBN 0 471 90901 7

British Library Cataloguing in Publication Data:

Ab initio methods in quantum chemistry
part II.—(Advances in chemical physics,
ISSN 0065-2385; V.69)
1. Quantum chemistry
I. Lawley, K. P. II. Series
541.2'8 QD462

ISBN 0 471 90901 7

Printed and bound in Great Britain

CONTRIBUTORS TO VOLUME LXIX

D. L. Cooper, Department of Inorganic, Physical and Industrial Chemistry, University of Liverpool, PO Box 147, Liverpool L69 3BX, UK

B. I. Dunlap, Code 6129, Naval Research Laboratory, Washington, DC 20375-5000, USA

J. Gerrat, Department of Theoretical Chemistry, School of Chemistry, University of Bristol, Bristol BS8 1TS, UK

J. Oddershede, Department of Chemistry, Odense University, DK-5230 Odense M, Denmark

P. Pulay, Department of Chemistry, University of Arkansas, Fayetteville, Arkansas 72701, USA

M. Raimondi, Departimento di Chimica Fisica ed Electrochimica, Universita di Milano, Via Golgi 19, 20133 Milano, Italy

B. O. Roos, Department of Theoretical Chemistry, Chemical Centre, PO Box 124, S-221 00 Lund, Sweden

D. R. Salahub, Département de Chimie, Université de Montréal, CP 6128, Succ. A, Montréal, Québec H3C 3J7, Canada

R. Shepard, Theoretical Chemistry Group, CHM, Argonne National Laboratory, Argonne, Illinois 60439, USA

F. B. van Duijneveldt, Rijkuniversiteit Utrecht, Vakgroep Theoretische Chemie, Padualaan 8, De Uithof, Utrecht, The Netherlands

J. G. C. M. van Duijneveldt-van de Rijdt, Rijkuniversiteit Utrecht, Vakgroep Theoretische Chemie, Padualaan 8, De Uithof, Utrecht, The Netherlands

J. H. van Lenthe, Rijkuniversiteit Utrecht, Vakgroep Theoretische Chemie, Padualaan 8, De Uithof, Utrecht, The Netherlands

H.-J. Werner, Institut für Physikalische and Theoretische Chemie, Johann Wolfgang Goethe-Universität Frankfurt am Main, Postfach 11 19 32, D-6000 Frankfurt, West Germany

INTRODUCTION

Few of us can any longer keep up with the flood of scientific literature, even in specialized subfields. Any attempt to do more, and be broadly educated with respect to a large domain of science, has the appearance of tilting at windmills. Yet the synthesis of ideas drawn from different subjects into new, powerful, general concepts is as valuable as eve, and the desire to remain educated persists in all scientists. This series, *Advances in Chemical Physics*, is devoted to helping the reader obtain general information about a wide variety of topics in chemical physics, which field we interpret very broadly. Our intent is to have experts present comprehensive analyses of subjects of interest and to encourage the expression of individual points of view. We hope that this approach to the presentation of an overview of a subject will both stimulate new research and serve as a personalized learning text for beginners in a field.

Ilya Prigogine

Stuart A. Rice

CONTENTS

MATRIX-FORMULATED DIRECT MULTICONFIGURATION SELF-CONSISTENT FIELD AND MULTICONFIGURATION REFERENCE CONFIGURATION-INTERACTION METHODS 1
H.-J. Werner

THE MULTICONFIGURATION SELF-CONSISTENT FIELD METHOD 63
R. Shepard

PROPAGATOR METHODS 201
J. Oddershede

ANALYTICAL DERIVATIVE METHODS IN QUANTUM CHEMISTRY 241
P. Pulay

SYMMETRY AND DEGENERACY IN $X\alpha$ AND DENSITY FUNCTIONAL THEORY 287
B. I. Dunlap

MODERN VALENCE BOND THEORY 319
D. L. Cooper, J. Gerratt and M. Raimondi

THE COMPLETE ACTIVE SPACE SELF-CONSISTENT FIELD METHOD AND ITS APPLICATIONS IN ELECTRONIC STRUCTURE CALCULATIONS 399
B. O. Roos

TRANSITION-METAL ATOMS AND DIMERS 447
D. R. Salahub

WEAKLY BONDED SYSTEMS 521
J. H. van Lenthe, J. G. C. M. van Duijneveldt-van de Rijdt and F. B. van Duijneveldt

AUTHOR INDEX 567

COMPOUND INDEX 583

SUBJECT INDEX 585

Ab Initio Methods in Quantum Chemistry—II
Edited by K. P. Lawley

MATRIX-FORMULATED DIRECT MULTICONFIGURATION SELF-CONSISTENT FIELD AND MULTICONFIGURATION REFERENCE CONFIGURATION-INTERACTION METHODS

HANS-JOACHIM WERNER

Institut für Physikalische und Theoretische Chemie, Johann Wolfgang Goethe-Universität Frankfurt am Main, Postfach 11 19 32, D-6000 Frankfurt, West Germany

CONTENTS

I. Introduction 2
II. Second-order Direct Multiconfiguration Self-consistent Field Theory . 5
A. Definition of Orbitals, Density Matrices and Integral Matrices . . 5
B. The Newton–Raphson Method and Related Optimization Procedures 7
C. Second-order Energy Approximations 12
D. The Variational Conditions 14
E. Solution of the Non-linear Equations 16
F. Optimization of Internal Orbital Rotations 20
G. Treatment of Closed Shells 24
H. A Direct Configuration-interaction Method for Complete Active Space Calculations 27
I. Optimization of Energy Averages of Several States 32
III. The Internally Contracted Multiconfiguration Reference Self-consistent Electron-pairs Method 33
A. Introduction 33
B. The Internally Contracted Configuration Space 38
C. Orthogonalization and Normalization of the Configurations . . 41
D. Coefficient Matrices and Coefficient Vectors 44
E. The Hamilton Matrix Elements 45
F. Matrix Formulation of the Residual Vector 48
G. Treatment of Closed Shells 51
H. Evaluation of the Coupling Coefficients 54
I. Optimization of the Contraction Coefficients 57

IV. Summary 59
Acknowledgements 59
References 60

I. INTRODUCTION

During the last 10 years, remarkable progress has been made in devising efficient procedures for accurate electronic structure calculations. Not only have the size and complexity of the problems which can be handled been extended, but also the reliability of quantum-chemical results has been considerably improved. Accurate quantum-chemical predictions of properties for small molecules have proven that theoretical calculations are often a useful complement to modern experimental work[1].

The progress in the calculation of highly correlated electronic wavefunctions is due both to the development of improved computational methods and to the rapidly increasing computing power available. In particular, the advent of vector computers has made it possible to perform much larger calculations than before in shorter times. In order to use such machines efficiently, it is essential to adjust the methods to the hardware available. Generally important is to remove all logic from the innermost loops and to perform as many simple vector or matrix operations as possible.

A central role among the available quantum-chemical tools is played by the multiconfiguration self-consistent field (MCSCF) and the multiconfiguration reference configuration-interaction (MR-CI) methods. The purpose of MCSCF calculations is to obtain electronic wavefunctions which represent the states under consideration at all investigated geometries at least qualitatively correctly. In order to obtain more accurate potential energy functions and to make reliable predictions for molecular properties, highly correlated electronic wavefunctions are necessary. It is usually impracticable to calculate such wavefunctions using the MCSCF method. Instead, one employs the MCSCF wavefunction as a zeroth-order approximation in an extended MR-CI calculation. In most MR-CI methods all single and double excitations (SD) relative to the MCSCF configurations are taken into account, and their coefficients are determined variationally. Since the length of such configuration expansions can be very large, conventional MR-CI methods[2–4] often require a configuration selection. However, during the last five years, efficient 'direct CI' methods have been developed which can handle very large configuration spaces.

The direct CI method was proposed in 1972 by Roos[5]. The idea of this method is to avoid the explicit construction and storage of the large Hamilton matrix. Instead, the eigenvectors are found iteratively. The basic operation in each iteration is to form the vector $\mathbf{g} = \mathbf{H} \cdot \mathbf{c}$ directly from the molecular integrals and the trial vector $\mathbf{c}$. The optimum algorithm to form this product depends on the structure of the wavefunction. For instance, in full CI calculations, in which all possible configurations within a given orbital basis are taken into account, it is virtually impossible to account explicitly for all different types of interactions. The necessary structure constants ('coupling coefficients') must then be calculated by a general method and stored on a formula file. This formula file is processed together with the molecular integrals in each iteration. On the other hand, if the wavefunction comprises all single and double excitations relative to a single Slater determinant (CI(SD)) only, the different configuration types and structure constants can be considered explicitly in a computer program. But even then, Roos and Siegbahn[6] distinguished more than 250 distinct cases for a closed-shell reference function. Their algorithm involved so much logic that it was rather inefficient and unsuitable for vectorization. A breakthrough for the closed-shell case was achieved by Meyer[7] in 1976 with his theory of 'self-consistent electron pairs' (SCEP). He showed that the calculation of $\mathbf{H}\cdot\mathbf{c}$ can be performed in terms of simple matrix operations, namely matrix multiplications, if the configurations are renormalized in a particular way. Particularly important in this development was that any dependence of the coupling coefficients on external orbital labels had been removed. Only very few and simple structure constants independent of the size of the basis set remained. An even more elegant formulation of the closed-shell SCEP method was recently presented by Pulay, Saebo and Meyer[8]. They were able to remove the coupling coefficients entirely and further reduce the computational effort. Because of its matrix structure, the SCEP procedure is optimally well suited for vectorization. It has been programmed and applied by Dykstra[9,10], Werner and Reinsch[11] and Ahlrichs[12].

It took a rather long time until similar matrix-formulated direct CI methods became available for more general wavefunctions. Dykstra generalized the SCEP theory for certain types of open-shell wavefunctions and for generalized valence bond (GVB) reference functions which consist only of closed-shell determinants[13,14]. In both cases, however, the important semi-internal configurations were not considered. Flesch and Meyer have developed an SCEP procedure for a spin-unrestricted Hartree–Fock (UHF) reference determinant[15]. In 1981 Chiles and Dykstra[16] presented a matrix formulation of Cicek's coupled-cluster theory. The first generalization of the SCEP method for arbitrary multiconfiguration reference wavefunctions was achieved by Werner and Reinsch[17,18] in 1981. They showed that even in this most general case the vector $\mathbf{H}\cdot\mathbf{c}$ can be obtained by performing a sequence of matrix multiplications and that all required coupling coefficients depend only on internal orbitals. In several applications the high efficiency of the method was demonstrated. Closely related matrix formulations of the direct MR-CI (SD) method were also given by Ahlrichs[19] and Saunders and van Lenthe[20].

Other multiconfiguration reference CI(SD) methods mostly used particular orthonormal spin-eigenfunction bases and employed group theoretical methods to evaluate the coupling coefficients. Most successful in this respect was the 'graphical unitary group approach' (GUGA) of Paldus[21,22] and Shavitt[23–25]. Siegbahn[26] was the first to succeed in developing a general direct MR-CI method using this technique. Other MR-CI methods were described by Buenker and Peyerimhoff[2,3], Brooks and Schaefer[27,28], Duch and Karwowski[29], Duch[30], Tavan and Schulten[31], Taylor[32], Liu and Yoshimine[33], Lischka *et al.*[34] and Saxe *et al.*[35]. Most of these methods used the fact that the external parts of the coupling coefficients are rather simple but did not eliminate them. In some of these methods, SCEP-like techniques were implemented later on (see, e.g., Ref. 36 and other articles in the same volume).

The MR-CI method usually yields most accurate results if the reference function has been fully optimized by an MCSCF procedure. In the MCSCF method not only the linear configuration coefficients but also the molecular orbitals are optimized. Owing to often strong couplings among the non-linear parameters describing changes of the molecular orbitals, early MCSCF strongly suffered with convergence difficulties. It would be beyond the scope of this chapter to review the numerous attempts to solve this problem. Considerable progress was made only quite recently with the development of second-order MCSCF methods[37–56] or approximate second-order methods[57–60]. In second-order MCSCF procedures the first and second derivatives of the energy with respect to all variational parameters, namely the orbital and configuration coefficients, are evaluated exactly. The energy is then approximated by a Taylor expansion, and the parameters are obtained by searching for a stationary point of this approximation. Close to the final solution, this method converges quadratically. Unfortunately, the radius of convergence is rather small. Therefore, many damping schemes and level-shift procedure schemes have been proposed with the aim of ensuring global convergence. This has been discussed in detail by Olsen, Yeager and Jørgensen in a previous volume of this series[53].

Another possibility to increase the radius of convergence is to include higher energy derivatives into the Taylor expansion[61,49]. The exact calculation of these derivatives is rather expensive, however. In 1980 it has been demonstrated by Werner and Meyer[42,43] that the radius of convergence can be improved considerably by treating the higher-order effects in an approximate manner. Only those terms were considered which account appropriately for the orthonormality condition of the orbitals. The extra effort to include these terms into a second-order MCSCF procedure is small. Recently, this method was further improved by Werner and Knowles[55,56], and a remarkable enhancement of convergence was achieved. Moreover, a new direct CI method devised by Knowles and Handy[62] was incorporated into the MCSCF procedure[56]. This allows one to optimize much longer configuration expansions than with previous methods.

The purpose of the present chapter is to describe in some detail the MCSCF and MCSCF-SCEP methods developed by the present author together with W. Meyer[42,43], E. A. Reinsch[17,18,63] and P. J. Knowles[55,56]. It is intended to require only little background. We hope that the rather explicit formulation will help the reader to understand various possible computational strategies, and give some insight into basic structures underlying both theories. In fact, our MCSCF and MCSCF-SCEP methods have many similarities. In both cases the integrals, the variational parameters and the coupling coefficients are ordered into matrices or vectors, and the quantities needed in each iteration are obtained by multiplying and linearly combining these matrices. Therefore, both methods are well able to be vectorized, which is important for efficient use of modern computer hardware. A reformulation of the SCEP theory in terms of non-orthogonal configurations is presented, and new techniques for efficient evaluation of the coupling coefficients are discussed. Since the emphasis of the present chapter is on the theoretical methods, only few examples for applications of our procedures are given. For a review of recent applications, the reader is referred to Ref. 64.

II. SECOND-ORDER DIRECT MULTICONFIGURATION SELF-CONSISTENT FIELD THEORY

A. Definition of Orbitals, Density Matrices and Integral Matrices

We consider a normalized N-electron wavefunction of the form

$$\Psi = \sum_I c_I \Phi_I \qquad \text{with} \qquad \langle \Phi_I | \Phi_J \rangle = \delta_{IJ} \qquad \sum_I c_I^2 = 1 \tag{1}$$

where $\{\Phi_I\}$ is a set of orthonormal configuration state functions (CSFs). Usually, the CSFs are symmetry adapted linear combinations of Slater determinants, but it is also possible to use the Slater determinants themselves as a basis. In the latter case one has to ensure in the optimization process that the wavefunction Ψ has the required symmetry. This will be discussed in more detail in Section II.H. The CSFs are constructed from the 'internal' subset of the orthonormal molecular orbitals $\{\phi_i\}$. Throughout this paper the internal orbitals will be labelled by the indices $i, j, k, \ldots$. The complementary space of external orbitals will be labelled $a, b, c \ldots$, and $r, s, t \ldots$ will denote any orbitals. The molecular orbitals (MOs) $\{\phi_r\}$ are approximated as linear combinations of atomic orbitals (AOs) or other suitable basis functions $\{\kappa_\mu\}$:

$$\phi_r = \sum_\mu X_{\mu r} \kappa_\mu \tag{2}$$

We assume that the orbitals are real and that the wavefunction is spin-restricted in the sense that each orbital can be occupied by two electrons with opposite spin. In terms of the expansion coefficients $\mathbf{X}$, the orthonormality condition of the orbitals takes the form

$$\langle r|s\rangle = (\mathbf{X}^{\dagger}\mathbf{S}\mathbf{X})_{rs} = \delta_{rs} \tag{3}$$

where

$$S_{\mu\nu} = \langle \kappa_{\mu}|\kappa_{\nu}\rangle \tag{4}$$

is the metric of the basis $\{\kappa_{\mu}\}$.

The energy expectation value of the wavefunction (1) can generally be written in the form

$$E^{(0)} = \sum_{IJ} c_I c_J \left(\sum_{ij} h_{ij}\gamma_{ij}^{IJ} + \tfrac{1}{2}\sum_{ijkl} (ij|kl)\Gamma_{ij,kl}^{IJ} \right) \tag{5}$$

where h_{ij} and $(ij|kl)$ are the one-electron and two-electron integrals in the MO basis, respectively, and

$$\gamma_{ij}^{IJ} = \langle \Phi_I|E_{ij}|\Phi_J\rangle \tag{6}$$

$$\Gamma_{ij,kl}^{IJ} = \langle \Phi_I|E_{ij,kl}|\Phi_J\rangle \tag{7}$$

are coupling coefficients. The one- and two-particle excitation operators E_{ij} and $E_{ij,kl}$ in Eqs (6) and (7) are defined as follows:

$$E_{ij} = \eta_i^{\alpha +}\eta_j^{\alpha} + \eta_i^{\beta +}\eta_j^{\beta} \tag{8}$$

$$E_{ij,kl} = \eta_k^{\alpha +}E_{ij}\eta_l^{\alpha} + \eta_k^{\beta +}E_{ij}\eta_l^{\beta} \tag{9}$$

Here η_i^{α} and $\eta_i^{\beta +}$ are the usual annihilation and creation operators for electrons with α and β spin, respectively. In order to satisfy the Pauli exclusion principle, they must obey the anticommutation relations

$$[\eta_i^{\rho}, \eta_j^{\sigma}]_+ = 0 \tag{10}$$

$$[\eta_i^{\rho +}, \eta_j^{\sigma +}]_+ = 0 \tag{11}$$

$$[\eta_i^{\rho +}, \eta_j^{\sigma}]_+ = \delta_{ij}\delta_{\rho\sigma} \qquad \text{with} \qquad \rho, \sigma = \{\alpha, \beta\} \tag{12}$$

The coupling coefficients γ_{ij}^{IJ} and $\Gamma_{ij,kl}^{IJ}$ depend only on the formal structure of the CSFs $\{\Phi_I\}$ but not on the particular form of the orbitals involved. They can, therefore, be calculated once and stored on a formula tape. However, as will be discussed later, it is in certain cases advantageous to re-evaluate them each time they are needed. From the coupling coefficients and CI coefficients, the first- and second-order density matrices can be obtained:

$$(\mathbf{D})_{rs} = \sum_{IJ} c_I c_J \gamma_{rs}^{IJ} \tag{13}$$

$$\Gamma_{rs,tu} = \sum_{IJ} c_I c_J \Gamma_{rs,tu}^{IJ} \tag{14}$$

For convenience in later expressions, we define the symmetrized density matrices

$$(\mathbf{P}^{kl})_{rs} = \tfrac{1}{2}(\Gamma_{rs,kl} + \Gamma_{sr,kl}) \tag{15}$$

$$(\mathbf{Q}^{kl})_{rs} = \tfrac{1}{2}(\Gamma_{rk,sl} + \Gamma_{kr,sl}) \tag{16}$$

Note that for MCSCF wavefunctions these matrices have non-vanishing elements only in the internal–internal block. According to the above definitions the following symmetry relations hold for real orbitals:

$$(\mathbf{D})_{ij} = (\mathbf{D})_{ji} \tag{17}$$

$$(\mathbf{P}^{kl})_{ij} = (\mathbf{P}^{kl})_{ji} = (\mathbf{P}^{ij})_{kl} \tag{18}$$

$$(\mathbf{Q}^{kl})_{ij} = (\mathbf{Q}^{lk})_{ji} = (\mathbf{Q}^{ij})_{kl} \tag{19}$$

We further define Coulomb and exchange matrices

$$(\mathbf{J}^{kl})_{rs} = (rs|kl) \tag{20}$$

$$(\mathbf{K}^{kl})_{rs} = (rk|ls) \tag{21}$$

which are ordered subsets of the two-electron integrals. This ordering of the integrals with at most two external orbitals is essential for the matrix formulation of both MCSCF and MCSCF-CI methods. As indicated by the parentheses in Eqs (18)–(21), superscripts denote different matrices, and subscripts their elements. This convention will be followed throughout this chapter and the parentheses will often be omitted. The energy expectation value can now be written in the simple form

$$E^{(0)} = \mathrm{tr}(\mathbf{hD}) + \tfrac{1}{2}\sum_{kl} \mathrm{tr}(\mathbf{J}^{kl}\mathbf{P}^{lk}) \tag{22}$$

where tr(**A**) denotes the trace of the matrix **A**.

B. The Newton–Raphson Method and Related Optimization Procedures

The purpose of the MCSCF method is to minimize the energy expectation value (Eq. (5)) with respect to the CI coefficients $\{c_I\}$ and the molecular-orbital coefficients $X_{\mu i}$ with the auxiliary orthonormality condition in Eq. (3). Since the energy expectation value is a fourth-order function of the orbitals, its direct minimization is impracticable. It is, therefore, necessary to employ an iterative procedure and minimize in each iteration an approximate energy functional. Provided this functional is a reasonably good approximation to the true energy as a function of the changes in the orbitals and CI coefficients, its minimization will yield improved orbitals and CI coefficients. These are used as a starting guess in the next iteration. If the approximate energy functional is accurate to second order in the changes of the orbitals and CI coefficients, the optimization will be quadratically convergent, i.e. when approaching the final solution the change of the energy in each iteration decreases quadratically. Far from the solution, however, the approximate second-order functional may not be a reasonable approximation. Special precautions are then necessary to ensure convergence. In fact, extensive work of several groups has been performed during the last few years in order to find stable and efficient algorithms which improve or even guarantee convergence in the non-local region[39–41,46,48,50–53].

The common approach in most methods is to describe the orbital changes by a unitary (orthogonal) transformation of the form[65,66]

$$|\tilde{i}> = \sum_r |r> U_{ri} \tag{23}$$

i.e.

$$\tilde{\mathbf{X}} = \mathbf{X}\mathbf{U} \tag{24}$$

with

$$\mathbf{U} = \exp(\mathbf{R}) = \mathbf{1} + \mathbf{R} + \tfrac{1}{2}\mathbf{R}\mathbf{R} + \cdots \tag{25}$$

where $\mathbf{R} = -\mathbf{R}^\dagger$ is an antisymmetric matrix. Since $\mathbf{U}$ remains unitary for any choice of this matrix, the elements $\{R_{ri}, r > i\}$ form a set of independent variational parameters. Some of the parameters R_{ij} may be redundant, i.e. they do not influence the energy to first order. This happens if the orbitals $|i\rangle$ and $|j\rangle$ have the same occupation number in all CSFs. An R_{ij} is also redundant if the same first-order energy change can be achieved by a variation of the CI coefficients. Since redundant variables do influence the energy in higher order, they must be set to zero in order to avoid convergence difficulties. The redundant variables can be determined automatically as described in Ref. 67.

In the Newton–Raphson method the energy is expanded up to second order in the variables R_{ri} and the changes of the CI coefficients $\{\Delta c_I\}$. Collecting these parameters into a vector $\mathbf{x}$, the stationary condition for the energy approximation

$$E^{(2)}(\mathbf{x}) = E^{(0)} + \mathbf{g}^\dagger\mathbf{x} + \tfrac{1}{2}\mathbf{x}^\dagger\mathbf{H}\mathbf{x} \tag{26}$$

takes the form of a system of inhomogeneous linear equations

$$\mathbf{H}\mathbf{x} + \mathbf{g} = 0 \tag{27}$$

where

$$g_i = (\partial E/\partial x_i)_{\mathbf{x}=\mathbf{0}} \tag{28}$$

is the energy gradient at the expansion point, and

$$H_{ij} = (\partial^2 E/\partial x_i \partial x_j)_{\mathbf{x}=\mathbf{0}} \tag{29}$$

is the Hessian matrix of second energy derivatives. Explicit expressions for these derivatives will be given in Section II.C. The solution of Eq. (27) yields the parameters $\mathbf{x}$, which are used to calculate new orbitals and a new CI vector. Experience has shown the radius of convergence of this method to be rather small. Far from the solution the Hessian often has many negative or very small eigenvalues, and convergence can then be achieved only by introducing a level shift which makes the Hessian positive definite and the step vector $\mathbf{x}$ sufficiently small:

$$(\mathbf{H} + \kappa\mathbf{1})\mathbf{x} + \mathbf{g} = 0 \tag{30}$$

If the level shift is chosen to be

$$\kappa = -\lambda\varepsilon \tag{31}$$

with

$$\varepsilon = \lambda\mathbf{g}^{\dagger}\mathbf{x} \tag{32}$$

the linear equations are transformed into an eigenvalue equation of the form:

$$\begin{pmatrix} -\varepsilon & \mathbf{g}^{\dagger} \\ \mathbf{g} & \mathbf{H}/\lambda - \varepsilon \end{pmatrix} \begin{pmatrix} 1/\lambda \\ \mathbf{x} \end{pmatrix} = \mathbf{0} \tag{33}$$

For $\lambda = 1$ this is known as the augmented Hessian (AH) method. It was first proposed by Lengsfield[39], and used with various modifications by several authors[44,46,51,52]. It can easily be proved that $\mathbf{H} - \varepsilon\mathbf{1}$ is always positive definite if ε is the eigenvalue obtained by solving Eq. (33). It can also be shown that the AH method is quadratically convergent[46]. A value $\lambda > 1$ has the effect of further reducing the step length $|\mathbf{x}|$. In fact, as shown by Fletcher[68] and

TABLE I
Convergence behaviour of step-restricted augmented Hessian calculations (Fletcher optimization).

	Energy difference		
Iter.	N_2[a]	CO[b]	CO[c]
1	– 0.019350483	– 0.016819128	– 0.005805628
2	– 0.016995391	– 0.017490633	– 0.015098631
3	– 0.016661071	– 0.013430376	– 0.027426813
4	– 0.002740679	– 0.015795238	– 0.026688321
5	– 0.000038990	– 0.003660917	– 0.009411258
6	– 0.000000013	– 0.001231696	– 0.018214694
7	– 0.000000000	– 0.000050933	– 0.016727086
8		– 0.000001714	– 0.001016208
9		– 0.000000003	– 0.000039643
10		– 0.000000000	– 0.000000210
11			– 0.000000000

[a]N_2 molecule; for details see Ref. 51; improved virtual orbitals (IVOs) were taken as starting guess.
[b]CO molecule; for details see Ref. 51; IVOs were used as starting guess.
[c]CO molecule; basis set and configurations as in footnote *b*, but canonical SCF orbitals used as starting guess; results from Ref. 55.

discussed in the context of the MCSCF problem by Jørgensen *et al.*[51,52,54], Eq. (33) can be derived by minimizing the second-order energy approximation (Eq. (26)) with the auxiliary condition $|\mathbf{x}| < s$. If the maximum step length s is updated automatically according to a particular scheme after each iteration, convergence can be guaranteed. This does not mean, however, that convergence is achieved with a small number of iterations as desired for a second-order scheme. Test calculations published by Jørgensen *et al.* have shown that typically 6–10 iterations are necessary unless a very good starting guess from a nearby geometry is available (cf. Table I). We have similar experiences using this method. Of these iterations, only the last two or three are in the local region and show quadratic convergence behaviour.

The origin of these difficulties is the orthonormality condition in Eq. (3). This causes the true energy to be periodic in individual orbital rotations. If only a single rotation between the orbitals $|i\rangle$ and $|j\rangle$ is considered, the unitary

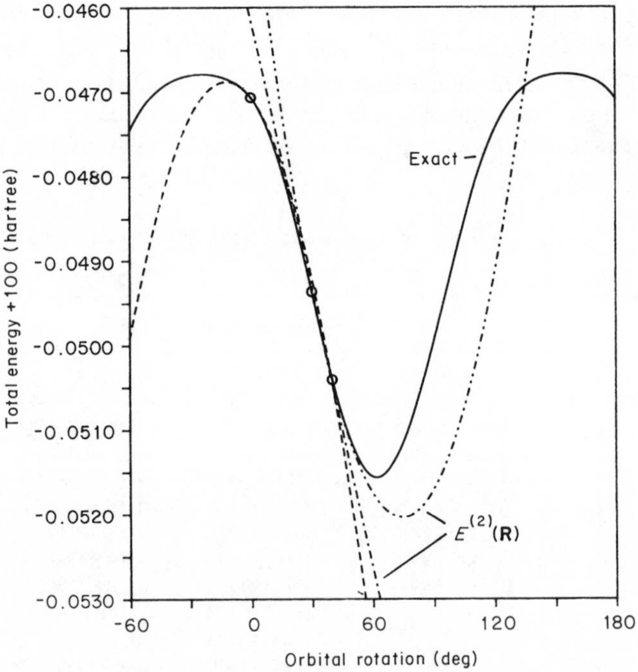

Fig. 1. The dependence of the exact energy and the second-order energy approximation $E^{(2)}(\mathbf{R})$ on the rotation 4σ–5σ for a three-configuration MCSCF calculation for the HF molecule. The configurations were: $1\sigma^2 2\sigma^2 3\sigma^2 1\pi^4$, $1\sigma^2 2\sigma^2 4\sigma^2 1\pi^4$ and $1\sigma^2 3\sigma^2 4\sigma^2 1\pi^4$. For other details see Ref. 55. The expansion point at zero degrees corresponds to canonical SCF orbitals. $E^{(2)}(\mathbf{R})$ is also shown for two other expansion points. In these cases all orbitals except 4σ and 5σ were canonical SCF orbitals.

matrix **U** can be written in the form

$$\mathbf{U} = \begin{pmatrix} \cos\alpha & \sin\alpha \\ -\sin\alpha & \cos\alpha \end{pmatrix} \tag{34}$$

where the rotation angle α equals R_{ij}. Clearly, as illustrated in Fig. 1, the second-order energy approximation does not describe this periodicity. Its minimization, therefore, predicts steps that are either too large or even of the wrong sign. Fig. 2 shows the effect of level shifts which restrict the step size to certain values. It is obvious that convergence will depend sensitively on the choice of the step size. Of course, Figs 1 and 2 are very idealized examples. In cases with many orbital rotations which influence each other, the situation is much more complicated, and a single level-shift parameter cannot be expected to be optimal for all orbital rotations.

From the above considerations it appears to be necessary to account more accurately, in the energy approximation, for the orthonormality of the orbitals. A straightforward extension of the Newton–Raphson method would be to expand the energy up to third or even higher order in **R** [49,53,61].

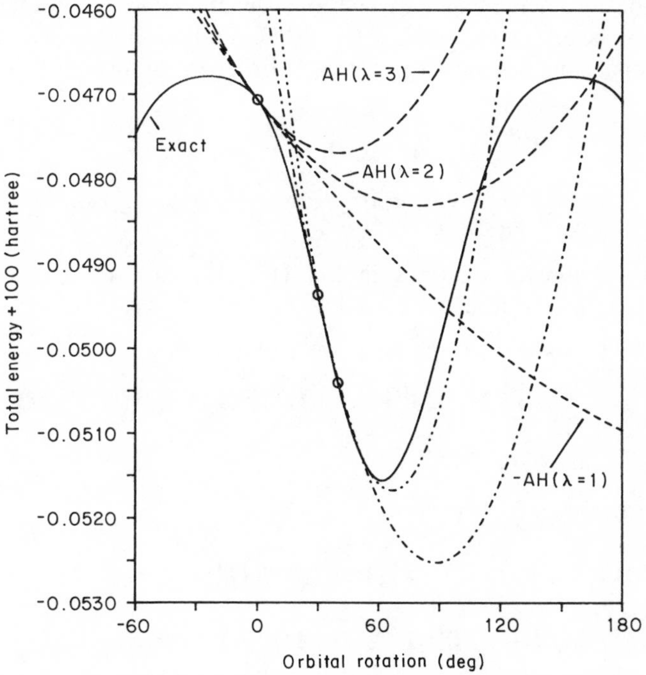

Fig. 2. Second-order energy approximations employed in step-restricted augmented Hessian calculations for the same model and the same expansion points as in Fig. 1. For the expansions at zero degrees, various level-shift parameters have been used.

However, this would be only a partial improvement, since the true orbital and energy changes are of infinite order in **R**. Furthermore, each iteration would require a considerably more expensive integral transformation than a second-order scheme. Fortunately, it is possible to account for the orthonormality of the orbitals appropriately in a rather simple way already in a second-order method. This will be explained in the next sections.

C. Second-order Energy Approximations

The energy expectation value (Eq. (5)) is a function of the CI coefficients $\{c_I\}$ and the orbital changes

$$|\Delta i\rangle = |\tilde{i}\rangle - |i\rangle = \sum_r |r\rangle T_{ri} \tag{35}$$

where

$$\mathbf{T} = \mathbf{U} - \mathbf{1} = \mathbf{R} + \tfrac{1}{2}\mathbf{RR} + \cdots \tag{36}$$

The dependence on the CI coefficients enters via the density matrices **D** and $\mathbf{P}^{kl}$ and will be considered explicitly later on. The exact energy is a fourth-order function of the orbital changes, i.e. $E = E^{(4)}(\mathbf{T})$, and of infinite order in **R**. If the energy expansion is truncated to second order in the orbital changes, one obtains

$$\begin{aligned} E^{(2)} = E^{(0)} &+ 2\sum_{ij} \langle \Delta i|h|j\rangle D_{ij} + \sum_{ij} \langle \Delta i|h|\Delta j\rangle D_{ij} \\ &+ \sum_{ijkl} [2(\Delta ij|kl)P_{ij}^{kl} + (\Delta i\Delta j|kl)P_{ij}^{kl} + 2(\Delta ik|l\Delta j)Q_{ij}^{kl}] \end{aligned} \tag{37}$$

where the symmetry relations in Eqs (17)–(19) have been used to sum equivalent terms. In terms of **U** and **T**, Eq. (37) takes the form

$$\begin{aligned} E^{(2)}(\mathbf{T}) = E^{(0)} &+ 2\,\mathrm{tr}(\mathbf{T}^\dagger \mathbf{hD}) + \mathrm{tr}(\mathbf{T}^\dagger \mathbf{hTD}) \\ &+ \sum_{kl} [2\,\mathrm{tr}(\mathbf{T}^\dagger \mathbf{J}^{kl}\mathbf{P}^{lk}) + \mathrm{tr}(\mathbf{T}^\dagger \mathbf{J}^{kl}\mathbf{TP}^{lk}) + 2\,\mathrm{tr}(\mathbf{T}^\dagger \mathbf{K}^{kl}\mathbf{TQ}^{lk})] \end{aligned} \tag{38}$$

Defining the matrices

$$\mathbf{A} = \mathbf{hD} + \sum_{kl} \mathbf{J}^{kl}\mathbf{P}^{lk} \tag{39}$$

$$\begin{aligned} \mathbf{B} &= \mathbf{A} + \mathbf{hTD} + \sum_{kl} (\mathbf{J}^{kl}\mathbf{TP}^{lk} + 2\mathbf{K}^{kl}\mathbf{TQ}^{lk}) \\ &= \mathbf{hUD} + \sum_{kl} (\mathbf{J}^{kl}\mathbf{UP}^{lk} + 2\mathbf{K}^{kl}\mathbf{TQ}^{lk}) \end{aligned} \tag{40}$$

and

$$\mathbf{G}^{ij} = \mathbf{h}D_{ij} + \sum_{kl} (\mathbf{J}^{kl} P_{ij}^{kl} + 2\mathbf{K}^{kl} Q_{ij}^{kl}) \tag{41}$$

the second-order energy can also be written in the more compact forms:

$$E^{(2)}(\mathbf{T}) = E^{(0)} + 2\,\mathrm{tr}(\mathbf{T}^{\dagger}\mathbf{A}) + \sum_{ij}(\mathbf{T}^{\dagger}\mathbf{G}^{ij}\mathbf{T})_{ij}$$
$$= E^{(0)} + \mathrm{tr}[\mathbf{T}^{\dagger}(\mathbf{A}+\mathbf{B})] \tag{42}$$

Note that, owing to the sparsity of the density matrices $\mathbf{D}$, $\mathbf{P}^{kl}$ and $\mathbf{Q}^{kl}$, all elements A_{ra} and B_{ra} ($|a\rangle$ external) vanish. Therefore, in a computer program only the rectangular blocks A_{ri} and B_{ri} have to be computed and stored. For the sake of compact expressions, however, it is advantageous to deal formally with the full square matrices.

As outlined in Section II.B, in the Newton–Raphson (NR) method the energy approximation in Eq. (42) is truncated to second order in $\mathbf{R}$. The explicit form of Eq. (26) for fixed CI coefficients is therefore

$$E^{(2)}(\mathbf{R}) = E^{(0)} + 2\,\mathrm{tr}(\mathbf{R}^{\dagger}\mathbf{A}) + \mathrm{tr}(\mathbf{RRA}) + \sum_{ij}(\mathbf{R}^{\dagger}\mathbf{G}^{ij}\mathbf{R})_{ij} \tag{43}$$

In contrast to this approximation, Eq. (42) contains terms up to infinite order in $\mathbf{R}$. These additional terms account appropriately for the orthonormality of the orbitals. In fact, as shown in Fig. 3 for the same model calculation as in Figs

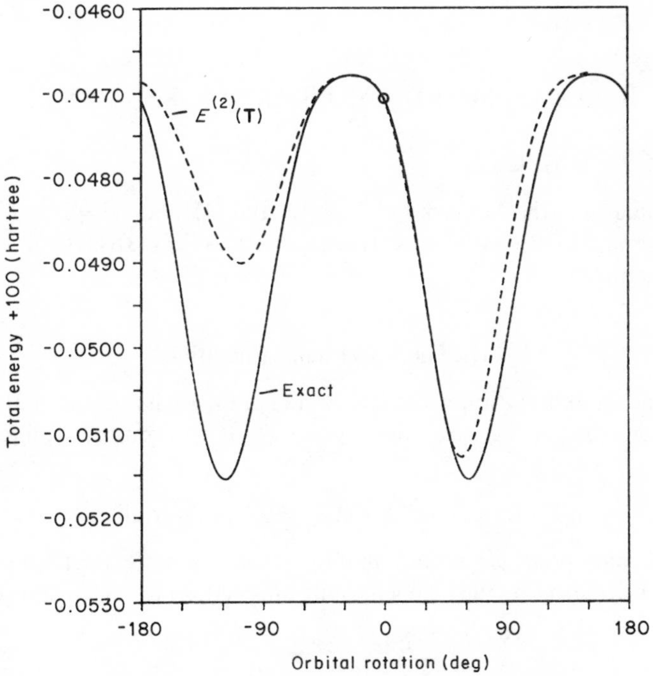

Fig. 3. The dependence of the second-order energy approximation $E^{(2)}(\mathbf{T})$ on the rotation 4σ–5σ for the HF molecule as in Fig. 1.

1 and 2, $E^{(2)}(\mathbf{T})$ is in close agreement with the true energy over a large range of rotation angles and predicts minima at nearly the correct angles. Thus, a much larger radius of convergence can be expected if $E^{(2)}(\mathbf{T})$ rather than $E^{(2)}(\mathbf{R})$ is used as approximate energy functional. It will be demonstrated also that the rate of convergence in the non-local region is much faster* than for a step-restricted augmented Hessian method.

We will now investigate how the second-order energy $E^{(2)}(\mathbf{T})$ at a particular point $\mathbf{T} = \mathbf{T}(\mathbf{R})$ changes if $\mathbf{T}$ undergoes a small variation. Such a change can be described by multiplying $\mathbf{U}$ with a second unitary transformation $\mathbf{U}(\Delta\mathbf{R})$:

$$\begin{aligned}\mathbf{U}(\mathbf{R}, \Delta\mathbf{R}) &= \mathbf{U}(\mathbf{R})\mathbf{U}(\Delta\mathbf{R}) \\ &= \mathbf{U} + \mathbf{U}(\Delta\mathbf{R} + \tfrac{1}{2}\Delta\mathbf{R}\Delta\mathbf{R} + \cdots)\end{aligned} \tag{44}$$

The antisymmetric matrix $\Delta\mathbf{R} = -\Delta\mathbf{R}^\dagger$ defines the change of $\mathbf{U}$. Note that $\mathbf{U}(\mathbf{R}, \Delta\mathbf{R}) \neq \mathbf{U}(\mathbf{R} + \Delta\mathbf{R})$ since $\mathbf{R}$ and $\Delta\mathbf{R}$ do not commute. Inserting this into Eq. (38) yields, up to second order in $\Delta\mathbf{R}$,

$$\begin{aligned}E^{(2)}(\mathbf{T}, \Delta\mathbf{R}) &= E^{(2)}(\mathbf{T}) + 2\,\mathrm{tr}(\Delta\mathbf{R}^\dagger\tilde{\mathbf{A}}) + \mathrm{tr}(\Delta\mathbf{R}\Delta\mathbf{R}\tilde{\mathbf{A}}) + \sum_{ij}(\Delta\mathbf{R}^\dagger\tilde{\mathbf{G}}^{ij}\Delta\mathbf{R})_{ij} \\ &= E^{(2)}(\mathbf{T}) + \mathrm{tr}[\Delta\mathbf{R}^\dagger(\tilde{\mathbf{A}} + \tilde{\mathbf{B}})] + \mathrm{tr}(\Delta\mathbf{R}\Delta\mathbf{R}\tilde{\mathbf{A}})\end{aligned} \tag{45}$$

where

$$\tilde{\mathbf{A}} = \mathbf{U}^\dagger\mathbf{B} \tag{46}$$

$$\tilde{\mathbf{B}} = \tilde{\mathbf{A}} + \mathbf{U}^\dagger\left(\mathbf{hU}\Delta\mathbf{RD} + \sum_{kl}(\mathbf{J}^{kl}\mathbf{U}\Delta\mathbf{RP}^{lk} + 2\mathbf{K}^{kl}\mathbf{U}\Delta\mathbf{RQ}^{lk})\right) \tag{47}$$

$$\tilde{\mathbf{G}}^{ij} = \mathbf{U}^\dagger\mathbf{G}^{ij}\mathbf{U} \tag{48}$$

One should note the similarity of Eqs (40) and (47). For $\mathbf{U} = \mathbf{1}$ and $\mathbf{T} = \mathbf{0}$ we have $\tilde{\mathbf{A}} = \mathbf{A}$ and $\tilde{\mathbf{G}}^{ij} = \mathbf{G}^{ij}$. Eq. (45) then reduces to Eq. (43) (Newton–Raphson approximation).

D. The Variational Conditions

The energy expectation value $E = E^{(4)}(\mathbf{T})$ in Eq. (5) has a stationary point if the first derivatives at the expansion point $\mathbf{T} = \mathbf{0}$ with respect to all R_{ri} vanish, i.e. if

$$(\partial E/\partial R_{ri})_{\mathbf{R}=\mathbf{0}} = 2(\mathbf{A} - \mathbf{A}^\dagger)_{ri} = 0 \qquad \text{for all } r > i \tag{49}$$

The stationary point is a minimum if the Hessian matrix (Eq. (29)) is positive definite. Furthermore, the CI coefficients must satisfy the eigenvalue equation

$$\sum_J (H_{IJ} - E\delta_{IJ})c_J = 0 \qquad \text{for all } I \tag{50}$$

*This may not be true for rotations between strongly occupied orbitals, cf. Section II. F.

This implies that for the electronic ground state E is the lowest eigenvalue of the Hamilton matrix

$$H_{IJ} = \langle \Phi_I | H | \Phi_J \rangle = \sum_{ij} h_{ij}\gamma_{ij}^{IJ} + \tfrac{1}{2}\sum_{ijkl}(ij|kl)\Gamma_{ij,kl}^{IJ} \quad (51)$$

If Eqs (49) and (50) are satisfied simultaneously, convergence of the MCSCF procedure is reached.

Similarly, the energy approximation $E^{(2)}(\mathbf{T})$ has a stationary point with respect to variations of $\mathbf{T}$ if the first derivatives

$$\left(\frac{\partial E^{(2)}(\mathbf{T}, \Delta\mathbf{R})}{\partial \Delta R_{ri}}\right)_{\Delta\mathbf{R}=\mathbf{0}} = 2(\tilde{\mathbf{A}} - \tilde{\mathbf{A}}^{\dagger})_{ri} \quad (52)$$

vanish for all $r > i$. These conditions are summarized in the matrix equation[42,43,55]

$$\mathbf{U}^{\dagger}\mathbf{B} - \mathbf{B}^{\dagger}\mathbf{U} = \mathbf{0} \quad (53)$$

The stationary point is a minimum if the matrix of second derivatives

$$\left(\frac{\partial^2 E^{(2)}(\mathbf{T}, \Delta\mathbf{R})}{\partial \Delta R_{ri}\partial \Delta R_{sj}}\right)_{\Delta\mathbf{R}=\mathbf{0}} = (1-\tau_{ri})(1-\tau_{sj})(2\tilde{\mathbf{G}}^{ij} - \delta_{ij}(\tilde{\mathbf{A}} + \tilde{\mathbf{A}}^{\dagger}))_{rs} \quad (54)$$

is positive definite. The operator τ_{ri} in Eq. (54) permutes the indices r and i. For $\mathbf{T} = \mathbf{0}$ Eqs (52) and (54) reduce to the explicit formulae for the energy derivatives used in the NR or AH methods (Eqs (28) and (29)).

In order to minimize the second-order energy approximation $E^{(2)}(\mathbf{T}, \mathbf{c})$ with respect to the CI coefficients, it can be written in the form

$$E^{(2)}(\mathbf{T}, \mathbf{c}) = \mathbf{c}^{\dagger}\mathbf{H}^{(2)}\mathbf{c}/\mathbf{c}^{\dagger}\mathbf{c} \quad (55)$$

where the second-order Hamiltonian $\mathbf{H}^{(2)}$ is defined as

$$H_{IJ}^{(2)} = \sum_{ij}(\mathbf{U}^{\dagger}\mathbf{h}\mathbf{U})_{ij}\gamma_{ij}^{IJ} + \tfrac{1}{2}\sum_{ijkl}(ij|kl)^{(2)}\Gamma_{ij,kl}^{IJ} \quad (56)$$

The integrals $(ij|kl)^{(2)}$ are the second-order approximations to the exact two-electron integrals as a function of $\mathbf{T}$:

$$(ij|kl)^{(2)} = -(ij|kl) + (\mathbf{U}^{\dagger}\mathbf{J}^{kl}\mathbf{U})_{ij} + (\mathbf{U}^{\dagger}\mathbf{J}^{ij}\mathbf{U})_{kl} + (1+\tau_{ij})(1+\tau_{kl})(\mathbf{T}^{\dagger}\mathbf{K}^{ik}\mathbf{T})_{jl} \quad (57)$$

According to this definition the second-order energy expressions in Eqs (38) and (55) are identical for a given set $\{\mathbf{c}, \mathbf{T}\}$. The minimization of the energy 'expectation value' (Eq. (55)) with respect to the $\{c_I\}$ yields the eigenvalue equation

$$(\mathbf{H}^{(2)} - E^{(2)}\mathbf{1})\mathbf{c} = 0 \quad (58)$$

The minimum of $E^{(2)}(\mathbf{T}, \mathbf{c})$ with respect to $\mathbf{T}$ and $\mathbf{c}$ is reached if the coupled non-linear equations (53) and (58) are satisfied simultaneously with the same $\mathbf{T}$ and $\mathbf{c}$. In this case the energy eigenvalue $E^{(2)}$ in Eq. (58) becomes identical with the expectation values in Eqs (38) and (55).

E. Solution of the Non-linear Equations

In order to make the optimization procedure outlined in Sections II.C and II.D practicable, a stable algorithm to solve the coupled non-linear equations (53) and (58) is necessary. The method should avoid the explicit construction and storage of large Hessian or Hamilton matrices in order to be flexible with respect to the number of orbitals and configuration state functions. Hence, for the optimization of the CI coefficients it is advantageous to employ a 'direct CI' procedure. In a direct CI method the desired eigenvector is obtained iteratively. In each iteration the 'residual vector' $\mathbf{y} = (\mathbf{H} - E\mathbf{1})\mathbf{c}$ is calculated directly from the one- and two-electron integrals, a trial vector $\mathbf{c}$ and the coupling coefficients. The residual vector is then used to improve the trial vector $\mathbf{c}$. Very similar techniques can be employed to solve iteratively large systems of linear equations or the non-linear equations (53) ('direct MCSCF'). The iterations needed to solve Eqs (53) and (58) are called 'micro-iterations'. After convergence of the micro-iterations, the final matrix $\mathbf{U}$ is used to transform the orbital coefficients according to Eq. (24). Then a new set of operators $\mathbf{J}^{kl}$, $\mathbf{K}^{kl}$ is evaluated. Efficient algorithms for this partial four-index transformation have been described by several authors[20,42,57]. The calculation of these operators and a variational energy initializes the next 'macro-iteration'.

In complete active space self-consistent field (CASSCF) calculations[57,59,69] with long configuration expansions the most expensive part is often the optimization of the CI coefficients. It is, therefore, particularly important to minimize the number of CI iterations. In conventional direct second-order MCSCF procedures[44,52,70], the CI coefficients are updated together with the orbital parameters in each micro-iteration. Since the optimization requires typically 100–150 micro-iterations, such calculations with many configurations can be rather expensive. A possible remedy to this problem is to decouple the orbital and CI optimizations[59], but this causes the loss of quadratic convergence. The following method allows one to update the CI coefficients much fewer times than the orbital parameters. This saves considerable time without loss of the quadratic convergence behaviour.

In order to minimize the second-order energy approximation $E^{(2)}(\mathbf{T})$ for fixed CI coefficients a step-restricted augmented Hessian method as outlined in Section II.B (Eqs (30)–(33)) is used. While in other MCSCF methods this technique is employed to minimize the exact energy, it is used here to minimize an approximate energy functional. The parameter vector $\mathbf{x}$ is made up of the non-redundant elements ΔR_{ri} $(r > i)$. The successive expansion points are defined by the matrix $\mathbf{T} = \mathbf{U} - \mathbf{1}$, which is updated according to Eq. (44) each time the eigenvalue equation (33) has been solved using Davidson's technique[71]. In each micro-iteration, the residual vector

$$\mathbf{y} = \mathbf{g} + (\mathbf{H} - \lambda\varepsilon)\mathbf{x} \tag{59}$$

has to be evaluated, where the elements of the gradient $\mathbf{g}$ and the Hessian $\mathbf{H}$ are given in Eqs (52) and (54), respectively. The damping parameter λ is determined automatically as the Davidson iteration proceeds such that the step length $|\mathbf{x}|$ remains smaller than a prescribed threshold (e.g. 0.5). The explicit form of the residual vector is obtained by deriving Eq. (45) with respect to all ΔR_{ri}. It can be written in matrix form as

$$\begin{aligned} &\mathbf{Y} = 2(\tilde{\mathbf{B}} - \tilde{\mathbf{B}}^{\dagger}) - (\tilde{\mathbf{A}} + \tilde{\mathbf{A}}^{\dagger})\Delta\mathbf{R} + \Delta\mathbf{R}(\tilde{\mathbf{A}} + \tilde{\mathbf{A}}^{\dagger}) - \lambda\varepsilon\Delta\mathbf{R} \\ &\text{with} \qquad \varepsilon = 2\,\mathrm{tr}(\Delta\mathbf{R}^{\dagger}\tilde{\mathbf{A}}) \end{aligned} \tag{60}$$

The matrices $\tilde{\mathbf{A}}$ and $\tilde{\mathbf{B}}$ have been defined in Eqs (46) and (47). In the Davidson procedure, the matrix $\Delta\mathbf{R}$ and the residual $\mathbf{Y}$ are obtained as linear combinations

$$\Delta\mathbf{R} = \sum_m \alpha_m \mathbf{S}^m \tag{61}$$

$$\mathbf{Y} = \sum_m \alpha_m \mathbf{Y}^m \tag{62}$$

where $\mathbf{S}^m$ are orthonormalized expansion vectors.* The $\mathbf{Y}^m$ are calculated according to Eq. (60) with the $\mathbf{S}^m$ instead of $\Delta\mathbf{R}$. The optimum parameters α_m are determined by solving a small eigenvalue problem[71]. A new expansion vector is then obtained as

$$S_{ri}^{m+1} = -\,Y_{ri}/(D_{ri} - \varepsilon) \tag{63}$$

and subsequently orthonormalized to all previous $\mathbf{S}^m$. The diagonal elements D_{ri} of the Hessian matrix (Eq. (54)) used in the update formula (63) are given by

$$D_{ai} = [2\mathbf{U}^{\dagger}\mathbf{G}^{ii}\mathbf{U} - (\mathbf{U}^{\dagger}\mathbf{B} + \mathbf{B}^{\dagger}\mathbf{U})]_{a\sigma} \tag{64}$$

$$\begin{aligned} D_{ji} = {}& 2[(\mathbf{U}^{\dagger}\mathbf{G}^{ii}\mathbf{U})_{jj} + (\mathbf{U}^{\dagger}\mathbf{G}^{jj}\mathbf{U})_{ii} - 2(\mathbf{U}^{\dagger}\mathbf{G}^{ji}\mathbf{U})_{ij}] \\ & - (\mathbf{U}^{\dagger}\mathbf{B} + \mathbf{B}^{\dagger}\mathbf{U})_{ii} - (\mathbf{U}^{\dagger}\mathbf{B} + \mathbf{B}^{\dagger}\mathbf{U})_{jj} \end{aligned} \tag{65}$$

Usually it is sufficient to calculate these elements only once per macro-iteration with $\mathbf{U} = \mathbf{1}$, such that the operators $\mathbf{G}^{ij}$ need not be transformed each time $\mathbf{U}$ is updated. If the Davidson procedure has converged (i.e. $\mathbf{Y}$ is smaller than a certain threshold) the unitary matrix $\mathbf{U}$ is updated according to Eq. (44), and a new matrix $\mathbf{B}$ is calculated. The process is repeated until Eq. (53) is

*In the orthonormalization process the non-redundant elements of the matrices $\mathbf{S}^m$ form a vector.

satisfied to the desired accuracy. One should note that no expensive integral transformation is necessary after updating **U**.

After an update of **U** it is possible to perform a direct CI step in order to improve the CI vector and the density matrices. In this case the calculation of **B** is done in two steps. First, the one-index transformations

$$(\tilde{\mathbf{h}})_{rj} = (\mathbf{hU})_{rj} \tag{66}$$

$$(\tilde{\mathbf{J}}^{kl})_{rj} = (\mathbf{J}^{kl}\mathbf{U})_{rj} \tag{67}$$

$$(\tilde{\mathbf{K}}^{kl})_{rj} = (\mathbf{K}^{kl}\mathbf{T})_{rj} \tag{68}$$

are performed, and, at the same time, the second-order integrals $(ij|kl)^{(2)}$ are evaluated by performing the second half transformations

$$(\mathbf{U}^{\dagger}\mathbf{hU})_{ij} = (\mathbf{U}^{\dagger}\tilde{\mathbf{h}})_{ij} \tag{69}$$

$$(ij|kl)^{(2)} \leftarrow (\mathbf{U}^{\dagger}\tilde{\mathbf{J}}^{kl})_{ij} \tag{70}$$

$$(ik|jl)^{(2)} \leftarrow (\mathbf{T}^{\dagger}\tilde{\mathbf{K}}^{kl})_{ij} \tag{71}$$

Since only the internal blocks of these transformed matrices are needed, the latter step is much cheaper than the first half of the transformation (Eqs (66)–(68)). The half-transformed operators $\tilde{\mathbf{h}}$, $\tilde{\mathbf{J}}^{kl}$ and $\tilde{\mathbf{K}}^{kl}$ are stored on disc, while the integrals $(ij|kl)^{(2)}$ are kept in high-speed memory and employed in the subsequent direct CI step. The direct CI procedure will be described in Section II.H. The improved CI vector is used to evaluate new density matrices **D**, $\mathbf{P}^{kl}$ and $\mathbf{Q}^{kl}$. Finally, the new matrix **B** is calculated according to

$$\mathbf{B} = \tilde{\mathbf{h}}\mathbf{D} + \sum_{kl}(\tilde{\mathbf{J}}^{kl}\mathbf{P}^{lk} + 2\tilde{\mathbf{K}}^{kl}\mathbf{Q}^{lk}) \tag{72}$$

As compared with other direct second-order MCSCF methods[44,52,70], the above procedure has the following advantages: (i) A CI step is only performed if the transformation matrix **U** has converged to a sufficiently stable value as indicated by a small step size $\Delta\mathbf{R}$ in the previous augmented Hessian iteration. This avoids unnecessary oscillations of the CI coefficients and minimizes the number of CI steps. (ii) Each update of the CI vector requires the cost of only two direct CI iterations, one for the calculation of the residual vector and one for the evaluation of the density matrices. If the orbitals and CI coefficients are optimized by a coupled Newton–Raphson or augmented Hessian procedure, each CI update should be about three times as expensive as a simple direct CI step. In practice, even a factor of 5 has been reported[70]. (iii) Since the density matrices are recalculated exactly, the effect of the change of the CI vector on the orbitals is taken into account more accurately than in the Newton–Raphson method. This considerably improves convergence, particularly in the region far from the solution.

In the (step-restricted) AH method as proposed by Lengsfield[39,44,47] and

TABLE II
Convergence of CASSCF calculations for formaldehyde[a].

Iter.	Energy difference: Without coupling[b]	With coupling[c]	With coupling[d]
1	−0.095301120	−0.106133147	−0.105835733
2	−0.009714002	−0.001264146	−0.001561529
3	−0.001696277	−0.000000014	−0.000000044
4	−0.000446948		
5	−0.000141131		
6	−0.000052973		
7	−0.000022911		
8	−0.000010824		
9	−0.000005361		

[a]Active space: $3a_1$–$7a_1$, $1b_2$–$3b_2$, $1b_1$–$2b_1$, 3644 configurations; basis set and other details, see Ref. 55.
[b]In each iteration, Eq. (54) is solved with fixed CI coefficients. The CI coefficients are optimized with $\mathbf{R} = \mathbf{0}$ at the beginning of each iteration.
[c]$E^{(2)}(\mathbf{T}, \mathbf{c})$ fully optimized with respect to $\mathbf{T}$ and $\mathbf{c}$.
[d]Same as footnote *c*, but only five updates of the CI coefficients in the first iteration.

used in various modifications by Shepard *et al.*[46], Golab *et al.*[52] and Jensen and Agren[70] the residual $\mathbf{Y}$ is calculated according to Eq. (60) for the special case $\mathbf{U} = \mathbf{1}$ only. (If the orbital CI coupling is included, further terms have to be added to $\mathbf{Y}$.) After having obtained the solution $\mathbf{Y} = \mathbf{0}$ with $\mathbf{U} = \mathbf{1}$ in these methods, the next macro-iteration is started, i.e. a four-index transformation to obtain new operators $\mathbf{J}^{kl}$ and $\mathbf{K}^{kl}$ is necessary. In our method this transformation is only performed after several applications of the AH method to the energy approximation $E^{(2)}(\mathbf{T})$. This greatly reduces the number of four-index transformations and the overall effort. Tables II and III demonstrate the fast convergence for some CASSCF calculations. It is seen that convergence is reached in only 2–3 macro-iterations. Similar fast convergence behaviour has been observed in many other applications.

The total number of micro-iterations needed for solving the non-linear equations (53) is often fairly large. It is, therefore, important to make them as efficient as possible. In order to minimize the I/O time, the operators $\mathbf{J}^{kl}$ and $\mathbf{K}^{kl}$ should be kept in high-speed memory whenever possible, since their recovery from disc may be more expensive than their use in the calculation of $\mathbf{Y}$. It is worth while to mention that we often found it advantageous to evaluate the operators $\mathbf{G}^{ij}$ as intermediate quantities. These operators only change if the CI coefficients are updated, and their calculation is particularly helpful if many micro-iterations are performed between CI updates. The matrix $\tilde{\mathbf{B}}$ is

TABLE III
Convergence of CASSCF calculations for various states of NO[a].

	Energy difference				
Iter.	$X^2\Pi$	$a^4\Pi$	$A^2\Sigma^+$	$B^2\Pi^b$	$b^4\Sigma^-$
1	−0.09099300	−0.06561940	−0.03737295	−0.06733878	−0.07296502
2	−0.00135266	−0.00930193	−0.00004712	−0.00282646	−0.00229034
3	−0.00000001	−0.00000786	−0.00000000	−0.00000012	−0.00000013

[a]Active space: 2σ–6σ, 1π–2π; 1σ, 2σ orbitals frozen. The starting orbitals were canonical SCF orbitals for the $^2\Sigma^+$ state in all cases. $R = 2.1$ bohr. Basis set:
Huzinaga[72] 11s, 7p (innermost 5s, 3p contracted), augmented on each atom by 2s, 1p, 2d functions with the following exponents:
N: s (0.051, 0.020); p (0.042); d (0.88, 0.22);
O: s (0.069, 0.027); p (0.053); d (1.2, 0.3);
Final energies are: −129.40697524, −129.12317041, −129.20572150, −129.08492834 and −129.14238817 hartree for the first to last columns, respectively.
[b]Second state of this symmetry optimized.

then obtained as

$$(\tilde{\mathbf{B}})_{ri} = (\tilde{\mathbf{A}})_{ri} + \sum_s U_{sr} \sum_j (\mathbf{G}^{ij}\mathbf{U}\Delta\mathbf{R})_{sj} \tag{73}$$

which requires about M^2N^2 operations. This is cheaper by a factor of $\frac{3}{2}M$ per micro-iteration than the application of Eq. (47). This saving often outweighs the additional cost for the calculation of the $\mathbf{G}^{ij}$ (Eq. (41), $\frac{3}{4}N^2M^4$ operations). The advantages are even greater if molecular symmetry can be employed, since only those blocks $(\mathbf{G}^{ij})_{rs}$ are needed in which (r, i) and (s, j) correspond to orbitals of the same symmetry. The evaluation of the operators $\mathbf{G}^{ij}$ is particularly efficient on vector computers, because it can be performed in terms of matrix multiplications with long vector lengths. In this case the elements G^{ij}_{rs}, P^{ij}_{kl} and Q^{ij}_{kl} (fixed ij) form vectors, and the operators J^{kl}_{rs} and K^{kl}_{rs} form supermatrices.

F. Optimization of Internal Orbital Rotations

The optimization method outlined in Sections II.C–E shows very fast and stable convergence behaviour when applied to CASSCF wavefunctions, in which all orbital rotations between occupied orbitals are redundant. However, experience has shown that convergence is often much slower when orbital rotations between strongly occupied valence orbitals have to be optimized. It has been shown that this is due to the fact that the energy approximation $E^{(2)}(\mathbf{T})$ is not invariant with respect to a unitary transformation between two doubly occupied orbitals[55]. If the 2×2 transformation in Eq. (34) is applied to a single Slater determinant with just two doubly occupied orbitals, the second-order energy $E^{(2)}(\mathbf{T})$ takes the form ($\alpha = R_{12}$, $T_{11} = T_{22} = \cos\alpha - 1$, $T_{12} = -T_{21} = \sin\alpha$)

$$E^{(2)}(\alpha) = e_1 + e_2(5 - 8\cos\alpha + 4\cos^2\alpha) \tag{74}$$

where

$$e_1 = 2(h_{11} + h_{22}) \tag{75}$$

$$e_2 = (11|11) + (22|22) + 4(11|22) - 2(12|12) \tag{76}$$

For small α, Eq. (74) can be approximated by

$$E^{(2)}(\alpha) = e_1 + e_2(1 + \tfrac{2}{3}\alpha^4 + \cdots) \tag{77}$$

Hence, $E^{(2)}(\mathbf{T})$ is not invariant with respect to α. In the Newton–Raphson approximation only terms up to second order in α are taken into account, and Eq. (77) shows that $E^{(2)}(\mathbf{R})$ has the correct invariance property. The implications of these findings are demonstrated in Fig. 4. This figure illustrates for a simple MCSCF wavefunction how the exact energy and the second-order energy $E^{(2)}(\mathbf{T})$ change as a function of the rotation angle α between two strongly occupied orbitals. Since one of these orbitals is correlated, the exact energy is not invariant with respect to α, but it is very flat. The energy approximation $E^{(2)}(\mathbf{T})$ shows the expected α^4 dependence. Therefore, the rotation angle predicted by minimization of $E^{(2)}(\mathbf{T})$ is much too small. In fact, too small step sizes for internal orbital rotations have been observed in many

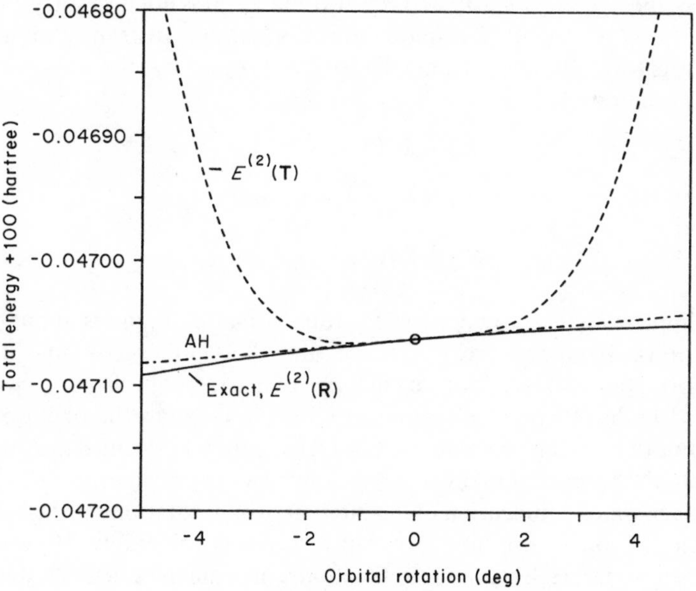

Fig. 4. The dependence of various energy approximations on the rotation 2σ–3σ for the HF molecule as in Fig. 1.

applications. The Newton–Raphson or undamped augmented Hessian methods predict steps which are of wrong direction or too large ($+61°$ and $-542°$, respectively; the optimum angle is $-53°$). Therefore, the use of the NR or AH method often does not remedy convergence difficulties for such rotations.

The fact that the rotations between occupied orbitals cause most difficulties suggests a special treatment of these rotations. This is possible at rather modest expense, since for any unitary matrix **U** which transforms the internal orbitals only among themselves, the transformation of the internal one- and two-electron integrals h_{ij} and $(ij|kl)$ is very cheap and can be performed in high-speed memory. Hence, we can start each macro-iteration with an optimization of the internal–internal orbital rotations and the CI coefficients only. As for the minimization of $E^{(2)}(\mathbf{T},\mathbf{c})$ an uncoupled step-restricted augmented Hessian method can be used for this optimization. Since the number of internal-internal orbital rotations is small, the augmented orbital Hessian can be constructed and diagonalized explicitly. A CI update is done after 1–3 orbital optimization steps, which converge quadratically. Usually a total of 2–3 CI updates is sufficient. Since in each step the one- and two-electron integrals are transformed exactly, the process yields a variational energy which is fully optimized with respect to the parameter subspace considered.

Owing to the fact that the internal orbitals change in this optimization process, the operators h_{rs}, J^{kl}_{rs} and K^{kl}_{rs} also change. It is not necessary, however, to perform a second four-index transformation. Instead, the modified operators are obtained from the original ones by the much cheaper transformations

$$h_{rs} \rightarrow (\mathbf{U}^\dagger \mathbf{h} \mathbf{U})_{rs} \tag{78}$$

$$J^{kl}_{rs} \rightarrow \sum_{ij} (\mathbf{U}^\dagger \mathbf{J}^{ij} \mathbf{U})_{rs} U_{ik} U_{jl} \tag{79}$$

$$K^{kl}_{rs} \rightarrow \sum_{ij} (\mathbf{U}^\dagger \mathbf{K}^{ij} \mathbf{U})_{rs} U_{ik} U_{jl} \tag{80}$$

The simplest method to perform this transformation requires about $\frac{3}{4}N^2M^4$ operations. Symmetry greatly reduces the effort. On vector machines the transformation can be performed in terms of matrix multiplications with long vector lengths (all $U_{ik}U_{jl}$ for a given k, l form a vector, the operators form supermatrices) and is therefore very fast. Using the transformed operators and orbitals, the optimization process is continued as described in Sections II.C–E.

The separate optimization of the internal orbital rotations at the beginning of each macro-iteration improves convergence considerably. However, this treatment so far neglects the coupling to the internal–external rotations. This coupling creates additional rotations R_{ij} between the internal orbitals when the non-linear equations (53) are solved. Convergence can be further improved

TABLE IV

Convergence of CASSCF calculations for the electronic ground state of CS_2[a].

	Energy difference		
Iter.	Without int. opt.[b]	With int. opt.[c]	With int. opt. and abs.[d]
1	−0.077490937	−0.085838896	−0.087484601
2	−0.005439995	−0.004644181	−0.003059391
3	−0.003146803	−0.000066851	−0.000006018
4	−0.002194213	−0.000000082	−0.000000000
5	−0.001326677		
6	−0.000661122		
7	−0.000240696		
8	−0.000047226		
9	−0.000002333		
10	−0.000000007		

[a]Geometry: $R = 3.1$ bohr, $\alpha = 105°$. Basis set:
C: Huzinaga[72] 8s, 4p, innermost 5s, 3p contracted; in addition one s (0.05), one p (0.05), and one d (0.5).
S: Huzinaga[72] 11s, 7p, innermost 6s, 4p contracted; in addition one s (0.05), one p (0.05), and one d (0.5).
Active space: $7a_1$–$10a_1$, $6b_2$–$8b_2$, $2b_1$–$3b_1$, $2a_2$ (3564 CSFs, 11 100 determinants). The $1a_1$–$5a_1$, $1b_2$–$4b_2$, $1b_1$ and $1a_2$ orbitals were canonical SCF orbitals and frozen; the $6a_1$ and $5b_2$ orbitals are doubly occupied and optimized.
[b]No extra optimization of internal–internal orbital rotations.
[c]Internal–internal orbital rotations fully optimized in the beginning of the second to fourth iteration.
[d]As in footnote *c*, but in addition one absorption step of elements R_{ij} in each iteration (see text).

TABLE V

Convergence of test calculations for the N_2 and CO molecules.

	Energy difference			
Iter.	N_2[a]	N_2[b]	CO[c]	CO[d]
1	−0.052436691	−0.053795845	−0.056168288	−0.114267390
2	−0.003387148	−0.002049568	−0.002605968	−0.006158512
3	−0.000023449	−0.000001875	−0.000001306	−0.000002589
4	−0.000000000	−0.000000000	−0.000000000	−0.000000000

[a]N_2 molecule; using IVOs as starting guess; basis set and configurations as in Ref. 51 and Table I. One absorption of elements R_{ij} per iteration.
[b]As in footnote *a*, but two absorptions of elements R_{ij} per iteration.
[c]CO molecule; using IVOs as starting guess; basis set and configurations as in Ref. 51, and Table I. One absorption of elements R_{ij} per iteration.
[d]As in footnote *c*, but canonical SCF orbitals used as starting guess.

by an iterative 'absorption' of these parameters R_{ij} into the orbital basis. This can be accomplished as follows. After solving the non-linear equations a unitary transformation matrix is formed from the elements $R_{ij} \simeq \frac{1}{2}(\mathbf{U} - \mathbf{U}^{\dagger})_{ij}$ (i, j internal only), and the transformations in Eqs (78)–(80) are repeated. The internal orbitals are transformed correspondingly. Then, some additional micro-iterations are necessary to solve the non-linear equations (53) and (58) with the new operators. In this case a good starting approximation is available by forming a unitary matrix from the previous R_{ai} and $R_{ij} = 0$. This process can be iterated until all R_{ij} remain zero when solving Eq. (53). In that case the internal orbital rotations have been treated to highest possible order with the operators $\mathbf{J}^{kl}$ and $\mathbf{K}^{kl}$ of the present macro-iteration. Since this is also true for the optimization of the CI coefficients, about the same convergence behaviour is expected for CASSCF (no internal–internal orbital rotations) and more general MCSCF calculations. In practice, we found it sufficient to absorb the internal–internal elements R_{ij} into the present orbital basis only once per macro-iteration. Furthermore, the initial optimization of the internal–internal orbital rotations is not necessary in the first macro-iteration.

Tables IV and V demonstrate the convergence improvements due to the internal optimization procedure. The first column of Table IV shows a calculation without the extra optimization of the internal orbital rotations. In the second column of Table IV the internal–internal rotations are optimized only at the beginning of the second and subsequent macro-iterations, but no absorptions of the elements R_{ij} have been performed. In the third column one absorption step has been done in each macro-iteration. The drastic reduction of the number of four-index integral transformations justifies the additional effort needed for the transformations in Eqs (78)–(80). The wavefunctions optimized in Table V are the same as those in Table I, and a comparison of these tables clearly demonstrates the convergence acceleration achieved with our method. The observed convergence behaviour is better than quadratic immediately from the first iteration even if bad starting orbitals are used. In the calculation for CO with SCF orbitals as a starting guess (last column in Table V), the initial orbital Hessian matrix had 21 negative diagonal elements, and probably a greater number of negative eigenvalues. In this case the initial 6σ and 2π orbitals were of Rydberg rather than of antibonding character. For N_2 it is shown in Table V that more than one absorption step in each iteration somewhat improves convergence. However, because it is usually not possible to save an iteration in this way, it is not recommended to perform more than one absorption step.

G. Treatment of Closed Shells

In many applications several orbitals are doubly occupied in all configurations of the MCSCF wavefunction. For such orbitals it is possible to simplify the algorithm, since the sub-blocks of the density matrices which involve closed-shell orbitals are of very simple structure. As shown below, they can be eliminated completely from the formalism.

The simplest way to deal with energetically low-lying closed-shell core orbitals is to take them directly from a preceding SCF calculation without further optimization. In this case one has to eliminate all rows and columns corresponding to core orbitals from the matrices $\mathbf{P}^{kl}, \mathbf{Q}^{kl}, \mathbf{R}, \mathbf{A}, \mathbf{B}$, etc., and replace the one-electron Hamiltonian $\mathbf{h}$ by a core Fock operator $\mathbf{F}^{c}$. This operator is calculated in the AO basis* according to

$$\tilde{\mathbf{F}}^{c} = \tilde{\mathbf{h}} + \tilde{\mathbf{G}}(\tilde{\mathbf{D}}^{c}) \tag{81}$$

where

$$(\tilde{\mathbf{D}}^{c})_{\mu\nu} = 2 \sum_{i(\text{closed})} X_{\mu i} X_{\nu i} \tag{82}$$

is the core first-order density matrix in the AO basis, and $\tilde{\mathbf{G}}(\tilde{\mathbf{D}})$ is defined as

$$\tilde{\mathbf{G}}(\tilde{\mathbf{D}})_{\mu\nu} = \sum_{\rho\sigma} \tilde{D}_{\rho\sigma}[(\mu\nu|\rho\sigma) - \tfrac{1}{2}(\mu\rho|\sigma\nu)] \tag{83}$$

The operator $\tilde{\mathbf{F}}^{c}$ has to be evaluated only once. It is transformed into the current MO basis at the beginning of each macro-iteration, i.e.

$$\mathbf{F}^{c} = \mathbf{X}^{\dagger}\tilde{\mathbf{F}}^{c}\mathbf{X} \tag{84}$$

Whenever the operator $\mathbf{G}(\mathbf{D})$ is used in the following, it will be assumed that it has been transformed into the MO basis similarly.

The freezing of core orbitals is usually a very good approximation if they are energetically well separated from the valence orbitals. However, the full optimization of all orbitals is sometimes desirable. This is the case, for instance, when the MCSCF calculation is followed by the evaluation of energy gradients with respect to the nuclear coordinates. It is, therefore, useful to consider explicitly the simplifications which are possible for closed-shell orbitals.

Using the anticommutation relations in Eqs (10)–(12) it is straightforward to derive the following expressions for the case that $|i\rangle$ is a closed-shell orbital:

$$D_{ij} = 2\delta_{ij}$$

$$P^{kl}_{ij} = 2\delta_{ij}D_{kl} - \tfrac{1}{2}(\delta_{il}D_{jk} + \delta_{ik}D_{jl}) \tag{85}$$

$$Q^{kl}_{ij} = 2\delta_{ik}D_{jl} - \tfrac{1}{2}(\delta_{il}D_{jk} + \delta_{ij}D_{kl}) \tag{86}$$

Using these relations, we obtain ($|i\rangle$ closed shell, $|j\rangle$ closed or open shell):

$$A_{ri} = 2(\mathbf{G}^{c})_{ri} \tag{87}$$

$$\mathbf{G}^{ij} = 2\delta_{ij}\mathbf{G}^{c} + \sum_{k} D_{jk}\mathbf{L}^{ik} \tag{88}$$

*In this section all quantities in the AO basis are marked with a tilde.

where

$$\mathbf{G}^c = \mathbf{F}^c + \sum_{kl(\text{open})} D_{kl}(\mathbf{J}^{kl} - \tfrac{1}{2}\mathbf{K}^{kl}) \tag{89}$$

and

$$\mathbf{L}^{ik} = 4\mathbf{K}^{ik} - \mathbf{K}^{ki} - \mathbf{J}^{ik} \tag{90}$$

The summations in Eq. (89) run over open-shell orbitals only. For the case that $|i\rangle$ and $|j\rangle$ are open-shell orbitals, the same quantities are given by

$$A_{ri} = \sum_{j(\text{open})} \left(F^c_{rj} D_{ji} + \sum_{kl(\text{open})} J^{kl}_{rj} P^{lk}_{ji} \right) \tag{91}$$

$$\mathbf{G}^{ij} = \mathbf{F}^c D_{ij} + \sum_{kl(\text{open})} (\mathbf{J}^{kl} P^{ij}_{kl} + 2\mathbf{K}^{kl} Q^{ij}_{kl}) \tag{92}$$

These formulae differ from Eqs (39) and (41) only in the use of $\mathbf{F}^c$ instead of $\mathbf{h}$ and the restrictions of the summations. Hence, all second-order density matrix elements involving closed-shell orbitals have been eliminated. However, all operators $\mathbf{J}^{kl}$ and $\mathbf{K}^{kl}$ are still needed in Eq. (88). Since the computational effort for their evaluation depends strongly on the number of optimized orbitals, it would also be useful to eliminate the operators $\mathbf{J}^{kl}$ and $\mathbf{K}^{kl}$ involving any closed-shell orbitals. As shown in the following, this is possible in a direct MCSCF procedure.

In each micro-iteration, we have to evaluate

$$B_{ri} = A_{ri} + \sum_j (\mathbf{G}^{ij}\,\mathbf{T})_{rj} \tag{93}$$

or the similar quantity $\tilde{\mathbf{B}}$. For the case that $|i\rangle$ is a closed-shell orbital, we obtain

$$B_{ri} = 2(\mathbf{G}^c\mathbf{U})_{ri} + \sum_j (\mathbf{L}^{ij}\mathbf{T}\mathbf{D})_{rj} \tag{94}$$

Defining the first-order change of the density matrix $\mathbf{D}$ as

$$\Delta\mathbf{D} = \mathbf{T}\mathbf{D} + \mathbf{D}\mathbf{T}^\dagger \tag{95}$$

Eq. (94) can be rewritten as

$$B_{ri} = 2(\mathbf{G}^c\mathbf{U} + \mathbf{G}(\Delta\mathbf{D}))_{ri} \tag{96}$$

The operator $\mathbf{G}(\Delta\mathbf{D})$ describes the first-order change of $\mathbf{G}^c$. It can be obtained directly from the two-electron integrals in the AO basis (cf. Eqs (83) and (84)). This requires transforming $\Delta\mathbf{D}$ into the AO basis:

$$\Delta\tilde{\mathbf{D}} = \mathbf{X}\Delta\mathbf{D}\mathbf{X}^\dagger \tag{97}$$

From Eq. (96) the relation of our method to a closed-shell Hartree–Fock procedure is apparent. For the case that there are only closed-shell orbitals, the variational conditions for $E^{(2)}(\mathbf{T})$ (Eq. (53)) take the form

$$(\mathbf{U}^\dagger\mathbf{B})_{ai} = 2(\mathbf{U}^\dagger\mathbf{F}^c\mathbf{U} + \mathbf{U}^\dagger\mathbf{G}(\Delta\mathbf{D}))_{ai} = 0 \qquad \text{for all } a, i \tag{98}$$

In the usual first-order SCF procedure the second term accounting for the change of the Fock operator is neglected.

Next we have to consider the columns of **B** which correspond to open-shell orbitals ($|i\rangle$ open shell). These columns are given by

$$B_{ri} = \sum_{j(\text{open})} (\mathbf{F}^{c}\mathbf{U})_{rj} D_{ji} + \sum_{jkl(\text{open})} [(\mathbf{J}^{kl}\mathbf{U})_{rj} P^{lk}_{ji} + 2(\mathbf{K}^{kl}\mathbf{T})_{rj} Q^{lk}_{ji}]$$
$$+ \sum_{j(\text{closed})} \sum_{k(\text{open})} D_{ik}(\mathbf{L}^{kj}\mathbf{T})_{rj} \quad (99)$$

The first three terms on the right-hand side of Eq. (99) differ from Eq. (40) only by the restrictions on the summations and the replacement of **h** by $\mathbf{F}^{c}$. The last term accounts for the change of the Fock operator $\mathbf{F}^{c}$ caused by a variation of the closed-shell orbitals. It can be brought into the form

$$\sum_{j(\text{closed})} \sum_{k(\text{open})} D_{ik}(\mathbf{L}^{kj}\mathbf{T})_{rj} = (\mathbf{G}(\Delta\mathbf{D}^{c})\mathbf{D})_{ri} \quad (100)$$

where

$$(\Delta\mathbf{D}^{c})_{rs} = (\mathbf{T}\mathbf{D}^{c} + \mathbf{D}^{c}\mathbf{T}^{\dagger})_{rs}$$
$$= 2 \sum_{j(\text{closed})} (T_{rj}\delta_{js} + T_{sj}\delta_{rj}) \quad (101)$$

is the first-order change of the closed-shell density matrix in the MO basis.

It follows from the above that the price one must pay for the elimination of the operators $\mathbf{J}^{kl}$ and $\mathbf{K}^{kl}$ involving closed-shell orbitals is the calculation of the two operators $\mathbf{G}(\Delta\mathbf{D})$ and $\mathbf{G}(\Delta\mathbf{D}^{c})$ from the two-electron integrals in the AO basis in each micro-iteration. If a large number of micro-iterations are necessary to solve the non-linear equations (53), this might be rather expensive. However, it is expected that energetically low-lying 'core orbitals' do not depend much on the changes of the valence orbitals and converge rapidly. It should, therefore, be possible to freeze all parameters R_{ri} which involve core orbitals in intermediate micro-iterations, and evaluate the operators $\mathbf{G}(\Delta\mathbf{D})$ and $\mathbf{G}(\Delta\mathbf{D}^{c})$ only a few times. Such an approximation is not appropriate, however, for closed-shell valence orbitals. For these orbitals the operators $\mathbf{L}^{ij}$ should be calculated explicitly.

In order to update the core orbitals the denominators D_{ri} in Eq. (63) are needed. Neglecting terms arising from the change of the operator $\mathbf{G}^{c}$, these can be approximated by

$$D_{ri} = 2(G^{c}_{rr} - G^{c}_{ii}) \quad (102)$$

Again, this approximation is not appropriate if $|i\rangle$ is a valence orbital.

H. A Direct Configuration-interaction Method for Complete Active Space Calculations

In the wavefunction optimization procedure outlined in the previous sections the coefficients $\{c_I\}$ can be optimized by any available CI procedure.

For short CI expansions one could simply construct and diagonalize the Hamiltonian matrices $\mathbf{H}^{(2)}$. For longer CI expansions a direct CI procedure must be employed. Here the desired eigenvectors are obtained iteratively without explicitly calculating and storing the Hamilton matrix. The basic operation in a direct CI iteration is the evaluation of the residual vectors*

$$\mathbf{y}^n = (\mathbf{H} - E^n)\mathbf{c}^n \tag{103}$$

for the required electronic states n. In terms of the molecular integrals and the coupling coefficients $\mathbf{y}$ can be written as

$$y_I = \sum_{ij} h_{ij} \sum_J \langle \Phi_I | E_{ij} | \Phi_J \rangle c_J + \tfrac{1}{2} \sum_{ijkl} (ij|kl) \sum_J \langle \Phi_I | E_{ij,kl} | \Phi_J \rangle c_J - Ec_I \tag{104}$$

The coupling coefficients $\langle \Phi_I | E_{ij} | \Phi_J \rangle$ and $\langle \Phi_I | E_{ij,kl} | \Phi_J \rangle$ in principle need only be constructed once if they are stored on a formula tape. In our program this step can be performed by a graphical unitary group approach (GUGA)[27]. However, for large CASSCF calculations, the formula tape becomes exceedingly long. This limits the length of the configuration expansion which can be handled by this method to about 10^4 configurations. For longer CASSCF expansions the only remedy is to recalculate the coupling coefficients each time they are needed. Clearly, this requires a particularly efficient algorithm.

Recently, Siegbahn[73] proposed the use of the factorization[74–76]

$$\langle \Phi_I | E_{ij,kl} | \Phi_J \rangle = \sum_K \langle \Phi_I | E_{ij} | \Phi_K \rangle \langle \Phi_K | E_{kl} | \Phi_J \rangle - \delta_{jk} \langle \Phi_I | E_{il} | \Phi_J \rangle \tag{105}$$

where the summation runs over the full spin-eigenfunction basis. This factorization follows from the anticommutation relations in Eqs (10)–(12). The formula tape then need only contain the one-particle matrix elements $\langle \Phi_I | E_{ij} | \Phi_K \rangle$, ordered after the intermediate state label K. Even though this greatly reduces the number of stored coupling coefficients, the length of the formula tape can still be rather large. Hence, sorting and processing of the coupling coefficients may take considerable time.

A further development was made by Knowles and Handy[62]. They proposed to use Slater determinants instead of spin eigenfunctions as a basis $\{\Phi_I\}$. In this case the coupling coefficients $\langle \Phi_I | E_{ij} | \Phi_J \rangle$ take only the values ± 1 or 0, and can rapidly be recalculated each time they are required. With a suitable 'canonical' addressing scheme for the determinants and CI coefficients, the construction and use of the coupling coefficients can be vectorized. This makes it possible to use modern vector processors very efficiently.

The calculation of the residual vector is done in the following steps:

$$A_{kl}^K = \sum_J \gamma_{kl}^{KJ} c_J \tag{106}$$

*Here and in the following we omit the superscripts indicating a second-order approximation.

$$B_{ij}^{K} = \sum_{kl} (ij|kl) A_{kl}^{K} \tag{107}$$

$$y_I = \sum_{K} \sum_{ij} (A_{ij}^{K} c_K \tilde{h}_{ij} + \tfrac{1}{2} \gamma_{ij}^{IK} B_{ij}^{K}) - E c_I \tag{108}$$

where

$$\tilde{h}_{ij} = h_{ij} - \tfrac{1}{2} \sum_{k} (ik|kj) \tag{109}$$

are modified one-electron integrals. This modification is necessary to account for the last term in Eq. (105). The density matrices are obtained as

$$D_{kl} = \sum_{K} A_{kl}^{K} c_K \tag{110}$$

$$\Gamma_{ij,kl} = \sum_{K} A_{ij}^{K} A_{kl}^{K} - \delta_{jk} D_{il} \tag{111}$$

In order to vectorize these steps efficiently, many matrices $\mathbf{A}^K$ and $\mathbf{B}^K$ must fit simultaneously in high-speed memory. For further computational details we refer the reader to the original paper of Knowles and Handy[62].

A disadvantage of using Slater determinants rather than spin eigenfunctions is an increased memory requirement. The length of the vectors $\mathbf{c}$ and $\mathbf{y}$, which must reside simultaneously in high-speed memory, is typically increased by a factor of 2. This factor is largest for singlet wavefunctions and becomes smaller with increasing spin multiplicity. Furthermore, one has to ensure that the wavefunction has the required spin symmetry. This can be achieved by using in the update formula

$$\Delta c_I = -y_i/(\bar{H}_{II} - E) \tag{112}$$

average values $\bar{H}_{II}$ for all Slater determinants which differ only in the spin function. Since the Hamilton operator used is spin-free, the improved CI vector $\mathbf{c} + \Delta\mathbf{c}$ will then represent a spin eigenfunction, provided the trial function is a spin eigenfunction.

The simple update formula in Eq. (112) is based on first-order perturbation theory. Therefore, it can only be applied if one Slater determinant strongly dominates the total wavefunction. In this case the coefficient of the leading determinant has to be kept constant in order to eliminate the redundancy arising from the normalization constraint. In cases in which several determinants are (nearly) degenerate or excited states are optimized, a modified procedure has to be applied. The method we use is based on the partitioning of the configuration space into a small primary (P) and a large secondary (Q) set. The P space contains all configurations whose energies H_{II} lie below a certain threshold. The total wavefunction can then be written as

$$\Psi^n = \sum_{P} c_P^n \Phi_P + \sum_{m} \alpha_m^n \Psi_Q^m \tag{113}$$

where

$$\Psi_Q^m = \sum_{Q} c_Q^m \Phi_Q \tag{114}$$

are contracted Q-space functions for each state m. These functions are automatically orthogonal to all P-space configurations and can be orthonormalized among themselves. The optimum P-space coefficients $\{\tilde{c}_P^n\}$ and the $\{\alpha_m^n\}$ are found by explicitly constructing and diagonalizing the Hamilton matrix in the basis of the P-space configurations and the contracted functions Ψ_Q^m (for more details see the appendix in Ref. 55). With the coefficients $\{\alpha_m^n\}$, modified Q-space vectors

$$\tilde{c}_Q^n = \sum_m \alpha_m^n c_Q^m \tag{115}$$

are evaluated. The resulting CI vectors $\tilde{\mathbf{c}}^n$ and the corresponding energy eigenvalues E^n are then used to calculate the residual vector $\mathbf{y}$. Owing to the variational determination of the P-space coefficients the elements y_P vanish. This automatically eliminates the redundancies which are due to the orthonormality constraints $\langle \Psi^n | \Psi^m \rangle = \delta_{nm}$, provided the P space contains at least as many configurations as there are states. The Q-space coefficients are updated using Eq. (112). Finally, the resulting CI vectors are re-orthonormalized. If more than one CI iteration is done with a particular set of integrals, convergence of the procedure can be improved by including the Q-space vectors of all previous iterations into the second summation of Eq. (113) (Davidson procedure[71]).

If Slater determinants are used as a basis, some further remarks concerning the above procedure are appropriate. In order to ensure that a proper spin eigenfunction is obtained, the P space should consist of full sets of determinants. Each set comprises all possible Slater determinants which can be obtained by assigning a given number of α and β spins to the orbitals. For each set of determinants we construct a complete set of spin eigenfunctions with the desired spin eigenvalues. The P-space Hamiltonian is transformed to this reduced basis before it is diagonalized. The resulting eigenvectors are transformed back correspondingly. If desired, spatial symmetry restrictions can be imposed on the wavefunction in the same way.

The Slater determinant-based direct CI procedure of Knowles and Handy has been implemented into our MCSCF procedure[56]. This enabled us to optimize fully much longer CASSCF wavefunctions than before. Some examples are shown in Tables VI and VII. Table VI compares the performance of our method with a large-scale augmented Hessian calculation published recently by Jensen and Agren[70]. It is seen that our method converged in only two iterations to 10^{-9} a.u. in the energy, while the AH calculation needed 11 steps. Even more important, however, is the reduction of the number of CI updates. Table VI shows that only 14–17 updates (depending on the desired accuracy) are needed. Each of these updates required only 4.9 s CRAY-1S CPU time for calculating the residual vector and 4.8 s for evaluating the new density matrices. This demonstrates that on modern computers calculations of

TABLE VI
Comparison of CASSCF calculations for CH_2[a].

Iter.	Energy difference		
	AH[b]	New method	New Method
1	−0.0051658625	−0.0313651501 (5)	−0.0313651501 (5)
2	−0.0034718425	−0.0001464216 (10)	−0.0001464107 (7)
3	−0.0016936113	−0.0000000005 (2)	−0.0000000058 (2)
4	−0.0004422620		
5	−0.0007721421		
6	−0.0013761648		
7	−0.0006363246		
8	−0.0001715291		
9	−0.0000286400		
10	−0.0000014239		
11	−0.0000000225		
12	−0.0000000001		

[a]Active space: $2a_1$–$9a_1$, $1b_2$–$4b_2$, $1b_1$–$2b_1$, $1a_2$, 24 156 configurations, 35 612 Slater determinants; for other details see Ref. 70.
[b]Augmented Hessian method with step-length control; results from Ref. 70. Start with orbitals of smaller MCSCF. Final energy −39.0278826738 hartree.
[c]Minimization of $E^{(2)}(\mathbf{T}, \mathbf{c})$ as described in text. Numbers in parentheses are numbers of CI updates performed in each iteration. Canonical SCF orbitals were used as starting guess. $E^{(2)}$ has been converged to within 10^{-10} a.u. in second and third iterations. Final energy −39.0278827366 hartree. Results from Ref. 56.
[d]As in footnote *c*, but E^2 converged only to 10^{-8} a.u. in second and third iterations. Final energy −39.0278827310 hartree. Results from Ref. 56.

TABLE VII
Convergence of CASSCF calculations for the $^5\Delta$ state of FeO[a].

Iter.	Energy difference		
	260 CSFs[b]	49 140 CSFs[c]	178 910 CSFs[d]
1	−0.0649079463 (5)	−0.1210861923 (5)	−0.0156927882 (3)
2	−0.0032670987 (15)	−0.0017511885 (15)	−0.0001465216 (14)
3	−0.0000005016 (6)	−0.0000001055 (8)	−0.0000000969 (6)

[a]For details, see Ref. 56; number of CI updates per iteration in parentheses.
[b]Active space: 8σ–10σ, 3π–4π, 1δ; start with canonical SCF orbitals.
[c]Active space: 8σ–11σ, 3π–5π, 1δ; start with canonical SCF orbitals.
[d]Active space: 8σ–11σ, 3π–5π, 2δ; start with orbitals from 49 140 CSF calculation.

this size can now be performed routinely. Table VII shows even larger test calculations for the FeO molecule. The longest CASSCF wavefunction optimized to date comprised 230 045 Slater determinants (178 910 CSFs). In this case the calculation of the residual vector took 36 s and the evaluation of the density matrices 17 s on a CRAY-1S.

I. Optimization of Energy Averages of Several States

In MCSCF calculations of excited electronic states in a given symmetry, one frequently encounters convergence problems due to the so-called 'root flipping problem'[40,43,77-79]. This is caused by the fact that during the optimization of the excited state the orbitals may become worse for the ground state. This can lead to a situation in which the energy eigenvalues of two states suddenly exchange their order. Then, the state under consideration no longer corresponds to the desired root of the CI matrix, but to a saddle point in orbital space[43,53]. Such solutions are generally considered to be undesirable, since the upper-bound property of the energy is lost. This problem can be avoided if an energy average of the states is optimized[43,47,55-57,77]. One then obtains a single set of molecular orbitals which is a compromise for all states.

The state-averaging procedure is very easily implemented into the MCSCF procedure described in the previous sections. One has only to replace the density matrices **D** and **Γ** by their state-averaged analogues, i.e.

$$D_{ij} = \sum_{n} W_n \sum_{IJ} c_I^n c_J^n \gamma_{ij}^{IJ} \tag{116}$$

$$\Gamma_{ij,kl} = \sum_{n} W_n \sum_{IJ} c_I^n c_J^n \Gamma_{ij,kl}^{IJ} \tag{117}$$

where W_n are arbitrary weight factors for the states n. If the states in question have different symmetries, their CI vectors can be calculated independently. Otherwise, a multistate direct CI treatment as outlined in Section II.H is necessary. Tables VIII and IX demonstrate that in state-averaged calculations convergence is as fast as in single-state optimizations. It is noted that the implementation of a state-averaging procedure into a Newton–Raphson or

TABLE VIII

Convergence of state-averaged and excited-state CASSCF calculations for the lowest two $^2A''$ states of the vinoxy radical[a].

	Energy difference	
Iter.	State averaged[b]	Excited state[c]
1	−0.0131063204 (5)	−0.0024425889 (5)
2	−0.0005477965 (12)	−0.0000677259 (11)
3	−0.0000000822 (6)	−0.0000000101 (5)

[a] 51 128 CSFs; for all details, see Ref. 56. Number of CI updates per iteration in parentheses.
[b] Start with optimized orbitals of the ground state. Final energy −152.31338897 hartree.
[c] Start with orbitals from state-averaged calculation. Final energy −152.31567256 hartree.

TABLE IX
Convergence of state-averaged CASSCF calculations for the X and A states of NH_3[a].

Iter.	Energy diff.[b]	Step length[c]
1	−0.12761811	1.9
2	−0.02034759	0.19
3	−0.00000004	0.0008

[a]Near-equilibrium geometry of A state: D_{3h} symmetry, $R = 2.0$ bohr. Basis set:
N: Huzinaga[72] 11s, 7p (first five s and three p contracted) augmented by two s (0.066, 0.033), one p (0.04) and two d (0.8, 0.2).
H: Huzinaga[72] 6s (first three contracted) augmented by two s (0.03, 0.011) and one p (0.2).
Start with canonical SCF orbitals of electronic ground state.
Active space (in C_{2v} symmetry): $2a_1$–$6a_1$, $1b_1$–$2b_1$, $1b_2$–$2b_2$, $1a_1$ orbital frozen.
[b]Change of average energy in successive iterations. Final energies: $E_1 = -56.30463570$ a.u., $E_2 = -56.11737479$ a.u.
[c]$(\sum_{r>i} R_{ri}^2)^{\frac{1}{2}}$.

augmented Hessian optimization method is much more complicated (see, e.g., Ref. 47). Therefore, state-averaged calculations of the size shown in Tables VIII and IX were previously not possible.

A typical example for the application of the state-averaging procedure is a calculation of the potential energy and transition moment surfaces for the $\tilde{X}$ and $\tilde{A}$ electronic states of NH_3 [80]. These states have a conical intersection which is shown in Fig. 5. The optimization of the excited-state wavefunctions in the vicinity of the crossing point would not have been possible in a single-state treatment.

III. THE INTERNALLY CONTRACTED MULTICONFIGURATION REFERENCE SELF-CONSISTENT ELECTRON-PAIR METHOD

A. Introduction

One of the most powerful tools presently available for accurate electronic structure calculations is the multiconfiguration reference CI(SD) method. In MR-CI(SD) wavefunctions, all configurations that are singly or doubly excited relative to any of the reference configurations are taken into account, and their coefficients are determined variationally. The reference wavefunctions are usually optimized by the MCSCF method. They should properly describe the dissociation of bonds and near-degeneracy effects. If the reference wavefunction includes the most important double excitations from the

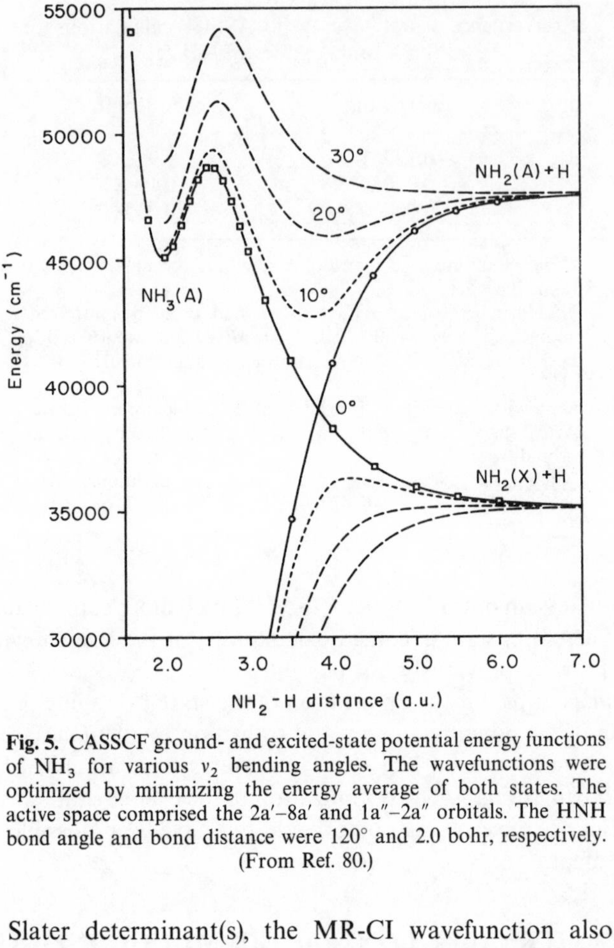

Fig. 5. CASSCF ground- and excited-state potential energy functions of NH_3 for various ν_2 bending angles. The wavefunctions were optimized by minimizing the energy average of both states. The active space comprised the 2a′–8a′ and 1a″–2a″ orbitals. The HNH bond angle and bond distance were 120° and 2.0 bohr, respectively. (From Ref. 80.)

dominant Slater determinant(s), the MR-CI wavefunction also contains important triple and quadruple excitations. It has been found in many applications that such configurations are essential to obtain accurate molecular properties.

The main drawback of the MR-CI method is the fact that the number of configurations and coupling coefficients increases rapidly with the number of reference configurations. It is, therefore, usually necessary to select a relatively small number of reference configurations. The choice of suitable reference configurations can be a difficult and time-consuming task. This is due to the fact that the importance of particular configurations may vary strongly with the internuclear geometry. Particular problems arise if there is an interaction of several states, e.g. near avoided crossings. It is then essential to describe these states in a balanced way. If the reference function is biased towards one of the states, this will also be the case, though to a lesser extent, for the MR-CI

TABLE X
Comparison of the results for the $^5\Delta$ state of FeO at $R = 3.0538$ bohr[a].

Method	CSFs	Energy	Dipole moment[b] (a.u.)	Correlation[c] (%)
SCF	1	−1337.137524	−3.524	0
CASSCF	260	−1337.220934	−1.286	20
CASSCF	49140	−1337.300164	−1.826	40
CASSCF	178910	−1337.323532	−2.005	45
CI(SD)	45366	−1337.533194	−2.978	96
MCSCF-CI[d]	207212	−1337.548794	−1.322	100

[a]For details, see Ref. 56.
[b]Negative sign corresponds to polarity Fe^+O^-.
[c]Relative to MR-CI calculation.
[d]Four reference configurations; see Ref. 56.

wavefunction. For instance, if the interacting states differ strongly in their polarity, this can lead to large errors in computed properties such as dipole moments. An extreme example[56] is shown in Table X for the $^5\Delta$ state of FeO. In this case the Hartree–Fock wavefunction is strongly polar (Fe^+O^-). A full valence CASSCF calculation (260 CSFs) showed, however, that there is a considerable $p \rightarrow d$ back-donation, which reduces the dipole moment. This effect is not properly accounted for in a CI(SD) calculation based solely on the restricted Hartree–Fock (RHF) reference wavefunction. If the six most important MCSCF configurations are selected as reference states, the MR-CI calculation yields a similar dipole moment as the CASSCF(260) calculation. Since in this calculation ionic and neutral structures are treated in a balanced way, it is likely that the MR-CI dipole moment is accurate. Also shown in Table X are some larger CASSCF calculations. It is seen that even the largest calculation with nearly 180 000 CSFs only yielded about 45% of the correlation energy obtained in the MR-CI calculation. Furthermore, the dipole moment is much too large. This is caused by the fact that the three additional orbitals included in the second CASSCF calculation become in the optimization process essentially oxygen 3p orbitals and correlate O^- but not Fe. Therefore, this CASSCF wavefunction is biased towards the ionic structure. Such effects are frequently encountered in MCSCF or CASSCF calculations, in particular if one includes more than the valence orbitals into the active space. A balanced treatment of dynamic correlation effects usually requires a larger number of orbitals than can be optimized with the MCSCF method. Therefore, these effects should be accounted for in a subsequent MR-CI calculation.

The computational effort in MR-CI calculations can be reduced by contracting certain classes of configurations with fixed coefficients. So far, two different contraction schemes have been applied. The first one is called 'external contraction' and was proposed by Siegbahn[81,82]. In this case all configurations which differ only in their external part, but not in their internal part, are contracted. The contraction coefficients are obtained perturbationally. This reduces the number of variational parameters by 2–3 orders of magnitude. The effort in an externally contracted MR-CI calculation is somewhat larger than that in one direct CI iteration for the uncontracted wavefunction. It still depends strongly on the number of reference configurations. Typically, 2–4% of the correlation energy is lost by the contraction. Errors of one-electron properties can be much larger, however.

The second possible contraction scheme was first proposed by Meyer[83], and discussed in the context of the direct CI method by Siegbahn[84]. In this case all configurations which have the same external but different internal parts are contracted, and the scheme is therefore called the 'internal contraction'. The internally contracted configurations are generated by applying pair excitation operators to the complete MCSCF reference function. Therefore, the number of contracted configurations and variational parameters is independent of the number of reference configurations. It only depends on the number of correlated internal orbitals and the size of the basis set.*

The price one must pay for the internal contraction is a complicated structure of the contracted configurations. Therefore, the evaluation of the coupling coefficients is more difficult than for uncontracted wavefunctions, and techniques like GUGA are difficult to apply. The structure of the coupling coefficients has been discussed for the special case of configurations with two electrons in external orbitals by Meyer[83] and Siegbahn[84]. Both authors did not treat the more difficult case of semi-internal configurations. A general formulation was first given by Werner and Reinsch[17,18,63], who also implemented the method and tested its accuracy. Following the work of Meyer[83] they showed that all coupling coefficients can be written as transition density matrices between a set of internal 'core functions', which are obtained by applying spin-coupled pair annihilation operators to the reference function. The structure of the residual vector is very similar as in Meyer's closed-shell SCEP theory[7]. Many applications of this 'internally contracted MCSCF-SCEP' method have proven that the loss of accuracy as compared to uncontracted MCSCF-CI wavefunctions is negligible. The correlation energy is typically reduced by only 0.2–0.3% (cf. Table XI), and also the values of properties such as dipole moments or electronic transition moments are hardly affected (cf. Table XII).

In principle, the advantage of applying the internal contraction scheme

*In fact, a small dependence can arise from the fact that some of the generated configurations may be linearly dependent and must be eliminated (cf. Section III. C).

TABLE XI
Comparison of correlation energies[a] for internally contracted and uncontracted MCSCF-SCEP wavefunctions (from Ref. 63).

Molecule	Number of ref. conf.	Correlation energy: Contracted	Correlation energy: Uncontracted	Difference (%)
$OH(X^2\Pi)^b$	3	−0.2168	−0.2171	0.14
	7	−0.2214	−0.2218	0.18
$CH_2(^3B_1) + H_2{}^c$	5	−0.1688	−0.1691	0.18
$O_3(^1A_1)^b$	2	−0.6192	−0.6208	0.26
$H_2O(^1A_1)^b$	11	−0.2684	−0.2691	0.23

[a] $E_{corr} = E_{tot} - E_{SCF}$.
[b] Near-equilibrium geometry.
[c] At saddle point of the reaction $CH_2 + H_2 \rightarrow CH_3 + H$; see Ref. 84.

TABLE XII
Comparison of dipole and transition moments of $OH(X^2\Pi–A^2\Sigma^+)$ for internally contracted and uncontracted MCSCF-SCEP wavefunctions[a].

	$R = 1.8$ bohr			$R = 2.5$ bohr		
	μ_1 (a.u.)	μ_2 (a.u.)	μ_{12} (a.u.)	μ_1 (a.u.)	μ_2 (a.u.)	μ_{12} (a.u.)
Contracted	0.6505	0.6540	0.1405	0.6598	0.9941	0.02991
Uncontracted	0.6495	0.6518	0.1412	0.6585	0.9900	0.02981
Difference (%)	0.15	0.33	0.50	0.28	0.41	0.33

[a] Eight and 10 reference configurations for the X and A states, respectively. The orbitals used for both states have been obtained by minimizing the energy average of the two reference wavefunctions.

should increase with the number of reference configurations, and reach a maximum for CASSCF reference wavefunctions. So far, however, the evaluation of the coupling coefficients has been a bottleneck in internally contracted MCSCF-SCEP calculations. This has restricted the length of the reference wavefunction to 20–30 configurations. In view of the difficulties of selecting these configurations it seems desirable to develop an improved method which allows one to handle 10^3–10^4 reference configurations. In many applications for small molecules this would make it possible to employ CASSCF reference wavefunctions, and to replace the selection of the reference configurations by a choice of active internal orbitals. For instance, if the active space consists of all atomic valence orbitals, proper dissociation of the electronic ground state is automatically ensured. In most cases this leads to a balanced description of the whole potential energy surface. It should also be noted CASSCF-SCEP wavefunctions are invariant with respect to any unitary transformation among the internal orbitals, provided the CI coefficients are fully re-optimized for the transformed orbitals. Thus, in the CASSCF-SCEP method uncertainties arising from the choice of the molecular orbitals and the reference configurations are minimized.

It is clear that CASSCF wavefunctions often contain many unimportant configurations. One could argue that this makes a CASSCF-SCEP calculation unnecessarily expensive. However, it is our opinion that the advantage of having a well defined model and avoiding the effort and the uncertainties of configuration selection might outweight this disadvantage. Considering such a method, one should also take into account that very fast vector computers are now becoming widely available. As will be shown in the next sections, the main effect of increasing the number of reference configurations in an internally contracted MCSCF-SCEP calculation is to increase the length of vector loops. Hence, the method allows one to use vector machines very efficiently.

The number of variational parameters in internally contracted MCSCF-SCEP wavefunctions rarely exceeds 10^5 even if large basis sets and complex reference wavefunctions are employed. In contrast to the number of variational parameters, the number of coupling coefficients depends on the number of reference configurations, and can become very large if CASSCF references are used. The main problem is, therefore, the calculation and storage of the coupling coefficients. It would be very helpful if at least part of them could be recalculated each time they are needed. As will be discussed in Section III.I, this would also allow one to relax the contraction coefficients in each direct CI iteration, thereby improving the quality of the wavefunction.

In the following sections we present a reformulation of the internally contracted MCSCF-SCEP method. The key for evaluating the coupling coefficients more efficiently is to deal with non-orthogonal configurations. Then, the coupling coefficients are simple linear combinations of higher-order density matrix elements, which can be evaluated by similar methods as discussed in Section II.H for the case of CASSCF wavefunctions. The transformation to an orthogonalized basis is applied to the residual vector at the end of each iteration.

B. The Internally Contracted Configuration Space

According to simple first-order perturbation theory, the configurations considered in an MR-CI treatment should span the 'first-order interacting space'[85,86] relative to the reference wavefunction Ψ_0. This space comprises all configurations $\{\Psi_{ext}\}$ which have a non-vanishing matrix element $\langle\Psi_{ext}|H|\Psi_0\rangle$. According to the Slater–Condon rules it is obvious that this space is made up of configurations which have at most two electrons in external orbitals. However, if Ψ_0 contains open-shell Slater determinants, often a large fraction of all possible doubly external configurations is non-interacting with Ψ_0. This is due to the fact that in open-shell cases many spin couplings are possible which do not interact with Ψ_0 either because they are obtained from Ψ_0 by exciting more than two spin orbitals, or because the contributions of various determinants cancel. The non-interacting configurations usually give only very small contributions to the energy and to other properties, and can therefore safely be neglected. The definition of the interacting configurations given below closely follows the work of Meyer[83].

In order to construct the first-order interacting configuration space we consider the Hamiltonian in second quantization

$$H = \sum_{rs} h_{rs} E_{rs} + \tfrac{1}{2} \sum_{rstu} (rs|tu) E_{rs,tu} \tag{118}$$

For the definitions of the excitation operators E_{rs} and $E_{rs,tu}$ we refer to Section II.A. The Hamiltonian can be rewritten in the form

$$H = \sum_{rs} h_{rs} E_{rs} + \tfrac{1}{2} \sum_{r \geqslant s} \sum_{t \geqslant u} \sum_{p} [(rt|su) + p(ru|ts)] \cdot (1+\delta_{rs})^{-1}(1+\delta_{tu})^{-1}(E_{rt,su} + pE_{ru,ts}) \tag{119}$$

where the parity p can take the values $+1$ and -1. We first consider the configuration subspace with two electrons in external orbitals. The matrix element between an arbitrary doubly external configuration Ψ_{ext} and Ψ_0 now takes the form

$$\langle \Psi_{\text{ext}} | H | \Psi_0 \rangle = \sum_{i \geqslant j} \sum_{a \geqslant b} \sum_{p} [(ai|bj) + p(aj|bi)] \cdot (1+\delta_{ij})^{-1}(1+\delta_{ab})^{-1} \langle \Psi_{\text{ext}} | \Psi^{ab}_{ijp} \rangle \tag{120}$$

where we have used the definition

$$\Psi^{ab}_{ijp} = \tfrac{1}{2}(E_{ai,bj} + pE_{aj,bi}) |\Psi_0\rangle \tag{121}$$

The configurations Ψ^{ab}_{ijp} are automatically spin eigenfunctions provided this is the case for the reference function Ψ_0. This is due to the fact that the operators $E_{ai,bj}$ commute with the spin operators S^2 and S_z. For single determinant reference functions it is easy to show that all Ψ^{ab}_{ijp} with $i \geqslant j$, $a \geqslant b$, $p = \pm 1$ are orthogonal or vanish (cf. Section III.G). From Eq. (120) it is obvious that they exactly span the first-order interacting space. According to the definition in Eq. (121) the two external electrons are coupled either to a singlet ($p = +1$) or to a triplet ($p = -1$) pair. The definition of these configurations is equivalent to that used by Meyer[83] and in our previous work[18] except that the normalization is different. For more general multiconfiguration reference functions the Ψ^{ab}_{ijp} may be non-orthogonal and linearly dependent. Then any linearly independent subspace of all Ψ_{ijp}^{ab} spans the first-order interacting space.

The doubly excited configurations in Eq. (121) are obtained by applying two-electron excitation operators to the complete reference function Ψ_0. If Ψ_0 is a multiconfiguration wavefunction we can write

$$\Psi_0 = \sum_I c_I \Psi_I \tag{122}$$

$$\begin{aligned}\Psi_{ijp}^{ab} &= \sum_I c_I (E_{ai,bj} + pE_{aj,bi})\Psi_I \\ &= \sum_I c_I \Psi_{ijp,I}^{ab}\end{aligned} \tag{123}$$

This shows that the Ψ_{ijp}^{ab} are linear combinations of a larger configuration set $\{\Psi_{ijp,I}^{ab}\}$ which is obtained by applying the excitation operators to the individual reference configurations. A linearly independent subset of these configurations spans the first-order interacting space with respect to all individual reference configurations. These configurations can be used as a basis in 'uncontracted' MR-CI calculations.* As mentioned in Section III.A, the number of uncontracted configurations depends rather strongly on the number of reference configurations (though not linearly, since some of the $\Psi_{ijp,I}^{ab}$ may be linearly dependent or vanish). On the other hand, the number of 'internally contracted' configurations Ψ_{ijp}^{ab} is essentially independent of the length of Ψ_0. If M is the number of correlated internal orbitals, there are at most $M(M+1)/2$ singlet pairs $\{ij, +1\}$ and $M(M-1)/2$ triplet pairs $\{ij, -1\}$. It is clear from Eq. (123) that the internal structure of the contracted configurations is much more complicated than in the uncontracted case.

The contraction coefficients are usually obtained by performing a small CI calculation with just the reference configurations and then kept fixed. As will be discussed in Section III.I it is possible, however, to relax them during the direct CI procedure. This is an important option if the weight of the individual reference configurations changes strongly in the correlated wavefunction. This can happen, for instance, in the neighbourhood of avoided crossings. It is even possible to optimize the contraction coefficients variationally. This might be important for the calculation of energy gradients with respect to the nuclear coordinates.

In the internal configuration subspace all reference configurations are explicitly included in order to allow a full relaxation of their coefficients. The remaining internal configurations can in principle be generated as in Eq. (121) by applying the operators $E_{ij,kl}$ to the reference wavefunction. However, in this case this procedure is less useful than for the doubly external configurations. This is most easily seen for the case that a CASSCF reference function is used.

*Treatments based on particular spin coupling schemes, such as GUGA, often include also the non-interacting configurations or part of them.

The excitation operators $E_{ij,kl}$ then only create redundant linear combinations of the reference configurations. In cases with extended but not complete reference spaces, the operators $E_{ij,kl}$ generate either a complete or nearly complete space. The contracted configurations Ψ_{ij}^{kl} are of a very complicated structure and not orthogonal. It is then much easier to use a slightly larger complete set of orthonormal spin-adapted internal configurations. An exception is the important case that the reference function contains many closed-shell orbitals. This will be discussed in Section III.G. Here we assume that the internal configuration space is complete and orthonormal.

Internally contracted singly excited and semi-internal doubly excited configurations can be defined as

$$\Psi_i^a = E_{ai}\Psi_0 \tag{124}$$

$$\Psi_{ij}^{ak} = E_{ai}E_{kj}\Psi_0 \tag{125}$$

We note that the spin coupling in Eq. (125) is different from that in our previous work[18]. As for the internal configurations, it may be more advantageous not to contract the singly external configurations. A full set of uncontracted configurations can be obtained by applying the excitation operators $E_{ai}E_{kj}$ to all individual reference configurations. If the labels k and j correspond to orbitals of the same symmetry, the operator E_{kj} only creates a particular configuration (or a linear combination of some configurations) of the complete internal space. Therefore, these operators can be omitted. If k and j have different symmetries, internal N-electron states which do not have the symmetry of the reference wavefunction are generated. Hence, the full set of uncontracted singly excited configurations is given by

$$\Psi_{iK}^a = E_{ai}\Phi_K \tag{126}$$

where the index K runs over all possible N-electron functions with the desired spin eigenvalues and all required spatial symmetries.

C. Orthogonalization and Normalization of the Configurations

As already mentioned, the configuration sets $\{\Psi_{ijp}^{ab}\}$, $\{\Psi_{ij}^{ak}\}$ and $\{\Psi_{iK}^a\}$ may not be orthogonal and are generally not normalized. In order to derive the overlap matrices it is convenient to define the simpler configurations

$$\Psi_{ij}^{ab} = E_{ai,bj}\Psi_0 \qquad (i \geqslant j, \text{ all } a, b) \tag{127}$$

such that

$$\Psi_{ijp}^{ab} = \tfrac{1}{2}(\Psi_{ij}^{ab} + p\Psi_{ji}^{ab}) \tag{128}$$

Then we get

$$\langle\Psi_{ijp}^{ab}|\Psi_{klq}^{cd}\rangle = \tfrac{1}{4}(\langle\Psi_{ij}^{ab}|\Psi_{kl}^{cd}\rangle + p\langle\Psi_{ji}^{ab}|\Psi_{kl}^{cd}\rangle + q\langle\Psi_{ij}^{ab}|\Psi_{lk}^{cd}\rangle + pq\langle\Psi_{ji}^{ab}|\Psi_{lk}^{cd}\rangle) \tag{129}$$

and we have to consider explicitly only one of the four contributions:

$$\langle \Psi_{ij}^{ab} | \Psi_{kl}^{cd} \rangle = \langle 0 | E_{ai,bj}^{+} E_{ck,dl} | 0 \rangle$$
$$= \delta_{ac}\delta_{bd} \langle 0 | E_{ik,jl} | 0 \rangle + \delta_{ad}\delta_{bc} \langle 0 | E_{il,jk} | 0 \rangle \tag{130}$$

This relation is easily obtained by matching the external indices. For convenience, we use the short-hand notation $|0\rangle = |\Psi_0\rangle$. The elements of the overlap matrix for the doubly external configurations are given by

$$\langle \Psi_{ijp}^{ab} | \Psi_{klq}^{cd} \rangle = \tfrac{1}{2}\delta_{pq}(\delta_{ac}\delta_{bd} + p\delta_{ad}\delta_{bc}) \langle 0 | E_{ik,jl} + pE_{il,jk} | 0 \rangle \tag{131}$$

Hence, the overlap matrix is completely defined by the elements of the second-order reduced density matrix $\Gamma_{ik,jl} = \langle 0 | E_{ik,jl} | 0 \rangle$. The configurations are automatically orthonormal if they differ in any external orbital or in the spin coupling of the external orbitals. It should be noted that the configurations Ψ_{ij}^{ab} could also be used as a basis. In fact, as shown by Pulay *et al.*[8], this leads to certain savings in the direct CI procedure for the special case of closed-shell single determinant reference functions. In more general cases these savings are not possible and it is better to use the configurations Ψ_{ijp}^{ab} in which the external orbitals are spin-coupled. In the orthogonalization procedure this has the advantage that the overlap matrix blocks into submatrices for singlet and triplet configurations, while the effort for other parts of the calculation remains unchanged.

Defining the internal overlap matrices ($p = \pm 1$)

$$S_{ij,kl}^{(p)} = \Gamma_{ik,jl} + p\Gamma_{il,jk} \tag{132}$$

we can orthonormalize the configurations by the transformation

$$\Psi_P^{ab} = \sum_{i \geq j} U_{ij,P}^{(p)} \Psi_{ijp}^{ab} \tag{133}$$

Here, P denotes a generalized internal $(N-2)$-electron hole state and includes the spin variable. The transformation matrices $\mathbf{U}^{(p)}$ are defined independently for $p = +1$ and $p = -1$ by the conditions

$$\mathbf{U}^{(p)\dagger}\mathbf{S}^{(p)}\mathbf{U}^{(p)} = \mathbf{1} \tag{134}$$

Now we have

$$\langle \Psi_P^{ab} | \Psi_Q^{cd} \rangle = \tfrac{1}{2}\delta_{PQ}(\delta_{ac}\delta_{bd} + p\delta_{ad}\delta_{bc}) \tag{135}$$

which implies the normalization

$$\langle \Psi_P^{ab} | \Psi_P^{ab} \rangle = (2 - \delta_{ab})^{-1} \tag{136}$$

The different normalization of the diagonal configurations Ψ_P^{aa} and the off-diagonal configurations Ψ_P^{ab} is essential for the removal of the external coupling coefficients and the matrix formulation of the theory. It should be noted that linear dependences in the basis Ψ_{ijp}^{ab} must be eliminated. Redundant configurations are easily found by diagonalizing the $\mathbf{S}^{(p)}$ matrices. For each zero or very small eigenvalue (e.g. $< 10^{-3}$) one pair (ijp) has to be deleted. The redundant pairs (ijp) are selected according to the magnitude of their coefficients in the corresponding eigenvectors.

The singly external configurations in Eqs (123) and (125) can be orthonormalized in a similar way. Defining the N-electron functions

$$|\Psi_j^m\rangle = E_{mj}|0\rangle \tag{137}$$

we obtain for the elements of the overlap matrix

$$\langle \Psi_{ik}^{am}|\Psi_{jl}^{bn}\rangle = \delta_{ab}\langle \Psi_k^m|E_{ij}|\Psi_l^n\rangle \tag{138}$$

$$\langle \Psi_{ik}^{am}|\Psi_j^b\rangle = \delta_{ab}\langle \Psi_k^m|E_{ij}|0\rangle \tag{139}$$

$$\langle \Psi_i^a|\Psi_j^b\rangle = \delta_{ab}\langle 0|E_{ij}|0\rangle \tag{140}$$

Similarly, for the uncontracted functions $\{\Psi_{iK}^a\}$ we have

$$\langle \Psi_{iK}^a|\Psi_{jL}^b\rangle = \delta_{ab}\langle K|E_{ij}|L\rangle \tag{141}$$

where the short-hand notation $|L\rangle = |\Phi_L\rangle$ has been used. In the latter case the internal overlap matrix

$$S_{iK,jL} = \langle K|E_{ij}|L\rangle = \gamma_{ij}^{KL} \tag{142}$$

strongly blocks, and orthogonalization presents no problem. On the other hand, the overlap matrix elements of the contracted functions in Eqs (138)–(140) are linear combinations of density matrix elements, e.g.

$$\langle \Psi_k^m|E_{ij}|\Psi_l^n\rangle = \Gamma_{km,ij,nl} + \delta_{im}\Gamma_{kj,nl} + \delta_{jn}\Gamma_{km,il} + \delta_{mn}\Gamma_{ij,kl} + \delta_{im}\delta_{jn}\gamma_{kl} \tag{143}$$

Here, we have introduced the third-order density matrix of the reference function

$$\Gamma_{ij,kl,mn} = \langle 0|E_{ij,kl,mn}|0\rangle \tag{144}$$

where

$$E_{ij,kl,mn} = \eta_m^{\alpha\dagger}E_{ij,kl}\eta_n^\alpha + \eta_m^{\beta\dagger}E_{ij,kl}\eta_n^\beta \tag{145}$$

Fourth-order and higher-order excitation operators and density matrices can be defined analogously. For CASSCF reference functions the internal overlap matrix $\langle \Psi_k^m|E_{ij}|\Psi_l^n\rangle$ can be rather large and in general blocks only according to spatial symmetries. Therefore, orthogonalization is considerably more difficult than for the uncontracted functions $\{\Psi_{iK}^a\}$. It appears that a variant in which only the doubly external configurations are internally contracted has most computational advantages.

In analogy to the Ψ_P^{ab} and Ψ_Q^{cd}, where the internal $(N-2)$-electron states have been denoted by the labels P and Q, the orthogonalized $(N-1)$-electron states will be denoted by the labels S and T:

$$\Psi_S^a = \sum_{iK} U_{iK,S} \Psi_{iK}^a \tag{146}$$

Internal configurations will always be labelled I, J.

D. Coefficient Matrices and Coefficient Vectors

After having defined our configuration space we can write the MCSCF-SCEP wavefunction in the form

$$\Psi = \sum_I c_I \Psi_I + \sum_S \sum_a c_a^S \Psi_S^a + \sum_P \sum_{ab} C_{ab}^P \Psi_P^{ab} \tag{147}$$

where $\mathbf{c}^S$ and $\mathbf{C}^P$ are coefficient vectors and matrices, respectively. To be consistent with the notation in Section II, we have labelled different matrices by superscripts and their elements by subscripts. Note that the last summation in Eq. (147) runs over all a, b, even though the distinct configurations are defined only for $a \geqslant b$. This implies the symmetry relations

$$C_{ab}^P = pC_{ba}^P \qquad \text{or} \qquad \mathbf{C}^P = p\mathbf{C}^{P\dagger} \tag{148}$$

since

$$\Psi_P^{ab} = p\Psi_P^{ba} \qquad (p = \pm 1) \tag{149}$$

We can also represent the wavefunction in the non-orthogonal configuration basis $\{\Psi_{iK}^a, \Psi_{ijp}^{ab}\}$. In this case it reads

$$\Psi = \sum_I c_I \Psi_I + \sum_{iK} \sum_a c_a^{iK} \Psi_{iK}^a + \sum_{i \geqslant j} \sum_{ab} \sum_p C_{ab}^{ijp} \Psi_{ijp}^{ab} \tag{150}$$

It is assumed here and in the following that only non-redundant configurations Ψ_{iK}^a and Ψ_{ijp}^{ab} are included in the summations. Inserting Eqs (133) and (146) into Eq. (147) we find the relations between the coefficients corresponding to orthogonal and non-orthogonal configurations:

$$\mathbf{C}^{ijp} = \sum_P U_{ij,P}^{(p)} \mathbf{C}^P \tag{151}$$

$$\mathbf{c}^{iK} = \sum_S U_{iK,S} \mathbf{c}^S \tag{152}$$

These relations can be used to evaluate the residual vector in the non-orthogonal configuration basis. For convenience in later expressions we define the contracted functions

$$\Psi_{\text{int}} = \sum_I c_I \Psi_I \tag{153}$$

$$\Psi_{iK} = \sum_a c_a^{iK} \Psi_{iK}^a \tag{154}$$

$$\Psi_{ijp} = \sum_{ab} C_{ab}^{ijp} \Psi_{ijp}^{ab} \tag{155}$$

The Ψ_{ijp} are called 'pair functions' and describe the correlation of an electron pair with spin coupling p in the internal orbitals i, j.

E. The Hamilton Matrix Elements

In a direct CI procedure we have to evaluate in each iteration the residuals

$$(\mathbf{T}^P)_{ab} = \langle \Psi_P^{ab} | H - E | \Psi \rangle \qquad \text{for all } P, a \geqslant b \tag{156}$$

$$(\mathbf{t}^S)_a = \langle \Psi_S^a | H - E | \Psi \rangle \qquad \text{for all } S, a \tag{157}$$

$$t_I = \langle \Psi_I | H - E | \Psi \rangle \qquad \text{for all } I \tag{158}$$

The desired solution is reached if the matrices $\mathbf{T}^P$, the vectors $\mathbf{t}^S$ and all elements t_I vanish. The quantities $\mathbf{T}^P$ and $\mathbf{t}^S$ are most easily calculated first in the non-orthogonal configuration basis, and then transformed to the orthogonal basis:

$$\mathbf{T}^P = \sum_{i \geqslant j} U_{ij,P}^{(p)} \mathbf{T}^{ijp} \tag{159}$$

$$\mathbf{t}^S = \sum_{iK} U_{iK,S} \mathbf{t}^{iK} \tag{160}$$

The transformation to the orthogonal basis is necessary in order to update the CI vector. We first consider the matrices $\mathbf{T}^{ijp}$:

$$T_{ab}^{ijp} = \langle \Psi_{ijp}^{ab} | H | \Psi_{\text{int}} \rangle + \sum_{iK} \langle \Psi_{ijp}^{ab} | H | \Psi_{kK} \rangle$$
$$+ \sum_{k \geqslant l} \sum_{q} \langle \Psi_{ijp}^{ab} | H - E | \Psi_{klq} \rangle \tag{161}$$

The derivation of the individual matrix elements is exemplified for the one-electron contribution in the last term. This can be written in the form

$$\langle \Psi_{ijp}^{ab} | h | \Psi_{klq}^{cd} \rangle = \tfrac{1}{4}(1 + p\tau_{ij})(1 + q\tau_{kl}) \sum_{rs} h_{rs} \langle 0 | E_{ai,bj}^{+} E_{rs} E_{ck,dl} | 0 \rangle \tag{162}$$

where τ_{ij} permutes the indices i and j. Since the reference function $|0\rangle$ only contains internal orbitals, the matrix element on the right-hand side of Eq. (162) is non-zero only if the external indices (a, b) match with the indices (c, d) or with the integral labels r, s. Using the anticommutation rules in Eqs (6)–(8) one obtains:

$$\sum_{rs} h_{rs} \langle 0 | E_{jb,ia} E_{rs} E_{ck,dl} | 0 \rangle$$
$$= (h_{ac}\delta_{bd} + h_{bd}\delta_{ac}) \langle 0 | E_{ik,jl} | 0 \rangle$$
$$+ (h_{bc}\delta_{ad} + h_{ad}\delta_{bc}) \langle 0 | E_{il,jk} | 0 \rangle$$
$$+ \sum_{mn} h_{mn} (\delta_{ac}\delta_{bd} \langle 0 | E_{ik,jl,mn} | 0 \rangle + \delta_{bc}\delta_{ad} \langle 0 | E_{il,jk,mn} | 0 \rangle) \tag{163}$$

It is straightforward to derive all other matrix elements in a similar manner.

The coupling coefficients obtained are defined as follows:

$$\sigma_{mn}(I, ijp) = \langle 0 | E_{im,jn} + pE_{jm,in} | I \rangle \tag{164}$$

$$\sigma_n(mK, ijp) = \langle 0 | E_{in,jm} + pE_{jn,im} | K \rangle \tag{165}$$

$$\alpha_{nkl}(mK, ijp) = \langle 0 | E_{in,jm,kl} + pE_{jn,im,kl} | K \rangle \tag{166}$$

$$\alpha_{mn}(ijp, klq) = \delta_{pq} \langle 0 | E_{ik,jl,mn} + pE_{jk,il,mn} | 0 \rangle \tag{167}$$

$$\beta_{mn}(ijp, klq) = \langle 0 | E_{im,nk,jl} + pE_{jm,nk,il} + qE_{im,nl,jk} + pqE_{jm,nl,ik} | 0 \rangle \tag{168}$$

$$\gamma(ijp, klq) = \delta_{pq} \left(\sum_{mn} h_{mn} \langle 0 | E_{ik,jl,mn} + pE_{jk,il,mn} | 0 \rangle + \tfrac{1}{2} \sum_{mn} \sum_{op} (mn|op) \langle 0 | E_{ik,jl,mn,op} + pE_{jk,il,mn,op} | 0 \rangle \right) \tag{169}$$

$$\alpha_{mn}(iK, jL) = \langle K | E_{ij,mn} | L \rangle \tag{170}$$

$$\beta_{mn}(iK, jL) = \langle K | E_{im,nj} | L \rangle \tag{171}$$

$$\gamma(iK, jL) = \sum_{mn} h_{mn} \langle K | E_{ij,mn} | L \rangle + \tfrac{1}{2} \sum_{mn} \sum_{op} (mn|op) \langle K | E_{ij,mn,op} | L \rangle \tag{172}$$

All coupling coefficients involve only internal indices. They are either density matrix elements evaluated with the reference wavefunction or transition density matrix elements involving internal N-electron states. The indices in parentheses denote the configurations involved. The coefficients $\boldsymbol{\sigma}$, $\boldsymbol{\alpha}$ and $\boldsymbol{\beta}$ are vectors with the elements given as subscripts. The indices mn or nkl are treated as single labels. In our previous work[18] we have written the same coupling coefficients as transition density matrix elements between sets of N-electron, $(N-1)$-electron and $(N-2)$-electron core functions. These core functions are in general not spin eigenfunctions and have a rather complicated structure. A brute-force method was used to represent them as linear combinations of determinants and to evaluate the coupling coefficients. The above formulation suggests that it might be possible to calculate the coupling coefficients much more efficiently directly with the spin-coupled internal N-electron states. As will be discussed in Section III.H, a similar method as proposed by Siegbahn for CASSCF wavefunctions (cf. Section II.H) can be employed.

We now obtain the following results after multiplying with the appropriate CI coefficients and summing over external indices for the individual interactions.

1. Single–internal:

$$\langle \Psi^a_{iK} | H | \Psi_I \rangle = \sum_j h_{aj} \langle K | E_{ij} | I \rangle + \sum_{jkl} (kl|ja) \langle K | E_{ij,kl} | I \rangle \tag{173}$$

$$\langle \Psi_I | H | \Psi_{iK} \rangle = \sum_j \sum_a h_{aj} c^{iK}_a \langle K | E_{ij} | I \rangle + \sum_{jkl} \sum_a (kl|ja) c^{iK}_a \langle K | E_{ij,kl} | I \rangle \tag{174}$$

2. Single–single:

$$\langle \Psi^{a}_{iK}|H|\Psi_{jL}\rangle = \sum_{b}\left(h_{ab}c^{jL}_{b}S_{iK,jL} + \sum_{mn}(ab|mn)c^{jL}_{b}\alpha_{mn}(iK,jL) + \sum_{mn}(am|nb)c^{jL}_{b}\beta_{mn}(iK,jL)\right) + c^{jL}_{a}\gamma(iK,jL) \tag{175}$$

3. Pair–internal:

$$\langle \Psi^{ab}_{ijp}|H|\Psi_{I}\rangle = \tfrac{1}{2}\sum_{mn}\alpha_{mn}(I,ijp)(am|nb) \tag{176}$$

$$\langle \Psi_{I}|H|\Psi_{ijp}\rangle = \tfrac{1}{2}\sum_{mn}\sigma_{mn}(I,ijp)\sum_{ab}(am|nb)C^{ijp}_{ab} \tag{177}$$

4. Pair–single:

$$\langle \Psi^{ab}_{ijp}|H|\Psi_{mK}\rangle = \tfrac{1}{2}\sum_{n}(h_{an}c^{mK}_{b} + ph_{bn}c^{mK}_{a})\sigma_{n}(mK,ijp) + \tfrac{1}{2}\sum_{kln}[(kl|na)c^{mK}_{b} + p(kl|nb)c^{mK}_{a}]\alpha_{nkl}(mK,ijp) + \tfrac{1}{2}\sum_{nc}[(an|bc) + p(bn|ac)]c^{mK}_{c}\sigma_{n}(mK,ijp) \tag{178}$$

$$\langle \Psi^{c}_{mK}|H|\Psi_{ijp}\rangle = \sum_{n}\sum_{a}h_{na}C^{ijp}_{ac}\sigma_{n}(mK,ijp) + \sum_{kln}\sum_{a}(kl|na)C^{ijp}_{ac}\alpha_{nkl}(mK,ijp) + \sum_{n}\sum_{ab}(na|bc)C^{ijp}_{ab}\sigma_{n}(mK,ijp) \tag{179}$$

5. Pair–pair:

$$\langle \Psi^{ab}_{ijp}|H|\Psi_{klq}\rangle = \delta_{pq}S^{(p)}_{ij,kl}\left(\sum_{c}(h_{ac}C^{klq}_{cb} + ph_{bc}C^{klq}_{ca}) + \sum_{cd}(ac|db)C^{klq}_{cd}\right) + \delta_{pq}\sum_{mn}\sum_{c}[(mn|ac)C^{klq}_{cb} + p(mn|bc)C^{klq}_{ca}]\alpha_{mn}(ijp,klq) + \tfrac{1}{2}\sum_{mn}\sum_{c}[(am|nc)C^{klq}_{cb} + p(bm|nc)C^{klq}_{ca}]\beta_{mn}(ijp,klq) + \delta_{pq}C^{klq}_{ab}\gamma(ijp,klq) \tag{180}$$

In the above formulae it has been assumed that the singly external configurations are not contracted. Similar formulae for the contracted configurations are obtained by replacing the bras and kets $\langle K|$ and $|L\rangle$ by functions like $|\Psi^{m}_{n}\rangle = E_{mn}|0\rangle$ (cf. Eq. (137)). The resulting coupling coefficients involve density matrices up to fifth order, and are more difficult to evaluate. Moreover, in contrast to the coefficients $\langle K|E_{ij,kl}|L\rangle$, the matrix elements

$\langle \Psi_n^m | E_{ij,kl} | \Psi_p^o \rangle$ are in general not sparse. It is, therefore, unlikely that the contraction of the singly external configurations saves much time, even though there are more non-redundant states Ψ_{iK} than Ψ_{ij}^k. This is in line with our experience using this method with a moderate number of reference configurations.

F. Matrix Formulation of the Residual Vector

In order to obtain the residual in a compact matrix form, we define the following integral matrices:

$$h_{ij}^{\text{int}} = h_{ij} \tag{181}$$

$$h_{ab}^{\text{ext}} = h_{ab} \tag{182}$$

$$I_{a,mkl} = (am|kl) \tag{183}$$

$$J_{ab,kl} = \tilde{J}_{a,bkl} = (ab|kl) \tag{184}$$

$$K_{ab,kl} = \tilde{K}_{a,bkl} = (ak|bl) \tag{185}$$

The matrices $\mathbf{h}^{\text{int}}$ and $\mathbf{h}^{\text{ext}}$ are the internal and external blocks of the one-electron Hamiltonian. The supermatrices **J** and **K** comprise all two-electron integrals with two external orbitals. The columns (kl) of these supermatrices represent the external blocks of the operators $\mathbf{J}^{kl}$ and $\mathbf{K}^{kl}$ defined in Section II.A, respectively. The matrix **I** comprises the two-electron integrals with one external orbital. The matrices $\tilde{\mathbf{J}}$ and $\tilde{\mathbf{K}}$ are identical to **J** and **K** except that the rows and columns have formally been redefined. Here the label *bkl* is a single column index. Of course, in a computer program only **I**, **J** and **K** need to be stored.

We further define a set of matrices which are contractions of coupling coefficients and CI coefficients:

$$V_{ma}^{ijp} = \sum_{nL} \sigma_{mn}(nL, ijp) c_a^{nL} \tag{186}$$

$$W_{mkl,a}^{ijp} = \sum_{nL} \alpha_{mkl}(nL, ijp) c_a^{nL} \tag{187}$$

$$X_{ma}^{I} = \sum_{nL} \langle I | E_{mn} | L \rangle c_a^{nL} \tag{188}$$

$$Y_{mkl,a}^{I} = \sum_{nL} \langle I | E_{mn,kl} | L \rangle c_a^{nL} \tag{189}$$

$$\mathbf{D}^{ijp} = \sum_{k \geqslant l} S_{ij,kl}^{(p)} \mathbf{C}^{klp} \tag{190}$$

$$\mathbf{E}^{ijp} = \mathbf{V}^{ijp} + p(\mathbf{V}^{ijp})^{\dagger} \tag{191}$$

Note that the matrices $\mathbf{D}^{ijp}$ have non-zero elements only in the external–external block, while the matrices $\mathbf{E}^{ijp}$ have non-zero elements only in the internal–external blocks. Similarly, we define the vectors

$$s_n^{mK} = \sum_I \langle K | E_{mn} | I \rangle c_I \tag{192}$$

$$t_{nkl}^{mK} = \sum_I \langle K | E_{mn,kl} | I \rangle c_I \tag{193}$$

$$x_a^{mK} = \sum_{nL} S_{mK,nL} c_a^{nL} \tag{194}$$

$$y_{akl}^{mK} = \sum_{nL} \alpha_{kl}(mK, nL) c_a^{nL} \tag{195}$$

$$z_{akl}^{mK} = \sum_{nL} \beta_{kl}(mK, nL) c_a^{nL} \tag{196}$$

Note that the external parts of the vectors **x** and **y** are particular rows of the matrices **X** and **Y** for the case that $\langle K|$ has the same symmetry as the internal configurations $\langle I|$.

The matrix elements between pair functions can be expressed conveniently by defining the operators

$$\mathbf{F}^{ijp,klq} = 2\delta_{pq} S_{ij,kl}^{(p)} \mathbf{h}^{\text{ext}} + 2\mathbf{J}\cdot\boldsymbol{\alpha}(ijp, klq) + \mathbf{K}\cdot\boldsymbol{\beta}(ijp, klq) + \mathbf{1}\gamma(ijp, klq) \tag{197}$$

which are contractions of integrals and coupling coefficients. These operators are evaluated as intermediate quantities but not stored. Finally, we define 'external exchange operators'

$$K(\mathbf{D}^{ijp})_{rc} = \sum_{ab} D_{ab}^{ijp}(ra|bc) \tag{198}$$

$$K(\mathbf{E}^{ijp})_{ab} = \sum_{rs} E_{rs}^{ijp}(ra|bs) \tag{199}$$

These operators are the only quantities which depend on integrals with three and four external indices. It is possible to calculate them directly from the integrals in the AO basis by the following steps:

$$\tilde{\mathbf{C}} = \mathbf{X}\mathbf{C}\mathbf{X}^\dagger \tag{200}$$

$$(\tilde{\mathbf{K}}(\tilde{\mathbf{C}}))_{\mu\nu} = \sum_{\rho\sigma} \tilde{C}_{\rho\sigma}(\mu\rho|\sigma\nu) \tag{201}$$

$$\mathbf{K}(\mathbf{C}) = \mathbf{X}^\dagger \tilde{\mathbf{K}}(\tilde{\mathbf{C}})\mathbf{X} \tag{202}$$

where **X** is the matrix of molecular-orbital coefficients (Eq. (2)). Hence, a full integral transformation can be avoided.

The residual now takes the form

$$t_I = g_I - Ec_I \tag{203}$$

$$\mathbf{t}^{mK} = \mathbf{g}^{mK} - E\mathbf{x}^{mK} \tag{204}$$

$$\mathbf{T}^{ijp} = \tfrac{1}{2}[\mathbf{G}^{ijp} + p(\mathbf{G}^{ijp})^\dagger] - E\mathbf{D}^{ijp} \tag{205}$$

The matrix elements $g_I = \langle \Psi_I | H | \Psi \rangle$ are given by

$$g_I = \sum_J \langle I|H|J\rangle c_J + \mathrm{tr}(\mathbf{h}\cdot\mathbf{X}^I) + \mathrm{tr}(\mathbf{I}\cdot\mathbf{Y}^I) + \sum_{i\geqslant j}\sum_p \mathrm{tr}\,[\mathbf{C}^{ijp\dagger}\cdot\mathbf{K}\cdot\boldsymbol{\sigma}(I, ijp)] \tag{206}$$

Here, the trace denotes summations over the appropriate blocks only. The matrix elements $\langle \Psi^a_{mK}|H|\Psi\rangle = g^{mK}_a$ are obtained in the form of vectors

$$\mathbf{g}^{mK} = \mathbf{h}^{\mathrm{int}}\cdot\mathbf{s}^{mK} + \mathbf{I}\cdot\mathbf{t}^{mK} + \mathbf{h}^{\mathrm{ext}}\cdot\mathbf{x}^{mK} + \tilde{\mathbf{J}}\cdot\mathbf{y}^{mK} + \tilde{\mathbf{K}}\cdot\mathbf{z}^{mK} + \sum_{nL}\gamma(mK, nL)\mathbf{c}^{nL} + \sum_{i\geqslant j}\sum_p p\{[\mathbf{C}^{ijp}\cdot\mathbf{h} + \mathbf{K}(\mathbf{C}^{ijp})]\cdot\boldsymbol{\sigma}(mK, ijp) + \mathbf{C}^{ijp}\cdot\mathbf{I}\cdot\boldsymbol{\alpha}(mK, ijp)\} \tag{207}$$

The matrix elements $\langle \Psi^{ab}_{ijp}|H|\Psi\rangle$ are represented by the matrices $\mathbf{G}^{ijp}$:

$$\mathbf{G}^{ijp} = \sum_I \mathbf{K}\cdot\boldsymbol{\sigma}(I, ijp)c_I + \mathbf{K}(\mathbf{D}^{ijp}) + \mathbf{K}(\mathbf{E}^{ijp}) + \sum_{k\geqslant l}\sum_q \mathbf{F}^{ijp,klq}\cdot\mathbf{C}^{klq} + \mathbf{h}\cdot\mathbf{V}^{ijp} + \mathbf{I}\cdot\mathbf{W}^{ijp} \tag{208}$$

Using these quantities, the energy expectation value can be written as

$$E = \frac{1}{N}\left(\sum_I g_I c_I + \sum_{mK}\mathbf{c}^{mK\dagger}\cdot\mathbf{g}^{mK} + \sum_{i\geqslant j}\sum_p \mathrm{tr}(\mathbf{C}^{ijp\dagger}\cdot\mathbf{G}^{ijp})\right) \tag{209}$$

where

$$N = \sum_I c_I^2 + \sum_{mK}\mathbf{c}^{mK\dagger}\cdot\mathbf{x}^{mK} + \sum_{i\geqslant j}\sum_p \mathrm{tr}(\mathbf{C}^{ijp\dagger}\cdot\mathbf{D}^{ijp}) \tag{210}$$

is the norm of the wavefunction.

In each CI iteration the residual has to be transformed into the orthogonal configuration basis according to Eqs (159) and (160). This requires linearly combining the vectors $\mathbf{g}^{mK}$ and the matrices $\mathbf{G}^{ijp}$. An update of the CI vector in the orthogonal basis can then be obtained as

$$\Delta c_I = -t_I/\langle \Psi_I|H - E|\Psi_I\rangle \tag{211}$$

$$\Delta c^S_a = -t^S_a/\langle \Psi^a_S|H - E|\Psi^a_S\rangle \tag{212}$$

$$\Delta C^P_{ab} = -T^P_{ab}/\langle \Psi^{ab}_P|H - E|\Psi^{ab}_P\rangle \tag{213}$$

In order to calculate the next residual vector the $\mathbf{c}^S$ and $\mathbf{C}^P$ are transformed back into the non-orthogonal basis according to Eqs (151) and (152). Alternatively, one could transform the coupling coefficients into the orthogonal basis, e.g.

$$\alpha_{nm}(P, Q) = \sum_{i\geqslant j}\sum_{k\geqslant l} U^{(p)}_{ij,P}U^{(p)}_{kl,Q}\alpha_{nm}(ijp, klp) \tag{214}$$

and calculate the matrices $\mathbf{G}^P$ and the vectors $\mathbf{g}^S$ directly in the orthogonal basis. Owing to the large number of coupling coefficients this might be difficult, however, in particular for the singly external configurations. Moreover, the transformed coupling coefficients may be less sparse than the original ones. This would make the calculation of the residual vector more expensive.

On the other hand, the diagonal matrix elements $\langle \Psi_P^{ab} | H | \Psi_P^{ab} \rangle$ and $\langle \Psi_S^a | H | \Psi_S^a \rangle$ are most easily calculated immediately in the orthogonal basis. This requires transforming only a small subset of the coupling coefficients, which presents no problems since the transformed quantities $\alpha_{mn}(P, P)$, $\beta_{mn}(P, P)$, $\gamma(P, P)$, etc., can be kept in high-speed memory. In order to simplify the calculation of the energy denominators in Eq. (213) certain approximations can be made which have been discussed in detail in Refs 7 and 18. Such approximations might influence the speed of convergence but not the converged results.

The above equations show that all relevant quantities are obtained by performing a sequence of matrix multiplications. One of the most crucial parts is the calculation of the terms $\mathbf{F}^{ijp,klq} \cdot \mathbf{C}^{klq}$. The operators $\mathbf{F}^{ijp,klq}$ are linear combinations of Coulomb and exchange operators. For simple reference functions only very few of these operators contribute to each $\mathbf{F}^{ijp,klq}$ since the third-order density matrix is then sparse. This sparsity is lost if CASSCF wavefunctions are used as reference. Then, up to $M(M+1)/2$ operators $\mathbf{J}^{mn}$ $(m \geqslant n)$ and M^2 operators $\mathbf{K}^{mn}$ may contribute to each interaction. On vector computers the operators $\mathbf{F}^{ijp,klq}$ can be calculated very efficiently using the supermatrix formulation in Eq. (197). Since the vector length is N_{ext}^2 it should be possible to perform this operation at maximum speed, e.g. 150 Mflop on a CRAY-1S. This requires, however, to keep the **J** and **K** supermatrices in high-speed memory. In order to minimize I/O time it is also advantageous to keep all $\mathbf{C}^{klq}$ in core. Clearly, if large basis sets are used, this can be a bottleneck. However, extremely large memories are presently becoming available, and it is likely that these problems will be eliminated in the near future.

It should be noted that in an uncontracted MR-CI calculation the third-order density matrices $\langle 0 | E_{ik,jl,mn} | 0 \rangle$ are replaced by the coefficients $\langle I | E_{ik,jl,mn} | J \rangle$. Then, instead of simply combining the Coulomb and exchange operators linearly, a large number of matrix multiplications have to be performed. Furthermore, the external exchange operators are needed for all coefficient matrices $\mathbf{C}^{ijp,I}$. These facts make uncontracted MR-CI calculations much more expensive than internally contracted calculations.

G. Treatment of Closed Shells

In many cases the reference wavefunction contains orbitals which are doubly occupied in all configurations. If these orbitals are energetically low-lying, they can also be kept doubly occupied in all configurations of the MR-CI wavefunctions without much loss of accuracy. However, if the closed-shell orbitals belong to the valence shell it is important to correlate them. So far, we have not considered these cases in our treatment. Therefore, even when only the valence electrons are correlated, the summations involving integral labels run over all internal orbitals, and all corresponding density matrix elements are needed. As discussed in Section II.G for the MCSCF case it is possible, however, to take advantage of the fact that the closed-shell density matrix blocks are sparse and of simple structure. In fact, as for the MCSCF case, one can eliminate these blocks entirely from the formalism.

In the following we divide the internal orbital space into three subspaces: the 'core' orbitals, which are doubly occupied and not correlated; the 'closed-shell' orbitals, which are correlated; and the 'active' orbitals, which are only partially filled in the MCSCF wavefunction. We will show that it is sufficient to evaluate coupling coefficients only for the active subspace. For the special case that the reference wavefunction is a single closed-shell determinant, the algorithm then reduces to the closed-shell SCEP method described by Meyer[7], and no coupling coefficients have to be calculated explicitly.*

First we consider the core orbitals, which are not correlated. By partitioning the summations in the Hamiltonian (Eq. (118)) it is easy to show that in this case one has simply to replace the one-electron Hamiltonian $\mathbf{h}$ by the core Fock matrix $\mathbf{F}^{c}$ (Eq. (84)) and to add the core energy

$$E_{c} = \sum_{i(\text{core})} (h_{ii} + F^{c}_{ii}) \tag{215}$$

to the total energy. Furthermore, all summations then exclude the core orbitals. Hence, the operators $\mathbf{J}^{kl}$ and $\mathbf{K}^{kl}$ are only needed for the correlated orbitals.

Next we consider excitations from the valence orbitals. Here we can distinguish the cases in which zero, one, or two electrons are excited from the closed-shell orbitals. It would be rather lengthy to discuss all possible matrix elements between the internal, singly external and doubly external configurations. Therefore, we restrict ourselves to some examples for the doubly external configurations. It is straightforward to derive all other matrix elements in a similar way.

In the following we denote closed-shell orbitals by the labels i, j, k, l and active orbitals by the labels m, n unless otherwise noted. It is obvious that configurations which have different numbers of electrons in the closed shells are orthogonal. The overlap matrix elements for configurations with one hole in the closed shells are given by

$$\langle \Psi^{ab}_{imp} | \Psi^{cd}_{knq} \rangle = \tfrac{1}{2}\delta_{pq}(\delta_{ac}\delta_{bd} + p\delta_{ad}\delta_{bc})(2-p)\delta_{ik}\langle 0|E_{mn}|0\rangle \tag{216}$$

Note that these configurations are orthogonal if the first-order density matrix

*Meyer has formulated the theory in the AO basis. Therefore, his treatment contains additional matrix multiplications involving the overlap matrix $\mathbf{S}$. It is possible to formulate the MR case similarly.

$\langle 0|E_{mn}|0\rangle$ is diagonal. For CASSCF reference functions it is therefore advantageous to transform the active orbitals to natural orbitals. For the case that both excitations are from closed-shell orbitals, Eq. (216) further reduces to

$$\langle \Psi^{ab}_{ijp}|\Psi^{cd}_{klq}\rangle = \tfrac{1}{2}\delta_{pq}(\delta_{ac}\delta_{bd} + p\delta_{ad}\delta_{bc})2(2-p)(\delta_{ik}\delta_{jl} + p\delta_{jk}\delta_{il}) \qquad (217)$$

Hence, all configurations Ψ^{ab}_{ijp} for $i \geqslant j$, $a \geqslant b$ are orthogonal, and the normalization is

$$\langle \Psi^{ab}_{ijp}|\Psi^{ab}_{ijp}\rangle = (1 + p\delta_{ij})(1 + \delta_{ab})(2-p) \qquad (218)$$

As examples for the simplifications in the Hamilton matrix elements we consider the contributions of the doubly external integrals to the operators $\mathbf{F}^{ijp,klq}$ (Eq. (197)). Depending on the number of excitations from the closed shells one can distinguish six distinct cases. Using the relations $\eta_i^{\alpha+}|0\rangle = \eta_i^{\beta+}|0\rangle = 0$ and the anticommutation rules in Eqs (6)–(8) one obtains the following results (the all-internal integral contributions are omitted).

1. i, j, k, l closed shell:

$$\mathbf{F}^{ijp,klq} = 2(2-p)(1 + p\tau_{ij})(1 + q\tau_{kl})\delta_{jl}[\delta_{pq}(\delta_{ik}\mathbf{G}^{c} - 2\mathbf{J}^{ik}) + (2-q)\mathbf{K}^{ik}] \quad (219)$$

All contributions of the active orbitals are absorbed in the operator $\mathbf{G}^{c}$, which is defined as in Eq. (89). For the case that there are only closed-shell orbitals $\mathbf{G}^{c}$ reduces to $\mathbf{F}^{c}$. The corresponding formula in Meyer's closed-shell SCEP theory (Eq. (16) in Ref. 7) can then immediately be obtained by multiplying with the normalization factor $[4(2-p)(2-q)(1+p\delta_{ij})(1+q\delta_{kl})]^{-1/2}$. For the other cases we obtain:

2. i, k closed shell, j, l active:

$$\begin{aligned}\mathbf{F}^{ijp,klq} = {} & 2(2-p)\delta_{pq}\delta_{ik}\left(\langle 0|E_{jl}|0\rangle\mathbf{F}^{c} + \sum_{mn}\mathbf{J}^{mn}\langle 0|E_{jl,mn}|0\rangle\right) \\ & - \delta_{ik}\sum_{mn}\mathbf{K}^{mn}[\langle 0|E_{jl,nm} + (p+q-2pq)\langle 0|E_{jm,nl}|0\rangle] \\ & + \langle 0|E_{jl}|0\rangle[2(2-p)\delta_{pq}\mathbf{J}^{ik} + (2-p)(2-q)\mathbf{K}^{ik}] \end{aligned} \qquad (220)$$

3. i, j, k closed shell, l active:

$$\begin{aligned}\mathbf{F}^{ijp,klq} = {} & 2(2-p)\delta_{pq}\sum_{n}\langle 0|E_{nl}|0\rangle(\delta_{ik}\mathbf{J}^{jn} + p\delta_{jk}\mathbf{J}^{in}) \\ & - pq(2-p)(2-q)\sum_{n}\langle 0|E_{nl}|0\rangle(\delta_{ik}\mathbf{K}^{jn} + p\delta_{jk}\mathbf{K}^{in}) \end{aligned} \qquad (221)$$

4. i, j closed shell, k, l active:

$$\mathbf{F}^{ijp,klq} = 0 \qquad (222)$$

5. i closed shell, j, k, l active:

$$\mathbf{F}^{ijp,klq} = \sum_{n}\langle 0|E_{jl,nk} + gE_{jk,nl}|0\rangle[(2-p)\mathbf{K}^{in} - 2\delta_{pq}\mathbf{J}^{in}] \qquad (223)$$

6. i, j, k, l active:

$$\mathbf{F}^{ijp,klq} = 2\delta_{pq} S^{(p)}_{ij,kl} \mathbf{F}^{c} + \sum_{mn} [2\mathbf{J}^{mn} \alpha_{mn}(ijp, klq) + \mathbf{K}^{mn} \beta_{mn}(ijp, klq)] \tag{224}$$

In all cases the summations run over active orbitals only. Hence, all coupling coefficients involving closed-shell orbitals have been eliminated. These equations show that the explicit treatment of closed-shell orbitals strongly reduces the computational effort.

Some additional remarks are necessary about the internal and singly external configurations. So far, we have assumed that the internal N-electron and $(N-1)$-electron function spaces are not contracted. One of the reasons for using the uncontracted functions was that they are much easier to orthogonalize than the contracted functions. The overlap matrices of the contracted $(N-1)$-electron and N-electron states depend on elements of the third- and fourth-order density matrices, respectively. For simple reference functions these are sparse, and the overlap matrices are strongly blocked. For more complicated reference functions this blocking is in general lost and the diagonalization of the overlap matrices may become impossible.

An obvious disadvantage of using the uncontracted functions is that their number strongly depends on the number of reference configurations and correlated orbitals. However, if there are closed-shell orbitals one can drastically reduce the number of internal $(N-1)$-electron and N-electron states by contracting those which have one or two holes in the closed shells. The overlap matrix of the internally contracted singly external configurations with one hole in closed shells only depends on the second-order density matrix of the active space. Hence, for active spaces of moderate size, orthogonalization is no problem. The same applies for contracted internal configurations

$$\Psi^{nm}_{ij} = E_{ni,mj}|0\rangle \tag{225}$$

with two holes in the closed shells. However, the overlap matrix of contracted internal configurations with only one hole still involves the third-order density matrix. It may, therefore, in this case be easier to use uncontracted configurations defined as

$$\Psi^{m}_{iK} = E_{mi}|K\rangle \tag{226}$$

where $|K\rangle$ is the full configuration basis generated by the active orbital space. This definition is analogous to that of the uncontracted singly external configurations in Eq. (126).

H. Evaluation of the Coupling Coefficients

An advantage of using the non-orthogonal configuration basis is that the coupling coefficients are simple linear combinations of transition density matrix elements between internal N-electron states. These can in principle be calculated by any technique applicable to the chosen spin-eigenfunction basis. The density matrices are then obtained by contracting the coupling coefficients with the coefficients of the reference configurations, e.g.

$$\Gamma_{ik,jl,mn} = \sum_{IJ} c_I c_J \langle I | E_{ik,jl,mn} | J \rangle \tag{227}$$

A further advantage of our formulation is that many coupling coefficients are identical or derived from the same basic quantities. For example, the coefficients $\alpha_{mn}(ijp, klq)$ and $\beta_{mn}(ijp, klq)$ for a given $i \geqslant j$ are all obtained from the N^4_{act} matrix elements

$$G^{ij}_{kl,mn} = \Gamma_{ik,jl,mn} \tag{228}$$

$$\alpha_{mn}(ijp, klq) = \delta_{pq}(G^{ij}_{kl,mn} + pG^{ij}_{lk,mn}) \tag{229}$$

$$\beta_{mn}(ijp, klq) = G^{ij}_{ml,nk} + pG^{ij}_{lm,nk} + qG^{ij}_{mk,nl} + pqG^{ij}_{km,nl} \tag{230}$$

Hence, considerable effort can be saved by calculating many coupling coefficients at the same time, thereby taking full advantage of the symmetry in their structure.

For performing the direct CI procedure it is essential to obtain the coefficients in a particular order. For instance, above all third-order density matrix elements $\Gamma_{ik,jl,mn}$ were needed for a fixed $i \geqslant j$. Similarly, for calculating the vectors $\mathbf{x}^{iK}$ and $\mathbf{y}^{iK}$ one needs the coupling coefficients $\alpha_{mn}(iK, jL)$ and $\beta_{mn}(iK, jL)$ for a fixed $\{mn\}$ and $\{iK\}$ (see below). Analogous considerations apply for all other interactions. It is important that the algorithm used for calculating the coupling coefficients generates them in the desired order.

The (transition) density matrix structure of the coupling coefficients suggests that it should be possible to evaluate them in a similar way as proposed by Siegbahn[87] for CASSCF wavefunctions (cf. Section II.H) and to make use of factorizations such as

$$\begin{aligned} E_{ik,jl,mn} &= E_{ik,jl}E_{mn} - \delta_{ml}E_{ik,jn} - \delta_{mk}E_{in,jl} \\ &= E_{ik}E_{jl}E_{mn} - \delta_{jk}E_{il}E_{mn} - \delta_{ml}E_{ik}E_{jn} - \delta_{mk}E_{in}E_{jl} \\ &\quad + \delta_{jk}\delta_{ml}E_{in} + \delta_{km}\delta_{jn}E_{il} \end{aligned} \tag{231}$$

To date, we have programmed only part of all coupling coefficients using this technique. The timings of preliminary test calculations were very encouraging. For instance, for a case with about 700 reference configurations (nine active orbitals) the evaluation of all $S^{(p)}_{ij,kl}$, $\alpha_{nm}(ijp, klp)$, $\beta_{mn}(ijp, klq)$ and $\gamma(ijp, klp)$ took less than 5 s CPU time on a CRAY-1S. In the following we will discuss the calculation of some coupling coefficients in detail.

As an internal N-electron basis $\{|K\rangle\}$ we use genealogical spin eigenfunctions. The one-particle coupling coefficients $\gamma^{KL}_{ij} = \langle K | E_{ij} | L \rangle$ are evaluated using well known group theoretical methods[88] and stored on disc. We first consider the coefficients $G^{ij}_{kl,mn}$ needed to calculate the $\boldsymbol{\alpha}(ijp, klp)$ and $\boldsymbol{\beta}(ijp, klq)$

(cf. Eqs (228)–(230)). These coefficients are obtained in the following sequence:

$$A_{mn,L} = \sum_K C_K \gamma_{mn}^{KL} \qquad \text{for all } L \tag{232}$$

$$C_{kl,L}^{ij} = \sum_K A_{jl,K} \gamma_{ik}^{KL} - \delta_{il} A_{jk,L} \tag{233}$$

$$G_{kl,mn}^{ij} = \sum_L C_{kl,L}^{ij} A_{nm,L} - \delta_{km} \Gamma_{in,jl} - \delta_{lm} \Gamma_{ik,jn} \tag{234}$$

In order to perform these steps with a minimum of input–output operations one has to keep all matrices **A** in high-speed memory. Furthermore, the coupling coefficients γ_{ik}^{KL} should be ordered according to the label i. Then, if all γ_{ik}^{KL} for a given i can be kept in core, a single read of these coefficients is sufficient to calculate the $G_{kl,mn}^{ij}$ for all $i \geqslant j$. For applications with up to 8–10 active orbitals, the memory requirement to perform these steps is not exceedingly large, in particular if molecular symmetry can be used. Note that the crucial step is the matrix multiplication in Eq. (234) and this can be perfectly vectorized.

The calculation of the coefficients $\gamma(ijp, klp)$ involves the density matrices up to fourth order, and might therefore appear to be more difficult. However, it is possible to obtain these coefficients with little extra cost from the same quantities needed above:

$$\gamma(ijp, klp) = H_{kl}^{ij} + pH_{lk}^{ij} \tag{235}$$

$$H_{kl}^{ij} = \sum_L C_{kl,L}^{ij} g_L + \sum_{mn} \Gamma_{im,jn}(km|ln) - \sum_L \sum_m (C_{km,L}^{ij} B_{lm,L} + C_{ml,L}^{ij} B_{km,L})$$
$$- \sum_m (\Gamma_{ik,jm} \tilde{h}_{lm} + \Gamma_{im,jl} \tilde{h}_{km}) \tag{236}$$

where

$$g_L = \sum_K c_K \left(\sum_{ij} h_{ij} \langle L|E_{ij}|K\rangle + \tfrac{1}{2} \sum_{ijkl} (ij|kl) \langle L|E_{ij,kl}|K\rangle \right) \tag{237}$$

$$B_{ij,L} = \sum_{kl} (ij|kl) A_{kl,L} \tag{238}$$

$$\tilde{h}_{ij} = h_{ij} - \sum_k (ik|kj) \tag{239}$$

The vector **g** is obtained in a direct CI step with the reference configurations only. This direct CI has to be performed anyway in order to determine the contraction coefficients c_I.

Next, we discuss the calculation of the coefficients

$$\alpha_{mn}(iK, jL) = \sum_I \langle K|E_{ij}|I\rangle \langle I|E_{mn}|L\rangle - \delta_{jm} \langle K|E_{in}|L\rangle \tag{240}$$

$$\beta_{mn}(iK, jL) = \sum_I \langle K|E_{im}|I\rangle \langle I|E_{nj}|L\rangle - \delta_{mn} \langle K|E_{ij}|L\rangle \tag{241}$$

For the efficient calculation of the residual vector $\mathbf{g}^{iK}$ it is advantageous to process the operators $\mathbf{J}^{mn}$ and $\mathbf{K}^{mn}$ sequentially. Then, for a fixed operator label $\{mn\}$, one needs for a given $\{iK\}$ all $\{jL\}$ together in order to evaluate the vectors $\mathbf{y}^{iK}$ and $\mathbf{z}^{iK}$. Unfortunately, it seems impossible to obtain the $\alpha_{mn}(iK, jL)$ and $\beta_{mn}(iK, jL)$ in this order without repeatedly reading the γ_{ij}^{IJ} or keeping them in high-speed memory. However, the coupling coefficients $\langle K|E_{ij,mn}|L\rangle$ can easily be calculated in a random order if the coefficients $\langle K|E_{ij}|I\rangle$ are ordered according to the intermediate state label $|I\rangle$. Since the coefficients $\langle K|E_{ij,mn}|L\rangle$ are also needed for the internal–single interactions it seems to be advantageous to evaluate them only once and sort them into the required order ($\langle K|E_{ij,mn}+E_{ij,nm}|L\rangle$, $\langle K|E_{im,nj}|L\rangle$, $\langle K|E_{in,mj}|L\rangle$ for $m \geqslant n$ in the outermost loop, $\{iK\}$ in the next inner loop and $\{jL\}$ in the innermost loop). Note that only the labels $\{iK\}$ which correspond to non-redundant $(N-1)$-electron states are needed. This considerably reduces the length of this formula file.

Finally, we consider the most crucial coupling coefficients

$$\alpha_{nkl}(mK, ijp) = (1 + p\tau_{ij})\left(\sum_I C_{nm,I}^{ij}\gamma_{kl}^{IK} - \delta_{km}C_{nl,K}^{ij} - \delta_{kn}C_{lm,K}^{ij}\right) \tag{242}$$

These are needed in an order where $\{ij\}$ runs in the outermost loop and either $\{nkl\}$ or $\{mK\}$ in the innermost loop. For the calculation of these coefficients it seems unavoidable to read the one-particle coefficients γ_{kl}^{IK} once for each orbital pair $i \geqslant j$ unless all γ_{kl}^{IK} can be kept in core. It should be noted that the coefficients for singlet and triplet pairs $(p = \pm 1)$ are obtained together. Furthermore, only a subset of all γ_{kl}^{IK} is needed since not all possible N-electron states are required to generate a non-redundant set of $(N-1)$-electron functions $\{mK\}$.

An obvious disadvantage of the procedure outlined above is that a relatively large amount of memory is needed. It should also be noted that the method is most advantageous for complete reference functions. For simple reference wavefunctions the matrices $A_{ij,I}$, $C_{kl,I}^{ij}$, etc., become very sparse. This sparsity cannot be exploited fully in a vectorized computer code. It may, therefore, be more efficient to use other techniques in such cases.

I. Optimization of the Contraction Coefficients

In most applications the internal contraction of the configuration state functions is an excellent approximation. In certain cases, however, the relative weight of the reference configurations strongly changes due to their coupling with the singly and doubly external configurations. This can happen, for instance, near avoided crossings of states with different correlation energies. In such cases it may be desirable to relax the contraction coefficients during the direct CI procedure. Obviously, this makes it necessary to recalculate all coupling coefficients which refer to contracted configurations.

In the most straightforward procedure, one would employ as improved contraction coefficients the coefficients of the reference configurations in the total wavefunction. The vector of contraction coefficients should always be renormalized. However, this simple procedure does not necessarily yield the optimum contraction coefficients. For simplicity, in the following we assume that the reference wavefunction comprises all internal configurations Ψ_I, and we only consider the doubly external configurations. The wavefunction can then be written in the form

$$\begin{aligned}\Psi &= \sum_I c_I \Psi_I + \sum_I a_I \sum_{i \geqslant j} \sum_p \sum_{ab} C^{ijp}_{ab} \Psi^{ab}_{ijp,I} \\ &= \Psi_{\text{int}} + \sum_I a_I \sum_{i \geqslant j} \sum_p \Psi_{ijp,I} \\ &= \Psi_{\text{int}} + \sum_{i \geqslant j} \sum_p \Psi_{ijp} \end{aligned} \tag{243}$$

where a_I are the contraction coefficients. It is not assumed that these coefficients are identical to the coefficients c_I of the internal configurations, even though it would be possible to impose this restriction. The energy derivatives with respect to the a_I are given by

$$\frac{\partial E}{\partial a_I} = 2 \sum_{i \geqslant j} \sum_p \left(\langle \Psi_{ijp,I} | H | \Psi_{\text{int}} \rangle + \sum_{k \geqslant l} \sum_q \langle \Psi_{ijp,I} | H - E | \Psi_{klq} \rangle \right) \tag{244}$$

For the optimum contraction coefficients all these derivatives must vanish. The matrix elements in Eq. (244) are explicitly given by

$$\langle \Psi_{ijp,I} | H | \Psi_{\text{int}} \rangle = \tfrac{1}{2} \sum_{mn} \text{tr}(\mathbf{C}^{ijp\dagger} \mathbf{K}^{mn}) \langle I | E_{im,jn} + p E_{in,jm} | \Psi_{\text{int}} \rangle \tag{245}$$

$$\begin{aligned}\langle \Psi_{ijp,I} | H | \Psi_{klq} \rangle = {} & \delta_{pq} \langle I | E_{ik,jl} + p E_{jk,il} | 0 \rangle [\text{tr}(\mathbf{C}^{ijp\dagger} \mathbf{K}(\mathbf{C}^{klq})) \\ & + 2\, \text{tr}(\mathbf{C}^{ijp\dagger} \mathbf{h} \mathbf{C}^{klq})] + 2 \sum_{mn} \text{tr}(\mathbf{C}^{ijp\dagger} \mathbf{J}^{mn} \mathbf{C}^{klq}) \alpha^I_{mn}(ijp, klq) \\ & + \sum_{mn} \text{tr}(\mathbf{C}^{ijp\dagger} \mathbf{K}^{mn} \mathbf{C}^{klq}) \beta^I_{mn}(ijp, klq) + \text{tr}(\mathbf{C}^{ijp\dagger} \mathbf{C}^{klq}) \gamma^I(ijp, klq) \end{aligned} \tag{246}$$

The definition of the coupling coefficients $\alpha^I_{mn}(ijp, klq)$, $\beta^I_{mn}(ijp, klq)$ and $\gamma^I(ijp, klq)$ is similar to Eqs (167)–(169) except that the bras $\langle 0|$ are replaced by $\langle I|$. Since the traces do not depend on the index I the derivatives with respect to all I should be calculated simultaneously. The matrix products $\mathbf{J}^{mn}\mathbf{C}^{klq}$ and $\mathbf{K}^{mn}\mathbf{C}^{klq}$ can also be used to compute the matrices $\mathbf{G}^{ijp}$ at the same time. Provided the coupling coefficients are not sparse this requires a total of about $\frac{3}{2}M^4$ matrix multiplications. The standard procedure in which the operators $\mathbf{F}^{ijp,klq}$ are evaluated as intermediate quantities requires at most M^4 matrix multiplications. Hence, the extra effort to calculate the energy derivatives with respect to the contraction coefficients is not exceedingly large. The derivatives in Eq. (244) can be employed to optimize the contraction coefficients in the usual way. They are also needed for the calculation of analytical energy gradients.

IV. SUMMARY

Efficient new MCSCF and MCSCF-CI techniques have been described which are based entirely on matrix operations and are therefore well suited for use on modern vector processing facilities. Both methods are formulated in terms of integral, density and coefficient matrices and have a rather similar structure. In the MCSCF method all first and second energy derivatives with respect to the variational parameters are taken into account exactly, while higher derivatives are incorporated approximately. It has been shown that the inclusion of the higher-order terms increases the radius and the speed of convergence remarkably. In previous second-order MCSCF methods the coupling of the orbital changes and the configuration coefficients was accounted for by evaluating first-order corrections of the density matrices. In the present method the density matrices are recalculated exactly after updates of the CI coefficients. Not only does this improve convergence, but it is also simpler and more efficient. By means of a new direct CI technique which avoids the storage of the coupling coefficients, CASSCF calculations with more than 10^5 configurations can now be performed.

The internally contracted MCSCF-SCEP method has been formulated in a non-orthogonal configuration basis. This requires additional transformations of the CI coefficients and of the residual vector but has the advantage that the coupling coefficients are much easier to evaluate than in an orthogonalized basis. The coupling coefficients are higher-order (transition) density matrices of the internal spin-adapted N-electron basis. Simplified formulae for the case that the reference wavefunction contains closed-shell orbitals have been derived. It has been shown that all coupling coefficients needed involve only the internal open-shell orbitals. A new algorithm to evaluate the higher-order density matrices efficiently has been discussed. In contrast to previous methods the new procedure can be vectorized and appears to be efficient enough to allow a recalculation of all coupling coefficients which depend on the contraction coefficients in each iteration. This eliminates large formula files and also makes it possible to relax or even fully optimize the contraction coefficients.

Acknowledgements

The author wishes to thank P. J. Knowles, W. Meyer and E.-A. Reinsch for their support and cooperation. They contributed essentially to the work described in this paper. Special thanks are due to M. I. McCarthy and E.-A. Reinsch for critically reading the manuscript and for making many useful suggestions, and to J. Senekowitsch for preparing the figures. Financial support of the Deutsche Forschungsgemeinschaft and the Fonds der Chemischen Industrie is gratefully acknowledged.

References

1. For some recent reviews, see, e.g., Bartlett, R. J. (Ed.), *Comparison of Ab Initio Quantum Chemistry with Experiment for Small Molecules—The State of the Art*, Reidel, Dordrecht, 1985.
2. Buenker, R. J., and Peyerimhoff, S. D., *Theor. Chim. Acta*, **35**, 33 (1974).
3. Buenker, R. J., Peyerimhoff, S. D., and Butcher, W., *Mol. Phys.*, **35**, 771 (1978).
4. For a review, see Shavitt, I., in *Modern Theoretical Chemistry* (Ed. H. F. Schaefer III), Plenum, New York, 1977.
5. Roos, B. O., *Chem. Phys. Lett.*, **15**, 153 (1972).
6. Roos, B. O., and Siegbahn, P. E. M., in *Modern Theoretical Chemistry* (Ed. H. F. Schaefer III), Plenum, New York, 1977.
7. Meyer, W., *J. Chem. Phys.*, **64**, 2901 (1976).
8. Pulay, P., Saebo, S., and Meyer, W., *J. Chem. Phys.*, **81**, 1901 (1984).
9. Dykstra, C. E., Schaefer, H. F., III, and Meyer, W., *J. Chem. Phys.*, **65**, 2740 (1976).
10. Dykstra, C. E., Schaefer, H. F., III, and Meyer, W., *J. Chem. Phys.*, **65**, 5141 (1976).
11. Werner, H.-J., and Reinsch, E.-A., unpublished, 1978.
12. Ahlrichs, R., *Chem. Phys.*, **17**, 31 (1979).
13. Dykstra, C. E., *J. Chem. Phys.*, **67**, 4716 (1977).
14. Dykstra, C. E., *J. Chem. Phys.*, **72**, 2928 (1980).
15. Flesch, J., and Meyer, W., private communication.
16. Chiles, R. A., and Dykstra, C. E., *J. Chem. Phys.*, **74**, 4544 (1981).
17. Werner, H.-J., and Reinsch, E.-A., in *Proc. 5th Semin. on Computational Methods in Quantum Chemistry* (Eds T. H. van Duinen and W. C. Niewpoort), MPI Garching, München, 1981.
18. Werner, H.-J., and Reinsch, E.-A., *J. Chem. Phys.*, **76**, 3144 (1982).
19. Ahlrichs, R., in *Proc. 5th Semin. on Computational Methods in Quantum Chemistry* (Eds T. H. van Duinen and W. C. Niewpoort), MPI Garching, München, 1981.
20. Saunders, V. R., and van Lenthe, J. H., *Mol. Phys.*, **48**, 923 (1983).
21. Paldus, J., *J. Chem. Phys.*, **61**, 5321 (1974).
22. Paldus, J., *Phys. Rev. A*, **14**, 1620 (1976).
23. Shavitt, I., *Int. J. Quantum Chem. Symp.*, **11**, 131 (1977).
24. Shavitt, I., *Int. J. Quantum Chem. Symp.*, **12**, 5 (1978).
25. Shavitt, I., *Chem. Phys. Lett.*, **63**, 421 (1979).
26. Siegbahn, P. E. M., *J. Chem. Phys.*, **72**, 1647 (1980).
27. Brooks, B. R., and Schaefer, H. F., III, *J. Chem. Phys.*, **70**, 5092 (1979).
28. Brooks, B. R., Laidig, W. D., Saxe, P., Handy, N. C., and Schaefer, H. F., III, *Phys. Scr.*, **21**, 312 (1980).
29. Duch, W., and Karwowski, J., *Theor. Chim. Acta*, **51**, 175 (1979).
30. Duch, W., *Theor. Chim. Acta*, **57**, 299 (1980).
31. Tavan, P., and Schulten, K., *J. Chem. Phys.*, **72**, 3547 (1980).
32. Taylor, P. R., *J. Chem. Phys.*, **74**, 1256 (1981).
33. Liu, B., and Yoshimine, M., *J. Chem. Phys.*, **74**, 612 (1981).

34. Lischka, H., Shepard, R., Brown, F. B., and Shavitt, I., *Int. J. Quantum Chem. Symp.*, **15**, 91 (1981).
35. Saxe, P., Fox, D. J., Schaefer, H. F., III, and Handy, N. C., *J. Chem. Phys.*, **77**, 5584 (1982).
36. Meyer, W., Ahlrichs, R., and Dykstra, C. E., in *Advanced Theories and Computational Approaches to the Electronic Structure of Molecules* (Ed. C. E. Dykstra), Reidel, Dordrecht, 1984.
37. Dalgaard, E., and Jørgensen, P., *J. Chem. Phys.*, **69**, 3833 (1978).
38. Yeager, D. L., and Jørgensen, P., *J. Chem. Phys.*, **71**, 755 (1979).
39. Lengsfield, B. H., III, *J. Chem. Phys.*, **73**, 382 (1980).
40. Yeager, D. L., and Jørgensen, P., *Mol. Phys.*, **39**, 587 (1980).
41. Yeager, D. L., Albertsen, P., and Jørgensen, P., *J. Chem. Phys.*, **73**, 2811 (1980).
42. Werner, H.-J., and Meyer, W., *J. Chem. Phys.*, **73**, 2342 (1980).
43. Werner, H.-J., and Meyer, W., *J. Chem. Phys.*, **74**, 5794 (1981).
44. Lengsfield, B. H., and Liu, B., *J. Chem. Phys.*, **75**, 478 (1981).
45. Jørgensen, P., Olsen, J., and Yeager, D. L., *J. Chem. Phys.*, **75**, 5802 (1981).
46. Shepard, R., Shavitt, I., and Simons, J., *J. Chem. Phys.*, **76**, 543 (1982).
47. Lengsfield, B. H., III, *J. Chem. Phys.*, **77**, 4073 (1982).
48. Olsen, J., Jørgensen, P., and Yeager, D. L., *J. Chem. Phys.*, **76**, 527 (1982).
49. Olsen, J., Jørgensen, P., and Yeager, D. L., *J. Chem. Phys.*, **77**, 456 (1982).
50. Olsen, J., and Jørgensen, P., *J. Chem. Phys.*, **77**, 6109 (1982).
51. Jørgensen, P., Swanstrøm, P., and Yeager, D. L., *J. Chem. Phys.*, **78**, 347 (1983).
52. Golab, J. T., Yeager, D. L., and Jørgensen, P., *Chem. Phys.*, **78**, 175 (1983).
53. Olsen, J., Yeager, D. L., and Jørgensen, P., *Adv. Chem. Phys.*, **54**, 1 (1983).
54. Jensen, H. J. A., and Jørgensen, P., *J. Chem. Phys.*, **80**, 1204 (1984).
55. Werner, H.-J., and Knowles, P. J., *J. Chem. Phys.*, **82**, 5053 (1985).
56. Knowles, P. J., and Werner, H.-J., *Chem. Phys. Lett.*, **115**, 259 (1985).
57. Ruedenberg, K., Cheung, L. M., and Elbert, S. T., *Int. J. Quantum Chem.*, **16**, 1069 (1979).
58. Siegbahn, P. E. M., Heiberg, A., Roos, B. O., and Levy, B., *Phys. Scr.*, **21**, 323 (1980).
59. Siegbahn, P. E. M., Almlof, J., Heiberg, A., and Roos, B. O., *J. Chem. Phys.*, **74**, 2384 (1981).
60. Shepard, R., and Simons, J., *Int. J. Quantum Chem.*, **14**, 211 (1980).
61. Yaffe, L. G., and Goddard, W. A., III, *Phys. Rev. A*, **13**, 1682 (1976).
62. Knowles, P. J., and Handy, N. C., *Chem. Phys. Lett.*, **111**, 315 (1984).
63. Werner, H.-J., and Reinsch, E.-A., in *Advanced Theories and Computational Approaches to the Electronic Structure of Molecules* (Ed. C. E. Dykstra), Reidel, Dordrecht, 1984.
64. Werner, H.-J., and Rosmus, P., in *Comparison of Ab Initio Quantum Chemistry with Experiment for Small Molecules—The State of the Art* (Ed. R. J. Bartlett), Reidel, Dordrecht, 1985.
65. Thouless, D. J., *The Quantum Mechanics of Many Body Systems*, Academic Press, New York, 1961.
66. Levy, B., *Chem. Phys. Lett.*, **4**, 17 (1969).
67. Hoffmann, M. R., Fox, D. J., Gaw, J. F., Osamura, Y., Yamaguchi, Y., Grew, R. S., Fitzgerald, G., Schaefer, H. F., III, Knowles, P. J., and Handy, N. C., *J. Chem. Phys.*, **80**, 2660 (1984).
68. Fletcher, R., *Practical Methods of Optimization*, Vol. 1, Wiley, New York, 1980.
69. Roos, B., Taylor, P., and Siegbahn, P. E. M., *Chem. Phys.*, **48**, 157 (1980).
70. Jensen, H. J. A., and Agren, H., *Chem. Phys. Lett.*, **110**, 140 (1984).

71. Davidson, E. R., *J. Comput. Phys.*, **17**, 87 (1975).
72. Huzinaga, S., *Approximate Atomic Functions*, Technical Report, Department of Chemistry, University of Alberta, 1965.
73. Siegbahn, P. E. M., *Chem. Phys. Lett.*, **109**, 417 (1984).
74. Cooper, I. L., and McWeeny, R., *J. Chem. Phys.*, **45**, 226 (1966).
75. Salmon, W. I., and Ruedenberg, K., *J. Chem. Phys.*, **57**, 2776 (1972).
76. Wetmore, R. W., and Segal, G. A., *Chem. Phys. Lett.*, **36**, 478 (1975).
77. Docken, K. K., and Hinze, J., *J. Chem. Phys.*, **57**, 4928 (1972).
78. Das, G., *J. Chem. Phys.*, **58**, 5104 (1973).
79. Chang, T. C., and Schwarz, W. H. E., *Theor. Chim. Acta*, **44**, 45 (1977).
80. McCarthy, M. I., Rosmus, P., Werner, H.-J., Botschwina, P., and Vaida, V., *J. Chem. Phys.*, in press.
81. Siegbahn, P. E. M., *Chem. Phys.*, **25**, 197 (1977).
82. Siegbahn, P. E. M., in *Proc. 5th Semin. on Computational Methods in Quantum Chemistry* (Eds T. H. van Duinen and W. C. Niewpoort), MPI Garching, München, 1981.
83. Meyer, W., in *Modern Theoretical Chemistry* (Ed. H. F. Schaefer III), Plenum, New York, 1977.
84. Siegbahn, P. E. M., *Int. J. Quantum Chem.*, **18**, 1229 (1980).
85. Bunge, A., *J. Chem. Phys.*, **53**, 20 (1970).
86. McLean, A. D., and Liu, B., *J. Chem. Phys.*, **58**, 1066 (1973).
87. Siegbahn, P. E. M., *Chem. Phys. Lett.*, **109**, 417 (1984).
88. This program has been developed by Knowles, P. J., 1985.

Ab Initio Methods in Quantum Chemistry—II
Edited by K. P. Lawley

THE MULTICONFIGURATION SELF-CONSISTENT FIELD METHOD

RON SHEPARD

Theoretical Chemistry Group, Chemistry Division, Argonne National Laboratory, Argonne, IL 60439, USA

CONTENTS

I. Introduction 64
II. Formal Background 66
A. Linear Algebra 67
B. Matrix Partitioning Theory 74
1. The Bracketing Theorem 75
C. The Definition of the Multiconfiguration Self-consistent Field Method 77
1. The Excited-state MCSCF Method 78
D. N-Electron Expansion Space Representation 80
E. Second Quantization 82
1. Bra and Ket Expansion Terms 83
2. Creation and Annihilation Operators 84
3. Representation of Quantum-mechanical Operators . . . 85
F. Orbital Transformations 87
G. Spin Eigenfunctions and the Unitary Group Approach . . . 92
1. The Canonical N-electron Expansion Basis 93
H. Matrix Element Evaluation 98
1. The DRT Representation of the CSF Expansion Space 99
I. Summary 101
III. Multiconfiguration Self-consistent Field Equations 102
A. Energy Expressions for Multiconfiguration Self-consistent Field Wavefunction Optimization 102
B. Matrix Element Evaluation 111
C. Solution of the Multiconfiguration Self-consistent Field Optimization Equations 114
D. Newton–Raphson Optimization Methods 118
E. Variational Super-configuration-interaction Optimization Methods . 120
F. Rational Function Approximation Optimization Methods . . 121
1. Augmented Hessian Optimization Methods 122
G. Extended Micro-iterative Methods 124
1. Alternate Energy Approximations 124
2. Explicit Third-order Contributions to the Energy 127

IV. Configuration State Function Expansion Spaces for Multiconfiguration Self-consistent Field Wavefunctions 127
A. Empirical Selection Methods 128
B. *A Priori* Selection Methods 130
1. The OVC Wavefunction 131
2. The Full CI Expansion 132
3. Subspace Full CI Expansions 133
4. Direct Product Full CI Expansions 137
5. Restricted Group Product Expansions 141
6. Other Linear Direct Product Expansions 142
7. Nonlinear Wavefunction Expansions 144
C. Orbital Symmetry and the Configuration State Function Expansion Space 148
D. Summary 151
V. Redundant Variables 151
A. Redundant Variables and the Configuration State Function Expansion 152
1. The Two-electron Case 152
2. The General N-electron Case 159
B. Wavefunction-dependent Redundant Variables 162
C. Properties of Redundant Variables 165
VI. Implementation of Multiconfiguration Self-consistent Field Methods . 169
A. General Considerations of Computer Implementations 169
B. Second-order Local Convergence 172
C. Elimination of the Hessian Matrix Formula File 174
D. Density Matrix Sparseness. 176
E. Simplifications Due to Doubly Occupied Orbitals 177
F. Transition Density Matrix Construction 180
G. Elimination of the Coupling Coefficient List 182
H. Direct Solution of the Wavefunction Correction Equations . . 184
I. Iterative Solution of the Wavefunction Correction Equations . . 185
J. Scaling and Level-shifting Modifications 189
K. Approximate Hamiltonian Operator Methods 191
VII. Summary. 194
Acknowledgements 195
References 196

I. INTRODUCTION

In this chapter, the recent developments of the multiconfiguration self-consistent field (MCSCF) method of the past few years are reviewed. Several aspects of the MCSCF method have been recently reviewed by other authors. Olsen *et al.*[1] and McWeeney and Sutcliffe[2] have reviewed the formal methodology, Roos[3] and Siegbahn[4] have reviewed the fully optimized reaction space/complete active space self-consistent field (FORS/CASSCF) method and applications, Detrich and Wahl[5] have reviewed chemical applications of the MCSCF method, and Dalgaard and Jørgensen[6] have reviewed various optimization methods. Although it is not practicable for the contents of this chapter to be orthogonal to these other papers, an attempt has been made to emphasize some of the aspects of the MCSCF method that have not been included elsewhere. The inclusion, or the neglect, of some aspect of the MCSCF method in this review, therefore, should not be used as a measure of its importance. This review is concerned primarily with the recent formal developments and the computer implementation of these developments. Other contributions in this volume[7] are concerned with specific chemical applications using some of the methods discussed in this review.

The acronym (or more correctly the initialism) MCSCF stands for multi-configuration self-consistent field and implies that an effective one-particle potential is adjusted until self-consistency is obtained for all of the electrons of a molecular system and that this is done for a wavefunction that consists of the superposition of several electronic configurations. Early MCSCF methods[6,8,9] were in fact formulated in this way and the optimization methods were simple extensions of the single-configuration, or SCF, methods. Modern MCSCF methods are not based on this approach. The electronic energy is considered instead to be a function of the orbitals and the coefficients of the configurations employed in the superposition and the energy is straightforwardly minimized within this space of variations.

For the MCSCF wavefunction and its properties to be meaningful, it is necessary for the orbitals to be sufficiently flexible, for the configurations used in the superposition to be appropriate, and for the underlying mathematical model to be adequate. The adequacy of the mathematical model is discussed in Section II. The correspondence between the lowest roots of the matrix representation of the Hamiltonian operator within a subspace of N-electron expansion terms and the exact electronic energies is established using the bracketing theorem. This correspondence then allows the MCSCF method to be defined for both ground and excited states. (In this chapter the term 'ground state' generally refers to the lowest state of a particular symmetry type and excited states are the higher states of this same symmetry.) The formal problems associated with excited-state optimization are then discussed. The remaining subsections of Section II introduce the notation used to define the MCSCF optimization equations. This notation includes the method of second quantization and the construction and use of spin eigenfunctions. Much of Section II is common to other recent MCSCF reviews and may be quickly scanned by those familiar with these aspects of the MCSCF method.

Section III then introduces the various approximate energy expressions that are used to determine the wavefunction corrections within each iteration of the MCSCF optimization procedure. Although many of these approximate energy expressions are defined in terms of the same set of intermediate quantities (i.e. the gradient vector and Hessian matrix elements), these expressions have some important formal differences. These formal differences result in MCSCF methods that have qualitatively different convergence characteristics.

The choice of N-electron expansion terms for MCSCF wavefunctions is discussed in Section IV. Many earlier MCSCF methods employed empirical selection methods in which the individual expansion terms were carefully selected in order to minimize the computational effort. Modern methods are more flexible with respect to expansion length and most methods allow selection of expansion terms based on rather general orbital occupation restrictions. Several methods for the specification of these expansion terms based on chemical intuition and general principles are described in Section IV. Although this is a very important aspect of the MCSCF method, this expansion term specification has received relatively little attention in the last few years compared to some of the other aspects of the MCSCF method.

Section V consists of a detailed discussion of redundant variables. The special case of MCSCF wavefunction optimization for two-electron systems is discussed in some detail. The relation between the configuration expansion space and the orbital variation space is quite straightforward for this case and this simplicity may be used to advantage in understanding the generalization to arbitrary numbers of electrons. There are two aspects of redundant variables that are important in the MCSCF method. First, if redundant variables are allowed to remain in the wavefunction variation space, then the optimization procedure becomes undefined or at least numerically ill-conditioned. Secondly, if the redundant variables are known for a given wavefunction then this flexibility may be used to transform the wavefunction to a form that is qualitatively easier to understand. The qualitative interpretation of MCSCF wavefunctions is one of the assets of the MCSCF method.

The details of the computer implementation of various MCSCF methods are discussed in Section VI. Modern high-speed computers are biased in their ability to perform certain operations efficiently. MCSCF methods that involve simple vector and matrix operations have a distinct advantage on these types of computers. The reduction of unnecessary I/O (input/output to external storage) is also very important on these computers. This is because the capacity of these machines to perform arithmetic operations outpaces their I/O capacity. These considerations have had a significant impact on the choice of MCSCF wavefunction optimization methods and on the specific details of the implementation of these methods.

II. FORMAL BACKGROUND

The discussion of the various aspects of the MCSCF method requires the discussion of some background material. In this section, some of the elementary concepts of linear algebra are introduced. These concepts, which include the bracketing theorem for matrix eigenvalue equations, are used to define the MCSCF method and to discuss the MCSCF model for ground states and excited states. The details of N-electron expansion space representations are then discussed. The method of second quantization is introduced and some of the advantages of this notation over the more traditional wavefunction notation, when applied to the MCSCF method, are demonstrated. The exponential operator representation of orbital transformations is fundamental in this discussion. The construction of spin eigenfunctions within the second quantization method is discussed using the unitary group approach. The advantages of the unitary group approach over other spin-eigenfunction methods in the construction of matrix elements result from the use of the distinct row table, a compact representation of the spin-eigenfunction expansion space.

A. Linear Algebra

In the MCSCF method many of the operations required in the formal development and in the actual computational implementation involve the use of linear algebra. These manipulations of matrices and vectors are discussed in this section and some necessary background for later discussions is introduced. The first reason that the manipulation of matrices is important in the MCSCF method is that the molecular orbitals (MOs) used to define the wavefunction are expanded in an atomic orbital (AO) basis. The orbital expansion coefficients may be collected into the matrix **C** and the relation between the two orbital sets may be written as

$$\boldsymbol{\phi} = \boldsymbol{\chi}\mathbf{C} \tag{1}$$

The set of AO basis functions is collected into the row vector χ and a column of the matrix **C** is the set of MO expansion coefficients for a particular molecular orbital ϕ. The second occurrence of the use of matrix operations results from the expansion of the wavefunction $|0\rangle$ in a set of N-electron expansion functions

$$|0\rangle = \sum_m |m\rangle c_m \tag{2}$$

These expansion terms will subsequently be called configuration state functions (CSFs). The actual construction of these expansion terms is discussed later in this section. It is assumed here that such an expansion exists and some of the algebraic notation that will be used in this review is introduced. The expectation value of an operator may be written as

$$\langle A\rangle = \langle 0|A|0\rangle\langle 0|0\rangle^{-1} \tag{3}$$

$$= \sum_m \sum_{m'} c_m^* < m|A|m' > c_{m'} \left(\sum_m \sum_{m'} c_m^* < m|m'> c_{m'} \right)^{-1} \tag{4}$$

$$= \sum_m \sum_{m'} c_m^* A_{mm'} c_{m'} \left(\sum_m \sum_{m'} c_m^* S_{mm'} c_{m'} \right)^{-1} \tag{5}$$

$$= \mathbf{c}^\dagger\mathbf{A}\mathbf{c}(\mathbf{c}^\dagger\mathbf{S}\mathbf{c})^{-1} \tag{6}$$

$$= \mathbf{c}^\dagger\mathbf{A}\mathbf{c}(\mathbf{c}^\dagger\mathbf{c})^{-1} \tag{7}$$

$$= \mathbf{c}^\dagger\mathbf{A}\mathbf{c} \tag{8}$$

Eq. (4) is the most general definition of the expectation value an is defined in terms of the N-electron integrals written as $\langle m|A|m'\rangle$ and $\langle m|m'\rangle$. Eq. (5) assumes that these integrals and CSF expansion coefficients are collected into arrays indexed by the expansion CSF labels. Eq. (6) gives the expression for the expectation value using matrix notation. Eq. (7) gives the expectation value in the case of an orthonormal basis and Eq. (8) assumes further that the expansion coefficients are normalized, $(\mathbf{c}^\dagger\mathbf{c}) = 1$. Many of the formal manipulations, and resulting computational methods, required in the MCSCF method involve linear algebraic operations on the matrix representation of various operators, particularly of the N-electron Hamiltonian operator, and of operations involving transformations of the orbitals and the integrals over these orbitals.

There are several types of matrix operations that are used in the MCSCF method. The transpose of a matrix $\mathbf{A}$ is denoted $\mathbf{A}^t$ and is defined by $(\mathbf{A}^t)_{ij} = \mathbf{A}_{ji}$. The identity $(\mathbf{AB})^t = \mathbf{B}^t\mathbf{A}^t$ is sometimes useful where $\mathbf{AB}$ implies the usual definition of the product of matrices. A vector, specifically a column vector unless otherwise noted, is a special case of a matrix. A matrix–vector product, as in Eq. (5), is a special case of a matrix product. The conjugate of a matrix is written $\mathbf{A}^*$ and is defined by $(\mathbf{A}^*)_{ij} = (\mathbf{A}_{ij})^*$. The adjoint, written as $\mathbf{A}^\dagger$, is defined by $\mathbf{A}^\dagger = (\mathbf{A}^*)^t$. The inverse of a square matrix, written as $\mathbf{A}^{-1}$, satisfies the relation $\mathbf{A}(\mathbf{A}^{-1}) = \mathbf{1}$ where $\mathbf{1}_{ij} = \delta_{ij}$ is called the identity or unit matrix. The inverse of a matrix product satisfies the relation $(\mathbf{AB})^{-1} = \mathbf{B}^{-1}\mathbf{A}^{-1}$. A particular type of matrix is a diagonal matrix $\mathbf{D}$, where $\mathbf{D}_{ij} = d_i\delta_{ij}$, and is sometimes written $\mathbf{D} = \mathrm{diag}(d_1, d_2, \ldots)$ or as $\mathbf{D} = \mathrm{diag}(\mathbf{d})$. The unit matrix is an example of a diagonal matrix.

The scalar product of two vectors, $\mathbf{u}$ and $\mathbf{v}$, may be written as $\mathbf{u}^\dagger\mathbf{v}$. Two vectors are orthogonal if their scalar product is zero. A particularly important quantity involving the scalar product is the Euclidean norm of a vector defined by $|\mathbf{u}| = (\mathbf{u}^\dagger\mathbf{u})^{1/2}$. The Euclidean norm of a vector is non-negative and is zero only if the vector is zero.

Some types of matrices that appear frequently are Hermitian matrices, for which $\mathbf{A} = \mathbf{A}^\dagger$, anti-Hermitian matrices, for which $\mathbf{A} = -\mathbf{A}^\dagger$, and unitary matrices, for which $\mathbf{U}^\dagger = \mathbf{U}^{-1}$. The MCSCF methods discussed in most detail in this review will involve only operations of real matrices. In this case these matrix types reduce to symmetric matrices, for which $\mathbf{A} = \mathbf{A}^t$, antisymmetric matrices, for which $\mathbf{A} = -\mathbf{A}^t$, and orthogonal matrices, for which $\mathbf{U}^t = \mathbf{U}^{-1}$. A particular type of orthogonal matrix is called a rotation matrix and satisfies the relation $\mathrm{Det}(\mathbf{R}) = +1$, where $\mathrm{Det}(\mathbf{R})$ is the usual definition of a determinant of a matrix. A determinant of a matrix is a scalar function of the elements of the matrix. The determinant of a matrix product satisfies the relation $\mathrm{Det}(\mathbf{AB}) = \mathrm{Det}(\mathbf{A})\,\mathrm{Det}(\mathbf{B})$. A matrix trace, written $\mathrm{Tr}(\mathbf{A})$, is another scalar function and is defined as the sum of the diagonal elements of the matrix. The trace of a matrix product satisfies the relation $\mathrm{Tr}(\mathbf{AB}) = \mathrm{Tr}(\mathbf{BA})$.

A similarity transformation of a matrix $\mathbf{A}$ may be written as $\mathbf{B}^{-1}\mathbf{AB}$. A congruence transformation of the matrix $\mathbf{A}$ is written as $\mathbf{B}^{\dagger}\mathbf{AB}$. Both transformations are identical if $\mathbf{B}$ is unitary and the result is called a unitary transformation, or, in the case of real matrices, an orthogonal transformation. An important property of Hermitian matrices is that they may be brought to diagonal form with a unitary transformation[10]. The diagonal elements of this diagonal form are called the eigenvalues. The eigenvalues of a Hermitian matrix are real (as are the diagonal elements of any Hermitian matrix). This transformation may be written in the form $\mathbf{AU} = \mathbf{U}\boldsymbol{\lambda}$ where $\boldsymbol{\lambda}$ is diagonal. This equation is called a matrix eigenvalue equation and the solution of this equation, the determination of $\mathbf{U}$ and $\boldsymbol{\lambda}$, is called the diagonalization of the matrix $\mathbf{A}$. Each column of the unitary matrix is called an eigenvector and is associated with a particular eigenvalue. An eigenvector–eigenvalue pair satisfy the relation

$$\mathbf{Au}^i = \mathbf{u}^i \lambda_i \tag{9}$$

where $\mathbf{u}^i$ is the ith column of the unitary matrix $\mathbf{U}$ and λ_i is the associated eigenvalue. Two eigenvectors corresponding to two different eigenvalues are necessarily orthogonal. When several eigenvalues are equal they are said to be degenerate and the associated eigenvectors, which may also be chosen to be orthogonal, belong to a degenerate subspace. The determination of a subset of the vectors that satisfy Eq. (9) is also called matrix diagonalization, or more specifically, partial matrix diagonalization.

The expectation value of a matrix in terms of a vector may be considered to be a scalar function of the matrix elements and vector elements. This function is called the Rayleigh quotient and is sometimes written as $\rho(\mathbf{A}, \mathbf{x})$. The eigenvector–eigenvalue pair in Eq. (9) satisfy the equation $\partial\rho(\mathbf{A}, \mathbf{u}^i)/\partial(\mathbf{u}^i)_k = 0$ for all components of the vector and for all eigenvectors. The eigenvalues satisfy the relations $\lambda_i = \rho(\mathbf{A}, \mathbf{u}^i)$ and $\mathrm{Det}(\mathbf{A} - \lambda_i \mathbf{1}) = 0$. The latter relation is called a secular equation. The lowest eigenvalue is thereby seen to be associated with the vector that minimizes the Rayleigh quotient. The second-lowest eigenvalue is associated with the vector that minimizes the Rayleigh quotient within the vector space that is orthogonal to the lower vector, and so on for the higher eigenvalues and associated eigenvectors.

The inertia of a matrix is a triple of integers $(n_<, n_0, n_>)$ where $n_<$ is the number of negative eigenvalues of a matrix, n_0 is the number of zero eigenvalues and $n_>$ is the number of positive eigenvalues. The inertia is defined only for matrices with real eigenvalues. A matrix with zero eigenvalues is called a singular matrix. A similarity transformation of a matrix does not change the eigenvalues. A congruence transformation with a non-singular matrix mat change the eigenvalues but it does not change the inertia of a matrix.

The commutator of two square matrices is defined as $[\mathbf{A}, \mathbf{B}] = \mathbf{AB} - \mathbf{BA}$. If $[\mathbf{A}, \mathbf{B}] = \mathbf{0}$ the matrices **A** and **B** are said to commute. All diagonal matrices commute, every matrix commutes with itself, every matrix commutes with its inverse, and every matrix commutes with the identity matrix. If **A** and **B** are Hermitian, then $[\mathbf{A}, \mathbf{B}] = \mathbf{0}$ if and only if both matrices may be diagonalized by the same unitary matrix. This does not mean that every matrix that diagonalizes **A** will diagonalize **B** but that at least one such matrix exists that will diagonalize both. This relation may be used to determine the classes of matrices that can be diagonalized by a unitary transformation. Let **A** be an arbitrary matrix and define $\mathbf{A}_+ = (\mathbf{A} + \mathbf{A}^\dagger)/2$ and $\mathbf{A}_- = (\mathbf{A} - \mathbf{A}^\dagger)/2\mathrm{i}$. Then

$$\mathbf{A} = \mathbf{A}_+ + \mathrm{i}\mathbf{A}_- \tag{10}$$

showing that an arbitrary matrix may be written as a linear combination of two Hermitian matrices. In order for **A** to be diagonalized by a unitary transformation, it may be shown[11] that both $\mathbf{A}_+$ and $\mathbf{A}_-$ must be simultaneously diagonalized. This only occurs if $[\mathbf{A}_+, \mathbf{A}_-] = \mathbf{0}$ which in turn only occurs if $[\mathbf{A}, \mathbf{A}^\dagger] = \mathbf{0}$. Therefore a matrix may be diagonalized by a unitary transformation if and only if it commutes with its adjoint. Besides the trivial case of diagonal matrices, three such cases that occur in the MCSCF method are Hermitian matrices, anti-Hermitian matrices and unitary matrices.

Some functions of matrices are defined as expansions. If $f(x)$ is a function of a scalar variable x that has the expansion

$$f(x) = f_0 + f_1 x + (1/2) f_2 x^2 + \cdots + (1/n!) f_n x^n + \cdots \tag{11}$$

where f_n is a scalar expansion coefficient, then the matrix function $f(\mathbf{A})$ is defined as

$$f(\mathbf{A}) = f_0 \mathbf{1} + f_1 \mathbf{A} + (1/2) f_2 \mathbf{A}^2 + \cdots + (1/n!) f_n \mathbf{A}^n + \cdots \tag{12}$$

Such functions of general matrices must usually be approximated by truncating the expansion at some practicable order. Functions of matrices satisfying the relation $[\mathbf{A}, \mathbf{A}^\dagger] = \mathbf{0}$ may be evaluated exactly, however, by using the relation $\mathbf{A} = \mathbf{U}\boldsymbol{\lambda}\mathbf{U}^\dagger$ to give

$$f(\mathbf{A}) = \mathbf{U}[f_0 \mathbf{1} + f_1 \boldsymbol{\lambda} + (1/2) f_2 \boldsymbol{\lambda}^2 + \cdots + (1/n!) f_n \boldsymbol{\lambda}^n + \cdots]\mathbf{U}^\dagger \tag{13}$$

$$= \mathbf{U}\,\mathrm{diag}(f(\lambda_1), f(\lambda_2), \ldots)\mathbf{U}^\dagger \tag{14}$$

$$= \mathbf{U} f(\boldsymbol{\lambda}) \mathbf{U}^\dagger \tag{15}$$

where the elements of the diagonal matrix $f(\boldsymbol{\lambda})$ are determined from scalar function values. The use of this relation usually involves finding the eigenvalues and eigenvectors of the original matrix, but sometimes this representation may be used formally without this requirement. For example, consider the scalar function of a matrix, Det(exp(**A**)). The sequence of identities

$$\mathrm{Det}(\exp(\mathbf{A})) = \mathrm{Det}(\mathbf{U}\exp(\boldsymbol{\lambda})\mathbf{U}^{\dagger}) = \mathrm{Det}(\exp(\boldsymbol{\lambda}))\mathrm{Det}(\mathbf{U}^{\dagger}\mathbf{U}) \tag{16}$$

$$= \prod_m \exp(\lambda_m) = \exp\left(\sum_m \lambda_m\right) \tag{17}$$

$$= \exp[\mathrm{Tr}(\mathbf{A})] \tag{18}$$

leads to a useful result that does not require explicit diagonalization. Eq. (17) follows from Eq. (16) because the determinant of a diagonal matrix is simply the product of the diagonal elements. Eq. (18) results from the fact that the trace of a matrix is invariant to similarity transformations giving $\mathrm{Tr}(\boldsymbol{\lambda}) = \mathrm{Tr}(\mathbf{A})$.

It is useful to consider the relation between the eigenvalues of some specific matrix functions and the eigenvalues of the matrix arguments of these functions. If $\mathbf{U}^{\dagger}\mathbf{A}\mathbf{U} = \boldsymbol{\lambda}$, then it follows that $\mathbf{U}^{\dagger}(a\mathbf{A} + b\mathbf{1})\mathbf{U} = a\boldsymbol{\lambda} + b\mathbf{1} = \boldsymbol{\lambda}'$. This shows that the addition of a constant to the diagonal elements of a matrix results in shifting all of the eigenvalues of the matrix by that same constant, and that scaling a matrix by a constant results in scaling all of the eigenvalues by that same constant. The eigenvectors of the original matrix are also eigenvectors of the shifted and scaled matrix. This is also true for general matrix functions as shown above. Another useful relation is that if $\mathbf{U}^{\dagger}\mathbf{A}\mathbf{U} = \boldsymbol{\lambda}$, then $\mathbf{U}^{\dagger}\mathbf{A}^{-1}\mathbf{U} = \boldsymbol{\lambda}^{-1}$. This shows that any chosen set of eigenvectors of a particular matrix is also a set of eigenvectors of its inverse. The inverse of a matrix only exists if all of its eigenvalues are non-zero. If all of the eigenvalues of a matrix are positive, the matrix is said to be positive definite. If all of the eigenvalues of a matrix are non-negtive, the matrix is said to be positive semidefinite. Negative definite and negative semidefinite matrices are defined analogously. Matrices with both positive and negative eigenvalues are indefinite. The relation $\mathbf{U}^{\dagger}\mathbf{A}^{\dagger}\mathbf{A}\mathbf{U} = \mathbf{U}^{\dagger}\mathbf{A}^{\dagger}\mathbf{U}\mathbf{U}^{\dagger}\mathbf{A}\mathbf{U} = \boldsymbol{\lambda}^{\dagger}\boldsymbol{\lambda}$ shows that the matrix $\mathbf{A}^{\dagger}\mathbf{A}$ is positive semidefinite. It may also be shown that the eigenvalues of an anti-Hermitian matrix are purely imaginary. It follows that the square of an anti-Hermitian matrix is negative semidefinite.

It is sometimes useful to judge the accuracy of an approximate eigenvalue or eigenvector when it is not practicable to diagonalize the matrix exactly. Consider the Hermitian matrix **A** and an approximate eigenvector **x** where $|\mathbf{x}| = 1$. The residual vector **r** may be defined as

$$\mathbf{r} = (\mathbf{A} - \rho)\mathbf{x} \tag{19}$$

If **x** is an exact eigenvector and if ρ is its exact eigenvalue, then $\mathbf{r} = \mathbf{0}$. Otherwise the norm of the vector **r** may be used to determine the accuracy of these quantities. For a fixed trial vector **x**, the best choice of ρ is given[10,12,13] by

$\rho = \mathbf{x}^\dagger \mathbf{A}\mathbf{x}$, which is the Rayleigh quotient. The vector $\mathbf{x}$ may formally be expanded in the set of exact eigenvectors of $\mathbf{A}$ (these vectors may be chosen to form a unitary matrix and therefore are linearly independent and complete within the vector space). The residual vector may then be written as

$$\mathbf{r} = (\mathbf{A} - \rho)\sum_k \mathbf{u}^k c_k \tag{20}$$

$$= \sum_k (\lambda_k - \rho)\mathbf{u}^k c_k \tag{21}$$

where $c_k = (\mathbf{u}^k)^\dagger \mathbf{x} = (\mathbf{U}^\dagger \mathbf{x})_k$. The vector $\mathbf{c}$ is simply the vector $\mathbf{x}$ represented in the basis defined by the columns of $\mathbf{U}$. The square of the residual norm satisfies the relations

$$|\mathbf{r}|^2 = \sum_k |\lambda_k - \rho|^2 |c_k|^2 \tag{22}$$

$$\geqslant \sum_k |\lambda_{\min} - \rho|^2 |c_k|^2 = |\lambda_{\min} - \rho|^2 \tag{23}$$

where $\lambda_{\min}$ is the exact eigenvalue that is closest to the approximate eigenvalue. The inequality in Eq. (23) is valid because each of the non-negative terms in the summation of Eq. (22) is replaced by another non-negative term of smaller magnitude. The inequality, $|\mathbf{r}| \geqslant |\lambda_{\min} - \rho|$, may be arranged to give the result

$$\rho - |\mathbf{r}| \leqslant \lambda_{\min} \leqslant \rho + |\mathbf{r}| \tag{24}$$

This shows that an exact eigenvalue of the matrix lies within the closed ball centered at ρ with radius $|\mathbf{r}|$ (i.e. on the real line segment of length $2|\mathbf{r}|$ centered at ρ). The relation of Eq. (24) is called the Weinstein residual norm theorem and is often used in the iterative solution of eigenvector equations as a measure of the accuracy of trial eigenvectors[12]. Unfortunately, the eigenvalue error bounds given by Eq. (24) are quite conservative. As shown by Shavitt[13], this may be demonstrated by defining an error vector with the relation

$$\varepsilon \mathbf{\Delta} = \mathbf{x} - \mathbf{u} \tag{25}$$

where $|\mathbf{\Delta}| = 1$, $\mathbf{u}$ is the nearest exact eigenvector and the scalar ε is the root-mean-square vector error and is a first-order function of the residual norm $|\mathbf{r}|$. The equation

$$\varepsilon^2 \mathbf{\Delta}^\dagger (\mathbf{A} - \lambda \mathbf{1})\mathbf{\Delta} = \rho - \lambda \tag{26}$$

shows that

$$|\rho - \lambda| = O(\varepsilon^2) \approx O(|\mathbf{r}|^2) \tag{27}$$

for sufficiently small ε. This demonstrates that the error in the eigenvalue is second order in the error of the eigenvector. For example, if $\varepsilon \approx |\mathbf{r}| \approx 10^{-3}$ then $|\rho - \lambda| \approx 10^{-6}$ assuming the remaining factor from Eq. (26) is approximately of unit magnitude. (This relation between the vector error and the function error is generally true for extremization problems.) Both the approximate relation of Eq. (27) and the rigorous relation of Eq. (24) are useful within their own ranges of applicability.

If the Weinstein residual norm theorem is applied in the special case that the trial vector consists of a column of the unit matrix, then $x_k = \delta_{km}$, $\rho = A_{mm}$ and $r_k = A_{km} - \delta_{km}A_{mm}$. In this case $|\mathbf{r}|$ is simply the norm of the vector constructed from the off-diagonal elements of the mth column of the matrix $\mathbf{A}$. Each diagonal element of the matrix $\mathbf{A}$ determines the center of a closed ball with a radius given by the norm of the corresponding off-diagonal elements of the matrix. The eigenvalues of the matrix lie within these closed balls (i.e. line segments). This property of the eigenvalues is most useful when the off-diagonal elements of the matrix are small, or when they can be made small with an appropriate sequence of matrix transformations. Perturbation theory may also be applied in the case of small residual vector elements[10,12] to give an approximate eigenvalue

$$\lambda \approx A_{mm} - \sum_{k(\neq m)} |r_k|^2/(A_{kk} - A_{mm}) \tag{28}$$

This shows again that the error of the approximate eigenvalue is of second order in the error of the approximate eigenvector.

Consider next the representation of a Hermitian matrix $\mathbf{A}$ in the form $\mathbf{U}\boldsymbol{\lambda}\mathbf{U}^\dagger$. This matrix product may be written in the form

$$\mathbf{A} = \sum_k \lambda_k \mathbf{u}^k(\mathbf{u}^k)^\dagger \tag{29}$$

The matrix formed from the product of vectors, $\mathbf{P}^k = \mathbf{u}^k(\mathbf{u}^k)^\dagger$, is called a vector outer product. The expansion of a matrix in terms of these outer products is called the spectral resolution of the matrix. The matrix $\mathbf{P}^k$ satisfies the relation $\mathbf{P}^k\mathbf{P}^k = \mathbf{P}^k$ as do matrices of the more general form, $\mathbf{P} = \sum \mathbf{P}^k$, where the summation is over an arbitrary subset of outer product matrices constructed from orthonormal vectors. Matrices that satisfy the relation $\mathbf{P}^2 = \mathbf{P}$ are called projection operators or projection matrices[14]. If $\mathbf{P}$ is a projection matrix, then $(\mathbf{1} - \mathbf{P})$ is also a projection matrix. Projection matrices operate on arbitrary vectors, measure the components within a subspace (e.g. spanned by the vectors $\mathbf{u}^k$ used to define the projection matrix) and result in a vector within this subspace.

A matrix of the form $\mathbf{A} = (\mathbf{1} - 2\mathbf{x}\mathbf{x}^\dagger)$ where $|\mathbf{x}| = 1$ is another outer product matrix that is useful in the MCSCF method. This matrix is both unitary and Hermitian and is called an elementary Householder transformation matrix[10,12,15]. These transformation matrices are useful in bringing Hermitian matrices to tridiagonal form.

Another outer product matrix that is particularly useful is of the form

$$\mathbf{S} = \mathbf{y}\mathbf{x}^\dagger - \mathbf{x}\mathbf{y}^\dagger \tag{30}$$

where $|\mathbf{x}| = 1$ and $\mathbf{x}^\dagger\mathbf{y} = 0$. The vector $\mathbf{x}$ may be regarded as a reference vector while the vector $\mathbf{y}$ belongs to the orthogonal complement to the vector $\mathbf{x}$. The anti-Hermitian matrix $\mathbf{S}$ is useful because the identities

$$\mathbf{S}^{2n}\mathbf{x} = (-1)^n|\mathbf{y}|^{2n}\mathbf{x} \tag{31}$$

$$\mathbf{S}^{2n+1}\mathbf{x} = (-1)^n|\mathbf{y}|^{2n}\mathbf{y} \tag{32}$$

may be used to show[16,17]

$$\exp(\mathbf{S})\mathbf{x} = \cos(|\mathbf{y}|)\mathbf{x} + \sin(|\mathbf{y}|)|\mathbf{y}|^{-1}\mathbf{y} \tag{33}$$

Since the vector $(|\mathbf{y}|^{-1}\mathbf{y})$ is a unit vector pointing in some direction orthogonal to the reference vector $\mathbf{x}$, the matrix $\exp(\mathbf{S})$ gives a convenient representation of a rotation between the two vectors $\mathbf{x}$ and $\mathbf{y}$. Since $[\exp(\mathbf{S})]^\dagger = \exp(-\mathbf{S}) = [\exp(\mathbf{S})]^{-1}$, the transformation is unitary: when operating on an arbitrary normalized vector the result is also a normalized vector. The quantity $|\mathbf{y}|$ determines the amount of rotation within the plane defined by the two vectors. This parametrization of an arbitrary vector rotation is useful in the MCSCF method in describing CSF mixing coefficient changes.

B. Matrix Partitioning Theory

It is sometimes convenient to consider a matrix as consisting of sub-blocks, each of which is another matrix of smaller dimension. Consider the eigenvector equation, Eq. (9), for a Hermitian matrix that is partitioned into four sub-blocks

$$\begin{pmatrix} \mathbf{B} & \mathbf{C} \\ \mathbf{C}^\dagger & \mathbf{M} \end{pmatrix}\begin{pmatrix} \mathbf{u} \\ \mathbf{v} \end{pmatrix} = \begin{pmatrix} \mathbf{u} \\ \mathbf{v} \end{pmatrix}\lambda \tag{34}$$

As written here, the $\mathbf{B}$ and $\mathbf{M}$ matrices are square. The usual rules for matrix multiplication also apply to the partitioned expressions. For example, Eq. (34) is equivalent to the pair or coupled equations

$$\mathbf{Bu} + \mathbf{Cv} = \mathbf{u}\lambda \tag{35}$$

$$\mathbf{C}^\dagger\mathbf{u} + \mathbf{Mv} = \mathbf{v}\lambda \tag{36}$$

The second equation may be solved for $\mathbf{v}$ in terms of $\mathbf{u}$ giving

$$\mathbf{v} = -(\mathbf{M} - \lambda\mathbf{1})^{-1}\mathbf{C}^\dagger\mathbf{u} \tag{37}$$

which, when substituted into Eq. (35), gives

$$[\mathbf{B} - \mathbf{C}(\mathbf{M} - \lambda\mathbf{1})^{-1}\mathbf{C}^\dagger]\mathbf{u} = \mathbf{u}\lambda \tag{38}$$

This is an expression giving the eigenvalue of the original equation and the component of the vector corresponding to the first block of the full vector. Substitution of $\mathbf{u}$, determined from Eq. (38), into Eq. (37) gives the unnormalized full vector. This matrix partitioning approach applied to the eigenvalue problem has been used by Löwdin[18] in the study of perturbation theory. A few of the results of this analysis that are particularly important in the MCSCF method will be discussed here. The reader is referred to the literature[18,19] for more details.

Consider the left-hand side of Eq. (38) as a matrix function of the parameter λ. Since the matrix elements depend on λ, the eigenvalues of this matrix are also functions of this parameter. Let this set of eigenvalues be collected together as a multivalued function, written $\mathbf{L}(\lambda)$, and let a particular branch of this function be written as $L^k(\lambda)$. There are as many branches of this function as the dimension of the original eigenvector equation (for brevity the eigenvalues are assumed to be distinct and to have non-vanishing $\mathbf{u}$ components). If the function $R(\lambda) = \lambda$ is defined to correspond to the right-hand side of Eq. (38), then the eigenvalues of the original equation are given by the set of values of λ for which $L^k(\lambda) = R(\lambda)$ for all the branches.

1. The Bracketing Theorem

An important special case of this analysis occurs when the dimension of the **B** matrix is unity. Without loss of generality the matrix **M** may be considered to be diagonal. (It may always be brought to diagonal form with a suitable block-diagonal unitary transformation.) In this case $L(\lambda)$ is a single-valued function of the form

$$L(\lambda) = B - \sum_n C_n^* C_n/(M_n - \lambda) \tag{39}$$

A plot of $L(\lambda)$ and $R(\lambda)$ is shown in Fig. 1 for a representative case. It may be verified from Eq. (39) that the individual branches of $L(\lambda)$ are monotonically decreasing functions of λ (this is also true for the multi-valued case). The simple poles in this function occur at the eigenvalues of the matrix **M**. The horizontal asymptote is the value of B and is approached from below as $\lambda \to -\infty$ and from above as $\lambda \to +\infty$. Because of these characteristics of the function $L(\lambda)$, the lowest value of λ for which $L(\lambda) = R(\lambda)$ is in the region $\lambda \leqslant M_1$. This corresponds to the lowest eigenvalue of the full matrix. The second intersection occurs in the region $M_1 \leqslant \lambda \leqslant M_2$. Each of the regions separated by the eigenvalues of the matrix **M** will contain an eigenvalue of the full matrix. The eigenvalues of the full matrix are said to 'bracket' the eigenvalues of the sub-block matrix **M**. This bracketing theorem is summarized by the sequence of inequalities:

$$\lambda_1 \leqslant M_1 \leqslant \lambda_2 \leqslant M_2 \leqslant \cdots \leqslant \lambda_k \leqslant M_k \leqslant \lambda_{k+1} \leqslant \cdots.$$

This relation is also called the Hylleraas–Undheim–MacDonald theorem[20,21].

An important application of the bracketing theorem is the case where the **M** matrix corresponds to a matrix representation of the Hamiltonian operator in

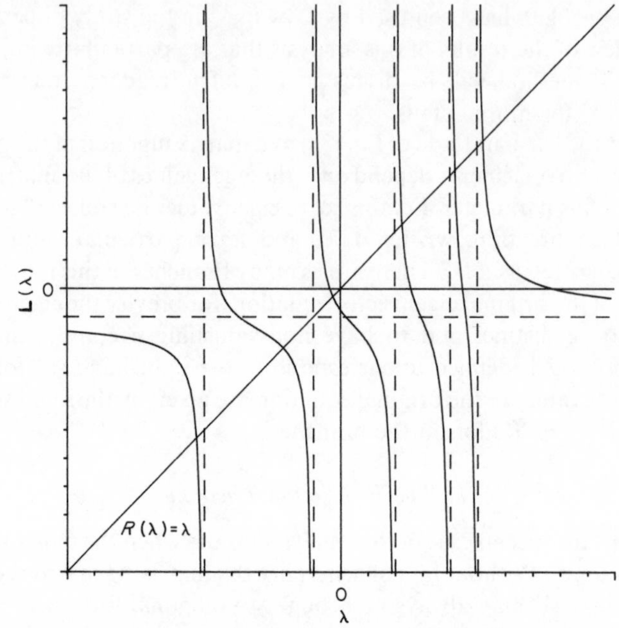

Fig. 1. Graphical example of the bracketing theorem. The vertical asymptotes are the eigenvalues of the matrix **M**. The horizontal asymptote is the diagonal element B of the full matrix. The intersections of the branches of the function $\mathbf{L}(\lambda)$ with the straight line $R(\lambda)$ are the eigenvalues of the full matrix. These intersections satisfy the bracketing theorem relations

$$\lambda_1 \leqslant M_1 \leqslant \lambda_2 \leqslant \cdots \leqslant \lambda_i \leqslant M_i \leqslant \lambda_{i+1} \leqslant \cdots \leqslant M_n \leqslant \lambda_{n+1}.$$

some limited CSF expansion space which may be chosen to be orthonormal. The eigenvalues of the **M** matrix are then the approximate energies, $E_1^{(n)} \leqslant E_2^{(n)} \leqslant \cdots \leqslant E_n^{(n)}$ where n is the subspace dimension. The addition of a new expansion term results in a new matrix representation of dimension $(n+1)$. The eigenvalues of this new representation satisfy the bracketing theorem ordering:

$$E_1^{(n+1)} \leqslant E_1^{(n)} \leqslant E_2^{(n+1)} \leqslant E_2^{(n)} \leqslant \cdots \leqslant E_k^{(n+1)} \leqslant E_k^{(n)} \leqslant E_{k+1}^{(n+1)} \leqslant \cdots \leqslant E_n^{(n)} \leqslant E_{n+1}^{(n+1)}.$$

If this process of CSF expansion term addition is continued, the approximate energies at each step form upper bounds to the approximate energies of the next step. In principle, this process may be continued until all possible expansion terms are added for the finite orbital expansions usually considered. This limiting case[13] is called the full CI expansion. A correspondence may then be established between the lowest roots of some limited expansion and the lowest roots of the full CI expansion. This process may be considered to be extended further with the addition of other orbitals and their corresponding

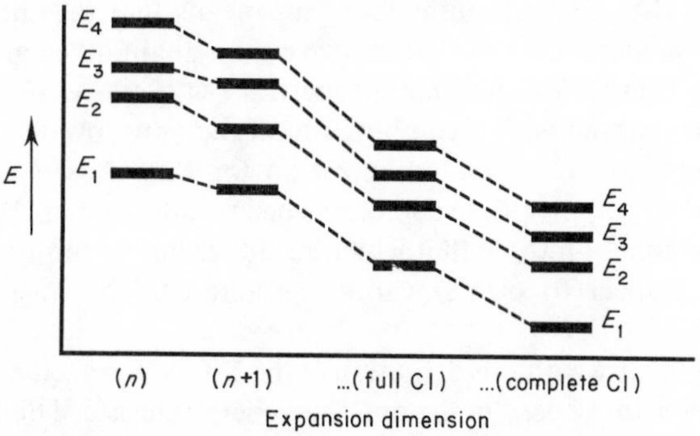

Fig. 2. Schematic representation of the lowest eigenvalues of the Hamiltonian matrix as a function of the wavefunction expansion length. As shown in Fig. 1, the eigenvalues for two successive dimensions satisfy the bracketing theorem ordering, given by $E_1^{(n+1)} \leqslant E_1^{(n)} \leqslant E_2^{(n+1)} \leqslant E_2^{(n)} \leqslant \cdots \leqslant E_i^{(n+1)} \leqslant E_i^{(n)} \leqslant E_{i+1}^{(n+1)} \leqslant \cdots$. The accuracy of the correspondence of the approximate eigenvalues and the exact eigenvalues depends on the CSF expansion space as discussed in the text.

CSFs into the wavefunction expansion. Although unattainable in molecular calculations, the second limiting case, corresponding to full CI for a complete orbital set, is called the complete CI expansion[13]. The eigenvalues of the complete CI expansion are the exact energies within the clamped-atomic-nucleus Born–Oppenheimer[14] approximation. A correspondence may then be established with the bracketing theorem between the lowest eigenvalues of a limited CSF expansion and those of the exact complete CI expansion. This is illustrated schematically in Fig. 2.

C. The Definition of the Multiconfiguration Self-consistent Field Method

This correspondence of approximate energy eigenvalues with the exact energy eigenvalues is the formal basis of the MCSCF method. The CSF expansion terms, which depend on the orbitals, determine the Hamiltonian matrix elements. The shape of the orbitals then affects the eigenvalues of the Hamiltonian matrix expanded in terms of a limited CSF basis. For a fixed CSF expansion, the shape of the orbitals may be varied, thereby producing different approximate energies. Since the bracketing theorem is valid for any or these orbital choices, the particular choice that gives the lowest approximate energy, and therefore the closest approximation to the exact energies, gives the best wavefunction for the given CSF expansion. It should be noted, however, that the minimization of the energy for some fixed CSF expansion set is not equivalent to the maximization of the overlap of the approximate wave function with the exact wavefunction. This means that the energy minimization condition along does not guarantee even a qualitative agreement of the approximate energy, or of other properties, with the exact results. This agreement is achieved with a combination of the appropriate choice of CSF expansion set and of orbital optimization for this chosen expansion set. Assuming such an appropriate choice has been made, we may then define the MCSCF wavefunction to be that which results from the minimization of the energy with respect to orbital variations and CSF expansion coefficient variations.

A formal difficulty with this definition of the MCSCF wavefunction is that it may lead to spurious cusps in the potential energy surface. This surface is the locus of approximate electronic energies as a function of the geometrical parameters of the molecule[14]. This cusp behavior results from the fact that, at a given geometry, different choices of orbitals for a particular CSF expansion may lead to identical energies. These different orbitals can correspond to qualitatively different descriptions of the electronic structure of the molecule. These two descriptions of the electronic structure result in two analytic energy surfaces that, in general, will not have the same geometrical dependence. These two approximate wavefunctions give energy curves that intersect and cross. The choice of the lowest-energy solution results in an approximate energy curve that is only piecewise analytic and has a cusp, a discontinuity of the first derivative, at these crossing points. (Although the full CI polyatomic potential energy surface may also possess similar cusp behavior[22] in the vicinity of allowed surface crossings, this spurious cusp behavior of the MCSCF wavefunction does not necessarily correspond to the exact cusp behavior. For example, an exact diatomic potential curve does not have cusps because there is only one geometrical degree of freedom but two independent conditions that must be satisfied. However, an approximate MCSCF diatomic curve may have cusps, which are of the spurious type discussed here.) It should be emphasized, however, that this multiple-solution problem is not unique to the MCSCF method. Any method designed to approximate the full CI wavefunction or energies may have this formal problem, the solution to which is to improve the approximate description until these artifacts disappear. The conceptual simplicity and interpretability of the MCSCF method aids in the recognition and solution of these problems.

1. The Excited-state MCSCF Method

This energy minimization definition of the MCSCF wavefuncton is applicable to both the lowest-energy eigenvalue, the ground state, and to the higher-energy eigenvalues, the excited states. The bracketing theorem places a restriction on the energy lowering possible for a particular excited state as a new expansion CSF is added. For example, $E_1^{(n)} \leqslant E_2^{(n+1)}$, which shows that no matter how well the $(n+1)$th expansion CSF is chosen, the excited-state eigenvalue cannot be improved beyond a particular point. In general this limit is determined by how well the lower-energy states are described.

a. The ideal excited-state method

These bracketing theorem considerations allow an ideal MCSCF method to be defined for excited states. This ideal method[23] is first to define the lower-energy states to be optimal as determined by separate MCSCF calculation on these states, and then to define the energy of the excited state to be the appropriate Hamiltonian matrix eigenvalue obtained from a wavefunction expansion that includes these lower-energy states as expansion terms.

For example, suppose that the lowest-energy eigenvalue has been optimized with the MCSCF method. That is, within some suitable CSF expansion space the orbitals have been chosen to minimize the lowest eigenvalue of the Hamiltonian matrix. Let this wavefunction be written as $|mc_1\rangle$. The expansion CSF space for the excited state is now constructed by adding expansion terms to the previously optimized state $|mc_1\rangle$. If the same expansion set is chosen for both the ground and excited states, this set is obviously linearly dependent as long as the orbitals are unchanged since $|mc_1\rangle$ is expanded with this set in the first place. However, as the orbitals are varied to minimize the energy of the excited state, the wavefunction $|mc_1\rangle$, for which the orbitals are not allowed to vary, becomes linearly independent. Since there is additional flexibility compared to the initial ground-state calculation, the ground-state energy may become lower but is cannot become higher (because of the bracketing theorem). The ground-state description is therefore improved by the addition of the excited-state expansion terms. The excited-state eigenvalues obtained in this expansion will all be upper bounds to the exact eigenvalues. This allows the excited-state orbitals to be determined, just as in the ground-state case, as those orbitals that minimize the appropriate eigenvalue. The computational difficulty with this ideal method is that the construction of the Hamiltonian matrix and overlap matrix is quite difficult because of the different orbital sets used to describe the individual wavefunction expansion terms. (A non-orthogonal orbital MCSCF method has been applied in an approximate form for small molecular systems by Ruttink and van Lenthe[24] and by van Lenthe and Balint-Kurti[25].) It is conceptually useful to define the ideal MCSCF model to aid in the analysis of other approximate MCSCF methods.

b. The usual excited-state method

The MCSCF method that is usually applied to excited states is an approximation of the ideal method just described. The approximation made is that the lower-energy states can be described in terms of the excited-state orbitals during the excited-state optimization process[23,26]. The use of a common set of orbitals during the optimization process results in a much simpler computational procedure. It is useful to consider the behavior of the lower-energy states as the orbitals are adjusted for the particular excited state. Consider the case where the ground state has been optimized, that is, where a set of orbitals has been determined that minimizes the energy of the lowest root of the Hamiltonian matrix. Now consider the optimization of the orbitals for the first excited state. As the orbitals change to give a better description of the second root of the Hamiltonian matrix, the lowest eigenvalue must increase away from its optimal value. (This behavior is in contrast to the ideal method in which the lowest eigenvalue actually is improved during the excited-state orbital optimization process.) In well behaved cases, the energy separation of the two roots is sufficiently large, and the CSF space is sufficiently flexible, so that both states may be described reasonably well. These conditions allow the excited state to be optimized with no difficulty.

In other cases however, the CSF space may not be sufficiently flexible, or the energy separation may be very small, so that the approximate ground-state energy increases until it approaches the excited-state energy. The excited-state energy is described essentially by determining when the two roots become degenerate, at which time the wavefunction becomes undefined within the degenerate space. Unless the two roots are degenerate in the full CI expansion case (e.g. at the polyatomic surface intersections discussed previously), this pseudo-degeneracy is simply an artifact of the MCSCF model and results in an unrealistic wavefunction and energy surface. This unrealistic behavior may be avoided, in principle, by expanding the CSF space so that both states are described adequately in terms of the excited-state orbitals. Thus, in the calculation of a general excited-state potential energy surface, the CSF expansion space must be sufficiently flexible for all of the lower-energy states to be adequately described in terms of the optimal excited-state orbitals. This aspects of excited-state wavefunction optimization is discussed in more detail in the following sections.

D. *N*-Electron Expansion Space Representation

The MCSCF wavefunctions discussed in this chapter consists of linear expansions in terms of N-electron basis functions. These N-electron functions depend on $4N$ variables and are spanned by the space of the Cartesian product of the spatial and spin coordinates for each electron. The total wavefunction must be antisymmetric with respect to interchange of the coordinates of any pair of electrons. This is because electrons are half-integer spin particles called fermions and all fermion wavefunctions must satisfy this property[14]. This is treated as a constraint on the class of admissible basis functions in the MCSCF method and in the CI method[12,13]. These admissible functions are further limited to be of the form of products of one-electron functions called orbitals, or more explicitly, because of their dependence on electron spin, spin orbitals. This is not the most effective expansion form for N-electron wavefunctions but, for computational reasons, it is the most popular. The simplest expansion functions of this form are called Slater determinants, and are the result of a particular projection operator, called the antisymmetrizer, operating on a simple product of singly occupied spin orbitals to produce an appropriately antisymmetrized function. These expansion terms may be written in the form of a determinant[14] where each row corresponds to a particular spin orbital and each column corresponds to the coordinates of a particular electron. The diagonal elements of such a determinantal representation may be used to specify the expansion term. For example, the determinant constructed from the first N spin orbitals may be written in the following forms:

$$|\phi_1(\mathbf{r}_1)\phi_2(\mathbf{r}_2)\cdots\phi_N(\mathbf{r}_N)| = \mathscr{A}[\phi_1(\mathbf{r}_1)\phi_2(\mathbf{r}_2)\cdots\phi_N(\mathbf{r}_N)] \tag{40}$$

$$= (N!)^{-1/2}\sum_{p}(-1)^{P_p}[\phi_1(\mathbf{r}_1)\phi_2(\mathbf{r}_2)\cdots\phi_N(\mathbf{r}_N)] \tag{41}$$

$$= (N!)^{-1/2}\begin{vmatrix} \phi_1(\mathbf{r}_1) & \phi_1(\mathbf{r}_2) & \cdots & \phi_1(\mathbf{r}_N) \\ \phi_2(\mathbf{r}_1) & \phi_2(\mathbf{r}_2) & \cdots & \phi_2(\mathbf{r}_N) \\ \vdots & & & \vdots \\ \phi_N(\mathbf{r}_1) & \phi_N(\mathbf{r}_2) & \cdots & \phi_N(\mathbf{r}_N) \end{vmatrix} \tag{42}$$

The $(N!)^{-1/2}$ factor included in these expressions ensures that the determinants are normalized when the orbitals are normalized. Eq. (41) gives an explicit representation of the antisymmetrizer. This summation is over the $N!$ permutations of electron coordinates for a fixed orbital order, or equivalently, over the permutations of spin-orbital labels for a fixed order of electrons. The exponent P_p is the number of interchanges required to bring a particular permuted order of electron coordinates, or of spin-orbital labels, back to the original order. Different expansion terms are generated when different spin orbitals are employed in the determinant. For convenience, we will choose this spin-orbital basis to be the direct product of the set of n spatial orbitals and the set of spin factors $\{\alpha, \beta\}$. A particular spin orbital of this form may be written as $\phi_{r\sigma}$, where r ($=1$ to n) labels the spatial orbital and σ labels the spin factor, or simply as ϕ_r, where the combined index r ($=1$ to $2n$) labels both the spatial and spin components. The notation used will be clear from the context.

The product form of the spin-orbital basis imposes no restriction on the space spanned by the full set of expansion determinants but it will impose restrictions on wavefunctions constructed from subsets of these expansion terms. For convenience, the spatial coordinates of the electrons are usually abbreviated to label only the electron, or even further, the electrons are assumed to occur in consecutive order and only the orbitals are specified. For example the determinant in Eq. (40) may be written as $|\phi_1(1)\phi_2(2)\cdots\phi_N(N)|$ or as $|\phi_1\phi_2\cdots\phi_N|$.

If the rows of a matrix become linearly dependent, the determinant of that matrix vanishes. This is also true of the determinantal expansion terms. The spin orbitals in the determinant must be linearly independent for the function to have non-zero values for some set of electron coordinates. For an orthonormal spin-orbital basis, this means that no spin orbital may be occupied more than once. The determinantal form of the expansion basis shows that a particular electron cannot be associated with a particular orbital or, accordingly, with a particular region of space. It is still convenient, however, to refer to the electrons in a particular orbital as if the wavefunction were a simple orbital product. The strict interpretation of these references is to the orbital, which is occupied in the determinant by all the electrons, and not to the individual electron which happens to occupy the orbital in some of the $N!$ terms of the determinant.

E. Second Quantization

The calculation of expectation values of operators over the wavefunction, expanded in terms of these determinants, involves the expansion of each determinant in terms of the $N!$ expansion terms followed by the spatial coordinate and spin integrations. This procedure is simplified when the spatial orbitals are chosen to be orthonormal. This results in the set of Slater–Condon rules[27] for the evaluation of one- and two-electron operators. A particularly compact representation of the algebra associated with the manipulation of determinantal expansions is the method of 'second quantization' or the 'occupation number representation'. This is discussed in detail in several textbooks and review articles[2,28,29], to which the reader is referred for more detail. An especially entertaining presentation of second quantization is given by Mattuck[30]. The usefulness of this approach is that it allows quite general algebraic manipulations to be performed on operator expressions. These formal manipulations are more cumbersome to perform in the wavefunction approach. It should be stressed, however, that these approaches are equivalent in content, if not in style, and lead to identical results and computational procedures.

In this chapter it is convenient to use both the second quantization approach and the traditional wavefunction approach, depending on the particular discussion. In the opinion of the author, some aspects of the MCSCF method are best understood in terms of a traditional wavefunction-oriented approach while other aspects are best understood in terms of the algebraic manipulations of operators for which the method of second quantization is best suited. This notation will be briefly reviewed so that the reader may confidently switch, when necessary, from one approach to the other.

1. *Bra and Ket Expansion Terms*

In the method of second quantization, a primitive expansion term is represented by a ket (or a bra, depending on how the expansion term is used). These expansion kets may be regarded as members of an abstract space, but, for our purposes here, they will be treated simply as an alternate representation of a determinant formed from orthonormal spin orbitals. An example of the correspondence between a determinant and a ket may be written for a three-electron determinant in a spin-orbital basis of dimension 4 as

$$|\phi_1\phi_2\phi_4| \rightarrow |1101\rangle \tag{43}$$

Without loss of generality, the electron coordinates may be assumed to occur in natural order and the spin orbitals permuted into their natural order in the determinantal expression to determine the sign of the ket. The occupation number representation of a general determinant simply has a '1' in positions corresponding to occupied orbitals and a '0' in positions corresponding to empty orbitals. There are as many positions in the ket as there are spin orbitals. It is useful to define a mapping, denoted by $[i]$, that gives the orbital index of the ith spin orbital in the determinant. In the example in Eq. (43), $[1]=1$, $[2]=2$ and $[3]=4$. A general term has the form

$$|m\rangle = |m_1 m_2 \cdots\rangle \tag{44}$$

where m_i corresponds to the occupation, 0 or 1, of spin orbital ϕ_i in the ket. Matrix elements between different determinants correspond to the integration over all the coordinates of the electrons. For orthonormal orbitals, the overlap between two determinants is zero unless the occupation pattern matches exactly

$$\langle m|k\rangle = \iint \cdots \int |\phi_{[1]}\phi_{[2]}\cdots\phi_{[N]}|_m^* |\phi_{[1]}\phi_{[2]}\cdots\phi_{[N]}|_k \, d\mathbf{r}_1 \, d\mathbf{r}_2 \cdots d\mathbf{r}_N \tag{45}$$

$$= \delta([1]_m, [1]_k)\delta([2]_m, [2]_k)\cdots\delta([N]_m, [N]_k) \tag{46}$$

$$= \langle m_1 m_2 \cdots | k_1 k_2 \cdots\rangle \tag{47}$$

$$= \delta(m_1, k_1)\delta(m_2, k_2)\cdots \tag{48}$$

$$= \delta_{mk} \tag{49}$$

Eqs (45) and (46) are the usual wavefunction approach expressions while Eqs (47) through (49) are the analogous expressions in the second quantized notation.

2. Creation and Annihilation Operators

It is convenient to introduce an electron annihilation operator, written as a_r, which destroys an electron in spin orbital ϕ_r to give an $(N-1)$-electron ket if spin orbital ϕ_r is occupied, and which gives zero if the orbital is unoccupied in the ket. For example

$$a_1|1101\rangle = |0101\rangle \tag{50}$$

$$a_3|1101\rangle = 0 \tag{51}$$

The general expression must include a sign factor to reflect the antisymmetry of the determinant with respect to spin-orbital interchange. The general expression for the result of an annihilation operator on an arbitrary ket is written as

$$a_r|m_1 m_2 \cdots m_r \cdots m_{2n}\rangle = m_r(-1)^{\theta_r}|m_1 m_2 \cdots 0_r \cdots m_{2n}\rangle \tag{52}$$

where the sign factor is determined by

$$\theta_r = \sum_{i=1}^{(r-1)} m_i \tag{53}$$

This gives a sign factor of $+1$ if there are an even number of electrons occurring before spin orbital ϕ_r and a factor of -1 for an odd number of electrons. It is also convenient to define an electron creation operator, written $a_r^\dagger$, which acts to create an electron in spin orbital ϕ_r or to give zero if such an electron already exists. The general expression for an arbitrary ket is

$$a_r^\dagger|n_1 n_2 \cdots n_r \cdots n_{2n}\rangle = \delta(n_r, 0)(-1)^{\theta_r}|n_1 n_2 \cdots 1_r \cdots n_{2n}\rangle \tag{54}$$

where the sign factor is again determined by Eq. (53). An important relation between the annihilation operators and the creation operators is the adjoint relation, $(a_r)^\dagger = a_r^\dagger$. This gives the correspondence between bras and kets involving these operators in the second quantized notation

$$(a_r^\dagger a_s|n\rangle)^\dagger = \langle n|a_s^\dagger a_r \tag{55}$$

From these definitions of electron creation and annihilation operators, the following anticommutator relations may be derived:

$$[a_r, a_s]_+ = 0 \tag{56}$$

$$[a_r^\dagger, a_s^\dagger]_+ = 0 \tag{57}$$

$$[a_r^\dagger, a_s]_+ = \delta_{rs} \tag{58}$$

where an anticommutator is defined as $[a, b]_+ = ab + ba$. These operator relations hold when operating on any ket involving any number of electrons. This allows formal manipulations of the creation and annihilation operators to be performed without regard to the set of bras and kets on which the final result will act. This is, of course, a very useful feature in deriving general expressions. Operators of the product form, $a_r^\dagger a_s$, $a_r^\dagger a_s^\dagger a_t a_u$, etc., are particularly important in the MCSCF method because they conserve the number of electrons. For example, if a ket contains an occupied spin orbital ϕ_s and an empty spin orbital ϕ_r, the effect of the operator $a_r^\dagger a_s$ is to 'excite' the electron from orbital ϕ_s to orbital ϕ_r. Operators of this form are therefore called single excitation operators

$$a_r^\dagger a_s |m_1 m_2 \cdots 0_r \cdots 1_s \cdots\rangle = (-1)^{\theta_{(r-s)}} |m_1 m_2 \cdots 1_r \cdots 0_s \cdots\rangle \tag{59}$$

The sign factor is determined from $\theta_{(r-s)} = \theta_r - \theta_s$ and may be interpreted as the number of occupied spin orbitals occurring between the locations corresponding to the spin orbitals ϕ_r and ϕ_s in the ket $|m\rangle$.

3. *Representation of Quantum-mechanical Operators*

The importance of these creation and annihilation operators results from the fact that quantum-mechanical operators may be written as an expansion in products of these operators. These products are then said to form an operator basis. For example, an arbitrary one-electron operator has the form

$$A = \sum_r \sum_s A_{rs} a_r^\dagger a_s \tag{60}$$

with the expansion coefficients A_{rs} given by the integral

$$A_{rs} = \int \phi_r^*(\mathbf{r}) A(\mathbf{r}) \phi_s(\mathbf{r})\, d\mathbf{r} \tag{61}$$

over the space and spin coordinates, where $A(\mathbf{r})$ is the usual definition of the quantum-mechanical operator[30] involving multiplication, differentiation with respect to $\mathbf{r}$, etc. An analogous result holds for two-electron operators, three-electron operators, etc. A one-electron operator may therefore be expanded in the single excitation operators basis, a two-electron operator may be expanded in the double excitation operator basis, and so on. The integration over the spin coordinate in Eq. (61) simplifies the general operator expansion of Eq. (60), if the operator is independent of spin, by forcing the integrals corresponding to different electron spins to be zero. Spatial symmetry simplifies the expression in a similar way since the integrand must contain the completely symmetric representation to be non-zero.

For an example of an operator representation, the electronic Hamiltonian operator may be written in this second quantized notation as

$$H = \sum_r \sum_s h_{rs} a_r^\dagger a_s + \tfrac{1}{2} \sum_r \sum_s \sum_t \sum_u g_{rstu} a_r^\dagger a_t^\dagger a_u a_s \tag{62}$$

where the one-electron coefficients h_{rs} include the electron kinetic energy and electron–nuclear attraction terms and the two-electron coefficients g_{rstu}

include the electron–electron repulsion

$$g_{rstu} = \iint \phi_r^*(\mathbf{r}_1)\phi_s(\mathbf{r}_1)|\mathbf{r}_1 - \mathbf{r}_2|^{-1}\phi_t^*(\mathbf{r}_2)\phi_u(\mathbf{r}_2)\,d\mathbf{r}_1\,d\mathbf{r}_2 \tag{63}$$

Spin and spatial symmetry also simplify the one- and two-electron expansion terms in Eq. (62) when the usual spin-independent Hamiltonian operator is used.

A unique feature of the occupation number representation is that the number of electrons does not appear in the definition of the Hamiltonian operator in this form as it does in the wavefunction form. This is because all of the occupation information resides in the bras and kets. This is true for any operator in second quantized form. This feature is used to advantage in theories that allow the number of particles to change, and to a more limited extent in the calculation of electron affinities and ionization potentials. It is less important to the MCSCF method but it is useful to remember that the bras and kets contain all of the occupation information. Other details of the wavefunction, such as the AO and MO basis set information, are included in the integrals that are used as expansion coefficients in the second quantized representation of the operator.

This occupation information may be determined for a particular ket using operators of the form $a_r^\dagger a_r$. It may be verified by inspection that this operator measures the occupation of spin orbital ϕ_r in a ket and leaves the occupation of the ket unchanged

$$a_r^\dagger a_r |m_1 m_2 \cdots m_r \cdots\rangle = m_r |m_1 m_2 \cdots m_r \cdots\rangle \tag{64}$$

Every primitive ket is an eigenfunction of an operator of this form and the matrix representation of this operator in the ket space is diagonal, i.e. $\langle m|a_r^\dagger a_r|k\rangle = m_r \delta_{mk}$. If this operator is summed over all spin orbitals, the resulting operator counts the number of electrons in the ket

$$\sum_r a_r^\dagger a_r |m_1 m_2 \cdots m_r \cdots\rangle = N|m_1 m_2 \cdots m_r \cdots\rangle \tag{65}$$

This operator, called the number operator[30], also has a diagonal matrix representation. If the summation in the number operator is limited to only α-type spin orbitals, the eigenvalue is the number of α-spin electrons instead of the total number of electrons. This, in addition to the analogous β-spin number operator, gives a convenient representation of the S_z operator

$$S_z = \tfrac{1}{2}\sum_r (a_{r\alpha}^\dagger a_{r\alpha} - a_{r\beta}^\dagger a_{r\beta}) \tag{66}$$

Every primitive ket is an eigenfunction of the S_z operator, a result of the direct product form of the spin-orbital basis. The ladder[31] operators S_+ and S_- defined as

$$S_+ = \sum_r a_{r\alpha}^\dagger a_{r\beta} \tag{67}$$

$$S_- = \sum_r a_{r\beta}^\dagger a_{r\alpha} \tag{68}$$

allow the S^2 operator to be defined in the usual way

$$S^2 = S_z^2 + \tfrac{1}{2}(S_+S_- + S_-S_+) \tag{69}$$

$$= S_z^2 + S_z + S_-S_+ \tag{70}$$

In general, an arbitrary ket is not an eigenfunction of the S^2 operator. It may be verified, however, that the three operators H, S_z and S^2 all commute with each other. This means that it is possible to construct wavefunctions that are simultaneous eigenfunctions of all three of these operators with suitable linear combinations of expansion kets.

The matrix representation of quantum-mechanical operators in the ket expansion space requires the evaluation of the matrix elements of the product operators $a_r^\dagger a_s$ for one-particle operators and of $a_r^\dagger a_t^\dagger a_u a_s$ for two-particle operators. These matrix elements may then be used in the evaluation of expectation values of these quantum-mechanical operators. For example, for the one-electron operator

$$\langle 0|A|0\rangle = \sum_r \sum_s A_{rs} \sum_m \sum_{m'} c_m^* \langle m|a_r^\dagger a_s|m'\rangle c_{m'} \tag{71}$$

The matrix elements $\langle m|a_r^\dagger a_s|m'\rangle$ and $\langle m|a_r^\dagger a_t^\dagger a_u a_s|m'\rangle$ are called the one- and two-particle coupling coefficients and for determinantal expansions take the values $-1, 0$ and $+1$. These coupling coefficients may then be used to form the one- and two-particle spin-orbital density matrices

$$D_{rs} = \sum_m \sum_{m'} c_m^* \langle m|a_r^\dagger a_s|m'\rangle c_{m'} \tag{72}$$

$$d_{rstu} = \sum_m \sum_{m'} c_m^* \langle m|a_r^\dagger a_t^\dagger a_u a_s|m'\rangle c_{m'} \tag{73}$$

Expectation values may be calculated using these quantities, without further explicit reference to the ket expansion space, as indicated in Eq. (71) for one-electron operators.

F. Orbital Transformations

The effect of a transformation of the orbital basis on the expansion kets using second quantization is important in the MCSCF method. An arbitrary ket may be written as an ordered product of creation operators corresponding to the occupied spin orbitals operating on a completely empty determinant[6]. For example

$$|1101\rangle = a_1^\dagger a_2^\dagger a_4^\dagger |0000\rangle \tag{74}$$

The completely empty determinant is called the true vacuum and is written as $|\text{vac}\rangle$. A determinant of the same occupation but defined with respect to a different orbital basis may be written as

$$|1101\rangle' = a_1'^{\dagger} a_2'^{\dagger} a_4'^{\dagger} |\text{vac}\rangle \tag{75}$$

where the relation between the two sets of orbitals is given by the unitary transformation

$$\boldsymbol{\phi}' = \boldsymbol{\phi}\mathbf{U} \tag{76}$$

The corresponding relation between the two sets of annihilation and creation operators is

$$a_s' = \sum_r a_r U_{rs}^* \tag{77}$$

$$a_s'^{\dagger} = \sum_r a_r^{\dagger} U_{rs} \tag{78}$$

This operator transformation gives the effect of an annihilation or creation operator, defined in terms of one spin-orbital basis, as it operates on kets defined with respect to another orbital basis. As shown by Thouless[28], and as demonstrated for the MCSCF method by Dalgaard and Jørgensen[6], this transformation of annihilation and creation operators may also be written as

$$a_s' = \exp(-\mathrm{i}\Lambda) a_s \exp(\mathrm{i}\Lambda) \tag{79}$$

$$a_s'^{\dagger} = \exp(-\mathrm{i}\Lambda) a_s^{\dagger} \exp(\mathrm{i}\Lambda) \tag{80}$$

where Λ is an Hermitian operator

$$\Lambda = \sum_r \sum_s \Lambda_{rs} a_r^{\dagger} a_s \tag{81}$$

constructed from the elements of the Hermitian matrix $\boldsymbol{\Lambda}$. The relation between the elements of $\boldsymbol{\Lambda}$ and the elements of $\mathbf{U}$ may be determined by using the commutator expansion of the exponential transformation

$$\exp(-\mathrm{i}\Lambda) a_s^{\dagger} \exp(\mathrm{i}\Lambda) = a_s^{\dagger} + [a_s^{\dagger}, \mathrm{i}\Lambda] + \tfrac{1}{2}[[a_s^{\dagger}, \mathrm{i}\Lambda], \mathrm{i}\Lambda] + \cdots \tag{82}$$

To simplify this expression, consider a commutator of the form $[a_s^{\dagger}, a_u^{\dagger} a_v]$. This appears to result in the difference of two terms each containing the product of three operators. However, the anticommutation relations of Eqs (57) and (58) may be used to produce the following sequence of identities:

$$[a_s^{\dagger}, a_u^{\dagger} a_v] = a_s^{\dagger} a_u^{\dagger} a_v - a_u^{\dagger} a_v a_s^{\dagger} \tag{83}$$

$$= -a_u^{\dagger} a_s^{\dagger} a_v - a_u^{\dagger} a_v a_s^{\dagger} \tag{84}$$

$$= -a_u^{\dagger}(\delta_{sv} - a_v a_s^{\dagger}) - a_u^{\dagger} a_v a_s^{\dagger} \tag{85}$$

$$= -a_u^{\dagger} \delta_{sv} \tag{86}$$

This shows, first that a single commutator of this form results in a single creation operator, and secondly, because of the nested commutator form of Eq. (82), that all such terms in this expression will reduce to a single creation operator. This operator rank reduction is a common occurrence with commutator expansions in the second quantized notation. Substitution of Eq. (86) into the first commutator term in the expansion gives

$$[a_s^\dagger, \mathrm{i}\Lambda] = \sum_u \sum_v \mathrm{i}\Lambda_{uv}[a_s^\dagger, a_u^\dagger a_v] = \sum_u \sum_v \mathrm{i}\Lambda_{uv}(-a_u^\dagger \delta_{sv}) \tag{87}$$

$$= \sum_u -a_u^\dagger \mathrm{i}\Lambda_{us} \tag{88}$$

Substitution of this expression into the second commutator term of Eq. (82) gives

$$[[a_s^\dagger, \mathrm{i}\Lambda], \mathrm{i}\Lambda] = \sum_{u'} \sum_{v'} \mathrm{i}\Lambda_{u'v'}[[a_s^\dagger, \mathrm{i}\Lambda], a_{u'}^\dagger a_{v'}] \tag{89}$$

$$= \sum_{u'} \sum_{v'} \sum_u \mathrm{i}\Lambda_{u'v'}(-\mathrm{i}\Lambda_{us})[a_u^\dagger, a_{u'}^\dagger a_{v'}] \tag{90}$$

$$= \sum_{u'} \sum_{v'} \sum_u \mathrm{i}\Lambda_{u'v'}(-\mathrm{i}\Lambda_{us})(-a_{u'}^\dagger \delta_{uv'}) \tag{91}$$

$$= \sum_{u'} a_{u'}^\dagger (\mathrm{i}\boldsymbol{\Lambda}\mathrm{i}\boldsymbol{\Lambda})_{u's} \tag{92}$$

The remaining higher-order commutators may be shown by induction to give the final result

$$\exp(-\mathrm{i}\Lambda)a_s^\dagger \exp(\mathrm{i}\Lambda) = \sum_u a_u^\dagger(\mathbf{1} - \mathrm{i}\Lambda + \tfrac{1}{2}(\mathrm{i}\boldsymbol{\Lambda}\mathrm{i}\boldsymbol{\Lambda}) + \cdots)_{us} \tag{93}$$

$$= \sum_u a_u^\dagger \exp(-\mathrm{i}\boldsymbol{\Lambda})_{us} \tag{94}$$

This allows the identification of the two matrices of Eqs (78) and (94)

$$\mathbf{U} = \exp(-\mathrm{i}\boldsymbol{\Lambda}) \tag{95}$$

It may be readily verified that the matrix $\exp(-\mathrm{i}\boldsymbol{\Lambda})$ is in fact unitary, provided the matrix $\boldsymbol{\Lambda}$ is Hermitian. The fact that the matrix $\exp(-\mathrm{i}\boldsymbol{\Lambda})$ is unitary also means that the operator $\exp(-\mathrm{i}\Lambda)$ is unitary and its matrix representation in the full ket expansion space, with matrix elements $\langle m|\exp(-\mathrm{i}\Lambda)|k\rangle$, is a unitary matrix. An analogous relation holds for transformations of the electron annihilation operators a_r but it is the creation operator expansion that is most important for the MCSCF method. Substitution of the operator transformation into the expression of an arbitrary determinant gives the relation

$$|m'\rangle = a_{[1]}'^\dagger a_{[2]}'^\dagger \cdots a_{[N]}'^\dagger |\mathrm{vac}\rangle \tag{96}$$

$$= \exp(-\mathrm{i}\Lambda)a_{[1]}^\dagger \exp(\mathrm{i}\Lambda)\exp(-\mathrm{i}\Lambda)a_{[2]}^\dagger \exp(\mathrm{i}\Lambda)\cdots$$
$$\times \exp(-\mathrm{i}\Lambda)a_{[N]}^\dagger \exp(\mathrm{i}\Lambda)|\mathrm{vac}\rangle \tag{97}$$

$$= \exp(-\mathrm{i}\Lambda)a_{[1]}^\dagger a_{[2]}^\dagger \cdots a_{[N]}^\dagger |\mathrm{vac}\rangle \tag{98}$$

$$= \exp(-\mathrm{i}\Lambda)|m\rangle \tag{99}$$

This shows that the effect of an orbital transformation, as in Eq. (76), on any expansion ket is exactly represented by the action of the exponential operator $\exp(-i\Lambda)$ on that ket. The important feature of the operator representaton is that it gives the effect of the orbital transformation expressed in the original orbital basis. Since Eq. (99) is valid for any ket, it also holds for any linear combination of kets and, therefore, for the MCSCF wavefunction. It may be noted that when $\mathbf{\Lambda} = \mathbf{0}$, this unitary operator reduces to the identity operator. For small values of the parameters Λ_{rs}, it is useful to consider the truncated expansion of the exponential operator

$$|m'\rangle = \exp(-i\Lambda)|m\rangle \tag{100}$$

$$= |m\rangle - \sum_r \sum_s i\Lambda_{rs}(a_r^\dagger a_s|m\rangle) + \tfrac{1}{2}\sum_r \sum_s \sum_t \sum_u i\Lambda_{rs} i\Lambda_{tu}(a_r^\dagger a_s a_t^\dagger a_u|m\rangle) + \cdots \tag{101}$$

The third term may be rearranged using the relation $a_r^\dagger a_s a_t^\dagger a_u = a_r^\dagger a_t^\dagger a_u a_s + \delta_{st} a_r^\dagger a_u$ into a combination of single and double excitation operators (this is called the normal order of the operator product[30]). This shows that the effect of the unitary operator on a ket is to produce an expansion consisting of the original ket, of single excitation kets with expansion coefficients determined by the elements of the matrix $\mathbf{\Lambda}$, of double (and single) excitation kets with expansion coefficients determined by the products of two elements of the matrix $\mathbf{\Lambda}$, and so on. Since the operator expansion is infinite, the usefulness of low-order expansions such as Eq. (101) is limited to cases involving small values of the expansion coefficients.

Before proceeding to the details of the MCSCF formalism, we first impose some restrictions on the matrices $\mathbf{\Lambda}$ and $\mathbf{U}$. We will later restrict the MCSCF wavefunction to be an eigenfunction of the spin operators S_z and S^2. The orbital transformations that are applied to this wavefunction should not destroy these symmetry properties. This condition will be satisfied if the operator Λ commutes with these spin operators. If the spin components of the operator are written explicitly

$$\Lambda = \sum_r \sum_s (\Lambda_{rs}^{\alpha\alpha} a_{r\alpha}^\dagger a_{s\alpha} + \Lambda_{rs}^{\alpha\beta} a_{r\alpha}^\dagger a_{s\beta} + \Lambda_{rs}^{\beta\alpha} a_{r\beta}^\dagger a_{s\alpha} + \Lambda_{rs}^{\beta\beta} a_{r\beta}^\dagger a_{s\beta}) \tag{102}$$

then the requirement that $[\Lambda, S_z] = 0$ using Eq. (66) leads to the conditions

$$\mathbf{\Lambda}^{\alpha\beta} = \mathbf{0} \tag{103}$$

$$\mathbf{\Lambda}^{\beta\alpha} = \mathbf{0} \tag{104}$$

and the requirement that $[\Lambda, S^2] = 0$ using Eq. (70) leads to the condition

$$\mathbf{\Lambda}^{\alpha\alpha} = \mathbf{\Lambda}^{\beta\beta} \tag{105}$$

The Λ operator may then be written in the simplified form

$$\Lambda = \sum_r \sum_s \Lambda_{rs}(a^\dagger_{r\alpha}a_{s\alpha} + a^\dagger_{r\beta}a_{s\beta}) \tag{106}$$

$$= \sum_r \sum_s \Lambda_{rs}E_{rs} \tag{107}$$

where the summations are over spatial orbitals only. Furthermore, we impose the restriction that only real orbital transformations will be allowed. This will be consistent with the later restriction that the one- and two-electron integrals of the Hamiltonian operator must be real. The more general case allowing complex orbital transformations is only slightly more complicated and has been considered in MCSCF response theory[32–35] and in complex coordinate rotation theory[36]. The Hermitian matrix $\mathbf{\Lambda}$ may be written as $\mathbf{\Lambda} = \boldsymbol{\lambda} + \mathrm{i}\mathbf{K}$ where $\boldsymbol{\lambda}$ is real symmetric and $\mathbf{K}$ is real antisymmetric. The elements of the matrix $\boldsymbol{\lambda}$ determine the complex contributions to the matrix $\exp(-\mathrm{i}\mathbf{\Lambda})$ and determine the phases of the individual columns. These contributions will hereafter be neglected and the Hermitian matrix $\mathbf{\Lambda}$ will be written as

$$\mathbf{\Lambda} = \mathrm{i}\mathbf{K} \tag{108}$$

resulting in the real orbital transformation matrix

$$\mathbf{U} = \exp(-\mathrm{i}\mathbf{\Lambda}) = \exp(\mathbf{K}) \tag{109}$$

and in the real operator

$$\exp(-\mathrm{i}\Lambda) = \exp(K) \tag{110}$$

The real transformation matrix $\mathbf{U}$ is actually a rotation matrix since $\mathrm{Det}(\mathbf{U}) = \mathrm{Det}(\exp(\mathbf{K})) = \exp(\mathrm{Tr}(\mathbf{K})) = \exp(0) = +1$. The orbital phases are not important in the MCSCF method so that this loss of generality, compared to more general orthogonal transformations, is not significant. There are two representations of the K operator that are useful. The first results directly from Eq. (108) and is given as

$$K = \sum_r \sum_s K_{rs}E_{rs} \tag{111}$$

and the second exploits the $\mathbf{K}$ matrix antisymmetry and is given as

$$K = \sum_{r>s} K_{rs}(E_{rs} - E_{sr}) = \sum_{r>s} K_{rs}T_{rs} \tag{112}$$

with $T_{rs} = (E_{rs} - E_{sr})$. It will sometimes be useful to collect the unique elements of the $\mathbf{K}$ matrix that are used in Eq. (112) into a vector that is denoted by $\boldsymbol{\kappa}$ such that $\kappa_{(rs)} = K_{rs}$ for $r > s$ and where (rs) is a combined vector index. The matrix $\exp(\mathbf{K})$ may be constructed without approximation[37], and using only real arithmetic, from the real eigenvectors and eigenvalues of the negative semidefinite symmetric matrix $\mathbf{K}^2$.

Besides the restrictions imposed on the orbital transformations to preserve spin symmetries, it is also useful to preserve spatial symmetry. This is done by allowing transformations only within sets of orbitals having the same symmetry properties and by not allowing these different sets of orbitals to mix. This restriction is accomplished by forcing the off-diagonal symmetry blocks of the **K** matrix, those labeled by spatial orbitals belonging to different symmetry types, to be zero. The notation required to label the symmetry species of the orbitals is somewhat cumbersome and will not be used except when explicitly required.

G. Spin Eigenfunctions and the Unitary Group Approach

The expression for the Hamiltonian operator may be simplified in the usual case in which the magnetic effects of electron spin are neglected. In this case the integrals of Eq. (62) take the form

$$h_{r\sigma,s\sigma'} = h_{rs}\delta_{\sigma\sigma'} \tag{113}$$

$$g_{r\sigma,s\sigma',t\sigma'',u\sigma'''} = g_{rstu}\delta_{\sigma\sigma'}\delta_{\sigma''\sigma'''} \tag{114}$$

where the spin factors are written explicitly and where the orbital labels specify the spatial orbital indices. This allows the Hamiltonian operator to be written in the form

$$H = \sum_r \sum_s h_{rs}E_{rs} + \tfrac{1}{2}\sum_r \sum_s \sum_t \sum_u g_{rstu}e_{rstu} \tag{115}$$

where the operator expansion coefficients are indexed by spatial orbital labels only. The operator $E_{rs} = a^\dagger_{r\alpha}a_{s\alpha} + a^\dagger_{r\beta}a_{s\beta}$ has already been introduced in Eq. (107) as the form of a single excitation operator that preserves the S_z and S^2 eigenvalues. The operator e_{rstu} may be written as

$$e_{rstu} = \sum_\sigma \sum_{\sigma'} a^\dagger_{r\sigma}a^\dagger_{t\sigma'}a_{u\sigma'}a_{s\sigma} \tag{116}$$

$$= E_{rs}E_{tu} - E_{ru}\delta_{ts} \tag{117}$$

The second form, the normal order of the generator product, shows that the operator e_{rstu} also preserves S_z and S^2 eigenvalues since it is constructed from operators that do so. The expansion of the Hamiltonian operator in this spin-preserving operator basis shows that the Hamiltonian operator itself must preserve the S_z and S^2 eigenvalues of the wavefunctions on which it acts. The definition of the operator E_{rs} results in the identities

$$E_{rs} = E^\dagger_{sr} \tag{118}$$

$$e_{rstu} = e_{turs} = e^\dagger_{utsr} = e^\dagger_{srut} \tag{119}$$

The anticommutation relations of the creation and annihilation operators of

Eqs (56) through (58) may be used to derive the following commutation relations of the operators E_{rs}

$$[E_{uv}, E_{rs}] = E_{us}\delta_{rv} - E_{rv}\delta_{us} \tag{120}$$

which may then be used to show

$$[e_{tuvw}, E_{rs}] = e_{tuvs}\delta_{rw} + e_{tsvw}\delta_{ru} - e_{turw}\delta_{vs} - e_{ruvw}\delta_{ts} \tag{121}$$

Eq. (120) is the same commutation relation as is satisfied by the generators of the unitary group $U(n)$, the group of all n-dimensional unitary matrices. For this reason, the operators E_{rs} are often referred to as generators[38–40]. The operator rank reductions shown in Eqs (120) and (121) prove to be important in some of the following derivations.

The commutator expressions, Eqs (120) and (121), are sufficient to derive expressions for the matrix elements required for the MCSCF optimization process. This results from the fact that both the orbital transformation and the Hamiltonian operator are written in terms of the generators and generator products of Eq. (117). Since all of these operators involve explicit references only to the spatial orbitals, and not to the spin orbitals, it would be possible to eliminate reference to the spin-orbitals entirely if the expansion kets could be represented in such a spin-free method and if matrix elements of these spin-free operators could be calculated without reference to the spin orbitals.

There are several approaches that may be used to achieve this goal. Many of these methods are discussed in the review of the CI method of Shavitt[13], in the review of MCSCF and CI methodology of McWeeny and Sutcliffe[2] and in the discussion of spin eigenfunctions by Pauncz[31]. The method with which the present author is most familiar is the graphical unitary group approach (GUGA) and this approach will be discussed briefly. For more details of this method, the reader is referred, in particular, to the contributions of Paldus[39] and of Shavitt[40] in the volume of *Lecture Notes in Chemistry* edited by J. Hinze.

1. *The Canonical N-electron Expansion Basis*

The expansion kets in the unitary group approach are constructed to be eigenfunctions of the operators S^2. These spin-eigenfunction expansion kets will be referred to as configuration state functions (CSFs). These spin eigenfunctions may be represented as a linear combination of the spin-orbital kets discussed in the previous section. The spin-orbitals kets, which correspond to determinants and which contribute to a particular CSF, all have the same spatial orbital occupancy but differ from each other in their spin assignments. It follows that each spin-adapted ket is an eigenfunction of operators of the type E_{rr} with eigenvalues of 0, 1 or 2, corresponding to empty, singly occupied and doubly occupied spatial orbitals. (A particular spatial orbital occupancy is sometimes called an electron configuration and is usually associated with several CSFs.) Since the Hamiltonian operator commutes with both S^2 and S_z, only determinants corresponding to a particular S_z eigenvalue are required in the expansion of a particular spin-adapted ket. If the matrix representation of the Hamiltonian operator, or any other spin-independent operator, were constructed in the full determinantal basis, it could be brought to block-diagonal form with the blocks labeled by S^2 eigenvalues (usually written as $S(S+1)$ with $S = 0, \frac{1}{2}, 1, \ldots$), where the transformation is independent of the particular values of the Hamiltonian matrix elements. Within each of these blocks, the matrix could further be block diagonalized according to the S_z eigenvalue, $M = -S, -S+1, \ldots, S-1, S$. All of these blocks, for a particular S^2 eigenvalue, would then be exactly identical and would result in sets of $(2S+1)$-degenerate eigenvalues of H. The construction of only one of these blocks, labeled by M, would be sufficient to determine the entire eigenvalue spectrum of the block labeled by S.

With only the M eigenvalue restriction imposed on the expansion determinants, the number of determinants is greater than the number of linearly independent spin eigenfunctions for a given number of singly occupied spatial orbitals. This is because these determinants span the space of spin eigenfunctions that correspond to several S^2 eigenvalues. The number of independent spin eigenfunctions for a particular orbital occupancy is given[31] by the expression

$$f(n^{[1]}, S) = B(n^{[1]}, \tfrac{1}{2}n^{[1]} - S) - B(n^{[1]}, \tfrac{1}{2}n^{[1]} - S - 1) \tag{122}$$

where $n^{[1]}$ is the number of singly occupied spatial orbitals and where the binomial coefficients are defined as usual as $B(r, s) = r!/[s!(r-s)!]$. The doubly occupied and empty orbitals do not contribute to this number. The number of independent spin eigenfunctions may also be determined from the 'branching diagram' recursion relation[31], $f(n^{[1]}, S) = f(n^{[1]} - 1, S + \frac{1}{2}) + f(n^{[1]} - 1, S - \frac{1}{2})$. The number of determinants required to span the space of these spin eigenfunctions is given by the expression

$$D(n^{[1]}, N_\sigma^{[1]}) = B(n^{[1]}, N_\sigma^{[1]}) \tag{123}$$

where $N_\sigma^{[1]}$ is either the number of α-spin or β-spin electrons in the singly occupied orbitals (Eq. (123) gives the same result in either case). The constraints $n^{[1]} = N_\alpha^{[1]} + N_\beta^{[1]}$ and $M = \frac{1}{2}(N_\alpha^{[1]} - N_\beta^{[1]})$ give the relation between these various quantities. Consider an example of the $M = 0$ determinants constructed from 12 singly occupied spatial orbitals: six α-spin electrons and six β-spin electrons. There are 924 determinants as determined by Eq. (123). These determinants span the space of 132 singlet CSFs with $S = 0$; 297 triplet CSFs with $S = 1$; 275 CSFs with $S = 2$; 154 CSFs with $S = 3$; 54 CSFs with $S = 4$; 11 CSFs with $S = 5$; and one CSF with $S = 6$. Clearly there is redundant information in the determinantal representation of spin eigenfunctions when

TABLE I
Comparison of determinantal and CSF expansion lengths. Comparison of the number of determinants and the number of CSFs for full CI and RCI wavefunction expansions. All wavefunctions are singlets with the number of orbitals equal to the number of electrons. Higher fractions of open-shell CSFs result in higher expansion length ratios for a particular wavefunction type. Expressions for these expansion lengths are given in the text.

	Full CI expansion			RCI expansion		
N	Determinant	CSF	Ratio	Determinant	CSF	Ratio
2	4	3	1.33	4	3	1.33
4	36	20	1.80	18	10	1.80
6	400	175	2.29	88	37	2.38
8	4 900	1 764	2.78	454	150	3.03
10	63 504	19 404	3.27	2 424	654	3.71
12	853 776	226 512	3.77	13 236	3 012	4.39
14	11 778 624	2 760 615	4.27	73 392	14 445	5.08
16	165 636 900	34 763 300	4.76	411 462	71 398	5.76
18	2 363 904 400	449 141 836	5.26	2 325 976	361 114	6.44
20	34 134 779 536	5 924 217 936	5.76	13 233 628	1 859 628	7.12

only those eigenfunctions corresponding to a particular S value are required.

Actual CSF space expansions include closed-shell terms, terms involving only a few singly occupied orbitals (for which the ratios of determinants to CSFs is relatively small) and expansion terms that have many singly occupied orbitals (for which these ratios are much larger). The overall ratios for the expansion space then depend on the relative numbers of these various types of terms. Table I shows the number of determinants and CSFs for two popular expansion forms of MCSCF wavefunctions. Both the full CI wavefunction and the restricted CI wavefunction are discussed in detail in Section IV. The expansion lengths in Table I correspond to singlet wavefunctions with the number of electrons equal to the number of orbitals. The large ratios of determinantal expansion lengths to CSF expansion lengths indicate that there are a large fraction of open-shell CSFs in these wavefunction expansion forms. These open-shell CSFs are required to describe the near-degeneracy and spin-recoupling effects of molecular systems as they dissociate to various fragments. These chemical effects are discussed in more detail in following sections.

The unitary group approach attempts to solve this redundancy in two ways. First, the expansion terms are constructed in such a way that only spin eigenfunctions of the correct S value are considered. Secondly, the generator and generator product matrix elements in the CSF expansion space are computed in such a way that no reference to determinants is made, either implicitly or explicitly. This leads to computational schemes that depend only on the CSF expansion length and have no component that depends on the equivalent determinantal representation of the wavefunction. Although this is not unique to the unitary group approach, it should be mentioned that not all spin-eigenfunction construction methods satisfy both of these properties[31]. Since many MCSCF wavefunction expansions contain a large percentage of open-shell CSFs, and since many of these CSFs have large numbers of open shells, it is particularly important to avoid reference to the determinantal expansion. In addition to this, all explicit reference to the M value is avoided in the unitary group approach[40], both in the specification of the CSF expansion space and in the resulting wavefunction optimization steps.

The CSFs may be represented in the unitary group approach with a 'step vector', denoted **d**, that specifies the occupation and spin coupling of the spatial orbitals. Each orbital corresponds to a 'level' in the unitary group approach. Each level of the step vector has associated with it a cumulative occupation and a cumulative spin. The cumulative spin at the ith level is denoted S_i and, consistent with the usual spin coupling of electrons, is allowed non-negative values of half-integer increments: $S_i = 0, \frac{1}{2}, 1, \frac{3}{2}, \ldots$ etc. To avoid the fractions, the intermediate spin is usually indicated with a b value where $b_i = 2S_i$. A step vector entry $d_i = 0$ indicates that the spatial orbital associated with the ith level is unoccupied in the CSF. The coupling of an empty orbital does not change the cumulative spin eigenvalue of the previously coupled electrons. This is indicated by $\Delta b_{i0} = 0$ with $\Delta b_i = b_i - b_{i-1}$ and where the second subscript indicates the step vector entry. A step vector entry $d_i = 3$ indicates that the orbital is doubly occupied. A doubly occupied orbital is necessarily a singlet giving $\Delta b_{i3} = 0$. A singly occupied orbital must change the cumulative spin eigenvalue of the previously coupled electrons. It may be coupled in such a way as to increase the cumulative spin by $\frac{1}{2}$ or to decrease the spin by $\frac{1}{2}$. These two possibilities are assigned step vector entries of $d_i = 1$ and $d_i = 2$ and correspondingly have $\Delta b_{i1} = +1$ and $\Delta b_{i2} = -1$. These four cases are the only allowed step vector entries.

For example, a CSF written as $|3300\rangle$ indicates that the first two orbitals are both doubly occupied. This CSF is a four-electron singlet. The ket $|3120\rangle$ indicates that the first orbital is doubly occupied; the second orbital is singly occupied and coupled to the lower-level orbitals to increase the spin, in this case giving a three-electron doublet. Finally the singly occupied third orbital is coupled to decrease the overall spin, in this case giving a four-electron singlet. The spin eigenvalue of any ket, denoted by S_n or by b_n, may be determined simply by the difference in the number of $d_i = 1$ step vector entries and the number of $d_i = 2$ step vector entries. In fact, the sum of the step vector entries is the same for all CSFs with the same total spin and number of electrons[40]. The CSF expansion space may easily be limited to include only those terms that contribute to the correct overall spin. An important property of CSFs constructed in this manner is that they form an orthonormal expansion basis.

It is, of course, possible to expand such a CSF in terms of determinants. Since only one electron is coupled at a time, these coefficients are determined simply from Clebsch–Gordan coefficients for each level[31]. The expansion coefficient of a determinant within a particular CSF is then given by the product of Clebsch–Gordan coefficients for each level in the CSF. The determinantal expansion space and the determinantal coefficients, of course, do depend on the M value chosen for the determinantal representation. We assume that the spin orbitals in the determinants are in the order $\phi_{1\alpha}, \phi_{1\beta}, \phi_{2\alpha}, \phi_{2\beta}, \ldots, \phi_{n\alpha}, \phi_{n\beta}$ and let M_i be the cumulative S_z eigenvalue of the determinant up through the ith level. With this notation, it is then straightforward to determine the Clebsch–Gordan coefficient associated with a particular level of a particular determinant. It is convenient to introduce an additional overall phase factor in the unitary group approach that is not usually included in the genealogical spin-eigenfunction scheme[31,40]. The possible Clebsch–Gordan coefficients with these phase factors are displayed in Table II.

The coefficient of a particular determinant within a CSF is given by the product of the factors f_k given in Table II for all the levels. The phase factors that are unique to the unitary group approach are those determined by b_k. These factors are determined by the CSF coupling and not by the individual determinants. Thus these phase factors result in the multiplication of the total CSF by some overall sign factor. Table III shows the determinantal expansion for the set of doublet CSFs consisting of five singly occupied orbitals, $\phi_1, \phi_2, \phi_4, \phi_5, \phi_6$, and one doubly occupied orbital, ϕ_3. The sparseness of the

TABLE II

Clebsch–Gordan coefficients with phase factors. The Clebsch–Gordan coefficients required in the genealogical construction of the CSFs in the unitary group approach. The phase factors depending only on b_k result in overall phase factors for the complete CSF. The determinantal expansion coefficients are given as products of the factors f_k for all the levels. These factors are also used to derive 'segment shape factors' that allow the evaluation of coupling coefficients directly from the stepvector entries or from the DRT representation of the CSF expansion space.

d_k	σ_k	f_k
0	–	1
1	α	$[(b_k + 2M_k)/2b_k]^{1/2}$
1	β	$[(b_k - 2M_k)/2b_k]^{1/2}$
2	α	$-(-1)^{b_k}[(b_k + 2 - 2M_k)/(2b_k + 4)]^{1/2}$
2	β	$(-1)^{b_k}[(b_k + 2 + 2M_k)/(2b_k + 4)]^{1/2}$
3	$\alpha\beta$	$(-1)^{b_k}$

TABLE III
Determinantal representation of CSFs. Expansion of CSFs in the unitary group approach in terms of spin-orbital determinants. The coefficients are determined as products of factors, f_k, determined from the Clebsch–Gordan coefficients and phase factors of Table II. The coefficient sparseness of the determinants is predictable and corresponds to the ‘allowed area’ principle.

	Step vector CSF representation				
Determinant	$\|113122\rangle$	$\|113212\rangle$	$\|123112\rangle$	$\|113221\rangle$	$\|123121\rangle$
$\|10\,10\,11\,10\,01\,01\rangle$	$-2^{-1/2}$	–	–	–	–
$\|10\,10\,11\,01\,10\,01\rangle$	$18^{-1/2}$	2/3	–	–	–
$\|10\,01\,11\,10\,10\,01\rangle$	$18^{-1/2}$	−1/3	$-3^{-1/2}$	–	–
$\|01\,10\,11\,10\,10\,01\rangle$	$18^{-1/2}$	−1/3	$3^{-1/2}$	–	–
$\|10\,10\,11\,01\,01\,10\rangle$	$18^{-1/2}$	−1/3	–	$-3^{-1/2}$	–
$\|10\,01\,11\,10\,01\,10\rangle$	$18^{-1/2}$	1/6	$12^{-1/2}$	$12^{-1/2}$	1/2
$\|01\,10\,11\,10\,01\,10\rangle$	$18^{-1/2}$	1/6	$-12^{-1/2}$	$12^{-1/2}$	−1/2
$\|10\,01\,11\,01\,10\,10\rangle$	$-18^{-1/2}$	−1/6	$12^{-1/2}$	$12^{-1/2}$	−1/2
$\|01\,10\,11\,01\,10\,10\rangle$	$-18^{-1/2}$	−1/6	$-12^{-1/2}$	$12^{-1/2}$	1/2
$\|01\,01\,11\,10\,10\,10\rangle$	$-18^{-1/2}$	1/3	–	$-3^{-1/2}$	–

determinantal coefficients is predictable. At any level, the M_i spin component of the determinant cannot be greater in magnitude than the S_i value of the CSF. This is called the ‘allowed area’ principle by Pauncz[31]. It may be verified from these determinantal expansions that the CSFs are orthonormal.

H. Matrix Element Evaluation

The optimization of the MCSCF wavefunction requires the construction of matrix elements of the Hamiltonian operator in the CSF expansion basis. The energy expectation value of the Hamiltonian operator of Eq. (115) may be written, assuming real normalized CSF expansion coefficients, in the forms

$$\langle 0|H|0\rangle = \sum_m \sum_{m'} c_m H_{mm'} c_{m'} \tag{124}$$

$$= \sum_m \sum_{m'} \sum_r \sum_s h_{rs} c_m \langle m|E_{rs}|m'\rangle c_{m'} + \tfrac{1}{2} \sum_m \sum_{m'} \sum_r \sum_s \sum_t \sum_u g_{rstu} c_m \langle m|e_{rstu}|m'\rangle c_{m'} \tag{125}$$

$$= \sum_r \sum_s h_{rs} D_{rs} + \tfrac{1}{2} \sum_r \sum_s \sum_t \sum_u g_{rstu} d'_{rstu} \tag{126}$$

$$= \sum_r \sum_s h_{rs} D_{rs} + \tfrac{1}{2} \sum_r \sum_s \sum_t \sum_u g_{rstu} d_{rstu} \tag{127}$$

$$= \mathrm{Tr}(\mathbf{hD}) + \tfrac{1}{2}\mathrm{Tr}(\mathbf{gd}) \tag{128}$$

The matrix elements of the generators and generator products, $\langle m|E_{rs}|m'\rangle$

and $\langle m|e_{rstu}|m'\rangle$, are called coupling coefficients. The contractions of these coupling coefficients with the CSF expansion coefficients indicated in Eq. (125) result in the symmetric matrix $\mathbf{D}$, called the one-particle density matrix, and the matrix $\mathbf{d}'$, which is called the two-particle density matrix. These matrices are the spin-traced versions of the spin-orbital density matrices of Eqs (72) and (73). The matrix $\mathbf{d}$ may be defined as

$$\mathbf{d}_{rstu} = \tfrac{1}{4}(\mathbf{d}'_{rstu} + \mathbf{d}'_{srtu} + \mathbf{d}'_{rsut} + \mathbf{d}'_{srut}) \tag{129}$$

resulting in the symmetric two-particle density matrix. This symmetric matrix has the same symmetry properties as the two-electron integrals. These density matrices may be used to calculate the expectation value of any operator by evaluating the matrix traces, as is shown in Eqs (127) and (128) for the Hamiltonian operator. The construction of the density matrices depends on the CSF expansion length but, once they are available, the calculation of expectation values depends only on the number of orbitals occupied in the CSF expansion. The density matrices may thus be regarded as convenient partial sums that will prove useful in the evaluation of various quantities required for the MCSCF wavefunction optimization.

The computation of the coupling coefficients is of utmost importance in the calculation of expectation values and of certain matrix–vector products. There are basically two different approaches that may be used in the evaluation of these terms. These terms may be computed once and stored as a separate file, which is read repeatedly when required, or they may be repeatedly computed and used as they are required. The first scheme has the advantage that any overhead associated with the repeated construction of these coupling coefficients is minimized. The second method has the advantage that no potentially large external files are required as in the first method. Both approaches have been used in MCSF calculations and the optimal approach is computer-dependent.

In either case, the unitary group approach provides an efficient method for the computation of the set of coupling coefficients corresponding to a fixed set of orbital labels[41]. An alternative 'CSF-driven' approach, proposed by Knowles *et al.*[42], involves the evaluation of the set of coupling coefficients that corresponds to a fixed expansion term $\langle m|$ for all kets and all orbital indices. Another method, called the 'loop-driven' approach[37,40,43], holds neither the CSF index nor the orbital indices fixed. The first method, which has some advantages in the MCSCF method, will be outlined here. This approach is called an 'index-driven' approach to differentiate it from the 'CSF-driven' and 'loop-driven' approaches.

1. The DRT Representation of the CSF Expansion Space

To describe the computation of the coupling coefficients with the formalism of the unitary group approach, the distinct row table (DRT) and its graphical representation, the Shavitt graph, is useful. A distinct row corresponds to a particular cumulative spin coupling and occupation at a particular level. The distinct row may be referenced by the three integers (i, N_i, b_i). For historical reasons[39,40], the quantity a_i is normally used to label the distinct row, instead of N_i, and is defined by $N_i = 2a_i + b_i$. The distinct rows correspond to vertices of a directed graph. Each vertex is connected to at most four vertices at the next higher level in the four different ways corresponding to the four different step vector entries. The step vector entries correspond to segments or 'arcs' that connect the vertices in the graphical representation. The connections of a vertex at some level to the vertices at the next higher level may be stored in an index array that accompanies the list of distinct rows. A particular ket may then either be represented with the step vector notation, or with the list of distinct rows that results from moving through the DRT from the lowest level to the highest level. A CSF may then be regarded as a 'walk' through the DRT.

Limitations are placed on the CSF space, relative to the full CI expansion, by deleting particular distinct rows or particular arcs connecting pairs of distinct rows. Although this does not lead to completely general CSF selection, it does lead to exactly the kind of occupation and spin-coupling restrictions that are most useful in the MCSCF method. This is discussed in more detail in Section IV.

The DRT representation of the CSF space is useful in two ways. First, the DRT representation proves to be a very compact representation of the CSF space. This is because the number of distinct rows is usually much smaller than the number of CSFs. Secondly, the coupling coefficients may be computed directly from the compact DRT representation. This coupling coefficient evaluation will now be examined.

A generator acting on a ket produces a linear combination of kets, in each of which the step vector entries outside the range of the generator indices remains unchanged[40]

$$E_{rs}|d_1 d_2 \cdots d_r d_{r+1} \cdots d_s d_{s+1} \cdots d_n\rangle = \sum_{d'} |d_1 d_2 \cdots d'_r d'_{r+1} \cdots d'_s d_{s+1} \cdots d_n\rangle \langle d'|E_{rs}|d\rangle \tag{130}$$

It is assumed in Eq. (130) that $r < s$. Not only are the step vectors unchanged outside of this range, but the entries from levels $(r+1)$ to $(s-1)$ must be such that the occupations of $|d\rangle$ and $|d'\rangle$ match identically. Additionally, the step vector entries must also be such that the occupations in the $|d'\rangle$ ket change by $+1$ at level r and by -1 at levels s relative to the $|d\rangle$ ket. These restrictions result in a very sparse set of coupling coefficients $\langle d'|E_{rs}|d\rangle$. The graphical representation of these restrictions results in the formation of a closed loop on the Shavitt graph. The coupling coefficient is given by a product of factors, called segment values. These segment values, one for each level, are unity outside the range r to s and depend only on the b_i values and the segment shapes of the levels within the range. This suggests that only the closed loop should be constructed during the coupling coefficient evaluation, followed by the subsequent addition of all possible 'upper walks' and 'lower walks' to specify the interacting CSFs completely. The definition of the segment values within the loop has been given by Shavitt[40,44]. These segment values were also derived by Paldus and Boyle using graphical spin algebra[39]. These segment values may be collected into tables that are easy to compute and store. This methodology results in the ability to compute coupling coefficients for arbitrary spin states and for arbitrary numbers of open shells. Both of these features are important in the MCSCF method.

The computation of coupling coefficients involving generator products is only slightly more complicated than for the case of a single generator. For these cases, two sets of products are generated simultaneously and added together to produce the required coupling coefficient. In most cases, these two sets of products may be combined in two different ways to produce two distinct sets of generator product matrix elements[40]. In the programs written by the author to compute these coupling coefficients, all possible loops are constructed that are consistent with a given set of levels. This is done by forming a 'template' for the loop type that is consistent with the level indices. An efficient tree search procedure then locates all consistent loops within the DRT. This particular search procedure allows the segment value products to be accumulated in two different arrays and it allows the intermediate cumulative products to be used for all loops that share common segment shapes. This is similar to the 'loop-driven' procedure of Brooks and Schaefer[43] but differs because the loop template restricts all the levels of the loop while the 'loop-driven' procedure only constraints one level to be fixed.

A feature of the template or 'index-driven' approach, which was first proposed by Shavitt for direct CI algorithms[40,44], is that all contributions to a particular density matrix element, D_{rs} or d_{rstu}, may be computed together. In the MCSCF procedure, it is also useful to compute quantities of the form $\langle 0|e_{rstu}|n\rangle$ for a fixed $(rstu)$ and for all possible $|n\rangle$. MCSCF procedures that do not use such a method must instead sort the list of coupling coefficients, which are computed in some arbitrary order, into an order that allows the orderly computation of these transition density matrix elements. The index-driven approach avoids this unnecessary sorting step in those cases where the coupling coefficients are explicitly written to an external file, and it allows the efficient computation of the required coefficients in those cases where they are used as they are computed. The relative merits of the index-driven and CSF-driven approaches are discussed further in Section VI.

I. Summary

The basic concepts necessary to the development of the MCSCF method have been introduced. These concepts include the bracketing theorem, which forms the formal basis of the MCSCF method. The N-electron expansion space used in the MCSCF method is introduced in the form of Slater determinants. The method of second quantization is used in the formal manipulations of the N-electron expansion space and of quantum-mechanical operators within this expansion space. Spin eigenfunctions are constructed within the second quantization method with the unitary group approach. This background material may now be used in the following sections to derive the MCSCF optimization equations, to discuss the selection of CSF expansion terms, to examine the relation between the orbital variation parameters and the CSF variations, and to discuss the implementation of various optimization procedures.

III. MULTICONFIGURATION SELF-CONSISTENT FIELD EQUATIONS

The formalism developed in the preceding sections may now be used to derive the working equations for MCSCF wavefunction optimization. It will be assumed that the wavefunction is expanded in a set of orthonormal CSFs, as described in Section II, that are appropriate eigenfunctions of the operator S^2. It will also be assumed that the one- and two-electron integrals of the Hamiltonian operator, the CSF expansion coefficients and the orbital transformation coefficients are all real. As discussed previously, this is not a limitation of the formalism, but rather is imposed for reasons of computational efficiency.

A. Energy Expressions for Multiconfiguration Self-consistent Field Wavefunction Optimization

The exponential operator parametrization of the orbital transformation of Eq. (99) may be used to write a trial wavefunction[6] in the form

$$|0'\rangle = \sum_m \exp(K)|m\rangle c_m \tag{131}$$

The trial wavefunction depends on the elements of the antisymmetric matrix **K**, which determine the orbital variations, and on the CSF expansion coefficients **c**. The Hamiltonian expectation value for this wavefunction then takes the form

$$E(\mathbf{K}, \mathbf{c}) = \sum_m \sum_{m'} c_m \langle m| \exp(-K) H \exp(K)|m'\rangle c_{m'} (\mathbf{c}^t\mathbf{c})^{-1} \tag{132}$$

This particular form of the energy expression emphasizes that the individual matrix elements of the Hamiltonian matrix depend on the parameters **K**.

Eq. (132) assumes an arbitrary normalization of the vector **c**. The expectation value is, of course, invariant to overall scaling of this vector and it is convenient to introduce this normalization constraint directly into the energy expression. This may be accomplished by using the exponential parametrization of Eq. (33) by assuming that a normalized trial vector **c** is available to define a reference wavefunction, $|0\rangle$. A CSF rotation operator may then be defined using the operator

$$S = |S\rangle\langle 0| - |0\rangle\langle S| \tag{133}$$

$$= \sum_{n(\neq 0)} p_n(|n\rangle\langle 0| - |0\rangle\langle n|) = \sum_{n(\neq 0)} p_n P_n \tag{134}$$

$$= \sum_n \tilde{p}_n(|\tilde{n}\rangle\langle 0| - |0\rangle\langle \tilde{n}|) = \sum_n \tilde{p}_n \tilde{P}_n \tag{135}$$

Eq. (133) is simply the operator version of the matrix expression in Eq. (30). In Eq. (133) the state $|S\rangle$ is orthogonal to the reference state $|0\rangle$ and the norm of $|S\rangle$ determines the angle of rotation between these two states. Eq. (134) assumes that a basis for the orthogonal complement to the reference vector exists and that the coefficients p_n are the expansion coefficients of the state $|S\rangle$ in this basis. Three bases have been used for this expansion. The Hamiltonian eigenvector basis[16,17,37,45,46], which satisfies $\langle n|H|n'\rangle = E_n\delta_{nn'}$, is linearly independent and orthonormal and is most convenient for formal manipulations. The projected basis[41,47,48], which is defined as $|n\rangle = (1 - |0\rangle\langle 0|)|\tilde{n}\rangle$ where $|\tilde{n}\rangle$ is an expansion CSF, is linearly dependent (since $\sum c_n|n\rangle = 0$) and non-orthogonal (since $\langle m|n\rangle = \delta_{mn} - c_m c_n$). The CSF basis itself[6] is orthonormal and overcomplete (since it also includes $|0\rangle$). In Eq. (135) it is assumed that the state $|S\rangle$ is expanded in the CSF basis. The vector $\tilde{\mathbf{p}}$ must be orthogonal to the vector **c** (i.e. $\tilde{\mathbf{p}}^t\mathbf{c} = 0$) in this expression. The CSF basis and the projected basis are closely related. In fact, the expansion coefficients in the two bases are identical for states $|S\rangle$ that are orthogonal to the reference wavefunction $|0\rangle$. The CSF basis or the projected basis are the most computationally convenient choices of orthogonal complement basis.

Using this exponential parametrization for both the orbital variations and the CSF variations allows the trial wavefunction to be written as

$$|0'\rangle = \exp(K)\exp(S)|0\rangle \tag{136}$$

and the energy expression to be written as

$$E(\mathbf{K}, \mathbf{p}) = \langle 0|\exp(-S)\exp(-K)H\exp(K)\exp(S)|0\rangle \tag{137}$$

(The trial wavefunction could also be written in the form $\exp(S')\exp(K')|0\rangle$ but this is not as useful because it eventually requires matrix elements of the form $\langle m|\exp(K')|n\rangle$ which are difficult to calculate.) This exponential parametrization of both the orbital and CSF variations was first employed by Yeager and Jørgensen[16] and by Dalgaard[17] and has been used in the formulation of several MCSCF methods[37,45,46,49]. The form of Eq. (137) emphasizes that the energy is a function of the parameters **K** and of the

$(n_{CSF} - 1)$ parameters required to specify the vector $\mathbf{p}$. The exponential parametrization of the wavefunction may now be used to advantage to expand the energy expression in various powers of the parameters $\mathbf{K}$ and $\mathbf{p}$:

$$E(\mathbf{K}, \mathbf{p}) = \langle 0|\exp(-S)(H + [H, K] + \tfrac{1}{2}[[H, K], K] + \cdots)\exp(S)|0\rangle \quad (138)$$

$$= \langle 0|H + [H, S] + \tfrac{1}{2}[[H, S], S] + \cdots + [H, K] + [[H, K], S] + \cdots + \tfrac{1}{2}[[H, K], K] + \cdots|0\rangle \quad (139)$$

It is important to notice that, because of the form of the wavefunction parametrization, all of the K-dependent commutators occur inside of the S-dependent commutators. All of the second-order terms are written explicitly in Eq. (139) and it is these terms that will be discussed in most detail in this review. Eq. (139) may also be written using matrix notation. Consider, for example, the fifth term of Eq. (139) where the parameters are factored from the operator basis terms

$$\langle 0|[[H, K], S]|0\rangle = \sum_{(rs)} \sum_{n} \kappa_{rs} \langle 0|[[H, T_{rs}], P_n]|0\rangle p_n \quad (140)$$

$$= \boldsymbol{\kappa}^{\mathrm{t}}\mathbf{C}\mathbf{p} \quad (141)$$

The representation of the K operator using the elements of the $\boldsymbol{\kappa}$ vector of Eq. (112) is most useful in this representation. The other terms of Eq. (139) may also be factored as in Eq. (141) to give the following set of equivalent matrix expressions of the truncated second-order energy expression:

$$E^{(2)}(\mathbf{K}, \mathbf{p}) = E(0) + \boldsymbol{\kappa}^{\mathrm{t}}\mathbf{w} + \mathbf{p}^{\mathrm{t}}\mathbf{v} + \boldsymbol{\kappa}^{\mathrm{t}}\mathbf{C}\mathbf{p} + \tfrac{1}{2}\boldsymbol{\kappa}^{\mathrm{t}}\mathbf{B}\boldsymbol{\kappa} + \tfrac{1}{2}\mathbf{p}^{\mathrm{t}}\mathbf{M}\mathbf{p} \quad (142)$$

$$= E(0) + (\boldsymbol{\kappa}^{\mathrm{t}} \;\; \mathbf{p}^{\mathrm{t}})\begin{pmatrix} \mathbf{w} \\ \mathbf{v} \end{pmatrix} + \tfrac{1}{2}(\boldsymbol{\kappa}^{\mathrm{t}} \;\; \mathbf{p}^{\mathrm{t}})\begin{pmatrix} \mathbf{B} & \mathbf{C} \\ \mathbf{C}^{\mathrm{t}} & \mathbf{M} \end{pmatrix}\begin{pmatrix} \boldsymbol{\kappa} \\ \mathbf{p} \end{pmatrix} \quad (143)$$

$$= E(0) + \tfrac{1}{2}(\boldsymbol{\kappa}^{\mathrm{t}} \;\; \mathbf{p}^{\mathrm{t}} \;\; 1)\begin{bmatrix} \mathbf{B} & \mathbf{C} & \mathbf{w} \\ \mathbf{C}^{\mathrm{t}} & \mathbf{M} & \mathbf{v} \\ \mathbf{w}^{\mathrm{t}} & \mathbf{v}^{\mathrm{t}} & 0 \end{bmatrix}\begin{bmatrix} \boldsymbol{\kappa} \\ \mathbf{p} \\ 1 \end{bmatrix} \quad (144)$$

In these matrix expressions, the vectors $\boldsymbol{\kappa}$ and $\mathbf{p}$ contain the orbital and CSF variation parameters respectively. The other matrix elements result from the commutator expressions of Eq. (139) and are given as

$$w_{rs} = \langle 0|[H, T_{rs}]|0\rangle \quad (145)$$

$$v_n = \langle 0|[H, P_n]|0\rangle = 2\langle 0|H|n\rangle \quad (146)$$

$$B_{rs,uv} = \tfrac{1}{2}\langle 0|[[H, T_{rs}], T_{uv}] + [[H, T_{uv}], T_{rs}]|0\rangle \quad (147)$$

$$C_{rs,n} = \langle 0|[[H, T_{rs}], P_n]|0\rangle = 2\langle 0|[H, T_{rs}]|n\rangle \quad (148)$$

$$M_{nn'} = \langle 0|[[H, P_n], P_{n'}]|0\rangle = 2\langle n|H - E(0)|n'\rangle \quad (149)$$

These matrix elements are the first and second derivatives of the energy with respect to orbital and state variations evaluated at the reference wavefunction. The vector $\mathbf{w}$ is called the orbital gradient vector and the vector $\mathbf{v}$ is called the state gradient vector. The combined vector $(\mathbf{w}^t \quad \mathbf{v}^t)^t$ is called the wavefunction gradient vector. The matrix $\mathbf{B}$ is called the orbital Hessian matrix. It is convenient to use the symmetric form of this matrix as defined in Eq. (147). The matrix $\mathbf{C}$ is called the orbital-state Hessian matrix or simply the coupling matrix. The matrix $\mathbf{M}$ is called the state Hessian matrix. Together, the four blocks of the Hessian matrix form the wavefunction Hessian matrix which is used in Eq. (143). The matrix in Eq. (144), consisting of both the wavefunction Hessian matrix and the wavefunction gradient vector, is called the augmented wavefunction Hessian matrix.

It is often the case during the iterative solution of the MCSCF wavefunction that the state gradient vector $\mathbf{v}$ is zero. This means that the mixing coefficients are optimal for the current set of orbitals and occurs when these coefficients are an exact eigenvector of the current Hamiltonian matrix. If it is further assumed that these mixing coefficients are to remain optimal with the orbital variations, then an implicit dependence of the vector $\mathbf{p}$ (or the expansion coefficients $\mathbf{c}$) on the orbital changes is established. The functional form of these CSF variations may be determined using perturbation theory[6] or alternately from one of the energy expressions of Eqs (142) through (144). The first-order response (giving the second-order contributions to the energy) may be written as

$$\mathbf{p} = -\mathbf{M}^{-1}\mathbf{C}^t\boldsymbol{\kappa} \tag{150}$$

using the matrix notation introduced above[16,19]. When substituted into the above energy expressions, this gives new energy expressions that explicitly depend only on the orbital variations

$$E^{(2)}(\mathbf{K}) = E(0) + \boldsymbol{\kappa}^t\mathbf{w} + \tfrac{1}{2}\boldsymbol{\kappa}^t(\mathbf{B} - \mathbf{C}\mathbf{M}^{-1}\mathbf{C}^t)\boldsymbol{\kappa} \tag{151}$$

$$= E(0) + \tfrac{1}{2}(\boldsymbol{\kappa}^t \quad 1)\begin{pmatrix} \mathbf{B} - \mathbf{C}\mathbf{M}^{-1}\mathbf{C}^t & \mathbf{w} \\ \mathbf{w}^t & 0 \end{pmatrix}\begin{pmatrix} \boldsymbol{\kappa} \\ 1 \end{pmatrix} \tag{152}$$

The matrix $(\mathbf{B} - \mathbf{C}\mathbf{M}^{-1}\mathbf{C}^t)$ is called the partitioned orbital Hessian matrix because of its connection to matrix partitioning theory. While the matrix $\mathbf{B}$ is the matrix of second derivatives with respect to orbital rotations when the CSF mixing coefficients are held constant, the partitioned orbital Hessian matrix is the matrix of second derivatives with respect to orbital rotations when the CSF coefficients relax optimally with the orbital changes. The matrix of Eq. (152), consisting of the partitioned orbital Hessian matrix and the orbital gradient vector, is called the augmented partitioned orbital Hessian matrix.

The MCSCF optimization process consists of finding the optimal set of orbital and CSF rotation parameters that minimize the energy. For excited states, the optimization process consists of finding the set of orbitals that minimize the appropriate eigenvalue of the Hamiltonian matrix. A set of necessary conditions for this to occur may be expressed using the equations

$$\left.\frac{\partial E}{\partial \boldsymbol{\kappa}}\right|_{\boldsymbol{\kappa}(\mathrm{opt}),\mathbf{p}(\mathrm{opt})} = \mathbf{0} \tag{153}$$

$$\left.\frac{\partial E}{\partial \mathbf{p}}\right|_{\boldsymbol{\kappa}(\mathrm{opt}),\mathbf{p}(\mathrm{opt})} = \mathbf{0} \tag{154}$$

It is not practicable to solve for the optimal orbitals and CSF coefficients directly from these equations. Instead, one of the truncated energy expressions, with possible modifications, is used to define the corrections appropriate for a given reference wavefunction. These corrections are then applied to the wavefunction, a new approximate energy expression is used to determine a new set of wavefunction corrections, and so on until convergence is obtained. In the following sections, these iterative methods will be discussed in detail. It is useful first to consider some of the properties of the gradient and Hessian matrices defined in the previous equations. These properties must be satisfied by the wavefunction obtained from any MCSCF iterative procedure and they may also be used during the optimization process to improve the convergence characteristics of particular iterative procedures.

Once a converged MCSCF wavefunction has been found, the stationary conditions of Eqs (153) and (154) reduce to the condition that the elements of the wavefunction gradient vector must be zero

$$w_{rs} = 0 \tag{155}$$

$$v_n = 0 \tag{156}$$

Eq. (155) is referred to as the generalized Brillouin theorem and as the Brillouin–Levy–Berthier theorem[51,52]. Eq. (156) implies that the converged CSF expansion coefficients are an eigenvector of the Hamiltonian matrix.

At convergence of the MCSCF iterative procedure, when the gradient terms vanish in the energy expression of Eq. (151), the most significant energy changes, with respect to wavefunction variations, result from the Hessian contributions. The MCSCF energy results from minimizing the appropriate eigenvalue of the Hamiltonian matrix with respect to orbital variations. For both ground-state and excited-state optimizations, this means that the partitioned orbital Hessian matrix must be positive semidefinite and that the converged CSF expansion coefficients must correspond to the appropriate eigenvector of the Hamiltonian matrix. In the case of zero eigenvalues of the partitioned orbital Hessian matrix, the higher derivatives of the energy must indicate that the energy is minimized within the subspace of wavefunction variations determined by the corresponding eigenvectors. For some wavefunction expansions, it occurs that all derivatives along certain directions are zero (i.e. the energy is invariant to certain wavefunction changes). These redundant variables are discussed in Section V where the relation between the orbital variations and the CSF mixing coefficient variations is considered in more detail.

Matrix partitioning methods may be used to show the relation between the eigenvalues of the partitioned orbital Hessian matrix and those of the wavefunction Hessian matrix[19]. If the wavefunction Hessian matrix is partitioned as in Eq. (34), it may be verified that the intersection of the branches of $\mathbf{L}(\lambda)$ with the vertical axis, $\lambda = 0$, are the eigenvalues of the partitioned orbital Hessian matrix. Fig. 3 shows three examples for a calculation of the first excited state. In this case it is clear from Eq. (149) that the state Hessian matrix $\mathbf{M}$ has exactly one negative eigenvalue. In general, the Nth excited state will have exactly N negative eigenvalues of the matrix $\mathbf{M}$ which result in N negative vertical asymptotes. The horizontal asymptotes of Fig. 3 are the eigenvalues of the orbital Hessian matrix $\mathbf{B}$. The three examples of Fig. 3 all have the same set of horizontal and vertical asymptotes.

The elements of the coupling matrix $\mathbf{C}$ determine the details of the shapes of the various branches. Equal coupling between all of the orbital and CSF variations is assumed in Fig. 3 (i.e. $C_{ij} = \text{constant}$ for all i, j). Fig. 3a shows an example of weak coupling between the orbital and CSF variations. This is characterized by small elements of the matrix $\mathbf{C}$. It is seen in this case that a negative eigenvalue of the partitioned orbital Hessian matrix exists and therefore the corresponding wavefunction is an inappropriate description of the first excited state. An orbital change in the direction corresponding to the eigenvector associated with this negative eigenvalue would decrease the Hamiltonian eigenvalue, thus violating the energy minimization condition. The wavefunction Hessian matrix corresponding to Fig. 3a has two negative eigenvalues, one corresponding to the negative vertical asymptote and one to the negative partitioned orbital Hessian matrix eigenvalue. This correspondence holds for general cases as may be verified from inspection. Each negative eigenvalue of the partitioned orbital Hessian matrix may be traced backward to a negative eigenvalue of the wavefunction Hessian matrix (of smaller magnitude since the individual branches are decreasing functions) and each negative vertical asymptote corresponds to a negative eigenvalue of the wavefunction Hessian matrix. One difficulty with iterative methods based directly on the wavefunction Hessian matrix is that some negative eigenvalues are unwanted in the final wavefunction while others are, in fact, necessary. It is difficult to distinguish between these two types of eigenvalues unless the partitioned orbital Hessian matrix is actually available for further examination.

Fig. 3b shows the case where the eigenvalue spectrum of the $\mathbf{B}$ and $\mathbf{M}$ matrices remains the same but with slightly stronger coupling between the orbital and CSF variations. In this case the lowest eigenvalue of the

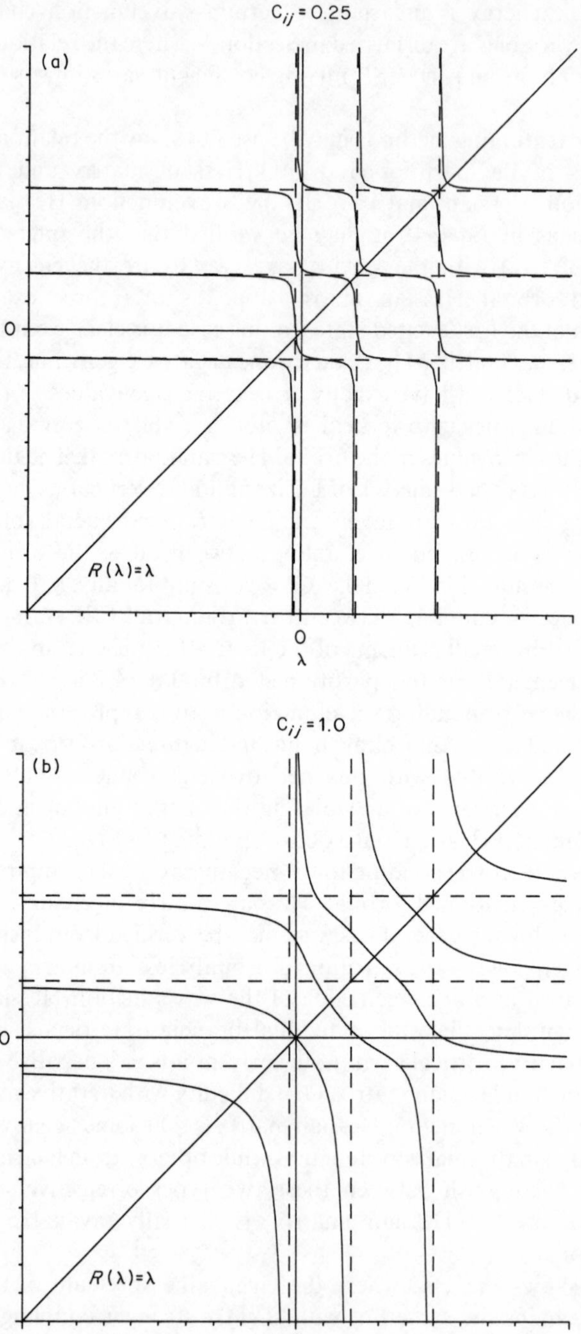
C_ij = 0.25
(a)
L(λ)
0
R(λ)=λ
0
λ
C_ij = 1.0
(b)
L(λ)
0
R(λ)=λ
0
λ

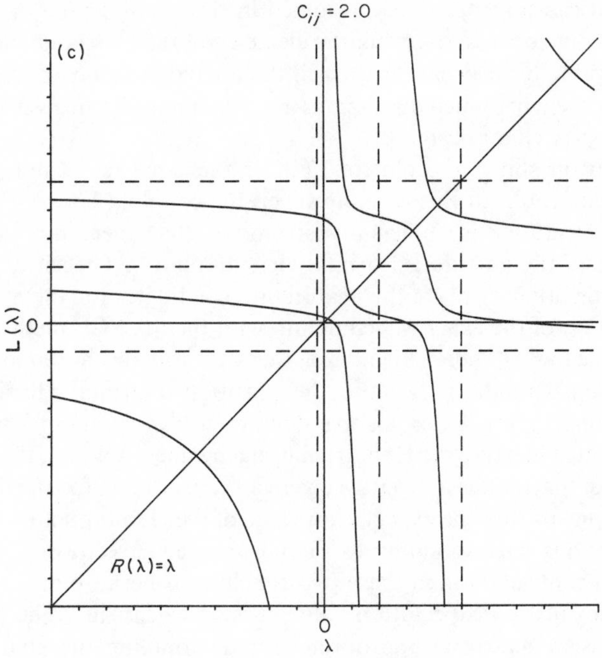

Fig. 3. Graphical representation of multidimensional matrix partitioning applied to the wavefunction Hessian matrix. The vertical asymptotes are the eigenvalues of the state Hessian matrix **M**. The horizontal asymptotes are the eigenvalues of the orbital Hessian matrix **B**. The intersections of the multivalued function $\mathbf{L}(\lambda)$ with the straight line $R(\lambda)$ occur at the eigenvalues of the wavefunction Hessian matrix. The intersections of $\mathbf{L}(\lambda)$ with the vertical line $\lambda = 0$ occur at the eigenvalues of the partitioned orbital Hessian matrix. These eigenvalues must be non-negative for acceptable MCSCF wavefunctions as discussed in the text. All three plots are for the first excited state for which **M** has one negative eigenvalue. The matrix **B** also has one negative eigenvalue. Fig. 3a corresponds to weak orbital–CSF coupling and results in an unacceptable approximation to the first excited state because of the negative eigenvalue of the partitioned orbital Hessian matrix. Fig. 3b corresponds to medium coupling and, coincidentally, to a spurious zero eigenvalue of the partitioned orbital Hessian matrix. Fig. 3c corresponds to strong coupling and to an acceptable approximate wavefunction. When the orbital–CSF coupling is strong, the branches of $\mathbf{L}(\lambda)$ avoid the intersections of the horizontal and vertical asymptotes by wider margins. This may be seen by comparing the shapes of the branches in the three figures.

partitioned orbital Hessian matrix has increased until it is equal to zero. The wavefunction Hessian matrix also has a zero eigenvalue in this case. This zero eigenvalue corresponds to a particular wavefunction variation for which the energy is invariant (at least to second order). This usually indicates that a redundant variable has not been removed from the wavefunction variation space but it can also occur, as presumed in this example, due to accidental numerical coincidences involving the elements of the Hessian matrix. Such a zero eigenvalue must occur, at least in the analytic regions of wavefunction variation space where such discussions are meaningful, as an eigenvalue of the Hessian matrix changes sign.

Fig. 3c again shows the plot of $\mathbf{L}(\lambda)$ for the same set of horizontal and vertical asymptotes. In this case, however, the coupling between the orbital and CSF variations is sufficiently strong so that there are no negative eigenvalues of the partitioned orbital Hessian matrix. The set of converged wavefunctions that displays this feature are candidates for the approximate representation of the exact wavefunction with the MCSCF method. At each geometry the lowest-energy member of this set is usually chosen to define this MCSCF approximation. Because the partitioned orbital Hessian matrix corresponding to Fig. 3c is positive definite, arbitrary infinitesimal orbital variations must increase the Hamiltonian eigenvalue. However, the wavefunction Hessian matrix has a negative eigenvalue, as it must for the first excited state. This means that the expectation value of the Hamiltonian will decrease when variations corresponding to the negative eigenvector are considered. There is no contradiction in these two conclusions because the excited-state energy is not just an expectation value, it is an eigenvalue. The partitioned orbital Hessian matrix is appropriate for determining the change in this Hamiltonian eigenvalue with respect to orbital variations while the wavefunction Hessian matrix is only appropriate for determining the change of the expectation value.

Fig. 3c also demonstrates another feature of MCSCF wavefunctions of excited states. For each negative eigenvalue of the matrix **M**, there may also be a negative eigenvalue of the orbital Hessian matrix **B**. For example, the second excited state may have zero, one or two negative eigenvalues of the **B** matrix. The second excited state must have exactly two negative eigenvalues of the **M** matrix and of the wavefunction Hessian matrix. It is not necessary for the matrix **B** to have any negative eigenvalues; in fact it usually does not, even for excited states, but clearly such cases appear to be possible. In the previous examples, corresponding to Fig. 3, it was argued that certain wavefunctions must be associated with strong coupling between the orbital and mixing coefficient variations before they satisfy the energy minimization condition. It may be just as easily argued that, under different conditions, acceptable wavefunctions must be characterized by weak coupling. This is because strong coupling between the orbital variations and higher-energy state variations may also result in negative eigenvalues of the partitioned orbital Hessian matrix that become positive as the coupling is reduced. In general, the stronger the coupling between the orbital and state variations, the larger will be the deviation away from the horizontal and vertical asymptote intersections by the branches of $\mathbf{L}(\lambda)$. This correlation may be observed in Fig. 3 where equal coupling is assumed for all the wavefunction variations. In general it may be concluded that MCSCF solutions that satisfy the eigenvalue minimization conditions must be accompanied with a correct balance of strong and weak coupling between the various orbital and state degrees of freedom. These considerations have important consequences for methods that explicitly use only the energy dependence on the orbital rotations in the computation of the wavefunction corrections[19].

There are two exceptions to these arguments relating the number of negative eigenvalues of the various Hessian matrix blocks. Olsen *et al.*[1,53] have argued that it may be necessary to relax the bracketing theorem upper-bound constraint on the MCSCF energy for excited state wavefunction optimization. These researchers have proposed instead that an approximation to the excitation energies from the reference function be used as a wavefunction selection criterion. This appears to be a weaker selection criterion than the bracketing theorem upper-bound condition. Certain wavefunctions have been found[54] that satisfy this criterion that are not consistent with the upper-bound condition while no wavefunctions have been found that result in upper bounds that do not also satisfy the approximate excitation energy criterion. More numerical experience and formal analysis is required to determine if such an approach is useful.

The second exception, noted by McLean *et al.*[55] and applicable to both ground and excited state wavefunction optimization, results from the fact that the imposition of spatial symmetry on a wavefunction may result in extraneous negative eigenvalues of the Hessian matrix. The wavefunction variations associated with these negative eigenvalues correspond, at symmetric molecular geometries, to the collapse of a symmetric wavefunction to an unsymmetric wavefunction which does not display the full molecular symmetry. The symmetric wavefunction, with the extraneous negative eigenvalues, smoothly deforms to an unsymmetrical wavefunction, which also has corresponding extraneous negative eigenvalues, when the molecule is distorted into an unsymmetrical geometry. The introduction of an energy hypersurface that depends on both the geometrical variables and the wavefunction variations is useful in this analysis[55]. In these cases the choice of the appropriate MCSCF representation may depend more on how well the approximate wavefunction mimics the behavior of the exact wavefunction (or at least that of a more accurate wavefunction) in the neighborhood of these symmetrical geometries than on the energy minimization conditions.

B. Matrix Element Evaluation

We now examine the elements of the gradient vector and Hessian matrix in more detail. The Hamiltonian operator of Eq. (115) and the generator commutation relations of Eqs (120) and (121) result in the following operator expression:

$$[H, T_{rs}] = (1 - P_{rs})\left(\sum_u h_{ru}(E_{us} + E_{su}) + \sum_t \sum_u \sum_v g_{rtuv}(e_{stuv} + e_{tsuv}) \right) \quad (157)$$

where P_{rs} acts to interchange the orbital indices r and s in the expression. This operator expression will be used in the evaluation of both the gradient and Hessian elements. The expectation value of the above operator gives directly the expression for the orbital gradient vector

$$w_{rs} = (1 - P_{rs})\left\langle 0 \left| \sum_u h_{ru}(E_{us} + E_{su}) + \sum_t \sum_u \sum_v g_{rtuv}(e_{stuv} + e_{tsuv}) \right| 0 \right\rangle \quad (158)$$

$$= 2(1 - P_{rs})\left(\sum_u h_{ru} D_{us} + \sum_t \sum_u \sum_v g_{rtuv} d_{stuv} \right) \quad (159)$$

$$= 2(1 - P_{rs})F_{rs} = 2(F_{rs} - F_{sr}) \quad (160)$$

The generalized Fock matrix **F**, defined above, is a convenient partial sum because it can also be used in the Hessian matrix element construction. This matrix is not symmetric before convergence is reached in the MCSCF iterative procedure, but becomes symmetric at convergence[56], resulting in vanishing orbital gradient vector elements.

Eq. (159) gives the expresion for the gradient vector elements in terms of the one- and two-particle density matrices defined in Eqs (126) and (129). It is important to note that these expressions result directly from the commutator relations of the generators and that the rank reduction is due to the exponential parametrization of the wavefunction variations of the orbitals. In fact the derivation of the gradient terms, the Hessian terms or higher derivative terms consists simply of index substitutions from Eqs (120) and (121). The examination of Eq. (159) shows that only a subset of the two-electron integrals are required to construct the gradient vector. Three indices of the integral are common with the indices of the density matrix element, the latter of which must all correspond to occupied orbitals in order for the density matrix element to be non-zero. For the usual type of MCSCF wavefunctions, only a fraction of the orbitals are occupied, as discussed in more detail in Section IV.

The gradient elements may also be written as

$$w_{rs} = 2\langle 0|HT_{rs}|0\rangle \quad (161)$$

and therefore are formally equivalent to matrix elements of the Hamiltonian operator involving the reference state $|0\rangle$ and the set of 'single excitation' states $\{T_{rs}|0\rangle : r > s\}$ where $T_{rs}|0\rangle = (E_{rs} - E_{sr})|0\rangle$. Evaluating **w** as this set of expectation values requires either the explicit construction of the single excitation states (e.g. in a determinantal representation) or the evaluation of three-particle density matrix type elements of the form $\langle 0|e_{tuvw}T_{rs}|0\rangle$. The first method becomes impracticable as the number of expansion CSFs increases; the second becomes impracticable as the size of the orbital basis set increases. Both of these schemes are less efficient than the method resulting from the straightforward commutator reduction of Eq. (159). The use of the one- and two-particle density matrices allows the gradient vector to be constructed with no explicit reference either to the CSF expansion or to the single excitation state expansion. Although this single excitation interpretation may be useful in understanding the effect of the operators T_{rs}, or of the operator $\exp(K)$, it has also been responsible for an unnecessary amount of confusion. Several researchers had initially discounted the effectiveness of the exponential operator parametrization of the orbital variations based on the single excitation interpretation without considering the rank reductions that result from the commutator expressions.

The expressions for the elements of the coupling matrix **C** also result directly from the operator expression of Eq. (157):

$$C_{rs,n} = 2(1 - P_{rs})\left\langle 0\left|\sum_u h_{ru}(E_{us} + E_{su}) + \sum_t\sum_u\sum_v g_{rtuv}(e_{stuv} + e_{tsuv})\right| n\right\rangle \quad (162)$$

$$= 4(1 - P_{rs})\left(\sum_u h_{ru}D^n_{su} + \sum_t\sum_u\sum_v g_{rtuv}d^n_{stuv}\right) \quad (163)$$

The density matrix elements of the form

$$D^n_{su} = \tfrac{1}{2}\langle 0|E_{us} + E_{su}|n\rangle \quad (164)$$

$$d^n_{stuv} = \tfrac{1}{4}\langle 0|e_{stuv} + e_{tsuv} + e_{stvu} + e_{tsvu}|n\rangle \quad (165)$$

are called symmetric transition density matrices and have the same orbital index symmetry properties as the density matrices **D** and **d** and as the integrals **h** and **g**. When the CSF basis is used to define the transition density matrices then, for a given set of orbital indices, it may also be observed that dot products of D^n_{su} or of d^n_{stuv} with the CSF coefficient vector **c** result in the usual density matrix elements D_{su} and d_{stuv}. When used with Eq. (163), this also implies that $\mathbf{Cc} = 2\mathbf{w}$. That is, the dot product of a row of the coupling matrix **C** with the vector **c** gives twice the corresponding orbital gradient vector element. Construction of the elements of **C** requires the same subset of the two-electron integrals as the gradient vector elements.

Elements of the orbital Hessian matrix may be derived from Eq. (157) by evaluating one more commutator expression. Because of the rank reduction properties of these commutator expressions, we already know that the form of the result will involve summations of generators and generator products and that the matrix elements of these operators will be constructed from the one- and two-particle density matrices and the one- and two-electron integrals. This is, in general, going to be true for higher-order commutator expressions. This means that not only will the first and second derivatives of the energy depend only on the one- and two-particle density matrix elements, but all higher-order derivatives will also depend only on these quantities. This results from the exponential parametrization of the MCSCF wavefunction and from the rank reductions that are inherent in the commutator expressions. The substitution of the commutator expressions of Eqs (120) and (121) and the operator of Eq. (157) results in the following expression for the orbital Hessian matrix elements:

$$B_{rs,uv} = (1 - P_{rs})(1 - P_{uv})\left((F_{rv} + F_{vr})\delta_{us} - 2h_{rv}D_{us} + \sum_w \sum_x (4g_{rwux}d_{swvx} + 2g_{ruwx}d_{svwx}) \right) \tag{166}$$

This expression is again written in an abbreviated form using the index permutation operators P_{uv} and P_{rs}. Since the symmetric combination of Fock matrix elements is required for the Hessian matrix and the antisymmetric combination for the gradient elements, the matrix **F** itself may be explicitly constructed and then its contributions included into the other matrix elements as needed. Eq. (166) shows that a larger subset of two-electron integrals are required for the Hessian elements than for the computation of the gradient elements (two unoccupied orbital indices instead of just one), but for typical MCSCF wavefunctions and orbital basis sets this is still only a small fraction of the total.

Finally the state Hessian matrix **M** is seen from Eq. (149) to be proportional to the representation of the Hamiltonian operator in the orthogonal complement basis, but with all the eigenvalues shifted by the constant amount $E(0)$. The dimension of the matrix **M** will be one less than the length of the CSF expansion unless it is constructed in the linearly dependent projected basis or the overcomplete CSF expansion set basis. Since the Hamiltonian matrix must usually be constructed in the CSF basis in the MCSCF method anyway, it is most convenient if **M** and **C** are also constructed in this basis. The transformation to the projected basis, if explicitly required, involves the projection matrix $(\mathbf{1} - \mathbf{c}\mathbf{c}^{\dagger})$. The matrix **M** only requires the two-electron integral subset that consists of all four orbital indices corresponding to occupied orbitals.

C. Solution of the Multiconfiguration Self-consistent Field Optimization Equations

The determination of the orbital corrections for each MCSCF iteration requires the manipulation of the gradient vector and Hessian matrix blocks. For relatively small orbital basis expansions and CSF expansions, the Hessian matrix elements may be constructed and explicitly stored. The solution of the

MCSCF equations, which are discussed in detail in the following sections, requires the solution of matrix equations involving the gradient vector and Hessian matrix. The solution to these matrix equations may be obtained using direct methods, such as Gaussian elimination or LU factorization, or iterative methods, such as the conjugate gradient or subspace expansion methods. For CSF expansion lengths exceeding a few hundred, it is impracticable to compute and store the **C** and **M** matrices. In these cases, it is still possible to solve the MCSCF equations using iterative methods that require the construction of the matrix–vector products **Br**, **Cs**, $\mathbf{C}^t\mathbf{r}$ and **Ms** where the vectors **r** and **s** are subspace expansion vectors in terms of which the correction vectors $\boldsymbol{\kappa}$ and **p** respectively are represented. The iterative solution of the MCSCF equations using the gradient vector and exact Hessian matrix was first implemented by Lengsfield and Liu[47] and is one of the more significant recent enhancements of MCSCF methodology. Because this iterative solution of the matrix equations is performed within an MCSCF iteration, these methods are usually called micro-iterative methods. Improvements in the MCSCF iterative procedures and in the actual construction of the required matrix–vector products have been proposed by several researchers[57–60]. The improvements in the construction of the matrix–vector products will be discussed in this section and will apply to the MCSCF methods discussed in the following sections.

Consider first the construction of the matrix–vector product **Ms**. For large CSF expansion lengths, this product is most efficiently computed in the CSF basis. Since the vector **s** is used as an expansion vector in the construction of the vector **p**, and since $\mathbf{p}^t\mathbf{c} = 0$, the vector **s** may also be chosen to be orthogonal to the Hamiltonian eigenvector **c**, i.e. $\mathbf{s}^t\mathbf{c} = 0$. Eq. (149) may then be used directly to give

$$\mathbf{Ms} = 2[\mathbf{H} - E(0)\mathbf{1}]\mathbf{s} = 2[\mathbf{Hs} - E(0)\mathbf{s}] \tag{167}$$

so that the required matrix–vector product may be computed with the same methodology as the matrix–vector products required in the iterative solution of the Hamiltonian matrix diagonalization. The unitary group approach is ideal for these products and other 'direct CI' approaches have also been used[42]. Eq. (167) may be modified for arbitrary reference wavefunctions by using the projection operator $(\mathbf{1} - \mathbf{cc}^t)$ to transform both the bra and ket expansion terms to the projected basis

$$\mathbf{Ms} = 2(\mathbf{1} - \mathbf{cc}^t)[\mathbf{H} - E(0)\mathbf{1}](\mathbf{1} - \mathbf{cc}^t)\mathbf{s} \tag{168}$$

If **c** is an eigenvector of **H**, Eqs (167) and (168) produce identical results. The above expression gives the representation of **M** in the linearly dependent projected basis. Although this projected basis was employed by the author in the direct solution of the optimization equations[41,58], it was first employed by Lengsfield and Liu[47] in the micro-iterative solution of these equations where its use actually extends the range of problems that may be approached with the MCSCF method. It is usually more efficient for the three matrices to operate successively, from right to left in Eq. (168), on the expansion vector rather than explicitly transforming the matrix $[\mathbf{H} - E(0)\mathbf{1}]$.

The matrix–vector product $\mathbf{Cs}$ is also performed most straightforwardly in the CSF basis. In this case, Eq. (157) may be used to show

$$(\mathbf{Cs})_{pq} = 4(1 - P_{pq})\sum_n \left(\sum_u h_{pu} D^n_{qu} + \sum_t \sum_u \sum_v g_{ptuv} d^n_{qtuv} \right) s_n \tag{169}$$

$$= 4(1 - P_{pq})\left(\sum_u h_{pu} D^s_{qu} + \sum_t \sum_u \sum_v g_{ptuv} d^s_{qtuv} \right) \tag{170}$$

$$= 4(1 - P_{pq}) F^s_{pq} = 4(F^s_{pq} - F^s_{qp}) \tag{171}$$

where, for a given trial vector $\mathbf{s}$, $\mathbf{D}^s$ and $\mathbf{d}^s$ may be computed similarly to the way $\mathbf{D}$ and $\mathbf{d}$ are computed. Once these transition density matrices are in hand for a given trial vector, the matrix–vector product $\mathbf{Cs}$ may then be computed by constructing the appropriate combinations of elements of the transtion Fock matrix $\mathbf{F}^s$. For trial vectors that are not constrained to be orthogonal to $\mathbf{c}$, the above equation may be replaced by

$$\mathbf{C}(\mathbf{1} - \mathbf{cc}^t)\mathbf{s} = \mathbf{Cs} - 2(\mathbf{c}^t\mathbf{s})\mathbf{w} \tag{172}$$

since, as shown previously, $\mathbf{Cc} = 2\mathbf{w}$. The above expression implicitly gives the transformation of the matrix $\mathbf{C}$ from the CSF basis to the linearly dependent projected basis. As indicated above, however, it is usually more efficient for the projection matrix to operate on the expansion vector than on the matrix $\mathbf{C}$.

Before consideration of the matrix–vector product $\mathbf{C}^t\mathbf{r}$, we first examine the following operator expression:

$$\sum_{(pq)} [H, T_{pq}] r_{pq} = \sum_p \sum_q [H, E_{pq}] R_{pq} \tag{173}$$

$$= \sum_u \sum_v h^R_{uv} E_{uv} + \tfrac{1}{2} \sum_u \sum_v \sum_w \sum_x g^R_{uvwx} e_{uvwx} \tag{174}$$

$$= H^R \tag{175}$$

The operator expansion coefficients are given by the effective integrals

$$h^R_{uv} = (\mathbf{hR} - \mathbf{Rh})_{uv} = (\mathbf{hR} + (\mathbf{hR})^t)_{uv} \tag{176}$$

$$= \sum_t (h_{tv} R_{tu} + h_{ut} R_{tv}) \tag{177}$$

$$g^R_{uvwx} = \sum_t (g_{tvwx} R_{tu} + g_{utwx} R_{tv} + g_{uvtx} R_{tw} + g_{uvwt} R_{tx}) \tag{178}$$

where the matrix $\mathbf{R}$ is the antisymmetric expansion of the parameters in the vector $\mathbf{r}$ (just as $\mathbf{K}$ is related to $\boldsymbol{\kappa}$). Both sets of effective integrals $\mathbf{h}^R$ and $\mathbf{g}^R$ may be constructed, as indicated in Eq. (176), with a one-index transformation of the original integrals, using the matrix **R** as the transformation matrix, followed by symmetrization. The effective integrals then have the same orbital index properties as the original integrals **h** and **g** and define an effective Hamiltonian operator H^R that has the same form as the original operator. Eq. (175) may now be used to simplify the matrix–vector product $\mathbf{C}^t\mathbf{r}$:

$$(\mathbf{C}^t\mathbf{r})_n = 2\langle n|H^R|0\rangle = 2\sum_m \langle n|H^R|\tilde{m}\rangle c_m \tag{179}$$

$$= 2(\mathbf{H}^R\mathbf{c})_n \tag{180}$$

$$= 2(\mathbf{1} - \mathbf{c}\mathbf{c}^t)(\mathbf{H}^R\mathbf{c})_n \tag{181}$$

The matrix–vector product indicated in Eq. (181) may be computed from the explicitly constructed matrix $\mathbf{H}^R$ or using a direct CI procedure along with the effective integrals. The result of this matrix–vector product must lie in the orthogonal complement space. In cases where the overcomplete CSF basis is used as the bra basis to define $\mathbf{H}^R$, the explicitly projected version of Eq. (181) should be used to ensure orthogonality to **c**.

The effective Hamiltonian operator of Eq. (175) may also be used to simplify the matrix–vector product **Br**. To this end, it is convenient to use the identity $[[A, B], C] - [[A, C], B] = [A, [B, C]]$ to write the symmetric **B** matrix in the form

$$B_{wx,uv} = \langle 0|[[H, T_{uv}], T_{wx}] + \tfrac{1}{2}[H, [T_{wx}, T_{uv}]]|0\rangle \tag{182}$$

The commutator relation

$$[T_{wx}, T_{uv}] = T_{wv}\delta_{xu} + T_{xu}\delta_{wv} + T_{uw}\delta_{vx} + T_{vx}\delta_{uw} \tag{183}$$

allows the second term to be reduced to a sum of gradient terms, which in turn may be defined in terms of Fock matrix elements. Using Eq. (145) and the effective operator of Eq. (175) for the first term in the above expression gives

$$(\mathbf{Br})_{wx} = \langle 0|[H^R, T_{wx}]|0\rangle + ([\mathbf{R}, \mathbf{F}] - [\mathbf{R}, \mathbf{F}]^t)_{wx} \tag{184}$$

$$= (\mathbf{w}^R)_{wx} + ([\mathbf{R}, \mathbf{F}] - [\mathbf{R}, \mathbf{F}]^t)_{wx} \tag{185}$$

In Eq. (185) the effective gradient vector $\mathbf{w}^R$ is computed in the same way as the usual gradient vector **w** except that the effective integrals of Eqs (176) and (178) are used instead of the usual integrals. (In practice, it may be most efficient if the effective integrals are used only for the construction of $(\mathbf{C}^t\mathbf{r})$ and the matrix–vector product $(\mathbf{Br})$ is instead constructed from the explicitly computed and stored **B** matrix.)

These relations show that it is possible to compute the matrix–vector products involving the Hessian matrix, which are required to solve the MCSCF equations, using a series of one-index transformations of the integrals and direct CI type matrix–vector products. For each expansion vector **r**, a one-index transformation of the integrals is required. For each expansion vector **s** the construction of a transition density matrix and the formation of a direct CI matrix–vector product is required. Various implementations of this approach are discussed in Section VI. With the formalism developed in the preceding sections it is now possible to introduce several MCSCF optimization methods and discuss their relative merits.

D. Newton–Raphson Optimization Methods

The most straightforward optimization equations for MCSCF wavefunctions result simply from imposing on one of the truncated energy expressions the condition that the energy is stable with respect to orbital and CSF variations. Given the approximate orbital and CSF corrections obtained from applying this condition to the truncated expression, a new reference wavefunction is defined in this new orbital basis. This requires transforming the one- and two-electron integrals from the old basis to the new basis. New Hessian matrix and gradient vector elements are computed in this new orbital and state basis and the process is repeated until convergence is obtained. This Newton–Raphson process converges quadratically in some neighborhood of a solution. The error in each iteration of both the wavefunction and the energy is the square of the error of the previous iteration. Applying Eqs (153) and (154) to the orbital and CSF variations at each iteration results in the correction vectors being determined from the linear equation system

$$\begin{pmatrix} \mathbf{B} & \mathbf{C} \\ \mathbf{C}^t & \mathbf{M} \end{pmatrix} \begin{pmatrix} \boldsymbol{\kappa} \\ \mathbf{p} \end{pmatrix} + \begin{pmatrix} \mathbf{w} \\ \mathbf{v} \end{pmatrix} = \begin{pmatrix} \mathbf{0} \\ \mathbf{0} \end{pmatrix} \tag{186}$$

if the truncated Eq. (143) is used, or from the linear equation

$$(\mathbf{B} - \mathbf{C}\mathbf{M}^{-1}\mathbf{C}^t)\boldsymbol{\kappa} + \mathbf{w} = \mathbf{0} \tag{187}$$

if the partitioned energy expression of Eq. (151) is used. If the latter equation is used to defined the orbital corrections, the CSF mixing coefficients must be determined for this new orbital basis in the next MCSCF iteration with the partial diagonalization of the Hamiltonian matrix. This MCSCF iterative method is called the partitioned orbital Hessian matrix Newton–Raphson (PNR) iterative procedure[6,16,37]. If Eq. (186) is used to define the orbital corrections for each iteration, the resulting method is called the wavefunction Hessian matrix Newton–Raphson (WNR) iterative procedure. If the CSF corrections are also determined from Eq. (186), the resulting method is called the one-step WNR procedure[16,49]. If, instead, the Hamiltonian matrix is diagonalized at the beginning of each MCSCF iteration, but Eq. (186) is used to define the orbital corrections, then the resulting method is called the two-step WNR procedure[16]. The PNR and two-step WNR methods result in identical orbital corrections for each MCSCF iteration. Both of these methods would be expected to be more stable than the one-step WNR method since the CSF coefficients are optimized more fully at each iteration, but all three of these Newton–Raphson methods display essentially the same local convergence properties.

All matrix elements in the Newton–Raphson methods may be constructed from the one- and two-particle density matrices and transition density matrices. The linear equation solutions may be found using either direct methods or iterative methods. For large CSF expansions, such micro-iterative procedures may be used to advantage. If a micro-iterative procedure is chosen that requires only matrix–vector products to be formed, expansion-vector-dependent effective Hamiltonian operators and transition density matrices may be constructed for the efficient computation of these products. Sufficient information is included in the Newton–Raphson optimization procedures, through the gradient and Hessian elements, to ensure second-order convergence in some neighborhood of the final solution.

There is, however, no guarantee that the Newton–Raphson procedure will converge to the correct MCSCF solution. In fact, there is no compelling reason even to expect convergence to the correct state with the WNR method. For the optimization of a given state, an iterative procedure should converge to a solution for which the appropriate eigenvalue of the Hamiltonian matrix is minimized with respect to orbital variations. The Newton–Raphson procedure only attempts to locate stationary points of the energy, only a small subset of which satisfy the eigenvalue minimization condition[19,61,62].

Numerical investigations of the Newton–Raphson method have shown that the neighborhood of convergence to the correct solution is often very small. By starting with a converged set of orbitals and systematically rotating the orbitals by varying amounts to generate different initial reference wavefunctions, the Newton–Raphson method has been shown to require starting orbitals within 5° to 10° of the converged orbitals even for well behaved cases[19,54,62]. Such well defined initial approximations are usually not available and, at any rate, more reliable procedures would be preferred to avoid convergence to any nearby spurious stationary points. The source of these convergence difficulties in the Newton–Raphson methods is that the truncated energy expression does not sufficiently mimic the true energy expression except in a small neighborhood of the reference wavefunction. Examination of Eq. (143) shows that as $|\boldsymbol{\kappa}|$ increases, the truncated energy expression is unbounded. If the wavefunction Hessian matrix has both negative and positive eigenvalues for the reference wavefunction, the truncated energy approximation may either increase or decrease without bound. In the following sections, other approximate energy expressions are introduced that do not display this incorrect and unphysical behavior.

E. Variational Super-configuration-interaction Optimization Methods

The first energy expression of the above type results from truncating the expansion of the wavefunction variations to include only the first-order changes

$$|0'\rangle \approx (1 + K + S)|0\rangle \tag{188}$$

This is called the first-order variational space which is spanned by the reference state, the orthogonal complement states and the single excitation states $\{|0\rangle, |n\rangle, T_{rs}|0\rangle\}$ or equivalently by the set of expansion CSFs and the single excitation states $\{|\tilde{n}\rangle, T_{rs}|0\rangle\}$. The parameters **K** and **p** may be determined by minimizing the expectation value of the Hamiltonian operator within this non-orthonormal basis. This results in the non-orthogonal matrix eigenvalue equation

$$\mathbf{H}^{\mathrm{SCI}}\mathbf{u} = \mathbf{S}^{\mathrm{SCI}}\mathbf{u}E^{\mathrm{SCI}} \tag{189}$$

where the eigenvector contains the wavefunction variation parameters, $\mathbf{u} = (\boldsymbol{\kappa}^{\mathrm{t}} \quad \mathbf{p}^{\mathrm{t}} \quad u_0)^{\mathrm{t}}$. These parameters may then be used, as in the Newton–Raphson method, to determine the new orbital basis. The approximate version of this method resulting from the deletion of the **p** parameters was first used by Grein and coworkers[23,63–65]. An alternate approach that uses the natural orbitals of the super-configuration-interaction (SCI) wavefunction to determine the orbital corrections has been proposed by Reudenberg and coworkers[66,67]. The above method was subsequently developed to improve the excited-state optimization characteristics by Chang and Schwarz[26,68]. The SCI method displays only first-order convergence in the neighborhood of the solution but otherwise it displays very good convergence properties. This convergence behavior results from the fact that the expectation value of the Hamiltonian in the SCI basis is bounded from below by the full CI energy and bounded from above by the current approximate MCSCF energy. Both of these bounds results from the bracketing theorem. The SCI energy of a particular iteration is approximately equal to the MCSCF energy of the next iteration. As the iterative procedure progresses toward convergence, both of these eigenvalues approach the final MCSCF energy.

Besides displaying only linear local convergence, another problem with the SCI method is that the construction of the $\mathbf{H}^{\mathrm{SCI}}$ matrix requires either the explicit construction of the single excitation states or the evaluation of four-particle density matrix type elements of the form $\langle 0| T_{pq} e_{rstu} T_{vw}|0\rangle$. In an implementation by the author[69,70] for example, the single excitation states were explicitly constructed (in a determinantal basis) and the construction of the $\mathbf{H}^{\mathrm{SCI}}$ matrix required an additional lengthy formula file. This places limitations on the orbital basis size and CSF expansion length[65,67,69,71] that are not imposed by the Newton–Raphson methods discussed previously.

These limitations have prompted several modifications to the matrix elements of the $\mathbf{H}^{SCI}$ matrix. Siegbahn, Roos and coworkers[72-74] have proposed that these matrix elements be approximated with Fock matrix elements which may be computed from the one- and two-particle density matrices. Chang and Schwarz[26] attempted to improve the linear convergence behavior by modifying the $\mathbf{H}^{SCI}$ matrix elements to include the effects of the second-order variations of the wavefunction. (However, the orbital transformation that was applied in this particular case was correct only to first order and the resulting MCSCF method only displayed first-order convergence.) All of these methods destroy the strict lower-bound relation between the SCI energy and the exact energy and replace it with an analogous but somewhat weaker lower-bound relation. Even with these modifications, however, the use of the expectation value type of energy expression results in stable convergence behavior.

F. Rational Function Approximation Optimization Methods

This stable behavior of these approximate super-CI methods suggests that the truncated energy expressions should be replaced with a modified energy expression that is of the expectation value form or, more generally, of the form of ratios of polynomial functions. Formal comparisons of the SCI methods with the Newton–Raphson methods[6,19,45,46,62,72] show the similarities between the individual matrix elements of the two methods, also suggesting such a functional form. One such set of approximations results from multiplying the augmented Hessian matrices of Eq. (144) or of Eq. (152) with a normalization type factor[19,61]. This results in an approximate energy expression of the same general form as the SCI energy expression and in the wavefunction corrections being determined from a matrix eigenvalue equation of the same form as Eq. (189).

Consider the multiplication of the augmented Hessian matrix of Eq. (152) by a factor of the form $(1 + \boldsymbol{\kappa}^t\mathbf{S}\boldsymbol{\kappa})^{-1}$ where the matrix $\mathbf{S}$ is some unspecified-metric matrix. Expansion of this factor in powers of $\boldsymbol{\kappa}$

$$(1 + \boldsymbol{\kappa}^t\mathbf{S}\boldsymbol{\kappa})^{-1} = 1 - \boldsymbol{\kappa}^t\mathbf{S}\boldsymbol{\kappa} + (\boldsymbol{\kappa}^t\mathbf{S}\boldsymbol{\kappa})^2 + \cdots \tag{190}$$

shows immediately that the first and second derivatives of the energy expression are unchanged. However, the new energy expression has third- and higher-order derivative terms that depend on the elements of the matrix $\mathbf{S}$ and on the gradient and Hessian matrix elements:

$$E^{(2)}(\mathbf{K}) = E(0) + \tfrac{1}{2}(\boldsymbol{\kappa}^t \quad 1)\begin{pmatrix} \mathbf{B} - \mathbf{C}\mathbf{M}^{-1}\mathbf{C}^t & \mathbf{w} \\ \mathbf{w}^t & 0 \end{pmatrix}\begin{pmatrix} \boldsymbol{\kappa} \\ 1 \end{pmatrix}(1 + \boldsymbol{\kappa}^t\mathbf{S}\boldsymbol{\kappa})^{-1} \tag{191}$$

$$= E(0) + \boldsymbol{\kappa}^t\mathbf{w} + \tfrac{1}{2}\boldsymbol{\kappa}^t(\mathbf{B} - \mathbf{C}\mathbf{M}^{-1}\mathbf{C}^t)\boldsymbol{\kappa} - \boldsymbol{\kappa}^t\mathbf{w}(\boldsymbol{\kappa}^t\mathbf{S}\boldsymbol{\kappa}) + \cdots \tag{192}$$

The superscript on this energy expression indicates that the approximation is correct through second order in the parameters **K**. The third- and higher-order terms in this expression (and in the analogous wavefunction Hessian matrix energy expression) do not in general correspond to the higher-order terms of the exact energy expression. The parameters of the matrix **S** could, in principle, be chosen to allow certain higher-order derivatives to be approximated. This could be done either analytically, using effective operator methods[75], or numerically, using an update procedure[76] that combines information from several MCSCF iterations. It should be emphasized that the expansion of Eq. (192) is used here only to demonstrate the correct local behavior of the approximate energy. The 'infinite-order' expression (Eq. (191)) is actually used to determine the wavefunction correction with this approach.

1. Augmented Hessian Optimization Methods

In all applications of the above approach so far, the matrix **S** has been chosen to be the unit matrix. In the case of the energy expression resulting from Eq. (144), the resulting iterative procedure has been called the augmented Hessian method[45], the non-variational super-CI method[61], the wavefunction Hessian super-CI (WSCI) method[19] and the norm-extended optimization (NEO) method[59]. In the case of the energy expression resulting from Eq. (152), this approach results in the partitioned orbital Hessian super-CI (PSCI) method[19]. These methods attempt to combine the advantages of the Newton–Raphson and the SCI iterative methods. They display second-order local convergence; all of the required matrix elements are defined in terms of commutator expansions and therefore may be constructed in terms of one- and two-particle density matrices; and they all have a much larger neighborhood of convergence than the analogous Newton–Raphson methods[19]. In the case of large CSF expansions, the effective operator methods discussed previously may be employed to form the matrix–vector products required within each micro-iteration which are used to solve for the orbital corrections at each MCSCF iteration[59]. In the wavefunction Hessian case, these orbital corrections result from the solution of the matrix eigenvalue equation

$$\begin{bmatrix} \mathbf{B}-\lambda\mathbf{1} & \mathbf{C} & \mathbf{w} \\ \mathbf{C}^{\mathrm{t}} & \mathbf{M}-\lambda\mathbf{1} & \mathbf{v} \\ \mathbf{w}^{\mathrm{t}} & \mathbf{v}^{\mathrm{t}} & -\lambda \end{bmatrix} \begin{bmatrix} \boldsymbol{\kappa} \\ \mathbf{p} \\ 1 \end{bmatrix} = \begin{bmatrix} \mathbf{0} \\ \mathbf{0} \\ 0 \end{bmatrix} \tag{193}$$

where the eigenvalue λ is twice the approximate energy lowering that will result from the wavefunction corrections. In the partitioned Hessian case, the orbital corrections result from the solution of the secular equation

$$\begin{pmatrix} (\mathbf{B}-\mathbf{C}\mathbf{M}^{-1}\mathbf{C}^{\mathrm{t}})-\lambda\mathbf{1} & \mathbf{w} \\ \mathbf{w}^{\mathrm{t}} & -\lambda \end{pmatrix} \begin{pmatrix} \boldsymbol{\kappa} \\ 1 \end{pmatrix} = \begin{pmatrix} \mathbf{0} \\ 0 \end{pmatrix} \tag{194}$$

For small CSF expansions, the matrix $\mathbf{M}$ may be explicitly constructed and inverted (or factored) to produce the $\mathbf{CM}^{-1}\mathbf{C}^{t}$ contributions to the first sub-block of this matrix. For larger CSF expansions[57,58] it is most efficient to solve the equivalent unfolded matrix equation

$$\begin{bmatrix} \mathbf{B}-\lambda\mathbf{1} & \mathbf{C} & \mathbf{w} \\ \mathbf{C}^{t} & \mathbf{M} & \mathbf{v} \\ \mathbf{w}^{t} & \mathbf{v}^{t} & -\lambda \end{bmatrix} \begin{bmatrix} \boldsymbol{\kappa} \\ \mathbf{p} \\ 1 \end{bmatrix} = \begin{bmatrix} \mathbf{0} \\ \mathbf{0} \\ 0 \end{bmatrix} \tag{195}$$

This equation is neither a pure matrix eigenvalue equation nor a pure linear equation. However, its solution may be found using iterative methods that are no more complicated than the iterative solution of the pure matrix eigenvalue equation of Eq. (193) or the pure linear equation of Eq. (186). The iterative solution of these equations may use the effective operator methods discussed previously.

Both of these methods display the same local convergence properties and approximately the same ground-state convergence properties. However, the iterative methods based on the partitioned orbital Hessian matrix have an advantage over the methods based on the wavefunction Hessian matrix when applied to excited-state optimization. In the case of the PSCI iterative method, the eigenvector corresponding to the lowest eigenvalue may always be chosen to define the orbital correction vector. Such a straightforward procedure is not available for methods based on the wavefunction Hessian matrix except for relatively simple and well behaved cases[19].

The relation between the wavefunction corrections determined by the WSCI and PSCI methods and the corresponding Newton–Raphson methods may be determined using the matrix partitioning approach. For example, the solution to the PCI matrix eigenvalue equation is equivalent to the solution of the linear equation

$$(\mathbf{B}-\mathbf{CM}^{-1}\mathbf{C}^{t}-\lambda\mathbf{1})\boldsymbol{\kappa}+\mathbf{w}=\mathbf{0} \tag{196}$$

which may be compared to the Newton–Raphson linear equation (Eq. (187)). A more complete comparison of the PSCI, WSCI and WNR methods is given in the literature[19]. The matrix in Eq. (196) is always positive definite when the lowest eigenvalue λ is chosen to define $\boldsymbol{\kappa}$ (a result of the bracketing theorem). The orbital correction vector determined from Eq. (194) (or from Eq. (195)) determines a direction along which the energy must decrease in some neighborhood of the current approximate wavefunction. In actual practice the magnitude of the step determined from Eq. (194) is usually, but not always, sufficient for convergence to be obtained.

There have been two proposed approaches to the problem of finding optimal step lengths when the straightforward solution of the matrix eigenvalue equation is inadequate. The first is a line search approach[77], which requires the evaluation of the exact energy at one or more points along the line determined by the vector $\boldsymbol{\kappa}$; and the second is a trust radius approach[78–80], which results from the solution of either the matrix eigenvalue equation or the linear equation when an additional level-shift parameter is included in the appropriate matrix equation. This level-shift parameter is chosen to force $|\boldsymbol{\kappa}|$ to be smaller than a tolerance which is dynamically adjusted during the iterative procedure. This tolerance is usually based on the comparison of the predicted energy lowering to the actual energy lowering of each MCSCF iteration. After it is determined that the step length of the previous MCSCF iteration was too large, a deficiency of both approaches is that most of the effort of the last integral transformation and Hamiltonian matrix eigenvalue equation solution is discarded. The line search methods at least use the energy determined from the subsequent iteration, along with the energy and line derivatives of the previous iteration (e.g. $\boldsymbol{\kappa}^t\mathbf{w}$ and $\boldsymbol{\kappa}^t(\mathbf{B} - \mathbf{CM}^{-1}\mathbf{C}^t)\boldsymbol{\kappa}$ for the partitioned Hessian matrix based methods) to estimate the optimal intermediate step along the line $\boldsymbol{\kappa}$. The trust radius approaches simply adjust the trust radius and waste all the effort of the current iteration by reverting to the approximate wavefunction of the previous iteration but with a more conservative trust radius. It is therefore advantageous to make conservative choices for step lengths since the penalty is so great for over optimistic choices of large step lengths. Such choices may involve overall scaling of the correction vector $\boldsymbol{\kappa}$ and of additional level shifts in the solution of the WSCI or PSCI matrix eigenvalue equation[19]. Several of these choices are discussed in Section VI.

G. Extended Micro-iterative Methods

In some cases the rational function approximation does not adequately mimic the behavior of the true energy expression. This is most likely to occur during the initial MCSCF iterations when the orbital and CSF corrections must be large. In this case the gradient and Hessian, which do provide adequate approximations in the local region, cannot by themselves be used to estimate the non-local behavior of the true energy. In these cases, it may be productive to use alternate energy approximations or to incorporate higher-order information into the approximate energy expression. Both of these approaches result in modifications of the micro-iterative procedures discussed previously.

1. Alternate Energy Approximations

Werner and Meyer[46] have proposed the use of an alternate energy approximation that does approximately describe non-local behavior of the energy. Consider the effect of the orbital transformation matrix $\mathbf{U}$ on the one- and two-electron integrals. In the new orbital basis, the integrals may be written exactly in terms of the original integrals as

$$h'_{p'q'} = (\mathbf{U}^t\mathbf{h}\mathbf{U})_{p'q'} \tag{197}$$

$$g'_{p'q'r's'} = \sum_p \sum_q \sum_r \sum_s g_{pqrs} U_{pp'} U_{qq'} U_{rr'} U_{ss'} \tag{198}$$

where the transformation matrix $\mathbf{U}$ must be orthogonal. This shows explicitly that the Hamiltonian operator, and therefore the energy for fixed CSF expansion coefficients, is a fourth-order function of the elements of $\mathbf{U}$. (The elements of $\mathbf{U}$ are not free to vary independently due to the orthonormalization constraint on the orbitals. The elements of the vector $\boldsymbol{\kappa}$ are free to vary independently but this choice of variational parameters results in the infinite-order energy dependence.) Consider applying the substitution $\mathbf{U} = \mathbf{1} + \mathbf{T}$ to the above two-electron integral expression. This results in an analogous fourth-order dependence on the elements of the matrix $\mathbf{T}$. Consider next the truncated expansion of this expression for the transformed integrals that contains terms through second order in $\mathbf{T}$. As shown by Knowles and Werner[81,82], this results in

$$g^{(2)}_{pqrs} = -g_{pqrs} + \sum_u \sum_v (g_{uvrs} U_{up} U_{vq} + g_{pquv} U_{ur} U_{vs})$$
$$+ (1 + P_{pq})(1 + P_{rs}) \sum_u \sum_v g_{purv} T_{uq} T_{vs} \tag{199}$$

where abbreviated index permutation expressions are used for the last set of summations. This approximate integral transformation involves only two-index transformations and requires less effort than the exact transformation. These approximate integrals may be used to define an approximate Hamiltonian operator

$$H^{(2)}(\mathbf{T}) = \sum_p \sum_q h'_{pq} E_{pq} + \tfrac{1}{2} \sum_p \sum_q \sum_r \sum_s g^{(2)}_{pqrs} e_{pqrs} \tag{200}$$

which is exact only when $\mathbf{T} = \mathbf{0}$. The expectation value of this operator, which is bounded by the exact energy only when $\mathbf{T} = \mathbf{0}$, results in an energy expression that, like the rational function approximations discussed earlier, contain approximately the effects of higher-order terms in the parameters $\mathbf{K}$. This approximate operator is useful when the parameters $\mathbf{K}$, which have been determined with some procedure, are too large to lie within the local region described with the truncated energy or rational function approximations. This $\mathbf{K}$ allows the matrices $\mathbf{U}$ and $\mathbf{T}$ to be constructed, which in turn are used to define $H^{(2)}$. This $H^{(2)}$ may then be used in he place of the exact Hamiltonian operator within a micro-iterative procedure to define a new orbital correction matrix $\mathbf{K}$. Also $H^{(2)}$ may be used to correct the CSF expansion coefficients $\mathbf{c}$ by solving[82] the approximate matrix eigenvalue equation $\mathbf{H}^{(2)}\mathbf{c}^{(2)} = \mathbf{c}^{(2)}E^{(2)}$. Since $H^{(2)}$ contains second-order information, $\mathbf{c}^{(2)}$ is correct through second order in $\mathbf{T}$ instead of just first order in $\mathbf{K}$ as in Eq. (150). This cannot improve the local convergence since $H^{(2)}$ and $E^{(2)}$ are only correct to second order, but such updates do appear to improve the non-local energy description and the global convergence characteristics[82]. Various implementations of this MCSCF method are discussed in Section VI.

Although this second-order operator approach and the rational function approach both effectively contain higher-order terms in the expansion of **K**, there are some important differences. The rational function approaches, including the WSCI and PSCI methods, result in an approximate energy expression that is defined for continuous variations of the independent variables of **K** and the exact energy derivatives at the reference wavefunction are always available, at least in those cases where the gradient and Hessian elements are constructed explicitly. The approximate effective operator approximation of Eq. (200), however, is available only for discrete values of the parameters **K** through the elements of **U** and through the approximate integral transformation step. The second significant difference of these approaches is that the form of the approximate transformation accounts for the cyclic dependence of the elements of **U** on the parameters **K**. At those values of the parameters **K** which produce a unit matrix, $\mathbf{U} = \mathbf{1}$, the effective operator again becomes exact. The fact that the approximate operator is exact for this infinite number of isolated points is probably the reason that this effective operator methods approximately describes the non-local energy behavior. In contrast, the truncated energy expansions and the rational function approximations are exact only at one point, $\mathbf{K} = \mathbf{0}$.

The advantages of this approach result from the fact that $H^{(2)}$ is a reasonably accurate approximation to the exact H operator, which in turn results in a reasonably accurate description of the non-local energy dependence on the orbital variations. Since so much effort is devoted to the convergence of the micro-iterations with this approach, the MCSCF iterations converge reliably and with greater stability. The neighborhood of second-order convergence is much larger than in the Newton–Raphson and rational function approximation approaches since the false starts and overshooting that sometimes occur in the initial iterations are largely corrected within the micro-iterative process. It should be noted that the subset of the exact integrals required to construct $H^{(2)}$ is the same as that required for the Hessian matrix construction.

The solution of the orbital corrections has usually been uncoupled from the CSF corrections within the micro-iterative procedure with this approach[81,82]. This usually causes no problems for ground-state calculations, but it is expected to be detrimental for excited states, particularly those with negative eigenvalues of the orbital Hessian matrix at convergence. This could be addressed by using, for example, the PSCI iterative method within a micro-iteration for a fixed operator $H^{(2)}$ to solve simultaneously for $\mathbf{K}^{(2)}$ and $\mathbf{p}^{(2)}$. Another disadvantage, for large numbers of virtual orbitals, is that the space required for integral storage is about one-and-a-half times that required for storage of the orbital Hessian matrix (this ratio is even larger when spatial symmetry is considered). In some cases, it would not be possible to hold these integrals in memory while it would be possible to hold the orbital Hessian matrix. Finally the approximate energy expression resulting from $H^{(2)}$ does not display the correct invariance to rotations between doubly occupied orbitals. For this reason, rotations between other highly occupied orbitals are also not described well with this approach[82].

2. *Explicit Third-order Contributions to the Energy*

Olsen *et al.*[1,75] have attempted to improve the non-local description of the approximate energy surface in the Newton–Raphson method by including explicitly third-order contributions of the energy. Instead of explicitly constructing the matrix elements corresponding to the third derivative terms, effective operator methods were used in micro-iterative procedures during the solution of the wavefunction corrections. (The reader is referred to the literature[1] for more details of the various methods that have been employed.) A disadvantage of this approach is that it requires a much larger subset of the integrals to be transformed. Since the approximate energy expression resulting from the truncation at third order is unbounded, as is the tuncated second-order expression, these methods would not be expected to be adequate outside of the local region, although they might be expected to have a somewhat larger neighborhood of convergence. The construction of a rational function approximation including the third-order terms might address the unphysical approximate energy behavior, as it does to a limited extent for the second-order expression, but it would still not be appropriate for a true global energy approximation.

IV. CONFIGURATION STATE FUNCTION EXPANSION SPACES FOR MULTICONFIGURATION SELF-CONSISTENT FIELD WAVEFUNCTIONS

There have been several CSF selection schemes proposed for MCSCF wavefunctions. Many of these have been enumerated in the review of Detrich and Wahl[5]. These schemes may be loosely grouped into two types: *a priori* and empirical. The *a priori* methods are those that include certain sets of CSFs based on general chemical principles, formal analysis, computational facility, or intuition. The empirical methods are those that are based on trial calculations on the molecular system being investigated. In actual practice, no selection scheme belongs in one of these categories entirely. Empirical selection methods include some amount of intuition and *a priori* selection methods must include empirical justification.

Depending on how the MCSCF wavefunction is to be used ultimately, there are two goals of the CSF selection procedure. The first goal, which is the more stringent, is that the MCSCF wavefunction itself must provide an accurate description of the exact wavefunction. The second goal, which is somewhat less demanding of the MCSCF procedure, occurs when the MCSCF wavefunction (or the MCSCF wavefunction expansion space) is used to construct a reference space which is then used in a more extensive calculation (e.g. multireference single and double excitation CI) designed to calculate the more difficult dynamical electron correlation effects. In this latter case, the MCSCF wavefunction is required only to describe qualitatively the more important features of the wavefunction such as the electron charge distribution and avoided crossings at the various molecular geometries. The discussions in this section will be, for the most part, from the point of view that the MCSCF wavefunction itself is to provide an accurate description of the wavefunction and that the potential energy surface is to be parallel to the full-CI energy surface. This rather difficult goal can only be achieved with sufficient flexibility of the CSF expansion space, with the appropriate choice of the individual expansion terms, and when a balanced description of the wavefunction has been obtained that is equally accurate at all relevant molecular geometries. However, the discussions in this section also remain relevant when the second, less demanding, goal is to be achieved. In this case the MCSCF wavefunction must deviate from the exact wavefunction in a continuous, slowly varying, and well understood manner that may be easily corrected in the subsequent, more extensive, calculations.

A few of these CSF selection methods will be described in this section. The empirical selection methods will be discussed first since these methods are somewhat direct in the attempt to attain these goals. The *a priori* methods will then be examined. Since it is rather difficult to categorize these selection methods systematically, they are introduced in a somewhat arbitrary order but an attempt is made to emphasize some of the significant similarities and differences among these methods. A few examples of non-linear wavefunction expansions are also given. Finally, the advantages and disadvantages of the use of symmetry orbitals are discussed.

A. Empirical Selection Methods

The empirical CSF selection method of Banerjee and Grein[65] will be discussed in detail. This method includes the essential features of other empirical selection methods[71] but is relatively general and has been applied to both ground and excited states. It is similar to the CSF selection approach used in conventional CI methods, as discussed by Shavitt[13], and to those used in the CI energy extrapolation methods, as discussed by Buenker and Peyerimhoff[83-85]. In this method, an initial CSF space is chosen, based perhaps on intuition and test calculations, that is believed to provide a minimal description of the relevant dissociation products and equilibrium geometries. In most cases this will be a restricted Hartree–Fock (RHF) type wavefunction with perhaps additional CSFs that allow dissociation to neutral fragments or to fragments of the correct spin state. A series of MCSCF calculations is then performed with this expansion space to generate an initial potential energy surface representation. This initial surface is used to locate approximately the geometries corresponding to energy minima and saddle points and to determine the molecular fragment geometries.

The fragment geometries are then examined in detail by performing CI calculations at these geometries. The expansion space for these CI calculations is based on the MCSCF orbitals and includes the MCSCF expansion CSFs, in addition to single and double excitations from the valence orbitals of these CSFs into a subset of the virtual orbitals. CSFs are selected from these CI calculations based on their energy or wavefunction contributions. Alternate methods may use the A_k or B_k approximate CI methods[13] instead of the small CI calculations. Of these possibilities, the B_k method is the most flexible and it allows the use of efficient direct CI algorithms[41]. If necessary, several such calculations are performed with different sets of virtual orbitals. If during these CI calculations any additional CSFs are identified as being important, they are added to the MCSCF set and new orbitals are determined. During this CSF selection step, a previously selected CSF may occasionally become unimportant as new terms are added and as the orbitals relax. When this occurs the expansion term may be deleted from the MCSCF set provided it is also known to be insignificant at the other fragment geometries. These steps are repeated, cyclically over the fragment geometries if necessary, until a satisfactory set of CSFs have been selected to describe all of the fragments adequately. If the initial CSF choice allowed an adequate description of all of the fragments, no new CSFs will have been added during this process. In some cases, however, additional CSFs are chosen that describe the near-degeneracy effects of the fragments. These steps will have eliminated any bias in the initial CSF set and will have produced an expansion set that should be equally accurate, or at least to the extent possible, at all of the fragment geometries as determined by the more extensive CI or B_k calculations.

Once an adequate description of the relevant fragments has been obtained, the CSF selection process is repeated at the other important geometries on the potential energy surface. These other geometries will include the stable minima on the surface and the transition-state geometries connecting them. The description of regions where avoided crossings occur is also important. These regions may be determined by monitoring the low-lying diagonal elements of the Hamiltonian matrix. For molecules with high-spin interacting fragments, there may be important regions where the qualitative description of the spin coupling of the molecule changes. The description of open-shell spin recoupling is often important in replacement reactions, abstraction reactions and distortions of multiple-bonded systems. Again, these cases may usually be identified by CSF crossings of the lower-energy expansion terms.

There is one important difference in the initial CSF selection at the fragment geometries and the final CSF selection at the molecular geometries. The CSFs that are added due to their molecular contributions are negligible at these geometries. This is because the accuracy of the fragments has already been determined in the initial selection step and only the CSFs that contribute primarily to the molecular correlation energy are sought during the final selection steps. This emphasizes that the goal of this selection procedure, particularly for excited-state calculations, is not primarily to obtain the lowest energy but rather to obtain a set of CSFs that describe the molecular wavefunction equally well at all geometries. This wavefunction would then produce, in ideal cases at least, a balanced wavefunction resulting in a potential energy surface that is parallel to the full CI surface.

Other empirical selection methods have been suggested that are similar to the one described here. The natural-orbital occupation numbers from B_k calculations[13], or from other approximate wavefunctions[37], may be used as a CSF selection guide by identifying important missing orbitals. Other properties may also be monitored, in addition to the energy lowering or wavefunction contribution, to determine the importance of the individual CSFs in the expansion. The disadvantage of the use of empirical selection methods is that a large amount of preparatory effort must precede the actual potential energy surface calculation. Although parts of these repetitive calculations may be automated and the use of small basis sets reduces the computational effort of the trial calculations, there are still a large number of calculations that must be performed to select a CSF space and verify its accuracy. Other problems also arise when there are several fragments to consider. Suppose a CSF contributes significantly at some molecular geometry, contributes only negligibly at some fragment geometries, but distorts other fragment discriptions. These difficulties have resulted in limited use of empirical CSF selection methods for MCSCF wavefunctions except for small diatomic systems[65,71].

B. *A Priori* Selection Methods

The use of the exponential operator MCSCF formalism, or more specifically the use of optimization methods that require only the density matrix instead of the coupling coefficients over the CSF expansion terms (or even worse, over the single excitation expansion terms), has allowed relatively large CSF expansion lengths to be used in MCSCF wavefunction optimization. These larger expansion lengths allow CSFs to be included based on formal analysis or computational facility with little or no penalty in those cases where some of these CSFs are unimportant at all geometries. This avoids the preparatory calculations of the empirical selection methods. Even with these improved wavefunction optimization methods, however, there is some advantage in keeping the expansion less than a few hundred CSFs in length. Expansion sets of this size may use the full second-order optimization methods described earlier and, as previously mentioned they may be used without further selection to define a reference space for more extensive calculations, such as MRSDCI, designed to calculate the dynamical electron correlation effects. In fact, the effort required in such large scale calculations is often seen to be proportional to the reference CSF expansion length, thereby providing even more incentive to find compact MCSCF expansions. In those cases where the full second-order optimization is not required for convergence to be obtained, or if micro-iterative procedures can be used effectively, or if the MCSCF wavefunction itself must provide a quantitatively accurate approximation, then the CSF expansion length may be, or may be required to be, much larger. The practical limit in these cases is probably less than 10^5 CSFs although larger calculations have been performed[81] and will likely continue to be in the future. This limit results from the fact that the Hamiltonian matrix diagonalization becomes the time-consuming bottleneck of the calculation. Recent work has been directed toward improving the computational efficiency of the direct CI step for these large expansions[4,81,86].

1. The OVC Wavefunction

One well known *a priori* CSF selection method is the optimized valence configuration (OVC) method of Das and Wahl[8,9]. This method is designed for the ground-state description to diatomic molecules. This method is mentioned first because its general approach to the CSF selection is similar to the empirical selection method previously described. The OVC method consists of forming an initial CSF space that describes the fragments, in this case the known atomic ground states, followed by the inclusion of terms that contribute to the interaction region but vanish at large internuclear distances. These CSFs describe the 'molecular extra correlation energy' and, in the OVC method, are assumed to be double excitations from the initial CSF space. To ensure that their contributions vanish at large internuclear distances, these CSF are constructed so that they reduce to simultaneous single excitations from each of the atoms, to charge transfer between the atoms, or to excited atomic states. At strongly interacting geometries, these double excitations describe the 'dissociative' behavior, the 'electron redistribution', the 'charge transfer' and the 'dispersive' interactions of the diatomic molecule being studied[9]. The problem with the OVC method is that is cannot be extended directly to other molecular systems and to excited states, particularly those involving single excitations from lower states.

2. *The Full CI Expansion*

The discussion of other MCSCF wavefunctions will be preceded by a brief discussion of the full CI expansion. The full CI expansion for a molecular system consists of all possible orbital occupations and all possible spin couplings consistent with the overall molecular spatial and spin symmetry. If the orbital basis set is complete (i.e. an arbitrary function of the three spatial coordinates may be represented exactly with the basis), the CSF expansion space is also complete and is called complete CI. Only finite, and therefore incomplete, basis sets are considered in this discussion. If the reductions due to spatial symmetry are ignored, the number of expansion terms is given by the Weyl[13] dimension formula

$$f(N, n, S) = B(n+1, \tfrac{1}{2}N - S)B(n+1, \tfrac{1}{2}N + S + 1)(2S+1)/(n+1) \qquad (201)$$

where S is the total spin (0 for singlet states, $\frac{1}{2}$ for doublets, etc.), n is the number of orbitals and N is the number of electrons. When spatial symmetry is present in the molecule and when symmetry orbitals are used to construct the CSFs, the total number of CSFs given in Eq. (201) may be partitioned into subsets corresponding to the different symmetry species. The number of CSF expansion terms belonging to a particular symmetry type may be evaluated for Abelian point groups by constructing the symmetry-dependent distinct row table (DRT). This recursive procedure[87] avoids any effort that is proportional to the number of expansion terms and depends instead only on the number of distinct rows in the table, a much smaller quantity. The set of expansion CSFs, or the resulting wavefunction depending on the context, will be written as $(\phi_1 \cdots \phi_n)^N$, or in shorthand form as $(\mathbf{n})^N$ if the individual orbitals need not be specified. This is simply an extension of the atomic-shell notation and emphasizes that all the orbitals are treated equivalently in the wavefunction expansion space.

An important feature of full CI expansions is that the resulting wavefunction, and therefore the energy, is independent of the choice of orbitals within a given orbital basis set. Of course, different orbital choices will result in different CSF expansion coefficients, giving different representations of the same wavefunction. This is considered in more detail in Section V when redundant variables are discussed. Since any orbital choice results in the same full CI wavefunction, the orbitals may be chosen to aid in the computation or the interpretation of the wavefunction. Useful choices include the natural orbitals (for which the one-particle density matrix is diagonal), the SCF canonical orbitals (which facilitate perturbation theory analysis of the wavefunction), or localized orbitals (which can result in more compact expansions for extended systems). Another computationally useful choice results from performing orbital updates during the iterative solution of the Hamiltonian eigenvalue equation to reduce the number of iterations required in this step. The calculation of the expansion coefficients for full CI wavefunctions could be considered as a special case of the MCSCF optimization procedure. Unfortunately, the length of these expansions is prohibitive for all but very small molecular systems and small basis sets. Such full CI calculations have been performed to determine the qualitative features of valence states of diatomic molecules[88] and to serve as benchmark calculations[58,86,89,90] to assess the accuracy of various approximate wavefunctions.

3. *Subspace Full CI Expansions*

One way to reduce the CSF expansion length is to allow only a subset of orbitals to be employed in generating the CSF list. In this case the wavefunction is invariant to transformations among the active orbitals (the orbitals potentially occupied in some CSF) and to transformations among the virtual orbitals (the orbitals that are not occupied in any CSF), but it depends on transformations that mix the orbitals of these subspaces with each other. The number of expansion terms in this subspace full CI is given by Eq. (201) where the number of active orbitals is used instead of the total number of orbitals.

For a given partitioning of the orbital space, the MCSCF optimization procedure for this wavefunction consists of determining the optimal CSF expansion coefficients and of determining the optimal mixing between these two orbital subspaces. The CSF expansion set for this type of wavefunction may be written as $(\phi_1 \cdots \phi_{n'})^N(\phi_{n'+1} \cdots \phi_n)^0$ to emphasize the direct product nature of this wavefunction. For convenience the virtual orbitals will not be written explicitly and this expression may be written in the abbreviated form $(\mathbf{n}')^N$. The quality of the final wavefunction depends on the partitioning of the orbitals into the two subsets; as the number of active orbitals approaches the total number of orbitals, the quality of the wavefunction may be expected to approach that of the full CI wavefunction. At the other extreme, the trivial example of this expansion is the closed-shell RHF wavefunction consisting of a single CSF, for which $n' = N/2$.

a. *Extended Independent-particle Model*

There are three expansions of the full CI form that are based on chemical intuition that will be briefly discussed. The first of these allows the same number of occupied orbitals as there are electrons. This wavefunction form is called the 'extended independent-particle model' by Ruedenberg *et al.*[66]. This CSF space, denoted $(\mathbf{N})^N$, is usually sufficiently flexible to describe molecular dissociation processes, including those involving various excited-state fragments, but unfortunately contains too many terms to be practicable for general molecular systems. There are also cases[91] where this choice does not result in a consistent molecular description (e.g. where fragment degeneracies or united-atom limits are not correctly described). This wavefunction form does define an unambiguous partitioning of the orbital set into the active and virtual orbital subspaces.

b. Conceptual Minimal Basis Full CI Wavefunction

The second, more restrictive, choice is to reduce the number of occupied orbitals to the conceptual minimal basis. Each carbon atom in a molecule, for example, would contribute five orbitals corresponding to the shells of its atomic Hartree–Fock configuration, $(\phi_{1s}^2\phi_{2s}^2\phi_{2p}^2)$. This usually results in fewer orbitals than in the previous case since the low-lying electron pairs contribute only one orbital instead of two to the active orbital space. The validity of this expansion depends on the approximation that the total energy contribution from the correlation effects of the core electrons and from orbitals outside the valence space is relatively constant over the molecular surface. It may be argued, however, that, as electrons are brought closer together at particular molecular geometries and the electron motions become more highly correlated, more correlating orbitals would be required to describe these dynamical correlation effects than are required at the molecular geometries where the electrons are spatially separated. These considerations place limits on the validity of this approximation of constant correlation energy neglect. The exclusion of these correlation energy contributions is approximated by deleting the correlating orbitals of the core electrons and the extra-valence orbitals from the expansion. This wavefunction form does include the limited core correlation effects described by allowing the core electrons to make use of the valence orbital space. However, this limited description of the core correlation would not be expected to be adequate for cases that require this correlation to be described accurately. The conceptual minimal orbital basis is invoked in the formulation of several semi-empirical electronic structure methods[14].

c. The FORS/CASSCF Wavefunction

The third approximate wavefunction of the full CI type results from neglecting all of the correlation of the core electrons[66,72]. This is accomplished by using an expansion space of the form $(\mathbf{n'/2})^{n'}(\mathbf{m})^{n''}$, where n' is the number of inactive electrons that are not correlated, n'' is the remaining number of active electrons and m is the number of active orbitals. This expansion may also be written as $(\phi_1^2\phi_2^2\cdots\phi_{n'/2}^2)(\mathbf{m})^{n''}$ to emphasize that the inactive orbitals are doubly occupied in all CSFs. The first form emphasizes the direct product nature of the expansion space and the full treatment of the electrons and orbitals within each subspace. As expected, a wavefunction constructed from this expansion space is invariant to transformations among the inactive orbitals, to transformations among the active orbitals and to transformations among the virtual orbitals. It does, of course, depend on the space spanned by these orbital subspaces and therefore is sensitive to transformations that mix the orbitals of these different subspaces with each other. Two simple examples of this wavefunction are the closed-shell RHF wavefunction, which contains no active orbitals, and the open-shell RHF wavefunction, for which only one active orbital occupation is consistent with the high-spin wavefunction. This is one of the most popular *a priori* CSF expansion spaces currently used for MCSCF wavefunctions. It was first proposed by Ruedenberg and co-workers[66,67] as the fully optimized reaction space (FORS) wavefunction and implemented within a variational super-CI optimization method. It was subsequently called the complete active space SCF (CASSCF) wavefunction and implemented within several density matrix based optimization methods[72–74,92]. This wavefunction form is described in more detail by Roos[7] in this volume.

The popularity of this wavefunction results from the fact that, in many cases, the choice of inactive and active electrons and the choice of inactive, active and virtual orbital partitioning is straightforward and produces small to moderate expansion lengths. It is slightly more flexible than the conceptual minimal basis wavefunction in that, if appropriate, electrons corresponding to closed-shell atomic orbitals may be included in the active space (e.g. the carbon 2s orbital). A representative table of expansion lengths for various molecules has been given by Camp and King[77]. In the FORS approach using this wavefunction form, the active orbitals are identified by first determining the important electrons and their corresponding orbitals that are involved in the molecular distortion process being described. This choice usually depends on both chemical intuition and test calculations. For example, in the dissociation of ethylene into two methylene fragments[67]

$$C_2H_4(^1A_1) \rightarrow 2CH_2(^3B_1) \tag{202}$$

the active electrons may be chosen to be the four electrons involved in the C=C double bond and the active orbitals are the four atomic orbitals involved in the molecular formation. This expansion may be written as $(\mathbf{6})^{12}(\mathbf{4})^4$. At molecular geometries, the four active orbitals describe the σ, σ^*, π and π^* orbitals. At large separations, these orbitals describe the two open-shell triplet-coupled electrons on each of the fragments. The inactive electrons are those corresponding to the C_{1s} cores and to the four σ_{C-H} bonds. If a more extensive treatment of the correlation is required, the eight electrons in these single bonds would be moved from the inactive space to the active space. The corresponding four inactive orbitals would be transferred from the inactive space to the active space, along with four correlating orbitals transferred from the virtual space to the active space. The expansion length would then increase, ignoring reductions resulting from spatial symmetry, from the 20 CSF expansion, $(\mathbf{6})^{12}(\mathbf{4})^{4}$, to the 226 512 CSF expansion, $(\mathbf{2})^{4}(\mathbf{12})^{12}$. In both cases, depending on the accuracy required, the orbital choice results from including the valence orbitals of the various possible fragments: the C_{2s}, C_{2p} and H_{1s} orbitals in this case.

In other cases, however, the choice of active electrons and orbitals is not so obvious[91]. Three frequently occurring cases involve the description of molecules containing lone-pair electrons, the description of avoided crossings where the two states require qualitatively different orbitals, and cases where the number of active orbitals required is greater than the number of active electrons. In the first case, consider the description of the bond deformation of a molecule containing a lone pair of electrons, :R—R. At large bond distances, the correlation of the bonding electrons required to describe the dissociation will take precedence over the correlation of the lone pair. The single correlating orbital localizes in the region of the bond and could be interpreted as a σ^* antibonding orbital. At other geometries, however, this correlation may become less important than the correlation of the electron lone pair. At these geometries, the correlating orbital will localize in the region of the lone pair to describe the radial correlation between these two electrons. At the geometries where these two effects compete equally, there are likely to be multiple MCSCF solutions resulting in spurious cusps in the potential energy surface if the lowest-energy solution is always chosen. The solution to this problem, once it is correctly identified, is to include the lone-pair electrons into the active space along with an appropriate correlating orbital.

The second case of active spaces that may be difficult to define is those for which the qualitative orbital description changes dramatically with small geometry changes. Consider the ground-state dissociation of a polar diatomic molecule A—B to neutral fragments A· and ·B. Close to the equilibrium geometry, the wavefunction adopts the charge-transfer character A^-B^+ where all of the orbitals localized primarily on A^- acquire the diffuse character appropriate to the description of an anion, and where all of the orbitals localized on B^+ acquire the contracted character appropriate to the description of a cation. At large displacements the orbitals on both fragments acquire the character of the neutral fragments. At some intermediate geometry, where this qualitative description changes from one form to the other, the wavefunction description must be sufficiently flexible to describe this change. One way that this change may be described is to rely on all of the orbitals to alter their form at the appropriate geometry. Unfortunately, this may not occur when the change must take place abruptly. The MCSCF wavefunction will instead have multiple solutions at geometries in this region. The alternative, and preferred, approach is to incorporate sufficient flexibility into the CSF expansion space to describe both qualitative forms of the wavefunction. The appropriate form will then be determined during the solution of the

Hamiltonian eigenvector equation. With this more flexible CSF expansion, this case may then be interpreted as a weakly avoided crossing. This solution, however, requires each valence orbital on the fragments to have two counterparts, a diffuse member describing the anion character of A^- or the neutral character of $\cdot B$, and a contracted member describing the neutral character of $A\cdot$ or the cation character of B^+. The requirement that each valence orbital must be represented twice is referred to as orbital doubling. This orbital doubling is also sometimes required to satisfy symmetry requirements of the wavefunction as discussed by McLean *et al.*[55]. Similar ambiguous active space choices may result from valence–Rydberg crossings. In this case orbital doubling occurs for the orbitals that must describe both the valence state and the Rydberg state, the latter of which may be thought of as a cation, requiring contracted orbitals, surrounded by an additional singly occupied diffuse orbital. The FORS/CASSCF wavefunction expansion for cases involving orbital doubling may suffer from impractically large CSF expansions.

The third case of unexpected active space requirements occurs when the number of active orbitals must be greater than the number of active electrons. This often occurs when two qualitatively different types of correlation effects for a particular electron pair must be described simultaneously. One dramatic example is the ground-state dissociation of Li_2 which requires a two-orbital active space to describe the left–right correlation necessary to dissociate correctly to doublet fragments. However, this wavefunction accounts[9,93] for only 44% of the bond energy of Li_2. In order to calculate the bond energy correctly, orbitals that correlate with the low-lying 2p orbitals of the atomic fragments must be included in the active space. Other diatomic molecule descriptions that require the addition of active orbitals outside of the valence space include N_2, as shown by Roos *et al.*[73,74], and O_2, as shown by Brown[94]. Siegbahn[95] has also argued that in some cases C_{3s}, C_{3p}, C_{3d} and Cl_{3d} type orbitals should be included in the active space of molecules containing these atoms. This is also supported by the ethylene calculations of Sunil *et al.*[96], which demonstrate that additional π and π^* orbitals must be included in the active space for an adequate description of the low-lying excited states.

4. *Direct Product Full CI Expansions*

The solution to the active orbital choice and the accompanying expansion length problems may be addressed in many cases by extending the direct product definition of the orbital space. The orbitals in the full CI expansion were partitioned into three types in the definition of the FORS/CASSCF wavefunction: inactive, active and virtual. Within each orbital type, the expansion may be considered to be full. That is, subject to the occupation restrictions on each subspace, all possible orbital occupations and spin couplings are allowed that are consistent with the requirements of the total wavefunction. In the more general direct product CSF expansions, the active electrons and their associated orbitals are further partitioned into subgroups.

The formal justification for this extension results from the fact that the most important electron correlation effects are those involving electrons, particularly electrons of opposite spin, confined to neighboring regions of space. Using localized orbitals, these neighboring regions often correspond to the usual chemical picture of hybridized orbitals resulting from interacting atomic fragments. In many cases, this allows the partitioning of the active orbitals to be performed based on intuition. In other cases, particularly for highly distorted geometries and for excited states, it is best to verify the orbital partitioning empirically with test calculations that include the effects of additional CSF expansion terms. In addition to the ability to mimic longer full CI expansions successfully, these direct product expansions are also capable of more consistent descriptions of the wavefunction. This is because the form of the expansion eliminates some of the multiple solution problems discussed previously[91].

The dissociation of ethylene may again be used as an example of this wavefunction type. Assume that the initial $(\mathbf{6})^{12}(\mathbf{4})^4$ expansion is not sufficient but that a more compact description than the $(\mathbf{2})^4(\mathbf{12})^{12}$ active space expansion is required. A reasonable intermediate wavefunction might be to partition the active space of the larger expansion into three sets: the initial four active orbitals describing the four electrons in the C=C double bond, an additional four orbitals describing the four electrons of the two adjacent C—H bonds of one end of the molecule, and the analogous four orbitals describing the four electrons on the other end of the molecule. This expansion space is written as $(\mathbf{2})^4(\mathbf{4};\mathrm{C}(1),\mathrm{H}(1),\mathrm{H}(2))^4(\mathbf{4};\mathrm{C}(2),\mathrm{H}(3),\mathrm{H}(4))^4(\mathbf{4};\mathrm{C}(1){=}\mathrm{C}(2))^4$ and consists of 25 656 CSFs. The orbital localization indicated in this expansion is not imposed during the wavefunction optimization, but rather occurs naturally because of the form of the CSF space. The CSF expansion terms that are neglected compared to the $(\mathbf{2})^4(\mathbf{12})^{12}$ expansion are those that describe the transfer of electrons from one orbital set to another. This wavefunction would not be expected to describe the excited states of ethylene that correspond to these types of charge-transfer states or the dissociation to charged methylene fragments involving these excitations but it would be expected to provide a flexible description of the ground state dissociation.

This active orbital partitioning many be continued by separating the electrons and orbitals involving the different hydrogen atoms into different subspaces. This expansion may be written as $(\mathbf{2})^4(\mathbf{2};\mathrm{C}(1){-}\mathrm{H}(1))^2(\mathbf{2};\mathrm{C}(1){-}\mathrm{H}(2))^2(\mathbf{2};\mathrm{C}(2){-}\mathrm{H}(3))^2(\mathbf{2};\mathrm{C}(2){-}\mathrm{H}(4))^2(\mathbf{4};\mathrm{C}(1){=}\mathrm{C}(2))^4$ and consists of 6144 CSFs. This wavefunction is even more restrictive than the previous example and neglects expansion terms that correspond to electron transfer between the various C—H bonds, including those bonds involving the same carbon atom.

Although this is not a serious restriction on a wavefunction designed to describe the ground-state dissociation of ethylene into two methylene fragments, it might limit the quality of a wavefunction that is also used to describe other dissociation processes such as to a hydrogen molecule and an acetylene molecule, $C_2H_4 \rightarrow H_2 + C_2H_2$.

a. The RCI Wavefunction

This partitioning may be continued until all active electron pairs are described by two localized orbitals. This expansion has the general form $(\mathbf{n'/2})^{n'}(\mathbf{2})^2 \cdots (\mathbf{2})^2$ where n' is the number of inactive electrons. In the ethylene example, this results in the expansion $(\mathbf{2})^4(\mathbf{2}; \mathrm{C}(1)\text{—}\mathrm{H}(1))^2(\mathbf{2}; \mathrm{C}(1)\text{—}\mathrm{H}(2))^2(\mathbf{2}; \mathrm{C}(2)\text{—}\mathrm{H}(3))^2(\mathbf{2}; \mathrm{C}(2)\text{—}\mathrm{H}(4))^2(\mathbf{2}; \mathrm{C}(1)\text{—}\mathrm{C}(2,\sigma))^2(\mathbf{2}; \mathrm{C}(1)\text{—}\mathrm{C}(2,\pi))^2$ which consists of 3012 CSFs. This is the most restrictive direct product full CI expansion space that describes the correlation of electron pairs. Any further subspace restrictions result in transferring active electrons into the inactive space. In this respect, it is also the most general wavefunction form that describes electron pairs with pairs of orbitals; other electron-pair expansions, which impose further occupation and spin restrictions, only span a subspace of this wavefunction, as is discussed in subsequent sections. This expansion space has been used by Harding and Goddard[97] and by Dunning *et al.*[98] (but with orbitals optimized for simpler MCSCF expansions) and is called the restricted CI (RCI) expansion. Goodgame and Goddard[99] have used this RCI expansion space with fully optimized orbitals to describe the dissociation of Cr_2. (The minimal description of this diatomic molecule consists of 12 active electrons and the 12 orbitals that correlate to the 4s and 3d atomic orbitals. The RCI expansion results in a considerable improvement over other more restrictive expansions for Cr_2 but unfortunately the results are still inadequate for this difficult diatomic molecule.) The expansion length for the RCI wavefunction for singlet states is given by the expression

$$R(n) = \sum_{k=0}^{n/2} 2^{n/2-k} B(n/2, k) B(2k, k)/(k+1) \tag{203}$$

where n is the number of active electrons. A list of representative expansions has already been given in Table I. The expression giving the determinantal expansion lengths for this wavefunction is identical to the above CSF expression except that the $1/(k+1)$ factor is dropped. This MCSCF wavefunction has also been called the full intra-pair spin-coupling (FIP) wavefunction by Chipman[100] in the study of the differential correlation effects of OH and OH^-.

The formal justification of this expansion form is analogous to that given by Hurley, Lennard–Jones and Pople[101] in the development of a correlated-pair extension of the closed-shell restricted Hartree–Fock (RHF) wavefunction.

Consider a closed-shell RHF wavefunction described in terms of localized orbitals. This localization may be assumed without loss of generality since this wavefunction is a full CI expansion in the doubly occupied orbital subspace. The antisymmetrized form of this wavefunction implicitly accounts for the correlation effects of electrons with the same spin. This is because the approximate wavefunction vanishes as two such electrons approach each other. As two electrons of opposite spin approach each other, the wavefunction should decrease in amplitude and possess a cusp (because of the $|\mathbf{r}_1 - \mathbf{r}_2|^{-1}$ repulsion). However, in the RHF approximation, the electrons only feel a spatially averaged repulsion and the RHF wavefunction does not adequately mimic the behavior of the true wavefunction in this respect. The electron interactions for which this is the worst approximation are those that are in the same regions of space and, of these, the worst offenders are the electron pairs occupying the same spatial orbital. This argues qualitatively that the most important electron correlation effects are those involving the description of the electron pairs and of the electrons in the neighboring electron pairs. This requires at least two orbitals for each electron pair. The most general direct product full CI wavefunction that describes these correlation effects but constrains each electron pair to occupy a pair of orthonormal orbitals is the RCI wavefunction. Harding and Goddard[97] have shown that this wavefunction includes both the dominant intra-pair correlation effects and the most important inter-pair correlation effects (e.g. the simultaneous charge polarization involving two nearby chemical bonds).

The notation $(\mathbf{n}; \text{RCI})^n$ will be used to designate the occupation restrictions for RCI wavefunctions. For high-spin cases the orbital pairing will be assumed

TABLE IV
The number of variational parameters of various wavefunction expansions. Singlet wavefunctions with the number of orbitals equal to the number of electrons. The number of variational parameters is one less than the number of CSFs in the linear expansion wavefunctions. The orbital rotation parameters are not included. See the text for descriptions of the wavefunction expansions.

N	Full CI	RCI	ERMC	PPMC	GP	APIG	PPGVB
2	2	2	1	1	1	1	1
4	19	9	5	3	4	3	2
6	174	36	19	7	9	5	3
8	1 763	149	69	15	16	7	4
10	19 403	653	251	31	25	9	5
12	226 511	3 011	923	63	36	11	6
14	2 760 614	14 444	3 431	127	49	13	7
16	34 763 299	71 397	12 869	255	64	15	8
18	449 141 835	361 113	48 619	511	81	17	9
20	5 924 217 935	1 859 627	184 755	1 023	100	19	10

for the first electron pairs followed by the singly occupied orbitals. This wavefunction form contains no further restrictions on the expansion CSFs; in particular, there are no spin restrictions imposed on subgroups of the orbitals.

Table IV shows the number of variational degrees of freedom for full CI expansions and for RCI expansions for singlet wavefunctions for which the number of active electrons is equal to the number of active orbitals. The number of variational degrees of freedom is one less than the CSF expansion length. This number is used in Table IV for the direct comparison of the various non-linear expansion wavefunctions discussed in the following sections. Expansion space reductions due to spatial symmetry are neglected in this list. Only the variational degrees of freedom due to the expansion coefficient variations are given since the remaining ones, resulting from orbital variations, depend not on the number of active orbitals but instead on the total number of basis functions. Expansion spaces for the RCI wavefunction and the more general direct product full CI expansions discussed previously, which explicitly include the effects of spatial symmetry, may be efficiently represented with the symmetry-dependent DRT.

5. *Restricted Group Product Expansions*

Instead of allowing all possible spin couplings as in the previous full CI direct product expansions, another possibility is to restrict the spin coupling within subgroups of the orbitals. For example, in the ethylene expansion $(\mathbf{2})^4(\mathbf{4};\mathrm{C}(1),\mathrm{H}(1),\mathrm{H}(2))^4(\mathbf{4};\mathrm{C}(2),\mathrm{H}(3),\mathrm{H}(4))^4(\mathbf{4})^4$ consisting of 25 656 CSFs described earlier, a more restrictive wavefunction could be constructed that consists of constraining the four electrons localized on the C(1)—H(1)—H(2) fragment to be singlet coupled and the analogous four electrons at the other end of the molecule to be singlet coupled. This spin-restricted expansion, written as $(\mathbf{2})^4(\mathbf{4};\mathrm{C}(1),\mathrm{H}(1),\mathrm{H}(2),\ S=0)^4(\mathbf{4};\mathrm{C}(2),\mathrm{H}(3),\mathrm{H}(4),\ S=0)^4(\mathbf{4}, S=0)^4$, consists of 8000 CSFs. Because the CSF expansion is full within each orbital subspace, this wavefunction has the same invariance properties as the more general expansion. (The more general spin-unrestricted expansion is a summation of such spin-restricted full CI expansions for each subspace. The invariance of the general expansion results from the invariance of each of the spin-restricted expansions.)

If the orbitals within each electron pair of the RCI expansion are restricted to be singlet coupled, the expansion may be written as $(\mathbf{n}'/\mathbf{2})^{n'}\ (\mathbf{2}; S=0)^2\cdots(\mathbf{2}; S=0)^2$ or as $(\mathbf{n}'/\mathbf{2})^{n'}\ (\mathbf{n}''; \mathrm{RCI}, S=0)^{n''}$ and, for singlet states, consists of $3^{n''/2}$ CSFs where n'' is the number of active electrons. In this expansion, each open-shell orbital occupation corresponds to a single CSF. For a single correlated electron pair, the open-shell CSF may be eliminated without restricting the wavefunction with an appropriate orbital choice. This is considered in more detail in the discussion of redundant variables in Section V. For more than one correlated pair, however, these open-shell expansion terms may not be eliminated without restricting the wavefunction flexibility. This expansion space has been called the singlet intra-pair spin-coupled (SIP) expansion by Chipman[100] and was found to describe the more important differential correlation effects of OH and OH^-. This spin-restricted RCI expansion for ethylene may be written as $(\mathbf{4})^2(\mathbf{12}; \text{RCI}, S=0)^{12}$ and consists of 792 CSFs. Of course there are numerous other expansions of this spin-restricted form that may be constructed. The validity of these spin-restricted wavefunction expansions depends on the particular molecular system being studied.

6. *Other Linear Direct Product Expansions*

In this section, other wavefunction expansions are considered that are not strictly of the form of direct products of full CI subspaces. These expansions are of direct product form but they include additional occupation and spin restrictions. Consider deleting all the terms from the orbital subspace full CI expansion $(\mathbf{n})^N$ containing singly occupied orbitals. This even replacement MC (ERMC) expansion may be written as $(\mathbf{n}; \text{ER})^N$ and consists of $B(n, N/2)$ CSFs. The one-particle density matrix resulting from this expansion is of diagonal form and the sparse two-particle density matrix consists only of direct and exchange type terms, i.e. the only non-zero terms are of the form d_{rrss} and d_{rsrs}. This expansion form has been advocated by Roothaan and coworkers[49,50] because this density matrix sparseness may be exploited computationally. Table IV gives the number of variational parameters for this expansion space for cases with the same number of electrons and active orbitals. It may be observed that the ERMC expansion lengths are reduced substantially from the corresponding full CI expansion length. However, the wavefunction is of much more restrictive form. In general, it would not be expected to describe spin-recoupling effects and it is not capable of describing general molecular dissociation. The even replacement approximation is appropriate for the description of limited electron-pair correlation effects, however, when used in conjunction with a general description of the more important active electrons. For example, the ethylene expansion $(\mathbf{2})^4(\mathbf{8}; \text{C—H})^8(\mathbf{4}; \text{C=C})^4$ consisting of 71 280 CSFs might be replaced with the expansion $(\mathbf{2})^4(\mathbf{8}; \text{C—H}, \text{ER})^8(\mathbf{4}; \text{C=C})^4$ consisting of 1400 CSFs to describe the dissociation to triplet methylene fragments. The wavefunction $(\mathbf{2})^4(\mathbf{12}; \text{ER})^{12}$ consisting of 924 CSFs cannot describe the dissociation to triplet fragments.

If the even replacement approximation is applied to the singlet RCI wavefunction, the result is the perfect pairing MC (PPMC) expansion. This expansion is written as $(\mathbf{n}'/\mathbf{2})^{n'}(\mathbf{2}; \text{ER})^2(\mathbf{2}; \text{ER})^2 \cdots (\mathbf{2}; \text{ER})^2$ or as $(\mathbf{n}'/\mathbf{2})^{n'}(\mathbf{n}''; \text{PP})^{n''}$ and consists of $2^{n''/2}$ CSFs, where n'' is the number of active electrons. Of course, not all the electrons must be described with this limited expansion form, but it may be appropriate for the description of the less important correlation effects in some fragments of the molecule. For example in the ethylene case, the C—H bonds could be described with this expansion while the four most important electrons could be described with a more general expansion form. This wavefunction is written as $(\mathbf{2})^4(\mathbf{8}; \mathrm{PP})^8(\mathbf{4}; \mathrm{C{=}C})^4$ and consists of 320 CSFs. This wavefunction not only describes the dissociation to methylene fragments but also describes the major correlation effects within the remaining single bonds. If all of the active electrons of ethylene are described in this manner, the resulting expansion is written as $(\mathbf{n}'/\mathbf{2})^{n'}(\mathbf{12}; \mathrm{PP})^{12}$ and consists of 64 CSFs. (Like the more general ERMC expansion, this wavefunction cannot describe the dissociation to triplet fragments.) In general, this expansion set is a subspace of the ERMC expansion: the ERMC expansion allows each electron pair to use all the orbitals in the expansion for correlating orbitals while the PPMC expansion only allows a single correlating orbital for each electron pair. The PPMC expansion is also a subset of the RCI expansion: the RCI expansion has the same orbital subspace restrictions as the PPMC expansion but allows all spin couplings and occupations while the PPMC expansion only contains the even replacement subset.

Table V gives a list of the various CSF expansion lengths for the ethylene molecule discussed in the previous sections. The expansion length reductions resulting from spatial symmetry are neglected in this list. This is because some of the expansions in Table V cannot exploit spatial symmetry. Table V thereby

TABLE V

CSF expansions for ethylene. CSF expansion lengths for the various ethylene wavefunctions discussed in the text. The number of distinct rows is typically much smaller than the number of CSFs indicating the efficiency of the DRT representation of these wavefunction expansions. The expansion types are discussed in the text.

Expansion	CSF expansion length	Number of distinct rows	Expansion type
$(\mathbf{2})^4(\mathbf{12})^{12}$	226 512	140	FORS
$(\mathbf{2})^4(\mathbf{8})^8(\mathbf{4})^4$	71 280	96	DPFCI
$(\mathbf{2})^4(\mathbf{4})^4(\mathbf{4})^4(\mathbf{4})^4$	25 656	72	DPFCI
$(\mathbf{2})^4(\mathbf{4}; S=0)^4(\mathbf{4}; S=0)^4(\mathbf{4}; S=0)^4$	8 000	40	Spin-restricted DPFCI
$(\mathbf{2})^4(\mathbf{8}; \mathrm{RCI})^8(\mathbf{4})^4$	6 144	56	DPFCI
$(\mathbf{2})^4(\mathbf{12}; \mathrm{RCI})^{12}$	3 012	52	RCI
$(\mathbf{2})^4(\mathbf{8}; \mathrm{RCI}, S=0)^8(\mathbf{4})^4$	1 620	30	Spin-restricted RCI
$(\mathbf{2})^4(\mathbf{8}; \mathrm{ER})^8(\mathbf{4})^4$	1 400	38	Restricted DP
$(\mathbf{2})^4(\mathbf{12}; \mathrm{ER})^{12}$	924	49	ERMC
$(\mathbf{2})^4(\mathbf{12}; \mathrm{RCI}. S=0)^{12}$	729	25	Spin-restricted RCI
$(\mathbf{2})^4(\mathbf{8}; \mathrm{PP})^8(\mathbf{4})^4$	320	26	Restricted DP
$(\mathbf{2})^4(\mathbf{12}; \mathrm{PP})^{12}$	64	19	PPMC
$(\mathbf{6})^{12}(\mathbf{4})^4$	20	14	FORS
$(\mathbf{8})^{16}$	1	9	RHF

gives a consistent comparison of the different CSF expansions that have been discussed. It should be mentioned that all of these expansions are constructed from orbitals that correspond to the valence orbitals of the carbon and hydrogen atoms. Important correlation effects which require orbitals from outside of this space cannot be described with these expansions. The choice of CSF expansion space for ethylene, or for any other system, would depend on the regions of the potential energy surface being studied, the accuracy required and the computational effort involved.

7. *Nonlinear Wavefunction Expansions*

The wavefunctions discussed in the previous sections have been linear expansions within the space of expansion CSFs. Another type of MCSCF wavefunction consists of non-linear expansions within this space. These expansions are characterized by some number of parameters which in turn determine the expansion coefficients in the CSF expansion space. The non-linear wavefunction expansion spans some subspace of a more general linear expansion space. The usefulness of this approach results from the fact that the number of these parameters is usually much smaller than the dimension of the underlying linear expansion space. For this to be exploited computationally, the effort required in the optimization of these non-linear parameters should depend only on the number of these parameters and not on the, presumably much larger, dimension of the underlying linear CSF space. All of the non-linear expansions discussed here will be of the simple product form. In general, such wavefunctions can be constructed by forming the appropriate antisymmetrized product of correlated molecular fragments. The number of parameters in these expansions is the total of the number of parameters of each fragment. This product expansion form, particularly when the fragments are limited to electron pairs, has been discussed in more detail elsewhere[93,102,103]. Only a few of the simpler product forms will be discussed here.

a. GVB Wavefunctions

One popular wavefunction of this form is the generalized valence bond (GVB) expansion[104–109]. With this method an N-electron molecule is described in terms of N non-orthogonal orbitals and an energy-optimized spin function as

$$\Psi = \mathscr{A}[\chi_1(1)\chi_2(2)\chi_3(3)\cdots\chi_N(N)\Theta(1,2,\ldots,N)] \tag{204}$$

where $\Theta(1,\ldots,N)$ is a general N-electron spin function. Because of the non-orthogonal orbitals, the optimization of this wavefunction is too difficult to be practicable for large molecular systems. However, calculations with this wavefunction form indicate that it is often reasonable to impose a strong orthogonality restriction on the orbitals[104,108]. Using this strongly orthogonal GVB (SOGVB) approximation, the orbitals describing an electron pair are allowed to remain non-orthogonal to each other, but the orbitals of different pairs are forced to be orthogonal. The two orbitals within an electron pair may then be described in terms of a pair of orthonormal orbitals. For example, the contribution to the wavefunction of the first pair of electrons may be written

$$\chi_1(1)\chi_2(2) = \phi_1^2(1,2) + \lambda_1\phi_2^2(1,2) + \lambda_1'[\phi_1(1)\phi_2(2) - \phi_2(1)\phi_1(2)] \quad (205)$$

where the last term must be triplet spin coupled. The total wavefunction is then given as a product of these geminals (two electron functions) for each of the electron pairs, of any remaining singly occupied orbitals and of any doubly occupied orbitals containing inactive electrons. The triplet-coupled terms of each pair are coupled to give the correct overall spin for the wavefunction. Since the last term in Eq. (205) is invariant with respect to rotations between the natural orbitals ϕ_1 and ϕ_2, the wavefunction dependence, and therefore the energy dependence, on this rotation is determined completely by the first two terms. The energy expression in terms of the parameters $\{\lambda_1, \lambda_1', \lambda_2, \lambda_2', \ldots\}$ along with various optimization procedures have been given by Bobrowicz[108,109] for several cases of the SOGVB wavefunction. It is this orthogonal orbital description that allows the GVB wavefunctions to be compared to the other MCSCF wavefunctions discussed in this review.

The SOGVB wavefunction consists of a product of electron-pair wavefunctions, each of which is described in terms of a pair of orthonormal orbitals. As a result of this product nature of the wavefunction, however, this wavefunction expansion is not as flexible as the underlying RCI expansion. Multiplication of the geminals of the various electron pairs shows that the SOGVB wavefunction spans a subspace of the RCI wavefunction. One restriction is that the CSF expansion coefficients are given by products of the parameters $\{\lambda\}$ and are not free to vary completely independently. Secondly, not all possible RCI expansion terms have non-zero coefficients. This is because the RCI wavefunction also includes the singlet spin couplings of the electron pairs and none of these terms arise in the product form of the SOGVB wavefunction. In fact, the product form of the SOGVB wavefunction allows these terms to be transformed away, without loss of generality, with a suitable orbital choice, leaving only the closed-shell terms and the triplet-coupled terms as shown in Eq. (205). Numerical comparisons by Harding and Goddard[97] and by Dunning *et al.*[98] of the SOGVB and RCI wavefunction forms shows that these singlet-coupled terms are sometimes important within the linear expansion. This is also supported by the calculations of Chipman[100], which demonstrate the importance of these singlet-coupled terms in the linear RCI expansion.

A further approximation to the GVB wavefunction is the deletion of the triplet term in Eq. (205) for each geminal. This results in the perfect pairing

GVB (PPGVB) wavefunction. This wavefunction form was first proposed by Hurley *et al.*[101], who derived the energy expression directly in terms of the parameters $\{\boldsymbol{\lambda}\}$. These authors arrived at this wavefunction form, however, as a special case of a geminal product wavefunction form. The PPGVB wavefunction form has been used extensively by Goddard and coworkers[97,98,104–109]. The wavefunction for N electrons (consisting of $N/2$ pairs) may be written as

$$\Psi = \mathscr{A}[(\phi_1^2 + \lambda_1\phi_2^2)(\phi_3^2 + \lambda_2\phi_4^2)\cdots(\phi_{N-1}^2 + \lambda_{N/2}\phi_N^2)\Theta_{\text{PP}}] \tag{206}$$

where the spin function Θ_{PP} is simply the product of $\alpha\beta$ terms for each electron pair. It is clear from this expression that the PPGVB wavefunction may be written as a linear combination of CSFs that contain doubly occupied orbitals only and where each orbital pair is constrained to contain exactly two electrons. This underlying CSF space is the PPMC space discussed in the previous section. The number of CSFs in this PPMC expansion is given by $2^{N/2}$. As with the SOGVB wavefunction, however, the expansion coefficients of these CSFs are not free to vary independently in the PPGVB wavefunction but are given as products of the $N/2$ parameters $\{\lambda_1, \lambda_2, \ldots, \lambda_{N/2}\}$. The PPGVB wavefunction is, of course, less flexible than the linear PPMC wavefunction. It is most accurate near equilibrium geometries where the parameters $\{\boldsymbol{\lambda}\}$ are small, for which the product nature of the wavefunction imposes no serious restriction, and for systems that consist of non-interacting electron pairs, for which the product nature of the wavefunction is exact. One feature of these perfect pairing wavefunctions is that the orbitals tend to localize into bonding and antibonding orbital pairs. This localization allows a straightforward chemical interpretation of these wavefunctions.

b. Geminal Product Wavefunctions

One limitation of the perfect pairing expansions is the constraint that an electron pair is restricted to a two-orbital subspace. As two electron pairs are forced into the same region of space during a molecular distortion, the correlation effects of the antibonding orbital of one electron pair are excluded for use by the other electron pair. The description of these correlation effects requires the transfer of the electrons from one localized orbital pair into the orbitals of the other localized electron pair. This restriction may be eliminated by allowing the geminals to be described by a set of correlating orbitals that are shared by all of the different electron pairs. The ith electron pair may be described as

$$g(2i-1, 2i) = \phi_i^2 + \sum_a T_{ai}\phi_a^2 \tag{207}$$

where the summation is over the entire set of correlating orbitals. The total wavefunction may then be written as the antisymmetrized product of such geminals, each of which is described with its own set of coefficients. This wavefunction will be referred to as the geminal product (GP) wavefunction and was first proposed by Hurley *et al.*[101]. The PPGVB wavefunction, discussed in the previous section, was treated by these authors as a special case of this GP wavefunction. The multiplication of these geminals shows that, like the perfect pairing wavefunctions previously described, this wavefunction is equivalent to an expansion involving CSFs with only doubly occupied orbitals. The linear expansion coefficients are products of the parameters T_{ai}. The number of parameters in this wavefunction expansion is the product of the number of electron pairs and the number of correlating orbitals. The underlying linear expansion space for the GP wavefunction is the ERMC expansion described earlier. Like the general ERMC expansion, the more restrictive GP, PPMC and PPGVB wavefunctions all result in a diagonal one-particle density matrix and in a sparse two-particle density matrix consisting of direct and exchange type terms only. This density matrix sparseness may be exploited computationally in all of these wavefunction forms.

Finally, one other non-linear wavefunction expansion will be described. If the geminals of the different electron pairs are further restricted to be identical for each electron pair, then the result is called the antisymmetrized product of identical geminals (APIG) wavefunction[110–114]. There are only $(n-1)$ parameters in the APIG wavefunction which spans a subspace of the GP wavefunction space. Because of the severely restrictive form of this wavefunction, it has not been used extensively for MCSCF calculations but it has been used as a reference function for propagator calculations[113,114] for which this wavefunction form has formal appeal.

In summary of the non-linear wavefunctions, the APIG wavefunction spans a subspace of the GP wavefunction which in turn spans a subspace of the linear ERMC wavefunction. The non-linear PPGVB wavefunction spans a subspace of the non-linear GP wavefunction, the non-linear SOGVB wavefunction and the linear PPMC wavefunction. The PPMC wavefunction in turn spans a subspace of the ERMC wavefunction and of the RCI wavefunction. The non-linear SOGVB wavefunction spans a subspace of the linear RCI wavefunction. All of these expansions, of course, span a subspace of the full CI expansion space. Table IV shows the number of variational parameters of these wavefunctions when the number of electrons is equal to the number of orbitals. It is clear from this table that wavefunctions expressed as non-linear contractions have an advantage over the more general linear expansion wavefunctions in reducing the number of variational parameters. The use of non-linear contractions, particularly of the GP and PPGVB type, has not been exploited in the recent MCSCF implementations. It is likely that this will change as more electrons are correlated in larger molecular systems. The advantages of these non-linear parametrizations will become more compelling, at least for subsets of the correlated electrons for which these wavefunction forms are expected to be most appropriate and when combined with more general expansion forms for the more important electrons.

C. Orbital Symmetry and the Configuration State Function Expansion Space

There are two effects of the imposition of orbital symmetry on the MCSCF wavefunction that should be discussed. Both of these effects may be considered by assuming that an initial MCSCF wavefunction has been obtained in terms of a set of localized orbitals. This wavefunction may then be symmetry projected to obtain a wavefunction that displays the full symmetry of the molecule. The first effect results from the case for which the symmetry projection does not change the orbital space but does induce changes in the CSF expansion coefficients so that new expansion terms are required when the CSFs are expressed in terms of the symmetry orbitals. The second effect results from the case for which the projected orbitals span a larger space than the localized orbitals.

The symmetry projection of the wavefunction is equivalent to a particular orbital transformation among the occupied orbitals of the wavefunction. If the CSF expansion is full within these sets of symmetry-related orbitals, no new CSFs will be generated by this orbital transformation. This type of wavefunction could have been computed directly in terms of symmetry orbitals with no loss of generality. (In fact, the CSF expansion expressed in terms of symmetry orbitals will usually result in fewer expansion terms because the symmetry blocking of the individual CSFs allows those of the incorrect symmetries to be deleted from the expansion.) However, if the CSF expansion is not full within these orbital sets, it is possible that the symmetry transformation of the orbitals will generate new CSF expansion terms. The coefficients of these new CSF expansion terms are determined by the old expansion coefficients and the symmetry transformation coefficients. For example, consider the case of two H_2 molecules, described in terms of localized orbitals, separated by a reflection plane. Assume that the localized description of the two H_2 molecules is of the form

$$\Psi_L = |\phi_{1\alpha}\phi_{1\beta}| + \lambda|\phi_{2\alpha}\phi_{2\beta}| \tag{208}$$

$$\Psi_R = |\phi_{3\alpha}\phi_{3\beta}| + \lambda|\phi_{4\alpha}\phi_{4\beta}| \tag{209}$$

for the left and right H_2 molecules. The total wavefunction may be written as the antisymmetrized product of these two fragments

$$\begin{aligned}\Psi = |\phi_{1\alpha}\phi_{1\beta}\phi_{3\alpha}\phi_{3\beta}| + \lambda^2|\phi_{2\alpha}\phi_{2\beta}\phi_{4\alpha}\phi_{4\beta}|\\ + \lambda|\phi_{2\alpha}\phi_{2\beta}\phi_{3\alpha}\phi_{3\beta}| + \lambda|\phi_{1\alpha}\phi_{1\beta}\phi_{4\alpha}\phi_{4\beta}|\end{aligned} \tag{210}$$

This four-CSF wavefunction contains the full molecular symmetry even though it is described in terms of localized orbitals and is not full within the symmetry-related orbital subsets. If the orbitals are subjected to a symmetry transformation to give the new orbitals

$$\phi_a = 2^{-1/2}(\phi_1 + \phi_3) \tag{211}$$

$$\phi_b = 2^{-1/2}(\phi_1 - \phi_3) \tag{212}$$

$$\phi_c = 2^{-1/2}(\phi_2 + \phi_4) \tag{213}$$

$$\phi_d = 2^{-1/2}(\phi_2 - \phi_4) \tag{214}$$

then the total wavefunction takes the form

$$\begin{aligned}\Psi = &|\phi_{a\alpha}\phi_{a\beta}\phi_{b\alpha}\phi_{b\beta}| + \lambda^2|\phi_{c\alpha}\phi_{c\beta}\phi_{d\alpha}\phi_{d\beta}| + \tfrac{1}{2}\lambda(|\phi_{a\alpha}\phi_{a\beta}\phi_{c\alpha}\phi_{c\beta}| \\ &+ |\phi_{a\alpha}\phi_{a\beta}\phi_{d\alpha}\phi_{d\beta}| + |\phi_{b\alpha}\phi_{b\beta}\phi_{c\alpha}\phi_{c\beta}| + |\phi_{b\alpha}\phi_{b\beta}\phi_{d\alpha}\phi_{d\beta}|) \\ &+ \tfrac{1}{2}\lambda(-|\phi_{a\alpha}\phi_{b\beta}\phi_{c\alpha}\phi_{d\beta}| + |\phi_{a\alpha}\phi_{b\beta}\phi_{c\beta}\phi_{d\alpha}| + |\phi_{a\beta}\phi_{b\alpha}\phi_{c\alpha}\phi_{d\beta}| \\ &- |\phi_{a\beta}\phi_{b\alpha}\phi_{c\beta}\phi_{d\alpha}|)\end{aligned} \tag{215}$$

It may be verified that the last term in parentheses may be written as $-\lambda|1212\rangle$, using the step vector notation with the symmetry-adapted orbital basis. Eq. (215) shows that the four-CSF expansion in terms of localized orbitals is expanded into the space of seven CSFs when symmetry orbitals are used. This increase of expansion length was expected since the original CSF expansion space was not invariant with respect to rotations between orbitals ϕ_1 and ϕ_3 and between orbitals ϕ_2 and ϕ_4.

This increase of CSF expansion length upon transformation to symmetry-adapted orbitals potentially affects any of the expansion forms that attempt to describe electron correlation in terms of localized orbitals and that are not invariant to transformations that mix the different localized orbitals. All of the product and direct product expansion forms (including the RCI, PPMC, PPGVB and SOGVB expansions) are potentially of this type. It often happens that these wavefunctions do have the full molecular symmetry even though they are described in terms of localized orbitals and not symmetry-adapted orbitals. The localized orbital description that results from these wavefunction optimizations is therefore both an asset and a liability; it aids the chemical interpretability and results in more compact CSF expansions but the computations must be performed in an orbital basis that does not possess the full molecular symmetry. This is computationally important since many steps of the MCSCF wavefunction optimization can exploit such orbital symmetry when it is present.

It the previous case it was assumed that the occupied orbital space did not change upon transformation to symmetry-adapted form. However, it sometimes happens that the occupied orbital space must be expanded to satisfy symmetry invariance properties. In contrast to the previous case, these MCSCF wavefunctions usually are not eigenfunctions of the symmetry operations of the molecule before symmetry projection. This aspect of spatial symmetry has been considered by several researchers[25,115,116]. One recent study is that of McLean *et al.*[55] who examined the symmetry-breaking problem of the formyloxyl radical HCO_2 and devised direct product CSF expansions that allowed the full molecular symmetry to be obtained. To understand the cause of this symmetry breaking, consider the RHF wavefunction of the He_2^+ ion at large internuclear separations[25,55]. The optimal orbitals for this three-electron wavefunction will consist of one that is localized on the left He atom and is doubly occupied, and one that is localized on the right atom and is singly occupied. This wavefunction is appropriate for the broken-symmetry ion $HeHe^+$. The spatial extents of these two orbitals are very different since one corresponds to a neutral atom while the other corresponds to a cation. Clearly these orbitals do not simply transform among themselves when subjected to reflection through the plane between the nuclear centers. There is of course an equivalent wavefunction that is the mirror image of this one. This wavefunction describes the He^+He ion. The orbitals required for this wavefunction are the mirror images of the previous ones and consist of a tight cation orbital on the left atom and doubly occupied orbital on the right atom. A symmetric wavefunction that has the correct mixture of both of these wavefunctions must include all four of these linearly independent orbitals and, as in the previous example, could be described either in terms of localized orbitals or symmetry orbitals. Other cases of orbital doubling that were caused by competing qualitative descriptions of the electronic structure have already been considered. It should be clear that this is another example of such a competing effect. The correlation energy resulting from mixing symmetry-related descriptions, such as the He_2^+ example discussed here, is usually called resonance energy. In the MCSCF method, this resonance energy lowering must compete with other types of correlation energy that are best described with symmetry-broken wavefunctions. Therefore, simply doubling the orbitals and computing the new MCSCF wavefunction still may not result in a symmetric wavefunction. It is only when all such competing correlation effects have been adequately described that the wavefunction will become symmetric and begin to provide an adequate description of the resonance energy.

These competing effects may be resolved by increasing the flexibility of the CSF space until all symmetry-related descriptions of the electronic structure are adequately described. This was done systematically in the case of HCO_2 until a wavefunction displaying the full molecular symmetry was obtained. The study of symmetry breaking in MCSCF wavefunctions leads to an important empirical generalization about the prediction of the behavior of potential energy surfaces in the vicinity of symmetric geometries[55]. An MCSCF calculation that displays a broken-symmetry solution at a symmetric geometry and indicates a broken-symmetry stable geometry cannot, by itself, be used to predict confidently that the molecule has such an unsymmetric geometry. An MCSCF calculation that displays a symmetric solution at symmetric geometries also cannot, by itself, be used to predict confidently that the molecule has an unsymmetric geometry. However, an MCSCF calculation that displays symmetric solutions at symmetric geometries (without imposed symmetry restrictions) and indicates a symmetric stable geometry may be used as compelling evidence that the molecule has such a symmetric geometry. This is because the resonance energy can be difficult to describe without imposing constraints on the expansion coefficients. Having a symmetric MCSCF solution is necessary, but not always sufficient, to describe this resonance energy adequately. If the resonance is inadequately described, the symmetric geometries tend to be local maxima between two lower-energy, symmetry-related, broken-symmetry geometries. However, if the MCSCF solution is symmetric, indicating that the resonance is at least partially accounted for, and the symmetric geometry is the lowest energy, then any improvements to the correlation energy description will tend to improve the resonance energy description even further, thereby enforcing the prediction of a stable geometry. This predictive ability of the MCSCF method is very useful and, when accompanied with a careful analysis as in the case of the HCO_2 radical[55], it becomes a very powerful tool.

D. Summary

In this section two approaches to the selection of CSF expansion spaces for MCSCF wavefunctions have been described. Most modern MCSCF methods are best suited to the *a priori* selection of CSFs based on orbital occupation and spin-coupling restrictions. These *a priori* selection approaches should be accompanied with empirical evidence (usually more extensive MCSCF or CI calculations at selected geometries) that a balanced description of the molecular electron correlation is achieved. A reasonable approach to use for molecular systems is to begin with an RCI expansion of all the valence electrons. The less important electrons may then optionally be described with the more restrictive, and more economical, PPMC expansion. The description of the more important electrons in the RCI expansion may be generalized, if necessary, to other direct product type expansions, thereby allowing a more thorough treatment of the correlation of these electrons.

V. REDUNDANT VARIABLES

The set of variables in the MCSCF optimization process determine the changes in the CSF mixing coefficients and the orbital expansion coefficients during each iteration. When particular choices of CSFs are employed in the wavefunction expansion or when particular relations between the CSF expansion coefficients are satisfied, redundant variables will occur within this set. These variables are such that the same final wavefunction would be obtained whether these variables are included explicitly in the wavefunction variation or deleted entirely. In the usual case redundant variable elimination may be performed by deleting particular CSFs from the expansion space or by performing only limited orbital optimizations. The connection between the CSF expansion space and the orbital optimization must be understood in order to ensure that the MCSCF optimization conditions may be satisfied when these restrictions are imposed on the optimization process. It will be shown that particular choices of methods for the elimination of these redundant variables may result in convergence difficulties even though these choices result in no loss of flexibility in the wavefunction. This is particularly true for excited states and for wavefunctions that must simultaneously describe several states of interest. Another aspect of redundant variables is that their occurrence may cause some of the intermediate equations that must be solved within an MCSCF iteration to become either ill-conditioned or undefined entirely. It is, therefore, of practical importance to understand the nature of redundant variables. The relation between the variables in the MCSCF optimization process and the CSF expansion space is first examined. Then the occurrence of wavefunction-dependent redundant variables is examined. Finally the effect of redundant variables on some of the MCSCF optimization equations is examined.

A. Redundant Variables and the Configuration State Function Expansion

As discussed earlier, many CSF expansion spaces are purposely chosen to be full with respect to some subset of the orbitals. It was mentioned that this resulted in the occurrence of redundant variables in the optimization procedure. In this section, the details of this relation between these redundant variables and the CSF expansion space are examined, first for the two-electron case and then for the general N-electron case.

1. The Two-electron Case

The general discussion of redundant variables will be preceded by a detailed discussion of the special case of two-electron wavefunctions. The formal manipulations are easier to understand for this case than in the general N-electron case. Additionally, the exact full-CI wavefunctions may be included in these formal manipulations. Finally, several features of the two-electron case may then be used in understanding the more general cases. For the discussion of this special case, the approach of McIver[117] is used to define a matrix of determinants of the form

$$\Phi_{ij} = |\phi_{i\alpha}(1)\phi_{j\beta}(2)| \tag{216}$$

for $i, j = 1$ to n. It may be verified that if the spatial orbital set $\boldsymbol{\phi}$ is subjected to an arbitrary transformation

$$\boldsymbol{\phi}' = \boldsymbol{\phi}\mathbf{T} \tag{217}$$

then the matrix $\boldsymbol{\Phi}$ transforms as

$$\boldsymbol{\Phi}' = \mathbf{T}^{t}\boldsymbol{\Phi}\mathbf{T} \tag{218}$$

It may be noted that Eq. (218) may be written in the alternate form

$$\Phi'_{ij} = \sum_{kl} \mathbb{T}_{ij,kl}\Phi_{kl} \tag{219}$$

where $\mathbb{T}_{ij,kl} = T_{ki}T_{lj}$. This shows that a linear transformation of the n-dimensional orbital space induces a linear transformation within the n^2-dimensional determinant space. The second form of this transformation, Eq. (219), may be extended to the general N-electron wavefunction case. It is, however, the first form, Eq. (218), that allows an explicit treatment of the two-electron case using familiar matrix operations and will be used in this section. For a given set of orbitals, an unnormalized two-electron wavefunction may be expanded in terms of this matrix of determinants as

$$\Psi = \sum_{j}\sum_{i} C_{ji}\Phi_{ij} = \mathrm{Tr}(\mathbf{C}\boldsymbol{\Phi}) \tag{220}$$

where the coefficient matrix $\mathbf{C}$ is symmetric for singlet states and antisymmetric for triplet states. The present analysis will concern the singlet states only. A CSF then consists of either a diagonal element of the matrix $\boldsymbol{\Phi}$ or the sum of two off-diagonal elements. Using Eq. (218) the dependence of the wavefunction on both the orbital transformation coefficients and the expansion coefficients may be written explicitly as

$$\Psi' = \mathrm{Tr}(\mathbf{C}\mathbf{T}^{t}\boldsymbol{\Phi}\mathbf{T}) = \mathrm{Tr}(\mathbf{T}\mathbf{C}\mathbf{T}^{t}\boldsymbol{\Phi}) \tag{221}$$

$$= \mathrm{Tr}(\mathbf{C}'\boldsymbol{\Phi}) \tag{222}$$

Eq. (221) shows that the wavefunction change induced by the orbital variations for a fixed set of expansion coefficients is equivalent to a transformation of the CSF expansion coefficient for a fixed set of orbitals. Eq. (222) shows that variations of the coefficients $\mathbf{C}$ and orbital transformation coefficients $\mathbf{T}$ do not need to be considered separately. Only the combined effect, expressed as variations of the coefficient matrix $\mathbf{C}'$, is required to allow an arbitrary two-electron wavefunction change. This occurs in this case because the wavefunction is expanded in the full CI set of CSFs. This demonstrates that a redundancy exists between the orbital coefficient variations and the CSF expansion coefficient variations and that this redundancy may be eliminated by considering only the CSF coefficient variations for some fixed set of orbitals. Other solutions to this redundancy will now be considered.

The following restrictions are hereafter imposed on the two-electron wavefunction variations. The CSF expansion coefficients **C** are assumed to be real and any transformation applied to these coefficients must be real. The orbital basis is allowed to be complex but any transformation applied to the orbitals must be real. These restrictions have no effect on the expectation values of real Hamiltonian operators. Finally, an orthonormal orbital basis is assumed and only orthogonal orbital transformations are allowed. This of course does place restrictions on some of the present discussions but is considered crucial for the extension of these results to the general N-electron case.

An orthogonal transformation matrix $\mathbf{U}'$ may be found to diagonalize the symmetric matrix **C** of Eq. (220). The matrix may be factored, as $\mathbf{U}' = \mathbf{Us}$, into the product of a rotation matrix **U** and a diagonal sign matrix **s** where $s_{ii} = \pm 1$. The matrix **U** may further be written as $\mathbf{U} = \exp(\mathbf{K})$ where **K** is antisymmetric as discussed in Section II. Eq. (18) shows $\mathrm{Det}(\mathbf{U}) = \exp(\mathrm{Tr}(\mathbf{K})) = +1$ as required for rotation matrices. An arbitrary two-electron wavefunction may then be written in the form

$$\Psi' = \mathrm{Tr}(\mathbf{C}\boldsymbol{\Phi}) = \mathrm{Tr}(\mathbf{U}'\mathbf{d}\mathbf{U}'^{\mathrm{t}}\boldsymbol{\Phi}) = \mathrm{Tr}(\mathbf{U}\mathbf{d}\mathbf{U}^{\mathrm{t}}\boldsymbol{\Phi}) \tag{223}$$

$$= \mathrm{Tr}(\mathbf{d}\mathbf{U}^{\mathrm{t}}\boldsymbol{\Phi}\mathbf{U}) = \mathrm{Tr}(\mathbf{d}\boldsymbol{\Phi}') \tag{224}$$

$$= \sum_i d_i \Phi'_{ii} \tag{225}$$

The fact that the diagonal matrices **d** and **s** commute is used in Eq. (223) to show that the wavefunction does not depend on the choice of sign factors and the matrix **s** will not be considered further. Eqs (224) and (225) demonstrate the well known fact that an arbitrary two-electron singlet wavefunction, including an energy-optimized full CI wavefunction, may be written in its natural-orbital form in terms of an n-term pair expansion consisting of doubly occupied determinants only[101–103,118,119].

In the full expansion case where all $n(n+1)/2$ unique elements of **C** are determined variationally, the orbitals do not need to be optimized simultaneously. This is, in fact, a general feature of full CI expansions and is not particular to the two-electron case. Eq. (224) shows that in principle it is possible to find an equivalent representation of this wavefunction that consists of the n-term pair expansion. To do so, however, requires knowledge of the matrix **U** and the pair-expansion coefficients **d**. Although Eq. (223) gives a prescription for finding **U** and **d** in terms of the full CI solution for the two-electron case, this is not practicable for the more general cases, so an alternative approach is required.

Such an approach is afforded in the MCSCF method by starting with the form of Eq. (224) as an ansatz, followed by the variational determination of the orbital transformation matrix **U** and pair coefficients **d**. The matrix **U** depends on the $n(n-1)/2$ independent parameters in the matrix $\mathbf{K}$. These variables, in addition to the n pair-expansion coefficients $\mathbf{d}$, bring the total number of variables to the same number as in the original full CI expansion, namely $n(n+1)/2$. (Of course in the usual N-electron case, the number of parameters in the full CI expansion will be much greater than the number of parameters in the orbital transformation matrix.) This trial wavefunction may be written in the form

$$\Psi' = \mathrm{Tr}(\mathbf{d}\exp(-\mathbf{K})\mathbf{\Phi}\exp(\mathbf{K})) \tag{226}$$

where the wavefunction dependence on the parameters $\mathbf{d}$ and $\mathbf{K}$ is shown explicitly. The variational determination of these parameters, however, does not guarantee the equivalence of the MCSCF wavefunction to the desired full CI wavefunction. Out of the total of $n(n+1)/2$ states in the full CI expansion, the MCSCF procedure could in principle converge to any of them in addition to possible spurious solutions that have no correspondence to energy optimized full CI states.

There is now a choice between two sets of variables for the MCSCF optimization of a general two-electron wavefunction. In the first choice, all $n(n-1)/2$ orbital rotation variations may be eliminated in favor of variations of the $n(n+1)/2$ CSF expansion coefficients. The second choice involves the elimination of the $n(n-1)/2$ off-diagonal CSF expansion coefficients in favor of variations of the n pair-expansion coefficients and $n(n-1)/2$ orbital rotation parameters. Both of these choices possess sufficient flexibility to allow the MCSCF procedure to converge to the same exact wavefunction. But an important difference in these two choices may be noted for excited-state wavefunction optimizations or for cases where more than one state needs to be adequately represented. All $n(n+1)/2$ states are described exactly in the full CI expansion, each state being associated with its own set of expansion coefficients. However, for the converged orbitals in the n-term pair expansion, only one state is described exactly and the other $(n-1)$ orthogonal complement terms only approximately correspond to other exact states. This is because the set of orbitals, which is used to describe all n states in the CSF expansion, has been chosen to optimize a particular state and the other states are only optimal with respect to the CSF expansion coefficients. The variation of the n pair-expansion coefficients alone, for fixed orbitals, is not sufficient for these other states to be described exactly. In principle then, it is possible to calculate an exact representation of some excited state but with the lower states being so poorly described that their approximate energies within the pair-expansion space are ordered incorrectly. This loss of correspondence of the exact states with the approximate MCSCF states, through the bracketing theorem, makes the identification of an excited MCSCF state difficult. With the loss of the correct ordering, it is not known whether the wavefunction corresponds to the exact wavefunction for the correct state, to an exact wavefunction for an incorrect state, or is simply some spurious solution (an artefact of the optimization process) representing perhaps a poor description of some state. Possible alternate wavefunction representations will now be considered in an attempt to allow the adequate simultaneous description of several states.

Since Eq. (224) shows that an arbitrary two-electron singlet wavefunction may be expanded in terms of only n CSFs, a possible solution to this problem may lie in the particular choice of these n expansion terms. If there are alternate n-term expansions for the general wavefunction, these might then be used to advantage in the simultaneous descriptions of several states since different CSF expansion spaces would produce different approximations to the other states. It is instructive to consider a possible n-term expansion in terms of Φ_{11} and its single excitations. This n-term wavefunction expansion consists of a symmetric $\mathbf{C}$ matrix of the form

$$\mathbf{C} = \begin{bmatrix} C_{11} & C_{12} & \cdots & C_{1n} \\ C_{21} & 0 & \cdots & 0 \\ \vdots & & & \vdots \\ C_{n1} & 0 & \cdots & 0 \end{bmatrix} \tag{227}$$

It may be verified by inspection that this matrix may be brought directly to tridiagonal form with a single elementary Householder transformation. The resulting matrix has the form

$$\mathbf{C}' = \begin{bmatrix} C_{11} & C'_{12} & 0 & \cdots & 0 \\ C'_{21} & 0 & & \cdots & 0 \\ 0 & 0\cdots & & \cdots & 0 \\ 0 & 0\cdots & & \cdots & 0 \end{bmatrix} \tag{228}$$

This shows that an n-term singles wavefunction is always equivalent to a two-term singles wavefunction. The $\mathbf{C}'$ matrix clearly has at most two non-zero eigenvalues and therefore this two-term singles wavefunction is equivalent to a two-term pair expansion. Furthermore the bracketing theorem, applied to the eigenvalues of the $\mathbf{C}'$ matrix, indicates that these two pair-expansion coefficients must have opposite signs. This leads to the conclusion that a two-term natural expansion with both coefficients having the same sign does not have an equivalent two-term singles representation (or an equivalent n-term singles representation). This example shows that a completely arbitrary choice of expansion CSFs, even one with the correct number of terms and with the correct number of occupied orbitals, may not possess sufficient flexibility to represent a particular two-electron wavefunction.

McIver has further analyzed the two-orbital case[117], which consists of three states in the full expansion. Because of the fact that the off-diagonal element of the Hamiltonian matrix in the two-term pair expansion is equal to an exchange integral and is therefore positive, the representation of the ground state has coefficients with opposite signs while both coefficients of the excited-state representation have the same sign. Therefore two-orbital ground states have equivalent two-term singles representations but the excited states may not always be represented in this form. In particular, assume that the highest-energy state is represented by a two-term singles expansion. This wavefunction must necessarily have a pair-expansion representation with coefficients of opposite signs. Since a higher-energy eigenvalue of the Hamiltonian matrix in the pair-expansion basis must exist for the general case of a non-zero exchange integral, there must correspondingly exist a higher-energy state of the full CI expansion. This contradicts the original assumption and shows that the highest-energy state cannot be represented with the two-term singles expansion. Numerical comparisons of all-singles MCSCF expansions and pair-excitation MCSCF expansions, including examples with more than two electrons, have also been performed for several atomic systems by Grein and Banerjee[64].

It does not appear that the general n-term pair expansion has an alternative n-term representation. In particular cases, however, it may be possible to replace a closed-shell determinant with an open-shell CSF. The previous discussion shows that this can be done only if the diagonal elements of the corresponding terms in the pair expansion have opposite signs. Instead of attempting to find an alternative n-term expansion for excited-state optimizations, a different approach is to begin with the n-term pair expansion and then selectively add open-shell determinants to the wavefunction to describe the lower-energy states adequately. These extra determinants have no effect on the state of interest since it may already be expanded in terms of the original pair-expansion terms; the effect of these additional determinants is solely to ensure the correct energy ordering within the other approximate states.

Suppose for example a trial wavefunction has the form of Eq. (226) and is written explicitly in terms of the n pair-expansion variables and the $n(n-1)/2$ variables in the matrix $\mathbf{K}$. The theory of the orthogonal rotation group (or of the more general unitary group) may be invoked to factor the orthogonal matrix $\exp(\mathbf{K})$ according to

$$\exp(\mathbf{K}) = \exp(\mathbf{K}'')\exp(\mathbf{K}') \tag{229}$$

where the antisymmetric matrix $\mathbf{K}'$ is zero everywhere except for the terms K'_{12} and K'_{21} (with $K'_{12} = K_{12}$) and where the antisymmetric matrix $\mathbf{K}''$ has zeros in those positions. This factorization is possible because the matrix $\exp(\mathbf{K}')$ is an element of a subgroup of the full rotation group that is parametrized, in this case, by the single parameter K'_{12}. Similarly, $\exp(\mathbf{K}'')$ belongs to a subgroup of the full group that is disjoint from the elements of $\exp(\mathbf{K}')$ except for the identity element. An arbitrary element of the full group, $\exp(\mathbf{K})$, may always be written in the left coset form[6,38] of Eq. (229). King *et al.*[120] have considered the representation of the elements of the matrix $\mathbf{K}''$ using commutator expansions. It may be verified that the matrix $\mathbf{U} = \exp(\mathbf{K}')$ is a unit matrix except for the elements $U_{11} = U_{22} = \cos(K'_{12})$ and $U_{12} = -U_{21} = \sin(K'_{12})$, i.e. the matrix $\mathbf{U}$ is an elementary plane rotation matrix. Substitution of Eq. (229) into the trial wavefunction of Eq. (224) gives

$$\Psi' = \mathrm{Tr}(\exp(\mathbf{K}')\mathbf{d}\exp(-\mathbf{K}')\exp(-\mathbf{K}'')\mathbf{\Phi}\exp(\mathbf{K}'')) \tag{230}$$

$$= \mathrm{Tr}(\mathbf{C}'\exp(-\mathbf{K}'')\mathbf{\Phi}\exp(\mathbf{K}'')) \tag{231}$$

The matrix $\mathbf{C}'$ consists of $(n+1)$ variables since the elementary plane rotation matrix transforms the diagonal matrix $\mathbf{d}$ into a symmetric matrix with only one pair of off-diagonal elements, $C'_{12} = C'_{21}$ in the example given here. The orbital rotations in Eq. (231) are parametrized in terms of the $n(n-1)/2 - 1$ independent variables remaining in the $\mathbf{K}''$ matrix. Since the $(n+1)$ variables in the matrix $\mathbf{C}'$ are to be variationally determined anyway, the explicit relation between $\mathbf{C}'$ and $\mathbf{d}$ does not have to be considered during the optimization. This shows explicitly for the two-electron case how a variable may be transferred from the orbital rotation space to the CSF coefficient space by increasing the length of the CSF expansion space. It is presumed that this new CSF expansion term may then be carefully chosen to improve the approximate description of the other states and to verify the correct correspondence of the MCSCF state with the exact state through the bracketing theorem.

There is a useful alternate interpretation of this example. The $(n+1)$ CSFs may be grouped into two types of terms: the first set consists of the three CSFs included in the full CI within the subspace of orbitals ϕ_1 and ϕ_2, and the second set consists of the remaining $(n-2)$ pair-expansion terms. Rotations between orbitals ϕ_1 and ϕ_2 simply transform the first group of three CSFs among themselves, since it is a full CI expansion in these orbitals, while the second group of CSFs is trivially invariant to these rotations. Of course, the choice of orbitals, ϕ_1 and ϕ_2 in this example, was completely arbitrary; any pair of orbitals could have been chosen.

Furthermore, another parameter could be moved from the orbital variation space into the CSF expansion space by factoring out another plane rotation matrix involving a pair of orbitals disjoint to the first pair. Suppose these orbitals are chosen to be ϕ_3 and ϕ_4. The orbital variations would then consist of $n(n-1)/2 - 2$ variational parameters and the CSF expansion space would include the n pair-expansion terms in addition to C_{12} and C_{34}, giving the same total of $n(n+1)/2$ MCSCF variables. This $(n+2)$-term CSF expansion set may then be partitioned into three subsets: the three CSFs that comprise the full CI expansion containing the orbitals ϕ_1 and ϕ_2, the analogous full CI expansion involving the orbitals ϕ_3 and ϕ_4, and the remaining $(n-4)$ pair-expansion terms. By the same reasoning as before, the redundant orbital rotations may be determined without recourse to Eq. (230) by recognizing that the spaces spanned by all three subsets of CSFs are each invariant to rotations between orbitals ϕ_1 and ϕ_2 and are each invariant to rotations between orbitals ϕ_3 and ϕ_4.

A further generalization of this approach is to allow full CI expansions in more than two orbitals at a time. For example, if all three of the orbital rotation parameters involving orbitals ϕ_1, ϕ_2 and ϕ_3 are factored out of the orbital space of the n-term pair expansion, they generate three new independent expansion terms in the CSF space. The resulting $(n + 3)$-term CSF space may then be partitioned into two subsets: those included in the full CI expansion of the three orbitals and those that do not contain these three orbitals. The redundant orbital rotation parameters may then be eliminated simply by recognizing that this CSF space partitioning exists. More general N-electron wavefunctions may also be constructed in such a way that they may be partitioned into subsets of CSFs, each of which displays invariance to particular orbital rotations. The recognition of this invariance may then be used to reduce the number of variables to be optimized by eliminating the redundant orbital rotations and thereby arriving at the same total number of independent MCSCF parameters as in the previous examples.

It may also be observed that the coset factorization of Eq. (229) is valid for any orbital pair regardless of the CSF space. However, the completely arbitrary transfer of parameters from the orbital variation space to the CSF space may not result in a CSF partitioning as in the above examples. For example, consider the $(n + 1)$-term expansion case considered previously in which C_{12} is allowed to be non-zero. The transfer of the orbital rotation parameter K_{13} from the orbital space into this CSF space requires the introduction of two new CSF expansion terms, C_{13} and C_{23}. However, these two terms may not be varied independently; one of the terms may be written as a function of the other terms in the expansion. Although the analysis of this transfer is straightforward in the two-electron case through Eq. (230), it is more difficult in the general case and this type of constrained CSF coefficient optimization will not be considered further.

2. *The General N-electron Case*

Many of the considerations of the two-electron case apply directly to the general N-electron case. Unfortunately, the analytic form of the equations that give the relationship between the orbital transformation, the CSF changes and the CSF expansion coefficient changes are not as simple as for the two-electron case. Another important difference is that the previous section was concerned with representing an exact state in some alternate representation in which approximations to the other states were sufficiently accurate to obtain the correct state ordering. In the usual N-electron case, the exact states are not known and the MCSCF optimization process is used to produce approximations to all of the states considered. This makes it even more difficult to obtain the correct state ordering in excited-state calculations because the CSF expansion terms that are added have an effect both on the state being optimized and on the orthogonal complement states. The best that can be achieved is a rapid convergence with respect to the expansion length so that the correct correspondence is achieved with minimal additional effort. Of course, the ability to perform accurate large-scale calculations based on the MCSCF wavefunction (Multireference configuration-interaction with single and double excitation, MRSDCI, for example) can give empirical evidence about the quality of the MCSCF approximation and can suggest the more important improvements to the wavefunction. With these differences noted, some of the results of the previous section will now be extended to the general N-electron case.

As in the previous section, the full CI case will be considered first. For the purpose of this discussion, a full CI expansion space is characterized by the relation

$$T_{rs}|n\rangle = \sum_{m} |m\rangle\langle m|T_{rs}|n\rangle \tag{232}$$

with $T_{rs} = E_{rs} - E_{sr}$ where the summation includes only the expansion CSFs, and where Eq. (232) is valid for any expansion term $|n\rangle$ and for any orbital pair, ϕ_r and ϕ_s. Eq. (232) is of course valid for any CSF space if the summation is allowed to range over states not included in the wavefunction expansion space. It is a property of a full CI expansion space that all potentially non-vanishing terms are already included in the expansion set. The recursive application of Eq. (232) may in fact be used to generate such a full CI expansion space.

Since an orbital transformation may be parametrized using the operator $\exp(K)$, the effect of an orbital transformation on a trial wavefunction may be written as

$$\exp(K)|0\rangle = \sum_{n} \exp(K)|n\rangle c_n \tag{233}$$

$$= \sum_{m}\sum_{n} |m\rangle\langle m|\exp(K)|n\rangle c_n \tag{234}$$

$$= \sum_{m}\sum_{n} |m\rangle \mathbb{T}_{mn} c_n \tag{235}$$

$$= \sum_{m} |m\rangle c'_m \tag{236}$$

Just as in the two-electron case of the full CI expansion, the simultaneous effect of both the orbital variations and the CSF expansion coefficient variations is redundant. An arbitrary wavefunction variation may be expressed by considering only the variations of the expansion coefficients for some fixed set of orthonormal orbitals. The transformation of Eq. (235) is equivalent to the transformation in Eq. (219) in the two-electron case and is the formal extension to the N-electron case. In both cases the effect of a linear transformation of the orbitals induces a linear transformation within either the space of expansion CSFs or the expansion coefficients. Unlike the relation in Eq. (218) for the two-electron case, however, there is no simple direct product representation for the CSF transformation matrix in Eq. (235) in the general N-electron case. This is due to the increased complexity of N-electron wavefunctions that results from spin symmetry and spatial symmetry restrictions imposed on the expansion CSFs.

Suppose that Eq. (232) is valid for some subset of operators $T_{r's'}$ but not for all such operators when the summation is limited to the CSF expansion terms. This will occur, for example, when the expansion space is full only with respect to a proper subset of the orbitals. In this case Eqs (233) through (236) will not hold for an arbitrary operator K but these relations will be valid for an operator K' that contains only the subset of operators $T_{r's'}$. An arbitrary operator $\exp(K)$ may then be factored

$$\exp(K) = \exp(K'')\exp(K') \tag{237}$$

where the K'' operator is a linear combination of T_{rs} operators which are disjoint from the $T_{r's'}$ operators included in the K' operator. This left coset representation of a full group element as a product of two subgroup elements is the operator version[6,38] of the matrix expression in Eq. (229). This factorization then gives a result analogous to Eq. (236).

$$\exp(K)|0\rangle = \sum_n \exp(K'')\exp(K')|n\rangle c_n \tag{238}$$

$$= \sum_m \sum_n \exp(K'')|m\rangle\langle m|\exp(K')|n\rangle c_n \tag{239}$$

$$= \sum_m \exp(K'')|m\rangle c'_m \tag{240}$$

This shows that any orbital variation induced by the components of the K operator involving the subset operators $T_{r's'}$ are redundant with the mixing coefficient variations and may be eliminated from the optimization process. Eq. (240) shows that the wavefunction does depend on the expansion coefficients and on the other orbital variations for which Eq. (232) is not satisfied.

The subset of operators satisfying Eq. (232) is usually known before the MCSCF optimization process is initiated. In fact, many CSF generation methods, including the graphical unitary group approach, allow the specification of the operators of Eq. (232) or the orbital subsets that define these rotation operators. The corresponding orbital rotation parameters may then be deleted from the MCSCF optimization process for any state and for any geometry for wavefunctions expanded in this CSF space. The direct product expansion spaces (including the important FORS/CASSCF and RCI expansions) are of this type where the redundant orbital rotations are known during the CSF space specification.

B. Wavefunction-dependent Redundant Variables

In the consideration of redundant variables so far, only those cases that are determined from the structure of the wavefunction, i.e. the CSF expansion space, have been considered. During the MCSCF optimization process and while performing MCSCF calculations at various points on a potential energy surface, there may occur additional redundant variables. For example, suppose that Eq. (232) is not satisfied for any orbital pair for the CSF expansion space. At some geometries, however, the wavefunction expansion may reduce to only a few terms and, within these non-vanishing expansion CSFs, Eq. (232) is satisfied for a set of orbital rotations. This occurs frequently when considering large bond lengths where the wavefunction reduces to a much simpler product of the fragment wavefunctions. Another common example occurs when computing the MCSCF wavefunction at points along some path which crosses a point of higher symmetry. At this particular geometry, a large fraction of the expansion coefficients may vanish because they contribute to states of the wrong symmetry.

In these cases, an active orbital may become doubly occupied, in which case the rotations between this orbital and other doubly occupied orbitals become redundant, or an active orbital may become unoccupied, in which case the rotations between this orbital and the other virtual orbitals become redundant. Another possibility is that the MCSCF wavefunction reduces to a high-spin RHF wavefunction and the orbital rotations among these high-spin singly occupied active orbitals become redundant. The dissociation of a singlet wavefunction into two triplet fragments is a common occurrence (e.g. the ethylene dissociation discussed previously). More extreme examples include the localized orbital description of the dissociation of N_2 into two $^4S(p^3)$ fragments or the dissociation of Cr_2 into two $^7S(s^1d^5)$ fragments (although in these cases spatial symmetry restrictions would eliminate the majority of these orbital rotation parameters anyway).

In these cases, the identification of the redundant variables resulting from the CSF expansion space alone does not guarantee that all convergence problems due to redundant variables will be avoided. It then becomes necessary to be able to determine the redundant variables for the approximate wavefunction during each iteration of the MCSCF optimization process. A general method of performing this redundant variable search has been proposed by Hoffmann *et al.*[121] and consists of numerically determining the linear dependences of the first-order variational space. This is equivalent to determining the infinitesimal operators that leave the wavefunction unchanged. Redundant variables of the type described by Eq (232) must result in linear dependences of the first-order variational space but there may also be other types of linear dependences. The first-order variational space of a trial wavefunction $|0\rangle$ is spanned by the union of the set of expansion CSFs $\{|n\rangle : n = 1, \ldots, n_{CSF}\}$ and the set of single excitation states $\{T_{rs}|0\rangle : r > s\}$. The actual wavefunction dependence (i.e. the CSF expansion coefficient dependence) enters from the last set of terms. This is, of course, the expansion space used in the variational super-CI method as described in Section III. A linear dependence in this space is given by the solutions to the equation

$$\sum_n |n\rangle c'_n + \sum_{r>s} T_{rs}|0\rangle \kappa'_{rs} = 0 \tag{241}$$

This equation is obviously satisfied by the operators that satisfy Eq. (232) and will also be satisfied by the redundant variables resulting from the wavefunction reductions discussed above. If the wavefunction reduces to a direct product of full CI expansions of orbital subsets, then the solution of Eq. (241) may be reduced to the simpler form of Eq. (232) involving only a single T_{rs} operator

$$T_{rs}|0\rangle = -\sum_n |n\rangle c'_n = \sum_n |n\rangle\langle n|T_{rs}|0\rangle \tag{242}$$

where the CSF summation involves only terms that contribute to the wavefunction. (In this case, the same wavefunction could have been computed within this reduced CSF expansion space initially.) Of course, the more general redundant operators that satisfy Eq. (232) also satisfy the wavefunction-dependent Eq. (242) and, therefore, satisfy the linear dependence condition, Eq. (241). Wavefunction-dependent orbital rotation operators that satisfy Eq. (242) may be factored according to Eqs (238) through (240). Wavefunctions involving an active orbital that becomes doubly occupied or unoccupied will result in a set of T_{rs} operators that produce zero when operating on the reference wavefunction. Wavefunctions involving high-spin states of fragments will result in a set of operators, those involving excitations among the high-spin orbitals, that produce zero when operating on the reference wavefunction. In all of these cases, the redundant orbital rotation operators may be eliminated from the MCSCF variational space.

For small wavefunction expansions where an explicit determinantal representation of the wavefunction may be constructed, it is straightforward to determine the solutions to Eq. (241) by constructing the overlap matrix of the first-order variational space. This is required in the solution of the variational super-CI equations for which these linear dependences must be explicitly identified and eliminated[26,64,66,69,71]. For larger CSF expansion lengths or large orbital basis sets, however, this step could easily dominate the entire iterative procedure, so alternate methods must be examined. The solutions to Eq. (241) may be determined[121] by first operating from the left with a CSF expansion term to give

$$c'_n = -\sum_{r>s} \langle n|T_{rs}|0\rangle \kappa'_{rs} \tag{243}$$

This equation is consistent with Eq. (242) when the above summation involves only a single term. Substitution of Eq. (243) into Eq. (241) eliminates the terms c'_n from the equation. The expansion coefficients κ'_{rs} may then be determined by operating from the left with the set of single excitation states to give

$$0 = \sum_{r>s}\left(\langle 0|T_{uv}T_{rs}|0\rangle - \sum_{n}\langle 0|T_{uv}|n\rangle\langle n|T_{rs}|0\rangle\right)\kappa'_{rs} \tag{244}$$

$$= \sum_{r>s} A_{uv,rs}\kappa'_{rs} \tag{245}$$

The matrix **A** in Eq. (245) is constructed from the two-particle density matrix contributions and the one-particle transition density matrix contributions for the current trial function $|0\rangle$. The dimension of this matrix is determined only by the number of active–active orbital rotations and is therefore relatively small for most problems. This is because the redundant variables resulting from an active orbital becoming either doubly occupied or empty may be determined simply by examining the one-particle density matrix. The determination of the set of degenerate eigenvectors corresponding to zero eigenvalues of the matrix **A** is straightforward using either direct or iterative methods. For each of these vectors $\boldsymbol{\kappa}'^{(i)}$, for $i = 1$ to n' (the rank of the degeneracy), there is a corresponding vector $\mathbf{c}'^{(i)}$ determined from Eq. (243).

The solution of equation systems that occur during the MCSCF iterative process may become ill-conditioned if redundant variables are allowed to be included in these equations. One example of this is the inversion of the Hessian matrix in the Newton–Raphson iterative method. If redundant variables are included in the optimization process, this Hessian matrix becomes singular and the method becomes undefined. This is discussed in detail in the next section. The solutions to Eq. (245) and to Eq. (243) may be used to overcome this problem by solving the Newton–Raphson equations in the orthogonal complement space to the vectors $\boldsymbol{\kappa}'^{(i)}$ and $\mathbf{c}'^{(i)}$. Equivalently, if the Newton–Raphson equations are solved using an iterative subspace expansion method, the trial vectors may be constrained to this orthogonal complement space. The imposition of this constraint requires only simple vector operations during the micro-iterative procedure to ensure orthogonality of the trial vectors to the vectors $\boldsymbol{\kappa}'^{(i)}$ and $\mathbf{c}'^{(i)}$. As long as the vectors are treated in this form, only the vector space spanned by $\boldsymbol{\kappa}'^{(i)}$ and $\mathbf{c}'^{(i)}$ is required. If information concerning the detailed structure of the redundancy of the wavefunction variations is required, these vectors spanning the degenerate subspace of the matrix **A** in Eq. (245) must be transformed among themselves to determine the solutions that have the form of Eq. (242).

In addition to the wavefunction reduction cases mentioned previously, another type of linear dependency may also be encountered. These are solutions to Eq. (241) that may not be cast in the form of Eq. (242). Consider a CSF expansion space that consists of the two CSFs, $|a\rangle$ and $|b\rangle$, with orbital occupations $\phi_1^2\phi_2\phi_3^2\phi_4$ and $\phi_1\phi_2^2\phi_3\phi_4^2$. It may be verified that $T_{12}|a\rangle + T_{34}|b\rangle = 0$ and that $T_{12}|b\rangle + T_{34}|a\rangle = 0$ but that neither of these orbital rotation operators alone satisfies Eq. (232) or the wavefunction-dependent Eq. (242). In other words, there is no structural redundancy in the orbital variation space. However, a solution to Eq. (241) exists if the two expansion coefficients are equal in magnitude, in which case

$$T_{12}|0\rangle \pm T_{34}|0\rangle = 0 \tag{246}$$

where the sign depends on the expansion coefficient signs. There is no simple coset factorization for this type of linear combination of orbital rotation generators. This implies that these types of solutions do not satisfy equations analogous to Eqs (238) through (240) in which either of these orbital rotations may be eliminated. Such accidental linear dependences within the first-order variational space may also cause convergence problems during the MCSCF iterative procedure.

In summary of the various types of redundant variables, the solutions to Eq. (241) are the most general that occur within the MCSCF method. Of the types of solutions to Eq. (241), some may be of the form of Eq. (232). This type of redundant variable is usually known prior to the optimization process and may be eliminated at that time from the wavefunction variation space. Other solutions to Eq. (241) may be cast in the form of Eq. (242). These redundant variables result from the simplification of the wavefunction. This simplification can occur at geometries corresponding to non-interacting fragments and at high-symmetry geometries. A particular case of such a wavefunction-dependent variable occurs when the right-hand side of Eq. (242) vanishes entirely for some set of orbital rotation parameters. All of these redundant variables result from the structure of the CSF expansion space (or the structure of the space of non-vanishing contributions to the wavefunction). The remaining redundant variables that satisfy Eq. (241) are determined numerically from Eq. (245). This latter type of variational parameter, which cannot be identified as one of the previous types, should be eliminated during the solution for the wavefunction correction vector within the particular MCSCF iteration in which it occurs but not for all subsequent iterations.

C. Properties of Redundant Variables

With a formal understanding of various types and causes of redundant variables, more of the properties of redundant variables may be determined. Consider first the redundant variables that satisfy Eq. (232) and the wavefunction reduction cases that satisfy Eq. (242). Substitution of Eq. (242) into the expression for the orbital gradient term gives

$$w_{rs} = 2\langle 0|HT_{rs}|0\rangle = \sum_n 2\langle 0|H|n\rangle\langle n|T_{rs}|0\rangle \tag{247}$$

$$= 2\langle 0|H|0\rangle\langle 0|T_{rs}|0\rangle + \sum_{n\neq 0} 2\langle 0|H|n\rangle\langle n|T_{rs}|0\rangle \tag{248}$$

$$= 0 \tag{249}$$

where the inner-projection basis is chosen as the MCSCF wavefunction and its orthogonal complement instead of the individual CSFs. The first term in Eq. (248) is zero because the one-particle density matrix is symmetric. The second term is zero if the current wavefunction expansion coefficients are an eigenvector of the current Hamiltonian matrix. If the expansion coefficients are not exactly an eigenvector of the Hamiltonian matrix, the magnitude of the gradient term w_{rs} (or equivalently the degree of symmetry of the Fock matrix elements F_{rs} and F_{sr}) may be used as a measure of the accuracy of these coefficients. The identification of zero gradient terms for an unconverged wavefunction (to within numerical accuracy) has been commonly used as a method of determining at least some of the redundant variables of an MCSCF calculation[37,45,46].

Next consider two rotation operators, T_{rs} and T_{uv}, that satisfy Eq. (242). Substitution of Eq. (242) into the expression for the blocks of the Hessian matrix, under the assumption that the current CSF expansion coefficients are an eigenvector of the current Hamiltonian matrix, gives

$$B_{rs,uv} = \sum_n 2(E^n - E^0)\langle n|T_{rs}|0\rangle\langle n|T_{uv}|0\rangle \tag{250}$$

$$C_{uv,n} = 2(E^n - E^0)\langle n|T_{uv}|0\rangle \tag{251}$$

$$C_{rs,n} = 2(E^n - E^0)\langle n|T_{rs}|0\rangle \tag{252}$$

$$M_{n,n'} = 2(E^n - E^0)\delta_{n,n'} \tag{253}$$

In these equations it is also assumed that the orthogonal complement states have been chosen to diagonalize the Hamiltonian matrix. This is merely a formal convenience and is not necessary for the validity of the results of this section. Substitution of these identities into the expression for the partitioned orbital Hessian matrix gives

$$(\mathbf{B} - \mathbf{C}\mathbf{M}^{-1}\mathbf{C}^{t})_{rs,uv} = 0 \tag{254}$$

This shows that all of the matrix elements of the partitioned orbital Hessian matrix involving redundant orbital rotation operators vanish. In particular the diagonal elements involving such operators vanish. This diagonal element relation, along with the gradient relation of Eq. (249), has been used by the author to identify redundant variables during the iterative MCSCF optimization procedure. (Note that Eqs (254) and (249) only demonstrate necessary conditions and are not sufficient to prove that a particular orbital rotation variable is redundant. It is highly unlikely, however, that both conditions would be satisfied simultaneously by mere numerical coincidence.) This method is only applicable to those cases where the partitioned orbital Hessian matrix is explicitly constructed. If the equation systems involving the partitioned orbital Hessian matrix are solved iteratively in an unfolded form, these matrix elements are not available for examination[57,58].

Consider next the elements of a row of the partitioned orbital Hessian matrix that corresponds to a redundant orbital rotation variable. In this case the orbital Hessian matrix elements may be written as

$$B_{rs,xy} = \tfrac{1}{2}\langle 0|[[H, T_{rs}], T_{xy}] + [[H, T_{xy}], T_{rs}]|0\rangle \tag{255}$$

$$= \langle 0|[[H, T_{xy}], T_{rs}] + \tfrac{1}{2}[H, [T_{rs}, T_{xy}]]|0\rangle \tag{256}$$

where the operator T_{rs} is redundant and the operator T_{xy} is an arbitrary operator. Substitution of Eq. (242) into the first commutator term of Eq. (256) results in the exact cancellation of the $\mathbf{CM}^{-1}\mathbf{C}^{t}$ terms. The second commutator term of Eq. (256) reduces to gradient elements, giving the elements of a row of the partitioned orbital Hessian matrix as

$$(\mathbf{B} - \mathbf{CM}^{-1}\mathbf{C}^{t})_{rs,xy} = w_{ry}\delta_{sx} + w_{sx}\delta_{ry} + w_{xr}\delta_{sy} + w_{ys}\delta_{rx} \tag{257}$$

This shows that the elements in the row are zero unless there is a coincidence of orbital labels in which case the Hessian elements reduce to a subset of the elements of the gradient vector. Eq. (257) reduces to Eq. (254) when the operator T_{xy} is also redundant. The Weinstein residual norm theorem applied to this row shows that the magnitude of an eigenvalue of the partitioned orbital Hessian matrix is bounded by the norm of the gradient vector. This eigenvalue is non-zero when the gradient vector elements are non-zero and approaches zero as the gradient vector elements approach zero during the MCSCF optimization process.

The matrix partitioning method may be used to show that the eigenvalues of the partitioned orbital Hessian matrix that are clustered around zero bound the corresponding eigenvalues of the wavefunction Hessian matrix. In particular, a small positive eigenvalue of the partitioned orbital Hessian matrix corresponds to a positive eigenvalue of the wavefunction Hessian matrix with smaller magnitude, and a small negative eigenvalue of the partitioned orbital Hessian matrix corresponds to a negative eigenvalue of the wavefunction Hessian matrix with smaller magnitude. This means that as an eigenvalue of the partitioned orbital Hessian matrix approaches zero, a corresponding eigenvalue of the wavefunction Hessian matrix also approaches zero and is smaller in magnitude. This bound property, which is also valid for eigenvalues that do not approach zero, may be the reason that iterative procedures based on the partitioned orbital Hessian matrix sometimes display better numerical stability than methods based on the wavefunction Hessian matrix. However, iterative methods based on either Hessian matrix may display convergence problems because of numerical instabilities resulting from these near-zero eigenvalues if the redundant variables causing them are not eliminated.

These results are also valid when the current CSF expansion coefficients are not an exact eigenvector of the Hamiltonian matrix, provided the error of these coefficients is less than the magnitude of the gradient vector elements. In this case, the elements of the partitioned orbital Hessian matrix and gradient vector that should be zero are instead small non-zero numbers. The magnitude of these elements is a measure of the CSF coefficient vector error. If this error is constrained to be less than the magnitude of the gradient vector elements during each MCSCF iteration, the eigenvalues of the partitioned orbital Hessian matrix and of the wavefunction Hessian matrix still approach zero. This analysis also shows that the orbital corrections calculated during each MCSCF iteration are only as accurate as the CSF expansion coefficients. When micro-iterative methods are used in both the CSF expansion coefficient evaluation and in the orbital transformation evaluation during each MCSCF iteration, these conditions determine the maximum error that should be allowed in each of these steps that will result in second-order convergence in the iterative procedure.

Consider next the solutions of the linear dependence condition, Eq. (241), that cannot be written in the form of Eq. (242). Operation from the left by $\langle 0|H$ shows that one of the gradient elements may be written as a linear combination of the other gradient elements. If Eq. (241) remains valid for each MCSCF iteration of a convergent sequence, the orbital rotation variable may be eliminated from the calculation and the corresponding gradient term will still approach zero because it is a linear combination of gradient terms that are approaching zero. It is possible, however, that Eq. (241) may not be valid for subsequent iterations, in which case the orbital rotation variable would be explicitly required for convergence. Consider the previous example in which Eq. (246) is satisfied for some MCSCF iteration. When this occurs, $w_{12} \pm w_{34} = 0$ for this iteration. Since T_{12} and T_{34} commute in this case, it may also be shown that $B_{12,12} = B_{34,34} = \pm B_{12,34}$ and $C_{12,n} = \pm C_{34,n}$. Since one row of the Hessian matrix is a multiple of another, the Hessian matrix has a zero eigenvalue and an infinite number of solutions exist for the Newton–Raphson equations. However, during subsequent iterations, the mixing coefficients may not have the same magnitude and none of these conditions will be repeated. An MCSCF iterative procedure must be sufficiently flexible to handle such accidental linear dependences while allowing the correct wavefunction solution to be found during subsequent iterations.

VI. IMPLEMENTATION OF MULTICONFIGURATION SELF-CONSISTENT FIELD METHODS

In this section, various features of the implementation of MCSCF optimization procedures are discussed. The relation between computer hardware and algorithm design is first discussed in general terms. The advantages of methods that allow a hierarchical program design with primitive matrix and vector operations at the lowest levels are emphasized. Some specific features of the implementation of the MCSCF optimization methods discussed in Section III are examined. The methods that use the explicitly constructed Hessian matrix are discussed first. It is for these methods that the solution of the various second-order MCSCF optimization procedures is most straightforward. Because of the dimensions of the blocks of the Hessian matrix, these methods are directly applicable to small and moderate size MCSCF calculations (e.g. less than about 50 MOs and less than about 500 CSFs). Then the methods that determine the wavefunction corrections using micro-iterative methods are discussed. These methods are applicable both to the efficient solution to moderate size MCSCF calculations and to MCSCF calculations involving larger CSF expansions (up to 10^5 CSFs) for which the various blocks of the Hessian matrix cannot be explicitly constructed and stored. The elimination of the formula files for gradient and Hessian construction, the elimination of the coupling coefficient list for CSF expansion vector optimization and density matrix construction, and the exploitation of density matrix sparseness are important considerations in the implementation of general second-order MCSCF computer programs. An attempt is made to present sufficient detail so that various approaches may be meaningfully compared.

A. General Considerations of Computer Implementations

Before the details of the implementation of MCSCF methods are discussed, it is useful to introduce a few general computer programming concepts. Modern computers may be classified in several ways[123] depending on size, cost, capabilities, or architecture. One such classification divides computers into scalar and vector machines. Scalar computers (e.g. the VAX 11/780) perform primitive arithmetic operations such as additions and multiplications on pairs of arguments. Vector computers (e.g. the CRAY X-MP and CYBER-205) have, in addition to scalar operations, vector instructions[124–129] which allow operations to be performed directly on vectors of floating-point arguments. A vector operation may be simulated with a series of scalar operations, but for sufficiently long vectors the use of the vector instruction is much more efficient than the sequence of scalar operations. Owing to the overheads associated with a vector instruction, longer vector operations are also typically more efficient than a series of shorter vector operations[123,124]. A particular type of scalar machine is a parallel–pipelined computer (e.g. the FPS-164). Such a computer is characterized[123,125] by the ability to perform several scalar operations simultaneously (parallel) and to initiate a new set of such operations before the results of a previous set have been completed (pipelined). One use of such an instruction set is to perform primitive vector operations. As with a vector computer, such vector operations may be performed more efficiently than the equivalent set of sequential–non-pipelined scalar operations. The considerations of algorithm design that are common to vector and parallel–pipelined scalar computers will be discussed in most detail in this section because the largest performance differences between well and poorly designed implementations will be observed for these types of computers[124].

It follows from this discussion not only that the optimal choice of MCSCF optimization procedure is machine-dependent, but that the details of the optimal implementation of that method would also be machine-dependent. Since the programming effort is substantial for the implementation of any MCSCF method, it is unlikely that an optimal implementation of an optimally chosen method will ever be realized. However, it is possible to achieve a near-optimal implementation of reasonable MCSCF methods on a variety of computers. This is achieved by recognizing that certain operations, such as the primitive vector operations discussed above, are efficiently performed on a wide class of computer types. A reasonable MCSCF method may then be chosen that may be implemented using this set of vector operations. One criterion of a reasonable MCSCF method would therefore be that its implementation has a hierarchical design, with the vast majority of the arithmetic operations required in the method being performed within a set of low-level machine-dependent subprograms. These low-level subprograms may then be assumed to be coded in an optimal way for each target machine. Other criteria for reasonable MCSCF methods involve the amount of memory required, the computer time and memory requirements as the orbital basis, CSF expansion length, or number of electrons increases, and the ratio of I/O (input/output to external storage) time to arithmetic processing time.

These considerations result in a set of fairly well accepted rules for computer implementations of computational algorithms that have been published in various forms by several researchers[123,124,126]. These guidelines include the use of a standard set of vector operations. One widely accepted standard set of vector operations is the basic linear algebra subprograms (BLAS), which are available in FORTRAN and are also available in efficient machine-dependent assembly code on many computers[130]. (Since FORTRAN is the most widely used language for scientific programming, it will be assumed when language dependence is discussed.) The subprograms SDOT (to compute the standard-precision dot product of two vectors) and SAXPY (to compute the standard-precision result of the multiplication of a scalar constant by a vector and its addition to another vector) are probably the most frequently used subprograms from this library. Both of these subprograms involve systematic memory references and they involve one multiplication and one addition for each vector element. These features allow these vector operations to be efficiently implemented on both parallel–pipelined scalar computers and vector computers (although the relative performance for these two operations will generally vary from machine to machine). When such standard vector operations are not suitable, it is important that the innermost DO loops should involve no logic (IF statements, GOTO statements, etc.) and should involve systematic memory references. (Indirect addressing should also be avoided where possible although it is generally preferable to the use of logical operations.) This increases the chance that the innermost DO loops may be implemented efficiently on various target machines[123–125].

There are many operations that involve matrices that may be formulated either as a series of inner product (using SDOT) or outer product (using SAXPY) vector operations. For example, matrix–vector products, matrix–matrix products and matrix factorization may be formulated using either method[130]. It follows that a series of vector operations may in turn be equivalent to such a matrix operation. Since it is desirable to perform these operations most efficiently, it is useful to design MCSCF methods that replace such series of vector operations with the equivalent matrix operation. This usually involves storing intermediate quantities, such as integrals or density matrix elements, in a particular order in the computer memory. The advantage of this hierarchical approach is that any machine-dependent code (possibly involving choices between SDOT and SAXPY formulations) may be assumed to be in the low-level library subprograms. The particular characteristics of these subprograms may be ignored in the formulation of the MCSCF method.

Many matrix operations in the MCSCF method involve matrices with dimensions determined by the size of the orbital basis. For an example of the efficiency improvements possible for such matrices, both the CRAY X-MP and the FPS-164 can compute such a matrix–matrix product using specially designed matrix library routines about four times faster than when the straightforward FORTRAN formulation is used. (On the FPS-164, the use of the library routine SDOT improves the timing over the straight FORTRAN inner product code by less than a factor of 2. On the CRAY X-MP, the straight FORTRAN outer product and library SAXPY perform equally, a consequence of the template recognition of the compiler. A technique called 'loop unrolling' improves the FORTRAN performance in both cases but is also machine-dependent[130].) The use of matrix-oriented routines allows algorithms to be designed without explicitly specifying the inner or outer product formulation. Of course, there may be other reasons to specify one formulation over the other. Some of these will be discussed in the following sections.

Another important rule, which is particularly applicable to implementations using modern supercomputers and attached processors, is to avoid unnecessary I/O. The time required to transfer each floating-point number is often equivalent to the time required to perform between 40 and 200 arithmetic operations[125,127]. This has important consequences on the choice of algorithms in the implementation of the MCSCF method. In most cases the elimination of I/O is accompanied by increased memory requirements and, in extreme cases, by an increased amount of CPU effort. The elimination of the storage of the coupling coefficient list in favor of the repeated computation of the coupling coefficients is a successful example of this trade-off that has already been mentioned. Another example is the 'direct SCF' approach[127] in which the integral I/O is eliminated in favor of the repeated integral evaluation during the orbital optimization procedure. It is not uncommon on modern computers to have several million words of available memory. Newer computers are becoming available (e.g. the CRAY-2 and the ETA-10 computers) that have several hundred million words of memory. The efficient use of such resources is a continuing challenge to computational chemists.

B. Second-order Local Convergence

The equations required for second-order local convergence for SCF wavefunctions and for MCSCF wavefunctions with fixed CSF expansion coefficients have been available in various forms for some time. Some of these optimization methods involved an exponential parametrization of the orbital rotation parameters[92,131–133] while others relied on the more traditional approach of introducing Lagrange multipliers[105,134,135] to enforce the orthogonality of the MO basis. However, the solution of these equations generally was not implemented. This was due in large part to the limited availability of computational resources and, in particular, to the larger memory requirements of the second-order methods. It was also due to the somewhat justified belief that the second-order solution of the orbital part of the problem, while completely neglecting the response of the CSF expansion coefficients to the orbital change, would result in an expensive overall first-order convergent process that would not compete favorably with other familiar first-order methods.

It was the more recent exponential operator formulation of the orbital variations of Dalgaard and Jørgensen[6] that provided the insight necessary to include correctly the orbital-state coupling terms required for true local second-order convergence of the MCSCF wavefunction. The initial implementations of this formalism, by Yeager and Jørgensen[16] and by Dalgaard[17], demonstrated that second-order local convergence could in fact be obtained. These initial implementations, which were necessarily somewhat inefficient and lacked generality, have provided the impetus for much of the recent activity in the development of the MCSCF method. Much effort has been devoted to the development of efficient, general-purpose, MCSCF programs that display second-order local convergence and that converge reasonably well in the non-local regions of wavefunction variation space. There are several features of this recent activity that are not obvious from examination of the equations of Section III, except perhaps in hindsight, that are discussed in the following sections.

The MCSCF optimization equations presented in Section III may be used in the implementation of either first-order[16,77,92] or second-order[16,17,37,45,46] convergent methods. The advantages of first-order procedures have been discussed previously[73,77]. However, there are several advantages that favor using second-order convergent methods. These include the ability to converge the wavefunction to very high accuracy with a reasonable amount of effort. Although the error of the MCSCF energy depends on the square of the wavefunction error, the error of other properties is directly proportional to the wavefunction error. The evaluation of many properties, such as the geometrical derivatives of the energy, are simplified if the gradients vanish exactly because the Hellmann–Feynman theorem is satisfied for such fully converged MCSCF wavefunctions[136]. Also important is the evaluation of response-dependent properties with finite difference methods. In these cases it is necessary to have highly converged results so that the small differences of nearly equal property values are sufficiently accurate. The evaluation of vibrational frequencies using finite differences of analytic gradients involves similar considerations. A second-order MCSCF method is capable of converging wavefunctions to high accuracy (e.g. 12–16 significant digits) with only one or two additional iterations beyond those required for normal chemical accuracy of the energy (equivalent to 3–4 significant digits in the wavefunction). Another advantage of using a second-order method, particularly one that allows the Hessian matrix (or its eigenvalue spectrum) to be examined, is that various convergence difficulties, convergence to spurious solutions, identification of multiple solutions and root-flipping difficulties are more readily recognized and corrected. Finally it should be mentioned that by changing computational algorithms from a slowly convergent first-order method to a rapidly convergent second-order method, it may be possible to reduce the overall amount of computer resources required to achieve convergence. Although the effort required during each iteration is increased, this may be more than compensated by the reduced number of iterations required. This is particularly true for supercomputer implementations if the amount of I/O is significantly decreased by the algorithm change.

C. Elimination of the Hessian Matrix Formula File

As discussed in Section III, the variational super-CI optimization methods are formally very similar to the second-order methods discussed in this review. A significant difference, however, is the fact that in the latter the gradient and Hessian elements may be constructed without a separate formula file (which contains the list of coupling coefficients, matrix element indices and integral indices) as is required in the construction of the super-CI Hamiltonian matrix. In the SCI implementation of Ruedenberg *et al.*[66], the SCI Hamiltonian matrix is constructed by applying a rectangular congruence transformation to a large Hamiltonian matrix constructed in the basis of single excitations from the individual CSFs. In an implementation of the variational SCI method by the present author[69], the SCI Hamiltonian matrix was constructed by implicitly performing this contraction during the construction step. In either case, this approach becomes impracticable as the number of expansion CSFs increases and as the dimension of the orbital basis increases. This limits the generality and the overall usefulness of the variational super-CI method[65,66,70,71].

The most efficient process that eliminates the formula tape must also use the one- and two-particle density matrix elements as indicated in the equations of Section III. The advantages resulting from the elimination of the Hessian matrix formula tape and the explicit use of the density matrix elements were recognized by Yaffe and Goddard[131] and by Siegbahn *et al.*[72] in their first-order convergent procedures but the initial implementations of general second-order MCSCF procedures did not fully exploit these possibilities. In an early implementation by the present author[37] of a second-order MCSCF method, which used a loop-driven unitary group approach, only the contributions to a density matrix element arising from a single loop of a Shavitt graph were summed and used in the Hessian construction step. This resulted in more arithmetic operations than necessary, although a significant improvement was realized over the earlier SCI approach[69] in this respect. This partial density matrix approach also eliminated the need for the separate formula file for the Hessian construction step[37].

In an independent second-order MCSCF implementation by Lengsfield[45], the density matrix was explicitly constructed and stored. It was correctly assumed that the number of occupied orbitals is sufficiently small for most MCSCF wavefunctions so that the two-particle density matrix may be stored in computer memory. This results in a significant improvement in reducing the number of arithmetic operations over the earlier SCI approaches. However, this program still employed a formula file in the construction of the Hessian matrix consisting of the list of integral indices and Hessian matrix indices associated with each density matrix element.

These initial shortcomings were quickly remedied and efficient general second-order MCSCF programs were subsequently used in several laboratories. In the case of the programs developed by the present author[19,41], the 'index-driven' unitary group approach, discussed in Section II, replaced the 'loop-driven' method, allowing efficient construction of the density matrix and transition density matrix elements.

To see how the formula tape may be eliminated in favor of an efficient computational method involving vector and matrix operations, consider the construction of the orbital gradient elments, or equivalently the construction of the Fock matrix **F**, as defined in Eq. (159). The one-electron terms are simply the result of the matrix–matrix product (**hD**) and will not be discussed in detail. The two-electron terms may be computed in several ways, as indicated in the following expressions:

$$F_{xs} = \sum_t \sum_u \sum_v g_{xtuv} d_{stuv} \tag{258}$$

$$= \sum_t \sum_{(uv)} (2 - \delta_{uv}) g_{xtuv} d_{stuv} \tag{259}$$

$$= \sum_{(uv)} (2 - \delta_{uv}) \left(\sum_t g_{xt}^{uv} d_{st}^{uv} \right) \tag{260}$$

These expressions, as written, all involve matrix–matrix multiplications. Eq. (258) may be considered to consist of the matrix **g** of dimension n by m^3 and the matrix **d** of dimension m by m^3 where n is the MO basis dimension and m is the number of occupied orbitals. Eq. (259) exploits the index symmetry of the integrals and density matrix to reduce the u and v indices to the unique subset (uv). The factor $(2 - \delta_{uv})$ is a redundancy factor to account for those terms that are not explicitly included relative to Eq. (258). Eq. (260) decomposes the matrix multiplication of Eqs (258) and (259) into a summation of matrix products of smaller dimension. The computation of the matrix **F** needs no formula file (which in this case might contain the list of integrals and Fock matrix elements associated with a particular density matrix element) since the memory references are well defined provided the integrals and density matrices are stored appropriately. The drawback of all of these expressions is that the integrals and density matrix elements are not usually stored in memory in such a way to allow any of these matrix multiplications to take place. In fact, to do so would increase the storage requirements of the matrix **d** by about a factor of 4 over that required when the index permutational symmetry is used to full advantage. These elements must be rearranged, and expanded, to take advantage of this matrix multiplication formulation. This rearrangement involves 'gather' and 'scatter' operations. It is therefore Eq. (260) that is usually implemented for the Fock matrix construction step since additional space for only two small matrices for each new (uv) combination is required. For each (uv), this rearrangement involves manipulation of nm elements of the matrix $\mathbf{g}^{uv}$ and m^2 elements of the matrix $\mathbf{d}^{uv}$

(ignoring the effects of spatial symmetry blocking of these arrays). The subsequent matrix multiplication involves $2nm^2$ arithmetic operations so that in most cases the $(nm + m^2)$ overhead associated with these rearrangements is insignificant compared to the actual computational work that is performed. Even with this small amount of overhead, this approach is much more efficient than a formula file approach which would be characterized by I/O and possibly some unpacking overhead associated with *each* arithmetic operation. Since there are $m(m+1)/2$ unique combinations of the combined index (uv), the Fock matrix construction requires roughly $nm^3(m+1)$ total arithmetic operations or, in other words, it is an m^5 process with the coefficient determined by the ratio n/m.

A common feature of both the gradient vector and Hessian matrix element construction is that the elimination of the formula file in favor of efficient matrix and vector operations must be accompanied by a particular arrangement of the integrals and density matrix elements. This sometimes requires an expansion of these quantities, with some elements stored redundantly, in order to simplify the memory references within these vector operations. Care must be taken during the computational algorithm design that this overhead does not become a significant fraction of the total effort.

D. Density Matrix Sparseness

Many of the wavefunction expansion forms discussed in Section IV result in sparse density matrices $\mathbf{D}$ and $\mathbf{d}$. This was discussed for the ERMC wavefunction and its subsets but it is also true for other direct product type expansion spaces. With the RCI expansion, for example, the matrix $\mathbf{D}$ consists of 2×2 blocks along the diagonal; the orbitals of each block are those associated with an electron pair. The matrix $\mathbf{d}$ is also sparse because of the orbital subspace occupation restrictions and the non-zero elements consist of those with two orbital indices belonging to one electron pair and with the other two indices corresponding to another (or the same) electron pair. It would be beneficial if this density matrix sparseness could be exploited when it exists. However, this must be done in such a way as to avoid any restrictions on the types of wavefunction expansions allowed.

A general way of exploiting this sparseness in both the gradient and the Hessian construction involves the use of 'outer product' algorithms to perform the matrix element assembly. In the case of the matrix multiplications required in the $\mathbf{F}$ matrix construction, this simply means that the innermost DO loop is over x in Eq. (260). (If t were the innermost DO loop, the result would be a series of dot products or an 'inner product' algorithm.) When an outer product algorithm is used, the magnitude of the density matrix elements may be tested and the innermost DO loop is only performed for non-zero d_{stuv} elements. (In the case of Hessian matrix construction, the test may occur outside of the two innermost DO loops.) In this way, the total number of arithmetic operations required to construct the matrix **F** is reduced by the fraction of non-zero density matrix indices. This often allows the **F** matrix to be constructed with much fewer (e.g. only about m^3 in the case of PPMC and RCI expansions) arithmetic operations with no loss of generality in the MCSCF program.

E. Simplifications Due to Doubly Occupied Orbitals

Another important feature in the construction of the gradient and Hessian elements is the elimination of the density matrix elements associated with the doubly occupied orbitals of the wavefunction. This is most important for transition-metal calculations, but it is also important for other calculations involving second- and third-row elements. There are two types of doubly occupied orbitals that occur in the MCSCF method: those that are frozen prior to the MCSCF step (e.g. at a simpler SCF level), and those that are optimized at the MCSCF level. Frozen orbitals may be eliminated completely from the MCSCF calculation by modifying the one-electron integrals and by including a frozen-core contribution to the diagonal elements of the Hamiltonian matrix (or equivalently by adjusting the nuclear repulsion term). This is a common practice in CI calculations and will not be discussed here. The elimination of the remaining density matrix elements involving the doubly occupied orbitals from the gradient and Hessian construction steps follows from the identities:

$$D_{ij} = 2\delta_{ij} \tag{261}$$

$$d_{ijpq} = 2D_{pq}\delta_{ij} \tag{262}$$

$$d_{ipjq} = -\tfrac{1}{2}D_{pq}\delta_{ij} \tag{263}$$

$$d_{ijkl} = 4\delta_{ij}\delta_{kl} - \delta_{ik}\delta_{jl} - \delta_{il}\delta_{jk} \tag{264}$$

In these expressions the indices i, j, k and l are used for doubly occupied orbital labels while p and q are used for active orbital labels. Other density matrix element types such as D_{ip}, d_{ipqr} and d_{ijkp} are all zero. The above expressions may also be used to simplify the transformation of the density matrices from one orbital basis to another as is required, for example, in the evaluation of MCSCF molecular properties in the AO basis. The use of these identities also eliminates the need for any coupling coefficients involving the doubly occupied orbitals.

Substitution of Eqs (261) through (264) into the equations for the gradient vector and Hessian matrix elements allows these expressions to be simplified. These simplifications may be maximally exploited if the gradient vector is partitioned into sub-blocks corresponding to active–double (ad), active–

active (aa), virtual–double (vd) and virtual–active (va) orbital combinations. The other gradient blocks, double–double and virtual–virtual, are identically zero by virtue of the fact that these rotations correspond to redundant variables. This also results in the partitioning of the coupling matrix **C** into the four corresponding blocks and in the partitioning of the orbital Hessian matrix into 10 separate blocks corresponding to adad, aaad, aaaa, vdad, vdaa, vdvd, vaad, vaaa, vavd and vava orbital labels. In the case that all active–active orbital rotations are redundant, the gradient vector contains only the remaining three non-zero blocks and only the six corresponding blocks of the orbital Hessian are required. This occurs for FORS/CASSCF wavefunctions as discussed in Section V.

A general MCSCF program must allow for the cases where some of the active–active orbital rotations are redundant and others are essential. The columns and rows of the Hessian matrix corresponding to these redundant variables do not need to be computed. This elimination may be performed with no loss of efficiency by storing the various blocks of the Hessian matrix in such a way that the most slowly varying indices are associated with active–active rotation variables both in the storage of these Hessian blocks and in their construction. This leaves the doubly occupied orbital and virtual orbital labels as the more rapidly varying array indices. For maximal efficiency, this Hessian block construction should also exploit the density matrix sparseness by using outer product assembly methods. The details of this matrix construction for all of the Hessian blocks have been discussed elsewhere[122,137] and will not be fully reproduced here.

It is useful to consider one example which demonstrates both the elimination of the Hessian matrix formula tape and the simplifications resulting from the elimination of the doubly occupied orbital components of the density matrices. Consider the vaad block of the orbital Hessian matrix. An element of this matrix block is the second derivative of the energy with respect to two orbital rotations, one of the virtual–active type and the other of the active–double type. The matrix elements are given by

$$B_{arip} = 2\tilde{h}_{ai}D_{pr} - F_{ai}\delta_{pr} - \sum_t 2(4g_{atip} - g_{aipt} - g_{apit})D_{rt} + \sum_u \sum_v (2g_{aiuv}d_{pruv} + 4g_{aviu}d_{purv}) \tag{265}$$

where the elements $\tilde{\mathbf{h}}$ are the modified one-electron Hamiltonian matrix elements

$$\tilde{h}_{xy} = h_{xy} + \sum_j (2g_{xyjj} - g_{xjyj}) \tag{266}$$

In Eq. (265) the indices p, r, t, u and v correspond to active orbitals, a to a virtual orbital and i to a doubly occupied orbital. In Eq. (266), the labels x and y correspond to general indices and the summation is over doubly occupied orbitals only. The combination of integrals $(4g_{atip} - g_{aipt} - g_{apit})$ occurs frequently in the MCSCF method and is called a P-supermatrix element. One example of an efficient construction algorithm results when the Hessian elements are stored in a four-index array in such a way that the virtual orbital label a is the most rapidly varying index, followed by the doubly occupied orbital label i, followed by the two active orbital labels r and p. Such a storage scheme is denoted by $(\mathbf{B}^{rp})_{ai} = B_{arip}$. The essential features of the present discussion may be understood by neglecting the complications in the array addressing that result from spatial symmetry. It may also be assumed that the two-electron integrals that contribute to this Hessian block are divided into two types, g_{aiuv} and g_{aviu}. These integrals may also be stored so that the indices a and i are the most rapidly varying, then followed by u and v. These arrays are called the 'direct' and the 'exchange' sorted integral lists and are denoted as $({}^{\mathrm{D}}\mathbf{g}^{(uv)})_{ai} = g_{aiuv}$ and $({}^{\mathrm{X}}\mathbf{g}^{vu})_{ai} = g_{aviu}$. The exchange list is roughly twice as long as the direct list because of the (uv) permutational symmetry present in the latter.

The P-supermatrix elements may be formed from the direct and exchange integral lists for each orbital pair, t and p, and stored as $({}^{\mathrm{P}}\mathbf{g}^{tp}) = 4({}^{\mathrm{X}}\mathbf{g}^{tp}) - ({}^{\mathrm{X}}\mathbf{g}^{pt}) - ({}^{\mathrm{D}}\mathbf{g}^{(tp)})$. A DO loop over r then gives a series of coefficients $2D_{rt}$, of this complete matrix as contiguous sequences of Hessian elements are updated with SAXPY operations, $\mathbf{B}^{rp} \leftarrow \mathbf{B}^{rp} - 2D_{rt}({}^{\mathrm{P}}\mathbf{g}^{tp})$. The direct and exchange integral lists are processed in a similar manner using the two-particle density matrix elements. Both of the loops over a and i are combined to give an effective vector length equal to the product of the number of virtual orbitals and the number of doubly occupied orbitals. The density matrix elements may be checked prior to this long vector operation in this outer product algorithm and the combined innermost DO loop may be skipped when appropriate. For the vaad Hessian block, it is most convenient if the ${}^{\mathrm{P}}\mathbf{g}$ elements are computed in blocks from the ${}^{\mathrm{D}}\mathbf{g}$ and ${}^{\mathrm{X}}\mathbf{g}$ sorted integral lists. In other cases, where ${}^{\mathrm{D}}\mathbf{g}$ and ${}^{\mathrm{X}}\mathbf{g}$ are not explicitly required, it is most efficient to compute ${}^{\mathrm{P}}\mathbf{g}$ during the integral sort step, thereby reducing the I/O and memory requirements during the Hessian construction step.

All 10 of the blocks of the orbital Hessian matrix may be constructed in a similar manner[122]. Each block should be examined individually to determine the best assembly strategy. This usually suggests the most appropriate storage scheme of the Hessian and of the contributing integrals. All but one of the blocks of the orbital Hessian may be constructed with outer product algorithms that allow the density matrix sparseness to be exploited. The only exception is the aaaa block. In this case, it is not possible to use an outer product routine and simultaneously avoid the use of indirect addressing because of the possibility of redundant active–active rotations. Fortunately, the aaaa block of the Hessian matrix is usually very small and its construction, using either an indirect addressing method or an inner product method, is not prohibitively inefficient.

F. Transition Density Matrix Construction

The construction of the blocks of the coupling matrix **C** is performed most straightforwardly in the CSF basis as discussed in Section III. The full list of transition density matrices $\mathbf{D}^n$ and $\mathbf{d}^n$ is quite large and, even for CSF expansions of only a few hundred in lengths, cannot be held in memory. There are two approaches that have been proposed for the construction of the coupling matrix **C**, both of which involve an orderly computation of the transition density matrices. The first such approach employs the 'index-driven' unitary group approach as discussed in Section III. With this approach the vectors of transition density matrices, for a fixed set of orbital indices, are computed for all of the CSFs simultaneously. For example, for the orbital indices p, q, r and s, three vectors are constructed simultaneously from the coupling coefficients with elements denoted as d_n^{pqrs}, d_n^{prqs}, and d_n^{psqr}. These vectors are then used to update the elements of the four blocks of the matrix **C**.

For the more restrictive CSF expansion spaces, such as PPMC and RCI expansions, it occurs that entire transition density vectors will vanish for particular combinations of orbital indices. It is most convenient if a logical flag is set during the construction of the three vectors, one flag for each vector, to indicate that it contains non-zero elements. This avoids the effort required to check each element individually for these zero vectors. The updates of the elements of the matrix **C** that result from a particular transition density vector involve two DO loops: one over the CSF index, which determines the second subscript of the matrix **C**, and the other over an orbital index, which, combined with a density matrix orbital index, is used to determine the first subscript of **C**. Either choice of the ordering of these two loops results in an outer product matrix assembly method.

With the first method, in which the innermost DO loop is over the CSF index, it is not convenient to check the magnitude of each individual element of the density vector. Only the sparseness indicated by the zero-vector flag is exploited with this approach. With the second method, in which the innermost DO loop is over the orbital label, the density vector sparseness may be fully exploited. However, this loop ordering may also require either indirect addressing or the computation of extraneous contributions to elements that correspond to redundant orbital rotation variables, and it may require that some integral blocks be expanded and stored redundantly in memory. The extraneous contributions are subsequently discarded but only after the additional arithmetic operations have been performed. The effective vector length in this case is n, the orbital basis dimension (or some subset of this basis), while with the first method the vector length is the CSF expansion length. The optimal construction strategy therefore depends on the relative sparseness of the density vectors, the occurrence of redundant orbital rotation variables, the lengths of the CSF expansion and of the orbital basis, and the amount of additional memory required for the expanded integral list. The optimal choice may vary depending on which block of the matrix **C** is being assembled since it is usually only the active–active block for which redundant variables must be eliminated.

Another approach to the **C** matrix construction is a 'CSF-driven' approach proposed by Knowles *et al.*[42]. With this approach, the density matrix elements d^n_{pqrs} are constructed for all combinations of orbital indices p, q, r and s, but for a fixed CSF labeled by n. Each column of the matrix **C** is constructed in the same way that the Fock matrix **F** is computed except that the arrays $\mathbf{D}^n$ and $\mathbf{d}^n$ are used instead of **D** and **d**. As with the **F** matrix construction described earlier, there are two choices for the ordering of the innermost DO loops. One choice results in an inner product assembly method while the other choice results in an outer product assembly method. The inner product choice, which does not allow the density matrix sparseness to be exploited, results in SDOT operations of length m or about m^3, depending on the integral storage scheme. The outer product choice, which does allow the density matrix sparseness to be exploited, has an effective vector length of n, the orbital basis dimension. However, like the second index-driven method described above, this may involve some extraneous effort associated with redundant orbital rotation variables in the active–active block of the **C** matrix.

It is useful to compare these approaches when applied to a wavefunction expansion that results in a sparse density matrix. For example with a PPMC expansion, each $\mathbf{d}^n$, with about $\frac{1}{8}m^4$ possible unique elements, contains only about $\frac{1}{8}m^2$ non-zero elements: $m(m+2)/8$ non-zero elements of the type d^n_{rrss} and $m/2$ non-zero elements of the type d^n_{rsrs}. For $m = 20$ the matrix $\mathbf{d}^n$ is only 0.29% non-zero. The inner product CSF-driven approach is clearly not suited for the sparse transition density matrix resulting from this type of wavefunction. The outer product CSF-driven approach does account for the density vector sparseness but the effective vector length is only n, the orbital basis dimension.

With the index-driven approach, m vectors will be processed of the type $\mathbf{d}^{rrrr}$ which are 50% non-zero, $m(m-2)/2$ vectors of the type $\mathbf{d}^{rrss}$ which are 25% non-zero, and $m/2$ vectors of the type $\mathbf{d}^{rsrs}$ which are 100% non-zero. For $m = 20$ the ratio of the number of non-zero elements processed to the total number of vector elements processed is approximately 0.31. Thus, even in the case of PPMC wavefunctions, which result in a very sparse set of transition density matrices, almost all of the sparseness is accounted for with the vector flag approach. All of this remaining sparseness may be accounted for with the second index-driven approach but, like the outer product CSF-driven approach, the effective vector length is only n. While the first index-driven approach does not account for as much of the sparseness, it has the advantage that the effective vector length is much larger (the CSF expansion length as opposed to the MO basis dimension) and, with more general CSF expansions, it avoids unnecessary operations associated with redundant orbital rotations. For large MO basis sets, dense transition density vectors and no redundant orbital rotations, all of these assembly methods should be comparable. For general wavefunction expansions, the index-driven approach is probably the best choice since it is flexible enough to allow the density vector sparseness to be exploited when necessary and it allows long vector operations when these vectors are predominantly non-zero while avoiding any additional effort associated with redundant orbital rotation variables.

G. Elimination of the Coupling Coefficient List

It is possible to eliminate the I/O associated with reading the list of coupling coefficients required in the MCSCF method if these elements are computed repeatedly as required during the optimization process. These coupling coefficients are required in basically two different places in the MCSCF method: the computaton of matrix–vector products involving the Hamiltonian matrix and a trial vector, and in the computation of density and transition density matrices for a given reference wavefunction. In the case of the index-driven unitary group approach, the coupling coefficients are computed from a shorter list of Shavitt loop values. Each loop value is then associated with the set of upper and lower walks which determine the set of CSF indices. For a particular Shavitt loop, the contributions to a matrix–vector product result in the following sequence of FORTRAN code:

```
      DO 1 LW = 1, NLW
        B0 = YB + R(Z + LW)
        K0 = YK + R(Z + LW)
          DO 1 UW = 1, NUW
1           HV(B0 + UW) = HV(B0 + UW) + T * V(K0 + UW)        (267)
```

The parameters NLW and NUW are the number of lower walks and upper walks for a particular Shavitt loop. (The mapping vector R(*) is loop-independent and gives the correspondence between the full set of upper and lower walks.) The walk offsets YB and YK are determined by the loop shape, and the coefficient T is a linear combination of the loop values and integrals associated with the Shavitt loop. Since only unique Shavitt loops are constructed, the innermost DO loop must be repeated with the bra and ket values interchanged, resulting in even more arithmetic operations for each Shavitt loop. The transition density matrix construction involves an analogous DO loop structure with the last statement replaced by

```
1   D(B0 + UW) = D(B0 + UW) + E * C(K0 + UW)        (268)
```

where D(*) is the transition density matrix vector associated with the loop value E and C(*) is the vector of CSF mixing coefficients. Since each loop is associated with one, two or three loop values, the transition density matrix construction requires somewhat more effort than that for a matrix–vector product. The construction of matrix–vector products and of the transition density matrix vectors are very similar and both involve SAXPY type vector operations. Although this is an efficient innermost DO loop structure, there are two shortcomings to this approach. First the construction of the loop values E is not particularly efficient for vector computers, although the recursive nature of this step does allow these scalar operations to be amortized over numerous Shavitt loops. Secondly, the number of upper and lower walks for each loop is often relatively small. Even the use of symmetry-dependent chaining indices[87,122], a significant improvement over the earlier symmetry-dependent distinct row approach[37], results in values typically in the range of 1 to 10. (Notable exceptions are direct product expansions involving several orbital subspaces which produce higher values for these quantities.) In order to approach maximal efficiency on vector computers, one or both of these features must be addressed in future implementations of the index-driven unitary group approach.

Siegbahn[4] has recently proposed a method that circumvents the explicit construction of the two-particle coupling coefficients. For expansions for which Eq. (232) is applicable, these coupling coefficients may be constructed from the one-particle coupling coefficients using Eq. (117):

$$\langle n|E_{pq}E_{rs}|m\rangle = \sum_k \langle n|E_{pq}|k\rangle\langle k|E_{rs}|m\rangle \tag{269}$$

(The remaining generator term in Eq. (117) is simpler to handle and is not discussed here.) This two-particle contribution to a matrix–vector product may be computed as

$$(\mathbf{Hv})_n = \tfrac{1}{2}\sum_k \left\{ \sum_p \sum_q \langle n|E_{pq}|k\rangle \left[\sum_r \sum_s g_{pqrs} \left(\sum_m \langle k|E_{rs}|m\rangle v_m \right) \right] \right\} \tag{270}$$

where the order of computation proposed by Siegbahn[4] is indicated by the bracketing. This approach requires all of the one-particle coupling coefficients $\langle n|E_{pq}|k\rangle$ to be available for fixed k and all p, q and n. For each k, this set of coefficients is then used twice, once in the sparse summation with the vector $\mathbf{v}$, and once in the sparse summation over p and q. These sparse summations which require indirect addressing are not particularly efficient on vector computers. The efficiency of this approach results from the intermediate summations, over r and s, which are dense and are computed efficiently on vector computers (as matrix-vector products) provided the integrals $\mathbf{g}$ are stored appropriately, and because the sparse summations are only computed for the one-particle coupling coefficient list. In the initial implementation of this approach, the required coupling coefficients were computed and stored on an external file. This list of one-particle coefficients is of course much shorter than the full set of one- and two-particle coefficients and results in a drastic reduction in the I/O requirements and in the number of sparse vector operations compared to the usual approach. Even this I/O could be eliminated by using a CSF-driven unitary group approach such as that proposed by Shavitt[40].

A similar approach has been implemented by Knowles and Werner[81]. This method uses determinantal expansion terms instead of CSFs. Although the expansion lengths are longer with this approach, as discussed in Section II, the calculation of the one-particle coupling coefficients is simplified so that these terms may be repeatedly computed as needed during the iterative method. All of the I/O associated with coupling coefficients is eliminated with this approach and the only sparse vector operations are those involving the one-particle coupling coefficients. Like the CSF approach of Siegbahn, however, this method is only efficient for full CI type expansions. If the expansion is not full within the active orbitals, or if spatial symmetry has been used to reduce the expansion length, then the expansion terms $|k\rangle$ of Eq. (269) must include terms outside of the expansion space. The full CI expansion lengths given in Eq. (201) or shown in Table I indicate the potential dependence of the dimension of the expansion space $|k\rangle$ on the number of active orbitals.

H. Direct Solution of the Wavefunction Correction Equations

For relatively small CSF expansion lengths and orbital basis sets, the blocks of the Hessian matrix and gradient vector may be explicitly computed and the appropriate equation may be solved directly to determine the wavefunction corrections for the subsequent MCSCF iteration. For example, the solution of the linear equations required for the WNR and PNR methods may use the stable matrix factorization methods found in the LINPACK library[130]. The routines from this library are often available in efficient machine-dependent assembly code for various computers. Even if this is not the case, the FORTRAN versions of these routines use the BLAS library, resulting in the efficient execution of the primitive vector operations. Similar routines are also available for the direct solution of the eigenvalue equations required for the WSCI and PSCI methods.

All of these direct solutions require a linearly independent basis to define the orthogonal complement space to the reference wavefunction. Initial implementations of second-order MCSCF methods[37,45,46] typically used the Hamiltonian eigenvectors for this expansion basis. This has the advantage of producing a diagonal state Hessian matrix **M** but it has the disadvantage that the explicit transformation of the **C** matrix (which is most efficiently computed in the CSF basis[37]) to this orthogonal complement basis is very time-consuming and that the complete diagonalization of the Hamiltonian matrix itself may also be quite time consuming. A simpler approach involves the projected basis as described in Section III. The problem with the use of this basis, with respect to the direct solution of the MCSCF wavefunction correction equations, is that it is linearly dependent and the matrix **M** is therefore singular in this basis. This linear dependence may be removed simply by deleting a term from this basis. An optimal choice of this expansion term is to delete the CSF that corresponds to the largest expansion coefficient. The deletion of a term corresponding to a small expansion coefficient might result in a numerical linear dependence. The rectangular transformation from the CSF basis to this modified projected basis is a matrix of the form $(\mathbf{1}' - \mathbf{c}'\mathbf{c}^{t})$ where $\mathbf{1}'$ is the unit matrix with the kth row deleted and $\mathbf{c}'$ is the mixing coefficient vector with the kth element deleted. Because of the simple form of this transformation matrix, the transformation of the matrices **M** and **C** to this modified projected basis only involves matrix–vector operations instead of matrix–matrix operations. Of course, with this basis, as opposed to the eigenvector basis, the matrix **M** is no longer diagonal. The factorization of the matrix **M**, required explicitly in the direct solution of the PNR and PSCI equations and implicitly in the direct solution of the WNR and WSCI equations, is more than compensated by the fact that only a single eigenvector of the Hamiltonian matrix is required.

I. Iterative Solution of the Wavefunction Correction Equations

As the dimension of the blocks of the Hessian matrix increases, it becomes more efficient to solve for the wavefunction corrections using iterative methods instead of direct methods. The most useful of these methods require a series of matrix–vector products. Since a square matrix–vector product may be computed in $2N^2$ arithmetic operations (where N is the matrix dimension), an iterative solution that requires only a few of these products is more efficient than a direct solution (which requires approximately N^3 floating-point operations). The most stable of these methods expand the solution vector in a subspace of trial vectors. During each iteration of this procedure, the dimension of this subspace is increased until some measure of the error indicates that sufficient accuracy has been achieved. Such iterative methods for both linear equations and matrix eigenvalue equations have been discussed in the literature[12,138,139].

A slight modification of these methods has also been used for the iterative solution of the PSCI equation[57,58]. As discussed in Section III, the PSCI iterative method has several advantages over the WNR and WSCI iterative methods, the latter of which may be implemented using the standard linear equation and matrix eigenvalue equation methods. It is most convenient to use the unfolded form of Eq. (195) for this iterative solution. It is assumed here that the CSF coefficient vector is an exact eigenvector of the Hamiltonian matrix. The extension to approximate eigenvectors is straightforward. Given an approximate solution, a residual vector is defined by

$$\begin{pmatrix} \mathbf{x}_\kappa \\ \mathbf{x}_p \end{pmatrix} = \begin{pmatrix} \mathbf{B} - \lambda\mathbf{1} & \mathbf{C} & \mathbf{w} \\ \mathbf{C}^t & \mathbf{M} & \mathbf{0} \end{pmatrix} \begin{pmatrix} \boldsymbol{\kappa} \\ \mathbf{p} \\ 1 \end{pmatrix} \tag{271}$$

where λ is the approximate PSCI eigenvalue. The components of the approximate solution $(\boldsymbol{\kappa}^t \quad \mathbf{p}^t \quad 1)^t$ are defined in terms of an expansion basis as

$$\boldsymbol{\kappa} = \mathbf{R}\mathsf{k} \tag{272}$$

$$\mathbf{p} = \mathbf{S}\mathsf{p} \tag{273}$$

where the columns of the matrices $\mathbf{R}$ and $\mathbf{S}$ are the individual expansion vectors for the orbital rotation correction vector and for the CSF correction vector, respectively. The optimal expansion coefficients in this subspace are determined from the subspace representation of the PSCI secular equation:

$$\begin{pmatrix} (\mathsf{B} - \mathsf{C}\mathsf{M}^{-1}\mathsf{C}^t) - \lambda\mathbf{1} & \mathsf{w} \\ \mathsf{w}^t & -\lambda \end{pmatrix} \begin{pmatrix} \mathsf{k} \\ 1 \end{pmatrix} = \begin{pmatrix} \mathbf{0} \\ 0 \end{pmatrix} \tag{274}$$

$$\mathsf{p} = -\mathsf{M}^{-1}\mathsf{C}^t\mathsf{k} \tag{275}$$

where the subspace representations of the various blocks of the Hessian matrix and gradient vector are defined as

$$\mathsf{w} = \mathbf{R}^t\mathbf{w} \tag{276}$$

$$\mathsf{B} = \mathbf{R}^t\mathbf{B}\mathbf{R} \tag{277}$$

$$\mathsf{C} = \mathbf{R}^t\mathbf{C}\mathbf{S} \tag{278}$$

$$\mathsf{M} = \mathbf{S}^t\mathbf{M}\mathbf{S} \tag{279}$$

The dimensions of these subspace representations are small enough so that Eqs (274) and (275) may be solved using direct linear equation an matrix eigenvalue equation methods. The computation of the residual vector uses the same matrix–vector products as is required for the construction of the subspace representations of the Hessian blocks. If either of the components of the residual vector are too large, a new expansion vector corresponding to the large residual component may be computed from one of the equations

$$\mathbf{r} = (\mathbf{B}_d - \lambda\mathbf{1})^{-1}\mathbf{x}_\kappa \tag{280}$$

$$\mathbf{s} = \mathbf{M}_d^{-1}\mathbf{x}_p \tag{281}$$

where $\mathbf{B}_d = \mathrm{diag}(B_{11}, B_{22}, \ldots)$ and $\mathbf{M}_d = \mathrm{diag}(M_{11}, M_{22}, \ldots)$. The new expansion vector $\mathbf{r}$ is then orthonormalized to the previous expansion vectors and appended to the matrix $\mathbf{R}$ and the new expansion vector $\mathbf{s}$ is orthonormalized to the previous expansion vectors and appended to the matrix **S**. It is also convenient at this stage to enforce any required orthonormalization constraints which result from redundant orbital rotation variables, particularly the wavefunction-dependent redundant variables discussed in Section V, that have not already been eliminated prior to the Hessian matrix construction step.

The micro-iterative procedure described here reduces to the standard iterative method proposed by Pople *et al.*[138] and by Purvis and Bartlett[139] for linear equation solutions if the λ-dependent first and third rows of Eq. (195) are neglected. This procedure also reduces to the standard Davidson iterative method[12] for matrix eigenvalue equation solutions if the λ-independent second row of Eq. (195) is neglected (and if intermediate normalization is used for the eigenvector). The convergence of the PSCI equation would therefore be expected to be similar to the convergence of these closely related methods. In particular, these methods converge only linearly. Although the MCSCF iterative procedure is a second-order convergent procedure, the micro-iterative solution required within each MCSCF iteration is only a first-order convergent process. The advantages of formulating the iterative procedure in terms of an approximate energy expression, instead of applying a first-order convergent method directly to the exact energy expression, are twofold. First, the approximate energy expressions are simpler than the exact infinite-order energy expression and the solution of the resulting equations is numerically better behaved. Additionally each micro-iteration of this procedure is much cheaper than an iteration of a first-order process on the exact energy expression. This is because only a small subset of integrals is required for a micro-iteration with the approximate energy expression while the complete integral list must be processed for the exact energy expression.

The matrix–vector products required in the PSCI method are the same as those required in the WNR and WSCI iterative methods. The matrix–vector products involving the **S** expansion vectors may be computed in any of the orthogonal complement bases discussed in Section III. Since the matrix **M** is no longer explicitly factored with the micro-iterative solution methods, it is no longer necessary to use a linearly independent orthogonal complement basis as described in the previous section. The simpler linearly dependent projected basis may be used instead with this method. An even simpler approach is to use the overcomplete CSF basis to represent the orthogonal complement space. The trial expansion vectors may then be constrained to lie within the orthogonal complement space during the micro-iterative procedure. The explicitly projected versions of the matrix–vector products, e.g. Eq. (181), should be used in these cases.

There are two additional features of the iterative solution of the PSCI equations that also have important implications for other MCSCF methods.

The first is that the vector **p** may be used as an expansion vector in the iterative solution of the Hamiltonian matrix diagonalization of the subsequent MCSCF iteration[57,58]. This vector is the first-order correction to the CSF expansion coefficients for the perturbaton induced in the Hamiltonian operator by the orbital change defined by $\boldsymbol{\kappa}$. This is particularly important for larger CSF expansions where the effort involved in the individual matrix–vector products during the diagonalization is a substantial fraction of the total MCSCF optimization. The second is that the partitioning of the Hessian matrix into blocks determined by orbital rotation parameters and by CSF variational parameters allows the dimension of the subspaces to vary independently. This is particularly useful for large CSF expansions where the addition of a new **s** vector requires much more computational effort than the addition of an **r** vector. The former requires the computation of **Ms** and **Cs**, which requires the coupling coefficient processing, while the latter only requires **Br** and $\mathbf{r}^t(\mathbf{CS})$ for the existing (**CS**). For a given dimension of the **S** matrix, the $\boldsymbol{\kappa}$ vector may be relaxed more fully before a new **s** expansion vector is computed. For large CSF expansions this partitioning results in significant improvements in computational efficiency[81].

For large CSF expansions, the matrix–vector products required in the iterative solution of the MCSCF equations may use the expansion-vector-dependent effective Hamiltonian and transition density matrix techniques discussed in Section III. The particular method used to compute the matrix–vector products, of course, has no effect on the convergence properties of the micro-iterative solution of the wavefunction correction vector. It becomes most importnt for large CSF expansions to minimize the number of direct CI type matrix–vector products. In addition to the partitioning of the wavefunction variation space described above, it may also be useful to perform multiple orbital transformations for fixed CSF expansion coefficients[82]. Since the density matrices depend only on the expansion coefficients, they may be used for several such iterations. These simpified iterations involve only the computation of new orbital Hessian and gradient vector elements. For excited states it may be necessary to compute at least a few expansion vectors **S** using the PSCI iterative method described above to prevent the orbitals form changing in such a way as to allow the wavefunction to acquire the character of the lower-energy states. This is an undesirable consequence of the incomplete incorporation of the coupling between the orbital and CSF variations into the iterative procedure and is remedied by the additional **S** expansion vectors. If the vectors **S** were to span the orthogonal complement space fully, all of the coupling of the PSCI method would be incorporated into the micro-iterative procedure. Further research is required before efficient methods are found that result in the optimal compromise between rigorous second-order convergence and the minimization of the number of expensive matrix–vector products for large CSF expansions.

J. Scaling and Level-shifting Modifications

The comparison of Eqs (196) and (187) shows that the solution of the PSCI equation may be regarded as a modified PNR solution in which the eigenvalues of the effective Hessian matrix have been subjected to a uniform shift relative to the exact Hessian matrix. The WSCI and WNR methods have an analogous relationship[19,61]. For ground-state calculations, in which the Hessian matrix is positive definite, this level shift serves to reduce the size of the step taken in the orbital variation space compared to the unshifted Newton–Raphson step size. This occurs because the eigenvalues are all shifted uniformly in the positive direction, increasing their magnitude. This curvature increase always moves the predicted energy minimum closer to the reference point in these cases.

For excited-state calculations and for the initial iterations for ground states, for which the Hessian matrix may have negative eigenvalues, this level shift may not always reduce the step size sufficiently to force convergence. The WSCI and PSCI approximate energy expressions are valid within some neighborhood of the reference wavefunction, but they would not be expected to provide an adequate representation of the energy outside of this local region. When such overshooting occurs, there are a few simple modifications that may be applied to the equations which are useful in producing smaller steps that are at least approximately in the correct direction.

One approach is to scale the solution vector prior to its use in defining the orbital transformation matrix. It may be shown[19,140] that the scaled vector

$$\boldsymbol{\kappa}' = \alpha \boldsymbol{\kappa} \tag{282}$$

will result in second-order local convergence provided $(1 - \alpha) = O(|\boldsymbol{\kappa}|^2)$. For small steps, this scaling will approach unity sufficiently rapidly so as not to hinder the local convergence properties. One such scaling choice is equivalent to using the usual eigenvector normalization instead of the intermediate normalization of Eqs (194) or (193). For the WSCI method, this results in the choice $\alpha = (1 + |\mathbf{p}|^2 + |\boldsymbol{\kappa}|^2)^{-1/2}$, and for the PSCI method it results in the choice $\alpha = (1 + |\boldsymbol{\kappa}|^2)^{-1/2}$. An even more conservative choice, suggested by Banerjee[140], is of the form $\alpha = (1 + |\boldsymbol{\kappa}|^2)^{-1}$. In general, the more conservative choices will result in smaller steps, in somewhat slower convergence and in less chance of divergence due to overshooting. It should be noted that the choice of scaling parameter has no effect on the direction of the wavefunction correction vector but changes only its magnitude.

Another simple approach is to include an additional level-shift parameter in the PSCI or WSCI solution[19,65]. For the modified PSCI equation, the orbital corrections are determined from the matrix eigenvalue equation

$$\begin{pmatrix} (\mathbf{B} - \mathbf{C}\mathbf{M}^{-1}\mathbf{C}^{t}) - \lambda \mathbf{1} & \mathbf{w} \\ \mathbf{w}^{t} & \Delta - \lambda \end{pmatrix} \begin{pmatrix} \boldsymbol{\kappa} \\ 1 \end{pmatrix} = \begin{pmatrix} \mathbf{0} \\ 0 \end{pmatrix} \tag{283}$$

instead of Eq. (194). The level-shift parameter Δ is assumed to be a small negative number. For both ground and excited states with the PSCI method, this parameter may always be chosen to be sufficiently large in magnitude to force $|\boldsymbol{\kappa}|$ to be small. This is because the partitioned orbital Hessian matrix must eventually become positive definite and even a small negative Δ usually suffices to produce a small $|\boldsymbol{\kappa}|$. For ground-state calculations, similar considerations apply to the WSCI method. For excited states, however, there may not exist a small Δ that results in a small $|\boldsymbol{\kappa}|$. For this reason the use of level-shift parameters is most appropriate for methods based on the partitioned orbital Hessian matrix.

It may be shown that if the level-shift parameter satisfies the condition $\Delta = O(|\mathbf{w}|)$, then second-order convergence will not be affected in the neighborhood of the final wavefunction[19]. A conservative iterative method could always use such a level-shift parameter to improve the non-local convergence characteristics without suffering any deterioration of the second-order local convergence. Such an approach has also been used by the author to improve the convergence of the micro-iterative procedure in those cases in which there are several small eigenvalues of the Hessian matrix. In contrast to the vector scaling described above, the use of a level-shifting parameter changes not only the magnitude of the wavefunction correction vector but also its direction.

One particularly interesting choice of this level-shifting parameter is based on the trust radius concept. A trust radius[78-80] is a measure of the size of the neighborhood about the reference wavefunction within which the exact energy expression is adequately represented by the approximate energy expression. During the micro-iterative solution of the wavefunction correction vector, the level-shifting parameter may be adjusted to produce a $|\boldsymbol{\kappa}|$ within this trust radius. As implemented by Golab *et al.*[80], the trust radius is adjusted in each MCSCF iteration. This adjustment is based on the ratio of the predicted energy lowering and the actual energy lowering at each MCSCF iteration. When this ratio is close to unity, the approximate energy expression is giving an adequate prediction of the exact energy dependence and the trust radius is either unchanged or increased. When this ratio is different from unity, indicating that the approximate energy expression is inadequate within the current trust radius, then the trust radius is decreased and, conditionally, the wavefunction correction vector of the previous iteration is recomputed with the smaller trust radius. By repeatedly reducing the trust radius when necessary, it has been shown that the sequence of MCSCF iterations must converge to a local minimum for ground-state calculations. Unfortunately, this process may involve many MCSCF iterations, many of which may have a small trust radius, and many integral transformations and energy evaluations that are essentially discarded[60]. This results in an inefficient, although stable, optimization procedure. The trust radius concept is useful, however, and it may be used in other parts of the MCSCF calculation such as within the micro-iterative and extended micro-iterative procedures and in conjunction with other scaling and level-shifting modifications of the iterative procedures[81].

K. Approximate Hamiltonian Operator Methods

One promising approach to the problem of effectively reducing the number of direct CI matrix–vector products is the approximate Hamiltonian operator method of Werner and coworkers[46,81,82] described in Section III. This is an extended micro-iterative method in which the Hamiltonian operator is allowed to be approximated during the solution of the wavefunction correction vector within an MCSCF iteration.

In the original procedure of Werner and Meyer[46] the orbital-state coupling is included using a partitioned Hessian matrix approach and, like the PSCI and PNR methods, the orbitals are determined from an energy expression that depends explicitly only on the orbital transformation. The orbital transformation is computed from a Fock matrix equation that is improved during the micro-iterative procedure to take account of the changes in the approximation of the Hamiltonian operator. The original micro-iterative method allowed the approximate Hamiltonian operator to be updated sequentially for every orbital rotation[46]. Using this micro-iterative method to solve the resulting non-linear optimization equation, only space for a few small matrices is required in memory but the integrals must be read in each micro-iteration (just as in a normal SCF iteration). Alternatively these integrals, the same subset as required for the Hessian matrix construction, may be held in memory during the micro-iterative process. The solution of the wavefunction corrections using this first-order convergent process within the MCSCF iteration sometimes requires 20–30 micro-iterations to converge the solution of the non-linear optimization equation[141].

The later procedure of Knowles and Werner[81] uses instead an augmented Hessian method to define the orbital corrections. An approximate Hamiltonian operator is constructed that accounts for the simultaneous change of the entire vector of orbital rotations. This procedure may be summarized by the following steps:

1. For the current MO basis, transform the required one- and two-electron integrals exactly. These integrals define the exact Hamiltonian operator H. For this operator, update the current CSF expansion vector $\mathbf{c}$.
2. For the current $\mathbf{c}$, construct the density matrices $\mathbf{D}$ and $\mathbf{d}$. From the current exact $H, \mathbf{D}$ and $\mathbf{d}$, construct $\mathbf{B}$ and $\mathbf{w}$ and solve for $\boldsymbol{\kappa}$ using an augmented orbital Hessian method. This orbital correction vector $\boldsymbol{\kappa}$ defines the transformation matrix $\mathbf{U}$.

3. Use **U** and **T** to define the approximate Hamiltonian operator $H^{(2)}$.
4. Use $H^{(2)}$ to construct an approximate Hessian matrix and gradient vector $\mathbf{B}^{(2)}$ and $\mathbf{w}^{(2)}$ and to solve for $\boldsymbol{\kappa}^{(2)}$. If $\boldsymbol{\kappa}^{(2)}$ is sufficiently large, then update **U** and cycle to step 3.
5. Use $H^{(2)}$ to update the vector **c**. If $\Delta\mathbf{c}$ is sufficiently large, then compute **D** and **d** and cycle to step 3.
6. If **U** is sufficiently different from a unit matrix, then cycle to step 1.

The micro-iterative procedure involves both the cycle between steps 4 and 3 and the cycle between steps 5 and 3. The micro-iterative procedures used to solve the augmented orbital Hessian matrix eigenvalue equations are not specifically indicated. The MCSCF iteration is the cycle between steps 6 and 1. In the micro-iterative cycles, the orbital corrections are calculated without the CSF coupling being included explicitly. This coupling is included in the cycle between steps 5 and 3 in which the density matrices are updated to reflect the changes of the CSF coefficients. At the beginning of step 6, a self-consistent orbital transformation matrix **U** and CSF vector **c** have been found for the sequence of approximate Hamiltonian operators $H^{(2)}$ generated at step 3. Since this operator is correct through second order in **U**, the MCSCF iterations display second-order convergence. Knowles and Werner[81] found it convenient to incorporate level shifting and trust radius constraints in steps 2 and 4 within the micro-iterative solution of the orbital correction vector. Since any overstepping is corrected within the micro-iterative procedure, the MCSCF iterations display improved stability.

There are two variations on the above iterative process that have been proposed. In the case of small CSF expansion lengths it is appropriate to perform several **c** updates within an MCSCF iteration[82]. This is because the expensive step of the MCSCF optimization process in this case is the two-electron integral transformation. The approximate transformations performed in step 3 of the micro-iterative procedure are less expensive than the exact transformations performed in step 1 of the MCSCF iteration. However, in the case of large CSF expansions, these updates should be avoided. This is because the exact transformation becomes a small part of the total iteration effort and it is preferable to perform the expensive CSF coefficient updates only with exact Hamiltonian operators. In this case step 5 and the cycle between steps 5 and 3 is not performed. Continuing this reasoning further, it may even be useful to perform several MCSCF iterations with the same CSF vector **c** to allow the orbitals to relax further for the exact Hamiltonian operators.

Werner and Knowles[82] have performed these exact transformations for a fixed CSF vector **c** but restricted to the active–active block of the orbital transformation matrix only. This transformation is performed using the same subset of integrals as required for the rest of the micro-iterative procedure, namely the integrals that contribute to the Hessian matrix. These additional transformation steps may then be performed without any additional integral I/O and may be included in step 3 of the above procedure instead of explicitly adding a new micro-iterative cycle. When the wavefunction is not invariant to these rotations, the explicit inclusion of these third- and fourth-order contributions to the approximate Hamiltonian operator allows a more accurate orbital transformation matrix to be determined. The approximate Hamiltonian operator is still only correct through second order and the resulting MCSCF iteration process displays second-order convergence. An important benefit of this approach is that the approximate energy expression regains the correct invariance to rotations between doubly occupied orbitals and the energy dependence is more correctly approximated for highly occupied orbitals than in the unmodified procedure involving the second-order approximate Hamiltonian operator of Eq. (200). This aspect of the approximate Hamiltonian operator method has been discussed by Werner and Knowles[82].

As discussed in Section III, the uncoupled solution of the orbital correction vector of steps 2 and 4 in the above procedure may not converge correctly for certain excited-state calculations (e.g. those states with negative orbital Hessian matrix eigenvalues but for which the energy is minimized with respect to orbital variations). To correct this deficiency, the orbital correction vector could be computed within the extended micro-iterative procedure using the PSCI or WSCI augmented Hessian methods. This coupled procedure would display convergence similar to the original Werner and Meyer procedure but presumably with much better convergence of the micro-iterative procedure[46]. The inclusion of appropriate trial vectors **S** in the PSCI iterative procedure allows convergence to MCSCF states with negative orbital Hessian matrix eigenvalues. As described here, this iterative method may be regarded as an extension of the PSCI or WSCI iterative methods in which an additional level has been added to the hierarchical structure of the micro-iterative procedure. This additional level involves the use of the approximate Hamiltonian operators. Alternately, it may be regarded as a modification of the original procedure of Werner and Meyer in which the slowly convergent sequential relaxation method[46] is replaced with the more reliable and more rapidly convergent PSCI method. The initial MCSCF iterations would include several approximate Hamiltonian operator construction steps while the later iterations would require only a single (exact) Hamiltonian operator definition and would therefore result in a usual PSCI iteration. Since the PSCI and WSCI methods display the correct energy dependence for doubly occupied orbitals when the exact Hamiltonian operator is used, the correct second-order convergence would be observed for this method in the later iterations without requiring the explicit third- and fourth-order contributions from the transformation as discussed above. Results of this method have not been reported in the literature.

VII. SUMMARY

The application of the MCSCF method to chemical problems requires an understanding of several aspects of electronic structure theory. Some of these have been discussed in this review and include the formal background of the various optimization methods, the CSF expansion spaces currently used for MCSCF wavefunctions, the relation of redundant variables to the CSF expansion space, the effect of redundant variables on the implementation of the MCSCF method, and the implementation of MCSCF methods on various types of computers. These aspects have been discussed in detail because they are relatively specific to the MCSCF method. This is not to imply that these discussions have no bearing whatsoever on other electronic structure methods, but rather that general principles from other electronic structure methods do not transfer directly to the MCSCF method in these respects.

For example, the quality of a large-scale CI calculation also depends on the choice of expansion CSFs, but this choice, which is dictated by the need to calculate quantitatively the dynamical electron correlation energy, is much different from that of a typical modern MCSCF calculation, which is more concerned with near-degeneracy, avoided-crossing and spin-recoupling effects.

There are other aspects of the application of the MCSCF method that have not been discussed in this review. The most notable of these probably is the lack of a discussion of orbital basis sets. Although the orbital basis set choice is very important in determining the quality of the MCSCF wavefunction, the general principles determined from other electronic structure methods also hold for the MCSCF method with very little change. For example, the description of Rydberg states requires diffuse basis functions in the MCSCF method just as any other method. The description of charge-transfer states requires a flexible description of the valence orbital space, triple or quadruple zeta quality, in the MCSCF method just as in other methods. Similarly, the efficient transformation of the two-electron integrals is crucial to the overall efficiency of the MCSCF optimization procedure. However, this is a relatively well understood problem (if not always well implemented) and has been described adequately in previous discussions of the MCSCF method and other electronic structure methods[37,46,66,77].

Finally it seems appropriate in this summary to attempt to extrapolate from the status of the current MCSCF methods reviewed here and to predict the direction of future developments. There are several forces at work in the developments of the last few years that are likely to remain in place in the future. The increased availability of computer resources to computational chemists will tend to fuel the quest for more accurate calculations and for the qualitative understanding of the results of ever more sophisticated calculations. The MCSCF method will be central to this development because it lends itself to qualitative interpretation and it forms the basis for other more accurate methods, particularly MRSDCI[41] and other closely related methods that are under development, such as multireference coupled-cluster methods[142,143] and quasi-degenerate perturbation theory methods[144,145].

As more accurate MCSCF calculations are required, non-linear wavefunction expansions will undoubtedly become more prominent. This will be particularly true when these methods are extended to larger systems and involve larger active orbital spaces. This may also be accompanied by the development of more sophisticated methods, based on these non-linear MCSCF wavefunctions, to compute quantitatively the dynamical correlation energy. Contracted CI methods have been developed for linear MCSCF expansions[146,147] and it seems only natural to extend these methods to non-linear reference functions.

Another promising development is the semi-empirical adjustment of Hamiltonian matrix elements to account for the effects of correlation of the electron core and for basis set limitations within this core[148–150]. As currently employed for diatomic molecules, the Hamiltonian matrix is constructed in a localized orbital basis. The diagonal elements of this Hamiltonian matrix correspond to products of configurations of the atomic fragments. The known errors of the excitation energies, ionization potentials and electron affinities of the atomic fragments for a given orbital basis set may then be used to modify the diagonal Hamiltonian matrix elements[149,150] or the integrals that contribute to these diagonal elements[148]. This semi-empirical adjustment has been applied to FORS/CASSCF wavefunctions by Reudenberg and coworkers[149,150] and to RCI type wavefunctions by Goodgame and Goddard[148].

Perhaps the most dramatic changes in the MCSCF method, and indeed in electronic structure methods in general, will result from the new computer architectures that are becoming available to computational chemists. Not only the supercomputers with about 10^8 words of memory and 10^9 arithmetic operations per second capability, but also the massively parallel computers with hundreds or thousands of computational units operating simultaneously may come to play important roles in modern computational chemistry.

Acknowledgements

The author wishes to acknowledge many informative discussions with Dr A. Banerjee, Dr F. B. Brown, Professor K. D. Jordan, Professor, I. Shavitt and Professor J. Simons concerning various aspects of this review. The author also wishes to thank his colleagues at Argonne National Laboratory, including Dr T. H. Dunning Jr, Dr L. B. Harding and Dr R. A. Bair for their numerous suggestions and comments during the preparation of this manuscript, and to thank Ms P. A. Davis for proofreading assistance.

This work was performed under the auspices of the Office of Basic Energy Sciences, Division of Chemical Sciences, US Department of Energy, under Contract W-31-109-Eng-38.

References

1. Olsen, J., Yeager, D. L., and Jørgensen, P., *Adv. Chem. Phys.*, **54**, 1 (1983).
2. McWeeny, R., and Sutcliffe, B. T., *Comput. Phys. Rep.*, **2**, 217 (1985).
3. Roos, B. O., in *Methods in Computational Molecular Physics* (Eds G. H. F. Diercksen and S. Wilson), p. 161, Reidel, Dordrecht, 1983.
4. Siegbahn, P. E. M., *Faraday Symp. Chem. Soc.*, **19**, 97 (1984).
5. Detrich, J. H., and Wahl, A. C., in *Recent Developments and Applications of Multiconfigurational Hartree–Fock Methods*, NRCC Proc. No. 10, p. 157, Report LBL-12157, Lawrence Berkeley Laboratory, University of California, Berkeley, CA, 1981.
6. Dalgaard, E., and Jørgensen, P., *J. Chem. Phys.*, **69**, 3833 (1978).
7. Roos, B. O., this volume; Werner, H.-J., this volume.
8. Das, G., and Wahl, A. C., *J. Chem. Phys.*, **56**, 1769 (1972).
9. Wahl, A. C., and Das, G. P., in *Methods of Electronic Structure Theory* (Ed. H. F. Schaefer III), p. 51, Plenum, New York, 1977.
10. Wilkinson, J. H., *The Algebraic Eigenvalue Problem*, Oxford University Press, London, 1965.
11. Mathews, J., and Walker, R. L., *Mathematical Methods of Physics*, Benjamin, Menlo Park, CA, 1964.
12. Davidson, E. R., in *Methods in Computational Molecular Physics* (Eds G. H. F. Diercksen and S. Wilson), p. 95, Reidel, Dordrecht, 1983.
13. Shavitt, I., in *Methods of Electronic Structure Theory* (Ed. H. F. Schaefer III), p. 189, Plenum, New York, 1977.
14. See, for example, Pilar, F. L., *Elementary Quantum Chemistry*, McGraw-Hill, New York, 1968; Levine, I. N., *Quantum Chemistry*, Allyn and Bacon, Boston, 1972; Schift, L. I., *Quantum Mechanics*, McGraw-Hill, New York, 1968; Messiah, A., *Quantum Mechanics*, Wiley, New York, 1958.
15. Hamming, R. W., *Numerical Methods for Scientists and Engineers*, McGraw-Hill, New York, 1962.
16. Yeager, D. L., and Jørgensen, P., *J. Chem. Phys.*, **71**, 755 (1979).
17. Dalgaard, E., *Chem. Phys. Lett.*, **65**, 559 (1979).
18. Löwdin, P. O., *in Perturbation Theory and its Applications in Quantum Mechanics* (Ed. C. H. Wilcox), p. 255, Wiley, New York, 1966.
19. Shepard, R., Shavitt, I., and Simons, J., *J. Chem. Phys.*, **76**, 543 (1982).
20. Hylleraas, E. A., and Undheim, B., *Z. Phys.*, **65**, 759 (1930).
21. MacDonald, J. K. L., *Phys. Rev.*, **43**, 830 (1933).
22. See, for example, Mead, C. A., and Truhlar, D. G., *J. Chem. Phys.*, **84**, 1055 (1986) and references therein.
23. Grein, F., and Banerjee, A., *Int. J. Quantum Chem.*, **S9**, 147 (1975).
24. Ruttink, P. J. A., and van Lenthe, J. H., *J. Chem. Phys.*, **74**, 5785 (1981).
25. van Lenthe, J. H., and Balint-Kurti, G. G., *J. Chem. Phys.*, **78**, 5699 (1983).
26. Chang, T. C., and Schwarz, W. H. E., *Theor. Chim. Acta*, **44**, 45 (1977).
27. Condon, E. U., and Shortley, G. H., *The Theory of Atomic Spectra*, Cambridge University Press, New York, 1957.
28. Thouless, D. J., *The Quantum Mechanics of Many-Body Systems*, Academic Press, New York, 1961.

29. Jørgensen, P., and Simons, J., *Second Quantization-Based Methods in Quantum Chemistry*, Academic Press, New York, 1981.
30. Mattuck, R. D., *A Guide to Feynman Diagrams in the Many-Body Problem*, McGraw-Hill, New York, 1976.
31. Pauncz, R., *Spin Eigenfunctions: Construction and Use*, Plenum, New York, 1979.
32. Yeager, D. L., and Jørgensen, P., *Chem. Phys. Lett.*, **65**, 77 (1979).
33. Dalgaard, E., *J. Chem. Phys.*, **72**, 816 (1980).
34. Dalgaard, E., in *Recent Developments and Applications of Multiconfigurational Hartree–Fock Methods*, NRCC Proc. No. 10, p. 136, Report LBL-12157, Lawrence Berkeley Laboratory, University of California, Berkeley, CA, 1981.
35. Nichols, J. A., Yeager, D. L., and Jørgensen, P., *J. Chem. Phys.*, **80**, 293 (1984).
36. Yabushita, S., and McCurdy, C. W., *J. Chem. Phys.*, **83**, 3547 (1985).
37. Shepard, R., and Simons, J., *Int. J. Quantum Chem.*, **S14**, 211 (1980).
38. Linderberg, J., and Öhrn, Y., *Int. J. Quantum Chem.*, **12**, 161 (1977).
39. Paldus, J., in *The Unitary Group for the Evaluation of Electronic Energy Matrix Elements* (Ed. J. Hinze), Lecture Notes in Chemistry, Vol. 22, p. 1, Springer, Berlin, 1981.
40. Shavitt, I., in *The Unitary Group for the Evaluation of Electronic Energy Matrix Elements* (Ed. J. Hinze), Lecture Notes in Chemistry, Vol. 22, p. 51, Springer, Berlin, 1981.
41. Lischka, H., Shepard, R., Brown, F. B., and Shavitt, I., *Int. J. Quantum Chem.*, **S15**, 91 (1981).
42. Knowles, P. J., Sexton, G. J., and Handy, N. C., *Chem. Phys. Lett.*, **72**, 337 (1982).
43. Brooks, B., and Schaefer, H. F., III, *J. Chem. Phys.*, **70**, 5092 (1979).
44. Shavitt, I., Annual Report to NASA Ames Research Center, June 1979.
45. Lengsfield, B. H., III, *J. Chem. Phys.*, **73**, 382 (1980).
46. Werner, H. -J., and Meyer, W., *J. Chem. Phys.*, **73**, 2342 (1980).
47. Lengsfield, B. H., III, and Liu, B., *J. Chem. Phys.*, **75**, 478 (1981).
48. Lengsfield, B. H., III, *J. Chem. Phys.*, **77**, 4073 (1982).
49. Roothaan, C. C. J., Detrich, J., and Hopper, D. G., *Int. J. Quantum Chem.*, **S13**, 93 (1979).
50. Roothaan, C. C. J., and Detrich, J. H., *Phys. Rev. A*, **27**, 29 (1983).
51. Levy, B., and Berthier, G., *Int. J. Quantum Chem.*, **2**, 307 (1968).
52. Bauche, J., and Klapisch, M., *J. Phys. B: At. Mol. Phys.*, **5**, 29 (1972).
53. Olsen, J., Jørgensen, P., and Yeager, D. L., *J. Chem. Phys.*, **76**, 527 (1982).
54. Golab, J., Yeager, D. L., and Jørgensen, P., *Chem. Phys.*, **93**, 83 (1984).
55. McLean A. D., Lengsfield, B. H., III, Pacansky, J., and Ellinger, Y., *J. Chem. Phys.*, **83**, 3567 (1985).
56. Golebiewski, A., Hinze, J., and Yurtsever, E., *J. Chem. Phys.*, **70**, 1101 (1979).
57. Shepard, R., 'Discussion of some multiconfiguration wave function optimization methods', 183rd ACS National Meeting, March 1982.
58. Brown, F. B., Shavitt, I., and Shepard, R., *Chem. Phys. Lett.*, **105**, 363 (1984).
59. Jensen, H. J. Aa., and Jørgensen, P., *J. Chem. Phys.*, **80**, 1204 (1984).
60. Jensen, H. J. Aa., and Ågren, H., *Chem. Phys. Lett.*, **110**, 140 (1984).
61. Shepard, R., in *Recent Developments and Applications of Multiconfigurational Hartree–Fock Methods*, NRCC Proc. No. 10, p. 117, Report LBL-12157, Lawrence Berkeley Laboratory, University of California, Berkeley, CA, 1981; Shepard, R., and Simons, J., *ibid.*, p. 121.
62. Shepard, R., Ph.D. Dissertation, University of Utah, 1980.
63. Grein, F., and Chang, T. C., *Chem. Phys. Lett.*, **12**, 44 (1971).
64. Grein, F., and Banerjee, A., *Chem. Phys. Lett.*, **25**, 255 (1974).
65. Banerjee, A., and Grein, F., *J. Chem. Phys.*, **66**, 1054 (1977).

66. Ruedenberg, K., Cheung, L. M., and Elbert, S. T., *Int. J. Quantum Chem.*, **16**, 1069 (1979).
67. Cheung, L. M., Sundberg, K. R., and Ruedenberg, K., *Int. J. Quantum Chem.*, **16**, 1103 (1979).
68. Chang, T. C., *J. Chin. Chem. Soc.*, **26**, 133 (1979).
69. Shepard, R., Banerjee, A., and Simons, J., *J. Am. Chem. Soc.*, **101**, 6174 (1979).
70. Shepard, R., and Simons, J., *Int. J. Quantum Chem.*, **S14**, 349 (1980).
71. Bauschlicher, C. W., Jr. and Yarkony, D. R., *J. Chem. Phys.*, **72**, 1138 (1980).
72. Siegbahn, P., Heiberg, A., Roos, B., and Levy, B., *Phys. Scr.*, **21**, 323 (1980).
73. Roos, B. O., *Int. J. Quantum Chem.*, **S14**, 175 (1980).
74. Roos, B. O., Taylor, P. R., and Siegbahn, P. E. M., *Chem. Phys.*, **48**, 157 (1980).
75. Olsen, J., Jørgensen, P., and Yeager, D. L., *J. Chem. Phys.*, **77**, 356 (1982).
76. Olsen, J., and Jørgensen. P., *J. Chem. Phys.*, **77**, 6109 (1982).
77. Camp, R. N., and King, H. F., *J. Chem. Phys.*, **77**, 3056 (1982).
78. Fletcher, R., *Practical Methods of Optimization*, Vol. 1, Wiley, New York, 1980.
79. Jørgensen, P., Swanstrøm, P., and Yeager, D. L., *J. Chem. Phys.*, **78**, 347 1983).
80. Golab, J., Yeager, D. L., and Jørgensen, P., *Chem. Phys.*, **78**, 175 (1983).
81. Knowles, P. J., and Werner, H.-J., *Chem. Phys. Lett.*, **115**, 259 (1985).
82. Werner, H.-J., and Knowles, P. J., *J. Chem. Phys.*, **82**, 5053 (1985).
83. Buenker, R. J., and Peyerimhoff, S. D., *Theor. Chem. Acta*, **35**, 33 (1974).
84. Buenkar, R. J., and Peyerimhoff, S. D., *Theor. Chim. Acta*, **39**, 217 (1975).
85. Jackels, C. F., and Shavitt, I., *Theor. Chim, Acta*, **58**, 81 (1981).
86. Handy, N. C., *Faraday Symp. Chem. Soc.* **19**, 17 (1984).
87. Shavitt, I., *Chem. Phys. Lett.*, **63**, 421 (1979).
88. Beebe, N. H. F., Thulstrup, E. W., and Andersen, A., *J. Chem. Phys.*, **64**, 2080 (1976).
89. Harrison, R. J., and Handy, N. C., *Chem. Phys. Lett.*, **95**, 386 (1983).
90. Purvis, G. D., Shepard. R., Brown, F. B., and Bartlett, R. J., *Int. J. Quantum Chem.*, **23**, 835 (1983).
91. Dunning, T. H., Jr. in *Advanced Theories and Computational Approaches to the Electronic Structure of Molecules* (Ed. C. E. Dykstra), p. 67, Reidel, Dordrecht, 1984.
92. Siegbahn, P. E. M., Almlöf, J., Heiberg, A., and Roos, B. O., *J. Chem. Phys.*, **74**, 2384 (1981).
93. Hurley, A. C., *Electron Correlation in Small Molecules*, Academic Press, London, 1976.
94. Brown, F. B., Ph.D. Dissertation, Ohio State University, 1982.
95. Siegbahn, P. E. M., *Chem. Phys. Lett.*, **119**, 515 (1985).
96. Sunil, K. K., Jordan, K. D., and Shepard, R., *Chem. Phys.*, **88**, 55 (1984).
97. Harding, L. B., and Goddard, W. A., III, *J. Am. Chem. Soc.*, **97**, 6293 (1975).
98. Dunning, T. H., Jr, Cartwright, D. C., Hunt, W. J., Hay, P. J., and Bobrowicz, F. W., *J. Chem. Phys.*, **64**, 4755 (1976).
99. Goodgame, M. M., and Goddard, W. A., III, *J. Phys. Chem.*, **85**, 215 (1981).
100. Chipman, D. M., *J. Chem. Phys.*, **84**, 1677 (1986).
101. Hurley, A. C., Lennard-Jones, J. and Pople, J. A., *Proc. R. Soc. A*, **220**, 446 (1953).
102. Kutzelnigg, W., in *Methods of Electronic Structure Theory* (Ed. H. F. Schaefer III), p. 129, Plenum, New York, 1977.
103. Meyer, W., in *Methods of Electronic Structure Theory* (Ed. H. F. Schaefer III), p. 413, Plenum, New York, 1977.
104. Goddard, W. A., III, and Ladner, R. C., *Int. J. Quantum Chem.*, **S3**, 63 (1969).
105. Hunt, W. J., Goddard, W. A., III, and Dunning, T. H., Jr, *Chem. Phys. Lett.*, **6**, 147 (1970).

106. Hay, P. J., Hunt, W. J., and Goddard, W. A., III, *J. Am. Chem. Soc.*, **94**, 8293 (1972).
107. Bobrowicz, F. W., and Goddard, W. A., III, in *Methods of Electronic Structure Theory* (Ed. H. F. Schaefer III), p. 79, Plenum, New York, 1977.
108. Bobrowicz, F. W., Ph.D. Dissertation, California Institute of Technology, 1974.
109. Bobrowicz, F. W., in *Recent Developments and Applications of Multiconfigurational Hartree–Fock Methods*, NRCC Proc. No. 10, p. 14, Report LBL-12157, Lawrence Berkeley Laboratory, University of California, Berkeley, CA, 1981.
110. Ortiz, J. V., Weiner, B., and Öhrn, Y., *Int. J. Quantum Chem.*, **S15**, 113 (1981).
111. Lozes, R. L., Weiner, B., and Öhrn, Y., *Int. J. Quantum Chem.*, **S15**, 129 (1981).
112. Sangfelt, E., Goscinski, O., Elander, N., and Kurtz, H., *Int. J. Quantum Chem.*, **S15**, 133 (1981).
113. Kurtz, H., Elander, N., Goscinski, O., and Sangfelt, E., *Int. J. Quantum Chem.*, **S15**, 143 (1981).
114. Weiner, B., and Öhrn, Y., *J. Chem. Phys.*, **83**, 2965 (1985).
115. Jackels, C. F., and Davidson, E. R., *J. Chem. Phys.*, **69**, 2908 (1976).
116. Martin, R. L., in *Recent Developments and Applications of Multiconfigurational Hartree–Fock Methods*, NRCC Proc. No. 10, p. 154, Report LBL-12157, Lawrence Berkeley Laboratory, University of California, Berkeley, CA, 1981.
117. McIver, J. W., private communication, May 1985.
118. Ahlrichs, R., Kutzelnigg, W., and Bingel, W. A., *Theor. Chim. Acta*, **5**, 289 (1966).
119. Ahlrichs, R., Kutzelnigg, W., and Bingel, W. A., *Theor. Chim. Acta*, **5**, 305 (1966).
120. King, H. F., Camp, R. N., and McIver, J. W., Jr, *J. Chem. Phys.*, **80**, 1171 (1984).
121. Hoffmann, M. R., Fox, D. J., Gaw, J. F., Osamura, Y., Yamaguchi, Y., Grev, R. S., Fitzgerald, G., Schaefer, H. F., III, Knowles, P. J., and Handy, N. C., *J. Chem. Phys.*, **80**, 2660 (1984).
122. Annual Report of the Theoretical Chemistry Group, October 1984 to September 1985, Argonne National Laboratory, 1985.
123. Shepard, R., Bair, R. A., Eades, R. A., Wagner, A. F., Davis, M. J., Harding, L. B., and Dunning T. H., Jr, *Int. J. Quantum Chem.*, **S17**, 613 (1983).
124. Ahlrichs, R., Böhm, H.-J., Ehrhardt, C., Scharf, P., Schiffer, H., Lischka, H., and Schindler, M., *J. Comput. Chem.*, **6**, 200 (1985).
125. Dunning, T. H., Jr, and Bair, R. A., in *Advanced Theories and Computational Approaches to the Electronic Structure of Molecules* (Ed. C. E. Dykstra), p. 1, Reidel, Dordrecht, 1984.
126. Bauschlicher, C. W., Jr, in *Advanced Theories and Computational Approaches to the Electronic Structure of Molecules* (Ed. C. E. Dykstra), p. 13, Reidel, Dordrecht, 1984.
127. Almlöf, J., and Taylor, P. R., in *Advanced Theories and Computational Approaches to the Electronic Structure of Molecules* (Ed. C. E. Dykstra), p. 107, Reidel, Dordrecht, 1984.
128. Dykstra, C. E., and Schaefer, H. F., III, in *Advanced Theories and Computational Approaches to the Electronic Structure of Molecules* (Ed. C. E. Dykstra), p. 197, Reidel, Dordrecht, 1984.
129. Binkley, J. S., in *Advanced Theories and Computational Approaches to the Electronic Structure of Molecules* (Ed. C. E. Dykstra), p. 209, Reidel, Dordrecht, 1984.
130. Dongarra, J. J., Moler, C. B., Bunch, J. R., and Stewart, G. W., *LINPACK User's Guide*, SIAM, Philadelphia, 1979.
131. Yaffe, L. G., and Goddard, W. A., III, *Phys. Rev. A*, **13**, 1682 (1976).
132. Douady, L. G. J., Ellinger, J., Subra, R., and Levy, B., *J. Chem. Phys.*, **72**, 1452 (1980).

133. Bacskay, G. B., *Chem. Phys.*, **65**, 383 (1982).
134. Hinze, J., and Roothaan, C. C. J., *Prog. Theor. Phys. Suppl.*, **40**, 37 (1967).
135. Das, G., *J. Chem. Phys.*, **74**, 5775 (1981).
136. See, for example, *Geometrical Derivatives of Energy Surfaces and Molecular Properties* (NATO ASI Series C, **166**) (Eds P. Jørgensen and J. Simons), Reidel, Dordrecht, 1986, in particular, Shepard, R., p. 193 and references therein.
137. Shepard, R., and Shavitt, I., manuscript in preparation.
138. Pople, J. A., Krishnan, R., Schlegel, H. B., and Binkley, J. S., *Int. J. Quantum Chem.*, **S13**, 225 (1979).
139. Purvis, G. D., and Bartlett, R. J., *J. Chem. Phys.*, **75**, 1284 (1981).
140. Banerjee, A., Adams, N., Simons, J., and Shepard, R., *J. Chem. Phys.*, **89**, 52 (1985).
141. Werner, H. J., private communication, November 1984.
142. Banerjee and Simons, J., *Int. J. Quantum Chem.*, **19**, 207 (1981).
143. Laidig, W. D., and Bartlett, R. J., *Chem. Phys. Lett.*, **104**, 424 (1984).
144. Davidson, E. R., and Bender, C. F., *Chem. Phys. Lett.*, **59**, 369 (1978).
145. Lee, Y. S., and Bartlett, R. J., *Int. J. Chem.*, **S17**, 347 (1983).
146. Werner, H.-J., and Reinsch, E. A., in *Advanced Theories and Computational Approaches to the Electronic Structure of Molecules* (Ed. C. E. Dykstra), p. 79, Reidel, Dordrecht, 1984.
147. Siegbahn, P. E. M., in *Methods in Computational Molecular Physics* (Eds G. H. F. Diercksen and S. Wilson), p. 189, Reidel, Dordrecht, 1983.
148. Goodgame, M. M., and Goddard, W. A., III, *Phys. Rev. Lett.*, **54**, 661 (1985).
149. Lam, B., Schmidt, M. W., and Ruedenberg, K., *J. Phys. Chem.*, **89**, 2221 (1985).
150. Schmidt, M. W., Lam, M. T. B., Elbert, S. T., and Ruedenberg, K., *Theor. Chim. Acta*, **68**, 69 (1985).

Ab Initio Methods in Quantum Chemistry—II
Edited by K. P. Lawley

PROPAGATOR METHODS

JENS ODDERSHEDE

Department of Chemistry, Odense University, DK-5230 Odense M, Denmark

CONTENTS

I. Introduction . . . 201
II. Definition of Propagators . . . 202
III. Relations to Response Functions . . . 204
IV. Equations of Motion . . . 210
V. Partitioning of the Equation of Motion . . . 214
VI. One-electron Propagators . . . 215
VII. Polarization Propagators . . . 218
A. The Random-phase Approximation . . . 218
B. The Multiconfiguration Random-phase Approximation . . . 223
C. The Antisymmetrized Geminal Power Method . . . 225
D. Perturbative Propagator Methods . . . 228
E. Coupled-cluster Propagator Methods . . . 234
F. Summary of Polarization Propagator Calculations . . . 235
References . . . 236

I. INTRODUCTION

Electronic structure calculations have over the years almost solely been performed using state approaches. The basic approach has been to develop better and better methods to evaluate the wavefunction and/or the energy and other properties of the individual state. To a large extent the efforts have concentrated on ground-state properties. Clearly these methods have also dominated the literature and even very recent reviews of electronic structure calculations (Schaefer, 1984) maintain the view that only state function methods have yet demonstrated their usefulness in electronic structure calculations. It is the purpose of the present review to describe a different approach, the propagator methods. In these methods we compute state energy differences, transition probabilities and response properties directly without knowing the wavefunctions of the individual states.

Propagators were introduced in molecular quantum mechanics about 20 years ago (Ball and McLachlan, 1964; McLachlan and Ball, 1964; Linderberg and Öhrn, 1965). However, they were used quite extensively in other areas of physics before that (Lindhard, 1954; Kubo, 1959; Zubarev, 1960; Thouless, 1961; Feynman and Hibbs, 1965; Hedin and Lundquist, 1969). Even though the term 'propagator' is not used explicitly in all of these references, they are in fact all dealing with the quantity that we here designate the propagator. The propagator formalism as it is presently used in quantum chemistry is akin to the formulations and methods of statistical physics. The development that is pursued in the present review owes much to authors like Zubarev and Kubo and an appropriate reference book is that of Zubarev (1974).

A propagator is not a single well defined quantity. First of all, the term is not used synonymously, and, secondly, even if we adhere to the definition that is used here, there are many different kinds of propagators (electron propagators, particle–hole propagators, polarization propagators, etc.). Propagator methods are not discussed in most textbooks of quantum chemistry. The only exceptions are those by Linderberg and Öhrn (1973), by Jørgensen and Simons (1981) and, to a certain extent, by Szabo and Ostlund (1982). There are reviews of propagator methods mainly concentrating on either the electron propagators (Öhrn, 1976; Simons, 1977; Öhrn and Born, 1981; Herman *et al.*, 1981), also called the Green's function method (Cederbaum and Domcke, 1977; Von Niessen *et al.*, 1984), or the polarization propagator method (Jørgensen, 1975; McCurdy *et al.*, 1977; Oddershede, 1978; Hansen and Bouman, 1980; Oddershede *et al.*, 1984). It is the intent in the present chapter to discuss several of the various propagators and propagator methods. However, it is unavoidable that the author's personal involvement in the development of the polarization propagator method will influence the 'flavour' of the review.

In a complete description of propagator methods there will inevitably appear rather lengthy mathematical derivations. In order to shorten the review these derivations are omitted in many places and references are given to textbooks or review articles. Some of the more common steps have also appeared so often in print that there is no need to repeat them here. Therefore, this review will focus more on principles than on derivations. For readers who are interested in an introduction to propagator theory where the steps leading to the working equations are explained in detail, I may refer to the textbooks or a set of lecture notes (Oddershede, 1983). Results of calculations using propagator methods that have appeared since my last compilation of propagator calculations (Oddershede, 1978) are reviewed in Sections VII.A and F.

II. DEFINITION OF PROPAGATORS

In this section we will discuss the propagators that are characterized by two operators $P(t)$ and $Q(t)$. These propagators are defined as

$$\langle\langle P(t); Q(t') \rangle\rangle = -\mathrm{i}\theta(t-t')\langle 0|P(t)Q(t')|0\rangle \pm \mathrm{i}\theta(t-t')\langle 0|Q(t')P(t)|0\rangle \quad (1)$$

where $\theta(t)$ is the Heaviside step function and the plus sign in front of the last term only applies when both P and Q are fermion (non-number-conserving) operators. Average values are taken with respect to $|0\rangle$, which we call the *reference state.* The propagator is thus defined for a particular state. This does not mean, as we will see later, that we must know $|0\rangle$ in order to calculate $\langle\langle P(t); Q(t) \rangle\rangle$, but it does mean that there are several propagators for a given system even if P and Q are fixed. Eq. (1) does not define all propagators. We will in Section III give examples of propagators that are defined for three and four one-electron operators. However, Eq. (1) defines the more common ones.

In order to see the usefulness of the propagator we consider the energy-dependent rather than the time-dependent propagator. The former is defined as

$$\langle\langle P; Q \rangle\rangle_E = \int_{-\infty}^{\infty} \mathrm{d}(t-t')\langle\langle P(t); Q(t') \rangle\rangle \mathrm{e}^{\mathrm{i}E(t-t')} \quad (2)$$

Using the interaction representation of the time-dependent operators $P(t)$ and $Q(t)$, we find (see e.g. Linderberg and Öhrn, 1973, p. 27)

$$\langle\langle P; Q \rangle\rangle_E = \sum_{m\neq 0}\left(\frac{\langle 0|P|m\rangle\langle m|Q|0\rangle}{E - E_m + E_0 + \mathrm{i}\eta} \pm \frac{\langle 0|Q|m\rangle\langle m|P|0\rangle}{E + E_m - E_0 - \mathrm{i}\eta}\right) \quad (3)$$

Here, η is a positive infinitesimal which ensures that the integrals in Eq. (2) are convergent (see Section III for a discussion of the physical reason for the appearance of this convergence factor). The summation in Eq. (2) extends over all states $|m\rangle$ different from the reference state $|0\rangle$. The real values of the (first-order) pole of $\langle\langle P; Q \rangle\rangle_E$ are

$$E = \pm(E_m - E_0) \quad (4)$$

and the residues are $\langle 0|P|m\rangle\langle m|Q|0\rangle$ and $\langle 0|Q|m\rangle\langle m|P|0\rangle$. The physical interpretation of the poles and residues thus depends on the nature of P and Q.

Consider first the case where P and Q are simple creation and annihilation operators, e.g. $P = a_r^+$ and $Q = a_\alpha$. The residue $\langle 0|P|m\rangle\langle m|Q|0\rangle$ is then non-vanishing only if $|m\rangle$ contains one electron *less* than $|0\rangle$ and the residue of the last term vanishes unless $|m\rangle$ contains one electron *more* than $|0\rangle$. The poles of the first term are thus the ionization potentials for state $|0\rangle$ while the poles of the second term are the electron affinities. The $\langle\langle a_r^+; a_\alpha \rangle\rangle_E$ propagator is called the *electron propagator* and will be discussed in more detail in Section VI.

Now assume that $P = a_r^+ a_s^+$ and $Q = a_\alpha a_\beta$. Following the same reasoning as above we see that $|m\rangle = |N+2\rangle$ if $|0\rangle = |N\rangle$ is an N-electron state. The poles and residues yield energy differences and overlap amplitudes of Auger spectra, respectively. Two-electron propagators of the type $\langle\langle a_r^+ a_s^+; a_\alpha a_\beta \rangle\rangle_E$ have in fact been used in calculations of Auger spectra of several molecules (Liegener, 1983; Ortiz, 1984).

We obtain a third kind of propagator if we assume that P and Q are number-conserving one-electron operators like

$$P = \sum_{rs} p_{rs} a_r^+ a_s \tag{5}$$

$$Q = \sum_{ij} q_{ij} a_i^+ a_j \tag{6}$$

The reference state $|0\rangle$ and $|m\rangle$ must then have the same number of electrons and the poles in Eq. (4) are energy differences between N-electron states of the systems. In all applications so far $|0\rangle$ has been the ground state and $|m\rangle$ are hence all excited states of the system. The poles of the propagator give energy differences between the ground state and an excited state but *not* between excited states. The residues are transition amplitudes between ground and excited states. This is seen by considering the most common choice of P and Q, namely $P = Q = \mathbf{r}$, the electric dipole operator. In this case the residues are $|\langle 0|\mathbf{r}|m\rangle|^2$. Other choices for P and Q give different transition moments. However, the poles of $\langle\langle P; Q\rangle\rangle_E$ are the same, irrespective of the form of p_{rs} and q_{rs}.

Both P and Q are sums of excitation operators (with weighting coefficients p_{rs} and q_{rs}). Thus, P and Q applied to $|0\rangle$ create a polarization of $|0\rangle$ and we call $\langle\langle P; Q\rangle\rangle_E$ a *polarization propagator*. In the special case where P and Q are both single particle–hole excitations, i.e. only one term in Eqs (5) and (6), we talk about the *particle–hole propagator*. It is important to note that only the residues of the polarization propagator and not of the particle–hole propagator determine transition moments (Oddershede, 1982). We must have the complete summations in Eqs (5) and (6) in order to represent the one-electron operator that induces the transition in question.

III. RELATIONS TO RESPONSE FUNCTIONS

Let us initially consider the $\langle\langle \mathbf{r}; \mathbf{r}\rangle\rangle_E$ polarization propagator as defined in Eq. (3). We find in the limit $\eta \to 0$

$$\begin{aligned}\mathrm{Tr}(\mathrm{Re}\langle\langle \mathbf{r}; \mathbf{r}\rangle\rangle_E) &= -2 \sum_{m \neq 0} \frac{\langle 0|\mathbf{r}|m\rangle \cdot \langle m|\mathbf{r}|0\rangle}{(E_m - E_0)^2 - E^2}(E_m - E_0) \\ &= -3 \sum_{m \neq 0} \frac{f_{0m}}{(E_m - E_0)^2 - E^2}\end{aligned} \tag{7}$$

where f_{0m} is the oscillator strength in the dipole length formulation

$$f_{0m} = \tfrac{2}{3}(|\langle 0|x|m\rangle|^2 + |\langle 0|y|m\rangle^2 + \langle 0|z|m\rangle|^2)(E_m - E_0) \tag{8}$$

Except for a factor of 3, the last part of Eq. (7) is readily recognized as the sum-over-states expression for the isotropic invariant of the energy-dependent polarizability of state $|0\rangle$, i.e.

$$\alpha(E) = -\tfrac{1}{3}\mathrm{Tr}\,(\mathrm{Re}\langle\langle \mathbf{r};\mathbf{r}\rangle\rangle_E) \tag{9}$$

The energy-dependent $-\langle\langle \mathbf{r};\mathbf{r}\rangle\rangle_E$ polarization propagator is thus the dynamic polarizability tensor, i.e. the induced electric dipole moment in units of an external electric field.

Let us analyse this relation in a little more detail. We assume that the external electric field $F(t)$ vanishes for $t \to -\infty$. Hence, for $\eta > 0$ we may write

$$\mathbf{F}(t) = \int_{-\infty}^{\infty} \mathrm{d}E\,\mathbf{F}_E \exp(-\mathrm{i}Et + \eta t) \tag{10}$$

The induced electric dipole moment is

$$\mu_{\text{ind}}(t) = \int_{-\infty}^{\infty} \boldsymbol{\alpha}(E)\mathbf{F}_E \exp(-\mathrm{i}Et + \eta t)\,\mathrm{d}E \tag{11}$$

where $\boldsymbol{\alpha}(E)$ is the polarizability tensor. Comparison of Eqs (9) and (11) suggests that we may express the dipole moment at time t as

$$\begin{aligned}\langle \mathbf{r}\rangle_t = \langle 0|\mathbf{r}|0\rangle &+ \int_{-\infty}^{\infty} \mathrm{d}E \langle\langle \mathbf{r};\mathbf{r}\rangle\rangle_E^r \mathbf{F}_E \exp(-\mathrm{i}Et + \eta t)\\ &+ \tfrac{1}{2}\int_{-\infty}^{\infty} \mathrm{d}E_1 \int_{-\infty}^{\infty} \mathrm{d}E_2 \langle\langle \mathbf{r};\mathbf{r};\mathbf{r}\rangle\rangle_{E,E_2} \mathbf{F}_{E_1}\mathbf{F}_{E_2} \exp[(-\mathrm{i}E_1 - \mathrm{i}E_2 + 2\eta)t]\\ &+ \tfrac{1}{6}\int_{-\infty}^{\infty} \mathrm{d}E_1 \int_{-\infty}^{\infty} \mathrm{d}E_1 \langle\langle \mathbf{r};\mathbf{r};\mathbf{r};\mathbf{r}\rangle\rangle_{E_1,E_2,E_3} \mathbf{F}_{E_1}\mathbf{F}_{E_2}\mathbf{F}_{E_3}\\ &\times \exp[(-\mathrm{i}E_1 - \mathrm{i}E_2 - \mathrm{i}E_3 + 3\eta)t] + \cdots\end{aligned} \tag{12}$$

In addition to the linear response function $\langle\langle \mathbf{r};\mathbf{r}\rangle\rangle_E$ we have also introduced the quadratic $\langle\langle \mathbf{r};\mathbf{r};\mathbf{r}\rangle\rangle_{E_1,E_2}$ and the cubic $\langle\langle \mathbf{r};\mathbf{r};\mathbf{r};\mathbf{r}\rangle\rangle_{E_1,E_2,E_3}$ response functions. The relation (9) only ensures that the real part of the second term is correct. In fact it turns out (Zubarev, 1974, Chap. 15; Oddershede *et al.*, 1984, Chaps 2.1 and 2.2) that $\langle\langle \mathbf{r};\mathbf{r}\rangle\rangle_E^r$ is the spectral representation of the *retarded* polarization propagator

$$\langle\langle \mathbf{r};\mathbf{r}\rangle\rangle_{E+\mathrm{i}\eta}^r = \sum_{m\neq 0}\left(\frac{|\langle 0|\mathbf{r}|m\rangle|^2}{E - E_m + E_0 + \mathrm{i}\eta} - \frac{|\langle 0|\mathbf{r}|m\rangle|^2}{E + E_m - E_0 + \mathrm{i}\eta}\right) \tag{13}$$

The real parts of $\langle\langle \mathbf{r};\mathbf{r}\rangle\rangle_E^r$ and the *causal* propagator defined in Eq. (3) are the same, whereas the imaginary parts are different. The real parts of the poles, i.e. the excitation energies, as well as the transition moments are also the same for the causal and the retarded propagators. Thus, for our purpose they are equally good. The infinitesimal η was introduced in Eq. (10) in order to make the applied external field and hence the response of the system vanish at $t \rightarrow -\infty$. This is the physical argument for the appearance of the convergence factor in the spectral representation of the propagator (Eq. (2)). Inverting Eq. (2) we find that the time-dependent *retarded polarization propagator* is defined as

$$\langle\langle P(t); Q(t') \rangle\rangle^r = -\mathrm{i}\theta(t-t')\langle 0[P(t), Q(t')]|0\rangle \tag{14}$$

The linear response function in Eq. (12) was identified from knowledge of the form of the polarizability. We also know (see e.g. Bloembergen, 1965) that the quadratic response function $\langle\langle \mathbf{r}; \mathbf{r}; \mathbf{r} \rangle\rangle_{E_1, E_2}$ is the energy-dependent dipole hyperpolarizability tensor and that the cubic response function is the second hyperpolarizability tensor. Expressions for the higher polarizabilities are given in many places and we could use them to find the actual forms of the quadratic and cubic polarization propagator. However, in order to determine more general response functions, i.e. not only response to homogeneous periodic electric fields, we may proceed as recently suggested by Olsen and Jørgensen (1985). They consider the time development of

$$A_{\mathrm{av}}(t) = \langle \bar{0} | A | \bar{0} \rangle \tag{15}$$

where A is a general time-independent operator in the Schrödinger picture (e.g. $\mathbf{r}$ in Eq. (12)) and the time evolution of $|\bar{0}\rangle$ is expressed as

$$|\bar{0}\rangle = \mathrm{e}^{\mathrm{i}P(t)}|0\rangle \tag{16}$$

By expanding $|\bar{0}\rangle$ in orders of the external perturbation V^t, we can identify the static, linear, etc., response of the system, that is the first, second, etc., terms in Eq. (12) with $\mathbf{r}$ replaced by A. We will outline the derivation.

The unperturbed system is described by a time-independent Hamiltonian H_0 and just as in Eq. (10) the interaction between the system and an external field $W(t)$ is written as

$$V^t = \int_{-\infty}^{\infty} \mathrm{d}E\, V^E \exp[(-\mathrm{i}E + \eta)t] \tag{17}$$

We expand the time evolution operator $P(t)$ in the eigenstates $\{|n\rangle\}$ of H_0

$$P(t) = \sum_n{}' [P_n(t)|n\rangle\langle 0| + P_n^*(t)|0\rangle\langle n|] + [P_0(t) + P_0^*(t)]|0\rangle\langle 0| \tag{18}$$

where the prime on the summation indicates that the term $n = 0$ is omitted. Only the real part of P_0 contributes to $P(t)$ and introducing the real and imaginary parts of P_n

$$P_n(t) = P_n^R(t) + \mathrm{i}P_n^I(t) \tag{19}$$

makes

$$P(t) = {}^{a}P(t) + {}^{b}P(t) \tag{20}$$

where

$${}^{a}P = \sum_{n}' P_n|n\rangle\langle 0| + P_n^*|0\rangle\langle n| \tag{21}$$

and

$${}^{b}P = 2P_0^{R}|0\rangle\langle 0| \tag{22}$$

Using Eq. (16) we obtain

$$\begin{aligned}|\bar{0}\rangle &= \exp[\mathrm{i}\,{}^{a}P(t)]|0\rangle \exp[2\mathrm{i}P_0^{R}(t)] \\ &= |\tilde{0}\rangle \exp[2\mathrm{i}P_0^{R}(t)]\end{aligned} \tag{23}$$

Thus, the ${}^{b}P(t)$ operator in Eq. (22) only introduces a time-dependent phase factor in the transformed wavefunction. By means of Ehrenfest's (1927) theorem we find

$$\langle\tilde{0}|\Lambda|\dot{\tilde{0}}\rangle + \langle\dot{\tilde{0}}|\Lambda|\tilde{0}\rangle = -\mathrm{i}\langle\tilde{0}|[\Lambda, H_0 + V^t]|\tilde{0}\rangle \tag{24}$$

where Λ is one of the operators $\Lambda_n = |n\rangle\langle 0|$ or $\Lambda_n^+ = |0\rangle\langle n|$. Note that it is $|\tilde{0}\rangle$ and not $|\bar{0}\rangle$ that appears in Eq. (24). The phase factors cancel in Ehrenfest's equation of motion.

By expressing the P_n coefficients in Eq. (18) in orders of the external perturbation V^t

$$P_n = P_n^{(1)} + P_n^{(2)} + \cdots \tag{25}$$

and also expanding $\exp(\mathrm{i}\,{}^{a}P)$ in Eq. (23) in a power series we can write

$$|\tilde{0}\rangle = |0^{(0)}\rangle + |0^{(1)}\rangle + |0^{(2)}\rangle + \cdots \tag{26}$$

where

$$|0^{(0)}\rangle = |0\rangle \tag{27}$$

$$|0^{(1)}\rangle = \mathrm{i}\sum_{n}'|n\rangle P_n^{(1)} \tag{28}$$

$$|0^{(2)}\rangle = -\tfrac{1}{2}|0\rangle\sum_{j}' P_j^{(1)}P_j^{*(1)} + \mathrm{i}\sum_{n}'|n\rangle P_n^{(2)} \tag{29}$$

etc. Inserting Eq. (26) in Eq. (24) gives a first-, second-, etc., order equation of motion that can be used to determine $P_n^{(1)}$, $P_n^{(2)}$, etc. For instance, the first-order equation of motion is (Olsen and Jørgensen, 1985)

$$\mathrm{i}\dot{P}_k^{(1)} - (E_k - E_0)P_k^{(1)} = -\mathrm{i}\langle k|V^t|0\rangle \tag{30}$$

a linear inhomogeneous differential equation that can be solved using standard techniques. Similar equations appear in higher orders and we can determine explicit expressions for the $P_k^{(1)}$, $P_k^{(2)}$, etc., coefficients and thus for $|0^{(1)}\rangle$, $|0^{(2)}\rangle$, etc., according to Eqs (27)–(29). Returning to Eq. (15) this means that

$$A_{\text{av}}(t) = \langle \bar{0} | A | \bar{0} \rangle = \langle \tilde{0} | A | \tilde{0} \rangle \tag{31}$$

can be expanded in orders of the external perturbation V^t as

$$\begin{aligned} A_{\text{av}}(t) = {} & \langle 0 | A | 0 \rangle + \langle 0^{(1)} | A | 0 \rangle + \langle 0 | A | 0^{(1)} \rangle \\ & + \langle 0^{(1)} | A | 0^{(1)} \rangle + \langle 0 | A | 0^{(2)} \rangle + \langle 0^{(2)} | A | 0 \rangle + \cdots \end{aligned} \tag{32}$$

Comparison of this equation and the generalization of Eq. (12)

$$\begin{aligned} A_{\text{av}}(t) = {} & \langle 0 | A | 0 \rangle + \int_{-\infty}^{\infty} \mathrm{d}E \langle\langle A; V^E \rangle\rangle_{E+\mathrm{i}\eta} \exp[(-\mathrm{i}E + \eta)t] \\ & + \tfrac{1}{2} \int_{-\infty}^{\infty} \mathrm{d}E_1 \int_{-\infty}^{\infty} \mathrm{d}E_2 \langle\langle A; V^{E_1}; V^{E_2} \rangle\rangle_{E_1+\mathrm{i}\eta, E_2+\mathrm{i}\eta} \\ & \times \exp[(-\mathrm{i}E_1 - \mathrm{i}E_2 + 2\eta)t] + \cdots \end{aligned} \tag{33}$$

and using the expressions for $P_k^{(i)}$ in terms of matrix elements of V^t leads to the following expressions for the (retarded) polarization propagators or response functions (Olsen and Jørgensen, 1985):

$$\langle\langle A; V^E \rangle\rangle_{E+\mathrm{i}\eta} = \sum_{k \neq 0} \left(\frac{\langle 0 | A | k \rangle \langle k | V^E | 0 \rangle}{E - E_k + E_0 + \mathrm{i}\eta} - \frac{\langle 0 | V^E | k \rangle \langle k | A | 0 \rangle}{E + E_k - E_0 + \mathrm{i}\eta} \right) \tag{34}$$

$$\begin{aligned} & \langle\langle A; V^{E_1}; V^{E_2} \rangle\rangle_{E_1+\mathrm{i}\eta, E_2+\mathrm{i}\eta} \\ & = (1 + P_{12}) \sum_{k,n \neq 0} \left(\frac{\langle 0 | A | k \rangle \langle k | V^{E_1} - \langle 0 | V^{E_1} | 0 \rangle | n \rangle \langle n | V^{E_2} | 0 \rangle}{(E_1 + E_2 - E_k + E_0 + 2\mathrm{i}\eta)(E_2 - E_n + E_0 + \mathrm{i}\eta)} \right. \\ & \quad + \frac{\langle 0 | V^{E_2} | n \rangle \langle n | V^{E_1} - \langle 0 | V^{E_1} | 0 \rangle | k \rangle \langle k | A | 0 \rangle}{(E_1 + E_2 + E_k - E_0 + 2\mathrm{i}\eta)(E_2 + E_n - E_0 + \mathrm{i}\eta)} \\ & \quad \left. - \frac{\langle 0 | V^{E_1} | k \rangle \langle k | A - \langle 0 | A | 0 \rangle | n \rangle \langle n | V^{E_2} | 0 \rangle}{(E_1 + E_k - E_0 + \mathrm{i}\eta)(E_2 - E_n + E_0 + \mathrm{i}\eta)} \right) \end{aligned} \tag{35}$$

The expression for the cubic response function is given in Eq. (2.60) of Olsen and Jørgensen (1985). All the propagators that are derived from response theory are retarded polarization propagators. The poles are placed in the lower complex plane. This is specified through the energy variables $E_1 + \mathrm{i}\eta$ and $E_2 + \mathrm{i}\eta$. The P_{12} operator in Eq. (35) permutes E_1 and E_2 and it is assumed that the $\eta \to 0$ limit must be taken of the response functions.

The form of V^E depends on the external perturbation $W(t)$. If $W(t)$ is a homogeneous electric field of frequency ω

$$\mathbf{F}(t) = \mathbf{F}^0\{\exp[(i\hbar\omega - \eta)t] + \exp[-(i\hbar\omega + \eta)t]\} \tag{36}$$

the interaction operators is

$$V^t = -\boldsymbol{\mu}\cdot\mathbf{F}(t) \tag{37}$$

where $\boldsymbol{\mu} = -e\mathbf{r}$ is the electric dipole operator. From Eqs (17), (36) and (37) it follows that

$$V^E = -\mathbf{F}_0\cdot\boldsymbol{\mu} \tag{38}$$

Since $-\mathbf{F}_0$ is a constant, it can be omitted in the response functions. When the external perturbation is an electric field, we wish to study the time development of the dipole moment, i.e. $A = \boldsymbol{\mu}$, and we see that Eq. (33) is identical to Eq. (13).

We have discussed the physical content of the linear response function in Section II. Let us now briefly consider the quadratic response function in Eq. (35). Let us again assume that $W(t)$ is an electric field, in which case the response function becomes

$$\begin{aligned}
&\langle\langle r^a; r^b; r^c \rangle\rangle_{E_1 + i\eta, E_2 + i\eta} \\
&= \sum_{k,n \neq 0} \left[\frac{\langle 0|r^a|k\rangle}{E_1 + E_2 - E_k + E_0 + 2i\eta} \left(\frac{\langle k|r_0^b|n\rangle\langle n|r^c|0\rangle}{E_2 - E_n + E_0 + i\eta} \right.\right. \\
&\quad \left. + \frac{\langle k|r_0^c|n\rangle\langle n|r^b|0\rangle}{E_1 - E_n + E_0 + i\eta} \right) + \left(\frac{\langle 0|r^c|n\rangle\langle n|r_0^b|k\rangle}{E_2 + E_n - E_0 + i\eta} \right. \\
&\quad \left. + \frac{\langle 0|r^b|n\rangle\langle n|r_0^c|k\rangle}{E_1 + E_n - E_0 + i\eta} \right) \frac{\langle k|r^a|0\rangle}{E_1 + E_2 + E_k - E_0 + 2i\eta} \\
&\quad - \langle k|r_0^a|n\rangle \left(\frac{\langle 0|r^b|k\rangle\langle n|r^c|0\rangle}{(E_1 + E_k - E_0 + i\eta)(E_2 - E_n + E_0 + i\eta)} \right. \\
&\quad \left.\left. + \frac{\langle 0|r^c|k\rangle\langle n|r^b|0\rangle}{(E_2 + E_k - E_0 + i\eta)(E_1 - E_n + E_0 + i\eta)} \right) \right]
\end{aligned} \tag{39}$$

where

$$r_0^a = r^a - \langle 0|r^a|0\rangle \tag{40}$$

From, for instance, Bloembergen (1965) and Stevens *et al.* (1963), we see that the quadratic response function is the r^a, r^b, r^cth component of the *dynamic* hyperpolarizability tensor, as it should be according to standard electric response theory. The residues of the quadratic response functions provide information about transition moments between two 'excited' states, i.e. neither of the two states is the reference state $|0\rangle$. More specifically we see from

Eq. (39) that

$$\lim_{E_1 \to E_0 - E_p - i\eta} \lim_{E_2 \to E_l - E_0 - i\eta} (E_1 + E_p - E_0 + i\eta)(E_2 - E_l + E_0 + i\eta)$$

$$\langle\langle r^a; r^b; r^c \rangle\rangle_{E_1 + i\eta, E_2 + i\eta}$$

$$= -\langle 0|r^b|p\rangle\langle p|r_0^a|l\rangle\langle l|r^c|0\rangle \tag{41}$$

Knowing the residues of the linear response function, $\langle 0|r^a|p\rangle$ and $\langle l|r^c|0\rangle$, we can find the transition moments between excited states $\langle p|r_0^a|l\rangle = \langle p|r^a|l\rangle$. The last equality follows from the definition in Eq. (40) when $|p\rangle$ and $|l\rangle$ are orthogonal excited states. In principle, the excited-state transition probabilities can also be obtained from the linear response function by using an excited state, say $|p\rangle$, as the reference state. This is a valid choice for the reference state. However, due to technical problems, such as the open-shell nature of most excited states, this has not been done yet and to obtain $\langle p|r^a|l\rangle$ from the residues of the quadratic response function is probably a better method.

The poles of the quadratic response function are the same as those of the linear response functions, i.e. the excitation energies of the system. This is also the case for the cubic response function which, furthermore, has the same kind of residues as the quadratic response function.

A whole range of different properties can be obtained by choosing other interaction operators $W(t)$ and thus other V^t and A operators. Examples are given by Olsen and Jørgensen (1985) and include a diversity of properties such as derivatives of the dipole polarizability, the B term in magnetic circular dichroism, the first anharmonicity of a potential energy surface, two-photon absorption cross sections and derivatives of the dynamic hyperpolarizability.

IV. EQUATIONS OF MOTION

Having analysed the form of the propagators and the information obtainable from the propagators, we now wish to determine the equations of motion. They are most conveniently obtained in the time representation. The time-dependent retarded linear response function is defined in Eq. (14) and was obtained from the energy-dependent propagator (Eq. (13)) through the transformation

$$\langle\langle P(t); Q(s)\rangle\rangle = \frac{1}{2\pi} \lim_{\eta\to 0^+} \int_{-\infty}^{\infty} \exp[-iE(t-s)]\langle\langle P; Q\rangle\rangle_E \, dE \tag{42}$$

This relation follows from insertion of Eq. (13) into Eq. (42) and

$$P(t) = \exp(iH_0 t)P(0)\exp(-iH_0 t) \tag{43}$$

We than recover Eq. (14). It should be pointed out that $P = P(0)$ is a time-independent Schrödinger operator, i.e. of the type we treated in Section III. In analogy we define

$$\langle\langle P(t); Q(s); R(v) \rangle\rangle$$
$$= \left(\frac{1}{2\pi}\right)^2 \lim_{\eta \to 0^+} \int_{-\infty}^{\infty} \mathrm{d}E_1 \exp[-\mathrm{i}E_1(t-s)] \int_{-\infty}^{\infty} \mathrm{d}E_2$$
$$\times \exp[-\mathrm{i}E_2(t-v)] \langle\langle P; Q; R \rangle\rangle_{E_1,E_2} \qquad (44)$$

Using the spectral from of the quadratic response function derived in Section III (Eq. (35)) we find (Dalgaard, 1982)

$$\langle\langle P(t); Q(s); R(v) \rangle\rangle$$
$$= -\langle 0|[[P(t), Q(s)], R(v)]|0\rangle \theta(t-s)\theta(s-v)$$
$$-\langle 0|[[P(t), R(v)], Q(s)]|0\rangle \theta(t-v)\theta(v-s) \qquad (45)$$

Dalgaard (1982) has a factor $\frac{1}{2}$ in front of the two terms on the right-hand side of this equation. He has included the factor coming from the Taylor series expansion of the response (see Eq. (33)) in the definition of the response functions. We follow the definitions used by Olsen and Jørgensen (1985).

The equations of motion for the linear and quadratic response functions are now obtained by taking the time derivatives and using the Heisenberg equation of motion

$$\mathrm{i}\frac{\mathrm{d}P(t)}{\mathrm{d}t} = [P(t), H_0] \qquad (46)$$

whereby we find from Eq. (14)

$$\mathrm{i}\frac{\mathrm{d}}{\mathrm{d}t}\langle\langle P(t); Q(s) \rangle\rangle = \langle\langle [P(t), H_0]; Q(s) \rangle\rangle + \delta(t-s)\langle 0|[P(t), Q(t)]|0\rangle \qquad (47)$$

and Eq. (45)

$$\mathrm{i}\frac{\mathrm{d}}{\mathrm{d}t}\langle\langle P(t); Q(s); R(v) \rangle\rangle = \langle\langle [P(t), H_0]; Q(s); R(v) \rangle\rangle$$
$$+ \delta(t-s)\langle\langle [P(t), Q(t)]; R(v) \rangle\rangle + \delta(t-v)\langle\langle [P(t), R(t)]; Q(s) \rangle\rangle \qquad (48)$$

Using the fact that

$$\delta(t-s) = \frac{1}{2\pi}\int_{-\infty}^{\infty} \exp[-\mathrm{i}E(t-s)]\,\mathrm{d}E \qquad (49)$$

and Eqs (42) and (44), respectively, combined with properties of Fourier integrals, we derive the equations

$$E\langle\langle P; Q \rangle\rangle_E = \langle 0|[P, Q]|0\rangle + \langle\langle [P, H_0]; Q \rangle\rangle_E \qquad (50)$$

$$(E+E')\langle\langle P;Q;R\rangle\rangle_{E,E'}$$
$$=\langle\langle [P,H_0];Q,R\rangle\rangle_{E,E'}+\langle\langle [P,Q];R\rangle\rangle_{E'}+\langle\langle [P,R];Q\rangle\rangle_{E} \quad (51)$$

At first sight these equations do not appear to be of any use since the simple response function is merely expressed in terms of a more complicated response function of the same kind involving $[P, H_0]$. However, it is possible to obtain a closed-form expression response function, as we shall see in the linear case. At the moment little is done to reformulate Eq. (51). Olsen and Jørgensen (1985) have shown how the quadratic and the cubic response function can be evaluated using a multiconfigurational self-consistent field (MCSCF) reference state.

In order to express Eq. (50) in a more compact form, it is useful to introduce the super-operator formalism (Pickup and Goscinski, 1973; Goscinski and Lukman, 1970). The time-independent operators are construed as elements in a super-operator space with a binary product

$$(P|Q)\equiv\langle 0|[P^+,Q]|\rangle \quad (52)$$

By means of the neutral element

$$\hat{I}P=P \quad (53)$$

the super-operator Hamiltonian

$$\hat{H}_0P=[H_0,P] \quad (54)$$

and the relation

$$\langle\langle [P,H_0];B\rangle\rangle_E=\langle\langle P;[H_0,B]\rangle\rangle_E \quad (55)$$

we can express the result of iterating on Eq. (50) as

$$\langle\langle P;Q\rangle\rangle_E=E^{-1}(P^+|\hat{I}Q)+E^{-2}(P^+|\hat{H}_0Q)+E^{-3}(P^+|\hat{H}_0^2Q)+\cdots$$
$$=(P^+|(E\hat{I}-\hat{H}_0)^{-1}|Q) \quad (56)$$

where the super-resolvent is defined through the series expansion

$$(E\hat{I}-\hat{H}_0)^{-1}=\frac{1}{E}\left[\hat{I}+\sum_{n=1}^{\infty}\left(\frac{\hat{H}_0}{E}\right)^n\right] \quad (57)$$

It is impracticable to work with an inverse operator, and using the inner projection technique (Löwdin, 1965) we can rewrite Eq. (56) as

$$\langle\langle P;Q\rangle\rangle_E=(P^+|\tilde{\mathbf{h}})(\mathbf{h}|E\hat{I}-\hat{H}_0|\tilde{\mathbf{h}})^{-1}(\mathbf{h}|Q) \quad (58)$$

Here $\mathbf{h}$ is a complete excitation operator manifold arranged as a column vector and $\tilde{\mathbf{h}}$ is the transposed row vector. Eq. (58) can also be derived from Eq. (56) using the identity (Simons, 1976)

$$\hat{I}=|\tilde{\mathbf{h}})(\mathbf{h}|\tilde{\mathbf{h}})^{-1}(\mathbf{h}| \quad (59)$$

Details of this derivation are given in Chapter 6.B.3 of Jørgensen and Simons (1981).

Eq. (58) represents the starting point for all approximate propagator methods. Even though in the derivation we only discussed the linear response functions or polarization propagators, a similar equation holds for the electron propagator. The equation for this propagator has the same form but there are differences in the choice of **h** and in the definition of the binary product (Eq. (52)), which for non-number-conserving, fermion-like operators should be

$$(P|Q) = \langle 0|P^{+}Q + QP^{+}|0\rangle = \langle 0|[P^{+}, Q]_{+}|0\rangle \tag{60}$$

The operator manifold **h** is traditionally chosen as

$$\mathbf{h} = \{\mathbf{h}_1, \mathbf{h}_3, \mathbf{h}_5, \ldots\} \tag{61}$$

with

$$\mathbf{h}_1 = \{a_m^{+}, a_\alpha\} \tag{62}$$

$$\mathbf{h}_3 = \{a_m^{+} a_n^{+} a_\beta, a_\alpha^{+} a_\beta^{+} a_n\} \qquad m > n \quad \text{and} \quad \alpha > \beta \tag{63}$$

etc., for the electron propagator and

$$\mathbf{h} = \{\mathbf{h}_2, \mathbf{h}_4, \ldots\} \tag{64}$$

with

$$\mathbf{h}_2 = \{a_m^{+} a_\alpha, a_\alpha^{+} a_n\} \tag{65}$$

$$\mathbf{h}_4 = \{a_m^{+} a_n^{+} a_\alpha a_\beta, a_\beta^{+} a_\alpha^{+} a_n a_m\} \qquad m > n \quad \text{and} \quad \alpha > \beta \tag{66}$$

etc., for the polarization propagator. Creation (a^{+}) and annihilation (a) operators with Roman indices refer to 'particles', i.e. spin orbitals that are unoccupied in the Hartree–Fock (HF) ground state, $|\text{HF}\rangle$, of the system, while unoccupied spin orbitals ('holes') are labelled with Greek indices. In the derivation of Eq. (58) we used that **h** operating on $|0\rangle$, the reference state, created a complete set of N-electron states for the polarization propagator and $(N \pm 1)$-electron states for the electron propagator. With the choices of **h** specified above, this statement is obvious if $|0\rangle$ were the HF ground state. However, it has been proved by Manne (1977) and Dalgaard (1979) that **h** applied to $|0\rangle$ also generates a complete set of states when $|0\rangle$ is an arbitrary correlated state, provided $|0\rangle$ is not orthogonal to $|\text{HF}\rangle$.

Using successively larger and larger operator manifolds we may define a hierarchy of approximate propagator methods based on Eq. (58). Also the choice of reference state influences Eq. (58) through the definition of the super-operator binary products in Eqs (52) and (60). It has been our experience that it is important to maintain a balance between the level of sophistication of the operator manifold and of the reference state. Better results are generally obtained this way, as we will see examples of in the subsequent sections.

V. PARTITIONING OF THE EQUATION OF MOTION

In matrix form Eq. (58) reads

$$\langle\langle P, Q \rangle\rangle_E = (\mathbf{P}_a^+, \mathbf{P}_b^+, \dots) \begin{pmatrix} \mathbf{M}_{aa} & \mathbf{M}_{ab} & \cdots \\ \mathbf{M}_{ba} & \mathbf{M}_{bb} & \cdots \\ \vdots & \vdots & \end{pmatrix}^{-1} \begin{pmatrix} \mathbf{Q}_a \\ \mathbf{Q}_b \\ \vdots \end{pmatrix} \tag{67}$$

where

$$\mathbf{P}_a = (P|\tilde{\mathbf{h}}_a) \tag{68}$$

and

$$\mathbf{M}_{ab} = (\mathbf{h}_a|E\hat{I} - \hat{H}_0|\tilde{\mathbf{h}}_b) \tag{69}$$

As shown in Section II, we wish to calculate the poles and residues of $\langle\langle P; Q \rangle\rangle_E$. However, even using moderately large operator manifolds, the inverse matrix becomes so large that we cannot evaluate all elements of it equally well. We therefore wish to treat one part of it better than the rest. Which part we choose will be directed by the physics of the problem. In order to do so it is convenient to partition (Löwdin, 1963) the inverse matrix, for instance in the following way (Nielsen *et al.*, 1980), letting $\mathbf{h}_b = \mathbf{h} - \mathbf{h}_a$

$$\begin{pmatrix} \mathbf{M}_{aa} & \mathbf{M}_{ab} \\ \mathbf{M}_{ba} & \mathbf{M}_{bb} \end{pmatrix}^{-1} \equiv \begin{pmatrix} \mathbf{T}_{aa} & \mathbf{T}_{ab} \\ \mathbf{T}_{ba} & \mathbf{T}_{bb} \end{pmatrix} \tag{70}$$

where

$$\mathbf{T}_{aa} = (\mathbf{M}_{aa} - \mathbf{M}_{ab}\mathbf{M}_{bb}^{-1}\mathbf{M}_{ba})^{-1} \tag{71}$$

$$\mathbf{T}_{ab} = -\mathbf{T}_{aa}\mathbf{M}_{ab}\mathbf{M}_{bb}^{-1} \tag{72}$$

$$\mathbf{T}_{ba} = -\mathbf{M}_{bb}^{-1}\mathbf{M}_{ba}\mathbf{T}_{aa} \tag{73}$$

$$\mathbf{T}_{bb} = \mathbf{M}_{bb}^{-1} + \mathbf{M}_{bb}^{-1}\mathbf{M}_{ba}\mathbf{T}_{aa}\mathbf{M}_{ab}\mathbf{M}_{bb}^{-1} \tag{74}$$

It should be mentioned that the partitioned matrix can be written in several other ways. The poles and residues of the partitioned propagator are exactly the same as those of the original propagator provided that we calculate all the individual $\mathbf{M}_{ab}$ matrices to the same level of accuracy. Inserting Eqs (70)–(74) in Eq. (67) gives

$$\langle\langle P; Q \rangle\rangle_E = (\mathbf{P}_a^+ - \mathbf{P}_b^+\mathbf{M}_{bb}^{-1}\mathbf{M}_{ba})\mathbf{T}_{aa}(\mathbf{Q}_a - \mathbf{M}_{ab}\mathbf{M}_{bb}^{-1}\mathbf{Q}_b) + \mathbf{P}_b^+\mathbf{M}_{bb}^{-1}\mathbf{Q}_b \tag{75}$$

The poles of $\langle\langle P; Q \rangle\rangle_E$ are also the poles of $\mathbf{T}_{aa}$ if we disregard the poles of $\mathbf{M}_{bb}$. Often the first non-vanishing terms in $\mathbf{M}_{ba}$ and $\mathbf{M}_{ab}$ are of one order higher than the leading term in the diagonal parts of the $\mathbf{M}$ matrix. This means that, in order to compute the poles of $\mathbf{T}_{aa}$ to a given order n, we need $\mathbf{M}_{aa}$ through order n but $\mathbf{M}_{bb}$ only through order (n–2). We thus choose $\mathbf{h}_a$ to contain the operators that describe the primary physical process we are interested in ($\mathbf{h}_1$

for ionization potentials/electron affinities, $\mathbf{h}_2$ for particle–hole excitations, etc.). The effects of, for example, 'shake-ups' on ionization potentials, i.e. $\mathbf{h}_3$, are then treated in a lower order. Had we been interested in computing a peak in the photoelectron spectrum that was predominantly of the shake-up type, we could just have inverted the definition of $\mathbf{h}_a$ $(=\mathbf{h}_3)$ and $\mathbf{h}_b$ $(=\mathbf{h}_1)$. The term *principal propagator* (Oddershede, 1982) is sometimes used for $\mathbf{T}_{aa}$ to indicate that it favours the description of the poles of primary interest.

The order concept is in most cases provided by perturbation theory and we thus compute $\mathbf{T}_{aa}$ (to find the poles) and the full $\langle\langle P;Q\rangle\rangle_E$ (to determine the residues as well) through a given order in perturbation theory by appropriate choices of $|0\rangle$ and $\mathbf{h}$, thereby obtaining the balanced approximations to Eq. (58) that was discussed in the last part of Section IV.

VI. ONE-ELECTRON PROPAGATORS

Historically the first applications of propagator theory were aimed at the evaluation of electron binding energies using the (one-)electron propagator (Pickup and Goscinski, 1973; Purvis and Öhrn, 1974), the equation-of-motion method (Simons and Smith, 1973) and the Green's function method (Cederbaum, 1975). Even though these methods might appear quite different, they are in fact equivalent. The electron propagator and the equation-of-motion methods lead to the same working equations (Herman *et al.*, 1978, 1981). The equivalence of the Green's function method and the electron propagator can probably best be demonstrated by means of Eq. (56). In both methods one starts out with a partitioning of H_0

$$H_0 = F_0 + V \tag{76}$$

where F_0 is the ground-state HF Hamiltonian and V thus becomes the fluctuation potential (electron repulsion minus the Fock potential). Since the commutator in Eq. (54) is linear

$$\hat{H}_0 = \hat{F}_0 + \hat{V} \tag{77}$$

and Eq. (56) reads

$$\mathbf{G}(E) = \langle\langle \mathbf{a}^+;\tilde{\mathbf{a}}\rangle\rangle_E = (\mathbf{a}|(E\hat{I} - \hat{F}_0 - \hat{V})^{-1}|\tilde{\mathbf{a}}) \tag{78}$$

where $\mathbf{G}(E)$ is the electron propagator matrix and $\mathbf{a} = \mathbf{h}_1$ is a column vector. The operators P and Q are chosen as annihilation and creation operators since we wish to calculate ionization potentials and electron affinities (see Section II). Using the identity

$$(A - B)^{-1} = A^{-1} + A^{-1}B(A - B)^{-1} \tag{79}$$

we write Eq. (78) as

$$\mathbf{G}(E) = \mathbf{G}_0(E) + (\mathbf{a}|(E\hat{I} - \hat{F}_0)^{-1}\hat{V}(E\hat{I} - \hat{F}_0 - \hat{V})^{-1}|\tilde{\mathbf{a}}) \tag{80}$$

where $\mathbf{G}_0(E)$ is defined as $\mathbf{G}(E)$ except that we omit V from the inverse operator. If we introduce the super-operator identity (Eq. (59)) and write $\mathbf{h}=\{\mathbf{a},\mathbf{x}\}$ with $\mathbf{x}$ being the orthogonal complement to $\mathbf{a}$, we find

$$\mathbf{G}(E)=\mathbf{G}_0(E)+\mathbf{G}_0(E)(\mathbf{a}|\hat{V}|\tilde{\mathbf{h}})(\mathbf{h}|(E\hat{I}-\hat{F}_0-\hat{V})^{-1}|\tilde{\mathbf{a}}) \tag{81}$$

Iterating this equation using Eq. (79) gives (Öhrn and Born, 1981, Section IV.B)

$$\begin{aligned}\mathbf{G}(E)=\mathbf{G}_0(E)+\mathbf{G}_0(E)[&(\mathbf{a}|\hat{V}|\tilde{\mathbf{x}})\\&+(\mathbf{a}|\hat{V}|\tilde{\mathbf{x}})(\mathbf{x}|(E\hat{I}-\hat{F}_0)^{-1}|\tilde{\mathbf{x}})(\mathbf{x}|\hat{V}|\tilde{\mathbf{a}})+\cdots]\mathbf{G}(E)\end{aligned} \tag{82}$$

or

$$\mathbf{G}(E)=\mathbf{G}_0(E)+\mathbf{G}_0(E)\mathbf{\Sigma}(E)\mathbf{G}(E) \tag{83}$$

where the self-energy $\mathbf{\Sigma}(E)$ is defined as

$$\mathbf{\Sigma}(E)=(\mathbf{a}|\hat{V}|\tilde{\mathbf{x}})+(\mathbf{a}|\hat{V}|\tilde{\mathbf{x}})(\mathbf{x}|(E\hat{I}-\hat{F}_0)^{-1}|\tilde{\mathbf{x}})(\mathbf{x}|\hat{V}|\tilde{\mathbf{a}})+\cdots \tag{84}$$

Eq. (83) is the Dyson equation (Abrikosov *et al.*, 1963) which is used in the Green's function method (Cederbaum and Domcke, 1977) and the two approximate schemes, the Green's function and the electron propagator methods, thus use the same starting equations.

The next step in the two methods consists of evaluating the electron propagator or the self-energy through a particular order in V defined in Eq. (76). Methods which include all terms through order 3 in perturbation theory have been derived and a term-by-term comparison shows that the third-order Green's function and electron propagator methods are identical (Öhrn and Born, 1981). Let us illustrate how the order analysis is carried out for the electron propagator. The operator manifolds in Eq. (70) have to be chosen in such a way that $\mathbf{h}_a=\mathbf{h}_1$ and $\mathbf{h}_b=\{\mathbf{h}_3,\mathbf{h}_5,\ldots\}$. Truncating $\mathbf{h}_b$ to only $\mathbf{h}_3$, the principal propagator of Eq. (71) becomes

$$\mathbf{T}_{11}(E)=(\mathbf{A}-\tilde{\mathbf{C}}\mathbf{D}^{-1}\mathbf{C})^{-1} \tag{85}$$

where

$$\begin{aligned}\mathbf{A}&=(\mathbf{h}_1|E\hat{I}-\hat{H}_0|\tilde{\mathbf{h}}_1)\\&=E\langle 0|[\mathbf{h}_1^+,\tilde{\mathbf{h}}_1]_+|0\rangle+\langle 0|[\mathbf{h}_1^+,[F_0,\tilde{\mathbf{h}}_1]]_+|0\rangle\\&\quad-\langle 0|[\mathbf{h}_1^+,[V,\tilde{\mathbf{h}}_1]]_+|0\rangle\end{aligned} \tag{86}$$

$$\mathbf{C}=(\mathbf{h}_3|E\hat{I}-\hat{H}_0|\tilde{\mathbf{h}}_1) \tag{87}$$

$$\mathbf{D}=(\mathbf{h}_3|E\hat{I}-\hat{H}_0|\tilde{\mathbf{h}}_3) \tag{88}$$

In Eq. (86) we have used the definitions of Eqs (54) and (60). Similar expanded expressions hold for the $\mathbf{C}$ and $\mathbf{D}$ matrices. We now wish to compute $\mathbf{T}_{11}(E)^{-1}$ to a particular order in perturbation theory. The order concepts is introduced in the references state through a Rayleigh–Schrödinger expansion

$$|0\rangle = K(|\mathrm{HF}\rangle + |0^{(1)}\rangle + 0^{(2)}\rangle + \cdots) \tag{89}$$

where $|0^{(1)}\rangle$ is the first-order correction to $|\mathrm{HF}\rangle$, i.e. it consists of doubly excited configurations relative to $|\mathrm{HF}\rangle$, $|0^{(2)}\rangle$ is the second-order correction to $|\mathbf{HF}\rangle$ (singles, doubles, triples, and quadruples) and K is a normalization constant. Zeroth-order contributions to **A** come from the first two terms in Eq. (86) with $|0\rangle = |\mathrm{HF}\rangle$, while in first order one of the reference states in the first two terms in Eq. (86) should be $|0^{(1)}\rangle$ and the other $|\mathrm{HF}\rangle$. Both $|0\rangle$ states must be equal to $|\mathrm{HF}\rangle$ in the last term to obtain a first-order contribution. Evaluation of the commutators shows that this first-order term is zero. In fact, through second order we obtain a rather simple expression for $\mathbf{T}_{11}(E)$ (Jørgensen and Simons, 1981, Chapter 6.D.2):

$$\mathbf{T}_{11}^{(2)}(E) = (\mathbf{A}_0 - \tilde{\mathbf{C}}_1\mathbf{D}_0^{-1}\mathbf{C}_1)^{-1} \tag{90}$$

where the subscripts on the **A**, **C** and **D** matrices specify the order of the matrices. We note also that $\mathbf{A}_2$ vanishes. In third order we find

$$\mathbf{T}_{11}^{(3)}(E) = [\mathbf{A}_0 + \mathbf{A}_3 - \tilde{\mathbf{C}}_1(\mathbf{D}_0 + \mathbf{D}_1)^{-1}\mathbf{C}_1 - \tilde{\mathbf{C}}_1\mathbf{D}_0^{-1}\mathbf{C}_2 - \tilde{\mathbf{C}}_2\mathbf{D}_0^{-1}\mathbf{C}_1]^{-1} \tag{91}$$

In order to evaluate the principal propagator, $\mathbf{T}_{11}(E)$, through third order it is not necessary to use the complete $|0^{(2)}\rangle$ wavefunction. Jørgensen and Simons (1975) have shown that only the singly excited part of $|0^{(2)}\rangle$ enters in $\mathbf{A}_3$ and none of the other matrices involve $|0^{(2)}\rangle$.

It should be noted that the strict order analysis applies to the *inverse* principal propagator. That means, as has always been carefully stated, that $\mathbf{T}_{11}$ itself is computed *through* a given order. There are several terms which are summed to infinite order. This can for instance be seen from iterating on Eq. (79):

$$(\mathbf{M}_0 - \mathbf{M}_1)^{-1} = \mathbf{M}_0^{-1} + \mathbf{M}_0^{-1}\mathbf{M}_1\mathbf{M}_0^{-1} + \mathbf{M}_0^{-1}\mathbf{M}_1\mathbf{M}_0^{-1}\mathbf{M}_1\mathbf{M}_0^{-1} + \cdots \tag{92}$$

where $\mathbf{M}_i$ is a matrix carrying an order (i). These infinite-order summations or renormalization terms (see e.g. Mattuck, 1976) are thus included in the electron propagator method. This was not the case in the original formulation of the Green's function method, that method now being referred to as the outer valence Green's function (OVGF) method (Cederbaum, 1975). However, another Green's function approach is now being used. That is the algebraic diagrammatic construction (ADC) and the basic idea there is to obtain the self-energy $\boldsymbol{\Sigma}(E)$ from the equation (Schirmer *et al.*, 1983)

$$\boldsymbol{\Sigma}_n(E) = \mathbf{U}^+(E\mathbf{1} - \mathbf{K}_0 - \mathbf{C}_n)^{-1}\mathbf{U} \tag{93}$$

where the indices on **K** and **C** again indicate the order of the matrices. A series of nth-order approximations to Eq. (93) is now obtained from comparison of the ADC form and a diagrammatic perturbation expansion, and requires that $\Sigma_n(E)$ is exact *through* order n

$$\Sigma_n(E) = \sum_{\nu=0}^{n} \Sigma^{(\nu)}(E) + O(n+1) \tag{94}$$

Obviously, those are the same considerations as we went through in order to obtain Eqs (90) and (91) and the electron propagator method and the ADC are thus equivalent methods. Using $n = 2$ in Eq. (93) we determine $\mathbf{T}_{11}^{(2)}$ and $n = 3$ gives $\mathbf{T}_{11}^{(3)}$. The $\mathbf{U}$ matrix in Eq. (93) corresponds to the transition matrix (cf. Eq. (75)). Both $\mathbf{T}_{11}^{(2)}$ and $\mathbf{T}_{11}^{(3)}$ only contain $\mathbf{C}$ and $\mathbf{D}$ terms (see Eqs (87), (88), (90) and (91), i.e. $\mathbf{h}_b = \mathbf{h}_3$ alone. From Eq. (63) we see that we may classify the operators in $\mathbf{h}_3$ as 2p–1h (two-particle, one-hole) and 2h–1p operators, and the $n = 3$ ADC approach, corresponding to the third-order electron propagator method, is therefore referred to as the extended 2p–1h Tamm–Dancoff approximation (TDA) (Walter and Schirmer, 1981). A fourth-order approximation to the ADC equations has also been described (Schirmer *et al.*, 1983) but not yet tested in actual applications.

Not all electron propagator calculations are based upon an order-by-order evaluation of the propagator. Redmon *et al.* (1975) included the $\mathbf{h}_5$ operator manifold, without computing all terms in $\mathbf{T}_{11}^{-1}$ in the next order of perturbation theory. They found that in some cases the effect of $\mathbf{h}_5$ was more important than that of $\mathbf{h}_3$. Also multiconfigurational electron propagator calculations have appeared in recent years, showing the need for a flexible representation of the reference state when we wish to calculate non p–h dominated peaks in the photoelectron spectra (Nichols *et al.*, 1984; Golab *et al.*, 1986).

The photoelectron spectra of several molecules have been calculated by means of the propagator and, in particular, the Green's function methods. A comprehensive listing of the atoms and molecules treated with these methods is given in Table 1 (161 entries) of Von Niessen *et al.* (1984).

VII. POLARIZATION PROPAGATORS

A. The Random-phase Approximation

In describing polarization propagator methods it is instructive to start out with the simplest consistent method of the kind, namely the random-phase approximation (RPA). Within the framework we use here, RPA is described as the approximation to the general equation of motion (Eq. (58)) in which we set $\mathbf{h} = \mathbf{h}_2$ and assume $|0\rangle = |\mathrm{HF}\rangle$, that is, use the simplest truncation in both Eqs (64) and (89). It is convenient to split $\mathbf{h}_2$ up into p–h and h–p excitation operators

$$\mathbf{h}_2 = \{\mathbf{q}^+, \mathbf{q}\} \tag{95}$$

where

$$q^+_{m\alpha} = a^+_m a_\alpha \tag{96}$$

$$q_{m\alpha} = a^+_\alpha a_m \tag{97}$$

Since $\mathbf{h}_a = \mathbf{h}_2$ and $\mathbf{h}_b = 0$, Eq. (67) reads

$$\left(\langle\langle P;Q\rangle\rangle_E = ((P^+|\tilde{\mathbf{q}}^+),(P^+|\tilde{\mathbf{q}}))\right)\begin{pmatrix} E\mathbf{1}-\mathbf{A} & -\mathbf{B}^* \\ -\mathbf{B} & -E\mathbf{1}-\mathbf{A}^* \end{pmatrix}^{-1}\begin{pmatrix} (\mathbf{q}^+|Q) \\ (\mathbf{q}|Q) \end{pmatrix} \tag{98}$$

where

$$\mathbf{A} = (\mathbf{q}^+|\hat{H}_0|\tilde{\mathbf{q}}^+) \tag{99}$$

$$\mathbf{B} = (\mathbf{q}|\hat{H}_0|\tilde{\mathbf{q}}^+) \tag{100}$$

and

$$(P^+|q^+_{m\alpha}) = P_{\alpha m} \tag{101}$$

In Eq. (101) we have used Eqs (5) and (52) and the properties of the $|\mathrm{HF}\rangle$ ground state.

We now wish to write Eq. (98) in a spectral form which readily allows identification of excitation energies and transition moments in the same way as we did for the exact propagator in Eq. (13), i.e. from the poles and residues. This is done by introducing the RPA eigenvector matrix (Oddershede *et al.*, 1984, Section 3.2)

$$\begin{pmatrix} -\mathbf{A} & -\mathbf{B} \\ -\mathbf{B} & -\mathbf{A} \end{pmatrix}\begin{pmatrix} \mathbf{Z} & \mathbf{Y} \\ \mathbf{Y} & \mathbf{Z} \end{pmatrix} = \begin{pmatrix} \mathbf{Z} & \mathbf{Y} \\ -\mathbf{Y} & -\mathbf{Z} \end{pmatrix}\begin{pmatrix} -\boldsymbol{\omega} & \mathbf{0} \\ \mathbf{0} & \boldsymbol{\omega} \end{pmatrix} \tag{102}$$

which fulfils the normalization condition

$$\begin{pmatrix} \tilde{\mathbf{Z}} & \tilde{\mathbf{Y}} \\ \tilde{\mathbf{Y}} & \tilde{\mathbf{Z}} \end{pmatrix}\begin{pmatrix} \mathbf{Z} & \mathbf{Y} \\ -\mathbf{Y} & -\mathbf{Z} \end{pmatrix} = \begin{pmatrix} \mathbf{1} & \mathbf{0} \\ \mathbf{0} & -\mathbf{1} \end{pmatrix} \tag{103}$$

Adding

$$\begin{pmatrix} E\mathbf{1} & \mathbf{0} \\ \mathbf{0} & -E\mathbf{1} \end{pmatrix}\begin{pmatrix} \mathbf{Z} & \mathbf{Y} \\ \mathbf{Y} & \mathbf{Z} \end{pmatrix}$$

to both sides of Eq. (102) and taking the inverse of the resulting equation we find, using Eq. (103),

$$\begin{pmatrix} E\mathbf{1}-\mathbf{A} & -\mathbf{B} \\ -\mathbf{B} & -E\mathbf{1}-\mathbf{A} \end{pmatrix}^{-1} = \begin{pmatrix} \mathbf{Z} & \mathbf{Y} \\ \mathbf{Y} & \mathbf{Z} \end{pmatrix}\begin{pmatrix} E\mathbf{1}-\boldsymbol{\omega} & \mathbf{0} \\ \mathbf{0} & E\mathbf{1}+\boldsymbol{\omega} \end{pmatrix}^{-1}\begin{pmatrix} \tilde{\mathbf{Z}} & \tilde{\mathbf{Y}} \\ -\tilde{\mathbf{Y}} & -\tilde{\mathbf{Z}} \end{pmatrix} \tag{104}$$

If we assume that **A** and **B** are real and use Eqs (98) and (104) we obtain

$$\langle\langle P; Q \rangle\rangle_E = \sum_\lambda \left(\frac{(P^+|O_\lambda^+)(O_\lambda^+|Q)}{E-\omega_\lambda} - \frac{(P^+|O_\lambda)(O_\lambda|Q)}{E+\omega_\lambda} \right) \tag{105}$$

where the RPA excitation operators are defined as

$$O_\lambda^+ = \tilde{\mathbf{q}}^+\mathbf{Z}_\lambda + \tilde{\mathbf{q}}\mathbf{Y}_\lambda \tag{106}$$

$$O_\lambda = \tilde{\mathbf{q}}^+\mathbf{Y}_\lambda + \tilde{\mathbf{q}}\mathbf{Z}_\lambda \tag{107}$$

with $\mathbf{Z}_\lambda$ and $\mathbf{Y}_\lambda$ being the eigenvector of the RPA eigenvalue ω_λ. We see from Eq. (105) that the RPA excitation energies are the eigenvalues of the (non-Hermitian) eigenvalue problem in Eq. (102) and that the transition moments

$$(P^+|O_\lambda^+) = \sum_{\alpha m} P_{\alpha m} Z_{\alpha m,\lambda} - P_{m\alpha} Y_{\alpha m,\lambda} \tag{108}$$

are given in term of the eigenvectors in Eq. (102) and the elementary properties integrals $P_{\alpha m}$. The computational steps in an RPA calculation are thus: perform an SCF calculation; transform the integrals to the SCF representation and evaluate the **A**, **B** and $P_{\alpha m}$ matrices; solve the eigenvalue problem in Eq. (102) and determine the excitation energies and transition moments as described above.

Explicit expressions for the **A**, **B**, etc., matrices are given in several places (e.g. Oddershede *et al.*, 1984, Appendix C). The RPA eigenvalue problem of Eq. (102) can be solved in many different ways (see e.g. Oddershede *et al.*, 1984, Appendix A). It is possible to cast it into two Hermitian eigenvalue problems for $\mathbf{A}+\mathbf{B}$ and $\mathbf{A}-\mathbf{B}$ (Jørgensen and Linderberg, 1970), thereby reducing the dimensionality of the problem in Eq. (102) by a factor of 2. However, that may not be sufficient in practical applications. The **A** and **B** matrices are of the dimension $N_{\text{ph}} \times N_{\text{ph}}$ where N_{ph} is the number of particle–hole excitations (see Eqs (96), (97), (99) and (100)). Even for moderately sized basis sets N_{ph} can easily be several hundred (Bouman *et al.*, 1983, 1985). Special methods are then used to determined the few lowest eigenvalues and eigenvectors of the RPA problem (Rettrup, 1982; Flament and Gervais, 1979; Hirao and Nakatsuji, 1982).

We saw in Section III that the polarization propagator is the linear response function. The linear response of a system to an external time-independent perturbation can also be obtained from the coupled Hartree–Fock (CHF) approximation provided the unperturbed state is the Hartree–Fock state of the system. Thus, RPA and CHF are the same approximation for time-independent perturbing fields, that is for properties such as spin–spin coupling constants and static polarizabilities. That we indeed obtain exactly the same set of equations in the two methods is demonstrated by Jørgensen and Simons (1981, Chapter 5.B). Frequency-dependent response properties in the

Hartree–Fock approximation are most conveniently calculated from the RPA response functions at the frequency of interest.

From the electronic transition moments and the excitation energies we may compute oscillator strengths. Using the residues of the $\langle\langle \mathbf{r};\mathbf{r}\rangle\rangle_E$ response function we obtain the dipole length oscillator strengths as (see Eq. (8))

$$f^{\mathrm{L}}_{0n} = \tfrac{2}{3}|\langle 0|\mathbf{r}|n\rangle|^2(E_n - E_0) \tag{109}$$

where transition moments and excitation energies are given in atomic units. From the residues of the $\langle\langle \mathbf{p};\mathbf{p}\rangle\rangle_E$ propagator the dipole velocity oscillator strengths

$$f^{\mathrm{V}}_{0n} = \tfrac{2}{3}|\langle 0|\mathbf{p}|n\rangle|^2/(E_n - E_0) \tag{110}$$

are obtained. Also the mixed (Hansen, 1967) oscillator strengths

$$f^{\mathrm{M}}_{0n} = \tfrac{2}{3}|\langle 0|\mathbf{r}|n\rangle\cdot\langle 0|\mathbf{p}|n\rangle| \tag{111}$$

can be computed. No matter which of the two propagators $\langle\langle \mathbf{r};\mathbf{r}\rangle\rangle_E$ or $\langle\langle \mathbf{p};\mathbf{p}\rangle\rangle_E$ we consider, the inverse matrix in Eq. (98), i.e. the principal propagator (Oddershede, 1982), is the same, so only inexpensive calculations of properties integrals are needed in order to compute the other propagator once one of them is known. It is thus possible at little extra cost to compute oscillator strengths in all three formulations. That is very useful since $f^{\mathrm{L}}_{0n} = f^{\mathrm{V}}_{0n} = f^{\mathrm{M}}_{0n}$ in RPA, *provided* a complete basis set is used. This is a rather unique feature of RPA (Harris, 1969; Jørgensen and Linderberg, 1970) which does not hold for most other approximate methods. This also gives us a way to judge the quality of the basis set. A necessary (but *not* sufficient) condition for a complete basis set is that it must give the same oscillator strengths in the dipole length, the dipole velocity and the mixed representation. In practical calculations we often use this condition to choose an appropriate basis set. It also proves useful to utilize the Thomas–Reiche–Kuhn sum rule in this aspect

$$\sum_n f_{0n} = N \tag{112}$$

since this relation holds in RPA as well.

The random-phase approximation is a very useful computational method for many properties, in particular for singlet response properties for which very reliable results are obtained. Triplet properties, both excitation energies and response properties such as most contributions to the indirect nuclear spin–spin coupling constants (Geertsen and Oddershede, 1984), can be quite incorrect in RPA due to near-instabilities of the SCF reference state in some cases. An instability of the SCF solution shows that there exists another state with a lower energy than the SCF state. The Hessians $\mathbf{A} + \mathbf{B}$ or $\mathbf{A} - \mathbf{B}$ have (at least) one negative eigenvalue (Golab *et al.*, 1983). This other state is generally of non-singlet spin multiplicity and singlet (but not triplet) properties are therefore well represented in RPA.

TABLE I
Review of molecular RPA calculations[a,b] using *ab initio* methods, 1977–1985[c].

System	Basis sets[d]	Properties[e]	References
H_2	37 STOs	$\Delta\varepsilon, f(q), S(n)$	Arrighini *et al.* (1980)
	[9s6p3d]	$S(n), I_\mu$	Geertsen *et al.* (1986)
He_2	[11s7p2d]	E_{disp}	Jaszunski and McWeeny (1985)
LiH	[9s7p2d/8s6p1d]	$S(0), \alpha(\omega), \gamma_{\alpha\beta}(\omega)$	Lazzeretti *et al.* (1983c)
FH	[9s6p4d/6s3p1d]	$\alpha(\omega), \gamma_{\alpha\beta}(\omega)\chi^{\mathrm{p}}, \sigma$	Lazzeretti *et al.* (1983a, b), Lazzeretti and Zamasi (1985)
N_2	[4s3p] + diffuse (32 CGTO)	$\Delta\varepsilon, f, S(n)$	Brattsev and Semenova (1980)
	50 STOs	$\alpha(\omega), \alpha'(\omega)$, Raman	Svendsen and Oddershede (1979)
H_2O	[15s8p4d/10s3p]	$S(n), I_\mu$	Geertsen *et al.* (1986)
CO_2	[3s2p]	$\Delta\varepsilon, f, S(n)$	Brattsev and Semenova (1980)
$(FH)_2$	[4s2p1d/3s]	E_{disp}	Jaszunski and McWeeny (1985)
CH_4	[5s3p2d/3s1p]	$\alpha'(\omega)$	Stroyer-Hansen and Svendsen (1984)
	[5s3p2d/3s1p]	$\alpha'(\omega), \alpha(0)$	Svendsen and Stroyer-Hansen (1985)
SiH_4	[6s5p2d/5s1p]	J_{AB}	Lazzeretti *et al.* (1984)
AlH_4^-	[6s5p2d/5s1p]	J_{AB}	Lazzeretti *et al.* (1984)
C_2H_2	[5s3p2d/3s1p]	$\alpha'(\omega), \alpha(0)$	Svendsen and Stroyer-Hansen (1985)
$CH_2{=}CH_2$	[5s3p/3s]	$\Delta\varepsilon, f, R$	Bouman *et al.* (1983)
	[5s4p1d/3s]	$\Delta\varepsilon, f$	Bouman and Hansen (1985)
	[5s3p2d/3s1p]	$\alpha'(\omega), \alpha(0)$	Svendsen and Stroyer-Hansen (1985)
C_2H_6	[5s3p2d/3s1p]	$\alpha'(\omega), \alpha(0)$	Svendsen and Stroyer-Hansen (1985)
$C_3H_3^+$	[6s4p1d/4s]	$\Delta\omega, f$	Eyler *et al.* (1984)
Propene	[5s4p1d/3s]	$\Delta\varepsilon, f$	Bouman and Hansen (1985)
CH_3CSCH_3	4/31G[f], 44/31G[f] + diffuse	$\Delta\varepsilon, f, R$	Lightner *et al.* (1984)
2-Butene	[5s4p1d/3s]	$\Delta\varepsilon, f$	Bouman and Hansen (1985)
$C_2H_5CSCH_3$	4/31G[f], 44/31G[f] + diffuse	$\Delta\varepsilon, f, R$	Lightner *et al.* (1984)
$C_2H_5COC_2H_5$	STO-4G[f]	$\Delta\varepsilon, f, R$	Bouman *et al.* (1979)
3-Methyl cyclopentene	[5s4p/2s]	$\Delta\varepsilon, f, R$	Hansen and Bouman (1985)
5-Methyl-1, 3-cyclohexadiene	STO-4G[f]	$\Delta\varepsilon, f, R$	Lightner *et al.* (1981a)
trans-Cyclooctene	[5s4p/2s]	$\Delta\varepsilon, f, R$	Hansen and Bouman (1985)

TABLE I (*contd.*)

System	Basis sets[d]	Properties[e]	References
7-Norbornenone	STO-4G[f]	$\Delta\varepsilon$	Lightner *et al.* (1981b)
3-Hydroxychromone	STO-5G[f]	$\Delta\varepsilon$	Bouman *et al.* (1985)

[a]Coupled Hartree–Fock calculations are not included (see text).
[b]Several of the references included in Table III give RPA results in addition to the ones listed in this table.
[c]This table is an extension of Table A. I in Oddershede (1978).
[d]Both Slater-type (STO) and Gaussian ([···]) basis sets are used. If more than one basis set was applied, only the largest set is given in the table. (CGTO = Contracted Gaussian type orbitals).
[e]The list of properties include.

$\Delta\varepsilon$	excitation energies,
f	oscillator strengths (often in several formulations),
$f(q)$	generalized oscillator strengths,
$S(n)$	$\sum_k f_{0k}\Delta\varepsilon_{0k}^n$,
I_μ	mean excitation energies,
$\alpha(\omega)$	frequency-dependent polarizabilities,
$\gamma_{\alpha\beta}(\omega)$	frequency-dependent nuclear electric shielding,
$\alpha'(\omega)$	derivatives of $\alpha(\omega)$ with respect to normal coordinates,
E_{disp}	dispersion energies,
J_{AB}	indirect nuclear spin–spin coupling constants,
R	optical rotatory strengths.

[f]The built-in basis set of the GAUSSIAN program system.

χ^p	paramagnetic susceptibility
σ	nuclear magnetic shielding

Because of its computational simplicity and other obvious qualities the random-phase approximation has been used in many calculations. Reviews of RPA calculations include one on chiroptical properties by Hansen and Bouman (1980), one on the equation-of-motion formulation of RPA (McCurdy *et al.*, 1977) and my own review of the literature through 1977 (Oddershede, 1978, Appendix B). *Ab initio* molecular RPA calculations in the intervening period are reviewed in Table I. Coupled Hartree–Fock calculations have not been included in the table. Only calculations which require diagonalization of both $\mathbf{A}+\mathbf{B}$ and $\mathbf{A}-\mathbf{B}$ and thus may give frequency-dependent response properties *and* excitation spectra are included. In CHF we only need to evaluate either $(\mathbf{A}+\mathbf{B})^{-1}$ or $(\mathbf{A}-\mathbf{B})^{-1}$ in order to determine the (static) response properties.

B. The Multiconfiguration Random-Phase Approximation

An improvement of the RPA would be obtained if we could remove the possibility of instabilities from the reference state $|0\rangle$, i.e. make sure that the Hessian is positive definite. The multiconfigurational SCF (MCSCF) wavefunction for the ground state fulfils this requirement. Furthermore, MCSCF introduces a much higher level of electronic correlation into the propagator method. Also, away from equilibrium (bond breaking, etc.), where the SCF reference state is very poor or perhaps even gives an incorrect description of physical reality, we would expect the MCSCF state

$$|0\rangle = \sum_i C_{i0}|\Phi_i\rangle \tag{113}$$

to be a good choice for the reference state. In Eq. (113) $\{|\Phi_i\rangle\}$ is a set of configuration state functions, the selection of which must be done with care and often using chemical intuition if reliable results are to be obtained. The selection of configuration probably represents the greatest challenge in multiconfiguration RPA (MCRPA) calculations.

The equation of motion for the MCRPA propagator is derived from the general equation of motion (Eq. (58)) using $|0\rangle$ in Eq. (113) as the reference state and a projection manifold $\mathbf{h}$ consisting of

$$\mathbf{h} = \{\mathbf{q}^+, \mathbf{R}^+, \mathbf{q}, \mathbf{R}\} \tag{114}$$

where $\{\mathbf{q}^+, \mathbf{q}\}$ are defined in Eqs (95)–(97) and the *state transfer operators* $\mathbf{R}^+ = \{R_n\}$ are

$$R_n^+ = |n\rangle\langle 0| \tag{115}$$

Here $\{|n\rangle\}$ are all the states except $|0\rangle$ that can be expanded in the configuration state functions $\{|\Phi_i\rangle\}$ in Eq. (113), i.e. the orthogonal complement of $|0\rangle$. The inclusion of $\mathbf{R}^+$ and $\mathbf{R}$ in the projection manifold introduces, in addition to the p–h and h–p excitations, the double, triple, etc., excitations that were considered to be important in the MCSCF calculation of $|0\rangle$. Sometimes additional transfer operators are included in the projection manifold in order to improve the performance of MCRPA, in particular with respect to the description of excitation to states of symmetries other than that of $|0\rangle$ (Yeager *et al.*, 1981; Albertsen *et al.*, 1980a; Lynch *et al.*, 1982).

From Eqs (67) and (114) it follows that the MCSCF polarization propagator is (for a detailed derivation see Jørgensen and Simons, 1981, Chapter 6.E.3))

$$\langle\langle P; Q\rangle\rangle_E = (\mathbf{P}, -\mathbf{P})\begin{pmatrix} E\mathbf{S} - \mathbf{A} & E\boldsymbol{\Delta} - \mathbf{B} \\ -E\boldsymbol{\Delta} - \mathbf{B} & -E\mathbf{S} - \mathbf{A} \end{pmatrix}^{-1}\begin{pmatrix} \mathbf{Q} \\ -\mathbf{Q} \end{pmatrix} \tag{116}$$

if we define

$$\mathbf{A} = \begin{pmatrix} \langle 0|[\mathbf{q}, [H_0, \tilde{\mathbf{q}}^+]]|0\rangle & \langle 0|[\mathbf{R}^+, [H_0, \tilde{\mathbf{q}}]]|0\rangle \\ \langle 0|[\mathbf{R}, [H_0, \tilde{\mathbf{q}}^+]]|0\rangle & \langle 0|[\mathbf{R}, [H_0, \tilde{\mathbf{R}}^+]]|0\rangle \end{pmatrix} \tag{117}$$

$$\mathbf{S} = \begin{pmatrix} \langle 0|[\mathbf{q}, \tilde{\mathbf{q}}^+]|0\rangle & \langle 0|[\mathbf{q}, \tilde{\mathbf{R}}^+]|0\rangle \\ \langle 0|[\mathbf{R}, \tilde{\mathbf{q}}^+]|0\rangle & \langle 0|[\mathbf{R}, \tilde{\mathbf{R}}^+]|0\rangle \end{pmatrix} \tag{118}$$

$$\mathbf{P} = (\langle 0|[P^+, \tilde{\mathbf{q}}^+]|0\rangle, \langle 0|[P^+, \tilde{\mathbf{R}}^+]|0\rangle) \tag{119}$$

and $\mathbf{B}$ and $\boldsymbol{\Delta}$ are obtained from $\mathbf{A}$ and $\mathbf{S}$ by replacing $\mathbf{q}$ by $\mathbf{q}^+$ and $\mathbf{R}$ by $\mathbf{R}^+$ as the first operators in the commutators. The inverse principal propagator in Eq. (116) is a 4×4 block matrix due to the choice in (114). We note that the top left blocks in **A** and **B**, except for the different choice of $|0\rangle$, are the same as the RPA matrices in Eqs (99) and (100). However, the extension of the manifold compared to Eq. (95) has introduced two new parts to **A** and **B**, namely the off-diagonal blocks describing interactions between the orbital space and the configuration space and a part which accounts for interactions within the configuration space. The matrices $\mathbf{A} - \mathbf{B}$ and $\mathbf{A} + \mathbf{B}$ are the Hessian matrices of the second-order MCSCF procedure (Dalgaard and Jørgensen, 1978), for real and imaginary variations, respectively. Thus, when a second-order MCSCF calculation for the reference state has been performed, a substantial part of the computational steps in an MCRPA calculation are done. Many of the MCSCF results can be re-used in the MCRPA calculation.

The formulation of MCRPA that is outlined here was first suggested by Yeager and Jørgensen (1979). A careful examination of the derivation leading to Eqs (116)–(118) would reveal that they forced the **A** matrix to be Hermitian. In the top right block of **A** in Eq. (117) we have interchanged $\mathbf{R}^+$ and $\mathbf{q}$ compared to the form it would have had in a straightforward derivation. This symmetrization is necessary in order to obtain real excitation energies. It also means that we retain many of the properties of RPA in MCRPA. For instance, the equivalence of the various formulations for the oscillator strengths (cf. Eqs (109)–(111) in a complete orbital basis set is maintained in MCRPA (Dalgaard, 1980; Albertsen *et al.*, 1980a). The same holds for the Thomas–Reiche–Kuhn sum rule, Eq. (112) (Dalgaard, 1980; Yeager *et al.*, 1981). It should also be mentioned that the coupled MCSCF equations are obtained in the $E = 0$ limit of MCRPA (Jørgensen and Simons, 1981, Chapter 5.B) and that MCRPA is invariant to rotations among the orbitals created by the redundant set of orbital excitation operators (Albertsen *et al.*, 1980b).

The excitation energies and transition moments of MCRPA can be determined as in RPA. Except for the dimensions of the **A** and **B** matrices and a different metric matrix $(\mathbf{S}, \mathbf{\Delta})$ the propagators in MCRPA and RPA have the same form and we can thus express $\langle\langle P; Q\rangle\rangle_E$ in Eq. (116) in a spectral form using the transformations described in Eqs (102)–(105). The solution of the eigenvalue problem that arises has to be handled a little differently due to the non-unit metric matrix (see e.g. Oddershede *et al.*, 1984, Appendix A).

The computational experience with the MCRPA method is still limited to a few systems. These calculations are reviewed later in Table III. However, the available results so far have been very encouraging and have demonstrated that MCRPA is a reliable method for the calculation of spectral properties.

C. The Antisymmetrized Geminal Power Method

This method is named after the reference state used

$$|\mathrm{AGP}\rangle = \mathscr{A} g(1,2) g(3,4) \cdots g(2n-1),(2n) \tag{120}$$

that is, an antisymmetrized ($\mathscr{A}$) geminal power (AGP) wavefunction constructed from n identical geminals expressed in two-electron determinants

$$g(1,2) = \sum_{k=1}^{M} g_k |u_k u_{k+M}| \tag{121}$$

where $\{u_k\}$ is a set of either $2M$ spin orbitals or natural spin orbitals (Jensen *et al.*, 1982). The interest in this form for the reference state wave was spurred by the findings of Linderberg and Öhrn (1977) and Öhrn and Linderberg (1979) who showed that $|\text{AGP}\rangle$ represents the natural RPA ground state. Natural in this context means the $|\text{AGP}\rangle$ under certain conditions (Linderberg, 1980; Weiner and Goscinski, 1980) fulfils the 'killer' condition

$$Q_\lambda |\text{AGP}\rangle = 0 \tag{122}$$

and simultaneously obeys

$$Q_\lambda^+ |\text{AGP}\rangle = |\lambda\rangle \tag{123}$$

where $|\lambda\rangle$ is an excited state and Q_λ and Q_λ^+ are the AGP de-excitation and excitation operators, respectively.

Fulfilment of Eqs (122) and (123) is a necessary condition for $|\text{AGP}\rangle$ to be the RPA ground state and $|\lambda\rangle$ an excited state. One may except that $|\text{SCF}\rangle$ would serve this purpose since we used this reference state to construct the RPA solutions. However, applying the RPA de-excitation operator of Eq. (107) to $|\text{SCF}\rangle$ we find

$$Q_\lambda |\text{SCF}\rangle = \tilde{\mathbf{q}}^+ |\text{SCF}\rangle \mathbf{Y}_\lambda \tag{124}$$

which is non-zero (but small, since it is proportional to the 'small' component $\mathbf{Y}_\lambda$ of the RPA eigenvector). This, inconsistency is referred to as *the ambiguity of RPA*: the state used to construct the RPA matrices is not the RPA 'vacuum' state. It is also sometimes expressed as '$|\text{SCF}\rangle$ is not the (self-) consistent RPA ground state' (Linderberg *et al.*, 1972). However, the fact that we do not know the exact form of the RPA ground state is not important for the determination of transition moments and excitation energies. We saw in Section VII.A (Eq. (108)) that the RPA transition moment is given as a *commutator*

$$\langle 0|\mathbf{r}|\lambda\rangle = \langle \text{HF}|[\mathbf{r}, O_\lambda^+]|\text{HF}\rangle \tag{125}$$

but *not* as

$$\langle 0|\mathbf{r}|\lambda\rangle = \langle \text{RPA}|\mathbf{r} O_\lambda^+ |\text{RPA}\rangle \tag{126}$$

as it would if the killer condition were obeyed for the state $|\text{RPA}\rangle = |0\rangle$. Knowing the form of the states involved in an excitation gives extra information in addition to that obtained from the poles and residues of the propagator, so clearly fulfilment of the killer condition is a desirable property of an approximate propagator calculation. Also, the question of N-

representability (Coleman, 1963) of the reference and the excited states cannot be answered definitely unless we know the states.

From the initial studies of Linderberg and Öhrn previously mentioned, it is expected that $|\mathrm{AGP}\rangle$ is a very good approximation to the consistent RPA ground state. However, Eq. (122) is not obeyed *exactly* unless the RPA **B** matrix is set to zero. Expressed a little differently, it is a consequence of the AGP assumption and Eq. (122) that the **B** matrix must be zero (Sangfelt *et al.*, 1984). Instead of evaluating the (presumably small) **B** matrix, most AGP propagator calculations (Weiner *et al.*, 1984; Sangfelt *et al.*, 1984; Goscinski and Weiner, 1980; Weiner and Goscinski, 1983) consider a model problem in which the exact Hamiltonian H_0 is replaced by an outer projection (Löwdin, 1965)

$$H_0^P = PH_0P \tag{127}$$

where

$$P = |\mathrm{AGP}\rangle\langle\mathrm{AGP}| + \sum_m \sum_{i<j} q_{ij}^+(m)|\mathrm{AGP}\rangle\langle\mathrm{AGP}|q_{ij}(m) \tag{128}$$

Using H_0^P instead of H_0 in Eq. (100) makes the $\mathbf{B}^P$ matrix zero and from Eq. (98) we obtain the AGP propagator

$$\langle\langle P;Q\rangle\rangle_E = ((P^+|\tilde{\mathbf{q}}^+),(P^+|\tilde{\mathbf{q}}))\begin{pmatrix} E\mathbf{1}-\mathbf{A}^P & \mathbf{0} \\ \mathbf{0} & -E\mathbf{1}-\mathbf{A}^P \end{pmatrix}^{-1}\begin{pmatrix} (\mathbf{q}|Q) \\ (\mathbf{q}|Q) \end{pmatrix} \tag{129}$$

In all applications of AGP propagator methods—also referred to as the self-consistent polarization propagator (SCPP) method (Goscinski *et al.*, 1982)—it is assumed that $|\mathrm{AGP}\rangle$ is a spin singlet constructed from pure singlet (pair) geminals

$$g(1,2) = \sum_{k=1}^{M} g_k|u_k u_{\bar{k}}| \tag{130}$$

(For an exception, see Sangfelt and Goscinski (1985).) The form of the AGP singlet $q_{ij}(1)$ and triplet $q_{ij}(3)$ excitation operators in Eqs (128) and (129) are given by Weiner *et al.* (1984).

From Eq. (129) it appears that the AGP excitation energies are simply the eigenvalues of $\mathbf{A}^P$ and Eqs (99) and (122) imply that we retrieve the mono-excited CI or Tamm–Dancoff approximation (TDA) within the set of AGP excitation operators. In more recent literature (Kurtz *et al.*, 1985) the level of approximation defined in Eq. (129) is therefore called AGPTDA to distinguish it from standard TDA based on an SCF reference state. Very recently Weiner and Öhrn (1985) and Sangfelt *et al.* (1985) have started to include a calculation of the AGP **B** matrix, thereby obtaining the AGPRPA method. They have tested the magnitude of **B** for systems like Be, Be_2 and CH^+ and found that it was indeed small compared to **A** and that inclusion of **B** had little effect on the excitation energies and transition moments.

TABLE II

Vertical excitation energies (eV) from the ground state of Li_2 and N_2.

System[a,b]	Final state	TDA	RPA	AGP[c]	SOPPA[d]	Experiment[e]
Li_2^a	$A^1\Sigma_u^+$	2.20	2.07	1.84	2.12	1.95
	$B^1\Pi_u$	2.71	2.55	2.71	2.58	2.58
	$^1\Pi_g$	3.46	3.88	3.05	3.41	–
	$a^3\Sigma_u^+$	0.69	unst.[f]	1.40	0.80	–
	$b^3\Pi_u$	0.78	unst.[f]	1.29	0.87	–
N_2^b	$o_3^1\Pi_u$	15.65	15.39	15.89	13.83	13.4
	$b^1\Sigma_u^+$	15.31	14.49	15.36	14.30	14.4
	$C^3\Pi_u$	11.85	11.42	12.4	11.10	11.1
	$W^1\Delta_u$	9.14	8.86	10.5	10.54	10.3
	$W^3\Delta_u$	7.41	5.98	8.0	9.01	8.9
	$a'^1\Sigma_u^-$	8.58	8.02	9.3	10.05	9.9
	$B'^3\Sigma_u^-$	8.58	8.02	10.1	9.99	9.7

[a] At $R = 5.04$ a.u. for Li_2, using a [4s2p1d], 32 CGTO basis set (Sangfelt *et al.*, 1984).
[b] At $R = 2.067$ a.u. for N_2, using a [6s5p2d], 66 CGTO basis set (Weiner *et al.*, 1984).
[c] Also referred to as AGPTDA. The results are taken from Sangfelt *et al.* (1984) and Weiner *et al.* (1984).
[d] The second-order polarization propagator approximation (see Section VII. D).
[e] Huber and Herzberg (1979).
[f] RPA is triplet unstable.

A summary of AGP calculations is part of Table III (see later). Inspection of these references will show that the AGP method works optimally for systems which are well described in a 'pair picture' but perhaps not so well for more complicated systems. In order to compare the performance of AGPTDA with other propagator methods we have used the same basis sets as Sangfelt *et al.* (1984) for Li_2 and Weiner *et al.* (1984) for N_2 to compute TDA, RPA and SOPPA (see Section VII.D) vertical excitation energies for these systems. The results are given in Table II. We see that AGPTDA gives excellent excitation energies for Li_2 whereas for some of the states of N_2 it gives results which are closer to TDA than to the more correlated (and experimental) results.

One of the advantages of AGP which is not apparent from Table II, and which it shares with MCRPA, is the ability to give proper dissociation products, i.e. good excitation energies at large internuclear separations. This is not the case with perturbative methods like, for example, SOPPA. This property of AGP has been demonstrated for Li_2, CH^+ (Elander *et al.*, 1983), Li_2 (Sangfelt *et al.*, 1984) and LiH (Kurtz *et al.*, 1985).

D. Perturbative Propagator Methods

In perturbative methods we evaluate the propagator through a given order n in the fluctuation potential (Eq. (76)). We make sure that the excitation energies, the transition moments and also the propagator itself, i.e. the response properties (see Section III), are correct through that order in perturbation theory. From Eq. (67) we see that this means evaluating the principal propagator (the inverse matrix) through that order to determine the poles. Since the principal propagator always contains zeroth-order terms, it is also necessary to calculate the transition moments, $\mathbf{P}_a$, etc., through the same order n. This is of course impossible for the full, infinite-dimensional, principal propagator. We must partition the propagator matrix and treat part of it better that the rest, in very much the same way as we did for the electron propagator (Section VI).

Most low-lying electronic states are predominantly described as single excitation out of a reference state. We will therefore chose $\mathbf{h}_a = \mathbf{h}_2$ in the partitioning of the propagator matrix (see Section V) and the polarization propagator is given in Eq. (75). The question then arises: how large must $\mathbf{h}_b = \{\mathbf{h}_4, \mathbf{h}_6, \ldots\}$ be taken in order to ensure that $\langle\langle P; Q \rangle\rangle_E$ is consistent (in the meaning discussed above) through order n? It was shown by Oddershede and Jørgensen (1977) that, to obtain the *excitation energies* correct through order 3, i.e. $\mathbf{T}_{aa}$ in Eq. (71) through order 3, it was only necessary to take $\mathbf{h}_b = \mathbf{h}_4$. The same conclusion holds for transition moments and response properties through *second* order (Nielsen *et al.*, 1980), whereas a third-order calculation of these quantities will require incorporation of higher projection manifolds.

The resulting second-order polarization propagator approximation (SOPPA) was first described in its present form by Nielsen *et al.* (1980). Splitting $\mathbf{h}_4$ up into 2p–2h and 2h–2p excitation operators as was done for $\mathbf{h}_2$ in Eqs (95)–(97)

$$\mathbf{h}_4 = \{\mathbf{q}^+\mathbf{q}^+, \mathbf{q}\mathbf{q}\} \tag{131}$$

we find from Eq. (75) (see e.g. Jørgensen and Simons, 1981, Chapter 6.E.4) that

$$\langle\langle P; Q \rangle\rangle_E = (\mathbf{t}(P, E), \mp \mathbf{t}(P, -E))\mathbf{P}^{-1}(E)\begin{pmatrix} \tilde{\mathbf{t}}(Q, E) \\ \mp \tilde{\mathbf{t}}(Q, -E) \end{pmatrix} + W_4(E) \tag{132}$$

where

$$\mathbf{t}(P, E) = (P^+ | \tilde{\mathbf{q}}^+)_{0,2} + (P^+ | \tilde{\mathbf{q}}^+ \tilde{\mathbf{q}}^+)_1 (E\mathbf{1} - \mathbf{D}_0)^{-1} \mathbf{C}_1 \tag{133}$$

$$W_4(E) = (P^+ | \tilde{\mathbf{q}}^+ \tilde{\mathbf{q}}^+)_1 [(E\mathbf{1} - \mathbf{D}_0)^{-1} + (-E\mathbf{1} - \mathbf{D}_0)^{-1}] (\mathbf{q}^+ \mathbf{q}^+ | Q)_1 \tag{134}$$

and

$$\mathbf{P}(E) = \begin{pmatrix} E\mathbf{S}_{0,2} - \mathbf{A}_{0,1,2} - \tilde{\mathbf{C}}_1 (E\mathbf{1} - \mathbf{D}_0)^{-1} \mathbf{C}_1 & \mathbf{B}_{1,2} \\ \mathbf{B}_{1,2} & -E\mathbf{S}_{0,2} - \mathbf{A}_{0,1,2} - \tilde{\mathbf{C}}_1 (-E\mathbf{1} - \mathbf{D}_0)^{-1} \mathbf{C}_1 \end{pmatrix} \tag{135}$$

The + and − signs in Eq. (132) refer to real and imaginary operators P and Q, respectively. The matrices **A** and **B** are formally defined as in RPA (Eqs (99) and 100)). However, we must use the Rayleigh–Schrödinger expansion of the reference state in Eq. (89) in order to obtain the second-order terms (as in Section VI, the subscripts of the matrices specify their order in V). The $\mathbf{A}_2$ and $\mathbf{B}_2$ matrices originate from $|0^{(1)}\rangle$ in Eq. (89). The other matrices in the second-order approach are

$$\mathbf{D}_0 = (\mathbf{q}^+\mathbf{q}^+|\hat{F}|\tilde{\mathbf{q}}^+\tilde{\mathbf{q}}^+) \tag{136}$$

$$\mathbf{C}_1 = (\mathbf{q}^+\mathbf{q}^+|\hat{V}|\tilde{\mathbf{q}}^+) \tag{137}$$

and

$$\mathbf{S}_{0,2} = (\mathbf{q}^+|\tilde{\mathbf{q}}^+) \tag{138}$$

Explicit expressions in terms of SCF two-electron integrals and orbital energies are given for all the matrices of the second-order theory in Oddershede *et al.* (1984, Appendix C). From this it appears that the only second-order term we need in the expansion of the reference state in Eq. (98) is the singly excited part of $|0^{(2)}\rangle$ and that this term only contributes to $(P|\tilde{\mathbf{q}}^+)_2$ in Eq. (133). Thus, we need not use the full second-order reference state to compute the polarization propagator through second order in the fluctuation potential. Note that the propagator is evaluated *through* second order. As for the electron propagator (see the discussion following Eq. (91)), this means that, in addition to all zeroth-, first- and second-order terms, we have included several terms summed to infinite order (Oddershede and Jørgensen, 1977).

The polarization propagator defined in Eqs (132) and (135) looks like the RPA propagator (Eq. (98)). The main difference is that $(P^+|\tilde{\mathbf{q}}^+)$ is replaced by an energy-dependent transition matrix, $\mathbf{t}(P, E)$, and that we now have an energy-dependent $\mathbf{A}$ matrix

$$\mathbf{A}(E) = \mathbf{A}_{0,1,2} + \tilde{\mathbf{C}}_1(E\mathbf{1} - \mathbf{D}_0)^{-1}\mathbf{C}_1 \tag{139}$$

The $\mathbf{S}$ matrix can also be treated as a matrix which simply redefines $\mathbf{A}$, $\mathbf{B}$ and $\mathbf{t}$ (Nielsen *et al.*, 1980). Since we are only concerned with p–h dominated excitation energies we can disregard poles from $\mathbf{t}(P, E)$ and $W_4(E)$ and the poles of $\mathbf{P}(E)$ in Eq. (135) are thus the SOPPA excitation energies. These can be calculated by means of the RPA diagonalization procedure described in Section VII.A. We must, however, apply an iterative procedure because $\mathbf{A}$ is now energy-dependent: a guess at the excitation energy E_0, e.g. the RPA excitation energy, is inserted in the 2p–2h term (the last term in Eq. (139)) and the $\mathbf{A}(E_0)$ matrix is used in the RPA fashion to evaluate a new excitation energy E_1 which is then used to compute $\mathbf{A}(E_1)$, etc. This procedure is continued until self-consistency, typically 3–4 iterations for about 0.01 eV accuracy of the excitation energy.

The second-order polarization propagator approximation is closely related to the equation-of-motion (EOM) method (McCurdy *et al.*, 1977). The equations that determine the *excitation energies* are the same up through the

TABLE III
Compilation of molecular *ab initio* polarization propagator calculations that go beyond RPA, 1977–1985[a].

System	Methods[b]	Basis sets[c]	Properties[d]	References
H_2	MCRPA	$8\sigma 5\pi 2\delta$	J_{AB}^{FC}	Albertsen *et al.* (1980c)
	MCRPA	[6s4p1d]	$\alpha(\omega)$	Jaszunski and McWeeny (1982)
	SOPPA	[6s4p2d]	J_{AB}	Geertsen (1985)
LiH	AGPTDA	31 CGTO	$\Delta\varepsilon(R), \tau, M(R)$	Kurtz *et al.* (1985)
He_2	MCRPA	[10s7p2d]	C_6	Jaszunski and McWeeny (1982)
Li_2	AGPTDA	[4s2p1d]	$\Delta\varepsilon(R), \tau, M(R)$, SpC	Sangfelt *et al.* (1984)
CH^+	CCDPPA[e]	[6s5p2d/4s3p]	$\Delta\varepsilon, f, M, \tau$	Geertsen and Oddershede (1986a)
Be_2	AGPTDA[f]	[6s3p2d]	$\Delta\varepsilon(R), \tau, M(R)$, SpC	Weiner and Öhrn (1985)
FH	MCRPA	$12\sigma 6\pi 2\delta/4\sigma 2\pi 1\delta$	J_{AB}^{FC}	Albertsen *et al.* (1980c)
N_2	SOPPA[e]	$9\sigma 6\pi 2\delta$	$\alpha(\omega), \alpha'(\omega)$, Raman	Oddershede and Svendsen (1982)
	AGPTDA	[6s5p2d]	τ	Weiner and Öhrn (1984)
	AGPTDA	[6s5p2d]	$\Delta\varepsilon(R), M(R), \tau, \alpha(0), f(R), S(0)$, SpC	Weiner *et al.* (1984)
	ADC(2)	[7s5p1d]	$\Delta\varepsilon, f$	Barth and Schirmer (1985)
	SOPPA[e]	$[4s3p] + p_B + d_B^g$	$\Delta\varepsilon, f$	Oddershede *et al.* (1985b)
	SOPPA	[5s3p2d]	$\alpha(\omega), \alpha'(\omega)$, Raman	Stroyer-Hansen and Svendsen (1986)
	SOPPA[e]	[5s4p2d]	$\Delta\varepsilon, \tau$	Dahl and Oddershede (1986)
CO	SOPPA	$10\sigma 5\pi 2\delta$	τ	Oddershede (1979)
	SOPPA	$10\sigma 5\pi 2\delta$	$\Delta\varepsilon, f, \alpha(\omega)$	Nielsen *et al.* (1980)
	MCRPA	[4s3p2d]	$\Delta\varepsilon, M, f$	Lynch *et al.* (1982)
	SOPPA[e]	$10\sigma 5\pi 2\delta$	$\alpha(\omega), \alpha'(\omega)$, Raman	Oddershede and Svendsen (1982)
	SOPPA[e]	$10\sigma 5\pi 2\delta$	$\Delta\varepsilon$	Oddershede and Sabin (1983)
	ADC(2)	$[7s5p1d] + d_B^g$	$\Delta\varepsilon, f$	Barth and Schirmer (1985)
CN^-	HRPA	6311G**[h]	J_{AB}	Galasso (1985)

TABLE III (*contd.*)

System	Methods[b]	Basis sets[c]	Properties[d]	References
O_2	MCRPA	$11\sigma 5\pi 2\delta^i$	$\Delta\varepsilon, f$	Albertsen *et al.* (1980a)
	MCRPA[e]	$11\sigma 5\pi 2\delta^i$	$\alpha(\omega)$	Albertsen *et al.* (1980b)
	MCRPA	$11\sigma 5\pi 2\delta^i$	$\Delta\varepsilon, f, \alpha(\omega), S(0)$	Yeager *et al.* (1981)
HCl	SOPPA[e]	$14\sigma 8\pi 3\delta/6\sigma 3\pi 1\delta$	$\alpha(\omega), \alpha'(\omega)$, Raman	Oddershede and Svendsen (1982)
Cl_2	SOPPA[e]	$9\sigma 6\pi 2\delta$	$\alpha(\omega), \alpha'(\omega)$, Raman	Oddershede and Svendsen (1982)
	SOPPA[e]	$9\sigma 6\pi 2\delta$	$M(R), \alpha(\omega)$, Raman	Ghandour *et al.* (1983)
HeH_2	MCRPA	[10s7p2d/6s4p1d]	C_6	Jaszunski and McWeeny (1982)
CH_2	MCRPA	[4s3p1d/2s1p]	$\Delta\varepsilon, f, \alpha(\omega)$	Nichols and Yeager (1981)
	AGPTDA	[4s3p1d/2s1p]	$\Delta\varepsilon(R)$	Jensen *et al.* (1983)
H_2O	SOPPA[e]	[6s5p1d/3s1p]	J_{AB}	Geertsen and Oddershede (1984)
HCN				
HNC	HRPA	6311G**[h]	J_{AB}	Galasso (1985)
CO_2	SOPPA[e]	[4s3p1d]	$\Delta\varepsilon, f$	Oddershede *et al.* (1985a)
SiC_2	SOPPA[e]	[5s4p1d/4s4p1d]	$\Delta\varepsilon, f, \tau$	Oddershede *et al.* (1985b)
Si_2C				
Si_3	SOPPA	[5s4p1d/4s4p1d]	$\Delta\varepsilon, f$	Sabin *et al.* (1986)
$(H_2)_2$	MCRPA	[6s4p1d]	C_6	Jaszunski and McWeeny (1982)
C_2H_2	HRPA	6311G**[h]	J_{AB}	Galasso (1985)
	SOPPA	[4s3p1d/3s1p]	$\Delta\varepsilon, J_{AB}$	Geertsen and Oddershede (1986b)
$HCNH^+$	HRPA	6-311G**[h]	J_{AB}	Galasso (1985)
SiHCH				
Si_2H_2	HRPA	6-31G**[h]	J_{AB}	Fronzoni and Galasso (1986)

CH_4 CH_2CH_2	HRPA	6311G**[h]	J_{AB}	Galasso (1985)
CH_2SiH_2 SiH_2SiH_2 SiH_3BH_2 SiH_3NH_2	HRPA	6-31G**[h]	J_{AB}	Fronzoni and Galasso (1986)
CH_3CH_3	HRPA	6311G**[h]	J_{AB}	Galasso (1985)
CH_3SiH_3 SiH_3SiH_3	HRPA	6-31G**[h]	J_{AB}	Fronzoni and Galasso (1986)

[a] This table is an update of Table A. I in Oddershede (1978).
[b] The methods include:

ADC (2)	algebraic diagrammatic construction (second order) (Sections VI and VII. D),
AGPTDA	antisymmetrized geminal power Tamm–Dancoff approximation (Section VII. C),
CCDPPA	coupled-cluster doubles polarization propagator approximation (Section VII. E),
HRPA	higher RPA (Section VII. D),
MCRPA	multiconfigurational RPA (Section VII. B),
SOPPA	second-order polarization propagator approximation (Section VII. D).

[c] Slater-type '$x\sigma y\pi z\delta$' and Gaussian '[*xsypzd*]' basis sets are given. If more than one basis set was applied, only the largest is given in the table.
[d] In addition to the list of properties described in footnote *e* of Table I, this table also includes:

J_{AB}	the Fermi contact, (FC) the orbital paramagnetic and diamagnetic and the spin dipolar contributions to the coupling constant,
$\Delta\varepsilon(R)$	potential energy curves,
τ	radiative lifetimes,
$M(R)$	transition moments as a function of the internuclear coordinates,
C_6	Van der Waals dispersion coefficients,
SpC	spectroscopic constants,
M	transition moment at one geometry,
$f(R)$	oscillator strengths as a function of the internuclear coordinates.

[e] RPA results are also given.
[f] The excitation energies are computed in AGPRPA as well.
[g] Bonded functions (s_B, p_B and d_B).
[h] The built-in basis set of the GAUSSIAN program system.
[i] Variable decomposition of the 50 STOs were used.

level referred to as higher RPA (HRPA) which is SOPPA without the 2p–2h corrections (Shibuya and McKoy, 1970). The 2p–2h corrections are sometimes treated slightly differently in the two methods (Rose *et al.*, 1973). However, a recent comparative study (Oddershede *et al.*, 19785a) of the two methods shows that the excitation energies are very nearly the same. On the other hand, the oscillator strengths were found to be quite different for several of the excitations studied for N_2 and CO_2.

Also the algebraic diagrammatic construction (ADC) method that was discussed in Section VI has been applied to the polarization propagator (Schirmer, 1982). Diagrammatic rules, rather than the analytic derivation used in SOPPA, are applied to formulate the second-order ADC(2) and basically the same approximation is obtained.

The second-order polarization propagator approximation has been applied to a range of molecules and properties and its characteristics are by now rather well understood. The performance of the method is discussed in some detail in Oddershede (1978) and the computational aspects are treated by Diercksen *et al.* (1983) and in the review by Oddershede *et al.* (1984). We refer the reader to these references for details. Here we will only mention some of the invariance properties of SOPPA and see how they relate to those of RPA and MCRPA. The hypervirial theorem, i.e. the equivalence between different formulations for the oscillator strengths (see Eqs (109–(111)), is only obeyed through second order in the fluctuation potential and so is the Thomas–Reiche–Kuhn sum rule (Jørgensen and Oddershede, 1983). The method is not invariant to separate unitary transformations among the occupied and the virtual orbitals (Oddershede and Sabin, 1983) so it is clear that SOPPA does not have the formal beauty of RPA and MCRPA. However, for most molecules both singlet and non-singlet properties around the equilibrium geometry come out quite well in SOPPA. A compilation of calculations is included in Table III.

E. Coupled-cluster Propagator Methods

Very recently we have formulated and tested a new coupled-cluster polarization propagator method (Geertsen and Oddershede, 1986a). We still use a perturbation expansion of the reference state but we replace the Rayleigh–Schrödinger correlation coefficients ($i = 1, 2$)

$$\kappa^{mn}_{\alpha\beta}(i) = \frac{(m\alpha|n\beta) - (-1)^i(m\beta|n\alpha)}{\varepsilon_\alpha + \varepsilon_\beta - \varepsilon_n - \varepsilon_m}(2i-1)^{1/2} \tag{140}$$

by the cluster amplitudes $t^{mn}_{\alpha\beta}(i)$ from the coupled-cluster doubles (Čižek, 1966; Paldus, 1977; Bartlett, 1981)

$$T_2 = \tfrac{1}{4}\sum_{i=1}^{2}\sum_{\substack{mn\\ \alpha\beta}} t^{\alpha\beta}_{mn}(i)S^+(i)_{m\alpha n\beta} \tag{141}$$

$$|0\rangle = e^{T_2}|\text{SCF}\rangle \tag{142}$$

We find that it is convenient to work with the spin-adapted form of the coupled-cluster doubles (CCD) equations. The spin-adapted double excitation operators $S^+(i), i = 1, 2$, are given, for example, in Oddershede *et al.* (1984, Appendix C).

The CCD polarization propagator approximation (CCDPPA) thus has an equation of motion which is identical in form to Eq. (132). The difference lies in the definition of the $\mathbf{A}_2, \mathbf{B}_2, \mathbf{S}_2, (P^+|\tilde{\mathbf{q}}^+)_2$ and $(P^+|\tilde{\mathbf{q}}^+\tilde{\mathbf{q}}^+)_1$ matrices where we simply replace $\kappa_{\alpha\beta}^{mn}(i)$ by $t_{\alpha\beta}^{mn}(i)$ in the formulae given in Oddershede *et al.* (1984, Appendix C).

We have tested the method for CH^+ and we find (Geertsen and Oddershede, 1986a) a substantial increase in $t_{\alpha\beta}^{mn}$ compared to $\kappa_{\alpha\beta}^{mn}$ for excitations out of the highest occupied molecular orbital (3σ). This increase has a positive effect on the excitation spectra. The triplet instability which we had in both RPA and SOPPA disappears and the singlet excitation energy of the $A^1\Pi$ state is improved by nearly 0.5 eV and lies very close to the experimental results. The same holds for the radiative lifetime.

F. Summary of Polarization Propagator Calculations

Table III gives a complication of the *ab initio* molecular polarization propagator calculations that have appeared since the previous survey of the literature up through 1977 (Oddershede, 1978). Only calculations which go beyond RPA are included in Table III, and to get the full picture of the activity within the field of polarization propagator calculations Table III must be seen in conjunction with Table I. If we compare with Table A.I. of Oddershede (1978) and consider only calculations of the same kind, i.e. exclude atomic calculations and calculations using the CHF method from the previous review, it is evident that the number of polarization propagator calculations, like other kinds of electronic structure calculations, have increased substantially within the last decade and by now is becoming a useful tool in computational quantum chemistry.

On the formal side there is a clear distinction between the state-specific approaches, such as MCSCF and configuration interaction, and propagator methods. The methods described in this review are all aimed at calculating 'differences' and response properties. We do not get information about the individual states. This makes it easier to calculate excited-state properties but at the same time we lose detailed knowledge about the states. Since the equations of motion and the state functions that are used in state-specific and propagator methods are so different it is often impossible to 'translate' from one picture to the other. This means that many of the methods we use in propagator calculations have no equivalent in the state methods. It then becomes difficult to get a feeling for (and perhaps a trust of) propagator methods since these cannot be translated into concepts that most quantum chemists are familiar with. It is the hope that the reader of the present review will overcome this barrier and thus be better equipped to understand propagator methods.

References

Abrikosov, A. A., Gorkov, L. P., and Dzyaloskhiniskii, I. E. (1963). *Methods of Quantum Field Theory in Statistical Physics*, Prentice-Hall, Englewood Cliffs, NJ.

Albertsen, P., Jørgensen, P., and Yeager, D. L. (1980a). *Int. J. Quantum Chem. Symp.*, **14**, 249.

Albertsen, P., Jørgensen, P., and Yeager, D. L. (1980b). *Mol. Phys.*, **41**, 409.

Albertsen, P., Jørgensen, P., and Yeager, D. L. (1980c). *Chem. Phys. Lett.*, **76**, 354.

Arrighini, G. P., Biondi, F., Guidotti, C., Biagi, A., and Marinelli, F. (1980). *Chem. Phys.*, **52**, 133.

Ball, M. A., and Lachlan, A. D. (1964). *Mol. Phys.*, **7**, 501.

Barth, A., and Schirmer, J. (1985). *J. Phys. B: At. Mol. Phys.*, **18**, 867.

Bartlett, R. J. (1981). *Annu. Rev. Phys. Chem.*, **32**, 359.

Bloembergen, N. (1965). *Non-linear Optics*, Benjamin, Reading, MA.

Bouman, T. D., and Hansen, Aa. E. (1985). *Chem. Phys. Lett.*, **117**, 461.

Bouman, T. D., Hansen, Aa. E., Voigt, B., and Rettrup, S. (1983). *Int. J. Quantum Chem.*, **23**, 595.

Bouman, T. D., Knobeloch, M. A., and Bohan, S. (1985). *J. Phys. Chem.*, **89**, 4460.

Bouman, T. D., Voigt, B., and Hansen, Aa. E. (1979). *J. Am. Chem. Soc.*, **101**, 550.

Brattsev, V. F., and Semenova, L. I. (1980). *Opt. Spectrosc. (USSR)*, **49**, 44.

Cederbaum, L. S. (1975). *J. Phys. B: At. Mol. Phys.*, **8**, 290.

Cederbaum, L. S., and Domcke, W. (1977). *Adv. Chem. Phys.*, **36**, 205.

Čižek, J. (1966). *J. Chem. Phys.*, **45**, 4256.

Coleman, A. J. (1963). *Rev. Mod. Phys.*, **35**, 668.

Dahl, F., and Oddershede, J. (1986). *Phys. Scr.*, **33**, 135.

Dalgaard, E. (1979). *Int. J. Quantum Chem.*, **15**, 169.

Dalgaard, E. (1980). *J. Chem. Phys.*, **72**, 816.

Dalgaard, E. (1982). *Phys. Rev.* A, **25**, 42.

Dalgaard, E., and Jørgensen, P. (1978). *J. Chem. Phys.*, **69**, 3833.

Diercksen, G. H. F., Grüner, N. E., and Oddershede, J. (1983). *Comput. Phys. Commun.*, **30**, 349.

Ehrenfest, P. (1927). *Z. Phys.*, **45**, 455.

Elander, N., Sangfelt, E., Kurtz, H., and Goscinski, O. (1983). *Int. J. Quantum Chem.*, **23**, 1047.

Eyler, J. R., Oddershede, J., Sabin, J. R., Diercksen, G. H. F., and Grüner, N. E. (1984). *J. Phys. Chem.*, **88**, 3121.

Feynman, R. P., and Hibbs, A. R. (1965). *Quantum Mechanics and Path Integrals*, McGraw-Hill, New York.

Flament, J. P., and Gervais (1979). *Int. J. Quantum Chem.*, **16**, 1347.

Fronzoni, G., and Galasso, V. (1986). *Chem. Phys.*, **103**, 29.

Galasso, V. (1985). *J. Chem. Phys.*, **82**, 899.

Geertsen, J. (1985). *Chem. Phys. Lett.*, **116**, 89.

Geertsen, J., and Oddershede, J. (1984). *Chem. Phys.*, **90**, 301.

Geertsen, J., and Oddershede, J. (1986a). *J. Chem. Phys.*, in press.

Geertsen, J., and Oddershede, J. (1986b). *Chem. Phys.*, in press.

Geertsen, J., Oddershede, J., and Sabin, J. R. (1986). *Phys. Rev.* A, in press.
Ghandour, F., Jacon, M., Svendsen, E. N., and Oddershede, J. (1983). *J. Chem. Phys.*, **79**, 2150.
Golab, J. T., Thies, B. S., Yeager, D. L., and Nichols, J. A. (1986). *J. Chem. Phys.*, **84**, 284.
Golab, J. T., Yeager, D. L., and Jørgensen, P. (1983). *Chem. Phys.*, **78**, 175.
Goscinski, O., and Lukman, B. (1970). *Chem. Phys. Lett.*, **7**, 573.
Goscinski, O., and Weiner, B. (1980). *Phys. Scr.*, **21**, 385.
Goscinski, O., Weiner, B., and Elander, N. (1982). *J. Chem. Phys.*, **77**, 2445.
Hansen, Aa. E. (1967). *Mol. Phys.*, **13**, 425.
Hansen, Aa. E., and Bouman, T. D. (1980). *Adv. Chem. Phys.*, **44**, 545.
Hansen, Aa. E., and Bouman, T. D. (1985). *J. Am. Chem. Soc.*, **107**, 4828.
Harris, R. A. (1969). *J. Chem. Phys.*, **50**, 3947.
Hedin, L., and Lundquist, S. (1969). *Solid State Phys.*, **23**, 1.
Herman, M. F., Freed, K. F., and Yeager, D. L. (1981). *Adv. Chem. Phys.*, **48**, 1.
Herman, M. F., Yeager, D. L., and Freed, K. F. (1978). *Chem. Phys.*, **29**, 77.
Hirao, K., and Nakatsuji, H. (1982). *J. Comput. Phys.*, **45**, 246.
Huber, K. P., and Herzberg, G. (1979). *Molecular Spectra and Molecular Structure*, Vol. IV, *Constants of Diatomic Molecules*, Van Nostrand-Reinhold, New York.
Jaszunski, M., and McWeeny, R. (1982). *Mol. Phys.*, **46**, 863.
Jaszunski, M., and McWeeny, R. (1985). *Mol. Phys.*, **55**, 1725.
Jensen, H. J. Aa., Weiner, B., and Öhrn, Y. (1983). *Int. J. Quantum Chem. Symp.*, **23**, 65.
Jensen, H. J. Aa., Weiner, B., Ortiz, J. V., and Öhrn, Y. (1982). *Int. J. Quantum Chem. Symp.*, **16**, 615.
Jørgensen, P. (1975). *Annu. Rev. Phys. Chem.*, **26**, 359.
Jørgensen, P., and Linderberg, J. (1970). *Int. J. Quantum Chem.*, **4**, 587.
Jørgensen, P., and Oddershede, J. (1983). *J. Chem. Phys.*, **78**, 1898.
Jørgensen, P., and Simons, J. (1975). *J. Chem. Phys.*, **63**, 5302.
Jørgensen, P., and Simons, J. (1981). *Second Quantization-Based Methods in Quantum Chemistry*, Academic Press, New York.
Kubo, R. (1959). *J. Phys. Soc. Japan*, **12**, 570.
Kurtz, H. A., Weiner, B., and Öhrn, Y. (1985). In *Comparison of Ab Initio Calculations with Experiment: State of the Art* (Ed. R. J. Bartlett), Reidel, Dordrecht.
Lazzeretti, P., Rossi, E., and Zanasi, R. (1983a). *Phys. Rev.* A, **27**, 1301.
Lazzeretti, P., Rossi, E., and Zanasi, R. (1983b). *J. Mol. Struct., Theochem*, **93**, 207.
Lazzeretti, P., Rossi, E., and Zanasi, R. (1983c). *J. Chem. Phys.*, **79**, 889.
Lazzeretti, P., Rossi, E., and Zanasi, R. (1984). *J. Chem. Phys.*, **80**, 315.
Liegener, C. M. (1983). *J. Chem. Phys.*, **75**, 1267.
Lightner, D. A., Bouman, T. D., Gawronski, J. K., Gawronski, K., Chappuis, J. L., Crist, B. V., and Hansen, Aa. E. (1981a). *J. Am. Chem. Soc.*, **103**, 5314.
Lightner, D. A., Bouman, T. D., Wijekoon, W. M. D., and Hansen, Aa. E. (1984). *J. Am. Chem. Soc.*, **106**, 934.
Lightner, D. A., Gawronski, J. K., Hansen, Aa. E., and Bouman, T. D. (1981b). *J. Am. Chem. Soc.*, **103**, 4291.
Linderberg, J. (1980). *Phys. Scr.*, **21**, 373.
Linderberg, J., Jørgensen, P., Oddershede, J., and Ratner, M. (1972). *J. Chem. Phys.*, **56**, 6213.
Linderberg, J., and Öhrn, Y. (1965). *Proc. R. Soc.* A, **285**, 445.
Linderberg, J., and Öhrn, Y. (1973). *Propagators in Quantum Chemistry*, Academic Press, London.
Linderberg, J., and Öhrn, Y. (1977). *Int. J. Quantum Chem.*, **12**, 161.

Lindhard, J. (1954). *K. Danske Vidensk. Selsk., Mat. Fys. Meddr.*, **28**, No. 8.
Löwdin, P.-O. (1963). *J. Mol. Spectrosc.*, **10**, 12.
Löwdin, P.-O. (1965). *Phys. Rev.* A, **139**, 357.
Lynch, D., Herman, M. F., and Yeager, D. L. (1982). *Chem. Phys.*, **64**, 69.
McCurdy, C. W., Rescigno, T., Yeager, D. L., and McKoy, V. (1977). In *Methods of Electronic Structures* (Ed. H. F. Schaefer, III), p. 339, Plenum, New York.
McLachlan, A. D., and Ball, M. A. (1964). *Rev. Mod. Phys.*, **36**, 84.
Manne, R. (1977). *Chem. Phys. Lett.*, **45**, 470.
Mattuck, R. D. (1976). *A Guide to Feynman Diagrams in the Manybody Problem*, 2nd Edn, McGraw-Hill, New York.
Nichols, J. A., and Yeager, D. L. (1981). *Chem. Phys. Lett.*, **84**, 77.
Nichols, J. A., Yeager, D. L., and Jørgensen, P. (1984). *J. Chem. Phys.*, **80**, 293.
Nielsen, E. S., Jørgensen, P., and Odderhede, J. (1980). *J. Chem. Phys.*, **73**, 6238.
Oddershede, J. (1978). *Adv. Quantum Chem.*, **11**, 275.
Oddershede, J. (1979). *Phys. Scr.*, **20**, 587.
Oddershede, J. (1982). *Int. J. Quantum Chem. Symp.*, **16**, 583.
Oddershede, J. (1983). In *Methods of Computational Molecular Physics* (Eds G. H. F. Diercksen and S. Wilson), p. 249, Reidel, Dordrecht.
Oddershede, J., Grüner, N. E., and Diercksen, G. H. F. (1985a). *Chem. Phys.*, **97**, 303.
Oddershede, J., and Jørgensen, P. (1977). *J. Chem. Phys.*, **66**, 1541.
Oddershede, J., Jørgensen, P., and Yeager, D. L. (1984). *Comput. Phys. Rep.*, **2**, 33.
Oddershede, J., and Sabin, J. R. (1983) *J. Chem. Phys.*, **79**, 2295.
Oddershede, J., Sabin, J. R., Diercksen, G. H. F., and Grüner, N. E. (1985b). *J. Chem. Phys.*, **83**, 1702.
Oddershede, J., and Svendsen, E. N. (1982). *Chem. Phys.*, **64**, 359.
Öhrn, Y. (1976). In *The World of Quantum Chemistry* (Eds B. Pullman and R. Parr), p. 57, Reidel, Dordrecht.
Öhrn, Y., and Born, G. (1981). *Adv. Quantum Chem.*, **13**, 1.
Öhrn, Y., and Linderberg, J. (1979). *Int. J. Quantum Chem.*, **15**, 343.
Olsen, J., and Jørgensen, P. (1985). *J. Chem. Phys.*, **82**, 3235.
Ortiz, J. V. (1984). *J. Chem. Phys.*, **81**, 5873.
Paldus, J. (1977). *J. Chem. Phys.*, **67**, 303.
Pickup, B. T., and Goscinski, O. (1973). *Mol. Phys.*, **26**, 149.
Purvis, G. D., and Öhrn, Y. (1974). *J. Chem. Phys.*, **60**, 4063.
Redmon, L. T., Purvis, G. D., and Öhrn, Y. (1975). *J. Chem. Phys.*, **63**, 5011.
Rettrup, S. (1982). *J. Comput. Phys.*, **45**, 100.
Rose, J. B., Shibuya, T., and McKoy, V. (1973). *J. Chem. Phys.*, **58**, 74.
Sabin, J. R., Oddershede, J., Diercksen, G. H. F., and Grüner, N. E. (1986). *J. Chem. Phys.*, **84**, 354.
Sangfelt, E., and Goscinski (1985). *J. Chem. Phys.*, **82**, 4187.
Sangfelt, E., Kurtz, H. A., Elander, N., and Goscinski, O. (1984). *J. Chem. Phys.*, **81**, 3976.
Sangfelt, E., Weiner, B., and Öhrn, Y. (1985). *Int. J. Quantum Chem. Symp.*, **25**, in press.
Schaefer, H. F., III (1984). *Quantum Chemistry, The Development of Ab Initio Methods in Molecular Electronic Structure Theory*, Clarendon, Oxford.
Schirmer, J. (1982). *Phys. Rev.* A, **26**, 2395.
Schirmer, J., Cederbaum, L. S., and Walter, O. (1983). *Phys. Rev.* A, **28**, 1237.
Shibuya, T., and McKoy, V. (1970). *Phys. Rev.* A, **2**, 2208.
Simons, J. (1976). *J. Chem. Phys.*, **64**, 4541.
Simons, J. (1977). *Annu. Rev. Phys. Chem.*, **28**, 1.
Simons, J., and Smith, W. D. (1973). *J. Chem. Phys.*, **58**, 4899.

Stevens, R. M., Pitzer, R. M., and Lipscomb, W. N. (1963). *J. Chem. Phys.*, **38**, 550.
Stroyer-Hansen, T., and Svendsen, E. N. (1984). *Int. J. Quantum Chem. Symp.*, **18**, 519.
Stroyer-Hansen, T., and Svendsen, E. N. (1986). *J. Chem. Phys.*, **84**, 1950.
Svendsen, E. N., and Oddershede, J. (1979). *J. Chem. Phys.*, **71**, 3000.
Svendsen, E. N., and Stroyer-Hansen, T. (1985). *Mol. Phys.*, **56**, 1025.
Szabo, A., and Ostlund, N. S. (1982). *Modern Quantum Chemistry*, MacMillan, New York.
Thouless, D. J. (1961). *Quantum Mechanics of Many Body Systems*, Academic Press, New York.
Von Niessen, W., Schirmer, J., and Cederbaum, L. S. (1984). *Comput. Phys. Rep.*, **1**, 57.
Walter, O., and Schirmer, J. (1981). *J. Phys. B: At. Mol. Phys.*, **14**, 3805.
Weiner, B., and Goscinski, O. (1980). *Int. J. Quantum Chem.*, **18**, 1109.
Weiner, B., and Goscinski, O. (1983). *Phys. Rev.* A, **27**, 57.
Weiner, B., Jensen, H. J. Aa., and Öhrn, Y. (1984). *J. Chem. Phys.*, **80**, 2009.
Weiner, B., and Öhrn, Y. (1984). *J. Chem. Phys.*, **80**, 5866.
Weiner, B., and Öhrn, Y. (1985). *J. Chem. Phys.*, **83**, 2965.
Yeager, D. L., and Jørgensen, P. (1979). *Chem. Phys. Lett.*, **65**, 77.
Yeager, D. L., Olsen, J., and Jørgensen, P. (1981). *Int. J. Quantum Chem. Symp.*, **15**, 151.
Zubarev, D. N. (1960). *Usp. Fiz. Nauk.*, **71**, 71; *Sov. Phys. Usp.* (Eng. Transl.), **3**, 320.
Zubarev, D. N. (1974). *Nonequilibrium Statistical Thermodynamics*, Consultants Bureau (Plenum), New York.

Ab Initio Methods in Quantum Chemistry—II
Edited by K. P. Lawley

ANALYTICAL DERIVATIVE METHODS IN QUANTUM CHEMISTRY

PETER PULAY

Department of Chemistry, University of Arkansas, Fayetteville, Arkansas 72701, USA

CONTENTS

I. Introduction 241
II. General Derivative Formulas 245
A. Parameter Dependence of the Wavefunction 245
B. General Derivative Expressions 247
C. Hellmann–Feynman Forces 254
III. Derivatives of Self-consistent Field Wavefunctions 256
A. Overview 256
B. First Derivatives 258
C. Second Derivatives 261
D. Higher Derivatives 268
E. The Evaluation of Integral Derivatives 269
IV. Derivatives for Dynamical Electron Correlation 274
A. Introduction 274
B. Gradients 276
C. Higher Derivatives 279
V. Property Derivatives 280
Acknowledgements 282
References 282

I. INTRODUCTION

Since their introduction in the late 1960s, gradient methods, or more properly analytical derivative methods, have become one of the most vigorously developing topics on modern quantum chemistry. They have also acquired considerable significance for the solution of practical chemical problems. The first review on this subject was written in 1974–75, although it was published much later (Pulay, 1977); a short chapter, limited to first derivatives,

*Dedicated to Professor Roger B. Bost, MD (Little Rock, Arkansas).

appeared somewhat later (Pulay, 1981). Schlegel's (1981) review is the only to treat the important developments of the late 1970s in which he played a major role. In the intervening five years, there has been much development in this field. Computational technology is progressing rapidly on two fronts: inexpensive computing using mass-produced components; and unprecedented, but unfortunately not cheap, computing power in the form of supercomputers. All this is changing the outlook in quantum chemistry. The purpose of the present article is to give an overview about direct derivative techniques in quantum chemistry. Unlike in my earlier reviews, applications cannot be covered here because of the breadth of the topic. In a chapter in the earlier volume, Schlegel (1986) summarizes the application of derivative techniques to the determination of stationary points on energy surfaces. Another large area of application, the determination of force constants, was recently covered in two review articles (Fogarasi and Pulay, 1984, 1985) although the more exhaustive of the two (1985) also suffered a long delay in publication. A review on the theoretical aspects is in preparation by Handy (1986) and, given his mastery of the subject, should be excellent reading. With my limited resources, I could not attempt a full coverage of the considerable literature in the field. Like in all new fields, a number of publications consist mainly in re-formulating the existing theory, perhaps in a pedagogically more satisfactory form. A sizeable fraction of the literature in the field consists of formal derivations, without considering the practicality of actual computer implementation. As gradient techniques are essentially a practical subject, such papers were not included in the present review.

One of the most important characteristics of molecular systems is their behavior as a function of the nuclear coordinates. The most important molecular property is total energy of the system which, as a function of the nuclear coordinates, is called the potential energy (hyper)surface, an obvious generalization of the potential energy curve in diatomics. Other expectation values as functions of the nuclear coordinates are frequently called property surfaces. The notion of the total molecular energy in a given electronic state, which depends only parametrically on the nuclear coordinates, is based on the fixed-nuclei approximation. In most cases (e.g. closed-shell molecules in the ground electronic state and in a low vibrational state) this is an excellent approximation. Even when it breaks down, the most convenient treatment is based on the fixed-nuclei picture, i.e. on the assumption that the nuclear mass is infinite compared with the electronic mass.

In the present review we shall concentrate on potential and property surfaces of general polyatomic molecules; techniques applicable only to diatomics will not be discussed. The complete potential energy surfaces of even a simple polyatomic molecule may be so complicated that its complete characterization is virtually impossible. Fortunately, in many problems the nuclear motion takes place in the vicinity of a reference nuclear configuration

$\mathbf{R}_0$, and the surface can be adequately characterized by its Taylor series expansion

$$E(\mathbf{R}) = E(\mathbf{R}_0) + \sum_a E^a \Delta R_a + \tfrac{1}{2}\sum_{a,b} E^{ab}\Delta R_a \Delta R_b + \tfrac{1}{6}\sum_{a,b,c} E^{abc}\Delta R_a \Delta R_b \Delta R_c + \cdots \quad (1)$$

where E^a, E^{ab}, etc., denote the first, second, etc., derivatives of the energy with respect to the nuclear coordinates R_a, $R_b, \ldots$ *at the reference geometry* $\mathbf{R}_0$, and $\Delta R_a = (\mathbf{R} - \mathbf{R}_0)_a$. These derivatives are important on their own, in that in some simplified treatments they can be directly related to physical observables. While this is not possible in more exact theories, the derivatives at cardinal points provide a concise characterization of potential surfaces, and greatly facilitate the reconstruction of complete surfaces.

The subject of the present review is the direct analytical calculation of energy and property derivatives from a single wavefunction. This is contrasted with the earlier pointwise method in which derivatives are evaluated by calculating the surface at a number of points and fitting an analytical function to the points. Direct calculation of the first derivatives is now usually called the gradient method. Originally (Pulay, 1969) this was called the force method, as the negative gradient with respect to nuclear coordinates is the force exerted on the nuclei by the presence of electrons and the other nuclei; the word 'force' has perhaps a bit more physical appeal than the word 'gradient'.

Direct calculation of the derivatives offers two obvious advantages: increased computational efficiency and increased numerical precision. In a molecule with N nuclear degrees of freedom, there are, in addition to the energy, N gradient components, $N(N+1)/2$ second derivatives and $\binom{N+k-1}{k}$ kth derivatives. For a molecule of modest size, say a dozen degrees of freedom, this means 12 first, 78 second and 364 third derivatives, contrasted with a single energy value. The usefulness of the gradient method depends on the fact that this large increase in information about the surface is not accompanied by a similar increase in computational effort. Indeed, as a rule of thumb, the effort needed to evaluate all first derivatives is equal, within a factor of 2–3, to the effort needed to calculate the wavefunction, for a wide range of wavefunctions and programs. As the information provided by the gradient calculation is equivalent to $(N+1)$ energy calculations in the pointwise method, the gradient method is about $(N+1)/2$ times more efficient than the traditional procedure. The advantage is thus particularly pronounced in large molecules. The recent surge of interest in gradient methods at the *ab initio* level is partly due to the fact that it is now possible to carry out *ab initio* calculations on fairly large molecules. The source of the increased efficiency in derivative methods is the utilization of much common information during the calculation of gradient components. This way, the subtle redundancy present in calculations using finite displacements is avoided.

A second advantage of derivative methods is increased numerical accuracy. This is perhaps less important in the present era of highly accurate computers. Nevertheless, the numerical evaluation of higher energy derivatives is still a difficult problem. Indeed, Hartree (1968) observes that 'the differentiation of a function specified only by a table of values... is a notoriously unsatisfactory process, particularly if higher derivatives than the first are required' (citation taken from Gerratt and Mills (1968)).

To be fair, we have to point out the disadvantages of the analytical gradient method. The foremost of these is that these methods provide basically the same information as the pointwise methods, and require significant extra programming effort. Secondly, computational times for a single run are longer than for an energy calculation, particularly for higher derivatives; this may be a limiting factor in some computing environments. Program suspension and restart features can partly solve the last problem but the necessity of keeping long files puts a limitation on this technique. Thus it seems that the direct

TABLE I
Correlation of energy derivatives with observables.

Quantity[a]	Observable
$\partial E/\partial R$	Forces on the nuclei; molecular geometry; saddle points
$\partial^2 E/\partial R_i \partial R_j$	Force constants; fundamental vibrational frequencies; infrared and Raman spectra; vibrational amplitudes; vibration–rotation coupling constants
$\partial^3 E/\partial R_i \partial R_j \partial R_k$	Cubic force constant; part of the anharmonic contribution to vibrational frequencies; anharmonic contribution to vibrationally averaged structures
$\partial^4 E/\partial R_i \partial R_j \partial R_k \partial R_l$	Quartic force constants; anharmonic corrections to vibrational frequencies
$\partial^2 E/\partial R_i \partial F_\alpha$	Dipole moment derivatives; infrared intensities in the harmonic approximation
$\partial^3 E/\partial R_i \partial F_\alpha \partial F_\beta$	Polarizability derivatives; Raman intensities in the harmonic approximation
$\partial^3 E/\partial R_i \partial R_j \partial F_\alpha$	Electrical anharmonicity; contributions to observed infrared intensities, mainly for overtones and combination bands; contributions to vibrationally averaged dipole moments, and the deflection of state-selected molecular beams in electric field
$\partial^4 E/\partial R_i \partial R_j \partial F_\alpha \partial F_\beta$	Raman intensities of overtone and combination bands; vibrationally averaged polarizabilities; deflection of state-selected molecular beams in inhomogeneous electric field
$\partial^3 E/\partial R_i \partial F_\alpha \partial H_\beta$	Infrared optical rotatory power

[a]Notation: E, total energy; R_i, nuclear coordinate; F_α, electric field component; H^β, magnetic field component.

calculation of geometry derivatives beyond the third or fourth is not practicable, although a recursive formalism has been developed for the calculation of arbitrarily high derivatives with respect to an external electric field by Dykstra and Jasien (1984). This case is, of course, considerably simpler than the case of geometrical derivatives, as the basis set is usually kept independent of the perturbation (see below).

In the previous paragraphs, the derivatives of the surface were treated as tools for the characterization of the energy surface in the vicinity of the reference configuration. However, derivatives at some cardinal points of the surface can be correlated, at least approximately, with physical observables. Thus the gradient, or with negative sign, the force on the nuclei, bears a direct relation to molecular geometry, or more generally to the location of stationary points on the surface, as the forces vanish at these points. The second derivatives of the molecular energy with respect to nuclear coordinates are the force constants; they determine the normal frequencies of the molecule in the harmonic approximation. Cubic, quartic and higher force constants are responsible for anharmonic effects in the vibrational spectrum. Mixed second derivatives with respect to a nuclear coordinate and an electric field component are dipole moment derivatives which determine the infrared intensities in the doubly (electrically and mechanically) harmonic approximation. Mixed third derivatives with respect to one nuclear coordinate and two electric field components are the polarizability derivatives, determining the Raman intensities. These, and a few less important combinations, are summarized in Table I.

II. GENERAL DERIVATIVE FORMULAS

A. Parameter Dependence of the Wavefunction

Nuclear coordinates as perturbation parameters play a role different from other common perturbations, e.g. weak external fields. The reason for this is the deep and singular potential well at the nucleus which leads to large charge density near the nucleus. In the usual basis set expansion method the utilization of the atomic nature of charge distribution leads to large savings, as the basis set can be limited to atomic functions. However, in order to maintain a uniform description over a range of nuclear coordinates, the basis functions must be coupled to the nuclei. This is a source of additional complexity, compared with ordinary perturbation theory: the latter yields derivatives with respect to perturbational parameters, but normally under the assumption that the expansion basis is independent of the perturbation. This aspect of geometrical perturbations has been stressed in my first publication (Pulay, 1969) dealing with the gradient method; the question of variable (perturbation-dependent) basis sets and derivative formulas is discussed for the case of self-consistent field (SCF) calculations in a fundamental, but unfortunately not widely known, paper by Moccia (1970). The question of the proper definition of the basis set parameters under geometrical perturbation has been clarified by Meyer and Pulay (1972) in response to some misunderstandings concerning the gradient method (Bishop and Macias, 1970). There are two ways to define the dependence of wavefunction parameters on the nuclear coordinates (Meyer and Pulay, 1972) (disregarding the trivial case of constant, i.e. perturbation-independent, parameters):

1. Optimize the parameter in question, i.e. choose the value of it which yields the minimum of the total energy expression, or, more generally, yields stationary energy.
2. Define a physically reasonable and simple, but to a certain extent arbitrary, functional relation between the wavefunction parameters and the perturbational parameters. This dependence should be chosen so that the quality of the description of the system is approximately the same with or without the perturbation.

The two kinds of parameters, defined by (1) or by (2), can be conveniently called variational and non-variational parameters. The distinction is important, as variational parameters allow simplification in the formalism, although their determination is inherently more difficult. Note that 'variational' in this context does not imply that the approximate energy is an upper bond for the true energy, as the energy expression itself may not be an exact expectation value.

Depending on the wavefunction and the parameter in question, one or the other definition may be more appropriate. Basis function parameters are almost always defined non-variationally nowadays, i.e. using definition (2). In the early days of quantum chemistry, however, extensive basis set optimization was quite common (see the comments on this in Pulay (1977, p. 160)). For basis functions, the following parameter definition is almost exclusively used: basis functions are assigned to atomic nuclei, and their centers move rigidly with the nucleus, while their exponents are fixed. Basis functions defined in this way describe properly the bulk of the change which takes place during displacement of the nuclei: the atomic part of the electron density follows the nuclei. The remaining finer changes are then taken into account by variation of other parameters, in particular orbital coefficients. This, of course, requires a basis set which is sufficiently flexible. The restriction caused by fixed orbital exponents can be relaxed by using several atomic functions of the same type but with different exponents, while rigid orbital following is compensated by using polarization functions.

It is instructive to consider some parameters in different types of wavefunctions. In SCF calculations, orbital coefficients, or more properly the independent unitary parameters defining the orbitals, are variational. In a configuration-interaction (CI) wavefunction built from SCF orbitals, however, the orbital parameters are non-variational, as they are not optimized for the energy expression in question. In the same way, CI coefficients are non-variational for a coupled-cluster calculation, although they are variational in a traditional CI calculation. For a further discussion, see Epstein and Sadlej (1979).

B. General Derivative Expressions

It is possible to derive general formulas for the derivatives of variational energy expressions. These formulas, through third derivatives, were given by Pulay (1983a). Recently, King and Komornicki (1986a) have extended these formulas up to fifth order, and introduced a convenient subscript and superscript notation, generalizing the superscript convention of Pulay (1977); we shall adopt their notation. Let us consider the energy functional

$$W = W(\mathbf{C}, \mathbf{R}) \tag{2}$$

where **R** represents a set of perturbation parameters, most importantly nuclear coordinates, and **C** is a set of variational parameters in the wavefunction, e.g. in the case of an SCF wavefunction, the C's are the SCF coefficients (for simplicity, we disregard the problem of non-independent parameters, see later). The parameters **C** are defined by the variational condition

$$\partial W/\partial C_i = 0 \tag{3}$$

which yields **C** as a function of the nuclear coordinates, $\mathbf{C} = \mathbf{C}(\mathbf{R})$. Substituted into Eq. (2), this gives the total energy

$$E = W(\mathbf{C}(\mathbf{R}), \mathbf{R}) \tag{4}$$

Expressions of this kind are frequently encountered in mathematical physics. A typical example (King and Komornicki, 1986a) is the energy profile along the reaction path; in this case **R** is the independent reaction coordinate and the C's are the remaining geometry parameters. More generally, energy expressions like this are characteristic of effective (or 'folded') Hamiltonians.

Our aim is to derive formulas for the derivatives of expression (4) with respect to the independent perturbation parameters **R**; the latter will be designated simply as nuclear coordinates in the following, although the results apply to other perturbations as well. Chain rule differentiation yields

$$\frac{\partial E}{\partial R_a} = \frac{\partial W}{\partial R_a} + \sum_i \left(\frac{\partial W}{\partial C_i}\right)\left(\frac{\partial C_i}{\partial R_a}\right) = \frac{\partial W}{\partial R_a}$$

or more simply

$$E^a = W^a + \sum_i W_i C_i^a = W^a \tag{5}$$

Here the superscript a denotes differentiation with respect to R_a, and the subscript i of W denotes differentiation with respect to C_i. The second equality follows from the variational condition (Eq. (3)), which in our notation can be written as

$$W_i = W_i(\mathbf{C}(\mathbf{R}), \mathbf{R}) = 0$$

The arguments of W are explicitly shown here to stress that this equation is valid only for the correct variational solution $\mathbf{C}(\mathbf{R})$. Eq. (5) shows the important fact that the first-order changes in the variational parameters are not needed for the evaluation of the gradients of a variational energy expression. This was already stressed by an early publication of Pulay (1969). The importance of this lies in the fact that the solution of the first-order response equations

$$\sum_j W_{ij} C_j^a = -W_i^a \tag{6}$$

for the perturbed coefficients C_j^a is time-consuming, owing to the large number of parameters even in simple wavefunctions. This linear system of equations (6) was obtained by differentiating the variational condition (Eq. (3)) with respect to R_a.

Generalization to higher derivatives (Pulay, 1983a) shows the clear analogy with the $(2n + 1)$ rule (Wigner, 1935; see also Hylleraas, 1930; Hirschfelder *et al.*, 1964): the derivatives of the variational parameters through order n suffice to determine the geometrical energy derivatives through order $(2n + 1)$. Jørgensen and Simons (1983) have extended this to fourth order, while King and Komornicki (1986a) gave a general procedure for arbitrary orders, and derived formulas explicitly through fifth order. For the special case of SCF wavefunctions, Dykstra and Jasien (1984) also derived a recursive procedure for arbitrary derivatives.

For the second derivatives, the first-order response of the wavefunction cannot be eliminated. Further differentiation of Eq. (5) yields

$$E^{ab} = W^{ab} + \sum_i W_i^a C_i^b \tag{7}$$

This expression is asymmetric in the superscripts a and b. It can be symmetrized using Eq. (6) (King and Komornicki, 1986a):

$$E^{ab} = W^{ab} - \sum_{i,j} W_{ij} C_i^a C_j^b$$

This form, however, does not seem to confer any advantage. On the contrary, it is possible to take advantage of the asymmetry if the perturbation parameters are of different nature (Pulay, 1983a); e.g. in the calculation of infrared intensities, mixed second derivatives with respect to the nuclear coordinates and the external electric field are needed. As there are only three electric field components but many nuclear coordinates, it is advantageous to choose the field components as coordinate b in the above expression, as only three response equations need to be solved in this case. An ingenious numerical procedure based on essentially the same idea was suggested by Komornicki and McIver (1979).

An alternative symmetrical form of the second derivative expression, also mentioned by King and Komornicki (1986a),

$$E^{ab} = W^{ab} + \sum_{i,j} W_{ij} C_i^a C_j^b + \sum_i W_i^b C_i^a + \sum_j W_j^a C_j^b \tag{8}$$

is advantageous in that it is a stationary point (normally a minimum) of the second derivative expression with respect to the perturbed parameters C_i^a (Sellers, 1986). Indeed, minimization of Eq. (8) with respect to C_j^b leads to the coupled perturbed equations (Eq. (6)) if the Hessian W_{ij} is positive definite, i.e. the wavefunction is stable. This quadratic form of E^{ab} is obviously much less sensitive to numerical inaccuracies in the perturbed parameters than the linear form (Eq. (7)). As the response equations are solved iteratively, this translates to fewer iterations. We may note in passing that this form of the second derivative is closely related to the Hylleraas (1930) functional form of the second-order perturbation theory. Like the latter, it can be generalized to fourth derivatives: the latter are minima with respect to the second-order perturbed coefficients C_i^{ab}.

Third derivatives are obtained from Eq. (7) by further differentiation:

$$E^{abc} = W^{abc} + \sum_i (W_i^{ac} C_i^b + W_i^{ab} C_i^c + W_i^a C_i^{bc}) + \sum_{i,j} W_{ij}^a C_i^b C_j^c + \sum_i W_i^a C_i^{bc} \tag{9}$$

The second-order derivative parameters C_i^{bc}, which would require the solution of the second-order response equations, can be eliminated by using the same kind of manipulations needed to derive the $(2n+1)$ rule (Wigner, 1935). The second-order response equations, obtained by differentiating Eq. (6), give

$$\sum_j W_{ij} C_i^{ab} = -W_i^{ab} - \sum_j (W_{ij}^a C_i^b + W_{ij}^b C_j^a) + \sum_{j,k} W_{ijk} C_j^a C_k^b \tag{10}$$

If we multiply Eq. (10) by C_i^a, and Eq. (6) by C_i^{bc}, and subtract the two, we can express the second-order term in Eq. (9) by first-order quantities. The resulting third derivative formula is symmetrical:

$$\begin{aligned} E^{abc} = W^{abc} &+ \sum_i [W_i^{ab} C_i^c + (cab) + (bca)] \\ &+ \sum_{i,j} [W_{ij}^a C_i^b C_j^c + (cab) + (bca)] + \sum_{i,j,k} W_{ijk} C_i^a C_j^b C_k^c \end{aligned} \tag{11}$$

Here the notation (cab), etc., means that the indices (abc) in the preceding expression have to be permuted to (cab). The general fourth- and fifth-order equations are given by King and Komornicki (1986a).

In the derivation of these equations, we have assumed that the parameters **C** are independent, i.e. unconstrained. In some cases, notably in the case of SCF wavefunctions, the simplest parameters, orbital coefficients, are constrained by orthonormality. If constrained parameters are used, then appropriate Lagrangian multipliers must be introduced. Pulay (1983a) gives the general derivative formulas for this case, up to third order. Let us introduce the Lagrangian function W as

$$W(\mathbf{C}, \boldsymbol{\lambda}, \mathbf{R}) = W'(\mathbf{C}, \mathbf{R}) - \sum_m f_m(\mathbf{C}, \mathbf{R})\lambda_m$$

Here W' is the energy expression which assumes that the constraints $f_m = 0$ hold, and the λ's are as yet undetermined Lagrangian multipliers. The constrained variational equations are

$$W_i = 0 \qquad \text{and} \qquad f_m = 0 \tag{12}$$

and are thus sufficient to determine the unknowns C_i and λ_m. The energy E as a function of the perturbation parameters $\mathbf{R}$ is given by

$$E = W(\mathbf{C}(\mathbf{R}), \boldsymbol{\lambda}, \mathbf{R})$$

where the Lagrangian multipliers are arbitrary. The gradient is given by

$$E^a = W^a$$

since the contributions $(\partial W/\partial C_i)$ and $(\partial W/\partial \lambda)$ vanish, by virtue of Eq. (12). Note that, in order to preserve the notation of King and Komornicki (1986a), derivatives of W with respect to λ_m are always replaced by f_m. The second derivative is

$$E^{ab} = W^{ab} + \sum_i W_i^a C_i^b - \sum_m f_m^a \lambda_m^b \tag{13}$$

The perturbed parameters can be obtained from the generalization of Eq. (6):

$$\begin{aligned} \sum_j W_{ij} C_j^a - \sum_m f_{mi} \lambda_m^a &= -W_i^a \\ \sum_i f_{mi} C_i^a &= -f_m^a \end{aligned} \tag{14}$$

Here the subscript i in f_{mi} denotes differentiation with respect to C_i. By expressing the perturbed Lagrangians from Eq. (14) and substituting them into Eq. (13), one finds that the simple variational form of the second derivatives (Eq. (8)) is exactly valid in the constrained case, although other forms, like Eq. (7), are more complex in the constrained case than in the unconstrained one.

The third derivative formula in the constrained case (Pulay, 1983a) contains the following extra terms in addition to the ones in Eq. (11):

$$\begin{aligned} E^{abc} = \text{Eq. (11)} - \sum_m \Bigg[\sum_i \Bigg(\sum_j f_{mij} C_i^a C_j^b \lambda_m^c \\ + f_{mi}^a (C_i^b \lambda_m^c + C_i^c \lambda_m^b) + f^{ab} \lambda_m^c \Bigg] + (cab) + (bca) \end{aligned}$$

If is often possible to find a simple set of independent parameters; e.g. in the case of an SCF wavefunction, the unitary parameters of Levy (1970) can be used. It is largely a question of taste which formalism to use. The final working formulas are usually exactly the same. If independent parameters are used, then part of the dependence of the energy formula on the SCF coefficients is shifted from the C's to the direct dependence. There is perhaps a certain pedagogical advantage in separating the essential parameters, for which the solution of the response equations is difficult, from the less essential parameters which are often easy.

As the foregoing discussion shows, variational energy expressions allow a significant simplification of derivative formulas. Nevertheless, the importance of this is perhaps less than originally assumed, due to an important discovery by Handy and Schaefer (1984). This will be described here in general, as it is not restricted to a certain type of wavefunction. In the case of a non-variational energy expression, the full gradient formula

$$E^a = W^a + \sum_i W_i C_i^a \tag{15}$$

must be used. The parameters $\mathbf{C}$ are frequently determined by minimizing a simpler type of energy expression. For instance, orbitals used in CI calculations generally minimize the underlying SCF energy but not the CI energy. In a similar way (see later) correlation coefficients used in third-order Møller–Plesset (MP3) theory minimize the Hylleraas form of the second-order correlation energy but not the third-order one. The difficulty with the gradient formula (15) is that the response equations have to be solved now for the parameter derivatives $\mathbf{C}^a$. Note that the wavefunction usually contains other parameters, e.g. positions of the basis functions, which are not optimized. However, the difference between these parameters and the near-variational parameters $\mathbf{C}$ is that the functional dependence of the former is simple and explicit, and therefore no response equations need to be solved. The response equations for the near-variational parameters $\mathbf{C}$ are often analogous to Eq. (6) but are derived from a simpler energy functional $w(\mathbf{C}, \mathbf{R})$. They can be written as

$$\sum_j w_{ij} C_j^a = -w_i^a \tag{16}$$

or, in matrix form, as

$$\mathbf{w}\mathbf{C}^a = -\mathbf{b}^a$$

where $(\mathbf{w})_{ij} = w_{ij}$ and $(\mathbf{b}^a)_i = w_i^a$. In some cases, e.g. for coupled-cluster wavefunctions, the parameters of the wavefunction are not derived from a functional but by a set of linear equations

$$w_i = 0$$

The corresponding response equations have the form of Eq. (16) except that the subscript i cannot be regarded as differentiation with respect to the parameter C_i (although it is generally close). In order to include this case, it will not be assumed that the matrix $\mathbf{w}$ is symmetrical.

A single solution of the response equation (16) should cost about as much as the determination of the wavefunction at a single geometry. Indeed, Eq. (16) is nothing else than the equation for the wavefunction parameters at an infinitesimally displaced geometry. The formal solution of Eq. (16) is

$$\mathbf{C}^a = -\mathbf{w}^{-1}\mathbf{b}^a$$

Substituting this into the gradient formula we obtain

$$E^a = W^a - \sum_i W_i \left(\sum_j (\mathbf{w}^{-1})_{ij} b_j^a \right)$$

By interchanging the order of summation,

$$E^a = W^a - \sum_j \left(\sum_i (\mathbf{w}^{-1})_{ij} W^i \right) b_j^a$$

a formula is obtained which requires the solution of only *one* response-like equation for all perturbations,

$$\mathbf{w}^{\dagger}\mathbf{X} = \mathbf{W}'$$

where the vector $\mathbf{W}'$ holds the components of W_i. The complete gradient equation requires only scalar products:

$$E^a = W^a - \mathbf{X}^{\dagger}\mathbf{b}^a$$

The method of Handy and Schaefer (1984) was originally introduced to eliminate the repeated solution of the coupled perturbed Hartree–Fock equations in CI and Møller–Plesset derivative calculations. It is, however, generally applicable. One of the most significant applications of this method is to correlated wavefunctions which are non-variational in the CI coefficients, such as Møller–Plesset perturbation theory (Bartlett and Silver, 1975; Pople *et al.*, 1976) and coupled-cluster theory (Čižek, 1966). Using the example of the latter wavefunction, gradient evaluation via Eq. (11) requires the perturbed CI coefficients for each perturbational parameter. As the determination of these is computationally roughly equivalent to the evaluation of the energy, there is little advantage in the analytical derivative method this way. Using the Handy–Schaefer device, however, requires only the solution of a single additional linear system of equations, approximately doubling the computational work (Adamowicz *et al.*, 1984). Note that a variational method, such as variational CI, linearized coupled-cluster theory (Čižek and Paldus, 1971) or variational coupled electron-pair approximation (CEPA) (Pulay, 1983b;

Ahlrichs *et al.*, 1985; Pulay and Saebø, 1985), still has an edge over non-variational methods, but not as decisively as assumed earlier.

Handy and Schaefer (1984) also discuss the corresponding transformation of higher energy derivatives. Obviously, this device can always be applied when the perturbed coefficients appear linearly in the derivative expression, and their coefficients are independent of the perturbation. The second derivative of a non-variational energy expression, for instance, gives

$$E^{ab} = W^{ab} + \sum_i (W_i^a C_i^b + W_i^b C_i^a) + \sum_{i,j} W_{ij} C_i^a C_j^b + \sum_i W_i C_i^{ab} \tag{17}$$

Obviously, the second-order perturbed coefficients C_i^{ab} can be eliminated from this by a technique exactly analogous to the first derivative case. In order to do this, the coefficients C_i^{ab} are formally expressed from the second-order response equations

$$\sum_j w_{ij} C_j^{ab} = -\sum_{j,k} w_{ijk} C_j^a C_k^b - \sum_j (w_{ij}^a C_j^b + w_{ij}^b C_j^a) - w_i^{ab} \tag{18}$$

and substituted into the last term of Eq. (17). The latter can then be expressed as the scalar product of the vector $(\mathbf{w}^\dagger)^{-1}\mathbf{W}'$ and the right-hand side of Eq. (18); again only one solution of a large system of equations is required.

It is less obvious that the third derivatives can be determined by solving only $(2m+1)$ response equations or equivalents, where m is the number of perturbational parameters. The third derivative of a non-variational energy expression is given by

$$\begin{aligned} E^{abc} = W^{abc} &+ \Bigg(\sum_i W_i^{ab} C_i^c + \sum_{i,j} W_{ij}^a C_i^b C_j^c \\ &+ \sum_i W_i^a C_i^{bc} + \sum_{i,j} W_{ij} C_i^a C_j^{bc} + (bca) + (cab) \Bigg) \\ &+ \sum_{i,j,k} W_{ijk} C_i^a C_j^b C_k^c + \sum_i W_i C_i^{abc} \end{aligned} \tag{19}$$

Here and below the permutation symbols (bca) and (cab) refer to all terms in the large parentheses. We need the first-order perturbed coefficients C_i^a; the determination of these requires the solution of m sets of response equations. The third-order response coefficients can be eliminated by using the Handy–Schaefer device for the corresponding response equation

$$\begin{aligned} \sum_{i,j} w_{ij} C_j^{abc} = -w_i^{abc} - \Bigg(&\sum_j (w_{ij}^{ab} + w_{ij}^a C_j^{bc}) \\ &+ \sum_{j,k} (w_{ijk}^a C_j^b C_k^c + w_{ijk}^a C_j^b C_k^c + w_{ijk} C_j^{ab} C_k^c) \\ &+ (bca) + (cab) \Bigg) - \sum_{j,k,l} w_{ijkl} C_j^a C_k^b C_l^c \end{aligned} \tag{20}$$

The second-order perturbed coefficients in the original third derivative equation (19), and those introduced by the Handy–Schaefer device, appear in linear combinations like

$$\sum_i W_i^a C_i^{bc}$$

Therefore, m solutions of linear equations (with a perturbation-dependent vector like $\mathbf{W}^a$ on the right-hand side) can replace the $O(m^2)$ solutions for $\mathbf{C}^{bc}$. Note that the situation, though similar, is not completely analogous to the case of the $(2n+1)$ rule.

C. Hellmann–Feynman Forces

For the exact wavefunction, the Hellmann–Feynman theorem (Hellmann, 1937; Feynman, 1939) holds: the energy gradient is equal to the expectation value of the derivative of the Hamiltonian

$$E^a = \langle \Psi | \mathbf{H}^a \Psi \rangle \tag{21}$$

This can be easily shown by noting that for a normalized wavefunction Ψ the other terms arising from the differentiation of the energy expectation value vanish:

$$2\,\mathrm{Re}\langle \Psi^a | \mathbf{H}\Psi \rangle = 2E\,\mathrm{Re}\langle \Psi^a | \Psi \rangle = E\langle \Psi | \Psi \rangle^a = 0$$

This theorem is also valid for many variational wavefunctions, e.g. for the Hartree–Fock one, *if complete basis sets are used*. As only the one-electron part of the Hamiltonian depends on the nuclear coordinates, $\mathbf{H}^a$ is a one-electron operator, and the evaluation of the Hellmann–Feynman forces is simple. Because of this simplicity, there have been a number of early suggestions to use the Hellmann–Feynman forces for the study of potential surfaces. These attempts met with little success, and the discussion below will show the reason for this. It is perhaps fair to say that the main value of the Hellmann–Feynman theorem for geometrical derivatives is in the insight it provides, and that numerical applications do not appear promising. For other types of perturbations, e.g. for weak external fields, the theorem is widely used, however. For a survey, see a recent book (Deb, 1981).

For approximate wavefunctions, Eq. (21) is not valid in general. In this case, the gradient of a variational energy expression is much superior to the Hellmann–Feynman force, as can be seen from the following argument. Consider the error in the energy, $\delta E = E - E_0$, where E_0 is the exact energy, and E is the expectation value of the Hamiltonian with the approximate wavefunction $\Psi_0 + \delta\Psi$. Using the properties of the exact wavefunction, the error of the gradient can be written as

$$\delta E^a = 2\,\mathrm{Re}\langle \delta\Psi^a | \mathbf{H}\delta\Psi \rangle + \langle \delta\Psi | (\mathbf{H} - E_0)^a \delta\Psi \rangle$$

i.e. it is second order in the error of the wavefunction, assuming that $\delta\Psi^a$ is the same order as $\delta\Psi$. The latter assumption is essentially equivalent to assuming that the error in the wavefunction does not change very fast with the nuclear coordinates. By contrast, the Hellmann–Feynman force contains a term first order in $\delta\Psi$.

Given this result, it is natural to ask under what conditions will the Hellmann–Feynman force be equal to the energy gradient. The difference of the two is called the wavefunction force (Pulay, 1969):

$$2\,\mathrm{Re}\langle \Psi^a | \mathbf{H}\Psi \rangle = 2\,\mathrm{Re} \sum_t \langle \partial\Psi/\partial p_t | \mathbf{H}\Psi \rangle p_t^a$$

Here the p_t are the parameters of the wavefunction, most importantly the coordinates of orbital centers. The contribution of p_t to the wavefunction force vanishes if either p_t is independent of the perturbation, or if the matrix element in the above expression is zero. The latter is the derivative of the energy functional on the parameter p_t, and thus vanishes if the value of the parameter is optimized. These conditions were first recognized by Hurley (1954); energy expressions which contain only constant or optimized parameters are called stable under the perturbation (Hall, 1961). Unfortunately, stable wavefunctions are generally not practicable for potential surface calculations. Stability can be achieved either by fixing the orbitals in space or by optimizing the orbital centers. Orbital centers which follow the nuclei are essential to provide approximately constant accuracy over a range of nuclear coordinates. Optimization of the basis function centers ('floating functions') is possible using the same techniques as for gradient evaluation, but is, in general, not cost effective.

In linear variational problems, one way of satisfying Hurley's conditions is to make the basis set closed with respect to the differential operators $\partial/\partial p$. Such a basis set is in principle infinite. Practically, however, the Hellmann–Feynman theorem will be approximately satisfied if, for each significantly populated basis function χ, its derivatives with respect to the orbital centers, χ^x, χ^y, χ^z, are included in the basis set (Pulay, 1969). The use of augmented basis sets in conjunction with the Hellmann–Feynman theorem was considered by Pulay (1969, 1977) but dismissed as expensive. Recently, Nakatsuji *et al.* (1982) have recommended such a procedure. However, an analysis of their procedure (Pulay, 1983c; Nakatsuji *et al.*, 1983) reveals that it is not competitive with the traditional gradient technique. Much of the error in the Hellmann–Feynman forces is due to core orbitals. Therefore, methods based on the Hellmann–Feynman theorem presumably work better for effective core potential calculations. The present author is generally skeptical about the prospect of using the Hellmann–Feynman forces in actual calculations.

III. DERIVATIVES OF SELF-CONSISTENT FIELD WAVEFUNCTIONS

A. Overview

Direct derivative methods were first applied to closed-shell SCF wavefunctions, and this application is still the most important one. A remarkably complete theory of SCF first and second derivatives was presented very early by Bratoz (1958). This method was implemented and applications using single-center basis sets were presented for NH_3, H_2O and CH_4 (Bratoz and Allavena, 1962; Allavena and Bratoz, 1963; Allavena, 1966). Analytical gradients with a single-center basis set were also calculated independently for SiH_4 (Pulay and Török, 1964). For single-center expansions, the gradient expression reduces to the Hellmann–Feynman force. While this greatly simplifies the formalism and the programming, it leads to poor results. The early reluctance to calculate the derivatives of two-electron integrals is understandable: around 1960, before the introduction of modern integral evaluation techniques, integral calculation was the most time-consuming part of *ab initio* SCF programs. It is regrettable that the work of Bratoz remained practically unknown for a long period, and is not cited properly even now; its formalism, particularly the treatment of SCF second derivatives, is excellent.

Gradients for the optimization of basis function exponents were first used by Moccia (1967). Formulas for geometrical gradients were derived by Gerratt and Mills (1968) as a first step toward their force constant evaluation via the analytical derivatives of the Hellmann–Feynman forces. Although the latter method is obsolete, this paper became very important, and is still the standard reference for the coupled Hartree–Fock method. It appears, however, that the significance of gradients was not realized in this paper. This was first emphasized by Pulay (1969), based on an implementation for multicenter basis sets in the summer of 1968. The usefulness of gradients was also independently pointed out by McIver and Komornicki (1971) for semi-empirical methods. The first practical *ab initio* gradient applications were given by Pulay (1970, 1971), Meyer and Pulay (1972) and Pulay and Meyer (1971, 1972). Generalization to open-shell wavefunctions (restricted and unrestricted Hartree–Fock), with applications, was reported by Meyer and Pulay (1973); the first serious application was Meyer's (1973) optimization of various structures of the CH_4^+ ion. Gradient formalism for multiconfigurational SCF (MCSCF) wavefunctions was given in Pulay's (1977) review, following Meyer's unpublished work. The first implementations, however, are those of Goddard *et al.* (1979) and Kato and Morokuma (1979). Note that the claim of Goddard *et al.* for the first implementation of restricted Hartree–Fock (RHF) gradients cannot be supported; e.g. Pulay's (1977) review paper describes the theory and, in Table II, presents optimized RHF geometries for several states of the ethylene positive ion.

Because of its superb clarity, the paper of Gerratt and Mills (1968) had a large influence on the development of practical second derivative algorithms. This important development is due to Pople *et al.* (1979). Previous attempts (Thomsen and Swanstrøm, 1973) were disappointing from a practical point of view, leading to considerable, but as it turned out, unjustified, skepticism (Pulay, 1977) concerning the analytical calculation of these quantities. A very significant contribution for the practical implementation of higher derivative methods was the introduction of Gaussian quadrature integral evaluation algorithms, particularly the Rys polynomial method of Dupuis *et al.* (1976); a similar but less practicable method has been devised by Saunders (1975).

Further development on SCF derivative methods proceeded on three fronts: generalization to more complex wavefunctions, the calculation of higher derivatives, and improvements in the known algorithms. MCSCF gradients, at least for a small number of configurations, can be handled by the same methods as in the closed-shell case (Pulay, 1977). For higher derivatives, the coupled perturbed MCSCF equations are needed. For the limited case of pair-excitation MCSCF wavefunctions, these were first obtained by Jaszunski and Sadlej (1975, 1977). The general case was treated by Dupuis (1981); he omitted, however, some non-zero contributions. The full equations were derived by Osamura *et al.* (1982a, b), Pulay (1983a) and Jørgensen and Simons (1983), and briefly by Camp *et al.* (1983). A lucid treatment was given by Page *et al.* (1984); the papers by Helgaker and Almlöf (1984) and Almlöf and Taylor (1985) are also recommended. The first MCSCF second derivative implementation is due to Yamaguchi *et al.* (1983), following their implementation of the restricted open shell Hartree–Fock (ROHF) case (Osamura *et al.*, 1982b). An important step toward MCSCF second derivatives was the solution of the coupled perturbed MCSCF equations (Osamura *et al.*, 1982a). MCSCF second derivative programs have been reported by Page *et al.* (1984) and Jensen (1985). A remarkably complete theory of closed-shell Hartree–Fock third derivatives was presented early by Moccia (1970). Implementation was not seriously considered until much later (Pulay, 1983a); see also a note by Simons and Jørgensen (1983). The method was implemented in a remarkably short time (Gaw *et al.*, 1984), using an alternative grouping of terms. There are plans in several research groups to go up to fourth order; the latter is of importance because the treatment of molecular anharmonicities at the second-order level requires the quartic force field. This is probably as far as one wants to go.

Improvements in the algorithm concentrate on the efficient calculation of integral derivatives, as this is the main cost in SCF and small MCSCF derivative calculations. This topic will be discussed separately. An important contribution was the adaptation of gradient algorithms to the efficient GAUSSIAN70 (Hehre *et al.*, 1972) program package (Schlegel *et al.*, 1975; Komornicki *et al.*, 1977). Another fundamental improvement was the use of conjugate gradient methods in the solution of coupled perturbed Hartree–Fock equations (Pople *et al.*, 1979). Other recent developments include the efficient use of symmetry (Dupuis and King, 1978), the use of the invariance properties of the integrals (e.g. Kahn, 1981), and an alternative solution of the coupled perturbed Hartree–Fock equations without integral transformation (Pulay, 1983a; Osamura *et al.*, 1983).

B. First Derivatives

In the following, a general formalism is given which includes the closed-shell, the open-shell and the multiconfigurational cases. The energy functional W' in MCSCF theory can be written as

$$W' = \sum_{K,L} H_{KL} A_K A_L \tag{22}$$

where the Hamiltonian matrix elements are given by

$$H_{KL} = \sum_{i,j} q_{KL}^{ij} h_{ij} + \tfrac{1}{2} \sum_{i,j,k,l} Q_{KL}^{ijkl} (ij|kl)$$

with

$$h_{ij} = \int \phi_i \mathbf{h} \phi_j \, d\tau$$

$$(ij|kl) = \iint \phi_i(1)\phi_j(1)(1/\mathbf{r}_{12})\phi_k(2)\phi_l(2) \, d\tau_1 \, d\tau_2$$

The orbitals ϕ are assumed to be real and orthonormal, and are expressed as a linear combination of atomic orbitals (basis functions)

$$\phi_i = \sum_r \chi_r C_{ri}$$

The reduction coefficients q and Q are independent of both the CI coefficients A_K and the orbitals, and are determined by the form of the many-electron wavefunction. For simple wavefunctions they are very simple; e.g. in the closed-shell case, the configuration subscripts can be omitted, and $q^{ij} = 2\delta_{ij}$, $Q^{ijkl} = 4\delta_{ij}\delta_{kl} - 2\delta_{ik}\delta_{jl}$. It is useful to introduce the orbital density matrices γ_{ij} and Γ_{ijkl} as the weighted sum of the reduction coefficients

$$\gamma_{ij} = \sum_{K,L} q_{KL}^{ij} A_K A_L$$

$$\Gamma_{ijkl} = \sum_{K,L} Q_{KL}^{ijkl} A_K A_L$$

With the density matrix notation, the energy expression takes the form

$$W' = \sum_{i,j} \gamma_{ij} h_{ij} + \tfrac{1}{2} \sum_{i,j,k,l} \Gamma_{ijkl}(ij|kl) \tag{23}$$

We shall later need W' expressed by atomic-orbital (AO) basis (function) integrals. This is obtained by substituting into Eq. (23) the expression of the molecular-orbital (MO) integrals:

$$W' = \sum_{p,q} d_{pq} h_{pq} + \tfrac{1}{2} \sum_{p,q,r,s} D_{pqrs}(pq|rs) \tag{24}$$

where the AO density matrices are

$$d_{pq} = \sum_{i,j} C_{pi} C_{qj} \gamma_{ij} \tag{25}$$

$$D_{pqrs} = \sum_{i,j,k,l} C_{pi} C_{qj} C_{rk} C_{sl} \Gamma_{ijkl}$$

Let us consider an infinitesimal change in orbital n of the form

$$\delta\phi_n = \tfrac{1}{2}\phi_p \lambda_{pn} \tag{26}$$

We can *define* the Lagrangian elements as

$$\varepsilon_{pn} = \partial W'/\partial\lambda_{pn} = \sum_j \gamma_{nj} h_{pj} + 2 \sum_{j,k,l} \Gamma_{njkl}(pj|kl) \tag{27}$$

Even for the optimized MCSCF wavefunction, these derivatives are not zero because the energy functional W' is valid only for orthonormal orbitals, and transformation (26) violates orbital orthonormality. The energy must be stationary against an infinitesimal 2×2 rotation of the form

$$\delta\phi_n = \tfrac{1}{2}\phi_p x_{pn}$$

$$\delta\phi_p = -\tfrac{1}{2}\phi_n x_{pn}$$

as this transformation preserves orbital orthonormality. From this, the condition for stationary energy with respect to orbital optimization is

$$\varepsilon_{pn} - \varepsilon_{np} = 0 \tag{28}$$

The Lagrangian function corresponding to the energy expression (22) is

$$W = W' - \lambda\left(\sum_K A_K^2 - 1\right) - \mathrm{Tr}[\boldsymbol{\varepsilon}(\mathbf{C}^\dagger \mathbf{S} \mathbf{C} - \mathbf{I})] \tag{29}$$

The terms added to W' correspond to the condition of the overall normalization of the wavefunction, and the orthonormalization of the molecular orbitals:

$$\sum_K A_K^2 = 1 \tag{30}$$

$$\mathbf{C}^\dagger \mathbf{S} \mathbf{C} = \mathbf{I}$$

Here $\mathbf{S}$ is the overlap matrix in atomic-orbital basis. The solution of the SCF and MCSCF equations has a large literature and will not be discussed here. The diagonal elements of $\boldsymbol{\varepsilon}$ can be visualized as the product of an orbital energy and the corresponding occupation number, in analogy with the one-electron picture in which the energy is the derivative of the energy expectation value with respect to the squared norm of the wavefunction. It is useful to point out that the Lagrangian components ε_{pn} are defined in the whole orbital space; however, they obviously vanish unless n is an occupied orbital.

The gradient expression is, according to Section II, simply

$$E^a = W^a = W'^a - \mathrm{Tr}[\boldsymbol{\varepsilon}(\mathbf{C}^\dagger \mathbf{S}^a \mathbf{C})] \tag{31}$$

The first term here is W'^a, the derivative of the energy expression (22). This consists of two parts, as W' depends directly on the nuclear coordinates in two ways: through the Hamiltonian, and through the positions of the basis functions. The first dependence gives the Hellmann–Feynman force, while the second one gives the energy expression evaluated with the derivative integrals; the latter corresponds to the 'wavefunction force' (Pulay, 1969). The last term in Eq. (30) contains the derivatives of the constraint equations. Note that contributions from the CI coefficients A_K are absent because the overall normalization condition does not contain parameters which depend on the nuclear coordinates.

Computationally, the most demanding part of the gradient expression is the evaluation of the contribution of integral derivatives. Here we discuss only the general organization of the calculation; details will be considered in the next chapter. Efficiency requires that the coordinates R_a are Cartesian coordinates of the nuclei. This guarantees that the position of a basis function depends only on a single nuclear coordinate in the usual case, i.e. basis functions centered on the nuclei. The integral-driven scheme has proven the most efficient in practice. This is based on evaluating the integral derivatives, multiplying them with the corresponding density matrix elements, and adding the result directly to the appropriate gradient component. As discussed by Pulay (1977), it is advantageous to calculate first the energy derivatives with respect to the basis function centers, as this procedure avoids an extra indexing step. Moreover, these quantities yield useful information about basis set deficiencies: a large component in the valence shell indicates the need for polarization functions.

The integral-driven procedure indicated above is practicable only if the elements of the two-particle density matrix can be rapidly accessed. In the closed-shell Hartree–Fock case, the two-particle density matrix can be easily constructed from the one-particle density. The situation is similar for open-shell and small multiconfigurational SCF wavefunctions: the two-particle density matrix can be built up from a few compact matrices. In most open-shell Hartree–Fock theories (Roothaan, 1960), the energy expression (Eq. (23))

contains only the following non-vanishing orbital density matrix elements:

$$\gamma_{ii} = f_I \qquad \Gamma_{ii,jj} = f_I f_J a_{IJ} \qquad \Gamma_{ij,ij} = -f_I f_J b_{IJ}$$

where I and J denote groups of orbitals with equal occupancy; orbital i belongs to group I and orbital j belongs to J. Substituting this into the two-particle AO density matrix expression (25), the latter factorizes as

$$D_{pqrs} = \sum_{I,J} f_I f_J (a_{IJ} D^I_{pq} D^J_{rs} - b_{IJ} D^I_{pr} D^J_{qs}) \tag{32}$$

Various open-shell and simple multiconfigurational SCF wavefunctions are characterized by different values of f, a and b; in the MCSCF case, the latter depend on the CI coefficients as well. Note that some simple MCSCF energy expressions can also be written in the above form (Bobrowicz and Goddard, 1977).

For open-shell and small (e.g. two-configuration) MCSCF wavefunctions, the construction of the AO density matrix according to Eq. (32) is computationally negligible compared to the evaluation of the integral derivatives. The open-shell case is thus computationally identical to the closed-shell case.

For long MCSCF expansions, construction of the AO density matrix becomes a major computational task, and an alternative method, similar to CI derivative calculation, must be used. This method, outlined by Meyer (1976), and first used by Brooks *et al.* (1980), consists of the explicit transformation of the MO two-particle density matrix to AO basis, prior to gradient evaluation. Transformation of the density matrix is obviously superior to the transformation of the $3N$ gradient components of each integral.

The final form of the gradient is obtained by substituting the detailed form of the energy expression in Eq. (31):

$$E^a = \sum_{p,q} d_{pq} h^a_{pq} + \tfrac{1}{2} \sum_{p,q,r,s} D_{pqrs}(pq|rs)^a - \mathrm{Tr}[\boldsymbol{\varepsilon}(\mathbf{C}^\dagger \mathbf{S}^a \mathbf{C})] \tag{33}$$

Analytical gradients have also been formulated for infinite periodic systems in the Hartree–Fock approximation (Teramae *et al.*, 1983); in metals, a new term arises from the change of the Fermi surface with the geometry (Kertesz, 1984).

C. Second Derivatives

Analytical second derivatives for closed-shell (or unrestricted Hartree–Fock (UHF)) SCF wavefunctions are used routinely now. The extension to the MCSCF case is relatively new, however. In contrast to the first derivatives, the coupled perturbed SCF equations have to be solved in order to calculate the second and third energy derivatives. The closed-shell case is relatively straightforward, and will be discussed. The multiconfigurational formalism is too technical to be treated here in detail; the reader is referred to the papers of Jaszunski and Sadlej (1977), Dupuis (1981), Osamura *et al.* (1982a), Pulay (1983a), Camp *et al.* (1983), Jørgensen and Simons (1983), Hoffmann *et al.* (1984), Page *et al.* (1984), Simons *et al.* (1984), Lengsfield (1986) and Helgaker *et al.* (1986). In particular, the papers by Osamura *et al.*, Hoffmann *et al.* and Page *et al.* discuss the practical implementation of second derivatives and are thus highly recommended. Most of these papers use a Fock-matrix-based formalism. Simons and Jørgensen, however, use an orthonormal AO basis. This gives a very compact set of formulas which must be expanded, however, for computer implementation.

In atomic-orbital basis, the Lagrangian function for a closed-shell SCF wavefunction (omitting the nuclear repulsion) becomes

$$W = \mathrm{Tr}[(\mathbf{h} + \mathbf{F})\mathbf{D}] - 2\,\mathrm{Tr}[\boldsymbol{\varepsilon}(\mathbf{C}^{\dagger}\mathbf{S}\mathbf{C} - \mathbf{I})] \tag{34}$$

Here $\mathbf{h}$ is the one-electron Hamiltonian defined in Eq. (22), $\mathbf{I}$ is the unit matrix and $\boldsymbol{\varepsilon}$ is the matrix of the Lagrangian multipliers; we included a factor of 2 in the last term to make $\boldsymbol{\varepsilon}$ identical with the usual orbital energies. The Fock matrix and the density matrix are defined as

$$F_{pq} = h_{pq} + \sum_{r,s} G_{pqrs} D_{rs} \tag{35}$$

$$\mathbf{D} = \mathbf{C}\mathbf{C}^{\dagger}$$

and

$$G_{pqrs} = 2(pq|rs) - (pr|qs)$$

It is useful to introduce the convention that $\mathbf{F}^{a}, \mathbf{F}^{ab}$, etc., denote the Fock matrix (Eq. (35)) built from derivative AO integrals. This is of course not equal to the true derivative of the Fock matrix, as the latter also contains the derivatives of the SCF coefficients $\mathbf{C}$, or alternately those of the density matrix $\mathbf{D}$.

Implementations of the SCF second derivative technique use the a-symmetrical form of the second derivative expression (Eq. (13)):

$$E^{ab} = W^{ab} + \sum_{i} W_{i}^{a} C_{i}^{b} - \sum_{m} f_{m}^{a} \lambda_{m}^{b}$$

In the present case, the terms in this equation are given by

$$W^{ab} = \mathrm{Tr}[(\mathbf{h}^{ab} + \mathbf{F}^{ab})\mathbf{D} - 2\boldsymbol{\varepsilon}\,\mathbf{C}^{\dagger}\mathbf{S}^{ab}\mathbf{C}] \tag{36}$$

$$\sum_{i} W_{i}^{a} C_{i}^{b} = -2\,\mathrm{Tr}[\mathbf{S}^{a}(\mathbf{C}^{b}\boldsymbol{\varepsilon}\,\mathbf{C}^{\dagger} + \mathbf{C}\boldsymbol{\varepsilon}(\mathbf{C}^{b})^{\dagger})] \tag{37}$$

$$-\sum_{m} f_{m}^{a} \lambda_{m}^{b} = -2\,\mathrm{Tr}[\mathbf{C}^{\dagger}\mathbf{S}^{a}\mathbf{C}\boldsymbol{\varepsilon}^{b}] \tag{38}$$

The first term above is the SCF energy expression evaluated with the second derivative integrals. This is usually the major computational task, although this term has an asymptotic computational dependence of only $O(m^4)$ where m is the basis set size; for a constant basis set quality, this term should increase in proportion as the fourth power of the molecular size, or more slowly if small integrals can be neglected. As we shall see, the solution of the coupled perturbed Hartree–Fock equations are expected to show a fifth-power dependence; integral transformation, which is part of the algorithm in some implementations, is also a fifth-order procedure. Nevertheless, the experience of King and Komornicki (1986b) shows that, even for fairly large calculations, the integral derivative step dominates the calculation on a supercomputer.

As discussed in Section II, an alternate form of the second derivative expression (Eq. (8))

$$E^{ab} = W^{ab} + \sum_i W_i^a C_i^b + (ba) + \sum_{i,j} W_{ij} C_i^a C_j^b$$

is variational in the perturbed parameters and therefore needs less sharp convergence in the coupled perturbed Hartree–Fock (CPHF) procedure. The first two terms are identical with the previous formulation. The third term is similar to the second but the perturbation parameters are interchanged. As discussed before, this extra term is not significant computationally for force constant calculations because the CPHF equation has to be solved for each perturbation anyway. The asymmetric form is more advantageous for dipole moment derivative calculations, however. The last term in Eq. (8), reproduced above, is new in the alternate second derivative formulation, and requires the wavefunction Hessian W_{ij}. The latter, of course, is needed for the CPHF equations, too. Straightforward differentiation of the Lagrangian function gives

$$\tfrac{1}{4} W_{pi,qj} = \delta_{ij}(F_{pq} - S_{pq}\varepsilon_{ji}) + 2\sum_{r,s} G_{prqs} C_{sj} C_{ri}$$

The wavefunction Hessian above has four subscripts because the individual variational parameters have two subscripts each. With the above result, the extra term in the alternative second derivative expression becomes

$$4\sum_i [((\mathbf{C}^a)^\dagger \mathbf{F} \mathbf{C}^b)_{ii} - \varepsilon_{ii}((\mathbf{C}^a)^\dagger \mathbf{S} \mathbf{C}^b)_{ii}] + 2 \sum_{p,q,r,s} G_{pqrs} D^a_{pq} D^b_{rs}$$

where the density matrix derivatives are simply

$$\mathbf{D}^a = \mathbf{C}^a \mathbf{C}^\dagger + (\mathbf{C}^a)^\dagger \mathbf{C}$$

The evaluation of this extra term is not significant; the alternative form may thus be more advantageous than the form commonly used.

The CPHF equations can be easily formulated with the help of the wavefunction Hessian. The general equation is Eq. (14):

$$\sum_j W_{ij} C_j^a - \sum_m f_{mi} \lambda_m^a = - W_i^a$$

In the closed-shell SCF case, W_i^a is given by

$$\tfrac{1}{4}W_{pi}^a = (\mathbf{F}^a\mathbf{C} - \mathbf{S}^a\mathbf{C}\boldsymbol{\varepsilon})_{pi}$$

The above formulas give the CPHF equations in AO basis:

$$\mathbf{F}\mathbf{C}_i^a - \mathbf{S}\mathbf{C}_i^a\varepsilon_{ii} = -\mathbf{F}^a\mathbf{C}_i - \mathbf{S}^a\mathbf{C}_i\varepsilon_{ii} - \mathbf{S}\mathbf{C}_i\varepsilon_{ii}^a - \mathbf{G}(\mathbf{D}^a)\mathbf{C}_i \tag{39}$$

This equation has been written in a form which facilitates its solution, with the dominant part on the left-hand side. $\mathbf{C}_i$ is the ith column of the coefficient matrix $\mathbf{C}$, and

$$\mathbf{G}(\mathbf{X}) = \sum_{r,s} G_{pqrs}X_{rs}$$

Eq. (39) contains $(nm + n)$ unknowns ($\mathbf{C}^a$ and $\boldsymbol{\varepsilon}^a$) and it has to be augmented by the derivative constraint equations

$$(\mathbf{C}^a)^\dagger\mathbf{S}\mathbf{C} + \mathbf{C}^\dagger\mathbf{S}\mathbf{C}^a + \mathbf{C}^\dagger\mathbf{S}^a\mathbf{C} = \mathbf{0} \tag{40}$$

Owing to the presence of the terms $\mathbf{D}^a$ and $\boldsymbol{\varepsilon}^a$, which contain the unknown $\mathbf{C}^a$, Eq. (39) has to be solved iteratively. Diagonal dominance is guaranteed if this is done in canonical MO basis, i.e. by writing $\mathbf{C}_i^a$ as

$$\mathbf{C}_i^a = \sum_j^{\text{occ}} U_{ij}\mathbf{C}_j + \sum_v^{\text{virt}} U_{iv}\mathbf{C}_v$$

Eq. (40) yields the following equation in the occupied subspace for the new unknowns U:

$$U_{ij} + U_{ji} = -\tfrac{1}{2}\mathbf{C}_j^\dagger\mathbf{S}\mathbf{C}_i \tag{41}$$

As the closed-shell SCF energy is invariant against unitary transformations within the occupied or the virtual subspaces, the solution of the CPHF equations is not completely determined within these sub-blocks. For SCF derivatives, this arbitrariness cannot manifest itself in the final results. In some correlated theories, for example Møller–Plesset perturbation theory, or in CI with selected configurations, the derivatives of the true *canonical* molecular orbitals are needed. The determination of these has been discussed by Gerratt and Mills (1968) and Pople *et al.* (1979). The calculation of the derivatives of the true canonical MOs suffers from the problem that the latter become arbitrary in the degenerate case; this appears in the equations as a division by a zero orbital energy difference (Gaw and Handy, 1986). For nearly degenerate orbitals, the equations may become numerically unstable. For the case of Møller–Plesset perturbation theory, it is possible to rewrite the equations so that the derivatives of canonical MOs are not needed (Handy *et al.*, 1985; Pulay and Saebø, 1986a). For SCF derivatives, a simple solution of Eq. (41) is (Gerratt and Mills, 1968):

$$U_{ij} = -\tfrac{1}{2}\mathbf{C}_j^\dagger\mathbf{S}^a\mathbf{C}_i$$

which is symmetrical. Page *et al.* (1984) show, however, that a computationally more efficient code results if the occupied–occupied block of **U** is chosen triangular. The occupied–virtual block of **U** is obtained by multiplying Eq. (40) by $\mathbf{C}_v^\dagger$ from the left, and using the fact that $\mathbf{C}_v$ is an eigenvector of the Fock operator in canonical orbitals, i.e. $\mathbf{FC}_v = \mathbf{SC}_v\varepsilon_v$. This yields

$$U_{iv} = (\varepsilon_i - \varepsilon_v)^{-1}\mathbf{C}_v^\dagger[\mathbf{F}^a\mathbf{C}_i + \mathbf{S}^a\mathbf{C}_i\varepsilon_i - \mathbf{G}(\mathbf{D}^a)\mathbf{C}_i]. \tag{42}$$

Although this equation has been known for a long time (Bratoz, 1958; Gerratt and Mills, 1968), its large dimension makes its solution by direct methods inefficient; e.g. in a calculation with 20 MOs and 100 basis functions, the number of U_{iv} elements is $n_0 n_v = 1600$ so a direct solution needs $O(1600^2)$ storage locations and $O(1600^3) \simeq 4 \times 10^9$ arithmetic operations for each nuclear coordinate. It is easy to see that the direct method is a sixth-power procedure, against the fourth-power dependence of the SCF procedure itself. Considerations like this have discouraged the calculation of SCF second derivatives. It was a significant breakthrough when Pople *et al.* (1979) introduced an efficient iterative method for the solution of the CPHF equations. In retrospect, it is clear that this is exactly analogous to what happens in the iterative Hartree–Fock procedure itself.

The method of Pople *et al.* (1979) for the solution of the CPHF equations is based on the following idea. The solution of Eq. (42) by an iterative method is straightforward, as the storage requirements are proportional to $O(m^2)$, and no random access to the coefficients is required. A single iteration consists of calculating $\mathbf{D}^a$ from the previous set of U_{iv}, and substituting it into the right-hand side of Eq. (42). The computational effort for this is equivalent to a cycle of conventional SCF iteration. Unfortunately, Eq. (42) is only weakly diagonally dominant, and convergence is usually not very good. Pople *et al.* (1979), however, pointed out that the subspace generated by the consecutive values of **U** during the iteration converges rapidly to contain the correct solution. The correct solution is therefore sought in the iterative subspace by direct matrix inversion. The final solution **U** is written as a linear combination of the consecutive iterates $\mathbf{U}_i$:

$$\mathbf{U} = \sum_i \alpha_i \mathbf{U}_i$$

The coefficients α_i are determined from the condition that the projection of the residuum of Eq. (42) on the consecutive iterates $\mathbf{U}_1, \mathbf{U}_2$, etc., vanishes. For this Eq. (42) is rewritten as

$$\mathbf{U} - \mathbf{AU} - \mathbf{b} = 0 \tag{43}$$

in which the unknown elements of **U** are represented as a column vector. The residuum is the deviation of the right-hand side of Eq. (43) from the zero vector.

The above condition leads to a small system of linear equations

$$\sum_j B_{ij}\alpha_j = \mathbf{U}_i^\dagger \mathbf{b}$$

where

$$B_{ij} = \mathbf{U}_i^\dagger(\mathbf{1} - \mathbf{A})\mathbf{U}_j$$

The construction of the coefficient matrix **B** requires only scalar products between the **U** vectors and the residuum vectors. Methods similar in spirit to the procedure of Pople *et al.* (1979) have been used to determine selected eigenvalues of large matrices (Brändas and Goscinski, 1970; Bartlett and Brändas, 1972; Roos, 1972; Davidson, 1975), and have, in fact, been recommended to solve large systems of linear equations (Roos and Siegbahn, 1977). They are now recognized (Wormer *et al.*, 1982) as versions of the conjugate gradient technique (see, for instance, Hestenes, 1980). The use of the conjugate gradient property allows some simplification of the procedure: instead of storing all previous iterates and residuum vectors, only the last two and the current vectors are needed if the method is appropriately reformulated (Wormer *et al.*, 1982). The saving provided by the use of a three-term recurrence relation is more significant in large-scale CI methods than in the CPHF procedure.

The original algorithm of Pople *et al.* (1979) required the storage of the first derivative integrals. This is likely to be a bottleneck in large calculations. Subsequently, it was pointed out (Pulay, 1983a; Schlegel *et al.*, 1984) that this step can be eliminated as only the $3N$ derivative Fock matrices are required in the calculation. However, there must be enough fast memory to hold the derivative Fock matrices simultaneously. With the increasing availability of large memories, this does not appear to be a problem any more. The algorithm of Pople *et al.* follows Gerratt and Mills (1968) and uses two-electron integrals transformed to the MO basis, instead of the AO formulation given above. Pulay (1983a) and Osamura *et al.* (1983) have pointed out that the equations can be formulated directly in AO basis, as shown above. In another context this is a well-known result (Stevens *et al.*, 1963). There is a trade-off here, however, because only $O(n^2m^2)$ transformed integrals are needed, as compared with $O(m^4)$ AO integrals. Thus, once the partial transformation to MO basis has been completed, the MO formalism is faster, particularly for many perturbations (Amos, 1986). Integral transformation is particularly fast on supercomputers, and the experience of King and Komornicki (1986b) on the CRAY shows that, even for fairly large molecules, it is only a minor part of the calculation. On the other hand, for large molecules the AO integrals set is sparse while the MO set is not.

Further economy can be realize in the solution of the CPHF equations if all equations, or at least a number of them, are solved simultaneously. This saves much input–output, as the two-electron integrals are read into the fast storage only once. Moreover, the iterative method discussed above can be generalized so that the solution is sought in the subspace spanned by the iterated vectors for all perturbations, not only for the one under consideration.

If one assumes that the number of perturbation parameters increases linearly with the molecular size, as is the case for geometry derivatives, then the analytical calculation of SCF second derivatives contains a fifth-order step, namely the solution of the $3N$ CPHF equations. Its computational dependence is therefore asymptotically the same as the numerical differentiation of the gradient, a considerably simpler procedure. This observation no doubt contributed to the fact that practical SCF second derivatives were implemented a decade later than the gradients. Practically, however, these considerations do not appear to be important, as for practical calculations the computationally most significant term is the evaluation of the second derivatives of the AO integrals. For these terms, the analytical technique provides large savings, because the computational effort of the second integral derivatives is a constant times that of the first derivatives. In the algorithm given by Schlegel *et al.* (1984), this factor ranges between 4 and 20, with smaller values for polarized basis sets.

The unrestricted Hartree–Fock (UHF) case is completely analogous to the closed-shell one. New terms do appear in the open-shell SCF and the few-configuration case. Nevertheless, the preferred technique is quite similar to the closed-shell case. In particular, the two-particle density matrix can be constructed from compact matrices, and the solution of the derivative CI equations is very simple, due to the small dimension.

In the general multiconfigurational case, the formalism, particularly the coupled perturbed (CP) equations, are much more complex. Nevertheless, there are important reasons to try to calculate MCSCF higher derivatives, as the calculation of the MCSCF energy is quite expensive, and it is important to exhaust the information in the wavefunction. The formalism given by Page *et al.* (1984) is perhaps easiest to follow. These authors arrange the algorithm so that much of the data generated in the second-order MCSCF procedure is used directy in the solution of the CP MCSCF equations. This is not surprising as the same Hessian occurs in both (see Eq. (6)). Page *et al.* use an unconstrained formalism, i.e. independent exponential parameters $\mathbf{X}$. The change of metric is incorporated by writing the orbital derivative matrix $\mathbf{U}$ of Eq. (41) as

$$\mathbf{U} = \mathbf{X} + \mathbf{T}$$

where $\mathbf{X}$ is an antisymmetric matrix and

$$\mathbf{T} + \mathbf{T}^{\dagger} = -\mathbf{C}^{\dagger}\mathbf{S}^{a}\mathbf{C} \tag{44}$$

The perturbation variable a is not explicitly shown here. Page *et al.* (1984) chose $\mathbf{T}$ to be upper triangular; this has some advantage in the MCSCF case over the symmetrical choice of Eq. (41) (see Camp *et al.*, 1983).

The CP MCSCF equations can be written symbolically (Lengsfield, 1986) as

$$\begin{bmatrix} \partial^2 W/\partial X^2 & \partial^2 W/\partial X \partial A \\ \partial^2 W/\partial X \partial A & \partial^2 W/\partial A^2 \end{bmatrix} = \begin{bmatrix} -g_X^a - g_X(T^a) \\ -g_A^a - g_A(T^a) \end{bmatrix}$$

In this equation, X stands for the elements of the matrix **X**, A denotes the CI coefficients, and the elements of g on the right-hand side are the derivatives of the energy function W with respect to the MCSCF parameters X or A, and the nuclear coordinate a, W_i^a in our general notation (Eq. (6)). The gradients g have two components: the first is the gradient evaluated with the derivative integrals, and the second arises from the change of the metric, i.e. from the effect of the matrix **T**. These terms, or course, correspond to the derivative constraint terms in the constrained formulation (Eq. (14)). The solution of the above equation can be carried out by a code similar to the one used to determine the MCSCF wavefunction itself.

In a long MCSCF expansion, the orbital (**X**) part of the Hessian may not be diagonally dominant, and the iterative procedure may not be rapidly convergent. Another difficulty is that the transformation of the first derivative integrals to orbital basis cannot be easily avoided; this is an expensive part of the algorithm. The second derivatives of the AO integrals, however, can be summed into the force constants directly. Finally, redundant variables may arise in a general MCSCF expansion if an orbital rotation transforms a configuration into another one which is present in the MCSCF expansion. Hoffmann *et al.* (1984) discuss their elimination by locating the zero eigenvalues of the Hessian. Finally, Lengsfield *et al.* (1984) and Saxe *et al.* (1985) used techniques similar to MCSCF second derivatives calculation to evaluate non-adiabatic coupling terms. Although the terms treated are first derivatives with respect to the nuclear coordinates, the quantity sought is not variational and the coupled perturbed equations are required.

D. Higher Derivatives

The calculation of third and fourth derivatives is attractive for two reasons: these are the quantities needed in the lowest-order treatment of vibrational anharmonicities, and a quartic surface is the simplest to exhibit a double minimum, i.e. the simplest model of a reaction surface. Moccia (1970) did apparently first consider the SCF third derivative problem. A detailed derivation of SCF and MCSCF third derivatives was given by Pulay (1983a); independently, Simons and Jørgensen (1983) also considered the calculation of MCSCF third, and even fourth, derivatives in a short note. As pointed out in Section II, third derivatives of the energy require only the first derivatives of the coefficients, and are thus computationally attractive. By contrast, fourth derivatives require the solution of the second-order CP MCSCF equations. The only computer implementation so far is that of Gaw *et al.* (1984) for closed shells, although the detailed theory has been worked out for the MCSCF case

(Pulay, 1983a). The problem with the implementation of these higher derivatives is twofold: the massive coding problem, and the long runnng times for a single calculation. Note that the analytical calculation of higher derivatives must be more efficient than the numerical differentiation of gradients or second derivatives, but the requirement for a single large calculation may impose limitations; e.g. a third derivative calculation on formaldehyde with double-zeta basis set (Gaw and Handy, 1986) took 122 min on the IBM 4381, versus 4 min for the energy calculation alone. Another slight disadvantage of the analytic third derivative calculation is that is must be performed in Cartesian coordinates. As the individual Cartesian force constants have little physical significance in a molecule, the complete cubic force field has to be evaluated in every case. If numerical differentiation of the gradients is used, curvilinear valence coordinates may be used, which allows the calculation of selected cubic force constants. In spite of these difficulties, now that the remarkable first implementation has been completed successfully by the Berkeley group (Gaw *et al.*, 1984), the analytical third derivative method is here to stay.

The computationally most demanding part of the algorithm is the calculation of integral third derivatives, and future effort should concentrate on this part.

The third derivative formula for the closed-shell case is fairly simple (Moccia, 1970; Pulay, 1983a):

$$\begin{aligned} E^{abc} = {} & 2\,\mathrm{Tr}[(\mathbf{h}^{abc} + \tfrac{1}{2}\mathbf{G}^{abc}(\mathbf{D}))\mathbf{D} - \mathbf{S}^{abc}\mathbf{R}] \\ & + \{2\,\mathrm{Tr}[\mathbf{h}^{ab} + \mathbf{G}^{ab}(\mathbf{D})\mathbf{D}^{c} - \mathbf{S}^{ab}\mathbf{R}^{c} + \mathbf{G}^{a}(\mathbf{D}^{b})\mathbf{D}^{c}] + (bca) + (cab)\} \\ & + \{4\,\mathrm{Tr}[(\mathbf{C}^{b})^{\dagger}(\mathbf{h}^{a} + \mathbf{G}^{a}(\mathbf{D})) + \mathbf{G}(\mathbf{D}^{a})\mathbf{C}^{c} - (\mathbf{C}^{b})^{\dagger}\mathbf{S}^{a}\mathbf{C}^{c}\varepsilon \\ & - ((\mathbf{C}^{c})^{\dagger}\mathbf{S}^{a} + (\mathbf{C}^{a})^{\dagger}\mathbf{S}^{c})\mathbf{C}\varepsilon^{b} - (\mathbf{C}^{a})^{\dagger}\mathbf{S}\mathbf{C}^{c}\varepsilon^{b}] + (bca) + (cab)\} \end{aligned}$$

Here

$$\mathbf{R} = \mathbf{C}\varepsilon\mathbf{C}^{\dagger}$$

Note that only the same coefficient derivatives are required as in the SCF second derivative case; in particular, the exact derivatives of the canonical molecular orbitals are not required. The formulation of Gaw *et al.* (1984) does apparently use the latter. As discussed in Section III.C, this may lead to numerical difficulties in the case of degeneracies or quasi-degeneracies. Recently Gaw and Handy (1986) eliminated the canonical orbital derivatives from their program, in agreement with the results above. An extension of the closed-shell third derivative program to open-shell and MCSCF wavefunctions would be highly desirable for calculating reaction surfaces.

E. The Evaluation of Integral Derivatives

The most essential step in analytical derivative calculations is the evaluation of one- and two-electron AO integrals differentiated with respect to the coordinates of their centers. For polyatomic molecules, Gaussian basis sets are used almost exclusively, and therefore we shall consider this case only. The first *ab initio* gradient program (Pulay, 1969, 1970) used a Gaussian lobe basis (Preuss, 1956). The calculation of the gradient is very simple for such basis sets. However, integral evaluation is less efficient than for Cartesian Gaussians because the shell property (Hehre *et al.*, 1969) cannot be efficiently utilized. The differentiation of a Cartesian Gaussian yields a sum of two Cartesian Gaussians, with the angular quantum number incremented and decremented by unity:

$$\frac{\partial}{\partial A_x}\{(x-A_x)^k(y-A_y)^l(z-A_z)^m\exp[-\alpha(\mathbf{r}-\mathbf{A})^2]\}$$
$$=[-k(x-A_x)^{k-1}+2\alpha(x-A_x)^{k+1}](x-A_y)^l(z-A_z)^m\exp[-\alpha(\mathbf{r}-\mathbf{A})^2]$$

The derivative calculation can thus be reduced to the evaluation of higher (and lower) angular momentum functions. Several programs using straightforward differentiation have been reported in addition to the one mentioned above (Huber *et al.*, 1976; Komornicki *et al.*, 1977; Pulay, 1979; Saebø and Almlöf, 1979). However, the most efficient program of this kind, still unsurpassed at this time, is the s- and p-function gradient routine of Schlegel (1982) for the GAUSSIAN80 suite of programs.

For higher angular momentum wavefunctions, or for higher derivatives, a more systematic procedure is needed as the code is rapidly becoming very complex and, because of the logic involved, inefficient, particularly on modern vector and array processors. Recent reviews on integral evaluation techniques (Saunders, 1975, 1983; Hegarty and van der Velde, 1983) apply also to the closely related integral derivative calculation.

The deservedly most popular method for integral derivative evaluation is currently the Rys quadrature technique (Dupuis *et al.*, 1976; King and Dupuis, 1976; Rys *et al.*, 1983). It is based on the realization that the computationally most significant step in the evaluation of two-electron integrals over high angular momentum Gaussians is the integration over the polynomial part. This is unlike the case of low angular momentum functions where the evaluation of the integral over the exponential part of the integrand is the most significant.

In the Rys polynomial method, the first step is to use the Gaussian transform of the inverse electronic distance to separate the electron repulsion integral to x, y and z components (Cook, 1974) using the Gaussian transform

$$r_{12}^{-1}=(2/\pi^{1/2})\int_0^\infty \exp(-u^2r_{12}^2)\,du \tag{45}$$

The integrand in Eq. (45) separates to

$$\exp[-u^2(x_1-x_2)^2]\exp[-u^2(y_1-y_2)^2]\exp[-u^2(z_1-z_2)^2]$$

Substituting Eq. (45) into the expression of the two-electron integral, the latter can be written as

$$(AB|CD) = \int_0^\infty I_x(u)I_y(u)I_z(u)\,\mathrm{d}u$$

where I_x, I_y and I_z are two-dimensional integrals. Dupuis *et al.* (1976) have shown that the integration with respect to u can be performed (after transformation to a related variable) by Gaussian quadrature (see e.g. Stroud and Secrest, 1966). The advantage of this is that the integrand must be evaluated at only a few values of u; the number of points needed is linear in the total angular momentum of the functions involved, and it is 5 for four d shells, 3 for four p shells, etc. Moreover, many different integrals share the same factors I_x, I_y, I_z. The two-dimensional integrals I_x, etc., can also be evaluated by Gaussian quadrature though later implementations (Rys *et al.*, 1983) use more efficient recursion formulas.

Several techniques have been suggested for the calculation of integral derivatives using the Rys quadrature method. Dupuis and King (1978) use integrals over higher and lower angular momentum functions whereas Saxe *et al.* (1982) and Schlegel *et al.* (1984) differentiate the factors I_x, etc. The relative efficiency of these methods is still a matter of debate; the most recent implementations (King and Komornicki, 1986b) seem to favor the original implementation of Dupuis and King (1978).

Another leading technique for integral evaluation is that of McMurchie and Davidson (1978). According to Saunders (1985), the ultimate efficiency of this method is higher than that of the Rys quadrature method. It has not become as popular as the latter, perhaps because of its slightly more complex logic. Saunders (1983) recommends the combination of the two techniques; this method was used in the evaluation of third derivative integrals by Gaw *et al.* (1984).

We shall now discuss four methods which can be used to enhance the efficiency of integral derivative evaluation:

1. The use of translational and rotational invariance.
2. The use of symmetry.
3. The use of derivative closure.
4. The neglect of small gradient components.

A two-electron integral is invariant against a common translation of its four centers. This property was first explicitly described by Komornicki *et al.* (1977). Thus the 12 derivatives with respect to the coordinates of the four centers can be expressed in terms of the nine derivatives with respect to the coordinates of three centers; the gradient with respect to the fourth center can be determined from the condition that the sum of the x components of the forces on the four centers should vanish. Much of the discussion of the general case in the literature is marred by the incomplete understanding of covariant and contravariant quantities. The following treatment is perhaps the simplest. Let us introduce, instead of the four x coordinates, the new coordinates X:

$$\mathbf{x} = \mathbf{A}\mathbf{X}$$

where

$$\mathbf{A} = \begin{bmatrix} 1 & 0 & 0 & 1 \\ 0 & -1 & 0 & 1 \\ 0 & 0 & 1 & 1 \\ 0 & 0 & 0 & 1 \end{bmatrix}$$

An infinitesimal change in X_4 is equivalent to the simultaneous equal shift of all four x coordinates, and therefore all derivatives involving X_4 must vanish. We can express the derivatives with respect to $\mathbf{x}$ through the derivatives of $\mathbf{X}$; for simplicity, this is shown only for the x coordinates:

$$\partial/\partial\mathbf{X} = \mathbf{A}^{\dagger}\partial/\partial\mathbf{x}$$

$$\partial^2/\partial\mathbf{X}^2 = \mathbf{A}^{\dagger}\partial^2/\partial\mathbf{x}^2\mathbf{A}$$

and similarly for the higher derivatives. The vanishing of all derivatives which contain X_4 (and Y_4 and Z_4) yields the equations which allow the generation of the derivatives with respect to x_4 from those of x_1, x_2 and x_3. Dupuis and King (1978) suggested that the translational equations be applied to the contribution from a shell of functions, instead of from individual functions. All analytical derivative programs make use of the translational invariance conditions.

A number of papers discuss the use of rotational invariance (Dupuis and King, 1978; Kahn, 181; Vincent *et al.*, 1983; Pulay, 1983a; Banerjee *et al.*, 1985). Obviously, rotational invariance can only be used for isotropic basis sets. The problem with this technique is that the individual integrals are not invariant under rotation, as a basis function with non-zero angular momentum transforms into a linear combination of other functions on the same center under rotation. Another, but much more trivial, problem is that the choice of independent coordinates must depend on the nuclear coordinates. The only known implementation is that of Vincent *et al.* (1983) for second derivatives with s and p basis functions; they report substantial savings. However, even this relatively simple case required massive coding, and the third derivative program (Gaw *et al.*, 1984) does not use rotational invariance, although its usefulness increases for higher derivatives: there are nine first, 45 second and 165 third derivatives of a general two-electron integral if only translational invariance is used, but only six, 21 and 56 if the rotational invariance is also utilized.

A simpler method (Pulay and Hamilton, 1987) is to use rotational invariance for contributions from four shells. The total contribution must be rotationally invariant, and therefore the detailed equations for the transformation of basis functions under rotation, which is the source of complexity, is not needed. The principle is very simple for the SCF gradients. One must be careful, however, because the obvious contribution of a shell quartet to the gradients:

$$\sum_{p,q,r,s}^{\text{shell}} (pq|rs)^a (2D_{pq}D_{rs} - D_{pr}D_{qs})$$

is not rotationally invariant. Invariance is only achieved if the less obvious contribution of Eq. (31) is included. In the closed-shell case this is

$$-2\,\mathrm{Tr}(\mathbf{C}^\dagger \boldsymbol{\varepsilon} \mathbf{C} \mathbf{S}^a) = -2\,\mathrm{Tr}(\mathbf{F}\mathbf{D}\mathbf{S}^a\mathbf{D})$$

where the two-electron integrals contribute to the Fock matrix **F**, and $\mathbf{S}^a$ is the derivative overlap matrix within the shell quartet. The method consists of evaluating the above two contributions just for the six independent coordinates, and generate the remaining six gradient components using the translational and rotational invariance equations. This method has not yet been implemented; its generalization to higher derivatives would be particularly important.

Important savings are possible by using symmetry efficiently. A systematic procedure was given by King and coworkers (Dupuis and King, 1978; Takada *et al.*, 1981, 1983), who generalized the symmetrization method of Dacre (1970) to derivative calculations. The new problem is that the derivatives, in contrast to the energy, are not totally symmetrical.

Finally, we may mention the technique of Almlöf and Helgaker (1981). These authors note that the contribution of an atomic orbital to the wavefunction force vanishes if the basis set contains the derivatives of the orbital (see Section II. C on the Hellmann–Feynman forces). For instance, in an uncontracted sp shell, the contribution of the s orbital to the wavefunction force can be omitted.

The computer time ratio of gradient evaluation to integral evaluation ranges between 1:2 and 1:4 if one evaluates the derivatives of all integrals which are retained, and if a similar algorithm is used in both cases. Schlegel *et al.* (1984) point out that the density matrix is known at the time the gradients are evaluated, and therefore the test for a gradient contribution can be made on the product of the integral and density matrix element; in particular, integrals over whole shells can be omitted if the estimated contribution is below a threshold. This results in a spectacular enhancement in the speed of the gradient calculation for larger molecules with small basis sets, the ratio $T_{\text{int}}/T_{\text{grad}}$ being around unity; the timing of energy versus gradient is even more advantageous for the gradients. This procedure cannot be applied, however, if the gradients are to be differentiated further numerically. One may also remark that, in a systematic series of calculations, one has a good approximation for the density matrix, and thus the same device could be used for the energy calculation, increasing the relative cost of the gradient calculation.

The timings for integral second derivatives are usually about five times those for the first derivatives (Schlegel *et al.*, 1984; King and Komornicki, 1986b); this seems to apply also to the ratio of the total gradient calculation versus the second derivative calculation. With the best computer codes, the total time needed for second derivative calculations is 5–10 times the time needed to calculate the energy.

IV. DERIVATIVES FOR DYNAMICAL ELECTRON CORRELATION

A. Introduction

The standard method for describing dynamical electron correlation is by means of superposition of configurations. Dynamical correlation arises from numerous small contributions. It is thus not effective to optimize the orbitals involved, as in the MCSCF method. Rather, the orbitals are quite arbitrary, e.g. Hartree–Fock orbitals. Three major methods are in use; these differ in the way the coefficients of the configurations are determined.

1. Historically, variational configuration interaction (CI) was the first method to be used widely. This method, at a fixed substitution level, suffers from size inconsistency, i.e. incorrect scaling with the number of particles, and is for this reason not used as often as formerly. Another disadvantage of the CI method is that the inclusion of higher substitutions without random configuration selection is computationally almost impossible for large molecules. On the other hand, CI has two advantages: it provides a variational upper bound for the energy, and can be used for excited states without particular difficulties.
2. Many-body perturbation theory (MBPT) approximates the true Hamiltonian by an average-field one-electron Hamiltonian, and calculates the correlation energy in low (second, third or fourth) orders of perturbation theory. MBPT is size-consistent. A further significant advantage of this method, compared with the other two techniques, is that it is not iterative: for example, third-order MBPT is roughly equivalent to a single iteration of all-doubles CI. For this reason, MBPT is significantly more efficient (up to an order of magnitude) than the other two methods. Its disadvantages are that the results depend on the somewhat arbitrary choice of the zeroth-order Hamiltonian, and that a unique generalization to open-shell and excited states is difficult. For closed-shell ground-state molecules, the Møller–Plesset (MP) (Møller and Plesset, 1934) partitioning of the Hamiltonian is the best choice. Although promising applications of MBPT were

reported in the late 1960s and early 1970s, many quantum chemists looked upon these methods with some suspicion because of their non-variational nature. MBPT became a generally accepted method in the mid-1970s (Bartlett and Silver, 1975; Pople *et al.*, 1976). Pople's group contributed much to the acceptance of Møller–Plesset theory by using a fixed hierarchy of methods, by reporting many applications and by making computer programs available. For open-shell systems, MP theory based on unrestricted Hartree–Fock (UHF) wavefunctions is generally used. This method does not yield a spin eigenfunction, however, and is thus liable to artefacts.

3. Coupled-pair and coupled-cluster methods are the size-consistent generalization of the CI method. The basic method, coupled-cluster doubles (CCD), or coupled-pair many electron theory (CP-MET), was introduced in chemistry by Čižek (1966). In this method, the CI wavefunction $(\mathbf{1}+\mathbf{T})\Psi_0$ is replaced by $\Psi = \exp(\mathbf{T})\Psi_0$. Here $\mathbf{T}$ is a doubles excitation operator. The resulting wavefunction contains higher-order substitutions, which precludes the determination of the CI coefficients ('excitation amplitudes') by energy minimization. Instead, the latter are determined from the condition that projection of the residuum vector $(\mathbf{H}-E)\Psi$ on the space of the double substitutions should vanish. In coupled-cluster methods, the number of CI coefficients is not larger than in the corresponding CI methods, although higher substitutions, which are the products of lower-order ones, are implicitly included in the wavefunction. Approximate versions of the coupled-cluster theory, e.g. the coupled electron-pair approximation (CEPA) methods of Meyer (1971, 1973), are simpler than the full theory and were implemented much earlier. From the perspective of perturbation theory, the major objection to coupled-cluster methods is their inclusion of high-order contributions from double substitutions, which results in an iterative, and therefore expensive, algorithm. For a review of MBPT and coupled-cluster methods, see Bartlett (1981).

The application of gradient techniques to dynamical correlation methods (with the possible exception of the second-order Møller–Plesset theory (MP2)) is not yet as widespread as their application to SCF and MCSCF wavefunctions. One of the reasons is that these methods have not yet been perfected to the degree to which SCF methods have been. The major reason is, however, that most of these methods are difficult to apply to larger molecules because their computational work scales steeply with the molecular size. However, as we have seen, the advantages of analytical derivative methods are most significant for larger molecules. In this light, the most important task appears to be to develop efficient derivative techniques to the more routinely applicable methods, such as MP theory or local correlation theory (Otto and Ladik, 1982; Laidig *et al.*, 1982; Pulay, 1983d; Saebø and Pulay, 1985; Pulay and Saebø, 1986b).

B. Gradients

The fact that the gradient of the variational CI energy does not contain the derivatives of the CI coefficients has been pointed out early (Kumanova, 1972; Tachibana *et al.*, 1978); indeed, this is implicit in some early work. However, no computationally attractive algorithm was given, in particular for the solution of the coupled perturbed Hartree–Fock (CPHF) equations, which are required for CI gradients.

The introduction of an efficient method for the solution of the CPHF equations (Pople *et al.*, 1979) has solved this problem, and this publication, a definite breakthrough, already deals with the calculation of MP2 derivatives. This is still a very important application, as much of the correlation correction is obtained already at the second-order level, and MP2 can be applied routinely for relatively large molecules. The MP2 energy expression used by Pople is not variational in the CI coefficients, and therefore contains the explicit derivatives of the latter. This is not a serious problem in MP perturbation theory, however, as the coefficients in the usual canonical case can be obtained without the expensive iterative procedure characteristic of CI and coupled-cluster techniques. It may appear that the Hylleraas functional form offers substantial advantages, on account of the variational property (Section II). Although the functional form has advantages in formulating the gradient (Pulay and Saebø, 1986a), it turns out that the terms which are equivalent to the CI derivatives reappear in the functional formalism (Bartlett, 1986). The MP2 gradient method has been applied quite extensively recently (e.g. Raghavachari, 1984).

As pointed out by Handy *et al.* (1985) and Pulay and Saebø (1986a), several expensive or problematic steps can be eliminated from the MP2 gradient algorithm of Pople *et al.* (1979):

1. the storage of the derivative integrals (see the comments on the CPHF equations in Section III),
2. the solution of the CPHF equations for the $3N$ nuclear displacements, using the device of Handy and Schaefer (1984),
3. the determination of MO integrals with three external indices, and
4. the determination of the derivatives of the true canonical MOs.

The first three modifications save time or storage space. The last point, particularly stressed by Handy *et al.* (1985), eliminates the need for the solution of the CPHF equations in the occupied–occupied and virtual–virtual blocks, which, as discussed in Section III, may lead to numerical instabilities. It is expected that the full implementation of these features will improve the efficiency the MP2 gradient algorithm further, leading to a very routinely applicable program above the Hartree–Fock level.

Historically, gradients for variational CI wavefunctions were developed next (Krishnan *et al.*, 1980; Brooks *et al.*, 1980). The approach described by these two groups is slightly different. Krishnan *et al.* formulate the gradient specifically for a CI wavefunction with all double substitutions. Brooks *et al.*, in a more general formalism, introduce the AO two-particle density matrix, a quantity whose usefulness for CI gradient evaluation was already stressed by Meyer (1976). The two-particle density matrix formalism eliminates the need for the transformation of the $3N$ derivative integrals; instead, only a single density matrix has to be transformed. Brooks *et al.* (1980) use the graphical unitary group formalism to evaluate the two-particle density matrix; this very general formalism permits virtually any type of CI expansion.

The first CI gradient implementations were restricted to a closed-shell or a UHF reference function. Soon, however, Osamura *et al.* (1981, 1982a, b) generalized the CI gradient method to open-shell and MCSCF reference functions. More recently Page *et al.* (1984) have described an efficient multireference CI gradient program. CI gradients have obviously much in common with (MC)SCF second derivatives, as the solution of the CP MCSCF equations is a key step in both procedures. As pointed out recently by Handy and Schaefer (1984), the repeated solution of the CPHF equations can be eliminated from the CI gradient algorithm (see Section II); this should lead to an improvement in the efficiency of the method, although Simons (1986) argues that the evaluation of the necessary density matrix elements may be more expensive than the solution of the CPHF equations. For a further discussion of this subject, see Shepard (1986). To eliminate the need for the repeated solution of the CPHF equations, Meyer (1978) recommended the optimization of the orbitals in large-scale CI expansion. This idea has not yet been put to a practical test.

In spite of its great potential significance, the CI gradient method did not become very popular. Presumably, the expensiveness of the CI calculation for molecules large enough to make gradient methods really necessary prohibits the routine application of these methods on the present generations of computers.

In principle, gradients could be evaluated for randomly selected CI expansions, although no such calculations have been reported. The formalism is somewhat more complicated than in the usual case where it can be assumed that the energy is invariant against unitary transformations within the occupied or virtual orbitals; e.g. if canonical orbitals are used, the exact derivatives of these must be determined. This problem already arises for frozen-core CI calculations (Rice *et al.*, 1986). The situation is similar for limited CI with localized orbitals (Pulay, 1983b), although the determination of the localized orbital derivatives does not suffer from the small denominator problem. A problem which may render gradients calculated with randomly selected configurations less useful is that the contribution of the omitted configurations may be an oscillatory function of the nuclear coordinates unless a very strict selection criterion is used. Thus the gradients may be significantly in error, even though the overall surface is well approximated.

Third-order Møller–Plesset (MP3) gradients have not been considered until very recently, in spite of their potential practical importance. Fitzgerald *et al.* (1985) have reported a first brute-force implementation of MP3 gradients. This implementation, and an earlier formulation by Jørgensen and Simons (1983), use derivative integrals in MO basis, and are thus not expected to be very efficient. An improved implementation (Fitzgerald *et al.*, 1986) was announced recently (Bartlett, 1986). This implementation presumably uses the AO density matrix of the MP3 wavefunction, dispensing thus with the transformation of the derivative integrals to MO basis. The MP3 energy expression is not variational in the CI coefficients. Nevertheless, determination of the CI response is not difficult, as in canonical MO basis the coefficients can be directly determined, without an iterative procedure. It is, however, doubtful whether the Handy and Schaefer (1984) device for the orbital response can be efficiently implemented for the MP3 energy. Fitzgerald *et al.* (1985) also calculate MP4 gradients restricted to double substitutions (D-MP4). As shown by Bartlett and Shavitt (1977), D-MP4 is computationally only insignificantly more expensive than MP3. Bartlett (1986) also discusses the determination of the gradient for the full MP4 energy expression.

Gradient evaluation for coupled-cluster (CC) wavefunctions has been formally derived by Jørgensen and Simons (1983). However, it would appear that these formulas do not lead to very significant savings over the numerical technique, and thus the considerable effort of the programming is not justified. Their method requires the determination of the derivative CI coefficients (often called 'amplitudes' in CC theory) for each degree of freedom. As this is essentially equivalent to $3N$ solutions of the CC equations, there are no savings in this step. The major source of savings thus remains the evaluation of the integrals. However, the latter is usually a minor component in the CC calculation for larger systems. To avoid this difficulty, Pulay (1983b) suggested the reformulation of coupled-pair theories, in particular CEPA-2 (Meyer, 1971, 1973), in a closely equivalent variational problem. Ahlrichs *et al.* (1985) suggest a similar functional; both their results and other tests (Pulay and Saebø, 1985) suggest that such a functional is a very close approximation to the coupled-pair method; moreover, it can be generalized to the full (coupled-cluster doubles) (CCD) model. Adamowicz *et al.* (1984) use an alternative technique which yields the exact CCD gradients. Their method is based on the device of Handy and Schaefer (1984), discussed in Section II, and replaces the $3N$ solutions of the derivative CCD equations by the solution of a single linear system. The method, as formulated, requires the derivative MO integrals, although the latter are not stored. In order to make this method truly efficient, a reformulation in AO form seems to be inevitable, in a way similar to the AO-based self-consistent electron-pair (SCEP) method (Meyer, 1976); the full CCD in spin-adapted AO matrix form has been presented recently (Pulay *et al.*, 1984). The remark made about CI gradients holds also for CC gradients: these methods seem to be too costly for routine application on the present generation of computers. For ground-state potential surfaces, the most urgent task seems to be the efficient calculation of the derivatives in many-body perturbation theory.

C. Higher Derivatives

Relatively little has been done for the calculation of higher derivatives of correlated wavefunctions which are not of the MCSCF type. In an impressive theoretical paper, Simons *et al.* (1984) have worked out general formulas for higher derivatives of the CI wavefunction up to fourth order. Using second quantization notation, and an orthonormal basis set, Simons *et al.* arrive at fairly compact formulas which must be, however, expanded considerably to make them directly programmable. It appears, however, that these methods are not practicable at the present time. Indeed, it is perhaps useful to remember that derivative methods are not a goal in themselves but a means to study potential surfaces. Unless a derivative method avoids some redundancy in a competing numerical scheme, it cannot be expected to be more efficient than a numerical method it replaces.

At the time of this writing, the only implementation of CI second derivatives is that of Fox *et al.* (1983). The superiority of this method to the numerical differentiation of the CI gradient is not clear to the present author. CI second derivatives require the perturbed CI vectors. The determination of the latter should cost about as much as the CI energy calculation. Therefore, $3N$ CI gradient calculations should be roughly equivalent to a CI second derivative calculation. Another difficulty in the CI second derivative algorithm is the need to store the derived CI coefficients. Hopefully, these difficulties will be eliminated in the future. CI third derivatives are attractive in principle but are not yet quite practicable.

Recently, Handy *et al.* (1985) have successfully implemented the second derivative algorithm for the second-order Møller–Plesset (MP2) energy. The MP2 method comes close to a routine correlation method, and an efficient MP2 second derivative program would contribute significantly to the solution of a number of problems.

Finally, we would like to mention Pople's (1986) recent work: this treats the derivatives of the (MP) correlation energy as a double perturbation problem, with respect to a physical perturbation (e.g. nuclear coordinate change) and a non-physical perturbation (electron correlation). This provides a unified theory for the treatment of geometry and property derivatives at the correlated level.

V. PROPERTY DERIVATIVES

The main subject of the present review is the direct evaluation of geometrical energy derivatives. The evaluation of derivatives of other properties is quite analogous, and in general much simpler because no perturbation-dependent basis functions are required in general. Field-dependent basis functions have been suggested for the calculation of electrical properties: dipole moments, dipole moment derivatives, polarizabilities and polarizability derivatives (Sadlej, 1977). However, their use requires a trade-off: although fewer basis functions are needed, the calculation of the properties becomes much more difficult. The situation is completely analogous to the difference between Hellmann–Feynman forces and energy gradient. However, whereas the energy gradient form (i.e. the use of perturbation-dependent basis functions) is obviously the method of choice, for electrical properties the advantages of field-dependent functions have not yet been convincingly demonstrated.

Field-dependent basis functions are used in Ditchfield's (1974) gauge-invariant formulation of nuclear magnetic resonance (NMR) chemical shifts. The chemical shift is a second-order property, the two perturbations being an external homogeneous magnetic field and the magnetic moment of the nucleus. In this method, the basis functions are multiplied by a complex exponential phase factor, the exponent of which is proportional to the external magnetic field. This restores the gauge invariance of the wavefunction; the situation is similar to the difference between the Hellmann–Feynman forces and the exact forces: the former, in contrast with the latter, may violate the translational and rotational invariance conditions. Indeed, the analogy between nucleus-coupled basis functions and guage-invariant functions is even deeper, as the latter can be considered as functions whose centers move along a purely imaginary vector proportional to the magnetic field. It is surprizing that no program for NMR shifts seems to exist which would incorporate the advances made in the last decade (Rys polynomial integral evaluation, modern methods for the solution of the coupled perturbed Hartree–Fock equations).

The most important electric properties are dipole moment and polarizability derivatives. The theory of dipole moment derivatives has been worked out by Bratoz (1958) and by Gerratt and Mills (1968). Both papers use the obvious definition of the dipole moment derivative: the change of the dipole moment with nuclear coordinates. Komornicki and McIver (1979) have pointed out, in an influential paper, that the alternative definition, namely the derivative of the geometrical force with respect to the electric field, is more useful, as there are only three field components versus $3N$ nuclear coordinates. Similarly, the polarizability derivatives can be defined as the second derivatives of the forces with respect to the field components. Komornicki and McIver (1979) suggest numerical differentiation with respect to the field components. This approach is viable (see Komornicki and McIver, 1979; Bacskay *et al.*, 1984; Fogarasi and Pulay, 1985). However, completely analytical versions have been programmed (Amos, 1986; Frisch *et al.*, 1986). The theory has been reviewed in several papers (Simons and Jørgensen, 1984; Helgaker and Almlöf, 1984).

Notes added in proof

Several authors have pointed out the minimum property of the second derivative expression, Eq. (8). At the time of the writing, only the paper of Sellers (1986) was known to me. Subsequently, this property has been recognized in the revised paper of King and Komornicki (1986a), and by Helgaker *et al.* (1986).

In an important recent contribution, Rice and Amos (1985) have simplified the gradient formalism for general correlated wave functions. For definiteness, let us consider a CI method based on an MC-SCF wave function. Rice and Amos (1985) point out that the two-electron integral derivative contributions to the gradients can be written in terms of an effective two-electron density matrix. This matrix is the sum of the familiar two-particle density matrix and another contribution which describes the orbital response to the perturbation. The latter contribution is a sum of two terms, one coming from the Lagrangian and one from the MC-SCF Hamiltonian. In earlier implementations, these orbital derivatives were explicitly calculated. Using the linearity of the response and the Handy–Schaefer (1984) device, Rice and Amos avoid the explicit calculation of the orbital derivatives. In essence, the effective two-particle density matrix elements are the derivatives of the total energy with respect to the two-electron integrals. The energy depends on the integrals both directly and through the orbitals. The former is the two-particle density matrix, and the total is the effective two-particle density matrix.

Harrison *et al.* (1986) have reported another implementation of MBPT second derivatives.

In a recent paper, Lee *et al.* (1986) report an improved implementation of CI second derivatives. Although analytical methods are certainly preferable to numerical ones, the present author still cannot see a very significant computational advantage in this method, as compared with the numerical differentiation of the gradients, in accordance with the discussion on p. 279. The timings show the expected 1:2 advantage of the analytical procedure over the numerical one *with central differences, i.e. double-sided displacements.* Considering the low anharmonicity of the correlation energy, one-sided displacements are adequate, leading to approximately equal computational requirements for both methods. Note that this program utilizes the translational and rotational conditions on the Cartesian derivatives (Page *et al.*, 1984). The latter are significant in a small molecule but not asymptotically [see the discussion of Pulay (1977), p. 164]. The calculation of second derivatives by the numerical differentiation of the forces in internal coordinates largely utilizes the savings connected with translational and rotational invariance.

Acknowledgements

This work has been partially supported by the National Science Foundation under Grant No. CHE-8500487. Acknowledgement is also made to the Donors of the Petroleum Research Fund, administered by the American Chemical Society, for partial support.

References

Adamowicz, L., Laidig, W. D., and Bartlett, R. J. (1984). *Int. J. Quantum Chem. Symp.*, **18**, 245.
Ahlrichs, R., Scharf, P., and Erhardt, C. (1985). *J. Chem. Phys.*, **82**, 890.
Allavena, M. (1966). *Theor. Chim. Acta*, **5**, 21.
Allavena, M., and Bratoz, S. (1963). *J. Chim. Phys.*, **60**, 1199.
Almlöf, J., and Helgaker, T. (1981). *Chem. Phys. Lett.*, **83**, 125.
Almlöf, J., and Taylor, P. R. (1985). *Int. J. Quantum Chem.*, **27**, 743.
Amos, R. D. (1986). In *Geometrical Derivatives of Energy Surfaces and Molecular Properties* (Eds P. Jørgensen and J. Simons), Proc. NATO Adv. Study Workshop, Reidel, Dordrecht, p. 135.
Bacskay, G. B., Saebø, S., and Taylor, P. R. (1984). *Chem. Phys.*, **90**, 215.
Banerjee, A., Jensen, J. O., and Simons, J. (1985). *J. Chem. Phys.*, **82**, 4566.
Bartlett, R. J. (1981). *Annu. Rev. Phys. Chem.*, **32**, 359.
Bartlett, R. J. (1986). In *Geometrical Derivatives of Energy Surfaces and Molecular Properties* (Eds P. Jørgensen and J. Simons), Proc. NATO Adv. Study Workshop, Reidel, Dordrecht, p. 35.
Bartlett, R. J., and Brändas, E. J. (1972). *J. Chem. Phys.*, **56**, 5467.
Bartlett, R. J., and Shavitt, I. (1977). *Chem. Phys. Lett.*, **50**, 190.
Bartlett, R. J., and Silver, M. D. (1975). *J. Chem. Phys.*, **62**, 325.
Bishop, D. M., and Macias, A. (1970). *J. Chem. Phys.*, **53**, 3515.
Bobrowicz, W., and Goddard, W. A., III, (1977). In *Methods of Electronic Structure Theory* (Ed. H. F. Schaefer, III), p. 79, Plenum, New York.
Brändas, E., and Goscinski, O. (1970). *Phys. Rev. A*, **1**, 552.
Bratoz, S. (1958). *Colloq. Int. C.N.R.S.*, **82**, 287.
Bratoz, S., and Allavena, M. (1962). *J. Chem. Phys.*, **37**, 2138.
Brooks, B. R., Laidig, W. D., Saxe, P., Goddard, J. D., Yamaguchi, Y., and Schaefer, H. F., III (1980). *J. Chem. Phys.*, **72**, 4652.
Camp, R. N., King, H. F., McIver, J. W., and Mullaly, D. (1983). *J. Chem. Phys.*, **79**, 1088.
Čižek, J. (1966). *J. Chem. Phys.*, **45**, 4256.
Čižek, J., and Paldus, J. (1971). *Int. J. Quantum Chem.*, **5**, 359.
Cook, D. B. (1974). *Ab Initio Valence Calculations in Chemistry*, Halsted, New York.
Dacre, P. D. (1970). *Chem. Phys. Lett.*, **7**, 47.
Davidson, E. R. (1970). *J. Comput. Phys.*, **17**, 87.
Deb, B. M. (1981). *The Force Concept in Chemistry*, Van Nostrand, New York.

Ditchfield, R. (1974). *Mol. Phys.*, **27**, 789.
Dupuis, M. (1981). *J. Chem. Phys.*, **74**, 5758.
Dupuis, M., and King, H. F. (1977). *Int. J. Quantum Chem.*, **11**, 613.
Dupuis, M., and King, H. F. (1978). *J. Chem. Phys.*, **68**, 3998.
Dupuis, M., Rys, J., and King, H. F. (1976). *J. Chem. Phys.*, **65**, 111.
Dykstra, C. E., and Jasien, P. G. (1984). *Chem. Phys. Lett.*, **109**, 388.
Epstein, S. T., and Sadlej, A. J. (1979). *Int. J. Quantum Chem.*, **15**, 147.
Feynman, R. P. (1939). *Phys. Rev.*, **56**, 340.
Fitzgerald, G., Harrison, R., Laidig, W. D., and Bartlett, R. J. (1985). *J. Chem. Phys.*, **82**, 4379.
Fitzgerald, G., Harrison, R., Laidig, W. D., and Bartlett, R. J. (1986). *Chem. Phys. Lett.*, **117**, 433.
Fogarasi, G., and Pulay, P. (1984). *Annu. Rev. Phys. Chem.*, **35**, 191.
Fogarasi, G., and Pulay, P. (1985). In *Vibrational Spectra and Vibrational Structure*, Vol. 14 (Ed. J. Durig), p. 125, Elsevier, Amsterdam.
Fox, D. J., Osamura, Y., Hoffmann, M. R., Gaw, J. F., Fitzgerald, G., Yamaguchi, Y., and Schaefer, H. F., III (1983). *Chem. Phys. Lett.*, **102**, 17.
Frisch, M. J., Yamaguchi, Y., Gaw, J. F., and Schaefer, H. F., III (1986). *J. Chem. Phys.*, **84**, 531.
Gaw, J. F., and Handy, N. C. (1986). In *Geometrical Derivatives of Energy Surfaces and Molecular Properties* (Eds P. Jørgensen and J. Simons), Proc. NATO Adv. Study Workshop, Reidel, Dordrecht. p. 79.
Gaw, J. F., Yamaguchi, Y., and Schaefer, H. F., III (1984). *J. Chem. Phys.*, **81**, 6395.
Gerratt, J., and Mills, I. M. (1968). *J. Chem. Phys.*, **49**, 1719, 1730.
Goddard, J. D., Handy, N. C., and Schaefer, H. F., III (1979). *J. Chem. Phys.*, **71**, 1259.
Hall, G. G. (1961). *Phil. Mag.*, 249.
Handy, N. C. (1986). To be published.
Handy, N. C., Amos, R. D., Gaw, J. F., Rice, J. E., and Simandiras, E. D. (1985). *Chem. Phys. Lett.*, **120**, 151.
Handy, N. C., and Schaefer, H. F., III (1984). *J. Chem. Phys.*, **81**, 5031.
Harrison, R. J., Fitzgerald, G. B., Laidig, W. D., and Bartlett, R. J. (1986). *Chem. Phys. Lett.*, **124**, 291.
Hartree, D. R. (1968). *Numerical Analysis*, Oxford University Press, Oxford.
Hegarty, D., and van der Velde, G. (1983). *Int. J. Quantum Chem.*, **23**, 1135.
Hehre, W. J., Lathan, W. A., Ditchfield, R., Newton, M. D., and Pople, J. A. (1982). GAUSSIAN70, Program No. 236, Quantum Chemistry Program Exchange, Indiana University, Bloomington.
Hehre, W. J., Stewart, R. F., and Pople, J. A. (1969). *J. Chem. Phys.*, **51**, 2657.
Helgaker, T. U., and Almlöf, J. (1984). *Int. J. Quantum Chem.*, **26**, 275.
Helgaker, T. U., Almlöf, J., Jensen, H. J. A., and Jørgensen, P. (1986). *J. Chem. Phys.*, in press.
Hellmann, J. (1937). *Einführung in die Quantenchemie*, Deuticke, Leipzig.
Hestenes, M. R. (1980). *Conjugate Direction Methods of Optimization*, Springer, New York.
Hirschfelder, J. O., Byers Brown, W., and Epstein, S. T. (1964). *Adv. Quantum Chem.*, **1**, 255.
Hoffmann, M. R., Fox, D. J., Gaw, J. F., Osamura, Y., Yamaguchi, Y., Grev, R. S., *et al.* (1984). *J. Chem. Phys.*, **80**, 2660.
Huber, H., Carsky, P., and Zahradnik, R. (1976). *Theor. Chim. Acta*, **41**, 217.
Hurley, A. C. (1954). *Proc. R. Soc. A*, **226**, 170.
Hylleraas, E. A. (1930). *Z. Phys.*, **65**, 209.
Jaszunski, M., and Sadlej, A. J. (1975). *Theor. Chim. Acta*, **40**, 167.

Jaszunski, M., and Sadlej, A. J. (1977). *Int. J. Quantum Chem.*, **11**, 233.
Jensen, H. J. A. (1985). Contribution at the 1985 Sanibel Symposium.
Jensen, H. J. A., and Jørgensen, P. (1984). *J. Chem. Phys.*, **80**, 1204.
Jørgensen, P. (1986). In *Geometrical Derivatives of Energy Surfaces and Molecular Properties* (Eds P. Jørgensen and J. Simons), Proc. NATO Adv. Study Workshop, Reidel, Dordrecht.
Jørgensen, P., and Simons, J. (1983). *J. Chem. Phys.*, **79**, 334.
Kahn, L. R. (1981). *J. Chem. Phys.*, **75**, 3962.
Kato, S., and Morokuma, K. (1979). *Chem. Phys. Lett.*, **65**, 19.
Kertesz, M. (1984). *Chem. Phys. Lett.*, **106**, 445.
King, H. F., Camp, R. N., and McIver, J. W. (1984). *J. Chem. Phys.*, **80**, 1171.
King, H. F., and Dupuis, M. (1976). *J. Comput. Phys.*, **21**, 144.
King, H. F., and Komornicki, A. (1986a). *J. Chem. Phys.*, **84**, 5645.
King, H. F., and Komornicki, A. (1986b). In *Geometrical Derivatives of Energy Surfaces and Molecular Properties* (Eds P. Jørgensen and J. Simons), Proc. NATO Adv. Study Workshop, Reidel, Dordrecht, p. 207.
Knowles, P. J., Sexton, G. J., and Handy, N. C. (1982). *Chem. Phys.*, **72**, 337.
Komornicki, A., Ishida, K., Morokuma, K., Ditchfield, R., and Conrad, M. (1977). *Chem. Phys. Lett.*, **45**, 595.
Komornicki, A., and McIver, J. W. (1979). *J. Chem. Phys.*, **70**, 2014.
Krishnan, R., Schlegel, H. B., and Pople, J. A. (1980). *J. Chem. Phys.*, **72**, 4654.
Kumanova, M. D. (1972). *Mol. Phys.*, **23**, 407.
Laidig, W., Purvis, G. D., III, and Bartlett, R. J. (1982). *Int. J. Quantum Chem. Symp.*, **16**, 561.
Lee, T. J., Handy, N. C., Rice, J. B., Scheiner, A. C., and Schaefer, H. F. III (1986). *J. Chem. Phys.*, **85**, 3930.
Lengsfield, B. H. (1986). In *Geometrical Derivatives of Energy Surfaces and Molecular Properties* (Eds P. Jørgensen and J. Simons). Proc. NATO Adv. Study Workshop, Reidel, Dordrecht, p. 147.
Lengsfield, B. H., III, Saxe, P., and Yarkony, D. (1984). *J. Chem. Phys.*, **81**, 4549.
Levy, B. (1970). *Int. J. Quantum Chem.*, **4**, 297.
McIver, J. W., and Komornicki, A. (1971). *Chem. Phys. Lett.*, **10**, 303.
McMurchie, L. E., and Davidson, E. R. (1978). *J. Comput. Phys.*, **26**, 218.
Meyer, W. (1971). *Int. J. Quantum Chem. Symp.*, **5**, 341.
Meyer, W. (1973). *J. Chem. Phys.*, **58**, 1017.
Meyer, W. (1976). *J. Chem. Phys.*, **64**, 2901.
Meyer, W. (1978). Invited talk at the International Congress of Quantum Chemistry, Kyoto.
Meyer, W., and Pulay, P. (1972). *J. Chem. Phys.*, **56**, 2109.
Meyer, W., and Pulay, P. (1973). In *Proc. 2nd Semin. on Computational Problems in Quantum Chemistry* (Eds A. Veillard and G. H. F. Diercksen), p. 44, Max-Planck-Institute, Munich.
Moccia, R. (1967). *Theor. Chim. Acta*, **8**, 8.
Moccia, R. (1970). *Chim. Phys. Lett.*, **5**, 260.
Møller, C., and Plesset, M. S. (1934). *Phys. Rev.*, **46**, 618.
Nakatsuji, H. Kanda, K., Hada, M., and Yonezawa, T. (1982). *J. Chem. Phys.*, **77**, 3109.
Nakatsuji, H., Kanda, K., Hada, M., and Yonezawa, T. (1983). *J. Chem. Phys.*, **79**, 2493.
Osamura, Y., Yamaguchi, Y., Saxe, P., Fox, D., Vincent, M. A., and Schaefer, H. F., III (1983). *J. Mol. Struct.*, **103**, 183.
Osamura, Y., Yamaguchi, Y., Saxe, P., Vincent, M. A., Gaw, J. F., and Schaefer, H. F., III (1982a). *Chem. Phys.*, **72**, 131.

Osamura, Y., Yamaguchi, Y., and Schaefer, H. F., III (1981). *J. Chem. Phys.*, **75**, 2919.
Osamura, Y., Yamaguchi, Y., and Schaefer, H. F., III (1982b). *J. Chem. Phys.*, **77**, 383.
Otto, P., and Ladik, J. (1982). *Int. J. Quantum Chem.*, **22**, 169.
Page, M., Saxe, P., Adams, G. F., and Lengsfield, B. H. (1984). *Chem. Phys. Lett.*, **107**, 587.
Page, M., Saxe, P., Adams, G. F., and Lengsfield, B. H. (1984). *J. Chem. Phys.*, **81**, 434.
Pople, J. A. (1986). In *Geometrical Derivatives of Energy Surfaces and Molecular Properties* (Eds P. Jørgensen and J. Simons), Proc. NATO Adv. Study Workshop, Reidel, Dordrecht, p. 109.
Pople, J. A., Binkley, J. S., and Seeger, R. (1976). *Int. J. Quantum Chem. Symp.*, **10**, 1.
Pople, J. A., Raghavachari, K., Schlegel, H. B., and Binkley, J. S. (1979). *Int. J. Quantum Chem. Symp.*, **13**, 225.
Preuss, H. (1956). *Z. Naturf.*, **11**, 823.
Pulay, P. (1969). *Mol. Phys.*, **17**, 197.
Pulay, P. (1970). *Mol. Phys.*, **18**, 473.
Pulay, P. (1971). *Mol. Phys.*, **36**, 2470.
Pulay, P. (1977). In *Applications of Electronic Structure Theory* (Ed. H. F. Schaefer, III), p. 153, Plenum, New York.
Pulay, P. (1979). *Theor. Chim. Acta*, **50**, 299.
Pulay, P. (1981). In *The Force Method in Chemistry* (Ed. B. M. Deb), p. 449, Van Nostrand-Reinhold, New York.
Pulay, P. (1983a). *J. Chem. Phys.*, **78**, 5043.
Pulay, P. (1983b). *J. Mol. Struct. Theochem*, **103**, 57.
Pulay, P. (1983c). *J. Chem. Phys.*, **79, 2491.**
Pulay, P. (1983d). *Chem. Phys. Lett.*, **100**, 151.
Pulay, P., and Hamilton, T. P. (1987) to be published.
Pulay, P., and Meyer, W. (1971). *J. Mol. Spectrosc.*, **41**, 59.
Pulay, P., and Meyer, W. (1972). *J. Chem. Phys.*, **57**, 3337.
Pulay, P., and Saebø, S. (1985). *Chem. Phys. Lett.*, **117**, 37.
Pulay, P., and Saebø, S. (1986a). *Theor. Chim. Acta*, in press.
Pulay, P., and Saebø, S. (1986b). In *Geometrical Derivatives of Energy Surfaces and Molecular Properties* (Eds P. Jørgensen and J. Simons), Proc. NATO Adv. Study Workshop, Reidel, Dordrecht.
Pulay, P., Saebø, S., and Meyer, W. (1984). *J. Chem. Phys.*, **81**; 1901.
Pulay, P., and Török, F. (1964). *Acta Chim. (Budapest)*, **41**, 257.
Raghavachari, K. (1984). *J. Chem. Phys.*, **81**, 2717.
Rice, J. E., and Amos, R. D. (1985). *Chem. Phys. Lett.*, **122**, 585.
Rice, J. E., Amos, R. D., Handy, N. C., and Schaefer, H. F., III (1986). *J. Chem. Phys.*, **85**, 963.
Roos, B. (1972). *Chem. Phys. Lett.*, **15**, 153.
Roos, B., and Siegbahn, P. E. M. (1977). In *Methods of Electronic Structure Theory* (Ed. H. F. Schaefer, III), p. 277, Plenum, New York.
Roothaan, C. C. J. (1960). *Rev. Mod. Phys.*, **32**, 179.
Rys, J., Dupuis, M., and King, H. F. (1983). *J. Comput. Chem.*, **4**, 154.
Sadlej, A. J. (1977). *Chem. Phys. Lett.*, **47**, 50.
Saebø, S., and Almlöf, J. (1979). Program Description for MOLFORC, an *ab initio* gradient program, Department of Chemistry, University of Oslo.
Saebø, S., and Pulay, P. (1985). *Chem. Phys. Lett.*, **113**, 13.
Saunders, V. R. (1975). In *Computational Techniques in Quantum Chemistry and Molecular Physics* (Eds G. H. F. Diercksen, B. T. Sutcliffe and A. Veillard), p. 347, Reidel, Dordrecht.

Saunders, V. R. (1983). In *Methods of Computational Molecular Physics* (Eds G. H. F. Diercksen and S. Wilson), NATO ASI, Series C, Vol. 113, p. 1, Reidel, Dordrecht.
Saunders, V. R. (1985). Private communication.
Saxe, P., Lengsfield, H. B., III, and Yarkony, D. (1985). *Chem. Phys. Lett.*, 113, 159.
Saxe, P., Yamaguchi, Y., and Schaefer, H. F., III (1982). *J. Chem. Phys.*, **77**, 5647.
Schlegel, H. B. (1981). In *Computational Theoretical Organic Chemistry* (Eds I. G. Csizmadia and R. Daudel), p. 129, Reidel, Dordrecht.
Schlegel, H. B. (1982). *J. Chem. Phys.*, **77**, 3676.
Schlegel, H. B. (1987). *Adv. Chem. Phys.*, **67**, 249.
Schlegel, H. B., Binkley, J. S., and Pople, J. A. (1984). *J. Chem. Phys.*, **80**, 1976.
Schlegel, H. B., and Wolfe, S. (1975).
Schlegel, H. B., Wolfe, S., and Bernardi, F. (1975). *J. Chem. Phys.*, **63**, 3632.
Sellers, H. L. (1986). *Int. J. Quantum Chem.*, **30**, 433.
Shavitt, I. (1977). In *Methods of Electronic Structure Theory* (Ed. H. F. Schaefer, III), p. 189, Plenum, New York.
Shepard, R. (1986). In *Geometrical Derivatives of Energy Surfaces and Molecular Properties* (Eds P. Jørgensen and J. Simons), Proc. NATO Adv. Study Workshop, Reidel, Dordrecht, p. 193.
Siegbahn, P. E., Almlöf, J., Heiberg, A., and Roos, B. O. (1981). *J. Chem. Phys.*, **74**, 2384.
Simons, J. (1986). In *Geometrical Derivatives of Energy Surfaces and Molecular Properties* (Eds P. Jørgensen and J. Simons), Proc. NATO Adv. Study Workshop, Reidel, Dordrecht, p. 27.
Simons, J., and Jørgensen, P. (1983). *J. Chem. Phys.*, **79**, 3599.
Simons, J., and Jørgensen, P. (1984). *Int. J. Quantum Chem.*, **25**, 1135.
Simons, J., Jørgensen, P., and Helgaker, T. U. (1984). *Chem. Phys.*, **86**, 413.
Stevens, R. M., Pitzer, R. M., Lipscomb, W. N. (1963). *J. Chem. Phys.*, **38**, 550.
Stroud, A. H., and Secrest, D. (1966). *Gaussian Quadrature Formulas*, Prentice-Hall, Englewood Cliffs, NJ.
Tachibana, A., Yamashita, K., Yamabe, T., and Fukui, K. (1978). *Chem. Phys. Lett.*, **59**, 255.
Takada, T., Dupuis, M., and King, H. F. (1981). *J. Chem. Phys.*, **75**, 332.
Takada, T., Dupuis, M., and King, H. F. (1983). *J. Comput. Chem.*, **4**, 234.
Teramae, H., Yamabe, T., Satoko, C., and Imamura, A. (1983). *Chem. Phys. Lett.*, **101**, 149.
Thomsen, K., and Swanstrøm, P. (1973). *Mol. Phys.*, **26**, 735, 751.
Vincent, M., Saxe, P., and Schaefer, H. F., III (1983). *Chem. Phys. Lett.*, **94**, 351.
Wigner, E. (1935). *Math. Naturwiss. Anz.* (*Budapest*), **53**, 477.
Wormer, P. E. S., Visser, F., and Paldus, J. (1982). *J. Comput. Phys.*, **48**, 23.
Yamaguchi, Y., Osamura, Y., Fitzgerald, G., and Schaefer, H. F., III (1983). *J. Chem. Phys.*, **78**, 1607.

Ab Initio Methods in Quantum Chemistry—II
Edited by K. P. Lawley

SYMMETRY AND DEGENERACY IN $X\alpha$ AND DENSITY FUNCTIONAL THEORY

BRETT I. DUNLAP

Code 6129, Naval Research Laboratory, Washington, DC 20375–5000, USA

CONTENTS

I. Introduction 287
II. Density Functional Theory 287
III. Self-consistency 292
IV. Approximations to the Indirect Electronic Coulomb Repulsion . . 296
V. Orbitals and Fractional Occupation Numbers. 305
VI. Broken Spin Symmetry 309
VII. Broken Spatial Symmetry. 313
References. 316

I. INTRODUCTION

Applied density functional theory is a rapidly evolving field, which has its origins in fast but approximate quantum-chemical calculations. In anticipation of substantial progress to come, this review emphasizes theoretical motivation and interpretation rather than concentrating on the present level of agreement with experiment. Fundamental issues that distinguish the Hohenberg–Kohn theorem, the Kohn–Sham construction, $X\alpha$ and its variants are presented. $X\alpha$ is generalized to be invariant under rotations in spin space by applying Slater's 4/3 power of the charge density approximation to the exchange energy of generalized Hartree–Fock theory. The concepts of chemical potential, Fermi statistics, fractional occupational numbers, broken symmetry and non-uniform magnetism are described.

II. DENSITY FUNCTIONAL THEORY

In either the Born–Oppenheimer (fixed-nuclei) or jellium (smeared-nuclei) approximations, the electronic part of a finite-molecular or infinite-solid wavefunction separates from the nuclear wavefunction. In this case the non-relativistic total energy is the sum of the constant nuclear repulsions plus the electronic energy,

$$E = -\tfrac{1}{2}\int \delta(x_1, x_2)\nabla_1^2 \rho(x_1, x_2)\,\mathrm{d}x_1\,\mathrm{d}x_2$$
$$- \int [\rho_{\mathrm{nuc}}(\mathbf{r}_2)\rho(x_1, x_1)/r_{12}]\,\mathrm{d}x_1\,\mathrm{d}\mathbf{r}_2$$
$$+ \int [\rho_2(x_1, x_1', x_2, x_2')/2r_{12}]\,\mathrm{d}x_1\,\mathrm{d}x_2 \qquad (1)$$

This in turn is only a function of the first-order density matrix of the N electrons,

$$\rho(x_1, x_1') = N \int \Psi^*(x_1, x_2, \ldots, x_N)\Psi(x_1', x_2, \ldots, x_N)\,\mathrm{d}x_2 \ldots \mathrm{d}x_N \qquad (2)$$

the second-order density matrix,

$$\rho_2(x_1, x_1', x_2, x_2')$$
$$= \tfrac{1}{2}N(N-1)\int \Psi^*(x_1, x_2, x_3, \ldots, x_N)\Psi(x_1', x_2', x_3, \ldots, x_N)\,\mathrm{d}x_3 \ldots \mathrm{d}x_N \qquad (3)$$

and of course the nuclear charge density $\rho_{\mathrm{nuc}}(\mathbf{r})$. In these equations x represents the three spatial coordinates and one spin coordinate of an electron, $\mathbf{r}$ represents the three spatial coordinates only and the integrations are over both spin and spatial coordinates. The term 'density functional' is now used to describe quantum-chemical methods that use functional forms involving the electronic density,

$$\rho(x) = \rho(x, x) \qquad (4)$$

the diagonal part of the first-order density matrix in the position-space representation, to evaluate or approximate terms in Eq. (1). The term 'density functional' (DF) has come into vogue with the fame of the proof by Hohenberg and Kohn (1964) that the density is the only necessary variable in an exact quantum-chemical calculation of ground-state properties of non-relativistic electrons acted upon by arbitrary electric and other scalar potentials, provided only that the ground state is non-degenerate. The proof is straightforward, relying on the fact that the interaction Hamiltonian between any number of electrons and an external scalar potential involves only the electron density. To complete the proof, Hohenberg and Kohn needed only to establish, via Schrödinger's equation, a one-to-one relationship between ground-state densities of N electrons and applied external potentials that differ by more than a constant. The one-to-one aspect of the proof limits its applicability to non-degenerate ground states. DF theories that compute ground-state electronic densities without computing any intermediate quantities such as the first-order density matrix, e.g. Thomas–Fermi, have recently been reviewed by March (1983), Kohn and Vashista (1983) and Rajagopal (1980) and will not be reviewed here.

The basic problem with pure density functional theories is that the electron density is a largely classical object. Quantum effects on the density, such as atomic and molecular shell structure, are small, arising via Schrödinger's equation through the replacement of the classical momentum of each electron by the quantum operator $-i\hbar\nabla$. That is not to say that pure density functional methods cannot ultimately be successful in indirectly yielding an exact quantum density. For instance, electronic shell structure is significantly amplified by calculating the gradient of the density. While no exact gradient method is known, gradient methods could be used to sample and amplify the quantum-mechanically induced structure of the density in an attempt to set up self-consistency requirements that include quantum effects indirectly. For examples of pure density functional methods using gradients, see Csavinszky (1981) and Murphy (1981). To circumvent this quantum problem, Slater (1951) advocated computing the kinetic energy directly and using a DF expression for the Hartree–Fock exchange term only. In such a method, shell structure arises naturally from the one-electron orbitals that must be computed in order to determine the kinetic energy directly. These computationally more demanding methods such as $X\alpha$ (Slater, 1974) are the subject of this review and will be called $X\alpha$-like. However, only characteristics common to all $X\alpha$-like models will be exphasized. They are called $X\alpha$-like to emphasize their approximate nature in contrast to the Kohn–Sham construction (Kohn and Sham, 1965), which although presently incompletely defined might be able to embrace an exact $X\alpha$-like method. Such methods include $X\alpha$ and the methods of Janak *et al.* (1975), of Gunnarsson and Lundqvist (1976) (GL), of Vosko *et al.* (1980) and of Perdew and Zunger (1981).

Kohn and Sham (1965) refined the Hohenberg–Kohn (HK) density functional in a manner appropriate to $X\alpha$ calculations and other one-electron models. They replaced the exact kinetic energy with the kinetic energy of the same number of non-interacting electrons. In order to have the possibility of a unique value (see Section VII) for this artificial kinetic energy, the electrons were constrained to be in their non-interacting ground state while simultaneously being constrained via an external electric field to have the same density as the original interacting system. This specification of the artificial external electric field via the exact density is required in order still to have a density functional theory. The perspective taken in this review differs slightly from that of Kohn and Sham in that they divide this artificial field into two parts. The two parts are the true applied external field, which is the quantity mapped onto the density in the HK theorem, and the remainder, which is roughly the self-consistent field of the density. This change is made to reduce problems associated with describing a collection of electrons that feel the direct part of their self-Coulomb repulsion as non-interacting. Thus in the Kohn–Sham (KS) construction the electronic Hamiltonian becomes a sum of three terms,

$$E(\rho) = T(\rho) - \int \rho(x) V(\mathbf{r})\,\mathrm{d}x + G'(\rho) \tag{5}$$

where T is the kinetic energy of non-interacting electrons having the exact interacting density, V is the electric scalar potential that is required to make the non-interacting electrons assume the proper density, and G' is everything that is left over. Provided that the three terms of Eq. (5) are always defined uniquely, the three terms are universal density functionals by construction and their sum is the universal density functional by the HK theorem.

The beauty of the KS construction, which maps the exact wavefunction onto a non-interacting electron-gas wavefunction via a common density, is that we can exactly solve the latter problem. In fact its solution is to solve the particularly trivial one-electron equations

$$\varepsilon_i u_i(\mathbf{r}) = -\tfrac{1}{2}\nabla^2 u_i(\mathbf{r}) - V(\mathbf{r}) u_i(\mathbf{r}) \tag{6}$$

and to occupy them in order of increasing one-electron energy up to the Fermi level in order to obtain the non-interacting kinetic energy

$$T(\rho) = \sum_i n_i u_i^*(\mathbf{r})(-\tfrac{1}{2}\nabla^2) u_i(\mathbf{r}) \tag{7}$$

and the density

$$\rho(\mathbf{r}) = \sum_i n_i u_i^*(\mathbf{r}) u_i(\mathbf{r}) \tag{8}$$

The possibility of fractional occupation numbers (FON),

$$0 < n_i < 2 \tag{9}$$

is allowed for in these definitions following the standard DF prescription, but in finding the ground state of the non-interacting electron gas, FON are only allowed in the case of degenerate electron orbitals at the Fermi level. (If FON appeared for one-electron energies below the Fermi level, the electron gas would be in an excited state for the external field in question.) Thus given the relationship between V and the density, Kohn and Sham have constructed a rigorously exact one-electron theory.

The allure of and Slater's goal in deriving the $X\alpha$ equations is evident in Eq. (6). In contrast to Hartree–Fock (HF) and HF-based configuration-interaction (CI) approaches, these one-electron equations include a single potential V that can depend only on the total density and not on the orbital in question. Historically the adjective 'local' in local density functional (LDF) or in local exchange and correlation potentials signified this important distinction between $X\alpha$ and HF (Connolly, 1975). More recently, 'local' is used to signify the computationally less significant property of the $X\alpha$ exchange potential that at each point it only depends on the density at that point (Langreth and Mehl, 1983). 'Local' in this work will have the historical meaning. In particular, the potential that leads to the direct part of the electron–electron repulsion will be called 'local' even though Poisson's equation, which involves an integration over the entire charge density, must be solved to find that local potential at any given point. In this nomenclature, gradient methods would be called 'local'.

The G' in Eq. (5) is different from the G of Eq. (2.1) of Kohn and Sham (1965). The difference is that in Eq. (5) what Lieb (1979) calls the direct part of the electronic Coulomb interaction energy,

$$\int [\rho(\mathbf{r}_1)\rho(\mathbf{r}_2)/2r_{12}]\,\mathrm{d}\mathbf{r}_1\,\mathrm{d}\mathbf{r}_2 = \sum_{i,j} J_{ij}/2 \qquad \text{(for the special case of HF)} \quad (10)$$

is not included as a separate entity. The direct part of the electron–electron Coulomb interaction is a special quantity in DF theory because it is the largest part of the electron–electron interaction that can be treated simply. The remaining part of the electron–electron interaction, which Lieb calls indirect, is conceptually and computationally more troublesome. It is troublesome even in HF where, although it is a direct function of the density matrix, it can only be expressed indirectly as a function of the density. As was alluded to above, this change from G to G' in Eq. (5) is made for two reasons. First, whether or not V depends in any way on ρ determines whether or not the Kohn–Sham (KS) equations are non-linear. If the KS equations are non-linear then they can have pathological solutions (complex and/or broken-symmetry orbitals) such as occur in HF (Fukutome, 1981). In such pathological cases, constraining the orbitals to be real or to be irreducible representations of the symmetry of V_{ext} will yield higher-energy solutions that can violate 'Fermi statistics', i.e. have unoccupied orbitals lower in one-electron energy than occupied orbitals, or that violate quantum mechanics by having fractions of an electron assigned to stationary orbitals. Secondly, the symmetry properties of the $X\alpha$-like solutions are more transparent if V includes the electrostatic potential of non-interacting electron gas having the density of the real system. The symmetry properties of the one-electron eigenfunctions are determined by the invariances of V under spatial and spin symmetry operations. Thus the symmetry-related properties of the density are determined by the symmetries of V—not by the symmetries of the nuclear and other externally applied electric fields, V_{ext}. Of course, in order for the KS scheme to work, there must be a one-to-one relationship between V and the externally applied fields. (A prime was not added to the V of Eq. (6) because the V of KS is V_{ext}.)

What is called the 'V-representability' (really V_{ext}-representability herein) problem is the question of whether or not there exists a one-to-one mapping between all exact non-degenerate ground-state densities and unique external fields, which act on non-interacting ground-state electrons to yield ground-state interacting electronic densities. The term 'V-representability' is attributed to E. G. Larsen in Ref. 14 of Harriman (1983). The question is still open in practical situations. Kohn has shown that densities close enough to a V-representable ground-state density are also V-representable ground-state densities (Kohn, 1983). Clearly if V changes in a region of space where ρ is non-zero, then ρ must change in order for Schrödinger's equation to remain valid. But the reverse mapping is problematical, particularly when the ground state is required to be degenerate by symmetry. The only postulated counter-example of non-V-representable yet reasonable densities continuous and having continuous gradients in order to be derivable from wavefunctions that are continuous with continuous (first derivatives) uses the degeneracy due to open $L > 0$ shells in a non-interacting atom acted upon by a central V (Levy, 1982; Lieb, 1983). (Other singular examples are given by Englisch and Englisch (1983).) For real atoms in this situation, with other than filled and half-filled $L > 0$ shells, Hund's rules require a non-spherical ground-state density, i.e. non-spherical V; in such cases, however, the ground state is degenerate and not covered by the HK theorem.

The problem of non-degeneracy seems less significant. Clearly if the true system is degenerate then any mapping between the density and the energy cannot be one-to-one. But in that case the exact kinetic energy is also degenerate via the virial theorem (Slater, 1972). If the degeneracy of the non-interacting electron gas is the same as the exact energy degeneracy, then nothing pathological needs to happen for cases where non-degenerate solutions pass smoothly into degeneracy when a symmetry-breaking external field or variable such as bond distance continuously varies toward a critical value (Gunnarsson and Lundqvist, 1976). Nevertheless, the V-representability problem must be solved before the degeneracies of the real and non-interacting electron gas of the same density can be enumerated for comparison. Further discussion of the consequences of fractional occupation numbers and degeneracy can be found in subsequent sections.

III. SELF-CONSISTENCY

If we require the KS energy (Eq. (5)) to be a minimum and the one-electron equations (6) to hold, then self-consistency requires that

$$V(\mathbf{r}) = \frac{\partial}{\partial \rho(\mathbf{r})}\left(\int \rho(\mathbf{r}')V(\mathbf{r})\,d\mathbf{r} + G'(\rho) \right) \tag{11}$$

It is universally agreed in KS theory and practice that the electric field due to the $(N-1)$ remaining electrons must contribute to $V(\mathbf{r})$ in Eq. (5). For practical reasons (see the next section) the effect of the electric field of all N electrons on each electron is separated out as the zeroth-order approximation to the electron–electron Coulomb repulsion. Thus Kohn and Sham use $G(\rho)$ and the potential

$$V(\mathbf{r}_1) = V_{\text{ext}}(\mathbf{r}_1) + \int [\rho(\mathbf{r}_2)/r_{12}]\, d\mathbf{r}_2 + \frac{\partial}{\partial \rho(\mathbf{r})} G(\rho) \tag{12}$$

and the variational energy that follows from it:

$$E(\rho) = T(\rho) + \int V_{\text{ext}}(\mathbf{r})\rho(\mathbf{r})\, d\mathbf{r} + \int [\rho(\mathbf{r}_1)\rho(\mathbf{r}_2)/2r_{12}]\, d\mathbf{r}_1\, d\mathbf{r}_2 + G(\rho) \tag{13}$$

The second term on the right-hand side of Eq. (13) shows that the KS problem is non-linear, and strongly so. One manifestation of this non-linearity is that for open-shell systems V breaks the symmetry of V_{ext} in the absence of FONs. Another manifestation is a strong dependence of the eigenvalues (Eq. (6)) on occupation numbers, particularly for finite systems.

Slater (1974) has shown the meaning of the one-electron eigenvalues for $X\alpha$:

$$\varepsilon_i = \frac{\partial E(\rho)}{\partial n_i} \tag{14}$$

Eq. (14) is true in general provided Eq. (11) holds. (Note that the derivatives taken here and throughout this section correspond to adding or subtracting in infinitesimal part of an electron to the system. The more commonly considered problem of electronic excitations involves the transfer of electrons, i.e. corresponds to differences of eigenvalue derivatives, which are far less predictable.) Owing to the dominance of the direct part of the electron–electron repulsion, the second derivatives with respect to occupation numbers,

$$\frac{\partial \varepsilon_i}{\partial n_j} > 0 \tag{15}$$

are always positive for $X\alpha$ and $X\alpha$-like models in which the indirect electron–electron interaction falls off with interaction coordinate no less rapidly than the direct part. In the absence of exchange, the derivatives of Eq. (15) are the Coulomb repulsion between orbitals i and j. Owing to the smoothness of current density functional approximations to the indirect part of the electron–electron interaction, viewing these derivatives as the direct Coulomb repulsion between orbitals is a very good approximation when exchange and correlation are included. This dominance of the direct part of the electron–electron repulsion energy makes the total energy approximately quadratic in the orbital occupation numbers,

$$E(\rho) = E(\mathbf{n}^0) + \sum_i \varepsilon_i^0 (n_i - n_i^0) + \sum_{i,j} U_{ij}^0 (n_i - n_i^0)(n_j - n_j^0) \tag{16}$$

where $\mathbf{n}^0$ is the vector containing the occupation numbers of a reference state, which is indicated by superscript 0. The effect of the quadratic coefficients U_{ij} is modeled by the I parameter of Hubbard (1964). The U_{ij} are always positive in the Born–Oppenheimer approximation. Negative U can only arise though a coupling of nuclear motion with electronic configuration (Anderson, 1975). The largest elements of U are diagonal, and these diagonal elements of U are to a good approximation the self-Coulomb repulsion of each one-electron eigenfunction.

Since the self-Coulomb interaction of each orbital is canceled in HF, the behavior of the two methods is quite different upon iteration toward self-consistency. In HF there is a bias against unoccupied orbitals that see a Fock potential due to N electrons whereas occupied orbitals see a Fock potential due to $(N-1)$ electrons. In $X\alpha$-like methods there is a bias against occupied orbitals because both orbitals see the same Fock potential which contains a term proportional to U_{ii} only if i is occupied. Thus during convergence $X\alpha$-like methods are much more oscillatory than is the HF method. This problem with $X\alpha$-like methods can be addressed by equally occupying nearly degenerate level until the end of the iterative cycle, an approach which is elegantly extended in the DVM computer code (Baerends *et al.*, 1973) by altering the occupancy of the levels in the vicinity of the Fermi energy at each iteration via a Fermi function of an appropriate temperature (Koelling, 1985). Convergence to a particular state can be sped up by considering how changes in the Fock matrix can mix the highest occupied molecular orbital (HOMO) and the lowest unoccupied molecular orbital (LUMO) (Dunlap, 1982).

In contrast to HF, where full use of symmetry in evaluating two-electron integrals is limited to one-dimensional representations, it is possible to use the full symmetry of V in $X\alpha$-like methods. Assume V is invariant under the generators $\{R_i\}$ of a group G. Then we need only concern ourselves with wavefunctions $|\lambda\mu\rangle$ that transform as some irreducible representation λ and basis partner μ of G:

$$R_i|\lambda\mu\rangle = \sum_{\upsilon} \Gamma^{\lambda}_{\mu\upsilon}(R_i)|\lambda\upsilon\rangle \tag{17}$$

In this case the transformation properties of the density matrix

$$\begin{aligned} R_i \rho_{\lambda\mu}(x_1, x_1') &= \sum_{\upsilon\sigma} \Gamma^{\lambda}_{\mu\upsilon}(R_i)^* \Gamma^{\lambda}_{\mu\sigma}(R_i) \int \psi^*_{\lambda\upsilon}(x_1, x_2, \ldots, x_N) \\ &\qquad \cdot \psi_{\lambda\sigma}(x_1', x_2, \ldots, x_N)\, \mathrm{d}x_2 \ldots \mathrm{d}x_N \\ &\equiv \sum_{\upsilon\sigma} \Gamma^{\lambda}_{\mu\upsilon}(R_i)^* \Gamma^{\lambda}_{\mu\sigma}(R_i) \gamma_{\lambda\upsilon\sigma}(x_1, x_1') \end{aligned} \tag{18}$$

are complicated theoretically (Davidson, 1972). However, an invariant can be formed by summing the density matrix over all partners of an irreducible representation (Burns, 1977):

$$R_i \sum_{\mu} \rho_{\lambda\mu}(x_1, x_1') = \sum_{\mu\upsilon\sigma} \Gamma^{\lambda}_{\mu\upsilon}(R_i)^* \Gamma^{\lambda}_{\mu\sigma}(R_i) \gamma_{\lambda\upsilon\sigma}(x_1, x_1')$$

$$= \sum_{\upsilon\sigma} \delta_{\upsilon\sigma} \gamma_{\lambda\upsilon\sigma}(x_1, x_1')$$

$$= \sum_{\upsilon} \rho_{\lambda\upsilon}(x_1, x_1') \tag{19}$$

which is the result of Pitzer (1972). Thus matrix elements involving the density matrix are invariant under each group operation if and only if we can average over all the l_λ basis partners of each irreducible representation. Taking the average is inconsequential when evaluating the kinetic energy

$$T_{\lambda\mu} = -\tfrac{1}{2} \int \delta(x_1, x_2) \nabla_1^2 \rho_{\lambda\mu}(x_1, x_2)\, dx_1\, dx_2 = \frac{1}{l_\lambda} \sum_{\upsilon} T_{1\upsilon} \tag{20}$$

or any one-electron operator, including V. In general, however, we cannot take this average for any energetic expression that is non-linear in the density or density matrix without altering that energy. Such non-linear energetic expressions include the direct part of the electron–electron Coulomb repulsion, its indirect part and any known $X\alpha$-like approximation to its indirect part. Thus HF cannot easily use the full group symmetry. (Nevertheless, one often takes this average. In atomic problems this average is called the central-field approximation.)

We circumvent this problem in $X\alpha$-like methods by lumping the non-linear terms into the local V and determining its symmetry. Now we can use an expression like Eq. (20) to simplify matrix elements over atom-centered basis functions when there are many symmetry-related atoms. For a given symmetry-adapted basis function centered on atom type C we expand the basis function into terms centered on the N symmetry-equivalent atoms,

$$|\lambda\mu C\rangle = \sum_{L=1}^{N} |\lambda\mu CL\rangle \tag{21}$$

To compute the Fock matrix element we must evaluate the expression

$$\langle \lambda\mu C_1 | -\tfrac{1}{2}\nabla^2 + V | \lambda\mu C_2 \rangle$$

$$= \frac{1}{l_\lambda} \sum_{L_1, L_2, \mu} \langle \lambda\mu C_1 L_1 | -\tfrac{1}{2}\nabla^2 + V | \lambda\mu C_2 L_2 \rangle$$

$$= \frac{N_1}{l_\lambda} \sum_{L_2, \mu} \langle \lambda\mu C_1 1 | -\tfrac{1}{2}\nabla^2 + V | \lambda\mu C_2 L_2 \rangle$$

$$= \frac{N_1 N_{12}}{l_\lambda} \sum_{L_2, \mu}{}' \langle \lambda\mu C_1 1 | -\tfrac{1}{2}\nabla^2 + V | \lambda\mu C_2 L_2 \rangle \tag{22}$$

where now the sum over symmetry-related L_2 atomic sites is limited to the first atom of the N_{12} atoms which are generated by group operations that leave the site $L_1 = 1$ fixed. Eq. (22) is a considerable simplification if the N's are larger than one. For four atoms in tetrahedral symmetry, there are two terms on the right-hand side of Eq. (22) instead of 16 on the left-hand side.

Similar simplifications hold for the $X\alpha$ scattered-wave (SW) method (Johnson, 1973). Diamond (1973) symmetrized the $X\alpha$-SW free-electron Green's function. Case and Yang (1980) used the invariance of the Green's function to eliminate the summation over the set atomic centers in the bra (or ket) of the equivalent of Eq. (22) for $X\alpha$-SW.

IV. APPROXIMATIONS TO THE INDIRECT ELECTRONIC COULOMB REPULSION

If the indirect part of the electronic Coulomb repulsion is neglected, we do not get the Hartree approximation as might be expected. Instead, we get a less accurate method, which will be called the neglect of indirect Coulomb repulsion (NICR) method. If only the direct part of the Coulomb repulsion is included, the electron–electron repulsion is

$$\begin{aligned} U_{\text{ee}} &= \int \rho(\mathbf{r}) V(\mathbf{r})\, d\mathbf{r} + G'(\rho) \\ &\simeq \int [\rho(\mathbf{r}_1)\rho(\mathbf{r}_2)/2r_{12}]\, d\mathbf{r}_1\, d\mathbf{r}_2 \end{aligned} \tag{23}$$

(when V_{ext} is zero) whereas the Hartree approximation is more complicated:

$$\begin{aligned} U_{\text{ee}} \simeq &\int [\rho(\mathbf{r}_1)\rho(\mathbf{r}_2)/2r_{12}]\, d\mathbf{r}_1\, d\mathbf{r}_2 \\ &- \sum_i \int [u_i^*(\mathbf{r}_1)u_i(\mathbf{r}_1)u_i^*(\mathbf{r}_2)u_i(\mathbf{r}_2)/2r_{12}]\, d\mathbf{r}_1\, d\mathbf{r}_2 \end{aligned} \tag{24}$$

The reason for including the NICR method in this discussion is that it yields one-electron equations having a local, orbital-independent potential,

$$\varepsilon_i u_i(\mathbf{r}_1) = \left(-\tfrac{1}{2}\nabla^2 + \int [\rho(\mathbf{r}_2)/r_{12}]\, d\mathbf{r}_2 - V_{\text{ext}}(\mathbf{r}_1) \right) u_i(\mathbf{r}_1) \tag{25}$$

and $V_{\text{ext}}(\mathbf{r}_1)$ is the externally applied electric field. In contrast, the non-local Hartree one-electron equations are

$$\varepsilon_i u_i(\mathbf{r}_1) = \left(-\tfrac{1}{2}\nabla^2 + \int [\rho(\mathbf{r}_2) - u_i^*(\mathbf{r}_2)u_i(\mathbf{r}_2)/r_{12}]\, d\mathbf{r}_2 - V_{\text{ext}}(\mathbf{r}_1) \right) u_i(\mathbf{r}_1) \tag{26}$$

Thus the Hartree problem is much more difficult to solve than the NICR problem. But the latter problem is not used in practical calculations. Fermi and Amaldi (1934) have developed a method that goes further toward eliminating the Coulombic self-interaction of each electron,

$$U_{ee} \simeq \int [\rho(\mathbf{r}_1)\rho(\mathbf{r}_2)/2r_{12}]\,d\mathbf{r}_1\,d\mathbf{r}_2 - \frac{1}{N}\int [\rho(\mathbf{r}_1)\rho(\mathbf{r}_2)/2r_{12}]\,d\mathbf{r}_1\,d\mathbf{r}_2 \qquad (27)$$

Yet yields a local potential. However, for solids, $N \to \infty$ and the Fermi–Amaldi approximation goes to NICR.

Slater (1951) simply and significantly improved upon the Fermi–Amaldi approach by averaging the potential derived from the indirect part of U_{ee} in the HF method. That potential is the HF exchange potential, and its average is

$$V(\mathbf{r}_1) = \frac{\sum_{i,j}\int [u_i^*(\mathbf{r}_1)u_j(\mathbf{r}_1)u_j^*(\mathbf{r}_2)u_i(\mathbf{r}_2)/r_{12}]\,d\mathbf{r}_2}{\sum_i u_i^*(\mathbf{r}_1)u_i(\mathbf{r}_1)} \qquad (28)$$

in spin-restricted form. This average can be evaluated in the jellium model. The only variable in the jellium model is r_s, which is the radius of a sphere containing one unit of electronic or nuclear charge,

$$1/\rho = \tfrac{4}{3}\pi r_s^3 \qquad (29)$$

From dimensional arguments alone the average HF exchange potential must be proportional to the inverse of r_s, i.e. proportional to the one-third power of the density. By averaging the HF exchange potential over the plane-wave one-electron eigenfunctions of the jellium model up to the Fermi wavevector

$$k_F = (3\pi^2\rho)^{1/3} \qquad (30)$$

Slater obtained the approximate indirect (HF exchange) potential

$$V^I(r) = -3[3\rho(r)/8\pi]^{1/3} \qquad (31)$$

or

$$U^I = G'(\rho) - \tfrac{1}{2}\int [\rho(\mathbf{r}_1)\rho(\mathbf{r}_2)/r_{12}]\,d\mathbf{r}_1\,d\mathbf{r}_2$$

$$= -\frac{9}{4}\left(\frac{3}{8\pi}\right)^{1/3}\int \rho^{4/3}(\mathbf{r})\,d\mathbf{r} \qquad (32)$$

By taking a different weighting in the average, i.e. averaging over the exchange energy instead of the exchange potential, Gaspar (1954) and Kohn and Sham (1965) obtained a value two-thirds as large. This discrepancy led to the α-dependent $X\alpha$ exchange energy

$$U^I_{X\alpha} = \frac{9\alpha}{4}\left(\frac{3}{8\pi}\right)^{1/3}\int \rho^{4/3} \qquad (33)$$

TABLE I
The Coulomb repulsion energy in atomic units for STO wavefunctions and the value of α necessary for the $X\alpha$ exchange energy to cancel that energy. For hydrogenic atoms, the exchange energy must be multiplied by Z/n.

n	l	$\|m\|$	Exchange	α
1	0	0	0.6250000	0.7772544
2	1	0	0.3914063	0.7187712
2	1	1	0.3703125	0.7638011
3	2	0	0.2765160	0.6672039
3	2	1	0.2641431	0.7014715
3	2	2	0.2694630	0.7155994
4	3	0	0.2144187	0.6519366
4	3	1	0.2071769	0.6826811
4	3	2	0.2045609	0.6884583
4	3	3	0.2140407	0.6968088
5	4	0	0.1755367	0.6425422
5	4	1	0.1709100	0.6707044
5	4	2	0.1678615	0.6754918
5	4	3	0.1686176	0.6764502
5	4	4	0.1786014	0.6822574
6	5	0	0.1488394	0.6159715
6	5	1	0.1456709	0.6412059
6	5	2	0.1430386	0.6458447
6	5	3	0.1421408	0.6456772
6	5	4	0.1442244	0.6452331
6	5	5	0.1538289	0.6493278

where α is allowed to vary between 2/3 and 1. Various optimal, in the sense of yielding HF atomic energies, values of α for the elements up to niobium are given by Schwarz (1972). Table I compares the self-Coulomb energy of hydrogenic wavefunctions with the approximation (Eq. (33)) to obtain the orbital-dependent values of α that make the approximation exact. Since the approximation is dimensionally sound, the α values are independent of the nuclear charge, i.e. independent of the exponent of any Slater-type orbital (STO). That the approximation is quite good is demonstrated by the fact that all the α values are rather close to 0.7, which is used in the most accurate $X\alpha$ calculations (Becke, 1983; Laaksonen *et al.*, 1985) that do not partition the approximation to exchange within atomic cells. Not only is Eq. (33) a good approximation to the indirect part of the Coulomb repulsion, but Lieb and Oxford (1981) have shown that for $\alpha = 1.52$ that expression yields a rigorous lower bound for the indirect part of the electronic Coulomb repulsion energy.

TABLE II
Magnetic quantum number dependence of the energies of the lowest states of various symmetries for the hydrogen atom using the GL potential from Gunnarsson and Lundqvist (1976).

State n	l	$\|m\|$	Energy (eV) GL	Exact
1	0	0	−13.39	−13.61
2	1	0	−3.79	−3.40
2	1	1	−3.63	−3.40
3	2	0	−1.86	−1.51
3	2	1	−1.78	−1.51
3	2	2	−1.76	−1.51

Unfortunately, a problem that increasingly haunts $X\alpha$-like methods as we strive for higher accuracy is evident from Table I. The problem is that Eq. (33) depends on the magnetic quantum number m in a manner that differs from the true exchange, which it is to cancel from the direct part of the Coulomb interaction energy. The same incomplete cancellation is shown for the Gunnarsson and Lundqvist (GL) potential in Table II, which is constructed to quantify its effect on the total energy. While this m dependence is proportionally largest for one-electron systems, it persists in larger systems, with the practical consequence that the dissociation energy of molecules is not uniquely defined in $X\alpha$-like models at the several tenths of an electronvolt level when, as is often the case, at least one of the fragments has a degenerate ground state. This non-uniqueness arises because the calculated binding energy will depend on which of the physically degenerate fragment energies is calculated. This same m-dependent problem occurs in HF theory for two-electron and larger systems. The dependence on l and m is due to the fact that the complex spherical harmonics used to generate these tables each have l nodal lines on the surface of the unit sphere, $|m|$ of which coalesce at the poles, zeroing $|m| - l$ derivatives there (Gunnarsson and Jones, 1985). A decreasing α correlates with spherical harmonics that have increasingly alpine modulus squared. The fact that α varies with l and almost equally strongly with m suggests that the small remaining error in $X\alpha$ is not easily correctable.

This significant m variability is reduced but does not disappear if we switch to the real spherical harmonics, which are arguably more important chemically, particularly in minimal-basis-set applications. In the context of chemistry we almost never use the complex spherical harmonics but rather their real and imaginary parts. The real spherical harmonics each have l nodal circles (that never coalesce to points). This basis change is made because the chemically most important aspect of a bond is its nodal character along the bond, i.e. one distinguishes between σ, π, δ, etc., character in a bond even in non-linear molecules which do not possess axial symmetry. The principal components in such bonds are linear combinations of the real and imaginary parts of the $l, m = \pm 1$ spherical harmonics when the bond is oriented along the z axis. Thus central to chemistry are the maximally alpine real spherical harmonics: p_x and p_y in the p shell, d_{xy} and $d_{x^2-y^2}$ in the d shell, etc., for bonds along the z axis. The complete set of spherical harmonics for each shell is necessary, however, to orient these special functions along any given bond direction. But the real spherical harmonics are not always used.

We can think of the new and very accurate diatomic $X\alpha$-like computer codes (Becke, 1983; Laaksonen *et al.*, 1985) as either using complex spherical harmonics or as using an axial-field approximation. If we view them as using complex spherical harmonics, this m-dependence problem (in the form of real versus complex spherical harmonics) gives rise to questions concerning axial symmetry breaking. For concreteness, consider the case of two singlet-coupled π electrons in a diatomic molecule. There are three such $X\alpha$-like states to be considered. Owing to the diagonal dominance of the U matrix of Eq. (16) ($U_{ii} > U_{ij}$) the lowest singlet state has the two electrons of opposite spins occupying both π orbitals. The lowest-energy $X\alpha$-like calculation related to this state can only be described using the more localized real π_x and π_y orbitals. Furthermore, if we want to consider bending (reducing symmetry to just reflection), then only the real spherical harmonics yield continuous potential surfaces in the absence of allowing complex eigenfunction expansions. While these arguments should not be taken as proving that real spherical harmonics are superior to complex spherical harmonics due to problems associated with generating a pure singlet state (see Section VI) and the question of mixing orbitals with arbitrary complex phase factors (see Section VII), it does show that the maximally alpine real spherical harmonics are important in practical calculations. In any event, $X\alpha$-like methods are slightly prejudiced toward bonds of higher nodal character along the bonds, as is evident from the decreasing value of α with l in Table I. However, this nodal prejudice is reduced in practical calculations that, via muffin-tinning the potential or finite spherical harmonic fits or expansions of the potential, lessen the effects of rapid spatial variation.

Perdew and Zunger (1981), in the $X\alpha$-like equivalent of the Hartree approximation, advocate subtracting the total self-interaction of each electron in $X\alpha$-like models. This proposal would remove the m dependence of hydrogenic systems. Since the self-interaction of each electron (orthonormal orbital), as well as their sum, is not invariant under a unitary transformation among the orbitals, in contrast to the first-order density matrix and thus $X\alpha$-like models, Perdew and Zunger propose picking out a unitary transformation that maximizes their self-interaction correction (SIC). SIC leads to a non-local (in the historical sense) potential that can be handled with minor difficulty in molecular applications (Pederson *et al.*, 1984). However, SIC only postpones the problem of *m*-dependence and the problem of broken symmetry that follows to two-electron and larger systems. A perhaps less significant problem with SIC is that no way is known to implement it for FON or other treatments of degeneracy.

It is possible to take different averages of the indirect part of the Coulomb interaction to yield different local potentials that give exact solutions to the SIC problem (Norman and Koelling, 1984), to the HF problem (Talman and

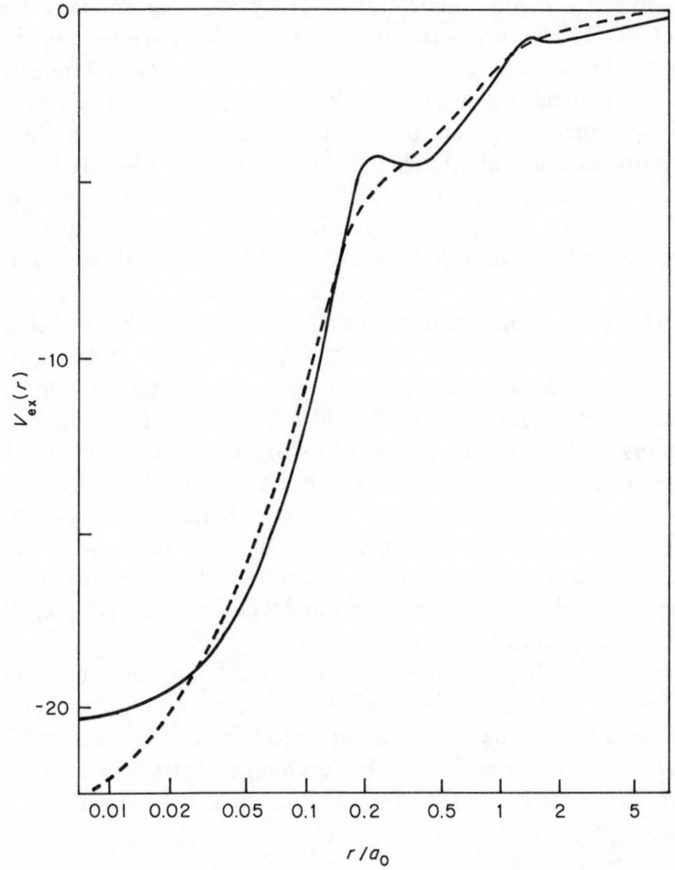

Fig. 1. The local potential as a function of radius that yields the HF energy and wavefunction for aluminum (full curve) is compared to the X_α potential (broken curve) for the α value that minimizes the HF energy, which is evaluated using the SCF X_α orbitals. (*From Talman and Shadwick (1976). Reproduced by permission of the American Physical Society.*)

Shadwick, 1976) and thus by extrapolation to the fully correlated problem via application to the multiconfiguration self-consistent field (MCSCF) problem. In spite of the fact that the mathematics of simultaneously averaging a non-local potential and proceeding to self-consistency is cumbersome at present, this approach seems a most promising avenue toward Slater's goal of using a local potential for orbital generation in CI calculations.

Fig. 1 compares the Talman and Shadwick (1976) local HF effective exchange potential and $X\alpha$ exchange potentials for aluminum, with α chosen to minimize the HF energy evaluated using SCF $X\alpha$ orbitals (Kmetko, 1970; Wood, 1970). (Owing to the extreme flatness of the HF energy using α-dependent orbitals, the Kmetko–Wood method for determining α has been superseded by the method of determining α by equating $X\alpha$ and HF energies (Schwarz, 1972).) The curves are quite similar and support the empirical result that $X\alpha$ and HF orbitals are interchangeable for all practical purposes. This practical equivalence holds even for the case of ozone (Salahub *et al.*, 1982), where HF incorrectly orders the orbitals while $X\alpha$ correctly orders them. Furthermore, this orbital interchangeability is visually displayed in the work of Bursten and Fenske (1977), in which the potential was muffin-tin-averaged in touching spheres. This latter result is remarkable in light of the fact that $X\alpha$ total energy surfaces are unreliable using the $X\alpha$-SW method with touching spheres; e.g. water is found to be linear in this approximation (Connolly and Sabin, 1972) while a non-muffin-tin $X\alpha$ calculation yields the correct geometry (Sambe and Felton, 1975). This problem with the $X\alpha$-SW total energies can be considerably reduced (for torsional and breathing modes) by using overlapping atomic spheres (Herman *et al.*, 1974; Casula and Herman, 1983; Case, 1982), particularly when the Norman (1976) criterion for choosing overlapping sphere radii is used. Thus it is important to distinguish comparisons between orbital eigenfunctions on the one hand, and between orbital eigenvalues or total energy expressions on the other. As a fortunate consequence of this equivalence between orbitals, basis sets optimized for HF are nearly optimal for $X\alpha$ calculation using STOs (Baerends *et al.*, 1973) and Gaussian-type orbitals (GTOs) (Dunlap *et al.*, 1979).

$X\alpha$-like models have been extended to treat magnetic (spin-polarized) systems (Slater, 1968), in which the electrons are required to be aligned either up or down along some quantization axis. For $X\alpha$ this extension follows trivially from the observation that HF exchange occurs between electrons of like spin. Thus Slater proposed the spin-polarized expression

$$U^{\mathrm{I}}_{X\alpha} = \frac{9\alpha}{4}\left(\frac{3}{4\pi}\right)^{1/3} \int \rho_{\uparrow}^{4/3} + \rho_{\downarrow}^{4/3}\,\mathbf{dr} \tag{34}$$

to approximate exchange in spin-unrestricted HF (UHF). The other $X\alpha$-like functionals usually interpolate between the completely spin-polarized (SP) and non-spin-polarized (NSP) extremes using the interpolation formula that has the same spin dependence (von Barth and Hedin, 1972):

$$u^{\mathrm{I}} = u^{\mathrm{I}}_{\mathrm{NSP}} + (u^{\mathrm{I}}_{\mathrm{SP}} - u^{\mathrm{I}}_{\mathrm{NSP}}) f(\mathbf{r}) \tag{35}$$

where lower-case u is used to denote an indirect Coulomb energy density, i.e. U is the integral over all space of u, and where

$$f(\mathbf{r}) = \frac{[\rho_{\uparrow}(\mathbf{r})/\rho(\mathbf{r})]^{4/3} + [\rho_{\downarrow}(\mathbf{r})/\rho(\mathbf{r})]^{4/3} - 2^{-1/3}}{1 - 2^{-1/3}} \tag{36}$$

But these expressions are incorrect if we want to mimic or approximate the spin dependence of UHF.

To see that Eq. (34) needs at least to be extended, consider a triplet wavefunction of two electrons. The three degenerate triplet components can be written as combinations of Slater determinants of orbitals $a(r)$ and $b(r)$, which for the triplet case can be assumed orthogonal without loss of generality:

$$\begin{aligned} |1,1\rangle &= \{a\alpha, b\alpha\} \\ |1,0\rangle &= [\{a\alpha, b\beta\} + \{a\beta, b\alpha\}]/\sqrt{2} \\ |1,-1\rangle &= \{a\beta, b\beta\} \end{aligned} \tag{37}$$

where curly brackets are used to denote Slater determinants of the spin orbitals. In $X\alpha$-like models the rule of Ziegler *et al.* (1977) states that only single determinants, i.e. $|1,1\rangle$ and $|1,-1\rangle$ and not $|1,0\rangle$, are to be used to determine $X\alpha$-like energies. Pick one of the good states, say $|1,1\rangle$, and consider a rotation in spin space through an angle of 90° about the Y axis. The resulting single determinant of generalized spin orbitals,

$$\{a(\alpha + \beta), b(\alpha + \beta)\}/2 \tag{38}$$

is the proper linear combination of the three triplet states as can be seen upon expansion:

$$(|1,1\rangle + \sqrt{2}|1,0\rangle + |1,-1\rangle)/2 \tag{39}$$

If we use single determinant (38) in Eq. (34) we get the wrong answer because it has equal population of spin up and spin down, just as does $|1,0\rangle$. This inconsistency is never a practical problem, at least for simple systems. The lack of invariance under rotations in spin space does, however, distinguish $X\alpha$-like models from HF (Fukutome, 1981), or to be more descriptive what Sykja and Calais (1982) call generalized Hartree–Fock (GHF). Any first-order density matrix can be decomposed into four mathematically equivalent pieces

$$\rho(x_1, x_2) = [N(\mathbf{r}_1, \mathbf{r}_2)I + \mathbf{S}(\mathbf{r}_1, \mathbf{r}_2)\cdot\sigma]/2 \tag{40}$$

using the Pauli spin matrices. For the wavefunctions (37) and (38), the spin rotationally invariant part is identical

$$N(\mathbf{r}_1, \mathbf{r}_2) = a^*(\mathbf{r}_1)a(\mathbf{r}_2) + b^*(\mathbf{r}_1)b(\mathbf{r}_2) \tag{41}$$

and its diagonal part integrates to the number of electrons. On the other hand, they differ in their non-zero parts of **S**,

$$|\mathbf{S}(\mathbf{r}_1, \mathbf{r}_2)| = a^*(\mathbf{r}_1)a(\mathbf{r}_2) + b^*(\mathbf{r}_1)b(\mathbf{r}_2) \tag{42}$$

in the z and x directions respectively. In GHF the indirect part of the Coulomb repulsion is the scalar expression

$$U^{\mathrm{I}}_{\mathrm{GHF}} = -\int [N^*(\mathbf{r}_1, \mathbf{r}_2)N(\mathbf{r}_1, \mathbf{r}_2)/4r_{12}]\,d\mathbf{r}_1\,d\mathbf{r}_2 - \int [\mathbf{S}^*(\mathbf{r}_1, \mathbf{r}_2)\cdot\mathbf{S}(\mathbf{r}_1, \mathbf{r}_2)/4r_{12}]\,d\mathbf{r}_1\,d\mathbf{r}_2 \tag{43}$$

(see for example Löwdin *et al.*, 1981). To the extent that Slater's $\rho^{1/3}$ approximation is valid in GHF, the indirect part of the Coulomb repulsion in the presence of spin polarization is

$$U^{\mathrm{I}}_{\mathrm{G}X\alpha} = -\frac{9}{4}\left(\frac{3}{8\pi}\right)^{1/3}\int \{\alpha\rho^{4/3}(\mathbf{r}) + \alpha_s[\mathbf{S}(\mathbf{r})\cdot\mathbf{S}(\mathbf{r})]^{2/3}\}\,d\mathbf{r} \tag{44}$$

where

$$\mathbf{S}(\mathbf{r}) = \mathbf{S}(\mathbf{r}, \mathbf{r}) \tag{45}$$

and where the two α's in Eq. (44) are allowed to differ in the spirit of $X\alpha$.

That Eq. (44) and Eq. (34) are different is due to the fact that GHF exchange is quadratic in the first-order density matrix, whereas Slater's approximation depends on the 4/3 power of the density. Only for the special case of a single determinantal wavefunction that has constant direction of **S** (i.e. for conventional but not generalized single determinants) can we regroup terms in Eq. (43) to yield Eq. (34). For this case N and **S** can be simplified as

$$\begin{aligned} N(\mathbf{r}_1, \mathbf{r}_2) &= \rho_{\uparrow\uparrow}(\mathbf{r}_1, \mathbf{r}_2) + \rho_{\downarrow\downarrow}(\mathbf{r}_1, \mathbf{r}_2) \\ \mathbf{S}(\mathbf{r}_1, \mathbf{r}_2) &= \hat{\mathbf{e}}_z[\rho_{\uparrow\uparrow}(\mathbf{r}_1, \mathbf{r}_2) - \rho_{\downarrow\downarrow}(\mathbf{r}_1, \mathbf{r}_2)] \end{aligned} \tag{46}$$

where z is the (constant) direction of **S**, and Eq. (43) can be transformed into the familiar expression

$$U^{\mathrm{I}}_{\mathrm{GHF}} = -\int \{[\rho^*_{\uparrow\uparrow}(\mathbf{r}_1, \mathbf{r}_2)\rho_{\uparrow\uparrow}(\mathbf{r}_1, \mathbf{r}_2) + \rho^*_{\downarrow\downarrow}(\mathbf{r}_1, \mathbf{r}_2)\rho_{\downarrow\downarrow}(\mathbf{r}_1, \mathbf{r}_2)]/4r_{12}\}\,d\mathbf{r}_1\,d\mathbf{r}_2 \tag{47}$$

from which Eq. (34) follows upon approximation.

In addition to extending $X\alpha$-like theory to admit Fukutome's (1981) eight classes of solutions, Eq. (44) clears up a number of problems such as what Gunnarsson and Jones (1980) call lack of particle–hole symmetry, which is the idea that for fixed orbitals the multiplet splittings of a number of electrons occupying various l shells should be identical to the multiplets for the situation in which an identical number of holes are identically distributed among these otherwise filled shells. (This particle–hole symmetry is only a property of HF.) Also Eq. (44) simplifies the $X\alpha$-like interpolation function of Eq. (36),

$$f(\mathbf{r}) = [|\mathbf{S}(\mathbf{r})|/\rho(\mathbf{r})]^{4/3} \tag{48}$$

To the degree that the spin dependence derived from Eq. (44) differs from that of conventional $X\alpha$, which is substantial, the spin dependence of conventional $X\alpha$ differs from the spin dependence of HF. However, without adjusting the two α's separately this change alone will not remove $X\alpha$'s bias toward high spin, which now comes from approximating GHF.

V. ORBITALS AND FRACTIONAL OCCUPATION NUMBERS

Löwdin (1955) has shown that the first-order density matrix can be diagonalized to yield a unique expression in terms of orbitals:

$$\rho(\mathbf{r}_1, \mathbf{r}_2) = \sum_i n_i u_i^*(\mathbf{r}_1) u_i(\mathbf{r}_2) \tag{49}$$

where the set of orbitals, $\{u_i\}$, are called natural orbitals. The occupation numbers are restricted to be between zero and one, and are all zero or one if and only if the wavefunction is a single determinant, a situation that never exactly occurs in practice. Since Eq. (49) deals with the exact wavefunction, the Schrödinger equation for which generates an independent second-order density matrix, we have no one-electron eigenvalues to associate with the occupation numbers. Thus we cannot ask if these occupation numbers satisfy Fermi statistics, in which case all non-integrally occupied orbitals would need to be degenerate. This situation would be necessary if it were possible to construct a KS method that yields the natural orbitals as one-electron orbitals. Thus we must give up precise adherence to one or two prevalent concepts in DF theory: that the KS orbitals are the natural orbitals or that the exact KS theory satisfies Fermi statistics. It is not hard to give up either.

First, consider Fermi statistics. Slater (1974) has shown for $X\alpha$ the proper interpretation of the one-electron eigenvalues

$$\varepsilon_i = \partial E/\partial n_i \tag{50}$$

which differs from the HF interpretation of them as unrelaxed one-electron excitation energies (Koopman's theorem). Eq. (50) also holds for any $X\alpha$-like method provided Eq. (11) holds. Thus minimizing the total energy subject to the constraint that the total number of electrons is fixed at N,

$$F(\mathbf{n}) = N - \sum_i n_i = 0 \tag{51}$$

requires that the occupation numbers be extremal (0 or 1) or else

$$\frac{\partial E}{\partial n_i} + \mu \frac{\partial F}{\partial n_i} = 0 \tag{52}$$

which implies the partially occupied levels are degenerate,

$$\varepsilon_i = \mu \tag{53}$$

This degeneracy occurs often in $X\alpha$-like theories due to crossing of levels as a function of internal specification (occupation numbers) or external specification (e.g. bond distances) of the state. In transition-metal and rare-earth atoms it occurs due to near-degeneracy and differing spatial extents of s and d or f electrons. If we erroneously seek the lowest-energy $X\alpha$-like solution in these cases of degeneracy via FONs then the positivity of the U_{ii} (the second derivatives of the energy with respect to occupation number) requires these orbitals be equally occupied. Parr and coworkers have constructed a theory of the thermodynamics of (fractional) charge transfer between chemical subsystems (Parr *et al.*, 1978; Ghosh *et al.*, 1984). In fact, μ in Eqs (26) and (27) is the chemical potential. Such considerations lead to fractional charges in the separated subsystem limit. A way to prohibit this unphysical result is to postulate a discontinuity with respect to occupation number in DF theory (Perdew *et al.*, 1982). Any such discontinuities would play havoc with the partial differentiations in the various Maxwell relations that are basic to modern thermodynamics (Reif, 1965).

A practical problem with fractional occupation numbers is that whenever they have been found to occur they can be removed by lowering (breaking) the symmetry of the externally applied fields; the broken-symmetry solution always lies lower in energy than the FON solution (Dunlap, 1984). The converse is not true. If a broken-symmetry solution exists, the lowest-energy higher-symmetry solution may have integral occupation numbers. The simplest such example is the hydrogen molecule. Beyond the Coulson and Fischer (1949) internuclear separation a broken-symmetry solution exists but in the symmetry-restricted case the σ_u orbital never lies lower than the σ_g orbital (Gunnarsson and Lundqvist, 1976). One FON example is the classic work of Slater *et al.* (1969). They showed that most transition-metal atoms, when treated as NSP, had fractional (n) s and ($n-1$) d orbital occupations, where n is the appropriate principal quantum number. Treated as SP, however, all transition-metal atoms have integral occupation numbers. Furthermore, broken spin symmetry (different orbitals for different spins) is well established and is necessary for current descriptions of ferromagnetism in solids (Slater, 1974).

FONs can occur most often for paramagnetic molecules. Ground-state examples are rare. Examples include dicarbon and the corresponding excited state of disilicon (Dunlap and Mei, 1983). However, lower-energy $X\alpha$-like solutions exist for these examples if antiferromagnetic (broken left–right symmetry) solutions are allowed (Dunlap, 1984). The most famous example of broken symmetry is dichromium. An early inversion-symmetric calculation of the $X\alpha$-like ground state erroneously (both within the $X\alpha$-like functional used and in comparison with experiment) found it to be high spin, $S = 6$ (Harris and Jones, 1978). However, even had that calculation been accurate in predicting a singlet ground state, any such symmetry-restricted $X\alpha$-like calculation of the singlet state will not dissociate properly (Goodgame and Goddard, 1982, 1985). Proper dissociation of Cr_2 can occur if inversion symmetry is relaxed. Good agreement with the experimental bond distance is found provided any $X\alpha$-like functional that favors ferromagnetism less than $X\alpha$ is used (Baykara *et al.*, 1984, and references therein). Such a distinction may be premature in light of questions as to whether or not some determinant or wavefunction should be symmetrized (see Section VI) and in light of Eq. (44). For other cases of greater chemical and biological interest in which broken symmetry is required, see Aizman and Case (1982), Noodleman *et al.* (1985) and Cook and Karplus (1985).

Thus it appears that we are stuck with broken symmetry and without FON (in the classic sense of degenerate eigenvalues) if we use conventional $X\alpha$-like methods. Unfortunately broken-symmetry calculations are much more demanding computationally, particularly for atoms, where the effects are small both in an absolute sense and compared to molecules where the incompletely paired electrons have greater space in which to separate (Janak and Williams, 1981). Also since broken symmetry can only result from a self-consistent calculation on a specific state, such considerations are ultimately incompatible with the various forms of multiplet averaging that are used to arrive at atomic one-electron promotion energies in discussions of which $X\alpha$-like method is best (Harris and Jones, 1978; von Barth, 1979; Wood, 1980). This is not a compelling indictment of broken symmetry since multiplet averaging is also inconsistent with the argument of Gunnarsson and Lundqvist (1976) that the HK theorem applies to the ground state of each symmetry, thus requiring separate self-consistent calculations. There is one positive outcome of this symmetry dilemma, which is conceptually identical to that of HF (Löwdin, 1969). When the first-order density matrix is diagonalized in symmetry-adapted orbitals, the occupation numbers turn out to be fractional as in a CI calculation rather than integral as in a restricted HF calculation. This configuration mixing can be significant despite the lack of orbital degeneracy due to core polarization.

A promising but less developed alternate approach to introducing natural-orbital configuration mixing is to follow the lead of Kohn and Sham to the ultimate conclusion that the $X\alpha$-like orbitals have no physical meaning. In extreme form, this point of view allows any one-electron orbitals whose moduli squared add up to the density. That there always exists at least one was proved by Harriman (1981). His explicit construction of orthonormal orbitals is particularly transparent in one dimension. For any function $F(x)$ whose derivative is proportional to the average density,

$$df/dx = 2\pi\rho(x)/N \tag{54}$$

any N of an infinite number of orbitals,

$$u_k = [\rho(x)/N]^{1/2} \exp[ikf(x)] \tag{55}$$

is a set of N orthonormal orbitals, the sum of whose modulis squared yields the density in question. Note that these orbitals each contribute one-Nth of the density everywhere. Thus they are radically different from conventional $X\alpha$-like orbitals, which for core electrons are particularly localized. The exponential phase factor in Eq. (55) is constructed so that the density, which appears as a common factor in binary products of such orbitals, is the appropriate factor to allow trivial integration of the various orbital products and to prove the orthonormality of these Harriman orbitals.

After realizing that there are kinetic energy contributions to G' in Eq. (3) in the KS construction, Levy (1979) introduced the important concept of a constrained search into DF theory. In doing so he limited the sets of orbitals to be considered in the KS construction to those that yield indistinguishable kinetic energy, i.e. to those orbitals that give a minimum kinetic energy as well as the density in question (see Levy and Perdew, 1985). Since these orbitals would be as smooth as possible, given the constraint that they yield the density, if we use the kinetic energy to form a continuum of orbitals yielding the density then Harriman-like and Levy-like will form the extremes with $X\alpha$-like somewhere in the middle. Furthermore, since these orbitals are as delocalized as possible, it is unlikely that they would break symmetry. However, the Levy construction will put severe conditions on G' to return the high kinetic energy associated with the core and other localization inherent in the exact solution of Schrödinger's equation. In any event, the concept of constrained search allows the extension of the HK theorem to degenerate cases (i.e. there could be more than one set of KS orbitals that yield the minimum kinetic energy) and replaces the V-representability problem with the problem of finding G'.

No matter how defined, restricting our $X\alpha$-like solutions to be the lowest-energy solutions, FONs never occur. Instead, for the case of near-degeneracy, such solutions will involve orbitals that are linear combinations of symmetry-distinct orbitals with complex coefficients. With such extensions we can use integral occupation numbers and employ finite differences to evaluate derivatives in Parr's exact HK-based thermodynamic theory of the electron gas. The HK variational problem

$$\delta[E(\rho, V_{\text{ext}}) - \mu N(\rho)] = 0 \tag{56}$$

contains the chemical potential, which may equal the μ of Eq. (53), through the Lagrange multiplier associated with the constraint that the density integrates to the number of electrons. Considering changes in the number of electrons and the external potential, Eq. (56) can be rewritten in differential form as

$$dE = \mu\, dN + \int \rho(\mathbf{r})\, dV_{\mathrm{ext}}(\mathbf{r})\, d\mathbf{r} \tag{57}$$

The Maxwell relation from Eq. (57) defines a function (Parr and Yang, 1984)

$$f(\mathbf{r}) = [\partial \rho(\mathbf{r})/\partial N]_{V_{\mathrm{ext}}} = [\partial \mu/\partial V_{\mathrm{ext}}(\mathbf{r})]_N \tag{58}$$

Taking the limit as N approaches the number of electrons from above and renormalizing yields an HK LUMO and the limit from below yields the HK HOMO, without recourse to the KS one-electron mapping. Without FON it should be possible to use powerful results such as Eq. (58) in non-singular $X\alpha$-like methods. However, we are left with questions concerning the interpretation of asymmetrical $f(\mathbf{r})$, which occur in cases of degeneracy.

VI. BROKEN SPIN SYMMETRY

The problem of two valence electrons that are singlet-coupled is richer than the problem of two electrons triplet-coupled because the two orbitals need not be orthogonal, i.e. for the triplet state the one-electron eigenvalue problem (Eq. (6)) returns two orthonormal orbitals. Consider the singlet ground state of the hydrogen molecule. In the approximate two-determinantal wavefunction of Heitler and London (1927),

$$\psi_{\mathrm{VB}}(\mathbf{r}_1, \mathbf{r}_2) = [\{u_{\mathrm{L}}\alpha, u_{\mathrm{R}}\beta\} - \{u_{\mathrm{L}}\beta, u_{\mathrm{R}}\alpha\}]/(2+2S)^{1/2} \tag{59}$$

the two singlet-coupled orbitals are centered each on separate atoms, labeled by L and R for left and right. This wavefunction, which correctly describes the dissociation limit of H_2, McWeeny (1979) calls valence bond (VB). Because the orbitals on separate atoms are not orthogonal (their overlap is the S of Eq. (59) computational difficulties arise that in part stimulated the molecular-orbital (MO) approach, where orbitals are constrained to be orthogonal, that is now used in most quantum-chemical method including $X\alpha$ and HF. In such methods this VB singlet coupling can only be treated in a post process such as CI. However, the Heitler–London approach survives in the generalized valence bond (GVB) method of Goddard and coworkers. In the GVB method Eq. (59) is generalized to allow the atom-centered orbitals to spread out over the molecule in MO-like fashion, yielding the GVB wavefunction for the hydrogen molecule (Bobrowicz and Goddard, 1977):

$$\begin{aligned}\psi_{\mathrm{GVB}}(\mathbf{r}_1, \mathbf{r}_2) = [&\{(u_{\mathrm{g}}\cos\omega + u_{\mathrm{u}}\sin\omega)\alpha, (u_{\mathrm{g}}\cos\omega - u_{\mathrm{u}}\sin\omega)\beta\} \\ &+ \{(u_{\mathrm{g}}\cos\omega - u_{\mathrm{u}}\sin\omega)\alpha, (u_{\mathrm{g}}\cos\omega + u_{\mathrm{u}}\sin\omega)\beta\}]/ \\ &\times (2 + 2\cos^2 2\omega)^{1/2}\end{aligned} \tag{60}$$

which is written using symmetrized orthogonal MO orbitals,

$$u_g = (u_L + u_R)/(2 + 2S)^{1/2}$$
$$u_u = (u_L - u_R)/(2 - 2S)^{1/2} \tag{61}$$

and mixing angle ω, but which avoids the orthogonality constraint of MO theory. In contrast, the best that can be done in HF theory near the dissociation limit is to use the UHF wavefunction

$$\psi_{UHF} = \{(u_g \cos\omega + u_u \sin\omega)\alpha, (u_g \cos\omega - u_u \sin\omega)\beta\} \tag{62}$$

The most dramatic difference between GVB and UHF is seen if the expectation value of spin along the z axis is computed using Eqs (61) and (62). For the GVB wavefunction the expectation value is zero everywhere whereas for UHF it is

$$(\rho_{\uparrow\uparrow} - \rho_{\downarrow\downarrow})_{UHF} = 2u_g u_u \sin 2\omega \tag{63}$$

The best or correct $X\alpha$-like treatment of H_2 is unresolved (Noodleman and Norman, 1979; Dunlap, 1984). The conventional SP treatment of H_2 near the separated-atom limit uses the orbitals of wavefunction (62). We should not use that wavefunction directly since Sabin and Trickey (1984) have correctly pointed out that $X\alpha$-like methods need not be thought of as wavefunction-based. Therefore, we must derive the spin density, which is central to DF theory, without recourse to wavefunction (62) using only orbitals and occupation numbers. In this rigorous procedure we use the two orbitals of wavefunction (62) each with unit occupation. The z component of the spin density from a conventional SP treatment of molecular hydrogen is Eq. (63), and we arrive at the same result that would follow from the shortcut of using the wavefunction (62). (Furthermore it is yet to be established that any meaningful density functional solutions cannot be associated with CI wavefunction (s).) But Eq. (63) is in error (beyond the Coulson–Fischer point where it is not zero) because that spin density will experience a torque in an inhomogeneous magnetic field in contrast to the true singlet ground state of H_2. Similar erroneous spin densities are found in the 'best' $X\alpha$-like calculations of singlet ground states of C_2 (Dunlap, 1984), Cr_2 (Baykara *et al.*, 1984) and models of 2-Fe ferredoxin (e.g. Noodleman and Baerends, 1984).

The only known ways to proceed to a better description of the antiferromagnetic systems in $X\alpha$-like methods use wavefunctions and the assumption that the $X\alpha$-like energies are those of single determinants, in direct contradiction to the position of Sabin and Trickey (1984). All known approaches in some fashion project out of single-determinantal wavefunctions having the proper symmetry. If two singlet-coupled orbitals are orthogonal we can project out of the antiferromagnetic single determinant the triplet component and use the fact that the exact Hamiltonian is diagonal for states of pure spin to obtain the singlet-state energy of Ziegler *et al.* (1977)

$$E_{\text{singlet}} = 2E(a\alpha, b\beta) - E(a\alpha, b\alpha) \tag{64}$$

where the quantities in parentheses indicate the spin orbitals that are occupied in the calculation of the unrestricted and triplet energies. Eq. (64) means that the singlet energy is not computed directly as it is not equal to the antiferromagnetic energy $E(a\alpha, b\beta)$. Rather, the antiferromagnetic energy lies halfway between the singlet and triplet energies. This result has been extended by Noodleman and coworkers (Noodleman and Norman, 1979; Noodleman, 1981) in two ways which follow directly from the single-determinant rule of Ziegler *et al.* (1977). The first comes from the consideration of two weakly interacting SP systems such as Fe^{3+} or Cr^{1+} atoms each having magnetically aligned half-filled 3d shells (approximated as having zero overlap of their wavefunctions with each other) and each having spin $s = 5/2$, which can couple to form six different states of total spin angular momentum S. In this situation two $X\alpha$-like calculations are easily performed. They are the aligned $S = 5$ calculation and the SP broken-symmetry calculation that has equal spin-up and spin-down electrons. The difference in energy is $4sJ$ where $2J$ is then the energy spacing between the $2s + 1$ eigenstates of spin S (Noodleman, 1981). Under this interpretation the singlet $S = 0$ energy is J lower than the antiferromagnetic state, reducing to Eq. (64) for $s = 1/2$.

This interpretation of $X\alpha$-like calculations would significantly increase the already too large binding energy of Cr_2 by J, where $10J$ is the spacing between the ferromagnetic and antiferromagnetic solutions labelled A and C respectively in Fig. 1 of Delley *et al.* (1983), which are many electronvolts apart near equilibrium. However, the assumptions of zero overlap and a common set of orbitals (with differing spins) in the two self-consistent calculations are less appropriate for Cr_2 than for transition-metal atoms separated by bridging ligands (Noodleman *et al.*, 1985).

Noodleman and coworkers' second extension of Eq. (64) involves no approximation but is computationally less tractable. In this extension the two magnetic subunits are no longer constrained or assumed to be orthogonal. Instead, pure spin eigenstates are obtained by projection; call this method energy projection. Energy projecting the pure singlet component out of the unrestricted wavefunction and using the rule of Ziegler *et al.*, Eq. (64) is slightly altered:

$$E_{\text{singlet}} = [2E(a\alpha, b\beta) - (1 - S^2)E(a\alpha, b\beta)]/(1 + S^2) \tag{65}$$

where S is the overlap between orbitals a and b. The result of this projection for H_2 beyond the Coulson–Fischer point for $X\alpha$ is shown in Fig. 2, which is from Dunlap (1984), and is labelled EQ1. The overlap S between the two orbitals is shown in Fig. 2a. Because the triplet component of the unrestricted solution

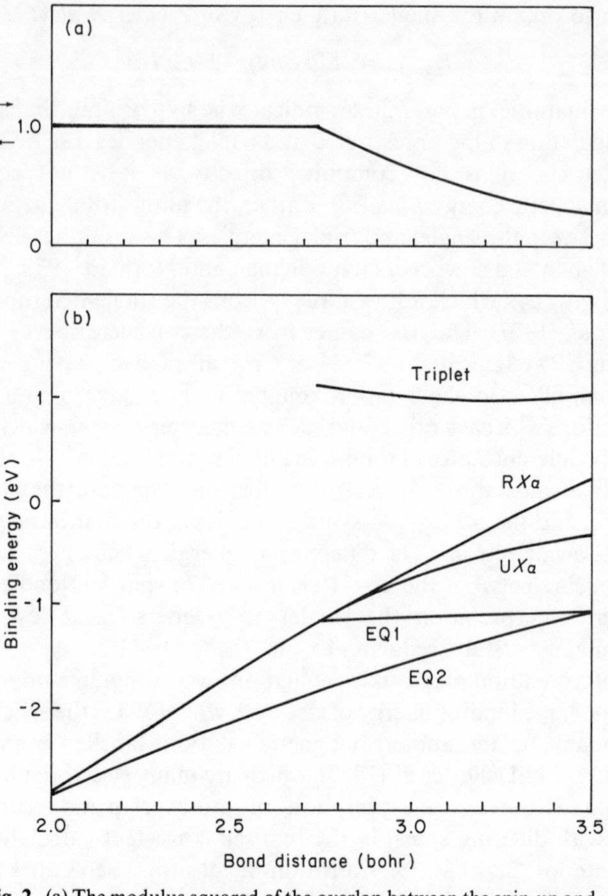

Fig. 2. (a) The modulus squared of the overlap between the spin-up and spin-down orbitals in a spin-unrestricted $X\alpha$ calculation on H_2 near the Coulson–Fischer point (b) The spin-polarized triplet, spin-restricted (R$X\alpha$), spin-unrestricted (U$X\alpha$), energy-projected singlet (EQ1) and self-consistently exchange-projected singlet (EO2) $X\alpha$ solutions for H_2 as a function of internuclear separation.

vanishes to the left of the Coulson–Fischer point, the energy-projected solution has a discontinuous derivative at that point. Thus, as is well known from HF theory, to get a well behaved solution we must project simultaneously with self-consistency, i.e. solve Eq. (65) self-consistently.

Dunlap (1984) advocates a computationally simpler procedure that does not involve solving an energy expression involving more than one set of orbital occupations simultaneously. This method is conceptually similar to Levy's constrained search over single determinants to find the one yielding the lowest kinetic energy. In this case we search over all single determinants to find the one yielding the lowest exchange energy that when properly symmetrized yields the density (and density matrix for kinetic energy evaluation) in question; call this alternative method exchange projection to contrast it with energy projection. Exchange projection yields the smooth curve labeled EQ2 in Fig. 2. (As was mentioned above, when self-consistently applied, energy projection will give a similar smooth curve.) This requirement of self-consistency invalidates the small apparent inconsistencies found by von Barth (1979), which arise from the use of a single radial part in the construction of wavefunctions for all the p^2 configurations of atomic carbon.

Unfortunately, if spatial symmetrization (in contrast to spin symmetrization considered above) is used, discontinuities will arise when spatial symmetry-breaking nuclear motion is considered. Thus broken spatial symmetry is best treated via the methods discussed in the next section that unfortunately do not yield states of pure spatial symmetry. (The alternative energy-projection method of Noodleman when self-consistently applied also becomes discontinuous during nuclear symmetry breaking.)

VII. BROKEN SPATIAL SYMMETRY

Spatial symmetry breaking can arise due to the crossing (and even near-degeneracy) of symmetry-distinct one-electron levels. The level crossing or degeneracy can only occur if the Fock matrix element between the states in question is identically zero, otherwise we will have an avoided crossing. Zero matrix elements arise for two reasons: either the Fock operator is zero in the region where the overlap between the states is non-zero, e.g. the states are somewhat spatially separated; or the positive and negative contributions to the matrix elements exactly cancel. Almost invariably the second situation occurs because under some symmetry operation the states in question are even and odd, i.e. they belong to different irreducible representations of some dynamical group. Such two-level crossing is generically depicted in Fig. 3, which is redrawn from Dunlap and Mei (1983) and refers to the crossing of the two π_u levels with a σ_u level as a function of carbon–carbon separation. The curves can cross because the orbitals are of different symmetry. They do cross because they are bonding and antibonding respectively. Such crossing becomes more prevalent as the number of valence electrons in the molecule increase. Portions of the curves correspond to other common cases of degeneracy that occur in atomic and molecular situations. The point of crossing can be thought of schematically as the degeneracy of the three p or five d levels of an atom (von Barth, 1979; Wood, 1980). One curve, the two degenerate π levels, can represent the two electron holes in the oxygen molecule (Cook, 1981).

Owing to the crossing of the two curves in Fig. 3, we cannot satisfy Fermi statistics in the vicinity of this crossing (Gunnarsson *et al.*, 1977). This problem

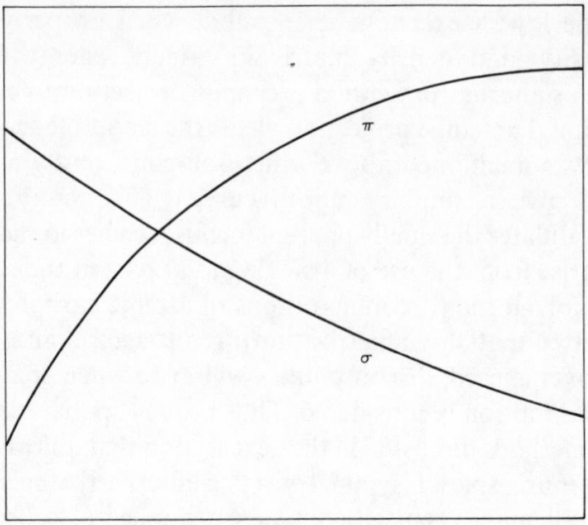

Fig. 3. A schematic representation of two-electron orbitals crossing as a function of nuclear coordinate. The curve is redrawn from Dunlap and Mei (1983) where it was the crossing of the doubly degenerate π_u and $2\sigma_u$ in the region of internuclear separation 1.0–4.5 bohr, and for occupations corresponding to the ${}^1\Sigma_g^+$ and ${}^1\Delta_g$ states which are degenerate in spatial symmetry-restricted $X\alpha$-like methods. The one-electron energy scale is from -4 to -14 eV.

is complicated due to the non-linearity of the $X\alpha$-like equations in that the exact position of the crossing of the two one-electron levels of C_2 depends on the occupation numbers (molecular state). Thus the point of degeneracy which is relevant to the HK theorem becomes a region of points in KS theory. The conventional $X\alpha$-like treatment of this region requires FONs. Computationally the molecular FON treatment is difficult due to the need to solve two levels of self-consistency: a self-consistent molecular calculation is required at each trial set of FONs until the FONs are found that are degenerate. Dunlap and Mei (1983) have proposed an excellent but approximate way around the double self-consistency problem based on the near-quadratic dependence of the $X\alpha$-like energy on occupation number. They quadratically interpolate between states of integral occupation numbers using the two energies and the two sets of energy derivatives with respect to occupation number (orbital eigenvalues). As was mentioned above, however, the FON solution is not the lowest-energy solution; rather we must use broken symmetry.

To compare FON and broken symmetry let us consider a simplified two-electron system corresponding to the crossing of the π_x and σ levels of Fig. 3. A general FON density can be thought of as arising from the complex single-determinantal wavefunction

$$\{(\sigma \cos \omega + i\pi_x \sin \omega)\alpha, (\sigma \cos \omega - i\pi_x \sin \omega)\beta\} \tag{66}$$

(The minus sign in the spin-down orbital is included to make the wavefunction odd under time reversal). The imaginary phase factor between the two contributions to each orbital results in the FON density

$$r = 2\rho_{\alpha\alpha} = 2\rho_{\beta\beta} = 2\sigma^2 \cos^2 \omega + 2\pi^2 \sin^2 \omega \tag{67}$$

Now we can ask under what conditions a variational calculation involving arbitrary phase factors will yield the FON imaginary phase factor. For simplicity we only consider phase variations, thus keeping the spin-up and spin-down densities equal using the wavefunction

$$\{(\sigma \cos \omega + \pi_x e^{i\delta})\alpha, (\sigma \cos \omega + \pi_x e^{-i\delta})\beta\} \tag{68}$$

Also, we keep the simplifying choice of π_x and σ orbitals which are symmetry distinct and thus have zero matrix elements involving any one-electron Fock matrix operator. Wavefunction (68) was considered for the HF case by Pople (1971). In this idealized problem we need only minimize the δ-dependent portion of the energy expression

$$8\cos^2 \omega \sin^2 \omega \cos^2 \delta \int [\sigma(\mathbf{r}_1)\pi_x(\mathbf{r}_1)\sigma(\mathbf{r}_2)\pi_x(\mathbf{r}_2)/r_{12}]\, d\mathbf{r}_1\, d\mathbf{r}_2$$

$$- C \int [2\sigma^2(\mathbf{r})\cos^2 \omega + 2\pi_x^2(\mathbf{r})\sin^2 \omega + 4\sigma(\mathbf{r})\pi_x(\mathbf{r})\cos \omega \sin \omega \cos \delta]^{4/3}\, d\mathbf{r}$$

$$\equiv D\cos^2 \delta - \int (A + B\cos \delta)^{4/3} \tag{69}$$

where we have specialized to $X\alpha$ and C is the combined numerical constants of Eq. (33). Since Eq. (69) only depends on $\cos \delta$, extrema always exist involving real combinations of these two orbitals. For D sufficiently large there does exist a FON solution,

$$\cos \delta = \frac{2}{3D} \int (A + B\cos \delta)^{1/3} B \tag{70}$$

However, if we had allowed more general variation in the two-electron, two-orbital problem, in particular permitting the spin-up and spin-down densities to differ by using two δ values in wavefunction (68), the FON solution would not lie lowest. Furthermore, for three or more quasi-degenerate levels (the true situation in C_2) non-magnetic FON solutions never occur.

If we seek the lowest-energy $X\alpha$-like solutions for the case of exact and near-degeneracy we face the problem of two levels of self-consistency—the familiar one and one involving the mixing of the real orbitals with complex coefficients. While such a computer code has been demonstrated for the case of H_2 (Dunlap, 1984) further progress awaits more accurate treatments of the non-analytic exchange and correlation energies of $X\alpha$-like local density functionals

(Becke, 1983; Laaksonen *et al.*, 1985; Dunlap and Cook, 1986) necessary to treat these complex and subtle effects.

References

Aizman, A., and Case, D. A. (1982). *J. Am. Chem. Soc.*, **104**, 3269–79.
Anderson, P. W. (1975). *Phys. Rev. Lett.*, **34**, 953–5.
Baerends, E. J., Ellis, D. E., and Ros, P. (1973). *Chem. Phys.*, **2**, 41–51.
Baykara, N. A., McMaster, B. N., and Salahub, D. R. (1984). *Mol. Phys.*, **52**, 891–905.
Becke, A. D. (1983). *J. Chem. Phys.*, **76**, 6037–45.
Bobrowicz, F. W., and Goddard, W. A., III (1977). In *Modern Theoretical Chemistry* (Ed. H. F. Schaefer), pp. 79–127, Plenum, New York.
Burns, G. (1977). *Introduction to Group Theory with Applications*, p. 63, Academic Press, New York.
Bursten, B. E., and Fenske, R. F. (1977). *J. Chem. Phys.*, **67**, 3138–45.
Case, D. A. (1982). *Annu. Rev. Phys. Chem.*, **33**, 151–71.
Case, D. A., and Yang, C. Y. (1980). *Int. J. Quantum Chem.*, **18**, 1091–9.
Casula, F., and Herman, F. (1983). *J. Chem. Phys.*, **78**, 858–75.
Connolly, J. W. D. (1975). In *Modern Theoretical Chemistry* (Ed. G. A. Segal), pp. 105–32, Plenum, New York.
Connolly, J. W. D., and Sabin, J. R. (1972). *J. Chem. Phys.*, **56**, 5529–32.
Cook, M. (1981). Ph.D. Thesis, Harvard University (1982 *Diss. Abstr. Int.*, **43**, 434-B (August), No. DA 8216185).
Cook, M., and Karplus, M. (1985). *J. Chem. Phys.*, **83**, 6344–6366.
Coulson, C. A., and Fischer, I. (1949). *Phil. Mag.*, **40**, 389–93.
Csavinszky, P. (1981). *Phys. Rev. A*, **24**, 1215–17.
Davidson, E. R. (1972). *Rev. Mod. Phys.*, **44**, 451–64.
Delley, B., Freeman, A. J., and Ellis, D. E. (1983). *Phys. Rev. Lett.*, **50**, 488–91.
Diamond, J. B. (1973). *Chem. Phys. Lett.*, **20**, 63–6.
Dunlap, B. I. (1982). *Phys. Rev. A*, **25**, 2847–9.
Dunlap, B. I. (1984). *Phys. Rev. A*, **29**, 2902–5.
Dunlap, B. I., Connolly, J. W. D., and Sabin, J. R. (1979). *J. Chem. Phys.*, **71**, 3696–402; 4993–9.
Dunlap, B. I., and Cook, M. (1986). *Int. J. Quantum Chem.*, **29**, 767–77.
Dunlap, B. I., and Mei, W. N. (1983). *J. Chem. Phys.*, **78**, 4997–5003.
Englisch, H., and Englisch, R. (1983). *Physica*, **121A**, 253–68.
Fermi, E., and Amaldi, E. (1934). *Acad. Ital. Rome*, **6**, 119.
Fukutome, H. (1981). *Int. J. Quantum Chem.*, **20**, 955–1065.
Gaspar, R. (1954). *Acta Phys.*, **3**, 263–86.
Ghosh, S. K., Berkowitz, M., and Parr, R. G. (1984). *Proc. Natl. Acad. Sci.*, **81**, 8028–31.
Goodgame, M. M., and Goddard, W. A., III (1982). *Phys. Rev. Lett.*, **48**, 135–8.
Goodgame, M. M., and Goddard, W. A., III (1985). *Phys. Rev. Lett.*, **54**, 661–4.
Gunnarsson, O., Harris, J., and Jones, R. O. (1977). *J. Chem. Phys.*, **67**, 3970–9.
Gunnarsson, O., and Jones, R. O. (1980). *J. Chem. Phys.*, **72**, 5357–62.
Gunnarsson, O., and Jones, R. O. (1985). *Phys. Rev. B*, **31**, 7588–602.
Gunnarsson, O., and Lundqvist, B. I. (1976). *Phys. Rev. B*, **13**, 4274–98.
Harriman, J.E. (1981). *Phys. Rev. A*, **24**, 680–2.
Harriman, J. E. (1983). *Phys. Rev. A*, **27**, 632–45.
Harris, J., and Jones, R. O. (1978). *J. Chem. Phys.*, **68**, 3316–17.
Harris, J., and Jones, R. O. (1979). *J. Chem. Phys.*, **70**, 830–41.
Heitler, W., and London, F. (1927). *Z. Phys.*, **44**, 455–72.
Herman, F., Williams, A. R., and Johnson, K. H. (1974). *J. Chem. Phys.*, **61**, 3508–22.

Hohenberg, P., and Kohn, W. (1964). *Phys. Rev.*, **136**, B864–71.
Hubbard, J. (1964). *Proc. R. Soc. A*, **281**, 401–19.
Janak, J. F., Moruzzi, V. L., and Williams, A. R. (1975). *Phys. Rev. B*, **12**, 1257–61.
Janak, J. F., and Williams, A. R. (1981). *Phys. Rev. B*, **23**, 6301–6.
Johnson, K. H. (1973). *Adv. Quantum Chem.*, **7**, 143–85.
Kmetko, E. A. (1970). *Phys. Rev. A*, **1**, 37–8.
Koelling, D. D. (1985). Unpublished.
Kohn, W. (1983). *Phys. Rev. Lett.*, **51**, 1596–8.
Kohn, W., and Sham, L. J. (1965). *Phys. Rev.*, **140**, A1133–8.
Kohn, W., and Vashista, P. (1983). In *Theory of the Inhomogeneous Electron Gas* (Eds S. Lundqvist and N. H. March), pp. 79–147, Plenum, New York.
Laaksonen, L., Sundholm, D., and Pyykko, P. (1985). *Int. J. Quantum Chem.*, **27**, 601–12.
Langreth, D. C., and Mehl, M. J. (1983). *Phys. Rev. B*, **28**, 1809–34.
Levy, M. (1979). *Proc. Natl. Acad. Sci.*, **76**, 6062–5.
Levy, M. (1982). *Phys. Rev. A*, **26**, 1200–8.
Levy, M., and Perdew, J. P. (1985). In *Density Functional Methods in Physics* (Eds R. M. Dreizler and J. da Providencia), pp. 11–30, Plenum, New York.
Lieb, E. H. (1979). *Phys. Lett.*, **70A**, 444–6.
Lieb, E. H. (1983). *Int. J. Quantum Chem.*, **24**, 243–77.
Lieb, E. H., and Oxford, S. (1981). *Int. J. Quantum Chem.*, **19**, 427–39.
Löwdin, P.-O. (1955). *Phys. Rev.*, **97**, 1474–89.
Löwdin, P.-O. (1969). *Adv. Chem. Phys.*, **14**, 283–340.
Löwdin, P.-O., Calais, J.-L., and Calazans, J. M. (1981). *Int. J. Quantum Chem.*, **20**, 1201–15.
McWeeny, R. (1979). *Coulson's Valence*, Oxford University Press, Oxford.
March, N. H. (1983). In *Theory of the Inhomogeneous Electron Gas* (Eds S. Lundqvist and N. H. March), pp. 1–77, Plenum, New York.
Murphy, D. R. (1981). *Phys. Rev. A*, **24**, 1682–8.
Noodleman, L. (1981). *J. Chem. Phys.*, **74**, 5737–43.
Noodleman, L., and Baerends, E. J. (1984). *J. Am. Chem. Soc.*, **106**, 2316–27.
Noodleman, L., and Norman, J. G., Jr (1979). *J. Chem. Phys.*, **70**, 4903–6.
Noodleman, L., Norman, J. G., Jr, Osborne, J. H., Aizman, A., and Case, D. A. (1985). *J. Am. Chem. Soc.*, **107**, 3418–26.
Norman, J. G., Jr (1976). *Mol. Phys.*, **31**, 1191–8.
Norman, M. R., and Koelling, D. D. (1984). *Phys. Rev. B*, **30**, 5530–40.
Parr, R. G., Donnelly, R. A., Levy, M., and Palke, W. E. (1978). *J. Chem. Phys.*, **68**, 3801–7.
Parr, R. G., and Yang, W. (1984). *J. Am. Chem. Soc.*, **106**, 4049–50.
Pederson, M. R., Heaton, R. A., and Lin, C. C. (1984). *J. Chem. Phys.*, **80**, 1972–5.
Perdew, J. P., Parr, R. G., Levy, M., and Balduz, J. L. (1982). *Phys. Rev. Lett.*, **49**, 1691–4.
Perdew, J. P., and Zunger, A. (1981). *Phys. Rev. B*, **23**, 5048–79.
Pitzer, R. M. (1972). *J. Chem. Phys.*, **58**, 3111–12.
Pople, J. A. (1971). *Int. J. Quantum Chem.*, **5**, 175–82.
Rajagopal, A. K. (1980). *Adv. Chem. Phys.*, **41**, 59–193.
Reif, F. (1965). *Fundamentals of Statistical and Thermal Physics*, McGraw-Hill, New York.
Sabin, J. R., and Trickey, S. B. (1984). In *Local Density Approximations in Quantum Chemistry and Solid State Physics* (Eds J. P. Dahl and J. Avery), pp. 333–52, Plenum, New York.
Salahub, D. R., Lampson, S. H., and Messmer, R. P. (1982). *Chem. Phys. Lett.*, **85**, 430–3.

Sambe, H., and Felton, R. H. (1975). *J. Chem. Phys.*, **62**, 1122–6.
Schwarz, K. (1972). *Phys. Rev. B*, **5**, 2466–8.
Slater, J. C. (1951). *Phys. Rev.*, **81**, 385–90.
Slater, J. C. (1968). *Phys. Rev.*, **165**, 658–69.
Slater, J. C. (1972). *J. Chem. Phys.*, **57**, 2389–96.
Slater, J. C. (1974). *Quantum Theory of Molecules and Solids*, Vol. 4, McGraw-Hill, New York.
Slater, J. C., Mann, J. B., Wilson, T M., and Wood, J. H. (1969). *Phys. Rev.*, **184**, 672–94.
Sykja, B., and Calais, J. -L. (1982). *J. Phys. C: Solid State Phys.*, **15**, 3079–92.
Talman, J. D., and Shadwick, W. F. (1976). *Phys. Rev. A*, **14**, 36–40.
von Barth, U. (1979). *Phys. Rev. A*, **20**, 1693–703.
von Barth, U., and Hedin, L. (1972). *J. Phys. C: Solid State Phys.*, **5**, 1629–41.
Vosko, S. H., Wilk, L., and Nusair, M. (1980). *Can. J. Phys.*, **58**, 1200–11.
Wood, J. H. (1970). *Int. J. Quantum Chem. Symp.*, **S3**, 747–55.
Wood, J. H. (1980). *J. Phys. B: At. Mol. Phys.*, **13**, 1–14.
Ziegler, T., Rauk, A., and Baerends, E. J. (1977). *Theor. Chim. Acta* (*Berl.*), **43**, 261–71.

Ab Initio Methods in Quantum Chemistry—II
Edited by K. P. Lawley

MODERN VALENCE BOND THEORY

D. L. COOPER

Department of Inorganic, Physical and Industrial Chemistry, University of Liverpool, PO Box 147, Liverpool L69 3BX, UK

J. GERRATT

Department of Theoretical Chemistry, School of Chemistry, University of Bristol, Bristol BS8 1TS, UK

and

M. RAIMONDI

Departimento di Chimica Fisica ed Elettrochimica, Universitá di Milano, Via Golgi 19, 20133 Milano, Italy

CONTENTS

I. Introduction . . . 320
II. Classical Valence Bond Theory . . . 320
III. The Spin-coupled Valence Bond Theory . . . 324
IV. Survey of Results of Spin-coupled Valence Bond Theory . . . 345
  A. $^1\Sigma^+$ States of CH^+ . . . 345
  B. Potential Surfaces of $(B\text{–}H_2)^+$ . . . 347
  C. Potential Energy Curves for $C^{3+}(2l) + H(1s) \rightarrow C^{2+}(nln'l';{}^1L) + H^+$ . . . 351
  D. The Electronic Structure of Benzene . . . 355
  E. Studies of Momentum-space Properties . . . 358
V. Calculation of Matrix Elements of the Hamiltonian . . . 360
VI. Recent Developments in Valence Bond Theory . . . 369
  A. The 'Diatomics-in-Molecules' Method, and Mixed Molecular Orbital–Valence Bond Methods . . . 369
  B. Semi-empirical Valence Bond Methods . . . 376
  C. Valence-shell Calculations . . . 378
VII. Intermolecular Forces . . . 378
  A. Introduction . . . 378
  B. Method of Calculation . . . 379
  C. Charge Transfer . . . 381
  D. The Intramolecular Correlation Energy . . . 382
  E. Intermolecular Perturbation Theory . . . 385

F. Pseudo-states 389
VIII. Conclusions 391
References 392

אֶבֶן מָאֲסוּ הַבּוֹנִים
הָיְתָה לְרֹאשׁ פִּנָּה

The stone rejected by the builders
Is become the chief cornerstone

I. INTRODUCTION

When atoms interact to form a molecule, the wavefunctions representing them overlap. Thus the physical phenomenon of bond formation is closely linked with the non-orthogonality of the electronic wavefunctions which describe the constituent fragments. The conceptual advantages of a formalism in which the wavefunction for the system is expanded in terms of wavefunctions of its constituents have long been recognized. The essential physical effects are incorporated at the outset, so that even a zeroth-order wavefunction in this framework provides a qualitatively correct description of the potential energy surfaces. However, not until the last few years has it proved possible for a general approach based upon these considerations to be fully developed.

The purpose of this review is to give an account of approaches of this type. That is to say we examine methods where non-orthogonal orbitals enter directly into the wavefunctions*. The fundamental prototype is of course the 'classical' valence bond (VB) theory[1], and accordingly we begin with a survey of the description it provides of molecular electronic structure, and of its important conceptual role in the description of many fundamental molecular processes.

The central theme of this chapter is to show how the many attractive physical concepts of qualitative VB theory can be incorporated into a more general framework (the spin-coupled valence bond theory, Section III) which leads to a computational procedure that provides quantitative descriptions of the ground-state and many excited-state potential energy surfaces of molecular systems.

II. CLASSICAL VALENCE BOND THEORY

The basic idea of valence bond (VB) theory is very simple: the wavefunctions for the electrons in a molecule are constructed directly from the wavefunctions of the constituent atoms. This implements in a very clear cut way a large part of the experience of chemistry. (For a review of classical valence bond theory, the reader should consult Ref. 1, for example.)

*This is to be clearly distinguished from the use of basis sets, which are almost always non-orthogonal, and which can be used in calculations to construct either orthonormal or non-orthogonal orbitals.

A fundamental requirement in this procedure is of course to ensure that the whole wavefunction is antisymmetric; in particular, that it is antisymmetric when electrons stemming from different atoms are interchanged. This gives rise to the exchange contribution to the total energy which is the key to understanding the strength of the covalent bond. In the simplest case of the H_2 molecule, this is given (without the normalization factor) by

$$K(1s_a, 1s_b) = \langle \psi_{1s_a}(1)\psi_{1s_b}(2) | V(1,2) | \psi_{1s_b}(1)\psi_{1s_a}(2) \rangle \tag{1}$$

The potential $V(1, 2)$ is the total interaction energy between the two atoms, and includes contributions from electron–nuclear attraction terms as well as from electron–electron repulsion effects. As a result $K(1s_a, 1s_b)$ is *negative*, and is primarily responsible for the stability of the H–H bond. Central to this description is the fact that the wavefunctions of the participating atoms, ψ_{1s_a} and ψ_{1s_b}, overlap. The magnitude of $K(1s_a, 1s_b)$, and consequently the strength of the covalent bond, is determined by the degree of non-orthogonality between the two orbitals.

The reason that $K(1s_a, 1s_b)$ contributes so heavily to the lowering of the energy as the two H atoms approach is that the spins of the two electrons making up the bond are coupled to form a net spin of zero. If the spins are antiparallel, as in the $^3\Sigma_u^+$ state, $K(1s_a, 1s_b)$ is multiplied by a factor of (-1) and consequently gives rise instead to a repulsive state.

These arguments can be extended in a consistent way to provide a coherent explanation for a whole range of chemical phenomena. For example, consider the formation of the ammonia molecule in its ground $^1A_1$ state from three hydrogen atoms and an N atom in its lowest state. The exchange interaction and hence the energy stabilization will be greatest if each H atom approaches along the axis of one of the three N 2p orbitals and also if its spin is coupled to give a net spin of zero with its N atom partner. Thus in a very direct way we predict that the shape of the NH_3 molecule is pyramidal. The resulting H–N–H bond angle is 90°, which differs from the experimentally observed value[2] of 106.47°. Higher-order effects in the form of configuration interaction (or hybridization) are needed to improve the agreement. However, the H–X–H angles in the ground states of PH_3, AsH_3 and SbH_3 are very closed to 90° (PH_3, 93.45°; AsH_3, 92°; SbH_3, 92.5°)[3], and in these cases configuration interaction (CI) plays practically no role in this respect. It is remarkable that such a simple zeroth-order theory leads at once to essentially the correct result, and provides at the same time a clear physical mechanism.

Another important concept which emerges from this example is that of spin recoupling. The ground state of $N(2p^3)$ is ^4S_u, in which the spins of the three valence electrons are coupled to give the maximum resultant, $S = \frac{3}{2}$. As the H atoms approach, these spins uncouple from each other and form three pair bonds. In many cases, this kind of recoupling has been found to occur over a very short range of internuclear distances (0.5–1 bohr), and is frequently associated with a maximum in the potential surface (see Section III).

Similar arguments based upon the magnitude and sign of exchange integrals can be extended to many other situations. Thus the exchange interaction between the electrons in a doubly occupied orbital on one atom or molecule and an unpaired electron on another system gives rise to a net repulsive interaction. This provides a clearcut explanation of the phenomenon of saturation of valence[4].

The existence of multiple bonds in a molecule such as ethylene and its consequent planarity and resistance to twisting also follows directly from this same kind of reasoning.

A particularly useful feature of this 'classical' VB theory is the concept of 'zeroth-order' potential curves which can be considered as arising from the interaction of specific atomic states. The states of the LiH molecule provide a simple example for which there is a good deal of experimental data as well as reliable theoretical calculations. Ground-state Li($1s^2 2s$; ^2S) and H(1s) give rise to weak covalent bonding. The same is true of Li(2p; ^2P) + H(1s). The major effect arises from the intervention of the ionic state $Li^+(^1S) + H^-(^1S)$. This lies asymptotically at 4.636 eV above the ground state. As R decreases, the energy of this state plunges downwards and gives rise to sharply avoided intersections with the Li(2p) + H(1s) and Li(2s) + H(1s) states. It is primarily responsible for the binding of LiH in its $^1\Sigma^+$ ground state. However, this analysis also predicts the existence of weakly bound excited $^1\Sigma^+$ states whose potential curves display unusual shapes with large flat portions ('iron bath' potentials)[6]. This has been confirmed by detailed calculations[7] and much experimental evidence[8].

This same analysis shows that at large separations the A state of LiH, for example, consists largely of the ion pair, Li^+H^-. The arguments given here are not special for LiH, and in general one may expect to find such ion-pair states in otherwise covalent molecules at a few electron volts above the ground state—though the situation will often be complicated by the intervention of Rydberg states. Recently, however, a whole series of such states have indeed been detected in I_2[9].

In physical terms, the formation of LiH in the ground state from its constituent atoms occurs by means of a transfer of an electron from the Li atom to H when the internuclear distance decreases below a critical separation R_c. This same concept underlies the 'harpoon' mechanism which is used to explain the very large cross-sections for reaction which are observed for such processes as $K + Br_2 \rightarrow KBr + Br$. As the reactants approach, the covalent $K + Br_2$ potential surface is intersected by an ionic $K^+ + Br_2^-$ surface. Accordingly, an electron transfers from K to Br_2. Subsequent production of KBr and Br is immediate. This model is also in accord with the observation in beam scattering experiments that the distribution of KBr product is strongly forward-peaked[10].

The Be_2 molecule provides an illuminating contrast between the numerical precision afforded in certain situations by 'state-of-the-art' CI methods and the physical insight provided by the qualitative VB theory under discussion here. Large-scale molecular-orbital configuration-interaction (MO-CI) calculations involving about 3×10^5 configurations show that the ground state of Be_2 possesses an unexpectedly deep well of 714 cm^{-1}, but give no hint of the mechanism for this. In VB terms it can be seen that the first two potential energy curves of $^1\Sigma_g^+$ symmetry, which arise from Be($2s^2$; ^1S) + Be($2s^2$; ^1S) and and Be(2s2p; ^1P) + Be($2s^2$; ^1S) respectively, can only be repulsive. However as shown by Gallup[12] the $^1\Sigma_g^+$ state arising from Be(2s2p; ^3P) + Be(2s2p; ^3P), whose asymptote lies 5.45 eV above the ground state, is strongly attractive and is essentially responsible for the weak bond. From this it follows that there is a direct connection between the computed well depth and the asymptotic splitting between the ground and second excited state: if Be(2s2p; ^3P) is well described, we may obtain a reliable ground-state potential curve. If it is not—and the (2s2p; ^3P) state is too high in energy—there will be no potential well. It is worth noting that similar arguments predict that there are a number of stable excited states of Be_2 of $^1\Sigma$, $^3\Sigma$, $^1\Pi$ and $^3\Pi$ (g and u) symmetries lying between 5 and 8 eV above the ground state.

This qualitative form of VB theory encounters difficulties in the well known case of O_2, where at first sight it seems to predict a $^1\Sigma_g^+$ ground state instead of $^3\Sigma_g^-$. However, it is worth reminding ourselves that detailed arguments given long ago by Wheland[13] in terms of signs of exchange integrals as above do in fact lead to $^3\Sigma_g^-$ as the ground state. Furthermore, this has recently been confirmed by actual computation[14].

Actual numerical applications of this classical VB theory show interesting and somewhat tantalizing features. Simple wavefunctions which consist of a few structures formed from just the most obvious participating atomic states—when combined with certain semi-empirical corrections—yield results which are often in remarkable agreement with experiment. The development of such semi-empirical approaches is a theme to which we return later in this review. Thus for example the ground state of CH is correctly found[15] to be $^2\Pi$ while self-consistent field (SCF) wavefunctions at anything less than the Hartree–Fock limit predict $^4\Sigma^-$. The same early VB calculation[16] on CH gives the $^4\Sigma^-$ state as lying 0.60 eV above the ground state. The correct value is now known to be 0.794 eV. An outstanding success of simple VB theory plus semi-empirical corrections was the definitive assignment of the binding energies of N_2[17] and CO[18].

However, the straightforward *ab initio* application of classical VB theory has so far not succeeded in attaining more than qualitatively correct results[19]. The rate of convergence from an initially promising start to a final approximately quantitative result is disconcertingly slow. The intervention of large numbers of ionic structures—even in pre-eminently covalent situations such as hydrocarbon fragments—is particularly distressing.

The basic reason for this slow convergence is that the VB wave function represents the molecular function as an expansion in terms of isolated atomic—or fragment—states. This representation does not take into account the essential *deformation* of the participating states on molecule formation, and it is the many ionic structures which attempt to provide for this phenomenon.

The situation is fundamentally altered if one uses orbitals which are not of fixed atomic-orbital (AO) type, but are expanded in a multicentred basis set of arbitrary size—much as in molecular-orbital (MO) theory. Such orbitals are able to account for the necessary distortions which occur as internuclear distances vary. However, the orbitals remain non-orthogonal, which confers on them characteristics entirely different to those of MO theory.

In the following section we present a general framework in which non-orthogonal orbitals are used to expand the exact wavefunction. This serves to explain the spin-coupled VB theory which is the basic *motif* of this chapter, and also to show how this reduces to classical VB theory on the one hand, and to the CI expansion on the other.

In Section IV results obtained so far by the spin-coupled VB theory are surveyed and in Section V we return somewhat briefly to classical VB theory.

One conclusion that emerges from this survey is that the most useful features of the classical VB theory are utilized not necessarily in *ab initio* work, but in providing a framework for semi-empirical theories. These last are proving to be of real value in the interpretation of results of beam scattering experiments, and in the provision for dynamical studies of potential surfaces which possess the correct general features.

The last sections are devoted to the application of VB theory to intermolecular interactions. Since van der Waals forces are so weak compared to 'chemical' forces, it is intuitively obvious that classical VB theory has much to offer in this context rather than a 'supermolecule' approach.

III. THE SPIN-COUPLED VALENCE BOND THEORY

We begin with some general considerations of perhaps lesser-known, but important, features of exact electronic wavefunctions. Our motive is to establish a theoretical framework together with a reasonably consistent notation in order to carry through the spin-coupled VB and other expansions of the total wavefunction. We consider an atomic or molecular system consisting of N electrons and A nuclei. We assume the Born–Oppenheimer separation[21] and write the Hamiltonian operator for the motion of the electrons in the form:

$$\mathscr{H} = \sum_{\mu=1}^{N}\left(-\tfrac{1}{2}\nabla_\mu^2 - \sum_{I=1}^{A} Z_I r_{\mu I}^{-1}\right) + \sum_{\mu>\nu=1}^{N} g_{\mu\nu}$$

$$= \sum_{\mu=1}^{N} h_\mu + \sum_{\mu>\nu=1}^{N} g_{\mu\nu} \qquad (2)$$

All equations are written in atomic units. With an origin chosen at one of the nuclei or at the centre of mass of the nuclei, we denote electronic coordinates by $\mathbf{r}_\mu$ and nuclear coordinates by $\mathbf{R}_I$. The scalar distances $|\mathbf{r}_\mu - \mathbf{R}_I|$ are denoted by $r_{\mu I}$ and $|\mathbf{r}_\mu - \mathbf{r}_\nu|$ by $r_{\mu\nu}$. The Coulomb repulsion between two electrons, $r_{\mu\nu}^{-1}$ is written as $g_{\mu\nu}$. The repulsion between the nuclei has been dropped from Eq. (2); it can be added later on.

The solutions Ψ of the Schrödinger equation

$$\mathscr{H}\Psi = E\Psi \tag{3}$$

are functions not only of the position coordinates of the electrons, $\mathbf{r}_1,\ldots,\mathbf{r}_\mu,\ldots,\mathbf{r}_N$, but also of the coordinates $\sigma_1,\ldots,\sigma_\mu,\ldots,\sigma_N$ of the spins of the electrons:

$$\Psi = \Psi(\mathbf{r}_1,\ldots,\mathbf{r}_N;\sigma_1,\ldots,\sigma_N;R) \tag{4}$$

(Note that the wavefunction also depends parametrically upon all the internuclear separations, which we have here denoted collectively by R. It will be convenient to drop this later on.) As is well known, solutions Ψ of Eq. (3) which correspond to actual states of the system must satisfy the Pauli principle, that is Ψ must be antisymmetric under any simultaneous permutation of spatial coordinates and spin coordinates:

$$P^r P^\sigma \Psi = \varepsilon_P \Psi \tag{5}$$

A permutation of the spatial coordinates $\mathbf{r}_1,\ldots,\mathbf{r}_N$ of Ψ is denoted by P^r and the same permutation of the spin coordinates by P^σ. The parity of this permutation, $+1$ or -1, is written as ε_P.

In addition we require the electronic wavefunction to be an eigenfunction of the operator for the square of the total spin, $\hat{\mathbf{S}}^2$, and also of the z component of the spin, $\hat{S}_z$:

$$\begin{aligned}\hat{\mathbf{S}}^2\Psi_{SM} &= S(S+1)\Psi_{SM}\\ \hat{S}_z\Psi_{SM} &= M\Psi_{SM}\end{aligned} \tag{6}$$

The corresponding quantum numbers are S and M respectively, and these are added to the notation for the wavefunction*.

As shown by Wigner[22], the most general form of Ψ_{SM} which satisfies requirements (5) and (6) is given by[23]

$$\Psi_{SM} = \left(\frac{1}{f_S^N}\right)^{1/2} \sum_{k=1}^{f_S^N} \Phi_{Sk}(\mathbf{r}_1,\ldots,\mathbf{r}_N)\Theta_{S,M;k}^N(\sigma_1,\ldots,\sigma_N) \tag{7}$$

The Φ_{Sk} are functions purely of spatial coordinates, and the $\Theta_{S,M;k}^N$ purely functions of spin coordinates. Eq. (7) expresses the fullest possible factoriz-

*In general we denote an operator corresponding to a physical quantity A with a circumflex over the letter: $\hat{A}$. The only exception to this is the Hamiltonian $\mathscr{H}$ which is defined as the operator corresponding to the energy E.

ation of Ψ_{SM} into functions of space and spin. These functions possess a fundamental permutational symmetry as follows:

$$P^r\Phi_{Sk} = \sum_{l=1}^{f_S^N} U_{lk}^{S,N}(P)\Phi_{Sl}$$
$$P^\sigma\Theta_{S,M;k}^N = \varepsilon_P \sum_{l=1}^{f_S^N} U_{lk}^{S,N}(P)\Theta_{S,M;l}^N \tag{8}$$

The Φ_{Sk} functions form a basis for an irreducible representation of the group $\mathscr{S}_N$ of permutations of N objects ('the symmetric group'). The particular representation is specified by S and N: the total spin of the system S and the total number of electron N. The matrices of the particular representation are denoted by $\mathbf{U}^{S,N}(P)$. They are of dimension f_S^N, where

$$f_S^N = \frac{(2S+1)N!}{(\frac{1}{2}N+S+1)!(\frac{1}{2}N-S)!} \tag{9}$$

Each individual function Φ_{Sk} is a solution of the Schrödinger equation:

$$\mathscr{H}\Phi_{Sk} = E\Phi_{Sk}$$

and from this it would appear that the energy level E is f_S^N-fold degenerate. However, this 'permutational degeneracy' cannot be removed by applying an external electric or magnetic field, since each component Φ_{Sk} is necessary to define a properly antisymmetric state function. The spin functions $\Theta_{S,M;k}^N$ are constructed as eigenfunctions of $\hat{\mathbf{S}}^2$ and $\hat{S}_z$, with eigenvalues S and M respectively. The significance of the index k will be explained shortly. They form a basis for the dual representation, whose matrices are just $\varepsilon_P\mathbf{U}^{S,N}(P)$.

A general discussion of the theory of the symmetric group would divert us too much from our main concern. The book by Kaplan[24] gives a particularly lucid account of the parts relevant to many-electron systems and several other detailed accounts are also available[22,25,26].

The U matrices play an important role in the theory of spin-coupled wavefunctions as will be seen. These matrices are always real, and since they are group representation matrices, they are also orthogonal:

$$U_{lk}^{S,N}(P) = [\mathbf{U}^{S,N}(P)]_{kl}^{-1} = U_{kl}^{S,N}(P^{-1}) \tag{10}$$

General programs for constructing the U matrices are available (see below). The functions Φ_{Sk} and $\Theta_{S,M;k}^N$ are also orthonormal:

$$\langle\Phi_{Sk}|\Phi_{Sl}\rangle = \delta_{kl}$$
$$\langle\Theta_{S,N;k}^N|\Theta_{S,M;l}^N\rangle = \delta_{kl} \tag{11}$$

The integration in the first of these equations is over all space, and in the second is over all values of the spin coordinates. Using property (10) in the general form for Ψ_{SM} in Eq. (7), it is not hard to show that the wavefunction does indeed satisfy the Pauli principle (5).

The most common way of constructing the spin functions $\Theta^N_{S,M;k}$ is simply by coupling individual electron spins successively using the normal rules of angular momentum theory. This builds up the spin functions as eigenfunctions of $\hat{\mathbf{S}}^2$ and $\hat{S}_z$ for $1, 2, \ldots, (N-1)$ and N electrons. The process is conveniently represented by means of the 'branching diagram' which is shown in Fig. 1[27]. Here the resultant spin S is plotted against the number of electrons. Each circle in the figure contains the number of ways of starting at $N = 1$, $S = \frac{1}{2}$ and arriving at the required resultant (N, S) by following possible rightward paths on the diagram. This is just the number f^N_S. Note that the relationship $f^N_S = f^{N-1}_{S+1} + f^{N-1}_{S-1}$ holds, as may be proved directly from (9). The significance of the index k is that it denotes the particular mode of coupling the individual spins to form the net spin. A more detailed specification of this index would be a series of partial resultant spins:

$$k \equiv (S_1 S_2 \cdots S_\mu \cdots S_{N-1})$$

where S_μ is the resultant spin after coupling μ electrons. Spin S_1 is always equal to $\frac{1}{2}$, and it is unnecessary to specify S_N since this is the same as the total spin S. We will occasionally make use of this more detailed notation.

The set of spin functions constructed in this way is commonly termed the 'standard' or 'Kotani' basis—though a more correct name would be 'the Young–Yamanouchi basis'[28]. There are, however, many other possible bases of spin functions (in general, there is an infinite number of choices) and we

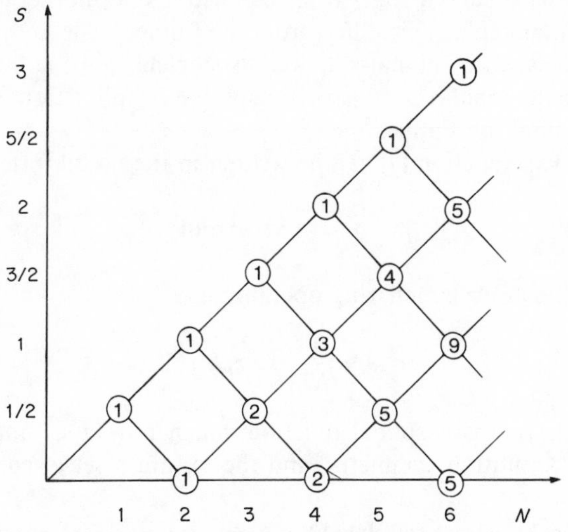

Fig. 1. The branching diagram.

mention briefly some of the more important ones. For certain problems it is useful to divide the electrons into two or more groups containing, say, N_1 and N_2 electrons ($N_1 + N_2 = N$), e.g. into groups corresponding to σ and π electrons in linear or conjugated systems, or into sets which correspond to fragment 1 and fragment 2 as a particular internuclear separation becomes large.

Corresponding to this we couple together Kotani spin functions from each subsystem to form complete functions of the form

$$\Theta^N_{S,M;S_1S_2k_1k_2} \equiv |S_1 S_2 k_1 k_2; SM\rangle$$

Here the index which denotes a particular spin function assumes a more complicated form since it is necessary to specify the net spin of each subsystem and the mode of coupling within the subsystems. The special properties of functions of this type are discussed in more detail in Refs 20 and 48.

The configuration-interaction stage of our calculations (to which we refer, not altogether accurately, as the 'VB stage') makes use of the traditional VB spin functions as first formulated by Rumer[30]. These are constructed by coupling together pairs of individual electron spins to form singlets in a series of specified ways. The physical significance of this basis is that is highlights the phenomenon of bond formation in molecular systems. The spin functions themselves are generated systematically by means of the 'leading term' algorithm (see Section V)*. While they are relatively simple to construct, the Rumer spin functions, however, are not orthogonal.

Finally it is worth mentioning the 'Serber basis' of spin functions in which pairs of electrons, 1 and 2, 3 and 4,..., etc., are coupled first to singlets or triplets, these pairs subsequently being coupled to one another to produce the required resultant spin. This differs from the Rumer basis in that triplet spin functions are used for the pairs as well as singlets, and the final set of spin functions is orthogonal. The construction of spin eigenfunctions is discussed in detail in the book by Pauncz[31].

The total wavefunction (7) can be written in the useful form[20]

$$\Psi_{SM} = \sum_{k=1}^{f_S^N} c_{Sk}(N!)^{1/2}\mathscr{A}(\Phi\Theta^N_{S,M;k}) \tag{12}$$

where $\mathscr{A}$ is the antisymmetrizing operator:

$$\mathscr{A} = \frac{1}{N!}\sum_{P\in\mathscr{S}_N} \varepsilon_P P^r P^\sigma \tag{13}$$

$\Phi = \Phi(\mathbf{r}_1, \mathbf{r}_2, \ldots, \mathbf{r}_N)$ is a spatial function which now does not possess any particular permutation symmetry, and the c_{Sk} are a set of coefficients. The

*It should perhaps be pointed out that the VB program is independent of any particular choice of basis of spin functions.

connection with (7) can be established by substituting (13) into (12) and making use of (8). We find that

$$\Phi_{Sl} = \sum_k c_{Sk}(\omega^S_{lk}\Phi) \tag{14}$$

where ω^S_{lk} is the projection operator given by[22,32]

$$\omega^S_{lk} = (f^N_S/N!)^{1/2} \sum_{P\in\mathscr{S}_N} U^{S,N}_{lk}(P)P^r$$

Thus beginning with an arbitrary spatial function Φ, we form from it a set of functions with the correct spatial symmetry according to (14), and finally the full antisymmetric function (7). This is all incorporated in expression (12) on which we now focus attention.

We now consider the expansion of the spatial function Φ in a set of orthonormal one-electron functions. For this purpose we can employ not a single orthonormal set but a different set for each electron coordinate:

$$\{\psi^{(i)}_\mu\} \qquad \mu = 1, 2, \ldots, N;\ i = 0, 1, 2, \ldots, \infty \tag{15}$$

The index μ denotes the particular electronic coordinate $\mathbf{r}_\mu$, and the superscript i the particular member of this set. While a given set of orbitals (15) is orthonormal, the different sets are not orthogonal to each other. Thus

$$\langle\psi^{(i)}_\mu|\psi^{(j)}_\nu\rangle = \begin{cases} \delta_{ij} & \text{if } \mu = \nu \\ \Delta^{ij}_{\mu\nu} & \text{otherwise} \end{cases} \tag{16}$$

Each set $\{\psi^{(i)}_\mu\}$ is in some sense an optimal expansion for that electronic coordinate. How this is achieved in practice is explained below. If we now expand Φ simultaneously in these N distinct sets, we obtain in fairly straightforward fashion the following expansion of the total wavefunction:

$$\Psi_{SM} = \{\psi^{(0)}_1\psi^{(0)}_2\cdots\psi^{(0)}_N\} + \sum_k \sum_{i_1,i_2,\ldots,i_N} c_{Sk}(i_1 i_2\cdots i_N)\{\psi^{(i_1)}_1\psi^{(i_2)}_2\cdots\psi^{(i_N)}_N\}_k \tag{17}$$

The first term in this expansion,

$$\{\psi^{(0)}_1\psi^{(0)}_2\cdots\psi^{(0)}_N\} \equiv \sum_{k=1}^{f^N_S} c_{Sk}(N!)^{1/2}\mathscr{A}\{\psi^{(0)}_1\psi^{(0)}_2\cdots\psi^{(0)}_N\Theta^N_{S,M;k}\} \tag{18}$$

is constructed from the orbitals $\psi^{(0)}_1\cdots\psi^{(0)}_N$ of lowest energy. It is termed the *spin-coupled wavefunction*[20]. We see that in each term of this wavefunction, the individual spins of the electrons are coupled to the required resultant S according to the coupling scheme k. The spin-coupled wavefunction is a linear combination of all the possible modes of coupling as shown in (18). The presence of these different spin couplings in the orbital wavefunction brings into sharp focus much of the physics and chemistry of the interactions between atoms and molecules, as we explain below. Because each orbital $\psi^{(0)}_\mu$ ($\mu = 1, 2, \ldots, N$) is a member of a different set, the N orbitals occurring in the spin-coupled wavefunction are all distinct and non-orthogonal; the double occupancy of orbitals, which is one of the hallmarks of MO theory, does not occur here.

The higher terms in Eq. (17):

$$\{\psi_1^{(i_1)}\psi_2^{(i_2)}\cdots\psi_N^{(i_N)}\}_k \equiv (N!)^{1/2}\mathscr{A}\{\psi_1^{(i_1)}\psi_2^{(i_2)}\cdots\psi_N^{(i_N)}\Theta_{S,M;k}^N\} \tag{19}$$

are obtained from the spin-coupled wavefunction by replacing one, two or more orbitals $\psi_\mu^{(0)}, \psi_\nu^{(0)}, \ldots$, by others $\psi_\mu^{(i_\mu)}, \psi_\nu^{(i_\nu)}, \ldots$, $(i_\mu, i_\nu > 0)$ in the different sets (15).[33]

If one uses only a single orthonormal set of orbitals ψ_i, say, then expansion (17) goes over to the MO-CI expansion for the total wavefunction. The first term of this will now contain doubly occupied orbitals:

$$\{\psi_1^2\psi_2^2\cdots\} \sim \mathscr{A}(\psi_1^2\psi_2^2\cdots\Theta_{S,M;f}^N) \tag{20}$$

Because of this, only one spin function is allowed: that corresponding to a series of pairs of spins coupled to give singlets. In the Kotani scheme of ordering, this is the last function in the set, $k = f_S^N$, as shown. If all the electrons are accommodated in doubly occupied orbitals, function (20) is nothing else but a single Slater determinant.

If on the other hand we identify the sets $\{\psi_\mu^{(i)}\}$ as atomic orbitals drawn from all the participating atoms, expansion (17) becomes the conventional VB expansion.

Intermediate between the spin-coupled expansion in N orthonormal sets and the MO expansion in a single set is an expansion in just two distinct sets. The first term of (17) now becomes the electron-pair wavefunction of Hurley, Lennard-Jones and Pople[34–36], in which pairs of orbitals (ψ_1, ψ_2), $(\psi_3, \psi_4), \ldots$, overlap with each other, but different pairs are (strongly) orthogonal. Much work has been carried out on this model by Goddard and coworkers under the nomenclature 'generalized valence bond (GVB) wavefunction'.[37,38]

Expansion (17) is therefore a generalization of that first carried out by Löwdin[39]. The convergence characteristics of the CI expansion are now well known, at least for ground-state wavefunctions. 'State-of-the-art' CI calculations currently involve 10^5–10^6 terms in such an expansion.

Because of the use of N distinct sets, we expect the spin-coupled VB expansion to converge much faster, and indeed the results so far show that this is the case. As we discuss in more detail below, the spin-coupled function (18) by itself possesses all the correct qualitative characteristics of the ground state of a molecular system, and 200–700 terms of expansion (17) are sufficient to attain chemical accuracy (~ 0.1 eV) for the first 10–15 eigenstates of a given symmetry, and spectroscopic accuracy ($\sim 100\ \text{cm}^{-1}$) for the ground state.[40,41]

We now turn to a more detailed discussion of the spin-coupled wavefunction (18), and for this purpose we drop the superscript (0) in the designation of the orbitals—though these will reappear when the multiconfiguration function (17) (the spin-coupled valence bond wavefunction) is discussed. We begin with a qualitative discussion of the wavefunction, before proceeding to a survey of the technical aspects of the implementation of this model.

The spin-coupled orbitals ψ_μ are expanded in a multicentred basis set,

$$\psi_\mu = \sum_{p=1}^{m} c_{\mu p}\chi_p \tag{21}$$

and the coefficients $c_{\mu p}$ are optimized simultaneously with the spin-coupling coefficients c_{Sk}[33,42]. The orbitals are generally highly localized and have the form of deformed atomic orbitals, the extent of the deformation varying with the internuclear separation. An example is shown in Fig. 2 for $CH(^2\Pi)$[43]. At large internuclear separations, the deformations tend to zero and the orbitals regain atomic form. At the same time the spin-coupling coefficients vary with nuclear geometry. When certain internuclear distances are large, the coefficients assume values characteristic of spin alignments in non-interacting fragments or reactants. As the subsystems approach and the interaction becomes strong (i.e. the exchange interaction between the systems becomes significant), the couplings change over to values associated with a reactive intermediate or with product molecules. In almost every case studied so far, this change occurs over a remarkably narrow range of internuclear distances (1 bohr or less), and is frequently associated with the existence of a potential barrier[44,45]. Otherwise the rapid variation in the spin-coupling coefficients takes place at the top of a potential well, usually at an internuclear separation between the interacting systems of 4–5 bohr. The behaviour of the spin-coupling coefficients is thus a very revealing indicator of the character of the ground-state electronic wavefunction, and shows for example that interacting molecules are essentially unchanged until they approach within about 5 bohr of each other, whereupon major changes occur. More detailed discussions of these aspects as applied to specific systems are given in the following section.

We turn now to a brief discussion of the symmetry properties of spin-coupled orbitals[20]. Because they almost always have the form of deformed atomic functions, the effect of a spatial symmetry operation upon the orbitals is to permute them amongst themselves. Thus each symmetry operation $\mathscr{R}$ corresponds to a certain permutation $P^r_{\mathscr{R}}$. From this it can be seen that the effect of $\mathscr{R}$ upon complete spin-coupled wavefunctions is to transform them amongst themselves as follows:

$$\begin{aligned}\mathscr{R}\mathscr{A}\{\psi_1\cdots\psi_N\Theta^N_{S,M;k}\} &= \mathscr{A}\{P^r_{\mathscr{R}}(\psi_1\cdots\psi_N)\Theta^N_{S,M;k}\}\\ &= \mathscr{A}\{\psi_1\cdots\psi_N P^{\sigma-1}_{\mathscr{R}}\Theta^N_{S,M;k}\}\\ &= \varepsilon_P\sum_l U^{S,N}_{kl}(P_{\mathscr{R}})\mathscr{A}\{\psi_1\cdots\psi_N\Theta^N_{S,M;l}\}\end{aligned} \tag{22}$$

(The use of Eq. (10) should be noted on going from the second to the third

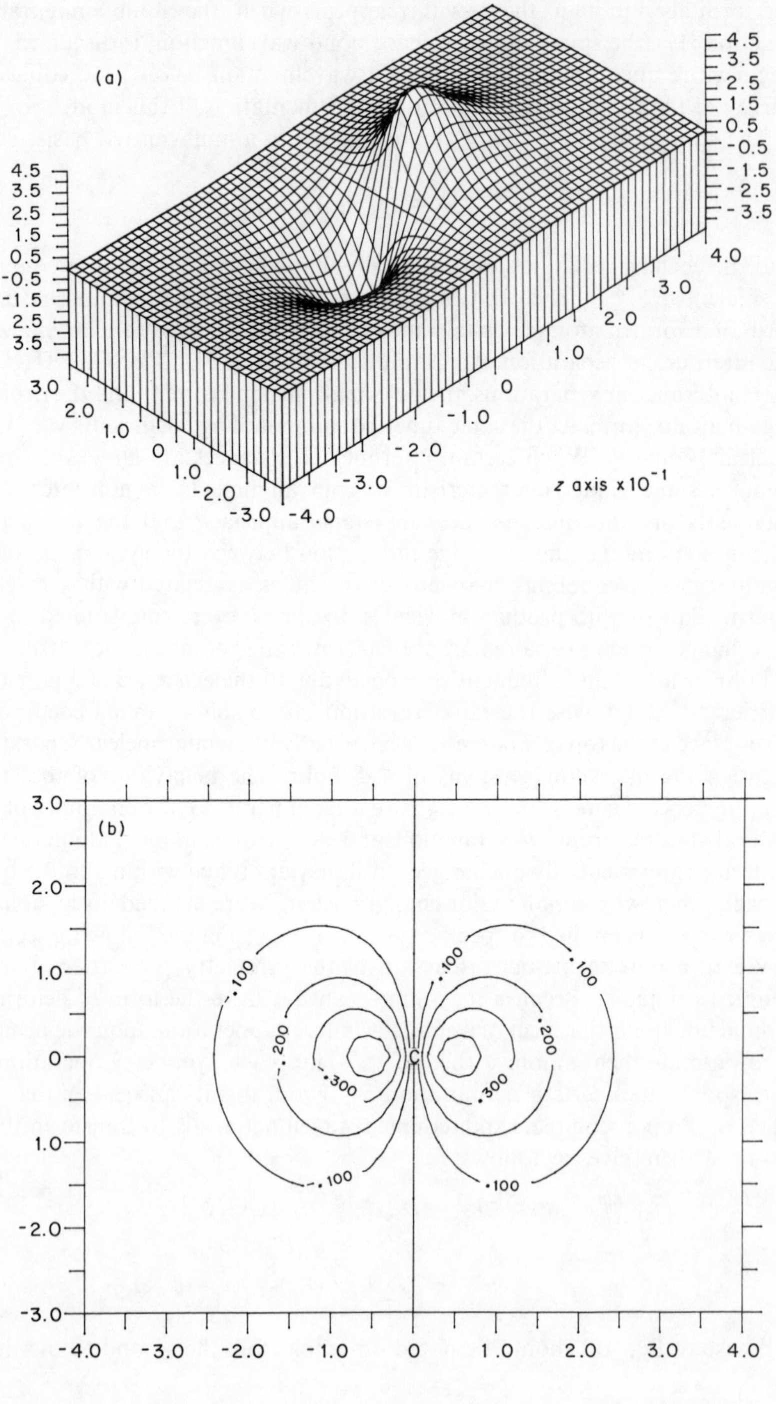
(a)
4.5
3.5
2.5
1.5
0.5
-0.5
-1.5
-2.5
-3.5
z axis x10^-1
-4.0
-3.0
-2.0
-1.0
0
1.0
2.0
3.0
4.0
(b)
.100
.200
.300
-3.0
-2.0
-1.0
0
1.0
2.0
3.0
4.0

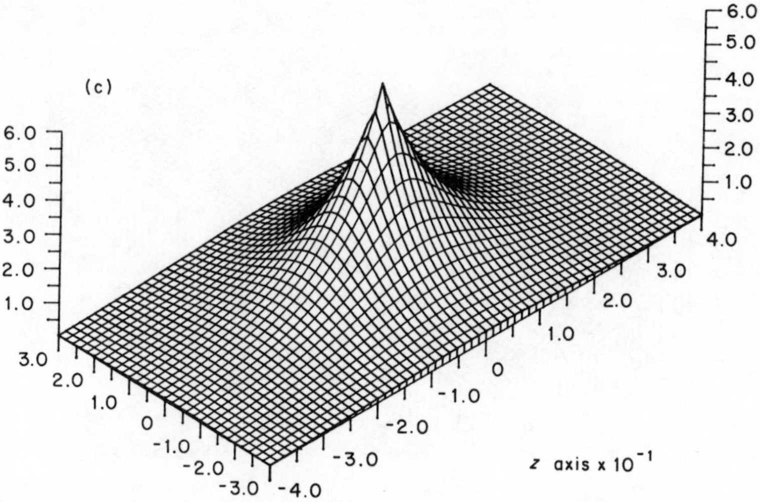
(c)
6.0
5.0
4.0
3.0
2.0
1.0
3.0
2.0
1.0
0
-1.0
-2.0
-3.0
-4.0
-3.0
-2.0
-1.0
0
1.0
2.0
3.0
4.0
z axis x 10^-1

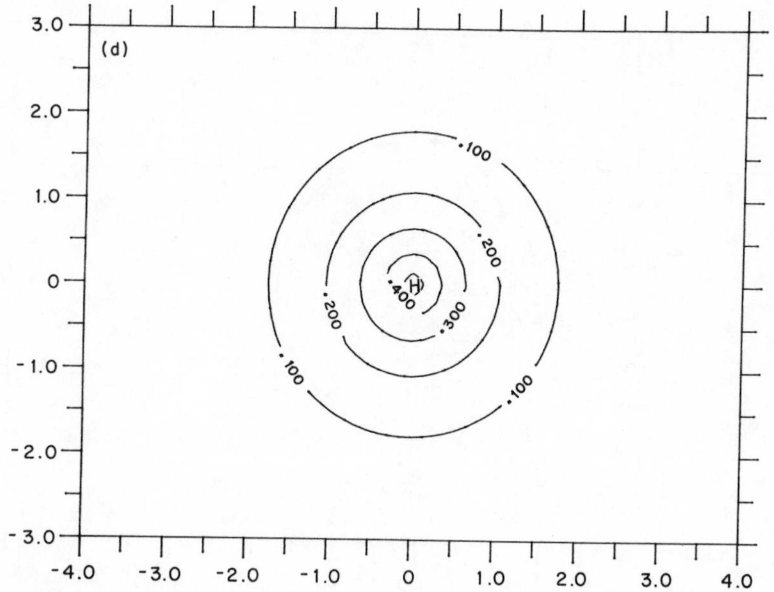
(d)
3.0
2.0
1.0
0
-1.0
-2.0
-3.0
-4.0
-3.0
-2.0
-1.0
0
1.0
2.0
3.0
4.0
.100
.200
.300
.400
H

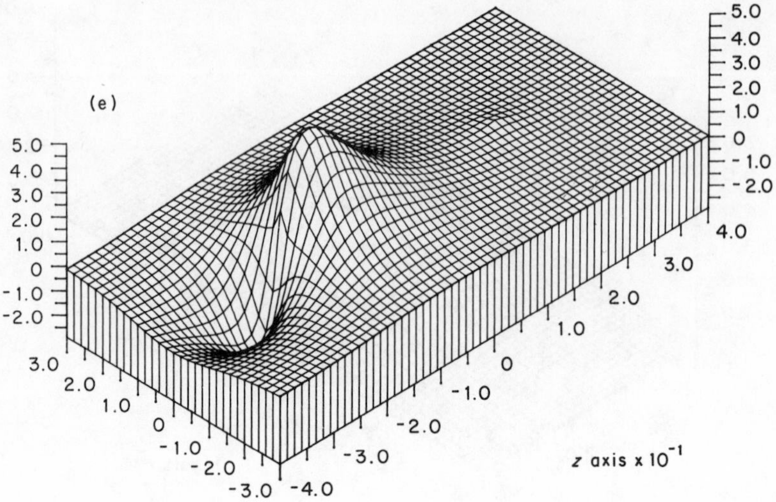
(e)
5.0
4.0
3.0
2.0
1.0
0
-1.0
-2.0
3.0
2.0
1.0
0
-1.0
-2.0
-3.0
-4.0
-3.0
-2.0
-1.0
0
1.0
2.0
3.0
4.0
z axis x 10^{-1}

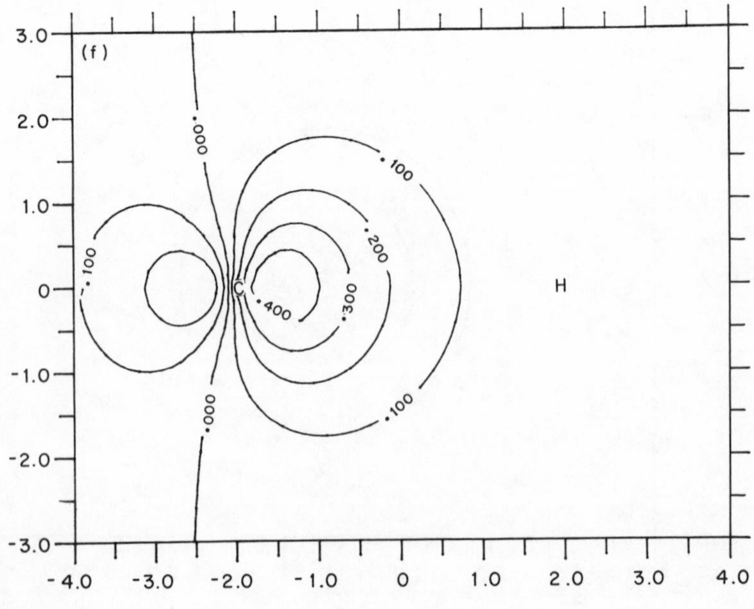
(f)
3.0
2.0
1.0
0
-1.0
-2.0
-3.0
-4.0
-3.0
-2.0
-1.0
0
1.0
2.0
3.0
4.0
.000
.100
.200
.300
.400
C
H

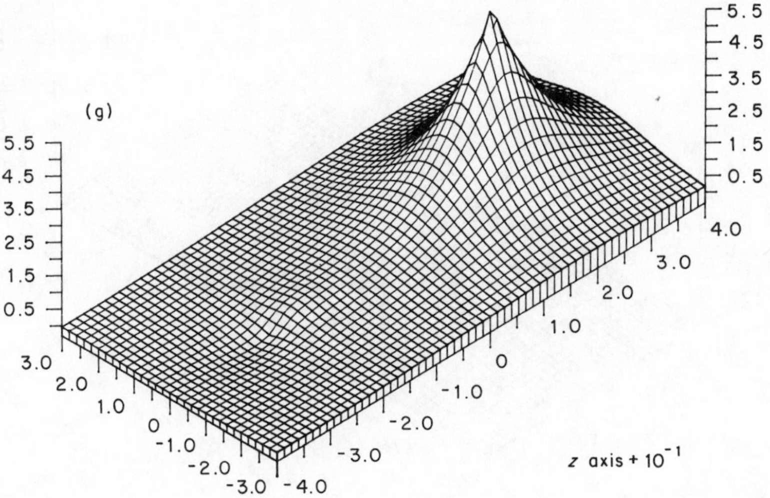
(g)
5.5
4.5
3.5
2.5
1.5
0.5
3.0
2.0
1.0
0
-1.0
-2.0
-3.0
-4.0
-3.0
-2.0
-1.0
0
1.0
2.0
3.0
4.0
z axis + 10^{-1}

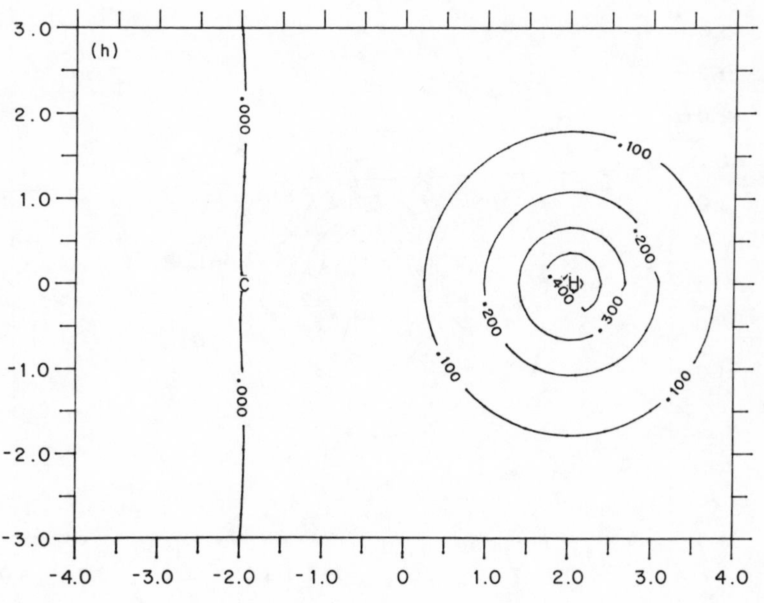
(h)
3.0
2.0
1.0
0
-1.0
-2.0
-3.0
-4.0
-3.0
-2.0
-1.0
0
1.0
2.0
3.0
4.0
.000
C
.100
.200
.300
.400
H

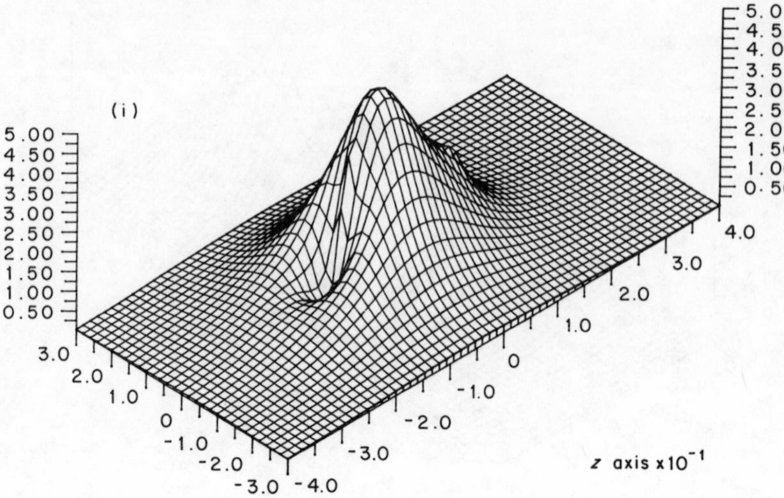
(i)
5.00
4.50
4.00
3.50
3.00
2.50
2.00
1.50
1.00
0.50
3.0
2.0
1.0
0
-1.0
-2.0
-3.0
-4.0
-3.0
-2.0
-1.0
0
1.0
2.0
3.0
4.0
z axis x10⁻¹

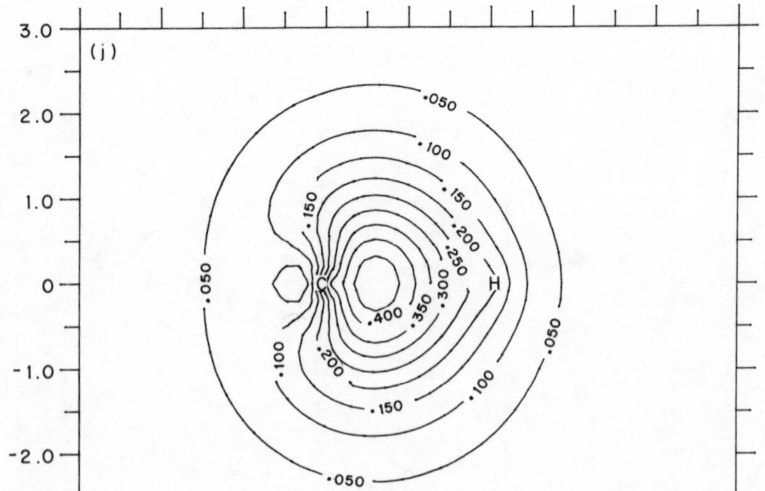
(j)
.050
.100
.150
.200
.250
.300
.350
.400
C
H
3.0
2.0
1.0
0
-1.0
-2.0
-3.0
-4.0
-3.0
-2.0
-1.0
0
1.0
2.0
3.0
4.0

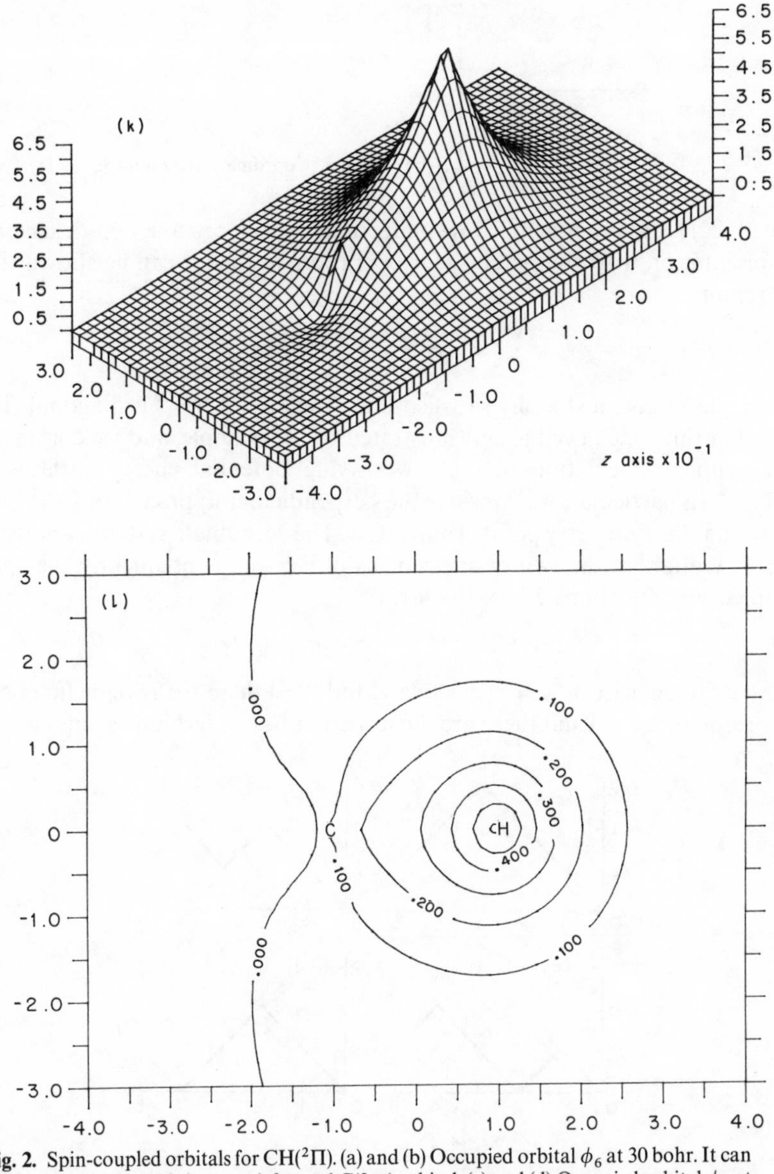

Fig. 2. Spin-coupled orbitals for $CH(^2\Pi)$. (a) and (b) Occupied orbital ϕ_6 at 30 bohr. It can be clearly seen that this is an undeformed $C(2p_z)$ orbital. (c) and (d) Occupied orbital ϕ_5 at 30 bohr. This is the bonding partner with ϕ_6. It is an undeformed H(1s) function. (e) and (f) Occupied orbital ϕ_6 at 4 bohr. Some deformation of the $C(2p_z)$ form due to the presence of the H nucleus can be seen in the perspective diagram. (g) and (h) Occupied orbital ϕ_5 at 4 bohr. Note the small amount of deformation of the 1s(H) that is now present. (i) and (j) Occupied orbital ϕ_6 at 2 bohr. Deformation of $C(2p_z)$ is now considerable, with some delocalization onto H(1s) nucleus. (k) and (l) Occupied orbital ϕ_5 at 2 bohr. The deformation from pure H(1s) character is obvious with some contribution from $C(2p_z)$.

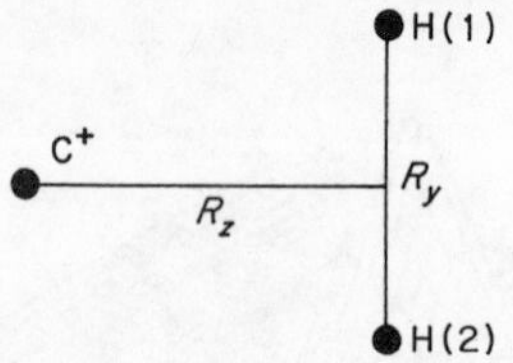

Fig. 3. Coordinate system for $C^+ + H_2$.

line of (22).) If the spin-coupled wavefunction (18) belongs to a non-degenerate representation with character $\zeta_{\mathscr{R}}$ for operation $\mathscr{R}$, then it can be shown that we require

$$\sum_l c_{Sl} U_{lk}^{S,N}(P_{\mathscr{R}}) = (\varepsilon_P \zeta_{\mathscr{R}}) c_{Sk}$$

This equation is most easily satisfied if the matrix $\mathbf{U}^{S,N}(P_{\mathscr{R}})$ is diagonal. The way that this is achieved is best illustrated by an example, and we consider a spin-coupled calculation of the lowest-lying potential energy surfaces of CH_2^+ [45]. In particular, we consider the perpendicular approach of $C^+(^2P)$ to H_2 so that the symmetry point group is C_{2v}. The coordinate system is shown in Fig. 3. We ignore the two core electrons of C^+ and consequently the spin-coupled wavefunctions are of the form

$$\Psi(CH_2^+; {}^2\Gamma) \equiv \{\sigma_1 \sigma_2 \sigma_3 \sigma_4 \psi_5\}$$

For the present case of $N = 5$ (effectively) and $S = \frac{1}{2}$ there are five spin functions according to Eq. (9) and these are illustrated in Fig. 4. Orbitals σ_1 and σ_2 are

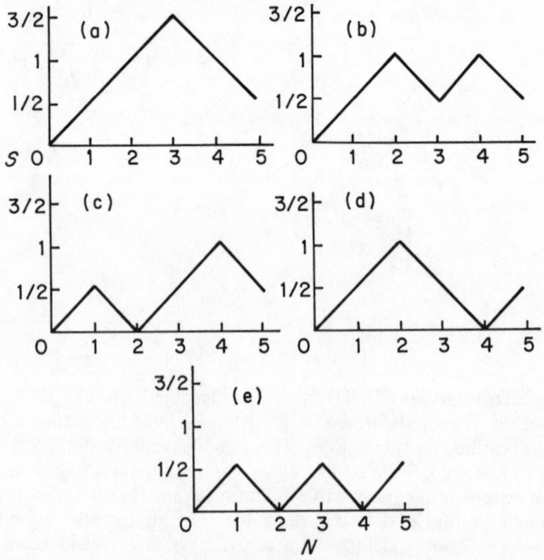

Fig. 4. Branching diagrams for $N = 5, S = \frac{1}{2}$.

two deformed 1s(H) functions, and the essential qualitative form of orbitals σ_3 and σ_4 is:

$$\sigma_3 \simeq \chi(C^+; 2s) + c_1\chi(C^+; 2p_y) + c_2\chi(C^+; 2p_z) + c_3\chi(H(1); 1s)$$
$$\sigma_4 \simeq \chi(C^+; 2s) - c_1\chi(C^+; 2p_y) + c_2\chi(C^+; 2p_z) + c_3\chi(H(2); 1s)$$

At short interatomic separations, orbital σ_3 possesses a maximum amplitude along the C^+–H(1) direction and overlaps strongly with σ_1. Similarly orbital σ_4 possesses a maximum amplitude along the C^+–H(2) direction and overlaps strongly with orbital σ_2. Orbitals σ_3, σ_4 and σ_1, σ_2 are interchanged by the operation of reflection in a plane σ_v passing through the C_2 axis and perpendicular to the molecular plane.

Orbital ψ_5 is very largely (C^+; 2p). We return to a more precise specification shortly.

Physically it is clear that the spin-coupled wavefunction $\Psi(CH_2^+; {}^2\Gamma)$ must remain unchanged when orbitals σ_1, σ_2 and σ_3, σ_4 are interchanged. The U matrix for the permutation $P_{12}P_{34}$ is diagonal and equal to the unit matrix if the pairs $(\sigma_1\sigma_2)$ and $(\sigma_3\sigma_4)$ are each coupled to give net singlet spins, or are each coupled to give net triplet spins. This condition is satisfied by all the spin functions displayed in Fig. 4 except Θ_3 which is therefore symmetry-forbidden*.

The actual representation Γ is then determined by ψ_5. For $\psi_5 \simeq \chi(C^+; 2p_z)$, $\Psi(CH_2^+; {}^2\Gamma)$ is ${}^2A_1$; for $\chi(C^+; 2p_y)$, ${}^2B_2$; and for $\chi(C^+; 2p_x)$, ${}^2B_1$.

This mode of coupling focuses attention on the behavior of the wavefunction as R_z becomes large, since the orbitals σ_1, σ_2 asymptotically describe the H_2 molecule. Singlet–singlet coupling of the two pairs gives rise asymptotically to $H_2(X^1\Sigma_g^+)$ and $C^+(^2P)$, whereas triplet–triplet coupling leads to H_2 in the repulsive ${}^3\Sigma_u^+$ state and $C^+(sp^2; {}^4P)$. Thus in this representation, the spin-coupling coefficients show directly the intervention of the ${}^3\Sigma_u^+$ state of H_2 and of $C^+(^4P)$ as the interaction between H_2 and C^+ increases.

It should be emphasized that, provided all the symmetry-allowed spin functions are included in the wavefunction, the actual ordering of the orbitals is immaterial and may be chosen to highlight different aspects of the same wavefunction. Thus in this present example we may equally well consider the orbital pairs (σ_1, σ_3) and (σ_2, σ_4) as singlet–singlet or triplet–triplet coupled. This is in a sense the natural representation to adopt for short R_z since it concentrates attention on the formation of C–H bonds. Some results for this system may be found in Ref. 46.

The occurrence of sets of degenerate orbitals requires a special type of spin coupling. We consider as an example the configuration $(\pi^4\ {}^1\Sigma^+)$ for a diatomic or linear system[42]. In order for the total wavefunction to remain invariant

*The Serber basis of spin functions is particularly useful here.

under any rotation about the axis of symmetry, this part of the spin-coupled wavefunction must be constructed as

$$(4!)^{1/2}\mathscr{A}(\pi_x\pi_y\pi'_x\pi'_y\Theta^4_{0,0,1}) \tag{23a}$$

assuming real orbitals π_x, π_y, π'_x, π'_y. Orbitals π_x and π_y are of course orthogonal by symmetry, as are π'_x and π'_y. However, π_x, π'_x and π_y, π'_y overlap. The spin function, number 1 in the Kotani numbering scheme, describes orbitals π_x, π_y as coupled to the maximum spin, $S = 1$, and similarly for π'_x and π'_y, with the two subsystems coupled to give a net spin of 0. This scheme may be represented as

$$((\pi_x\pi_y\,{}^3\Sigma^-)(\pi'_x\pi'_y\,{}^3\Sigma^-)\,{}^1\Sigma^+) \tag{23b}$$

Spin function $\Theta_{0,0,1}$ is referred to as the 'Hund's rule coupling'. Because of its construction, wavefunction (23) is unchanged by any linear transformation of the pairs of orbitals (π_x, π_y) amongst themselves, and we are free to use complex orbitals (π^+, π^-) if desired. The same is true of the pair (π'_x, π'_y).

It is worth noting that the degenerate pair π_x, π_y, when coupled to give a triplet spin, yields a function of symmetry ${}^3\Sigma^-$, while if coupled to a singlet spin, the resulting wavefunction is one component (xy) of a ${}^1\Delta$ state.

Thus it can be seen that there is a wide range of choice of bases for spin functions for use in spin-coupled wavefunctions, and this confers on the theory a great flexibility for describing in a compact and highly visual manner the various physical situations that occur in atoms and molecules. A very general spin-coupled program has been written which incorporates these features and is able to handle complex configurations of σ and π electrons in any basis of spin functions[47].

We now consider some of the technical aspects of the spin-coupled theory[23,42]. The total electronic energy given by the spin-coupled wavefunction (18) may be written as

$$E = \frac{1}{\Delta}\left(\sum_{\mu,\nu=1}^{N} D(\mu|\nu)\langle\psi_\mu|h|\psi_\nu\rangle + \tfrac{1}{2}\sum_{\mu,\nu,\sigma,\tau=1}^{N} D(\mu\nu|\sigma\tau)\langle\psi_\mu\psi_\nu|g|\psi_\sigma\varphi_\tau\rangle\right) \tag{24}$$

In this equation, $\langle\psi_\mu|h|\psi_\nu\rangle$ and $\langle\psi_\mu\psi_\nu|g|\psi_\sigma\varphi_\tau\rangle$ denote the usual one- and two-electron integrals over the indicated orbitals, $D(\mu|\nu)$ and $D(\mu\nu|\sigma\tau)$ are respectively elements of the one- and two-electron density matrices, and Δ is the normalization integral. In the notation for the density matrices it is understood that orbital indices μ, ν,... which occur together on the *same* side of a vertical bar may never be equal to one another. The various orders of density matrices are connected with one another, and this provides a highly efficient recurrence scheme for their computation. Thus,

$$\Delta = \sum_{\nu=1}^{N} D(\mu|\nu)\langle\psi_\mu|\psi_\nu\rangle \qquad (\mu = 1, 2, \ldots, \text{ or } N)$$

$$D(\mu|\nu) = \sum_{\tau=1}^{N} D(\mu\sigma|\nu\tau)\langle\psi_\sigma|\psi_\tau\rangle \qquad (\sigma = 1, 2, \ldots, \text{ or } N; \sigma \neq \mu) \quad (25)$$

$$D(\mu_1\mu_2|\nu_1\nu_2) = \sum_{\nu_3=1}^{N} D(\mu_1\mu_2\mu_3|\nu_1\nu_2\nu_3)\langle\psi_{\mu_3}|\psi_{\nu_3}\rangle \qquad (\mu_3 \neq \mu_1, \mu_2)$$

This procedure is particularly convenient since in order to optimize the energy, density matrices of order 1, 2, 3 and 4 are needed, and these are all produced simultaneously. Note that the normalization integral involves all N electrons, $D(\mu|\nu)$ involves $(N-1)$ electrons, $D(\mu\nu|\sigma\tau)$ involves $(N-2)$ electrons, etc. The final relationship in the series is

$$D(\mu_1\mu_2\cdots\mu_{N-1}|\nu_1\nu_2\cdots\nu_{N-1}) = \sum_{\nu_N} D(\mu_1\mu_2\cdots\mu_{N-1}\mu_N|\nu_1\cdots\nu_{N-1}\nu_N)\langle\mu_N|\nu_N\rangle$$

which connects the $(N-1)$- and N-electron density matrices. This last involves no electrons at all and is given by

$$D(\mu_1\cdots\mu_N|\nu_1\cdots\nu_N) = \sum_{k,l=1}^{f_S^N} c_{Sk}c_{Sl}\{U_{kl}^{S,N}(P) \oplus U_{lk}^{S,N}(P)\} \qquad (26)$$

In this equation, P is the permutation corresponding to

$$P \equiv \begin{pmatrix} \mu_1\mu_2\cdots\mu_N \\ \nu_1\nu_2\cdots\nu_N \end{pmatrix}$$

and the symbol $\oplus$ signifies the symmetrization of the U matrix: add only if $l \neq k$. A general program has been written which computes the U matrices in a standard sequence of permutations P, and is available through the CPC Program Library, Queen's University of Belfast[48].

In actual practice, the four-electron density matrices are computed directly from (26) by means of

$$D(\mu_1\cdots\mu_4|\nu_1\cdots\nu_4) = \sum_{\nu_5,\ldots,\nu_N} D(\mu_1\cdots\mu_N|\nu_1\cdots\nu_N)\langle\mu_5|\nu_5\rangle\cdots\langle\mu_N|\nu_N\rangle$$

and from these all the other required density matrices are determined by working up through Eqs (25).

The energy is optimized by means of the 'stabilized Newton–Raphson' procedure[33,42]. Let $\delta\mathbf{c}$ denote the vector of corrections to all the orbital and all the spin-coupling coefficients. It is determined by solving the equations

$$(\mathbf{G} + \alpha\mathbf{I})\delta\mathbf{c} = -\mathbf{g} \qquad (27)$$

In this equation $\mathbf{g}$ is the vector of gradients of the energy with respect to orbital and spin-coupling parameters:

$$\mathbf{g} = \begin{pmatrix} |g_\mu\rangle \equiv \partial E/\partial\psi_\mu^* \\ |g_{c_S}\rangle \equiv \partial E/\partial\mathbf{c}_S \end{pmatrix} \qquad (28)$$

and $\mathbf{G}$ is the matrix of second derivatives. This can be written in blocked form as follows

$$\mathbf{G} = \left[\begin{array}{c|c} \mathbf{G}(\mu, \lambda) & \mathbf{G}(\mu, c_S) \\ \hline \mathbf{G}(\mu, c_S)^+ & \mathbf{G}(c_S, c_S) \end{array}\right]$$

where $G(\mu, \lambda)$ represents the matrix of second derivatives of E with respect to the coefficients of orbitals ψ_μ and ψ_λ, $G(\mu, c_S)$ is the cross-product of derivatives with respect to orbital ψ_μ and the spin-coupling coefficients, and $G(c_S, c_S)$ is the matrix of second derivatives of the energy with respect to the spin-coupling coefficients. Explicit expressions for these and for the gradients are given in Refs 33 and 42.

The parameter α in Eq. (27) is given by

$$\alpha = -e_0 + R\langle g|g\rangle^{1/2}$$

where e_0 is the lowest eigenvalue of $\mathbf{G}$ (assumed negative), and the scale parameter R which multiplies the length of gradient vector $\langle g|g\rangle^{1/2}$ is given the initial value of 0.5. This value for α ensures that $(\mathbf{G} + \alpha\mathbf{I})$ is positive definite, so that the corrections $\delta\mathbf{c}$ are always reasonably small. As the iteration proceeds and e_0 becomes positive, α is put to zero since convergence of the pure Newton–Raphson method is then guaranteed.

Our experience using this method, even with the very large basis sets which are our standard usage for diatomic systems, has always been excellent. The tolerance on the elements of $\delta\mathbf{c}$ is normally set to 10^{-8} so that E is correct to about 10^{-16}. Convergence is achieved in 10–15 iterations at an initial point on a potential surface, and in 5–6 iterations at subsequent points. It is worth mentioning that simpler minimization methods using gradients only are at least an order of magnitude slower. In addition we have found methods using orbital equations similar to Hartree–Fock theory to be entirely unsuccessful.

When convergence is reached, all gradients are zero. The orbitals then satisfy equations which may be written as

$$\sum_{\nu=1}^{N} \hat{F}_{\mu\nu}\psi_\nu = \varepsilon_\mu\psi_\mu \qquad (\mu, 1, 2, \ldots, N) \tag{29}$$

The $\hat{F}_{\mu\nu}$ are a set of one-electron operators which depend upon one-, two- and three-electron density matrices. Explicit expressions for them are given in Refs 33 and 42. Eqs (29) may be rewritten in the form:

$$\hat{F}_\mu^{(\text{eff})}\psi_\mu = \varepsilon_\mu\psi_\mu \qquad (\mu = 1, 2, \ldots, N) \tag{30}$$

where

$$\hat{F}_\mu^{(\text{eff})} = \hat{F}_{\mu\mu} - \sum_{\substack{\lambda,\nu=1 \\ (\neq\mu)}}^{N} \hat{F}_{\mu\nu}(\hat{F}^{(\mu)} - \varepsilon^{(\mu)})_{\nu\lambda}^{-1} F_{\lambda\mu} \tag{31}$$

In this equation $\hat{F}^{(\mu)}$ stands for the supermatrix of $\hat{F}_{\mu\nu}$ operators with row and column μ missing, and similarly $\varepsilon^{(\mu)}$ represents a diagonal matrix of all the orbital energies except ε_μ. All the operators $\hat{F}_{\mu\nu}$ are constructed from $(N-1)$ orbitals, and consequently the effective operator $\hat{F}_\mu^{(\text{eff})}$ may be viewed as determining orbital ψ_μ in the field of the $(N-1)$ other electrons. Since $\hat{F}_\mu^{(\text{eff})}$ is in general Hermitian, solution of (30) gives rise to a whole set of orthonormal functions

$$\psi_\mu^{(0)}, \psi_\mu^{(1)}, \ldots, \equiv \{\psi_\mu^{(i)}\}$$

One of these solutions, $\psi_\mu^{(0)}$ say, corresponds to the occupied orbital already found, and the higher solutions $\psi_\mu^{(i)}$ to states which arise when the electron in orbital $\psi_\mu^{(0)}$ is excited.

There are clearly N different effective operators and each one gives rise to an orthonormal set of functions. We therefore generate N distinct sets of orbitals with properties as described by Eq. (16). Because of the construction of the $\hat{F}_\mu^{(\text{eff})}$ operators, each set indeed appears to be optimal for the expansion of that particular electron coordinate. Excited structures are obtained by replacing one, two,... or more occupied orbitals with higher solutions from their respective stacks, as described above.

The final spin-coupled valence bond wavefunction is of the form (17), but where the number of structures is of course now finite. The coefficients $c_{Sk}(i_1, i_2, \ldots, i_N)$ are determined by constructing the matrix of the Hamiltonian (2) over the chosen set of structures and diagonalizing.

This expansion of the total electronic wavefunction is very compact, and provides a great deal of physical and chemical visuality. The spin-coupled structure (18) by itself reproduces with very reasonable accuracy all the features of a ground-state molecular potential energy surface. For example the spin-coupled function typically yields 85% of the observed binding energy, and equilibrium internuclear separations are accurate to 0.01 Å. This function consequently dominates expansion (17) for all nuclear geometries. The various excited structures provide angular and other types of correlation as an extra quantitative refinement but do not alter the qualitative picture.

The same is true of excited states. A reasonably large set of virtual orbitals $\psi_\mu^{(i)}$ (six or seven from each stack, say) and 200–700 structures is sufficient to reproduce the first 10 states of a given symmetry to an accuracy of about 0.01 eV over an energy range of 40 eV. Reference spin-coupled configurations for the excited states are obtained by single replacements of occupied orbitals $\psi_\mu^{(0)}$ by appropriate virtual orbitals $\psi_\mu^{(i)}$. (Double replacements usually produce states whose energies are so high that they lie in the continuum of the positive ion, and strictly speaking are subject to autoionization.) These singly excited structures are refined by further one-, two-,..., etc., excitations from them, giving rise to single, double, triple and quadruple excitations from the ground-state function. Nevertheless, just as for the ground state, the excited states are dominated by very few structures, and these provide the essential physical interpretation.

The effect of the non-orthogonality of the orbitals in the spin-coupled wavefunctions occurs in the computation of the density matrices. If N_σ is the number of occupied orbitals of σ symmetry and N_π of π symmetry, present programs are capable of treating problems with $N_\sigma = 8$ and $N_\pi = 8$ for each (x and y) component. Doubly occupied orbitals in closed-shell core configurations of the form

$$\{\psi_1^2 \psi_2^2 \cdots \psi_{N_c}^2\} \tag{32}$$

play no role in this respect and any number of them may be included in addition to the $N_\sigma + N_\pi$ spin-coupled orbitals. These limits allow the theory to be applied to a wide range of atomic and molecular systems—as will be seen in the following section.

The current restrictions on N_σ and N_π arise essentially because of the length of the file of four-electron density matrices, $D^{(4)}(\mu_1 \cdots \mu_4 | \nu_1 \cdots \nu_4)$. The recurrence procedure itself requires a negligible amount of computing time. The $D^{(4)}$'s are in fact only required in one routine in the computation of the $G(\mu, \lambda)$ block of the second derivative matrix. Versions of the programs are under development in which the $D^{(4)}$ elements are computed *in situ*, obviating the need for such files—and so removing the present limitations on N_σ and N_π. In general it should be emphasized that the newer machines with an available 5–10 Mbyte of fast memory and extremely fast disk I/O (even without vector or parallel processing capability)—combined of course with highly efficient algorithms as outlined above—have reduced the 'non-orthogonality problem' to a very considerable extent.

The inclusion of closed-shell configurations in spin-coupled calculations is most easily accomplished by carrying out a self-consistent field (SCF) (or small multiconfiguration SCF (MCSCF)) step first. The integrals over the basis functions are transformed to the molecular-orbital representation and input to the spin-coupled program. The use of MOs as basis functions also shows directly which of them are significant in the spin-coupled wavefunction.

The valence bond or configuration-interaction stage of the computation is carried out by means of a flexible, cofactor-driven program which is based upon the Löwdin formulation of matrix elements of the Hamiltonian between Slater determinants composed of non-orthogonal orbitals. The computational work here is proportional to N^4, and the program is not subject to the current spin-coupled restrictions on N_σ and N_π. It is described in more detail in Section V.

The spin-coupled method was first applied to atoms by Kaldor[49]. He obtained remarkably accurate results for the spin densities at the nuclei, particularly in the case of Li(^{2}S) and Li(^{2}P). This is due directly to the involvement of the different spin-couplings. Spin functions in which two core orbitals, 1s and 1s′, are coupled to a triplet give a negligible contribution to the total energy, but are essential for obtaining accurate values of the Fermi contact term. The orbitals were optimized by an extension of Brillouin's theorem, which is equivalent to utilization of first derivatives of the energy with respect to the variational parameters.

Early work on molecules was carried out by Goddard and his coworkers[50], who used an orbital equation method to optimize the orbitals. The spin-coupling coefficients were separately optimized in pairs by a non-linear search procedure. This approach proved to be very inefficient and has since been abandoned.

IV. SURVEY OF RESULTS OF SPIN-COUPLED VALENCE BOND THEORY

The purpose of this section is to discuss in some detail several results of recent applications of the spin-coupled VB method. We have chosen a few representative examples in order to illustrate different aspects of the theory: (i) CH^+ ion; (ii) potential energy surfaces of the $(B–H_2)^+$ system; (iii) potential energy curves of the CH^{3+} system; (iv) the benzene molecule; and (v) studies of momentum-space properties. The CH^+ ion is the first full-scale application of the method to a system for which very high-quality MCSCF-CI wavefunctions exist[51]. The $(B–H_2)^+$ problem represents an application to a reactive system where there are many different modes of dissociation in many different electronic states[52]. The CH^{3+} system is characterized by many highly contrasting excited states stretching over a range of more than 40 eV[53]. The benzene molecule illustrates the new insights afforded by the spin-coupled VB description[54]. Finally we consider briefly the important advantages of spin-coupled wavefunctions for studies of momentum-space properties.

A. $^1\Sigma^+$ States of CH^+

The CH^+ ion is of considerable importance in interstellar chemisty, and has also been studied by MCSCF and CI methods[51]. It is therefore well suited as a full-scale demonstration of the spin-coupled VB procedure described above. The basis set used was of modest size ($18\sigma, 20\pi, 6\delta$ Slater orbitals), and is the same as that used by Green[55] except for omission of 4f functions. However, no diffuse 3s(C) or 3p(C) functions, which would be needed to describe any Rydberg character in excited states, were included.

The spin-coupled wavefunction is of the form

$$\{\sigma_1\sigma_2\sigma_3\sigma_4\sigma_5\} \tag{33}$$

Orbitals σ_1 and σ_2 are core orbitals and can be characterized simply as $1s(C^+)$ and $1s'(C^+)$. They change very little as R varies. Orbitals σ_3 and σ_4 are lone-pair orbitals. At large internuclear distances they are of the form $2s(C^+)$ and $2s'(C^+)$, but as the atoms approach they undergo considerable deformation. The same is true of orbital σ_5, the first member of the bonding pair σ_5, σ_6. At large values of R, σ_5 has the form of $2p_z(C^+)$ but changes greatly as R decreases with significant amounts of $2s(C^+)$ character. Orbital σ_6 is almost completely a 1s function on the H atom, but shows delocalization onto C^+ at shorter internuclear separations. Wavefunction (33) thus provides a qualitatively correct description of the whole ground-state potential curve of CH^+, yielding 85% of the observed binding energy as well as the correct dissociation products:

$$C^+(1s1s'\,2s2s'\,2p_z\,{}^2P) + H(1s)$$

Orbitals σ_1 and σ_2 were coupled to give a singlet spin throughout. Consequently there remain just two possible spin functions which give a net spin $S = 0$. If we denote these simply as Θ_1 and Θ_2, then Θ_1 corresponds to the coupling

$$((\sigma_3, \sigma_4)^3\Sigma^+, (\sigma_5, \sigma_6)^3\Sigma^+)^1\Sigma^+$$

and Θ_2 to

$$((\sigma_3, \sigma_4)^1\Sigma^+, (\sigma_5, \sigma_6)^1\Sigma^+)^1\Sigma^+$$

Although coupling Θ_2 is dominant over all R, it actually becomes *less* so at large separations, and there is considerable ^3S character in the (2s, 2s′) pair at

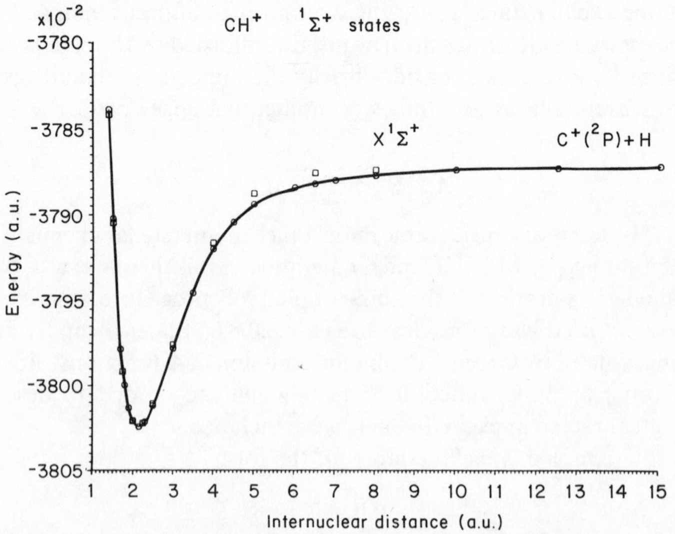

Fig. 5. Comparison of spin-coupled VB (○; 500 structures[51]) and MCSCF-CI calculations (□; Green *et al.*[55]) for the ground state of CH^+.

TABLE I
Spectroscopic properties of $CH^+(X^1\Sigma^+)$.

Wavefunction	E_{min}(a.u.)	R_e(Å)	ω_e(cm^{-1})	$\omega_e x_e$(cm^{-1})	D_e(eV)
Spin-coupled[51]	−37.95056	1.141	2761	60	3.63
Spin-coupled VB[51]	−38.02426	1.135	2845	69	4.14
MCSCF-CI[56]	−38.02233	1.128	2860	59	4.14
MCSCF-CI (incl. core correlation)[55]	−38.06064	1.130	2847	63	4.11
Experiment[57]		1.131	2858	59.3	4.26

$R = 15a_0\,(c_{01} \simeq 0.360, c_{02} \simeq 0.933)$. This provides additional correlation in the lone pair which diminishes at shorter R.

Spin-coupled VB calculations were carried out using a total of 26 orbitals: six occupied, six σ virtual and 14 π virtual orbitals. The final wavefunctions consisted of 500 structures of the type described in Eq. (17) formed from 286 distinct spatial configurations of Σ^+ symmetry. These consist of the spin-coupled reference function and $(1+2+3+4)$-fold excitations. No excitations from the (σ_1, σ_2) core were included. About half of these structures (single plus double replacements) contribute to the ground state, the remainder improves the description of the excited states.

In Fig. 5 the potential energy curve for the $X^1\Sigma^+$ state is compared to that obtained by Green *et al.*[55] using an MCSCF-CI method. This wavefunction includes core correlation and gives the lowest energy of any calculation on CH^+ in the literature. In Fig. 5 this potential curve has been shifted upwards in energy so as to coincide with the spin-coupled VB result at $R = 2.0$ bohr. The two potential curves are seen to be practically indistinguishable over a wide range of R. Computed spectroscopic constants for the spin-coupled VB function are shown in Table I where they are compared with other calculations and with experiment.

B. Potential Surfaces of $(B\text{–}H_2)^+$

The reaction of B^+ with H_2 gives rise to a variety of products which have been studied in ion beam experiments by several groups[58–61], for collision energies up to about 13 eV. These are summarized as follows:

$$B^+(2s^2\,{}^1S, 2s2p\,{}^3P) + H_2 \begin{cases} \rightarrow BH^+(X^2\Sigma^+, A^2\Pi, B^2\Sigma^+) + H \\ \rightarrow B(2s^22p\,{}^2P, 2s2p^2\,{}^2D, 2s^23s\,{}^2S) + H_2^+ \end{cases}$$

It is worth noting in this the occurrence of the previously unknown $B^2\Sigma^+$ state of BH^+ which is bound. It was originally identified on the basis of spin-

TABLE II
Calculated and experimental asymptotic energies in $(B\text{–}H_2)^+$ (eV).

Process	Present work[52]	MCSCF or CASSCF	Experiment[65]
$B^+(2s^2\,{}^1S) \rightarrow B^+(2s2p\,{}^3P)$	4.71	4.68[63]	4.66
$B(2s^22p\,{}^2P) \rightarrow B(2s^23s\,{}^2S)$	4.98	4.86[64]	4.96
$B^+(2s^2\,{}^1S) + H \rightarrow B(2s^22p\,{}^2P) + H^+$	5.60		5.30

coupled VB calculations on the BeH molecule[33]*. This state (together with H(1s)) is almost the exclusive reaction product of $B^+(2s^2\,{}^1S) + H_2$ at collision energies above 8 eV.

A good understanding of the mechanisms which give rise to these processes requires first of all a reliable picture of the potential surfaces which lie within 13 eV of the ground state, and this is the purpose of the spin-coupled VB calculations. These must include the states $BH(X^1\Sigma^+, A^1\Pi, B^1\Sigma^+) + H^+$, for although these products are not observed in reactions with H_2, they lie below the $B + H_2^+$ channels, and are in fact the preferred product in reactions of B^+ with hydrocarbons[62]. Very large MCSCF and complete active space SCF (CASSCF) computations on several states of BH^+ [63] and BH[64] are available for comparison in the asymptotic regions of the BH_2^+ surfaces—and it is therefore sensible to confine our attention initially to these regions.

The calculations were carried out using a fairly large Gaussian basis set, B(11s6p2d/9s5p2d) and H(6s2p1d/4s2p1d), though this is a little smaller than those used for the MO-based calculations on isolated BH and BH^+. However, it is sufficiently saturated to describe the different chemical situations which occur in the $(B\text{–}H\text{–}H)^+$ system, as can be seen in Table II. Only an extended Slater basis, such as that used in spin-coupled VB calculations on diatomic molecules, is capable of reducing the discrepancy of 0.3 eV between theory and experiment in the description of the charge-transfer process:

$$B^+(2s^2;\,{}^1S) + H \rightarrow B(2s^22p;\,{}^2P) + H^+$$

(see next section).

The spin-coupled VB calculations comprised one-, two- and three-fold excitations from the original spin-coupled structure, and give rise to 400 spatial configurations or 592 spin-coupled structures. In Fig. 6 are shown sections of the collinear path $B \cdots\cdots H(1) \cdots\cdots H(2)$ where H(2) remains fixed at 30 bohr from B, and the $B \cdots\cdots H(1)$ distance R_{BH} is varied from 1.6 to 15 bohr. The lowest state corresponds to $BH^+(X^2\Sigma^+) + H$, followed by $BH^+(B^2\Sigma^+)$

*Although this state was predicted long ago by Herzberg[6] to be bound, the only other calculation available at the time of the experiment showed this was apparently not the case.

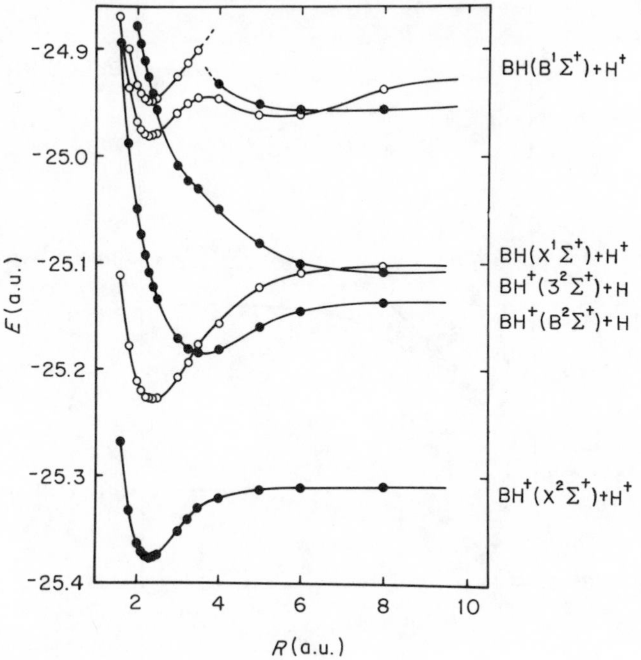

Fig. 6. Potential energy curves of $BH^+ + H$ (●) and of $BH + H^+$ (○).

$+ H$. These dissociate at large values of R_{BH} to give $B^+(2s^2\,{}^1S) + H + H$ and $B^+(2s2p\,{}^3P) + H + H$ respectively. The $BH^+(B^2\Sigma^+) + H$ state is intersected by the state corresponding to $BH(X^1\Sigma^+) + H^+$. It should be emphasized for clearity of presentation and analysis that the potential curves in Fig. 6 have been drawn adiabatically for the $BH^+ + H$ states and also for the $BH + H^+$ states, but *diabatically* for the interaction between $BH^+ + H$ and $BH + H^+$. In reality there are many avoided crossings between the two sets of states.

It is worth noting that the bound $BH(X^1\Sigma^+) + H^+$ and repulsive $BH^+(3^2\Sigma^+) + H$ states become degenerate at large values of R_{BH}. This is because both give $B(2s^2 2p\,{}^2P) + H + H^+$ as dissociation products. Otherwise it is obvious that these two states differ widely from one another. The chosen set of spin-coupled structures provides a correct description of both.

The state corresponding to $BH(B^1\Sigma^+) + H^+$ possesses a double minimum. It dissociates to give $B(2s^2 3s; {}^2S) + H + H^+$. At short values of R_{BH}, the wavefunction contains much $B^+ \cdots\cdots H^- \cdots\cdots H^+$ character. Consequently the inner well of this state may be regarded as arising from an intersection with a higher-lying ionic state.

In Fig. 7, the two lowest $BH^+ + H$ and $BH + H^+$ states are compared directly with MCSCF and CASSCF calculations on the isolated BH^+ and BH

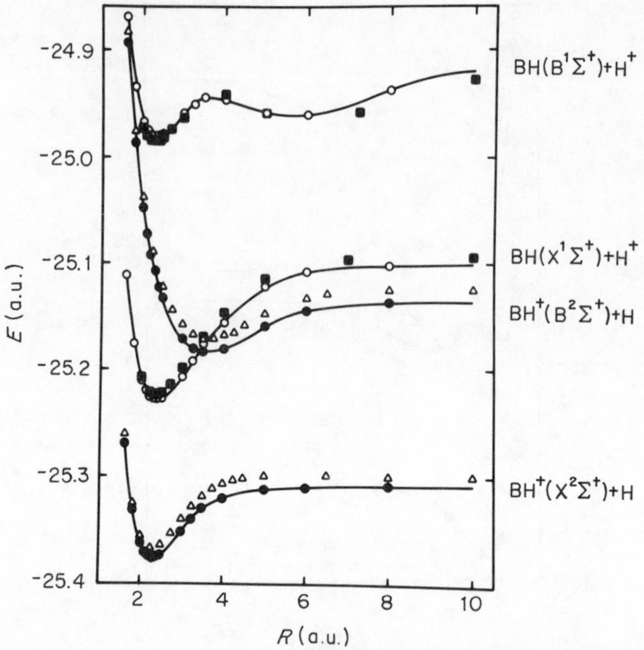

Fig. 7. Comparison between present results (●, ○, as in Fig. 6) and previous work (Δ, Ref. 63; ■, Ref. 64). See footnote to text.

species*. The MCSCF[63] calculation includes 60 configurations, and the CASSCF[64] work includes a total of 645 configurations. It can be seen that the present calculations compare well with these isolated-molecule computations. The total energies given by the spin-coupled VB method are significantly lower even though, at these nuclear geometries, less than 100 spin-coupled structures play any significant role. In particular it should be noted how well the $BH(B^1\Sigma^+) + H^+$ state is represented here. Because of intervening $BH^+ + H$ states in the VB calculation, this state occurs as the sixth or sometimes the *seventh* root.

In Fig. 8a the spin-coupled and final spin-coupled VB potentials are compared for the ground state. Fig. 8b shows the associated spin-coupling coefficients. It can be seen that the spin-coupled calculation by itself yields a potential energy curve with all the essential features of the final result. The spin-coupling coefficients display much of the essential chemistry of bond formation in BH^+. At large R_{BH} distances, coefficients are $c_2 \simeq 1$ and $c_1 \simeq 0$, which shows that the two 2s-like orbitals stemming from B^+ are coupled to a

*An extra 0.5 a.u. has been added to the MCSCF energies for the BH^+ states. This enables all results to be plotted on the same scale.

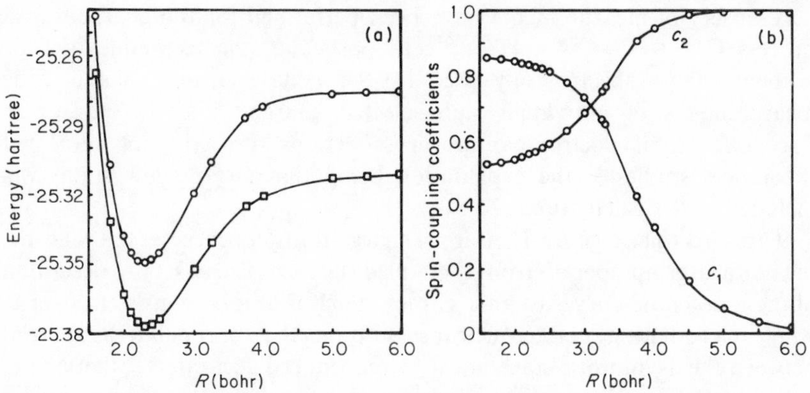

Fig. 8. (a) Spin-coupled and final spin-coupled VB potential energy curves for BH^+ ($X^2\Sigma^+$) + H. (b) Spin-coupling coefficients for the potential curves of (a).

singlet. Between 4.0 and 5.0 bohr, there is an abrupt change in the spin coupling as one of the B^+(2s) orbitals forms a singlet pair with H(1s). The theoretical limit for the coefficients for pure spin coupling of this type is $\sqrt{3}/2$ and 1/2, and in Fig. 8b it can be seen that this is approached at small R_{BH}. In other words, the BH^+ molecule at equilibrium in the ground state possesses much character essentially derived from $B^+(2s2p\,^3P)$, and conversely the $B^2\Sigma^+$ state much $B^+(2s^2\,^1S)$ character. In this light, the almost exclusive production of $BH^+(B^2\Sigma^+) + H$ from $B^+(2s^2\,^1S) + H_2$ above 8 eV collision energy immediately becomes understandable[61].

C. Potential Energy Curves for $C^{3+}(2l) + H(1s) \rightarrow C^{2+}(nln'l';\,^1L) + H^+$

Multicharged ions coexist with neutral species in plasmas produced by high-frequency photons, or when a current of neutral atoms is injected into a magnetically confined plasma. The degree of penetration of a neutral hydrogen beam is very dependent upon charge-transfer processes of the type

$$X^{(n+1)+} + H \rightarrow X^{n+} + H^+$$

The X^{n+} ions are frequently produced in excited states and the resulting radiation losses

$$(X^{n+})^* \rightarrow X^{n+} + h\nu$$

cool the plasma. These processes continue until X becomes electrically neutral or until it again becomes fully ionized—so that impurity ions X^{n+} are very efficient cooling agents. The ions of carbon are particularly important in this respect since they originate from the steel walls of the torus containing the plasma.

A variety of measurements have been performed for the charge-transfer process $C^{3+} + H \rightarrow C^{2+} + H^+$[53,66]. In particular, the technique of translational energy spectroscopy used by the Belfast group[67] has provided measurements of individual state-selected charge-transfer cross-sections. Except for total electron-capture cross-sections these do not agree with theoretical studies[68], the explanation being that far too few states were included in the calculations.

It appears to be a general feature of charge-transfer processes that one must include a large number of strongly coupled states. It is necessary to describe all of these potential curves to an accuracy which is at least consistent over the whole relevant range of internuclear separations R. In particular, the splittings between the asymptotic states must be reproduced accurately, otherwise the regions of strongly avoided crossings, where the charge transfer actually occurs, are displaced and the whole calculation is vitiated. The charge transfer is driven by radial coupling matrix elements of the form $\langle \Phi_1 | \partial/\partial R | \Phi_2 \rangle$ which are large in the regions of avoided crossings. These integrals are most easily computed numerically using finite differences. However, it is important to note that this imposes a requirement of a consistent amount of electron correlation in all the relevant states for all relevant values of R.

The aim of the spin-coupled VB calculations is to determine potential energy curves for sufficient states of the CH^{3+} system that are of sufficient accuracy and include a uniform amount of electron correlation. For this

TABLE III
Asymptotic energies[a] of the first 11 states of CH^{3+}.

State	Energy (eV)	
	Experiment[65]	Calculated
$C^{2+}(2s^2\ {}^1S) + H^+$	0	0
$C^{2+}(2s2p\ {}^1P) + H^+$	12.69	13.03
$C^{2+}(2p^2\ {}^1D) + H^+$	18.08	18.54
$C^{2+}(2p^2\ {}^1S) + H^+$	22.63	23.55
$C^{2+}(2s3s\ {}^1S) + H^+$	30.65	30.69
$C^{2+}(2s3p\ {}^1P) + H^+$	32.10	32.19
$C^{3+}(2s\ {}^2S) + H(1s)$	34.27	34.29
$C^{2+}(2s3d\ {}^1D) + H^+$	34.28	36.73
$C^{2+}(2p3s\ {}^1P) + H^+$	38.44	39.13
$C^{2+}(2p3p\ {}^1P) + H^+$	39.64	39.97
$C^{3+}(2p\ {}^2P) + H(1s)$	42.28	42.34

[a]The experimental values are based on Moore's[65] tables of atomic energy levels. The calculated values are based on results obtained at $R = 30$ bohr, but have been corrected for the Coulombic repulsion at this distance.

purpose the 'universal even-tempered' (UET) basis set was used[69]. This is a very large basis of Slater functions (36σ, $18\pi_x$, $18\pi_y$, $6\delta_{xy}$, $6\delta_{x^2-y^2}$) with exponents systematically generated as a geometric series. This basis is not biased towards any particular range of R, nor towards any one state. On each centre there are nine 1s, six 2p and three 3d functions. The s and p spaces are reasonably saturated and there are sufficient d orbitals to act as polarization functions. However, the basis may need extension for the accurate description of atomic d orbitals. The basis includes several diffuse functions and consequently it is able to describe the large variety of states, including those with Rydberg character, that one needs in CH^{3+}. We have employed this basis on several occasions in spin-coupled VB calculations.

The spin-coupled VB calculations include a total of 12σ and 16π orbitals. These were used to generate 127 spatial configurations which give rise to 228 spin-coupled configurations.

In Table III the computed splittings between the asymptotic states are compared with experimental values for the first 11 states. The agreement is very good indeed and is maintained over a range of 40 eV of energy. The single exception is the $C^{2+}(2s3d; ^1D) + H^+$ limit. Careful checks indicate that this can only be improved by extending the d space of the UET basis, as indicated above.

The lowest nine potential curves are shown in Fig. 9. Note that it is possible to trace diabatic paths through the figure for the $C^{3+}(2s) + H(1s)$ and $C^{3+}(2p) + H(1s)$ charge-transfer states, both of which are bound. Avoided crossings with repulsive C^{2+}/H^+ states lead to potential wells capable of supporting vibrational levels. For example, as a result of such an avoidance the $(5)^1\Sigma^+$ state possesses a potential maximum which lies nearly 4000 cm^{-1} above the minimum, and supports eight vibrational levels. The first few of these possess negligible widths due to tunnelling through the barrier. Consequently if the transition dipole moments to lower levels are large enough (as one might expect considering the change in polarity between C^{3+}/H and C^{2+}/H^+ states), it should be possible to observe photoemission from them.

In Fig. 10 the present potential energy curves for $C^{2+}(2p^2; ^1S, ^1D) + H^+$ and $C^{3+}(2s) + H(1s)$ are compared with those used by Bienstock *et al.* in their quantal scattering study[68]. The potentials employed in their work were obtained using CI procedures. Away from the crossing region the two sets of curves seem to be essentially parallel—the spin-coupled VB giving substantially lower energies for all states at all internuclear separations. The largest discrepancies occur in the important region of the avoided crossing. This is seen to be far less sharp in the spin-coupled VB case, and the minimum energy separation is significantly greater than in the MO-CI calculations. The computed cross-sections for charge transfer depend sensitively upon this separation, and this is almost certainly one reason for the lack of agreement between previous theory and experiment. The UET basis set, which is

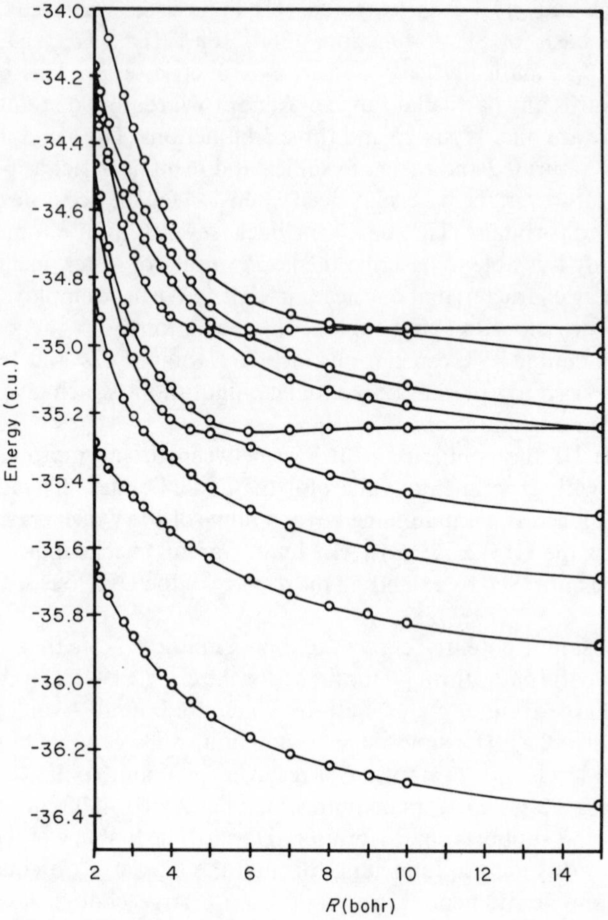

Fig. 9. Potential curves for the process $C^{3+}(2l) + H(1s) \rightarrow C^{2+}(nln'l'; {}^1L) + H^+$. The nine lowest states of CH^{3+} of ${}^1\Sigma^+$ symmetry are shown.

probably too large for MO-CI calculations, should describe the $C^{2+} + H^+$ and $C^{3+} + H$ states to equal accuracy.

The spin-coupled VB calculations show that the 10 lowest states of the CH^{3+} system, which are of highly contrasting character, can be determined to a useful accuracy by single and double excitations from the 'valence' orbitals of a single reference function. The resulting total wavefunctions are very compact and consist of a linear combination of just 228 spin-coupled configurations. The potential curves represent a significant improvement over previous work, particularly in the crucial regions of the avoided crossings.

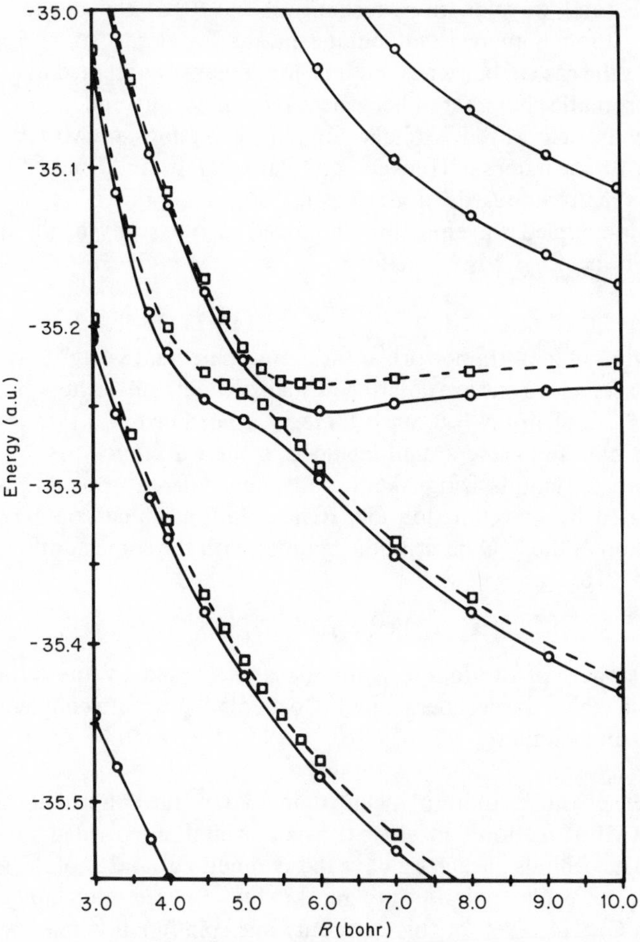

Fig. 10. Comparison of the present potential curves for $C^{2+}(2p^2; {}^1S, {}^1D) + H^+$ and for $C^{3+}(2s) + H(1s)$ with those used in the quantal scattering study of Bienstock *et al.*[68].

D. The Electronic Structure of Benzene

In the molecular orbital theory of electronic structure, the orbitals are all determined as eigenfunctions of a single operator $\hat{F}$, the Hartree–Fock operator. They are consequently orthogonal, and the state of lowest energy of the molecule is obtained when the individual MOs are doubly occupied. Generally speaking, as eigenfunctions of $\hat{F}$, the MOs form a basis for irreducible representations of the point group of the molecule (D_{6h} in the case of benzene), and are necessarily delocalized[70]. It is possible in certain instances, such as that of H_2O, to carry out a unitary transformation (which leaves the total wavefunction unchanged) such that the orbitals become localized—though there is no unique means for doing so[71]. This is not possible in the case of benzene, and it is now generally accepted at all levels[72] that the aromatic character of benzene is most naturally understood in terms of physically delocalized orbitals. Simplified versions of MO theory, associated with the names of Hückel[73] and Pariser, Parr and Pople[74] have been applied to many hundreds of π-electron systems.

The spin-coupled wavefunction employed in our study of the electronic structure of benzene[54] is of the form:

$$\{\psi_1^2\psi_2^2\cdots\psi_{18}^2\pi_1\pi_2\cdots\pi_6\} \tag{34}$$

The electrons of the σ framework are accommodated in 18 orthonormal MOs, while the six π electrons are described by six distinct non-orthogonal orbitals $\pi_\mu\,(\mu = 1, 2, \ldots, 6)$. For $N = 6$, $S = 0$, there are according to Eq. (9) a total of five spin functions, and these are all included in the wavefunction.

It is worth emphasizing the number of different possible solutions encompassed by wavefunction (34). One solution which could result is a modification of the MO description $\{a_{2u}^2 e_{1g}^4\}$ with the some additional radial correlation, i.e.

$$\{(a_{2u}a'_{2u})(e_{1g}^{(1)}e_{1g}^{(2)}, e_{1g}^{\prime(1)}e_{1g}^{\prime(2)})\}X^1A_{1g}$$

where the spins of the four e_{1g} orbitals are coupled by the 'Hund's rule function' in which the degenerate pair of orbitals $e_{1g}^{(1)}, e_{1g}^{(2)}$ are coupled to a spin of $S = 1$, and similarly for $e_{1g}^{\prime(1)}$, $e_{1g}^{\prime(2)}$. All these orbitals would then be delocalized.

Another possible solution included in (34) is the 'alternant molecular orbital' (AMO) function[75] in which the six π orbitals are divided into two sets of three. The orbitals in each set have the symmetry a_2'' and e'' of the subgroup D_{3h} of the molecular point group, and the two sets are interchanged by the operation C_6. However for this case, only one spin function may occur: that where the spins of each set are coupled to the maximum $S = \frac{3}{2}$, and the two subsystems coupled to give the final resultant $S = 0$ (i.e. the 'Hund's rule coupling' for $N = 6$, $S = 0$).

The actual spin-coupled wavefunction corresponds to none of these cases, but to six identical highly localized orbitals on each of the six carbon atoms. One of these is shown in Fig. 11, and it can be seen that it is essentially a $C(2p_z)$ orbital which is deformed by the two adjacent C atoms. The orbitals are permuted by the operation C_6. All five of the spin functions participate, giving a function which corresponds to a symmetric coupling of the spins around the ring.

As can be seen from Table IV, the total energy of the spin-coupled wavefunction is 0.0750 a.u. lower than that of the MO solution. This is a substantial energy lowering. A full CI calculation amongst these orbitals in

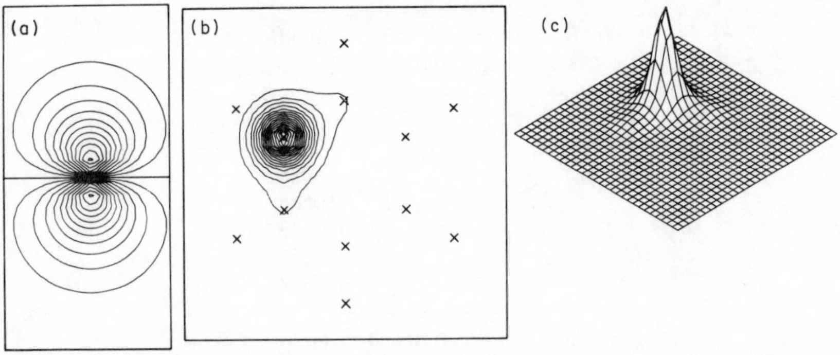

Fig. 11. One of the spin-coupled orbitals ϕ_i in benzene. The others are generated by successively C_6 rotations. (a) Contours of $\phi_i(\mathbf{r})$ in a σ_v mirror plane (i.e. perpendicular to the molecular frame and containing two C–H bonds). (b) Contours of $|\phi_i(\mathbf{r})|^2$ in a plane 1 bohr above the C_6H_6 plane (the nuclei below are denoted by X). (c) An isometric view corresponding to (b).

TABLE IV

Total energies for benzene (X^1A_{1g}) using [3s2p/2s] and [3s3p/2s] contracted Gaussian basis sets.

Calculation	Energy (a.u.)	
	[3s2p/2s]	[3s3p/2s]
SCF	– 229.99539	– 230.00150
Spin-coupled (one structure)	– 230.07043	– 230.07627
Full VB(CI) with all ionic structures	– 230.07739	– 230.08324

which all 175 possible structures were included (i.e. singly, doubly and triply polar structures[76,77]) only gives a further energy lowering of 0.007 a.u. This actually overestimates the importance of ionic structures; their contribution would have been even less if excited spin-coupled structures had been added. We have not at the time of writing included such excitations, but it should be strongly emphasized that in all our experience so far, addition of further correlation has not altered the physical picture afforded by the spin-coupled function above.

One of the most unacceptable features of classical VB theory is the proliferation of physically unreasonable ionic structures, whose role is to provide for deformation of the atomic functions when the molecule forms. As shown particularly clearly here, these are now unnecessary; the spin-coupled wavefunction for benzene incorporates all the deformation required.

The five spin functions used in the spin-coupled calculations are the

TABLE V
Occupation numbers of the spin functions in the Rumer basis[a].

Rumer function	Occupation number
1	0.4028
2	0.0648
3	0.0648
4	0.4028
5	0.0648

[a]Functions 1 and 4 correspond to the two Kekule structures, and 2, 3 and 5 to the Dewar structures.

orthonormal Kotani functions. The coefficients in the more traditional basis of Rumer functions have also been determined, i.e. two Kekulé and three Dewar structures, and the occupation numbers corresponding to these are shown in Table V. The values are astonishingly close to those given long ago by Coulson[78]. This very strongly suggests that the Kekulé description of benzene, as expressed in classical VB form, is in fact much closer to reality than a description in terms of delocalized molecular orbitals.

It is therefore virtually certain that the π electrons of benzene are essentially localized in deformed $C(2p_z)$ orbitals. The notable aromatic characteristics of this molecule arise from the symmetric coupling of the electrons around the carbon ring framework. This mode of description is not unfamiliar to solid-state physicists who use similar language to describe cooperative phenomena such as ferro- and antiferromagnetism. Benzene must now be considered in this light, and consequently aromaticity is a much more profoundly quantum phenomenon than we have hitherto realized.*

E. Studies of Momentum-space Properties

Spin-coupled wavefunctions have proved to be very useful in studies of momentum-space properties[79–81]. Except for very simple systems, it is rather difficult to solve the Schrödinger equation directly in the momentum representation; fortunately, the momentum-space wavefunction is also given by the Fourier transform of that in position space and this indirect approach proves to be much more tractable. The momentum-space formalism is particularly convenient for the interpretation of various scattering techniques[82] such as Compton scattering and binary (e, 2e) spectroscopy.

The (e, 2e) experiment is concerned with processes of the general type

$$M + e^- \rightarrow M^+ + e^- + e^-$$

***Note added in proof.** Since the completion of this Review, further results have been obtained for a whole series of heterocyclic systems such as pyridine, furan, thiophene, etc. In each case parallel results to those obtained for beutene have been obtained.

where a fast electron, with an energy in the range of a few hundred electronvolts to several kiloelectronvolts, collides with a target molecule M so as to knock out an electron and produce a molecular ion. The experiment is unique in that it is usually interpreted in terms of the momentum distributions for *individual* orbitals. Two electron detectors determine the energies and momenta of the (coincident) scattered and ejected electrons—and thus the recoil momentum of the ion. In the simplest approximation, this is equal in magnitude to the momentum the electron had at the moment of impact.

Central to a more detailed interpretation is the momentum-space electronic overlap integral $\langle \mathbf{p}\Psi^{N-1} | \Psi^N \rangle$ which involves integration over the $(N-1)$ common electrons of the target and ion wavefunctions; it is thus a function of the momentum $\mathbf{p}$ of the ejected electron. Allan *et al.*[79] have shown for spin-coupled wavefunctions that the orbital amplitude can be approximated by an expression that can be written

$$\langle \mathbf{p}\Psi^{N-1} | \Psi^N \rangle = \sum_{\mu} f_{\mu} \Phi_{\mu}(\mathbf{p})$$

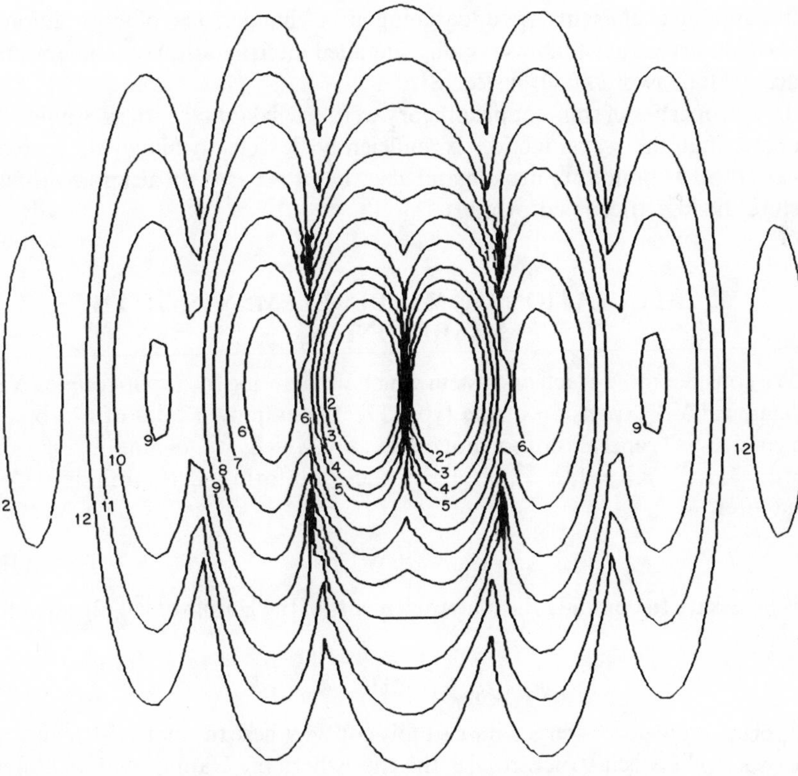

Fig. 12. Momentum-space spin-coupled orbital for BH^+.

where f_μ is a numeric quantity which is independent of $\mathbf{p}$ and $\Phi_\mu(\mathbf{p})$ is the Fourier transform of the orbital $\psi_\mu(\mathbf{r})$

$$\Phi_\mu(\mathbf{p}) = (2\pi)^{-3/2} \int \phi_\mu(\mathbf{r}) \exp(-i\mathbf{p}\cdot\mathbf{r})\, d\mathbf{r}$$

The measured differential cross-sections depend on the rotationally and vibrationally averaged values of $|\langle \mathbf{p}\Psi^{N-1}|\Psi^N\rangle|^2$. Allan *et al.* have discussed the interpretation of (e, 2e) experiments in terms of spin-coupled orbitals and have presented momentum-space electron density maps for individual orbitals for LiH, NH and the π electrons in $CH_2CHX(X = F, Cl)$. The non-orthogonality of the spatial orbitals allows them to assume properties very different from those of MO theory; this leads to striking differences in momentum space.

Spin-coupled wavefunctions have also recently been used in a critical appraisal of local density approximations in small molecules[80] and for a comparison of the effects of bond formation on electron densities in momentum space and in position space[81]. An example of a momentum space-orbital is shown in Fig. 12 for BH^+ at 3.75 bohr in the range of R where the spin-coupling coefficients are undergoing large changes. The obvious diffraction effects are greatest in this region; a detailed interpretation of momentum-space orbitals is presented in Ref. 81.

Two properties of spin-coupled theory were crucial for all of these studies—namely, that the wavefunction is sufficiently flexible to allow for correct dissociation as bonds are broken and also that it is based on a simple orbital picture of singly occupied orbitals.

V. CALCULATION OF MATRIX ELEMENTS OF THE HAMILTONIAN

We consider an N-electron system which we describe by a spin-coupled VB or classical VB expansion of the type (17). We employ a total of n orbitals which are in general non-orthogonal, and which we denote by ψ_μ or $\psi_{\mu_i} (\mu = 1, 2, \ldots, n;\ i = 1, 2, \ldots, N)$. From these we form a certain finite set of structures

$$\{\phi_{\mu_1}\phi_{\mu_2}\cdots\phi_{\mu_N}\}_k \tag{35}$$

It is necessary to compute the matrix elements of the Hamiltonian (2) between functions (35):

$$\langle\{\phi_{\mu_1}\phi_{\mu_2}\cdots\phi_{\mu_N}\}_k|\mathscr{H}|\{\phi_{\lambda_1}\phi_{\lambda_2}\cdots\phi_{\lambda_N}\}_l\rangle \tag{36}$$

Our present purpose is served if we simply consider here the matrix elements of the overlap between structures, i.e. the case where the Hamiltonian in (36) is replaced by the unit operator:

$$\Delta^{(kl)}_{\mu_1\cdots\mu_N;\lambda_1\cdots\lambda_N} = \langle\{\phi_{\mu_1}\phi_{\mu_2}\cdots\phi_{\mu_N}\}_k|\{\phi_{\lambda_2}\cdots\phi_{\lambda_N}\}_l\rangle \tag{37}$$

In addition to the chosen set of orbitals ϕ_μ, we introduce the 'dual basis'

$$\bar{\phi}_\lambda = \sum_{\mu=1}^{n} (\Delta^{-1})_{\lambda\mu}\phi_\mu \qquad (\lambda = 1, 2, \ldots, n) \tag{38}$$

where $\mathbf{\Delta}$ is the $n \times n$ matrix of overlaps between the orbitals. The dual orbitals have the property

$$\langle\bar{\phi}_\lambda|\phi_\mu\rangle = \delta_{\lambda\mu} \tag{39}$$

From them a dual set of structures

$$\{\bar{\phi}_{\lambda_1}\bar{\phi}_{\lambda_2}\cdots\bar{\phi}_{\lambda_N}\}_k \tag{40}$$

can also be formed. The relationship between the two sets (35) and (40) is given by

$$\{\phi_{\mu_1}\phi_{\mu_2}\cdots\phi_{\mu_N}\}_k = \sum_{\lambda_1,\lambda_2,\ldots,\lambda_N} \Delta_{\mu_1\lambda_1}\Delta_{\mu_2\lambda_2}\cdots\Delta_{\mu_N\lambda_N}\{\bar{\phi}_{\lambda_1}\bar{\phi}_{\lambda_2}\cdots\bar{\phi}_{\lambda_N}\}_k \tag{41}$$

Now, because of their construction, the spin-coupled structures form bases for irreducible representations of the *general linear group* GL(n)[83]. The particular irreducible representation is determined by N and S, i.e. the same two quantities which determine the irreducible representation of the symmetric group. In addition, if all the symmetry-allowed spin couplings k are included, the orbitals $\phi_{\mu_1}, \ldots, \phi_{\mu_N}$ occurring in a particular structure may be ordered in any way, e.g. in the canonical order

$$\mu_1 \geqslant \mu_2 \geqslant \cdots \geqslant \mu_N$$

Because of the Pauli principle, no more than two such indices may be equal. The indices occurring in the designation of an individual structure $(\mu_1, \mu_2, \ldots, \mu_N; k)$ may be combined into a single '*Gelfand tableau*'[84] $[m]$ or $[m_\mu]$. The relationship (41) may then be written as

$$\{\Phi\}_{[m_\mu]} = \sum_{[m_\lambda]} G^{N,S}_{[m_\lambda][m_\mu]}(\mathbf{\Delta})\{\bar{\Phi}\}_{[m_\lambda]} \tag{42}$$

In this equation, the notation $\{\Phi\}_{[m_\mu]}$ is used for the structure (35) in order to highlight its role as a basis function $[m_\mu]$ of GL(n), and similarly for the dual structures. The quantity

$$G^{N,S}_{[m_\lambda][m_\mu]}(\mathbf{\Delta})$$

is the $([m_\lambda], [m_\mu])$ element of the matrix $\mathbf{G}^{N,S}(\mathbf{\Delta})$ which is an *irreducible representation* of GL(n) corresponding to the linear transformation in an n-dimensional space induced by the matrix Δ of orbital overlaps. Replacing

Eq. (41) by (42) and using it in (37), we obtain

$$\Delta_{[m_\mu],[m_\lambda]} = G^{N,S}_{[m_\lambda][m_\mu]}(\boldsymbol{\Delta}) \tag{43}$$

This equation represents the general solution of the 'non-orthogonality problem', since the construction of the matrix $\mathbf{G}^{N,S}(\boldsymbol{\Delta})$ does not involve $N!$ operations. All the other density matrices needed in the evaluation of the matrix of the Hamiltonian (36) may be derived similarly. Thus, for example, the required one-electron density matrix elements are given by

$$D_{[m_\mu][m_\lambda]}(\mu_i|\lambda_i) = G^{N-1,S\pm\frac{1}{2}}_{[m'_\lambda][m'_\mu]}(\Delta[\mu_i,\lambda_i]) \tag{44}$$

In this equation, $\boldsymbol{\Delta}[\mu_i, \lambda_i]$ is an $(n-1)\times(n-1)$ submatrix of $\boldsymbol{\Delta}$ in which the row corresponding to orbital ψ_{μ_i} is missing and similarly for column λ_i. The irreducible representation $\mathbf{G}^{N-1,S\pm\frac{1}{2}}$ now refers to $\mathrm{GL}(n-1)$, and the particular element, $([m'_\lambda],[m'_\mu])$, is obtained by removing λ_i and μ_i from the Gelfand tableaux $[m_\lambda]$ and $[m_\mu]$ respectively[85].

Most VB methods rely on the reduction of structures of the form (35) to a linear combination of Slater determinants, and form the matrix of the Hamiltonian over the set of such determinants constructed from non-orthogonal orbitals. The exception to this is the method of Gallup[86] which has been reviewed recently. The alternative approach of orthogonalizing the orbitals, while indeed eliminating the overlap problem, suffers from conceptual difficulties. It has long been recognized[87] that the orthogonalization of the orbitals destroys the physical insight afforded by the original non-orthogonal basis. This was the subject of a debate in 1968 between Prosser and Hagstrom[87] on the one side and King and Stanton[88] on the other. The development of spin-coupled VB theory has given a further impetus to the direct use of non-orthogonal functions. Furthermore, in the determination of intermolecular interactions (Section VII), the use of orthogonalized orbitals gives rise to basis-set superposition errors. These do not occur when non-orthogonal functions are used.

The most general formulation of matrix elements between non-orthogonal determinants is due to Löwdin[39] and this provides the basis of the 'factorized cofactor-driven algorithm' which we now describe.

Consider two Slater determinants, U and V, constructed from a set of spin orbitals $u_1, u_2, \ldots, u_N$ and $v_1, v_2, \ldots, v_N$ respectively:

$$U = \det\{u_1 u_2 \cdots u_N\} \qquad V = \det\{v_1 v_2 \cdots v_N\} \tag{45}$$

Löwdin showed that

$$\langle U|V\rangle = D_{uv} \tag{46a}$$

$$\langle U|\mathscr{H}|V\rangle = \sum_{i,j=1}^{N} D_{uv}(i|j)\langle u_i|h|v_j\rangle + \tfrac{1}{2}\sum_{i,j,k,l=1}^{N} D_{uv}(ij|kl)\langle u_i u_j|g|v_k v_l\rangle \tag{46b}$$

In these expressions, D_{uv} is the determinant of the matrix of overlap integrals $\boldsymbol{\Delta}_{uv}$ associated with U and V. The elements of $\boldsymbol{\Delta}_{uv}$ are given by

$$\Delta_{uv}^{ij} = \langle u_i | v_j \rangle$$

and

$$D_{uv} = \det\{\Delta_{uv}^{11} \cdots \Delta_{uv}^{NN}\}$$

The $D_{uv}(i|j)$ and $D_{uv}(ij|kl)$ are respectively minors of order $(N-1)$ and $(N-2)$. The similarity of Eqs (46) with those arising in the spin-coupled theory (24) is worth noting.

Formulae (46) are much simpler in appearance than in practice. In addition to multiple summations, they involve the calculation of all the determinants D_{uv} and of all their cofactors $D_{uv}(i|j)$ and $D_{uv}(ij|kl)$. A straightforward evaluation of (46b) is prohibitive since it would involve an N^7 algorithm: each of the cofactors appearing in the fourfold summation requires about N^3 operations. However, by precomputing the cofactors, the N^7 process is broken down into an N^3 and an N^4 step. As pointed out long ago by Löwdin, cofactors of order $(N-1)$ and $(N-2)$ are proportional to the elements of the inverse of the matrix of overlaps, $(\boldsymbol{\Delta}_{uv})^{-1}$. Provided that $\boldsymbol{\Delta}_{uv}$ is non-singular, the cofactors required for a matrix element (46b) are all obtained from a single matrix inversion, which is an N^3 process.

In this case the cofactor method appears to be very efficient and superior to the method of corresponding orbitals of King *et al.*[89], which is one of the most efficient ways of orthogonalizing the orbitals. Prosser and Hagstrom[90] gave a more general validity to this approach by showing how cofactors for an arbitrary matrix $\boldsymbol{\Delta}_{uv}$ may be readily obtained by means of the biorthogonalization technique. In addition they hinted at the solution of the more general problem of precomputing all the unique cofactors which are needed for determining a wavefunction in the form of a linear combination of many non-orthogonal configurations. Raimondi, Campion and Karplus[91] further showed how to incorporate both spin and spatial symmetry to reduce the total number of non-redundant cofactors. These features are all included in a general factorized cofactor-driven program, known as MILANO. It has been applied to many systems within the framework of both classical[19,92,93] and spin-coupled VB[33,40,41,43,46,51,62,67]. The number of electrons which are explicitly included in the calculations, N, is usually small ($N \leqslant 12$), while the total number of electrons may be large (see e.g. the calculations on H_2S[93] or on benzene[77]).

The spin functions used by the program are the set of Rumer[30] (or bonded) functions. Besides highlighting the formation or disintegration of covalent bonds, this basis has the advantage that the structures in it are very easily decomposed into Slater determinants. The coefficients of the individual determinants are just ± 1. The Rumer basis of spin functions is not orthogonal, and a linearly independent set is systematically generated using the 'leading term' method. This is described in Ref. 19, and in the book by Pauncz[31].

The efficiency of the program is based on the assumption that the general VB expansion (17) involves a relatively small set of carefully optimized orbitals. This is the opposite of the CI strategy where a very large set of orbitals must be employed. Consequently the precomputation and storage of the cofactors does not present a severe problem. At the same time, the treatment of the two-electron integrals is simplified since they can be held in core. The rate-determining step is that of combining the cofactors with the two-electron integrals, and this is an N^4 process.

The method is now briefly described. For given values of N and of the quantum numbers S and M, the number of electrons with α and β spin, N_α and N_β, is determined ($N = N_\alpha + N_\beta$). Because of the orthogonality of the one-electron spin functions, the matrices $\mathbf{\Delta}_{uv}$ are block diagonal, and consequently the determinants factorize:

$$D_{uv} = D^{\alpha}_{uv} D^{\beta}_{uv}$$

where D^{α}_{uv} and D^{β}_{uv} are determinants of order N_α and N_β respectively. Similarly the cofactors of order $(N-1)$ and $(N-2)$ have the form:

$$D_{uv}(i|\,j) = \begin{cases} D^{\alpha}_{uv}(i|\,j) D^{\beta}_{uv} & (i, j \leqslant N_\alpha) \\ D^{\alpha}_{uv} D^{\beta}_{uv}(i|\,j) & (i, j \geqslant N_\alpha) \end{cases} \tag{47}$$

and

$$D_{uv}(ij|kl) = \begin{cases} D^{\alpha}_{uv}(ij|kl) D^{\beta}_{uv} & (i, j, k, l \leqslant N_\alpha) \\ D^{\alpha}_{uv}(i|k) D^{\beta}_{uv}(j|l) & (i, k \leqslant N_\alpha; j, l > N_\alpha) \\ D^{\alpha}_{uv} D^{\beta}_{uv}(ij|kl) & (i, j, k, l \geqslant N_\alpha) \end{cases} \tag{48}$$

TABLE VI
Order of the arrays containing the determinants D^{α}_{uv}, D^{β}_{uv} and D_{uv} equal to

$$\binom{N}{N_\alpha}, \binom{N}{N_\beta} \text{ and } \binom{N}{N_\alpha} \times \binom{N}{N_\beta}.$$

N	N_α	N_β	$\binom{N}{N_\alpha}$	$\binom{N}{N_\beta}$	$\binom{N}{N_\alpha} \times \binom{N}{N_\beta}$
4	2	2	6	6	36
5	3	2	10	10	100
6	3	3	20	20	400
7	4	3	35	35	1225
8	4	4	70	70	4900
9	5	4	126	126	15876
10	5	5	252	252	63504
11	6	5	462	462	213444
12	6	6	924	924	853776

where D^{α}_{uv}, D^{β}_{uv}, $D^{\alpha}_{uv}(i|j)$, $D^{\beta}_{uv}(i|j)$, $D^{\alpha}_{uv}(ij|kl)$ and $D^{\beta}_{uv}(ij|kl)$ are determinants of order N_α, N_β, $(N_\alpha - 1)$, $(N_\beta - 1)$, $(N_\alpha - 2)$ and $(N_\beta - 2)$ respectively. The spin symmetry reduces the number of cofactors very effectively, as shown in Table VI. The entries give the number of factorized determinants D^{α}_{uv} and D^{β}_{uv}, and also the total number of determinants D_{uv}. It should be noted that the number of factorized determinants does not vary when increasing a VB calculation from one in which covalent structures only are included (i.e. no doubly occupied orbitals; all $\mu_1, \mu_2, \ldots, \mu_N$ in Eq. (35) equal to 1), to a full VB calculation which includes all possible structures. The last column of Table VI gives the number of determinants D_{uv} which occur in the limiting case of a full VB calculation. This number varies with the degree of ionicity (i.e. the maximum number of doubly occupied orbitals) of the structures included in a calculation, its minimum value being the number of determinants D^{α}_{uv} or D^{β}_{uv}.

It can be seen that these numbers increase very rapidly with the number of orbitals included in a calculation, and this emphasizes the need for a set of well optimized functions in terms of which the VB expansion converges rapidly. Using Eqs (47) and (48), formula (46b) becomes (see also Refs 19 and 91)

$$\langle U|\mathscr{H}|V\rangle = D^{\beta}_{uv} \sum_{i,j=1}^{N_\alpha} D^{\alpha}_{uv}(i|j)\langle u_i|h|v_j\rangle + D^{\alpha}_{uv} \sum_{i,j=1}^{N_\beta} D^{\beta}_{uv}(i|j)\langle u_i|h|v_j\rangle \quad (49a)$$

$$+ \tfrac{1}{2}D^{\beta}_{uv} \sum_{i,j,k,l=1}^{N_\alpha} D^{\alpha}_{uv}(ij|kl)\langle u_i u_j|g|v_k v_l\rangle$$

$$+ \tfrac{1}{2}D^{\alpha}_{uv} \sum_{i,j,k,l=1}^{N_\beta} D^{\beta}_{uv}(ij|kl)\langle u_i u_j|g|v_k v_l\rangle \quad (49b)$$

$$+ \sum_{i,k=1}^{N_\alpha} \sum_{j,l=1}^{N_\beta} D^{\alpha}_{uv}(i|k) D^{\beta}_{uv}(j|l)\langle u_i u_j|g|v_k v_l\rangle \quad (49c)$$

All contributions in (49a) and (49c) are precomputed and stored, these being common to many matrix elements. Once this step is completed there is no longer any need to keep the cofactors $D^{\alpha}_{uv}(ij|kl)$ and $D^{\beta}_{uv}(ij|kl)$. Contribution (49c) is reduced by means of techniques similar to those employed in the four-index transformation of two-electron integrals. This can be accomplished in one, two, three or four passes, depending upon the amount of disk space and CPU time available. A two-pass program is probably a good compromise. In the one-pass case, contribution (49c) requires $(N_\alpha N_\beta)^2$ operations, corresponding to $N^4/16$ for closed-shell systems. For $N \leqslant 16$, this approximates to N^3. This is an important consideration, since the number of active electrons in accurate *ab initio* CI or spin-coupled VB calculations is usually less than 16. The two-pass program is based upon the following nesting of the summations in (49c):

$$\sum_{i,k=1}^{N_\alpha} \sum_{j,l=1}^{N_\beta} D^{\alpha}_{u_\alpha v_\alpha}(i|k) D^{\beta}_{u_\beta v_\beta}(j|l)\langle u_i u_j|g|v_k v_l\rangle$$

$$= \sum_{i,k=1}^{N_\alpha} G(u_\beta v_\beta; ik) D^{\alpha}_{u_\alpha v_\alpha}(i|k)$$

where

$$G(u_\beta v_\beta; ik) = \sum_{j,l=1}^{N_\beta} D^{\beta}_{u_\beta v_\beta}(j|l)\langle u_i u_j|g|v_k v_l\rangle$$

The notation $u \equiv (u_\alpha u_\beta)$, $v \equiv (v_\alpha v_\beta)$ has been used. These simplifications are easy to introduce in any method based upon the utilization of cofactors. This is true whether the cofactors are computed directly, or indirectly as elements of the adjoint* matrix of the overlap $\mathbf{\Delta}_{uv}$—provided only that the block diagonal form of $\mathbf{\Delta}_{uv}$ is fully exploited.

The case of intermolecular forces (see Section VII) deserves special attention. Here there are usually N^c_α and N^c_β closed-shell core electrons, and the core orbitals remain unchanged in all the Slater determinants. Consequently the contribution from (49c) for these,

$$\sum_{i,k=1}^{N^c_\alpha} \sum_{j,l=1}^{N^c_\beta} D^{\alpha}_{uv}(i|k) D_{uv}(j|l)\langle u_i u_j|g|v_k v_l\rangle \tag{50}$$

is computed once only. This saving increases with increasing N.

We now analyse briefly the problem of the precomputation of the determinants and their cofactors. First of all a list of non-redundant minors of order $(N_\alpha - 2)$ and $(N_\beta - 2)$ is generated and all the $D^{\alpha}_{uv}(ij|kl)$ and $D^{\beta}_{uv}(ij|kl)$ are computed. Normally this requires $(N_\alpha - 2)^3$ or $(N_\beta - 2)^3$ operations for each cofactor. In the special case of intermolecular forces, this step can be greatly reduced by incorporating the suggestion of Hayes and Stone[94]. We return to this shortly.

The next step is the computation of the non-redundant cofactors of order $(N_\alpha - 1)$ and $(N_\beta - 1)$ by means of the Laplace formula[95] using the cofactors of order $(N_\alpha - 2)$, $(N_\beta - 2)$ evaluated in the previous step, e.g.

$$D^{\alpha}_{uv}(i|j) = \sum_{l=1}^{N_\alpha} D^{\alpha}_{uv}(ik|jl)(\mathbf{\Delta}_{uv})_{kl}$$

(compare Eqs (25)). The formation of each $D^{\alpha}_{uv}(i|j)$ and $D^{\beta}_{uv}(i|j)$ thus requires only $(N_\alpha - 1)$ or $(N_\beta - 1)$ operations. All the unique D^{α}_{uv} and D^{β}_{uv} are similarly formed by means of N_α or N_β operations per determinant.

An alternative procedure is based on the general relations

$$\begin{aligned} D_{uv}(i|j) &= (\mathbf{\Delta}_{uv})^{-1}_{ij} D_{uv} \\ D_{uv}(ij|kl) &= D_{uv}(\mathbf{\Delta}_{uv})^{-1}_{lk}(\mathbf{\Delta}_{uv})^{-1}_{ji} - (\mathbf{\Delta}_{uv})^{-1}_{jk}(\mathbf{\Delta}_{uv})^{-1}_{li} \end{aligned} \tag{51}$$

These can also be expressed in terms of the adjoint matrix of $\mathbf{\Delta}$ as in the biorthogonalization technique of Prosser and Hagstrom[90]. A recently proposed method is a reformulation of this procedure and simply consists of

*I.e. the matrix formed from $\mathbf{\Delta}_{uv}$ by replacing every element by its cofactor. This is sometimes called the 'adjugate' or 'adjugate compound' matrix.

using (51) in the Löwdin formula (46b)[96]. The block diagonal form of $\mathbf{\Delta}_{uv}$ reduces the number of operations required for each inversion from N^3 to $N_\alpha^3 + N_\beta^3$. In this way all the cofactors needed for a given matrix element are obtained by a single inversion of the corresponding matrix $\mathbf{\Delta}_{uv}$, and can be carried out when evaluating (49a–c). The advantage of this procedure is that it can be programmed very easily and does not require any disk space. This implies that a given block of $\mathbf{\Delta}_{uv}$, which occurs in different overlap matrices, is inverted many times. However, if a list is generated of the non-redundant α and β blocks which constitute all the $\mathbf{\Delta}_{uv}$ matrices, the method gains in efficiency. Nevertheless, as shown by Gianinetti *et al.*[97], it still remains less efficient than the factorized cofactor-driven algorithm. In fact this is obvious if one considers that, in the latter procedure, only non-redundant cofactors of order $(N_\alpha - 1)$, $(N_\beta - 1)$, $(N_\alpha - 2)$ and $(N_\beta - 2)$ are determined and used to carry out the minimum amount of computation. In Table VII a comparison is shown of the total number of operations required for the precomputation of all the cofactors occurring in the matrix elements (49a–c), directly and according to Eqs (51)[97]. The particular VB wavefunction includes at least all covalent structures (i.e. all $\mu_i = 1$ in (35)), but can be extended to the full VB set without increasing the time for this step.

It is therefore likely that the factorized cofactor-driven algorithm, which is incorporated in the program MILANO, is optimal for a wavefunction of spin-coupled VB form (17) where the structures are derived by excitations from a reference configuration of N singly occupied orbitals.

The determination of intermolecular forces, however, presents a special situation. An approach which is proving to be very fruitful in this respect (see Section VII) is to describe the individual monomer molecules by means of a CI wavefunction using orthonormal orbitals, but, in computing the interaction between them, the non-orthogonality of the orbitals stemming from different monomers is taken directly into account. In this case there is a reference

TABLE VII
Number of operations required for the calculation of all the cofactors computed directly or by means of the inverse $(\mathbf{\Delta}_{uv})^{-1}$.

N	N_α	N_β	Direct calculation	Through $(\mathbf{\Delta}_{uv})^{-1}$
5	3	2	740	1485
6	3	3	870	5670
7	4	3	4872	40320
8	4	4	15540	159040
9	5	4	84051	1000125
10	5	5	313350	3984750
11	6	5	1518198	23101848
12	6	6	11990880	92307600

configuration which consists of the maximum possible number of doubly occupied orbitals. The expansion (17) is restricted to low-level excitations from a limited number of active orbitals in the reference configuration while the many core orbitals remain unexcited. Hayes and Stone[94] showed that the cofactors of the excited determinants can be expressed in terms of the cofactors of the reference determinant in a simple way. This provides a very great saving in computer time as the number of electrons increases. It should be emphasized that in the case of intermolecular forces the total number of Slater determinants can be made very small, and the above simplifications represent a significant reduction in computing time. The fourfold summations in Eq. (49c) can then be carried out, the simplifications contained in Eq. (50) becoming more important with increasing N.

Figari and Magnasco[98] simplify and generalize the work of Hayes and Stone[94] by providing a formulation which, in addition to saving computational time, remains valid even when the inverse of $\mathbf{\Delta}_{uv}$ does not exist because of a singularity (see, however, Gallup *et al.*[86] who has also shown how to avoid this problem). The approach is based upon the fact that

$$\det(\mathbf{\Delta}_{uv}) = \det(\mathbf{\Delta}_{u^\circ v^\circ}) \det \mathbf{D} \tag{52}$$

where $\mathbf{\Delta}_{u^\circ v^\circ}$ and $\mathbf{\Delta}_{uv}$ are respectively the matrices of the overlap integrals between the Slater determinants U° and V°, and between U and V. U° and V° are the unexcited reference Slater determinants, while U and V are determinants obtained from U° and V° by excitations of orbitals. Every spin orbital in U° is identical to the corresponding spin orbital in U° except for m_u terms, this being the degree of excitation. The same is true of V and V°, where there are m_v differences. In most applications U° and V° are identical, but this is not necessary. In Eq. (52), $\mathbf{\Delta}_{uv}$ and $\mathbf{\Delta}_{u^\circ v^\circ}$ are of order $N \times N$, while $\mathbf{D}$ is of order $n \times n$, where $n = m_u + m_v$: the total number of rows and columns where $D_{uv} = \det(\mathbf{\Delta}_{uv})$ differs from $D_{u^\circ v^\circ} = \det(\mathbf{\Delta}_{u^\circ v^\circ})$.

This technique can be incorporated in the factorized cofactor-driven algorithm and used when possible for the computation of the cofactors of order $(N_\alpha - 2)$ and $(N_\beta - 2)$. However, it does not afford any advantage in the general spin-coupled VB case, since the orbitals in the reference configuration are singly occupied. This is because here $n = m_u + m_v$ always increases up to N_α or N_β so that D^α_{uv} and D^β_{uv} become completely different from $D^\alpha_{u^\circ v^\circ}$ or $D^\beta_{u^\circ v^\circ}$.

In general, the precomputation of all the cofactors and contributions (49a) and (49b) is a small fraction of the overall time in the spin-coupled VB method. As an example we consider the computation of the states of $(\mathrm{B{-}H_2})^+$ discussed in Section IV. We take as the unit of time that taken to compute all the one- and two-electron integrals for the Gaussian basis set B(11s6p2d/9s5p2d) and H(6s2p1d/4s2p1d) on each hydrogen nucleus. The time required to determine a wavefunction consisting of 500 structures, which gives results for several states of the same accuracy as large MCSCF or CASSCF calculations (see

Section IV), with 10^6 matrix elements computed by Eq. (46), is 5 units. The time necessary to precompute all the determinants, their cofactors and carry out the summations (49a) and (49b) is 0.1 units, while the time to carry out the fourfold summation in (49c) is 3 units.

In the case of intermolecular forces, where the reference Slater determinants U° and $V^\circ \equiv U^\circ$ consist of doubly occupied orbitals, the distribution of the time changes. For the He + HF system, using an uncontracted Gaussian basis set of F(11s6p3d), H(6s2p) and He(10s3p), 140 VB structures and 8000 matrix elements between Slater determinants, the total time is 1.6 units. The first step of precomputing the determinants, their cofactors and the terms (49a) and (49b) takes 0.3 units, while the fourfold sums of (49c) take 0.1 units. The use of the suggestion of Hayes and Stone[94] or of Figari and Magnasco[98] would make the time for the first step negligible. In addition it would not increase for heavier systems such as He–HCl or for He–HBr, etc. This represents a very significant improvement of the factorized cofactor algorithm. The calculation of intermolecular interactions in heavier systems is possible because the wavefunctions can be expressed in terms of very few structures and the time needed for (49c) does not increase prohibitively.

An early method due to Moffitt for calculating matrix elements of the Hamiltonian between non-orthogonal Slater determinants is perhaps worth mentioning. In this, the determinants are expanded directly in terms of determinants composed of orthogonal orbitals. If the orbitals in this new set of determinants are obtained by the Schmidt procedure, Moffitt showed that the expansion coefficients of the determinants are just cofactors of the matrix of orthogonalization coefficients. The matrix elements of the Hamiltonian between the orthogonalized determinants are of course simple to evaluate. This procedure was used extensively in early VB calculations by Hurley and others. It involves a four-index transformation of the two-electron integrals from the original non-orthogonal set of orbitals to the Schmidt representation. This step alone involves at least N^5 operations and consequently the method is unlikely to be competitive.

VI. RECENT DEVELOPMENTS IN VALENCE BOND THEORY

A. The 'Diatomics-in-Molecules' Method, and Mixed Molecular Orbital–Valence Bond Methods

The data available from experiments such as molecular beam scattering are now becoming very detailed and include measurements of the number of product molecules in individual vibration–rotation states as a result of reactive encounters[99]. The first reasonably unambiguous resonance in reactive collisions (in the $F + H_2$ reaction) has recently been observed[100]. These phenomena can only be understood through dynamical studies of various kinds. For this purpose, potential surfaces are needed, often for several different electronic states, that cover the entire range of the relevant nuclear coordinates. The complete determinations of such surfaces by *ab initio* methods is necessarily accomplished pointwise, and is therefore an exceedingly tedious task, so that one turns naturally to the contemplation of semi-empirical procedures. Of these the currently most important is the 'diatomics-in-molecules' (DIM) method. Its prominence is fundamentally due to its use by experimental groups who utilize it in conjunction with classical trajectory calculations with the aim of providing some insight into at least the gross features of, say, observed product velocity flux distributions[101]. The method is based upon VB theory, and is a development of the earlier 'atoms-in-molecules' (AIM) approach founded by Moffitt[102].

The DIM method is essentially a technique for the approximate construction of the Hamiltonian matrix for a polyatomic system from smaller Hamiltonian matrices computed separately for the constituent atomic and diatomic fragments. It was first introduced by Ellison[103], and it has since been refined by a number of authors. Detailed reviews have been given by Tully[104] and by Kuntz[105]. A very clear derivation, essentially a projection operator reformulation, has been presented by Faist and Muckerman[106], who subsequently presented a real application to the $F + H_2$ surfaces (but with values that are uniformly too high for the important height of the barrier in the ground state), which illustrates both the practical considerations in constructing the DIM matrices, and also some of the complications which can arise. An elegant procedure for performing the spin decoupling of fragment states in DIM calculations, using systematic elimination rules based on second quantization techniques, has been described by Radlein[107].

We begin from a set of antisymmetrized products of atomic functions of the kind

$$\Psi_i = \mathscr{A}\{\zeta_{i_A}^{(A)}(1,\ldots,n_A)\zeta_{i_B}^{(B)}(n_A+1,\ldots,n_A+n_B) \cdot\zeta_{i_C}^{(C)}(n_A+n_B+1,\ldots,n_A+n_B+n_C)\cdots\} \quad (53)$$

The $\zeta_{i_k}^{(k)}$ are antisymmetric atomic functions pertaining to atom k in state i_k. The n_k space and spin coordinates of atom k are indicated as shown. The explicit form of the $\zeta_{i_k}^{(k)}$ will be generally as given in Eq. (12), but this will not be needed here. We do, however, assume in (53) that the individual spins of atoms A, B, C, ..., etc., namely $S_A, S_B, S_C, \ldots$, have been coupled to the required overall resultant S in a definite fashion. The total number of electrons is N, as before. Function Ψ_i is termed a 'polyatomic basis function' (PBF), the index i standing for the collection of individual indices $(i_A i_B i_C \cdots)$. The total wavefunction is expanded as a linear combination of the PBF:

$$\Psi = \sum_i C_i \Psi_i \quad (54)$$

Clearly if the complete set of the Ψ_i is used, this expansion is exact. Substitution of (54) into the Schrödinger equation (3) leads to a secular equation for determining the C_i of the form

$$(\mathbf{H} - E\mathbf{\Delta})\mathbf{C} \tag{55}$$

in which $\mathbf{H}$ is the Hamiltonian matrix with elements $\langle \Psi_i | \mathscr{H} | \Psi_j \rangle$ and $\mathbf{\Delta}$ is the overlap matrix with elements $\langle \Psi_i | \Psi_j \rangle$. The coefficients C_i occur as the column vector $\mathbf{C}$.

The partitioning of the Hamiltonian operator for the polyatomic system takes the form

$$\mathscr{H}\Psi_i = \left(\sum_{a>b} \mathscr{H}_{ab} - (N-2) \sum_a \mathscr{H}_a \right) \Psi_i \tag{56}$$

In this equation $\mathscr{H}_{ab}$ is the Hamiltonian operator for the diatomic fragment AB; it operates only upon the electronic coordinates pertaining to that fragment. Similarly $\mathscr{H}_a$ is the Hamiltonian operator for the isolated atom A. In general we denote an atomic or diatomic fragment by the index γ, i.e. (a) or (ab). The total antisymmetrizing operator $\mathscr{A}$ can always be factorized as

$$\mathscr{A} = \mathscr{A}_{\gamma\bar{\gamma}} \mathscr{A}_{\gamma} \mathscr{A}_{\bar{\gamma}} \tag{57}$$

in which $\mathscr{A}_\gamma$ antisymmetrizes only the electrons of the γ fragment, $\mathscr{A}_{\bar{\gamma}}$ similarly antisymmetrizes all the other electrons not belonging to γ, and $\mathscr{A}_{\gamma\bar{\gamma}}$ completes the identity.

Now if—and only if—the set of PBF is complete, the following operator identity holds true:

$$I = \sum_{i,j} |\Psi_i\rangle (\Delta^{-1})_{ij} \langle \Psi_j | \tag{58a}$$

By writing the $|\Psi_i\rangle$ (or Ψ_i) as a row vector $\mathbf{\Psi}$, and the $\langle \Psi_j|$ (or Ψ_j^*) as a column vector $\mathbf{\Psi}^\dagger$, this assumes the form:

$$I = \mathbf{\Psi}(\mathbf{\Delta}^{-1})\mathbf{\Psi}^\dagger \tag{58b}$$

From this we derive the identity

$$\begin{aligned}(\mathscr{H}\Psi_1, \ldots, \mathscr{H}\Psi_i, \ldots) &\equiv \mathscr{H}\mathbf{\Psi} \\ &= \mathbf{\Psi}(\mathbf{\Delta}^{-1})\mathbf{\Psi}^\dagger \mathscr{H} \mathbf{\Psi}\end{aligned}$$

i.e.

$$\mathscr{H}\mathbf{\Psi} = \mathbf{\Psi}(\mathbf{\Delta}^{-1})\mathbf{H} \tag{59}$$

In addition, by applying the partial antisymmetrizer $\mathscr{A}_\gamma \mathscr{A}_{\bar{\gamma}}$ to the product of atomic eigenfunctions of (53), we obtain a partially antisymmetrized PBF $\Phi_{\gamma i}$ in which there is no exchange of electrons between the sets of coordinates γ and $\bar{\gamma}$:

$$\Phi_{\gamma_i} = \mathscr{A}_\gamma \mathscr{A}_{\bar{\gamma}} \{\zeta_{i\mathrm{A}}^{(\mathrm{A})}(1, \ldots, n_\mathrm{A}) \zeta_{i\mathrm{B}}^{(\mathrm{B})}(n_\mathrm{A}+1, \ldots, n_\mathrm{A}+n_\mathrm{B}) \cdots \}$$

These functions too form a complete set (in this partially antisymmetrized space) and an identity parallel to (59) also holds:

$$\mathscr{H}_\gamma \mathbf{\Phi}_\gamma = \mathbf{\Phi}_\gamma(\mathbf{\Delta}_\gamma^{-1})\mathbf{H}_\gamma \tag{60}$$

in which the matrix elements of $\mathbf{H}_\gamma$ are given by

$$(\mathscr{H}_\gamma)_{ij} = \langle \Phi_{\gamma i} | \mathscr{H}_\gamma | \Phi_{\gamma j} \rangle$$

and $\mathbf{\Delta}_\gamma$ is the corresponding overlap matrix. With the help of these relations, we obtain for the Hamiltonian matrix **H** occurring in the secular equation (57):

$$\mathbf{H} = \mathbf{\Delta}\left(\sum_{a>b} (\mathbf{\Delta}_{ab}^{-1})\mathbf{H}_{ab} - (N-2)\sum_{a} (\mathbf{\Delta}_a^{-1})\mathbf{H}_a \right) \tag{61}$$

Thus the construction of the polyatomic Hamiltonian matrix is reduced in the *ab initio* DIM method to the construction of the corresponding matrices of the atomic and diatomic fragments.

An immediate concern is that although each matrix $\mathbf{H}_\gamma$ is Hermitian, the terms $\mathbf{\Delta}\mathbf{\Delta}_{ab}^{-1}\mathbf{H}_{ab}$ and $\mathbf{\Delta}\mathbf{\Delta}_a^{-1}\mathbf{H}_a$ are not, and nor are they fully independent. As a result, the DIM Hamiltonian matrix (61) is also non-Hermitian. This is a result of using truncated basis sets, so that the identities (59) and (60) are only approximate. These constitute the essential approximations of DIM theory. An obvious remedy would be simply to use the symmetric sum:

$$\mathbf{H} \rightarrow \tfrac{1}{2}(\mathbf{H} + \mathbf{H}^+)$$

However, much physical information is lost by this projection, and in addition this can be shown to produce instabilities. Otherwise one must simply use the non-Hermitian Hamiltonian matrix (61). The inclusion of sufficiently extensive basis sets will of course always reduce this problem. From a slightly different points of view, it can be said that the central purpose of the DIM approach is to reproduce the full Hamiltonian matrix—which contains important three- and four-centre contributions—by the expression (61), which involves no more than diatomic and atomic terms. This can only be done if the expansion (54) is essentially complete. Nonetheless, the DIM procedure presents a powerful approximation that has enjoyed fair success in a range of applications.

The DIM method is most commonly employed as a semi-empirical technique. The fragment Hamiltonian matrices are usually related to atomic and diatomic energies by making various approximations for the overlap matrices. Both the form of the DIM equation and the chosen set of PBF must be sufficient to account for all the qualitative features of the system being studied. Under such circumstances the approach may offer acceptable accuracy for modest computational effort. Given the input of experimental and accurate theoretical data for the fragments, it is not unreasonable to suppose that the method can yield results comparable to those from larger *ab initio* calculations. It is for this reason that the DIM approach has been particularly favoured by dynamicists as a means of producing entire surfaces and of characterizing interactions between them.

Rather than present a catalogue of DIM applications to small molecules, metal clusters, conjugated hydrocarbons, solids and so on, we have selected a few of those studies of potential surfaces for reactive systems that we have found particularly interesting or informative. For example, Meier *et al.*[108] found in a study of collinear He–H–He$^+$ that the DIM formalism can give good results for geometries and binding energies and can produce accurate surfaces. They caution that care is essential to make sure that a sufficient number of configurations is included. In some cases it is necessary to adjust the excited diatomic potential energy curves to obtain results of usable accuracy. A problem mentioned by a number of authors concerns the shifting or scaling of potential curves. This might seriously affect the nature and position of (avoided) crossings between different states and, for some studies, this is more important than obtaining correct asymptotic energies.

The collinear reaction Ca + HCl → CaCl + H exemplifies another problem: accurate potential curves are unlikely to be known for all the neutral and ionic diatomic fragments. This system has been studied by Isaacson and Muckerman[109], who were able to demonstrate that most of the prominent features of their results could not be attributed to the large uncertainties in some of the diatomic potentials.

The sensitivity of DIM calculations to the diatomic input has been very carefully considered by Kuntz and Roach[110] in a very extensive case study of the reaction Be + HF → BeF + H. They found good agreement between DIM and (conventional) VB *ab initio* calculations based on identical lists of structures. Some of the diatomic potentials employed were very poor, as a cursory comparison with more recent data demonstrates for the $^2\Pi$ states of BeF, the excited $^2\Sigma^+$ states of BeH, and so on. They caution that use of exact experimental energy curves may not compensate for the exclusion of structures, particularly those that contribute significantly to ground-state diatomic curves. Nonetheless, Kuntz and Roach were able to obtain very reasonable potentials from some of their calculations.

A simplified approach has been proposed by Olson and Garrison[111] who do not partition $\mathscr{A}$. They thus avoid the construction of all the $\frac{1}{2}N(N+1)$ DIM matrices for an N-atomic system. However, their approach is equivalent in the sense that it yields the same surfaces for identical input. We suggest that a possible drawback is that careful analysis of the DIM matrices for the fragments may be important in assessing the reliability of the calculations. Criteria for the selection of the DIM basis, and a practical systematic procedure for extracting fragment matrices from *ab initio* calculations, have been presented by Schreiber and Kuntz[112].

Finally, we mention an important extension called DIM-3C in which three-centre terms are added; these are necessary in the case of a heavy central atom. This approach has been applied successfully to the series of reactive surfaces $H + XY \rightarrow X + HY$ (with X, Y = F, Cl, Br, I) by Last and Baer[113] using a small common set of parameters. For an example of a use of DIM-3C potentials in reaction dynamics, see the study of Cl–HCl by Baer and Last[114].

Clearly the DIM approach will become increasingly important for the calculation of potential surfaces, especially those for chemical reactions. However, it must be used with caution: *quantitative* features are certainly very sensitive to poor input data. On the other hand, the use of experimental diatomic potentials does not ensure *qualitatively* correct surfaces if sufficiently careful analysis of the nature of the bonding and of the important structures has not been performed. The use of DIM as a 'black-box' procedure is certainly not to be recommended.

We conclude this section with a brief discussion of a closely related *ab initio* method, which always gives rise to Hermitian matrices at all stages, but requires correspondingly more computation. Consider a simple chemical reaction of the type

$$A + BC \rightarrow AB + C$$

which nevertheless covers many experimental situations under study. When atom A is far from the BC molecule, the system is well described by wavefunctions of the form

$$\Psi_i^{(\alpha)} \equiv \{\Phi_{i_A}^{(A)}\Phi_{i_{BC}}^{(BC)}\} \tag{62}$$

Here $\Phi_{i_A}^{(A)}$ stands for a Hartree–Fock or multiconfiguration wavefunction for atom A in state i_A; for a single configuration, it is identical to the function $\zeta_{i_A}^{(A)}$ of Eq. (53). Similarly $\Phi_{i_{BC}}^{(BC)}$ represents an MO or compact MCSCF wavefunction for molecule BC in state i_{BC}. The orbitals which make up $\Phi_{i_A}^{(A)}$ or $\Phi_{i_{BC}}^{(BC)}$ are in general orthonormal. The index i on $\Psi_i^{(\alpha)}$ stands for the pair i_A, i_{BC}, $i \equiv (i_A, i_{BC})$, and the superscript (α) indicates the particular *arrangement** of the system into A + BC. The brackets { } indicate multiplication by an N-electron spin function and complete antisymmetrization as in Section III. We assume that the spins of A and B are coupled to give the required overall resultant as before. Similar wavefunctions are constructed for the other possible arrangements, e.g.

$$\Psi_j^{(\beta)} \equiv \{\Phi_{j_C}^{(C)}\Phi_{j_{AB}}^{(AB)}\} \quad \text{and} \quad \Psi_k^{(\gamma)} \equiv \{\Phi_{k_B}^{(B)}\Phi_{k_{AC}}^{(AC)}\}$$

The total wavefunction for the ABC system is written as a superposition of the different possible situations as follows:

$$\Psi = \sum_i C_i^{(\alpha)}\Psi_i^{(\alpha)} + \sum_j C_j^{(\beta)}\Psi_j^{(\beta)} + \sum_k C_k^{(\gamma)}\Psi_k^{(\gamma)} \tag{63}$$

*This name corresponds to that used in the quantum theory of reactive scattering.

The total energies and coefficients $\mathbf{C} \equiv (\mathbf{C}^{(\alpha)}, \mathbf{C}^{(\beta)}, \mathbf{C}^{(\gamma)})$ are determined by solution of a secular equation of the form (55). It is clear that the wavefunctions $\Psi_i^{(\alpha)}$, $\Psi_j^{(\beta)}$, $\Psi_k^{(\gamma)}$ are non-orthogonal, since the constituent parts in each are formed from different sets of orthogonal orbitals. By construction, function (63) describes correctly any mode of disintegration of the ABC system into atom plus diatomic molecule. This would not be possible if a single set of orthogonal orbitals were used as in the MO-CI approach without including an excessive number of configurations. Wavefunction (63) also takes into account—at least to some extent—the central drawback of classical VB theory: the allowance for deformation as atom A, say, approaches BC. This is provided by the presence of functions from arrangements β and γ, whose coefficients will increase as the A–(BC) distances decrease.

The resulting Hamiltonian matrix possess the following block structure:

$$\mathbf{H} = \begin{bmatrix} \mathbf{H}^{\alpha\alpha} & \mathbf{H}^{\alpha\beta} & \mathbf{H}^{\alpha\gamma} \\ \mathbf{H}^{\beta\alpha} & \mathbf{H}^{\beta\beta} & \mathbf{H}^{\beta\gamma} \\ \mathbf{H}^{\gamma\alpha} & \mathbf{H}^{\gamma\beta} & \mathbf{H}^{\gamma\gamma} \end{bmatrix} \tag{64}$$

the indices α, β, etc., indicating the two arrangements connected by a particular block, e.g.

$$H_{ij}^{\alpha\beta} = \langle \Psi_i^{(\alpha)} | \mathscr{H} | \Psi_j^{(\beta)} \rangle \equiv \langle \{\Phi_{i_A}^{(A)} \Phi_{i_{BC}}^{(BC)}\} | \mathscr{H} | \{\Phi_{j_C}^{(C)} \Phi_{j_{AB}}^{(AB)}\} \rangle \tag{65}$$

As indicated above, the individual fragment wavefunctions, $\Phi_{i_A}^{(A)}$, $\Phi_{j_{AB}}^{(AB)}, \ldots,$ etc., are calculated separately beforehand by, say, MCSCF methods. However, matrix elements such as (65) are computed using a VB program.

Wavefunctions of the form (63) have been used in a study of the Rydberg states of H_2O, in which a total of just 60 structures drawn from two arrangements $\{\Phi^{(H(1))}\Phi_i^{(OH(2))}\}$ and $\{\Phi^{(H(2))}\Phi_j^{(OH(1))}\}$ were included to produce potential surfaces for a number of states to good accuracy[115]. The configurations which make up the states of the OH fragment were selected so as to reproduce correctly the known excitation energies T_e of OH. Multiphoton excitation of the lowest $^1B_1\,(^1A'')$ state of H_2O[116] predissociates to give OH + H, the OH fragments being found to be exclusively in the ground $(X^2\Pi)$ state[117]. The mixed MO-VB calculations show that, as a function of the H–OH distance R(H–OH), the $^1A''$ state, which arises from $OH(X^2\Pi)$ + H(1s), is strongly repulsive at large R(H–OH) distances. However, at shorter R(H–OH) separations it interacts strongly with a second $^1A''$ state which is highly attractive. This state stems from a Rydberg state of OH $(2\sigma^2 3\sigma_{ky}^2 3s; {}^4\Sigma^-)$ and H(1s), and as a result an inner well is formed in the $^1A''$ surface. The observed predissociation takes place from this well. The somewhat awkward concept of 'Rydbergization' of the H(1s) orbital (which is invoked in CI calculations of the similar process $NH_3^* \rightarrow NH_2 + H$) is entirely unnecessary[118].

Similar calculations on reactions such as $Be + F_2 \rightarrow BeF + F$ are in progress[115]. In this case additional functions corresponding to $Be^+ + F_2^-$ need to be included since these are actually occurring species.

B. Semi-empirical Valence Bond Methods

The VB approach has been particularly popular for studies of π-electron systems, mostly using semi-empirical techniques. We shall, however, concentrate on a few of the semi-empirical VB treatments that have been used to obtain potential surfaces for reactive systems.

Much work has been done in this field by Zeiri and Shapiro, and subsequently by others. For example, semi-empirical potential surfaces have been reported for the alkali atom (M = Li, Na, K, Rb) and halogen molecule ($XY = F_2, Cl_2, Br_2, I_2$) reactions[119]. In this model, the three valence orbitals were represented by Slater-type orbitals centred on M, X and Y. Three VB structures were used: a covalent structure based on the spatial configuration MXY and two ionic structures based on M^+X^-Y and M^+XY^-. The full Hamiltonian was written as a sum of diatomic and atomic terms, but, contrary to the DIM formalism, only the ground-state potentials were required. The approach of Zeiri and Shapiro is very cheap and can be applied to heavy systems.

Goldfield *et al.*[120] used a very similar model to study in more detail potential energy surfaces for the reactions of M (Li, Na, K, Rb, Cs) with Br_2. The three valence orbitals were represented as *ns* Slater-type functions and the three VB structures were written:

$$(1/\sqrt{2})\{|M(1)\overline{X(2)}Y(3)| - |M(1)X(2)\overline{Y(3)}|\} \qquad \text{covalent}$$

$$|X(1)\overline{X(2)}Y(3)| \qquad \text{ionic}$$

$$|X(1)\overline{Y(2)}Y(3)| \qquad \text{ionic}$$

These surfaces were subsequently used[121] to calculate reactive cross-sections and good agreement with experimental data was obtained at low energies. They found a vibrational barrier between $M + Br_2$ and $M^+ + Br_2^-$ which plays an important role in the reaction (vibrational capture). Goldfield *et al.*[120,121] claim that some of their surfaces are the most reliable currently available. It would be very interesting indeed to see how DIM potentials compare with these.

Zeiri *et al.* have reported an ambitious application of a semi-empirical VB formalism to dissociative electron transfer on ionic surfaces[122]. They studied electron transfer from a charged AgBr surface to Ag^+ ions and from Br^- ions to AgBr surface holes. With very few VB structures it is possible to describe such complex processes quite well.

As a simpler application we mention the work of Zeiri and Balint-Kurti[123] who were interested in the photodissociation of the alkali-metal halides. Semi-empirical potential curves obtained using a very similar model have been subsequently used to study mutual neutralization and chemi-ionization in collisions of alkali-metal and halogen atoms[124]. The Hamiltonian and overlap matrix elements were expressed in terms of ionization potentials, electron affinities, polarizabilities and a few two-centre atomic overlap integrals. The structures involved in the valence-shell covalent–ionic resonance model were

$$X_1 = |\mathrm{p}_z(\mathrm{X})\overline{\mathrm{p}_z(\mathrm{X})}| \qquad \text{ionic}$$
$$X_2 = (1/\sqrt{2})\{|\mathrm{s(M)}\overline{\mathrm{p}_z(\mathrm{X})}| - |\overline{\mathrm{s(M)}}\mathrm{p}_z(\mathrm{X})|\} \qquad \text{covalent}$$
$$X_3 = (1/\sqrt{2})\{|\mathrm{p}_z(\mathrm{M})\overline{\mathrm{p}_z(\mathrm{X})} - |\overline{\mathrm{p}_z(\mathrm{M})}\mathrm{p}_z(\mathrm{X})|\} \qquad \text{excited covalent}$$

and solution of the 3 × 3 secular equation gave 'valence' energies. For each value of the internuclear distance R, the core energy was estimated by constraining the diabatic ionic curve to be a T-Rittner potential with experimental parameters. The three (diabatic) potentials are shown schematically in Fig. 13. Features of the potentials that made them very suitable for the scattering calculations included: correct dissociation; 'sensible' short-range forms; a good description of the ionic curve; reasonable R values for the crossing points because the covalent curves are essentially flat and the ionic curve is essentially $-1/R$ at long range.

As mentioned above, a detailed comparison between such a semi-empirical VB method and a DIM calculation for a system such as $K + Br_2$ would be very worth while to ascertain the relative merits of these approaches.

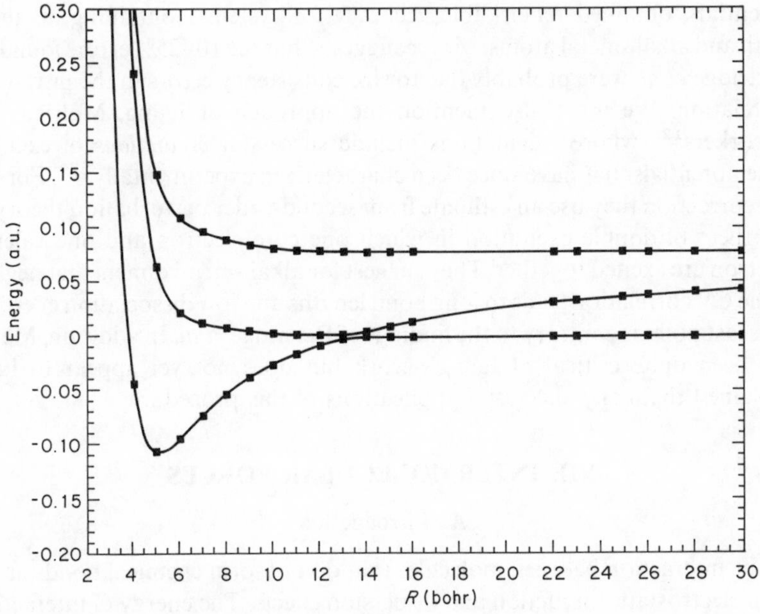

Fig. 13. Semi-empirical potential energy curves for the study of collisions in Na I.

C. Valence-shell Calculations

High-accuracy all-electron calculations on small molecules containing atoms beyond the first or second rows of the Periodic Table are difficult and expensive. All-electron calculations on larger molecules containing many such atoms are currently impossible, even at the SCF level. Techniques have been proposed for approximately accounting for the core electrons in a valence-electrons-only calculation. For example, pseudo-potentials are in widespread use, often including relativistic effects for very heavy atoms. Most of the applications of these have not been in VB calculations. Nonetheless, it is useful to consider briefly some of the methods or conclusions that may be relevant for achieving high accuracy in VB calculations on molecules containing heavier atoms.

The importance of core–valence (CV) correlation effects has long been recognized[125]. In approximate treatments, many workers have ignored terms that were present in the careful analysis of Bottcher and Dalgarno[125]—good results have been a reflection of cancellation of errors. The inclusion of CV correlation effects in *ab initio* CI calculations results in contraction of the valence shell; this contraction decreases along a row and increases down a column of the Periodic Table. The effect is particularly important for Group I and II elements. Excluding core–core correlation, Partridge *et al.*[126] suggested for alkali-metal dimers that CV effects reduce R_e but do not significantly affect D_e. Their dissociation energies were not, however, very reliable. Meyer and coworkers[127], who employed effective core polarization potentials, obtained remarkable accuracy for systems containing alkaline-earth and alkali-metal atoms. Meyer suggests that the 10–25% errors found by Partridge *et al.* were probably due to size consistency errors in the intra-core correlation. We must also mention the approach of Jeung, Malrieu and coworkers[128] whose calculations include successful *predictions* of excited-state potentials that have since been characterized experimentally[129]. For the CV correction they use an estimate from second-order perturbation theory of the effect of double excitation in which one core electron and one valence electron are excited together. They suggest for alkali-metal dimers that neglect of the CV correlation leads to long bond lengths and low dissociation energies. This last point is contrary to the findings of Partridge *et al.* In addition, Meyer has been quite critical of Jeung's work but does not yet appear to have explained the many successful applications of this procedure.

VII. INTERMOLECULAR FORCES

A. Introduction

The interaction between molecules that do not form chemical bonds arises from electrostatic, induction and dispersion effects. The energy of interaction is about 10^{-5} a.u., and in such situations a description based upon the role of the individual subsystems is obviously suitable. The use here of a 'supermolecule' approach—except possibly at very short interatomic separations—flies in the face of the physical situation.

In particular, the VB method allows one to calculate both the attractive and repulsive parts of the potential with a consistent formalism. The approach is similar to a perturbation treatment and yields a physically meaningful decomposition of the interaction energy in terms of electrostatic, inductive and dispersive contributions. The role of charge-transfer effects, and of correlation effects within the monomers, can be directly investigated[130,131]. The non-physical 'basis-set superposition error' (BSSE)—which is an unavoidable concomitant of supermolecule descriptions—can be excluded in a clear physical way[130–134]. Using structures carefully chosen according to criteria discussed below, intermolecular potentials determined in this manner are numerically reliable over the range of intermolecular separations sampled by collisions at thermal energies. Very short-range forces, which are important in high-energy scattering experiments, are indeed more conveniently studied by 'supermolecule' methods since the individual subsystems no longer play a significant role.

These features of the VB approach, however, are preserved only under certain conditions[130,131,133,134]. One of them is the direct treatment of the non-orthogonality, which becomes important when the separation between the systems decreases to the sum of the van der Waals radii and below. The approach is based upon the variation theorem, and consequently the interaction energy is computed as the difference between two large numbers. This difference is about 10^{-5}, and therefore very high accuracy is required in the computation in order not to lose all significant figures. However, under certain conditions this cancellation can be entirely avoided[134] (see below).

The study of charge-transfer effects must be carried out with particular care, since the straightforward inclusion of ionic structures leads to the appearance of BSSE—for which no satisfactory correction presently exists. Even when correction is attempted, the presence of BSSE leads to results which are physically unreliable and difficult to interpret[135].

The fact that the interaction energy is so small suggests strongly that this should be reflected in the calculational procedure, and several important initiatives have been made recently. We refer to these in due course.

B. Method of Calculation

The basic strategy of the method was proposed by Wormer and van der Avoird who first applied it to the case of the He–He interaction[132]. They showed that very simple VB wavefunctions provided the self-same interaction energies as fairly sophisticated MO-CI (supermolecule) calculations. Of particular interest is the demonstration that the BSSE disappears when ionic structures are excluded. This simple criterion is obviously reasonable for this case.

The Nijmegen group emphasized the connection of the VB method with perturbation theory and with the multipole expansion. The use of monomer MOs obtained from SCF calculations on the separate fragments led for the first time to the use of large basis sets in VB-type calculations. In addition to the He–He work[132], calculations were carried out on the He–H_2 [137] interaction, on the ethylene dimer[138], and more recently on the N_2 dimer[139], the Ne–H_2 [140] potential and the Heisenberg exchange interaction in O_2 [141]. Much insight into the nature and implications of the BSSE was also gained.

The computations were carried out using programs based upon the Reeves algorithm[142]. This requires as input orthonormal spatial orbitals, and for this purpose the Löwdin ($\Delta^{-1/2}$) procedure[143] was used, which minimizes the mixing of the basis orbitals and at the same time preserves their spatial symmetry. However, this orthogonalization is clearly unphysical and introduces an error which depends on overlap and distance. Although it did not significantly influence the result in the He–He case (which was a small calculation), this is not generally so. This error cannot be rectified exactly, and remains the chief drawback of the method.

The general procedure of these calculations is now briefly outlined. We consider two molecules A and B, and introduce orbitals $\phi_A^0, \phi_B^0, \phi'_A$ and ϕ'_B. Orbitals ϕ_A^0, ϕ_B^0 are SCF molecular orbitals which are occupied in the ground states of A and of B respectively, and ϕ'_A, ϕ'_B are 'excited' orbitals of the respective fragments satisfying Brillouin's theorem. In general, these last are not virtual SCF orbitals, but are instead functions suitable for describing correctly the properties of the individual molecules which are connected with intermolecular forces, such as polarizability. The wavefunctions Φ_A^0 and Φ_B^0 represent ground-state SCF wavefunctions for A or B, and are constructed out of the orbitals ϕ_A^0 or ϕ_B^0. Functions Φ'_A and Φ'_B are perturbed states of A or B. They are obtained from ground-state functions when one occupied molecular orbital, ϕ_A^0 say, is replaced by an excited orbital ϕ'_A. These represent 'local single excitations'. The Φ'_A and Φ'_B are generally not spectroscopic states, nor even approximations to such states, but are instead 'pseudo-states': Their matrix elements with the ground-state wavefunction determines the response of the subsystem A or B to an external perturbation, such as an electric field F.

The following kinds of structures are included in the VB calculations.

1. $\{\Phi_A^0\Phi_B^0\}$—in the long-range region these describe the electrostatic interactions.
2. $\{\Phi_A^0\Phi'_B\}, \{\Phi'_A\Phi_B^0\}$—these represent A in the ground state interacting with B in the perturbed state and vice versa and in the long-range limit they express polarization effects.

3. $\{\Phi'_A \Phi'_B\}$—these represent perturbed molecule A interacting with perturbed molecule B, and in the long-range region they describe the dispersion energy. It should be stressed that the full Hamiltonian is used without any multipole expansion. It is unfortunate that many of the obvious benefits of this procedure are dissipated by the orthogonalization of the orbitals.

A similar approach has been pursued by Daudey, Claverie and Malrieu[144], who used the same orthogonalization procedure for the orbitals but determined the interaction energy by means of classical perturbation theory.

The VB program[19,88] described in Section V has enabled the Milan University group to apply the procedure described above without resorting to orthogonalization of the basis sets centred on the interacting fragments. The method has been applied to the He–He, He–Li[130] and He–HF[133] interactions. The He–HF study was carried out using an extended basis set which results from its optimization of energy, dipole moment and polarizability. Its suitability was determined by computing the Coulomb, exchange, induction and dispersion contributions using basis sets that were systematically increased. The limiting values of each contribution were determined together with an estimate of the uncertainties in the anisotropy of the resulting potential. Pseudo-state orbitals ϕ'_{He}, ϕ'_{HF} were obtained by means of a finite field technique. The results are well in line with experimental results for the interaction of HF with rare-gas atoms other than He. However, there is disagreement with previous theoretical results[145] which were obtained using smaller basis sets, and with a 'supermolecule' (MO) approach. The results are also in disagreement with the 'Hartree–Fock and damped dispersion' (HFD) treatment of Rodwell *et al.*[146]. This is due to some inadequacy both in the HFD damping function used and in the HFD model itself. Charge-transfer effects were found to be negligible.

C. Charge Transfer

In order to assess the importance of charge transfer, it is necessary to include structures which describe ionic terms. In principle the set of structures $\{\Phi_A \Phi_B\}$ is complete*, and consequently it is possible to describe any state of the dimer system AB. However, when AB is more closely represented by an ionic wavefunction, $\{\Phi_{A^+} \Phi_{B^-}\}$ for example, expansion of the total wavefunction in terms of states $\{\Phi_A \Phi_B\}$ will converge very slowly. This can only be improved by including ionic structures (see Murrell *et al.*[147] for a more general discussion of this problem). But in addition to this difficulty, which is technical in nature, there is the more physical aspect concerning the nature of the electronic structure of the AB system. The question as to whether the complex

*It is in fact overcomplete: recall that the parentheses mean multiplication by a spin function and complete antisymmetrization.

must be described as AB or A^+B^- (or A^-B^+) is relevant particularly when A has a low ionization potential and B a high electron affinity.

For this purpose, we introduce orbitals of the type $\phi^0_{A^+}$, $\phi^0_{B^-}$ and ϕ'_{A^+}, ϕ'_{B^-} which are obtained from separate calculations on the isolated systems A^+ and B^-. The VB structures for A^+B^- are generated in the same way as for the AB system. It is clear that wavefunctions of the form $\{\Phi_A\Phi_B\}$ are optimal for describing AB, while $\{\Phi_{A^+}\Phi_{B^-}\}$ are similarly so for A^+B^-. In this last case, the wavefunctions describe a real state A^+B^- formed from parent ions. According to the variational principle, the actual wavefunction for the system may be expressed as a linear combination of these two types of structures. The consequent lowering of the energy and the resulting values of the coefficients provide a description of the true state of AB.

This procedure has been applied to the He–Li system[130]. The computed well depth compared very well with the experimental value, while a previous theoretical study, which used a much larger basis set in the framework of an MO treatment, gave a value for the well depth three times as large[148]. Charge-transfer effects were found to be negligible—again in disagreement with the MO-based calculation.

It should be stressed that, although the basis set used in the VB calculation is of good quality, it is smaller than that employed in the MO treatment. It follows from this that, if the method of including charge transfer had been incorrect, the contributions of He^-Li^+ structures would have been exaggerated in order to compensate for the lack of flexibility of the smaller basis set employed for He–Li. This is in fact the misleading result of methods which try to describe charge-transfer effects by means of straightforward excitations from occupied orbitals of A to virtual orbitals of B and vice versa.

The key to the elimination of the BSSE is therefore the use of structures $\{\Phi_A\Phi_B\}$, $\{\Phi_{A^+}\Phi_{B^-}\}$ and $\{\Phi_{A^-}\Phi_{B^+}\}$ formed from wavefunctions for actual states of A, B, A^+, B^-, A^- and B^+. This approach enables one to decide almost immediately whether or not charge-transfer effects are likely to be important, for it is necessary actually to determine wavefunctions of A^- and B^-. The rare gases do not form stable negative ions except as transitory resonance states in which the parent atom is excited. Hence wavefunctions corresponding to $He^-(1s^22s\,^2S)$ for example assume an increasing continuum character for the extra electron as the basis set is improved, and in actual fact no true He^- wavefunction can be determined for use in forming a structure $\{He^-Li^+\}$. Since the interaction energy in a dimer is so small, we can assume that this is also insufficient to stabilize an otherwise unstable negative ion.

D. The Intramolecular Correlation Energy

For systems which interact so weakly, one would not expect the correlation energy within a subsystem to vary significantly with intermolecular separations. The identity of the individual fragments always remains well defined and no electron spin recouplings occur. It follows that there will be no great difference in the intermolecular potential when the wavefunctions of the separate monomers are computed at the CI level instead of by SCF theory. However, if the interaction potential is required with very high accuracy, this effect must be included. The He–He case is typical in this respect.

In order to include this phenomenon, two procedures are possible: (i) the wavefunctions of A and B are represented by CI expansions; or (ii) the wavefunctions are determined at the spin-coupled VB level.

(i) The wavefunctions of A and B are represented by CI expansions. The orbitals and interacting configurations are determined by means of separate calculations on the monomers[131]. The structures $\{\Phi_A^0\Phi_B^0\}$ for the interacting system now become linear combinations of configurations whose coefficients are determined at each intermolecular separation. Since it is easy to include only those configurations which correspond to the AB system, no BSSE occurs in this step.

The same procedure can be followed for describing $\{\Phi_A^0\Phi_B'\}$, $\{\Phi_A'\Phi_B^0\}$ and $\{\Phi_A'\Phi_B'\}$, but the number of configurations rapidly becomes very large. Thus if Φ_A^0 and Φ_B^0 each consist of a Hartree–Fock function plus double excitations, then $\{\Phi_A^0\Phi_B^0\}$ includes double and quadruple excitations relative to the ground-state AB. In order to maintain a consistent description, the pseudo-states, Φ_A' for example, must consist of a set of singly excited functions plus further double replacements, i.e. one- and three-fold excitations relative to the Hartree–Fock ground state of A. Structures $\{\Phi_A'\Phi_B'\}$ for the dimer consequently include two-, four- and six-fold excitations. In essence, one requires that the AB wavefunctions fulfil the condition of size consistency.

However, the method has been applied to the He–He interaction[131] in order to determine accurately the contribution of the intra-atomic correlation energy to the repulsive Coulomb and exchange part of the interaction in the van der Waals region for which no pseudo-states are needed. The results are shown in Table VIII. A compact CI wavefunction was used for each He atom, consisting of an expansion in four natural-orbital (NO) configurations of type (ns^2), two of (np^2) and one of ($3d^2$). The contribution of configurations ($nsn's$ ^1S, ^3S) to the He–He interaction energy was found to be negligible. The basis set was of even-tempered type and increased systematically as shown. At the SCF level, the basis-set limit for the Coulomb and exchange part of the interaction energy corresponds to a temperature 9.69 K. For correlated wavefunctions, the s—only limit is 11.86 K, the limit being 10.70 K for s + p orbitals and 10.58 K for s + p + d orbitals. The net contribution of the intra-atomic correlation energy to this part of the interaction energy is 8% of the total.

(ii) A more general and promising approach, though applicable only to small systems, is based upon the determination of the wavefunction at the

TABLE VIII
He–He Coulomb and exchange interaction energies at 5.6 bohr[131].

	Atomic basis set	Interaction energy (K)
HF	12s	9.53
	14s	9.63
	16s	9.67
	18s	9.69
CI[a]	12s	11.60
	14s	11.75
	16s	11.84
	18s	11.86
	16s3p	10.67
	16s4p	10.70
	16s5p	10.70
	16s4p1d	10.59
	16s4p2d	10.58

[a]The CI wavefunction for He consists of four NOs of type (ns^2), $n = 1$–4; two NOs of type (np^2), $n = 2, 3$; and one NO of type ($3d^2$).

spin-coupled VB level of theory. In order to avoid the occurrence of BSSE, the spin-coupled orbitals—both occupied and virtual—of A or B are expanded in basis functions stemming respectively from either molecule A or B only. In this approach, the spin-coupled orbitals are allowed to 'relax' at each intermolecular separation. The advantage of this procedure is that the subsequent VB expansion converges very quickly. Work on He–He, He–H_2, He–Li, He–Be and He–LiH is in progress[149].

Both procedures (i) and (ii) share the important property of size consistency, so that at least in principle no corrections for lack of this should be needed.

A different VB strategy has been developed by Gallup *et al.*[86,136]. The approach is variational and the non-orthogonality between basis functions situated on different fragments is also treated directly. However, the list of structures includes direct transfer of electrons from orbitals centred on one system to orbitals centred on the other. Consequently, the BSSE makes its appearance. Gallup *et al.*[136] assert that it can be eliminated by selecting a specially optimized basis set. Their most accurate calculation for the He–He interaction was carried out with a basis set of the type (s, s′; p, p′, p″; d) at each centre. Each orbital was separately optimized as follows:

1. Orbital s—an SCF orbital contracted from 10 primitive Gaussian functions.
2. Orbital s′—an s orbital contracted from six primitive Gaussians which maximizes the radial correlation energy in the atom.

3. Orbital p—contracted from a set of six p primitive Gaussians which approximates a natural orbital of the atom; it is equivalent to an STO-6G with an exponent of $2.45a_0^{-1}$.
4. Orbital p′—another STO-6G function with exponent $2.05a_0^{-1}$ which optimizes the full CI energy of He in the basis s, s′, p, p′.
5. Orbital p″—STO-6G function with exponent $1.3a_0^{-1}$ which optimizes the dispersion energy of He–He at $R = 5.6a_0$.
6. Orbital d—an STO-3G function with exponent $1.75a_0^{-1}$ optimized as above. Several calculations were carried out which included up to 3694 Slater determinants. A total interaction energy of 10.05 K was obtained, compared to the experimental value of 10.74 K. In all of these calculations, the computed interaction energy is less than the experimental value. This gives support to the conjecture that once the basis set is carefully optimized, the calculated curve of the interaction energy always lies above the curve of the true interaction energy. However, it remains to be shown that the results are not due to a compensating effect and why, when working in the framework of VB theory, this method should be considered preferable to, or more convincing than, those which avoid BSSE by omitting the structures responsible for it[130–134].

E. Intermolecular Perturbation Theory

Closely related to the strictly variational VB method described so far, there have been a number of recent approaches which use perturbation theory. All of these are characterized by the use of non-orthogonal functions and fully antisymmetrized wavefunctions. In addition the full Hamiltonian is used without a multipole expansion.

Stone and Hayes[135] have developed a method based upon the SCF molecular orbitals of the separate fragments. Single and double excitations are included. The secular determinant (cf. Eq. (55))

$$\det(\mathbf{H} - E\boldsymbol{\Delta}) = 0 \tag{66}$$

is expanded to obtain an estimate of the lowest eigenvalue in the form:

$$E = H_{00} - \sum_p \frac{(H_{0p} - \Delta_{0p}H_{00})^2}{H_{pp} - H_{00}} \tag{67}$$

In this expression the subscript 0 denotes the zeroth-order function $\{\Phi_A^0\Phi_B^0\}$ where Φ_A^0 and Φ_B^0 are SCF wavefunctions (single Slater determinants in the closed-shell case) of the respective monomers[130,132]. The index p runs over all single and double excitations from the zero-order function. The same expression is also used by Magnasco and Musso[150] but in this case index p runs over singly excited configurations only; dispersion effects are not included in their treatment. In addition the unoccupied orbitals are expressed in the form of antibonding orbitals.

In both these approaches the excitations are not constrained to be localized on the same fragment. Consequently, basis-set superposition errors of various types occur, and there are difficulties in correcting for them. In fact, the corrections turn out to be dependent upon the basis set, and a number of techniques are proposed for the removal of these shortcomings. While the errors in the cases treated by this perturbation approach (see below) do not appear to be too serious, there is as yet no definitive procedure for their rectification[135,151,152].

Important simplifications for the case of intermolecular forces introduced by Hayes and Stone[94], and generalized by Figari and Magnasco[98], concern the calculation of the matrix elements of the Hamiltonian between Slater determinants constructed from non-orthogonal orbitals by means of the Löwdin formula. This has been described in Section V. The method has been applied to a number of interesting problems such as He–Be, Ar–HF, Ar–HCl, Ne–H_2 and Ne–HF[135,152] and recently to the dimer of H_2O[153]. 'Split valence' or more extended basis sets were employed, and charge-transfer effects were found to be significant. However, the previous discussion concerning the BSSE must be borne in mind, particularly as the method is prone to overestimate charge-transfer contributions originating in this way.

A variation perturbation approach has recently been given by Gallup and Gerratt[134]. We begin from the secular equation for the lowest eigenvalue E_0 of the dimer:

$$(\mathbf{H} - E_0\mathbf{\Delta})\mathbf{C}_0 = 0 \tag{68}$$

in which $\mathbf{C}_0$ is the eigenvector corresponding to E_0. This equation is satisfied for all separations between the interacting monomers. The Hamiltonian $\mathbf{H}$ is constructed from the sets of structures $\{\Phi_A^0\Phi_B^0\}$, $\{\Phi_A^0\Phi_B'\}$, $\{\Phi_A'\Phi_B^0\}$ and $\{\Phi_A'\Phi_B'\}$ discussed above. We introduce the matrix $\mathbf{M}$ which diagonalizes $\mathbf{H}$ and $\mathbf{\Delta}$ at infinite separations, and use it to transform Eq. (68) to

$$[\bar{\mathbf{H}} - V(R)\bar{\mathbf{\Delta}}]\mathbf{D}_0 = 0 \tag{69}$$

In this equation,

$$\bar{\mathbf{H}} = \mathbf{M}^\dagger(\mathbf{H} - E_\infty\mathbf{\Delta})\mathbf{M}$$
$$\bar{\mathbf{\Delta}} = \mathbf{M}^\dagger\mathbf{\Delta}\mathbf{M}$$
$$\mathbf{D}_0 = \mathbf{M}^{-1}\mathbf{C}_0$$

E_∞ is the value of E_0 at infinite distance, i.e. $E_\infty = E_A + E_B$, where E_A and E_B are the respective energies of the individual fragments A and B. Eq. (69) now contains the interaction energy V, shown with explicit dependence upon the monomer–monomer separation R, and is defined as

$$V(R) = E_0 - E_\infty \tag{70}$$

The matrices $\bar{\mathbf{H}}$ and $\bar{\mathbf{\Delta}}$ are partitioned in the form

$$\bar{\mathbf{H}} = \begin{bmatrix} \bar{\mathbf{H}}_{00} & \bar{\mathbf{H}}_{0e} \\ \bar{\mathbf{H}}_{e0} & \bar{\mathbf{H}}_{ee} \end{bmatrix} \qquad \bar{\mathbf{\Delta}} = \begin{bmatrix} \bar{\mathbf{\Delta}}_{00} & \bar{\mathbf{\Delta}}_{0e} \\ \bar{\mathbf{\Delta}}_{e0} & \bar{\mathbf{\Delta}}_{ee} \end{bmatrix}$$

The subscript 0 indicates the ground state of the dimer $\{\Phi_A^0\Phi_B^0\}$ as before, and e denotes all the excited states. Since $\bar{\mathbf{H}}_{00}$ and $\bar{\mathbf{\Delta}}_{00}$ are 1×1 matrices, we obtain from (5) the following implicit equation for V[134,154]:

$$V = \frac{1}{\Delta_{00}}[\bar{H}_{00} - (\bar{\mathbf{H}}_{0e} - V\bar{\mathbf{\Delta}}_{0e})(\bar{\mathbf{H}}_{ee} - V\bar{\mathbf{\Delta}}_{ee})^{-1}(\bar{\mathbf{H}}_{e0} - V\bar{\mathbf{\Delta}}_{e0})] \tag{71}$$

In general the terms involving V on the right-hand side are small compared to $\bar{H}_{00}$, so that V can be quickly determined by iteration.

It is important to stress that in the element $\bar{H}_{00}$ there is an *exact* cancellation with E_∞ so that only interaction terms are left[155]. This cancellation does not depend upon how accurately E_∞ represents the true values of the molecular energies at infinite separation, but depends only upon the VB form of the wavefunction. It means that V has the same percentage error e as the most approximate quantity used to determine e—usually the intersystem integrals (see below). Tests carried out on the Ne–HF system (see below) show conclusively that V is the same whether computed directly by Eq. (71) or by solving (68) and using (70), even when some two-electron integrals over Slater orbitals are accurate to no more than three decimal places. In the linear Ne–HF configuration with the Ne–HF (centre-of-mass) distance at $7a_0$, Eq. (71) gives $V = -0.000\,189\,6082$ a.u., while (68) gives $E_0 = -228.590\,811\,15$ a.u. and $E_\infty = -228.590\,621\,54$ a.u.

The same considerations go through if $\bar{H}_{00}$ is determined from a linear combination of wavefunctions Φ_A^0 and Φ_B^0 which provide correlation in the two fragments—as long as the coefficients in the linear combinations remain fixed at the isolated monomer values. That is, the potential V is obtained without any serious differencing effects only if the correlation within the ground-state wavefunctions is 'frozen'. Indeed the same is true of variational calculations carried out under the same conditions.

This development takes into explicit account the fact that intermolecular interactions are weak, i.e. that $V \sim 10^{-5}$ a.u. Within the same spirit, Gallup and Gerratt introduced a systematic series of further approximations[134]: The matrix elements $\bar{\mathbf{H}}_{00}$ and $\bar{\mathbf{H}}_{0e}$ (in which excited states interact directly with the ground state) are calculated exactly, and also the diagonal elements of $\mathbf{H}_{ee}$. However, the great bulk of the matrix elements which occur in the off-diagonal parts of $\bar{\mathbf{H}}_{ee}$ are obtained by neglecting the non-orthogonality between the orbitals of the two monomers. This means that, as the number of configurations increases, the number of matrix elements to be evaluated exactly increases only linearly with the order of the matrices, while the number treated approximately increases as the square of the order. This approximation has been checked in a number of instances, and it is found that the value of V is changed by about 1 cm^{-1} compared to evaluation without approximation. At the same time the computation time is reduced by about an order of magnitude.

Further approximations concern the multicentre intersystem integrals[134]. As the monomers approach, the 'tails' of the respective wavefunctions overlap and consequently the exchange interaction begins to contribute significantly. This causes the change-over from the long-range form of the interaction, where $V \sim \sum_n C_n R^{-n}$, to a form better fitted by functions with an exponential factor. The outer parts of the wavefunction are therefore responsible for the important 'damping' effect and hence essentially determine much of the shape of the intermolecular potential. Consequently it is crucial that the electronic wavefunctions possess the correct (exponential) behaviour at large distances from the nuclear framework. This can of course always be achieved by using a sufficient number of Gaussian functions, though in practice the temptation is to use less. However, Slater functions do possess the correct properties in this respect, and by using them a more compact representation is achieved in a directly physical way.

The absence of serious differencing effects (within the stated conditions) in the VB framework means that Slater functions may be safely used. However, for large systems, there are very many intersystem integrals whose computation may well dominate the calculation. Consequently it is necessary to implement simplifications that arise as a result of the relatively long distances involved in such integrals. The details may be found in Ref. 134, but are summarized as follows:

1. Integrals $[aa'|r_b^{-1}]$, $[bb'|r_a^{-1}]$ are determined by multipole expansions.
2. Integrals $[ab|r_a^{-1}]$, $[ab|r_b^{-1}]$, $[aa'|bb']$, $[aa'|a''b]$, $[bb'|ab'']$ and $[ab|a'b']$ are approximated using a Fourier transform method.
3. Intra-system integrals are computed without approximation. These are independent of the interaction between monomers and are computed once only.

The method has been applied to the Ne–HF interaction, in which the H–F distance was kept fixed at $1.7328a_0$[156]. The pseudo-states were determined by a method described below. The maximum well depth is 41.6 cm^{-1} and occurs for the linear Ne·····H–F configuration, with the Ne·····H distance approximately $5.5a_0$. A secondary minimum of depth 20 cm^{-1} was found for the H–F·····Ne linear configuration at a F·····Ne distance of about $7.0a_0$. The primary well depth of 41.6 cm^{-1} is in very close agreement with that found by Stone and Hayes[135]. There is, however, a factor of 4 discrepancy in the Ne–FH well (20 cm^{-1} compared to 5 cm^{-1}). Stone and Hayes used a considerably smaller basis set, and this is almost certainly the main reason for this disagreement. Additional errors may also be present as a result of correcting for BSSE.

F. Pseudo-states

'Pseudo-states' or 'effective states' were introduced above as a certain set of excited states whose interaction with the ground-state wavefunction through the appropriate operator determines the response of the system to an external perturbation. The concept of such states was introduced by Murrell and Shaw[157], who showed for the H–H interaction at long range that a single Slater 2p function centred on each H atom with exponent optimized to give the maximum contribution to the dispersion energy provided a value of the C_6 coefficient within 1% of the exact value. The same is true of the He–He interaction: a single configuration (1s2p′ ^{1}P) for each atom with a suitably optimized 2p′ orbital interacts with the SCF function (1s^2 ^{1}S) to yield a value of C_6 within 5% of the true value[158].

Generally speaking, the determination of orbitals ϕ'_A and ϕ'_B, which are best suited to describe the pseudo-states, represents a very important step that is crucial both for the convergence of the expansions of the perturbed states Φ'_A and Φ'_B and for a correct estimate of induction and dispersion contributions to the total interaction energy.

A fairly general approach for this purpose has been given by Gallup and Gerratt[134,159] and proceeds as follows. We consider for simplicity the case of an atom in a finite external electric field F, and form a set of orthonormal functions Φ_i consisting of the ground-state SCF wavefunction and a series of single, double, triple, etc., excitations. The Hamiltonian in the presence of F is given by

$$\mathscr{H} = \mathscr{H}^{(0)} - F\mu$$

where $\mathscr{H}^{(0)}$ is the field-free Hamiltonian, and $\mu = \sum_i z_i$. The secular equation for the lowest eigenvalue of $\mathscr{H}$, $E_0 \equiv E_0(F)$, can be written in the form:

$$\begin{aligned} &[H_{00}^{(0)} - E_0(F)]C_0 - F\sum_{j\neq 0} \mu_{0j}C_j = 0 \\ &\sum_{j\neq 0} [H_{ij}^{(0)} - E_0(F)\delta_{ij}]C_j - F\mu_{i0}C_0 = 0 \qquad (i \neq 0) \end{aligned} \tag{72}$$

The equations divide into two sets because of the parity of the operator μ: functions which contribute to the ground-state wavefunction give a zero contribution to the pseudo-state and vice versa. The pseudo-states Φ_p are thus given by

$$\Phi_p = \sum_{i\neq 0} C_i\Phi_i \tag{73}$$

The polarizability α of the atom is found by taking

$$\alpha = \lim_{F\to 0}\left(-\frac{\partial^2 E_0(F)}{\partial F^2}\right)$$

A very similar procedure was also used to determine the relevant pseudo-states for the He–HF interaction. It can be shown reasonably directly that this procedure is the same (to order F^4) as minimizing the functional

$$J(\Phi_p) = \langle \Phi_p | \mathscr{H}^{(0)} - E_0^{(0)} | \Phi_p \rangle - 2F\langle \Phi_p | \mu | \Phi_0^{(0)} \rangle \tag{74}$$

with respect to variations in the pseudo-state Φ_p, where $\Phi_0^{(0)}$ is the lowest solution of secular equation (72) when $F = 0$, $E_0^{(0)}$ is the corresponding energy and Φ_p is represented as in Eq. (73). Eq. (74) represents the application of the Hylleraas variation method[160] to the determination of 'second-order' properties, and has been extensively used by Burke and Robb[161] for determining atomic polarizabilities.

The representation of the ground state of a monomer by an SCF wavefunction Φ_A^0 or Φ_B^0, and the corresponding pseudo-states Φ'_A or Φ'_B as a sum of single excitations out of Φ_A^0 or Φ_B^0, represents a well defined and consistent description which we might refer to as the 'pseudo-state SCF limit'. Generally speaking, if a sufficiently large basis set is used, the polarizabilities and van der Waals coefficients $C_6, C_8, \ldots$ are given by this procedure to 5–10% of the accurate values (where known); although, particularly for the $C_8, C_{10}, \ldots$ coefficients, *very* large basis sets, which include high angular quantum numbers, are necessary.

However, as discussed briefly above, in order to produce highly accurate results, both the SCF functions and the pseudo-states require additional correlation. Thus both Φ_A^0 and Φ'_A, for example, must be augmented by double replacements. This generates such a large number of configurations that the

TABLE IX
Calculated values for the van der Waals C_6 coefficient.

Monomer A	Monomer B	C_6(uncorrected) (a.u.)	$\Delta_A + \Delta_B$ (a.u.)	C_6(corrected) (a.u.)	C_6(other) (a.u.)
He	He	1.451	0	1.451	1.459[b]
He	Ne	2.582	−0.150	2.755	3.0[c]
Ne	Ne	6.122	−0.300	6.708	6.882[d]
Ne	HF(∥)	7.114	−0.273[a]	7.905	–
Ne	HF(⊥)	4.599	−0.273[a]	5.065	

[a]These shifts were determined on the basis of computed values of $\alpha_\parallel$ and $\alpha_\perp$ for HF; see Ref. 162. The experimental value of $\bar{\alpha} = \frac{1}{3}(\alpha_\parallel + 2\alpha_\perp)$ is unknown, and these theoretical values of $\alpha_\perp$ and $\alpha_\parallel$ exactly match the known anisotropy[166].
[b]Ref. 163.
[c]Ref. 164.
[d]Ref. 165.

resulting monomer wavefunctions cannot be used in the dimer calculations. In order to avoid this situation, Gallup and Gerratt[159] introduced semi-empirical level shifts Δ_A and Δ_B into Eq. (72). These are defined so as to bring the computed polarizabilities into coincidence with the observed (or with any other accurate) value. These self-same shifts are then used in the VB calculation on the dimer. Their presence makes a significant improvement to the values of the van der Waals C_6 coefficient as shown in Table IX. In this way it is possible to use reasonably sized monomer wavefunctions in the dimer calculations while at the same time assuring a fairly accurate behaviour in the long-range region. It should be emphasized, however, that this procedure is only to be used with good-quality basis sets, e.g. at the pseudo-state SCF limit. If used in an attempt to remedy plainly inadequate wavefunctions, the results for the potential energy curves will be unpredictable.

VIII. CONCLUSIONS

Few groups apply VB methods to the study of intermolecular forces. One hopes that the obvious advantages as displayed here will encourage more to do so. Nevertheless, much progress has been achieved in making the method efficient and applicable to large systems. For those who already use non-orthogonal functions, it is a simple matter to modify their strategy so as to exclude the basis-set superposition error. The VB approach has been applied to many interesting systems, and it has been shown that the results are often superior to those obtained from MO-based methods[130,133]. The size of systems that have been studied, e.g. Ne–HF, Ar–HF, Ar–HCl, is at least as large as those treated by supermolecule approaches. In addition, the results obtained are physically meaningful and easy to interpret[131,133,134]. The role of the different components of the interaction energy, such as Coulomb, exchange, induction, dispersion and charge transfer, can be assessed. The effect of intra-system correlation can be studied accurately.

It is worth pointing out the advantages of this approach over MO-CI methods. In the supermolecule treatment, the problem of correcting for the BSSE and lack of size consistency is far from trivial. We stress this difficulty since it has been the subject of extensive recent study[167–169,173]. In particular, it is clear from the analysis of Fowler and Buckingham[168] that even at the SCF level there is no way of correcting exactly for this error by means of techniques based upon the introduction of 'ghost orbitals', as in the counterpoise method of Boys and Bernardi[170]—even when improved according to the suggestions of Fowler and Madden[167]. In addition, Wells and Wilson[169] have used even-tempered basis sets to study the behavior of the basis superposition as the size is increased. They develop extrapolation procedures whose aim is to provide an empirical estimate of the energy of the system as the basis tends to completeness. Unfortunately, these estimates are unreliable to an extent which renders them inapplicable for determining the superposition corrections in the limit of infinite bases: the size of the corrections necessary in MO-CI methods are often of the same order of magnitude as the well depths—even when reasonable-sized basis sets are employed[171,172]. The same point is made very forcibly by Schwenke and Truhlar[173], who calculated the interaction energy between two HF molecules at the SCF level. They used no less than 34 different basis sets in order to assess the reliability of the counterpoise correction for BSSE. The corrections are large in all of the configurations studied and, more seriously, large basis sets—for which the corrections are small—do not guarantee accurate results.

References

1. See, e.g., Gerratt, J., in *Theoretical Chemistry*, Vol. 4, Specialist Periodical Reports, Chemical Society, London, 1974.
2. Herzberg, G., *Infrared and Raman Spectra*, Van Nostrand-Reinhold, New York, 1945.
3. Townes, C. H., and Schawlow, A. L., *Microwave Spectroscopy*, McGraw-Hill, New York, 1955; Dover, New York, 1975.
4. Heitler, W., *Marx Handb. Radiol.*, **II**, 485 (1934).
5. Nordheim-Pöschl, G., *Ann. Phys.*, **26**, 258 (1936).
6. Herzberg, G., *Spectra of Diatomic Molecules*, Van Nostrand-Reinhold, New York, 1950.
7. Partridge, H., and Langhoff, S. R., *J. Chem. Phys.*, **74**, 2361 (1981).
8. Way, K. R., and Stwalley, W. C., *J. Chem. Phys.*, **59**, 5298 (1973).
   Stwalley, W. C., Zemke, W. T., Way, K. R., Li, K. C., and Proctor, T. R., *J. Chem. Phys.*, **66**, 5412 (1977) (erratum: *J. Chem. Phys.*, **67**, 4748 (1977)).
   Zemke, W. T., and Stwalley, W. C., *J. Chem. Phys.*, **68**, 4619 (1978).
   Zemke, W. T., Crooks, J. B., and Stwalley, W. C., *J. Chem. Phys.*, **68**, 4628 (1978).
   Zemke, W. T., Way, K. R., and Stwalley, W. C., *J. Chem. Phys.*, **69**, 409 (1978).
   Li, K. C., and Stwalley, W. C., *J. Chem. Phys.*, **70**, 1736 (1979).
   Orth, F. B., and Stwalley, W. C., *J. Mol. Spectrosc.*, **76**, 17 (1979).
9. Donovan, R. J., O'Grady, B. V., Shobatake, K., and Hiraya, A., *Chem. Phys. Lett.*, **122**, 612 (1985).
10. Bernstein, R. B., *Chemical Dynamics via Molecular Beam and Laser Techniques*, Clarendon Press, Oxford, 1982.
11. Lengsfield, B. H., McLean, A. D., Yoshimine, M., and Liu, B., *J. Chem. Phys.*, **79**, 1891 (1983).
    R. J. Harrison and N. C. Handy, *Chem. Phys. Letts.*, **98**, 97 (1983).
12. Gallup, G. A., and Collins, J. R., to be published.
13. Wheland, G. W., *Trans. Faraday Soc.*, **33**, 1499 (1937).
14. Kuntz, P. J., private communication.
15. Hurley, A. C., *Proc. R. Soc. A*, **248**, 119 (1958).
16. Hurley, A. C., *Proc. R. Soc. A*, **249**, 402 (1959).
17. Hurley, A. C., *Proc. R. Soc. A*, **69**, 767 (1956).
18. Hurley, A. C., *Rev. Mod. Phys.*, **32**, 400 (1960).
19. Raimondi, M., Simonetta, M., and Tantardini, G. F., *Comput. Phys. Rep.*, **2**, 171 (1985).
20. Gerratt, J., *Adv. At. Mol. Phys.*, **7**, 141 (1971).

21. Born, M., and Huang, K., *Dynamical Theory of Crystal Lattices*, Clarendon Press, London, 1954.
22. Wigner, E. P., *Group Theory*, Academic Press, New York, 1959.
23. Kotani, M., Amemiya, A., Ishiguro, E., and Kimura, T., *Tables of Molecular Integrals*, 2nd edn, Maruzen, Tokyo, 1963.
24. Kaplan, I. G., *Symmetry of Many-Electron Systems*, Academic Press, New York, 1975.
25. Rutherford, D. E., *Substitutional Analysis*, Edinburgh University Press, Edinburgh, 1948.
26. Hamermesh, M., *Group Theory and its Application to Physical Problems*, Addison-Wesley, Reading, MA, 1967.
27. van Vleck, J. H., *The Theory of Electric and Magnetic Susceptibilities*, Oxford University Press, London and New York, 1932.
28. Yamanouchi, T., *Proc. Phys.-Math. Soc. Japan*, **18**, 623 (1936); *ibid.*, **19**, 436 (1936).
29. Corson, R. M., *Perturbation Methods in the Quantum Mechanics of n-Electron Systems*, Blackie, London, 1951.
30. Rumer, G., *Göttingen Nachr.*, 377 (1932).
31. Pauncz, R., *Spin Eigenfunctions*, Plenum, New York and London, 1979.
32. Jahn, H. A., *Phys. Rev.*, **96**, 989 (1954).
33. Gerratt, J., and Raimondi, M., *Proc. R. Soc. A*, **371**, 525 (1980).
34. Hurley, A. C., Lennard-Jones, J., and Pople, J. A., *Proc. R. Soc. A*, **220**, 446 (1953).
35. Gerratt, J., and Lipscomb, W. M., *Proc. Natl. Acad. Sci. USA*, **59**, 332 (1968).
36. Parks, J. M., and Parr, R. G., *J. Chem. Phys.*, **28**, 335 (1958).
37. Goddard, W. A., Dunning, T. H., Hunt, W. J., and Hay, P. J., *Acc. Chem. Res.*, **6**, 368 (1973).
38. Goddard, W. A., and Harding, L. B., *Annu. Rev. Phys. Chem.*, **29**, 363 (1978).
39. Lowdin, P. O., *Phys. Rev.*, **97**, 1474 (1955).
40. Cooper, D. L., Gerratt, J., and Raimondi, M., *Mol. Phys.*, **56**, 611 (1985).
41. Cooper, D. L., Gerratt, J., and Raimondi, M., *Chem. Phys. Lett.*, **118**, 580 (1985).
42. Pyper, N. C., and Gerratt, J., *Proc. R. Soc. A*, **355**, 402 (1977).
43. Penotti, F., Cooper, D. L., Gerratt, J., and Raimondi, M., to be published.
44. Walters, S. G., Ph.D. Thesis, University of Bristol, 1984.
Walters, S. G., Gerratt, J., and Raimondi, M., to be submitted.
45. Walters, S. G., Penotti, F., Gerratt, J., and Raimondi, M., to be submitted.
46. Cooper, D. L., Gerratt, J., and Raimondi, M., *Faraday Symp., Chem. Soc.*, **19**, 149 (1984).
47. Manley, J. C., Ph.D. Thesis, University of Bristol, 1982.
48. Manley, J. C., and Gerratt, J., *Comput. Phys. Commun.*, **31**, 75 (1984).
Gerratt, J., *Mol. Phys.*, **33**, 1199 (1979).
49. Kaldor, U., and Harris, F. E., *Phys. Rev.*, **183**, 1 (1969).
Kaldor, U., *Phys. Rev. A*, **1**, 1586 (1970).
Kaldor, U., *Phys. Rev. A*, **2**, 1267 (1970).
50. Goddard, W. A., *Phys. Rev.*, **157**, 81 (1967).
Palke, W. E., and Goddard, W. A., *J. Chem. Phys.*, **50**, 4524 (1969).
Melius, C. F., and Goddard, W. A., *J. Chem. Phys.*, **56**, 3348 (1972).
Blint, R. J., and Goddard, W. A., *J. Chem. Phys.*, **57**, 5296 (1972).
Ladner, R. C., and Goddard, W. A., *J. Chem. Phys.*, **51**, 1073 (1969).
51. Gerratt, J., Manley, J. C., and Raimondi, M., *J. Chem. Phys.*, **82**, 2014 (1985).
52. Cooper, D. L., Gerratt, J., and Raimondi, M., *Chem. Phys. Lett.*, **127**, 600 (1986).
53. Cooper, D. L., Ford, M. J., Gerratt, J., and Raimondi, M., *Phys. Rev. A*, in press
54. Cooper, D. L., Gerratt, J., and Raimondi, M., *Nature*, **323**, 699 (1986).

55. Green, S., Bagus, P. S., Liu, B., McLean, A. D., and Yoshimine, M., *Phys. Rev. A*, **5**, 1614 (1972).
56. Saxon, R. P., Kirby, K., and Liu, B., *J. Chem. Phys.*, **73**, 1873 (1980).
57. Huber, K. P., and Herzberg, G., *Constants of Diatomic Molecules*, Van Nostrand-Reinhold, New York, 1979.
58. Sondergaard, N. A., Sauers, J., Jones, A. C., Kaufman, J. J., and Koski, W. S., *J. Chem. Phys.*, **71**, 2229 (1979).
59. Friedrich, B., and Herman, Z., *Chem. Phys.*, **69**, 433 (1982).
60. Ottinger, Ch., and Reichmuth, J., *J. Chem. Phys.*, **74**, 928 (1981).
61. Ottinger, Ch., in *Gas Phase Ion Chemistry* (Ed. M. Bowers), Vol. 3, Academic Press, New York, 1983.
62. Reichmuth, J., Diplomarbeit, Max-Planck-Institut für Strömforschung, Göttingen, 1981.
63. Klein, R., Rosmus, P., and Werner, H. J., *J. Chem. Phys.*, **77**, 3559 (1982).
64. Jaszunski, M., Roos, B. O., and Widmark, P. O., *J. Chem. Phys.*, **75**, 306 (1981).
65. Moore, C. E., *Atomic Energy Levels*, National Bureau of Standards Reference Data System, US Dept of Commerce, Washington, DC, 1971.
66. Gardner, L. D., Bayfield, J. E., Koch, P. M., Sellini, I. A., in press.
Pegg, D. J., Peterson, R. S., and Crandall, D. H., *Phys. Rev. A*, **21**, 1397 (1980).
Crandall, D. H., Paneuf, R. A., and Meyer, F. W., *Phys. Rev. A*, **19**, 504 (1979).
Ciric, D., Brazuk, A., Dijkkamp, D., de Heer, P. J., and Winter, H., *J. Phys. B: At. Mol. Phys.*, **18**, 3629 (1985).
67. McCullough, R. W., Wilkie, F. G., and Gilbody, H. B., *J. Phys. B: At. Mol. Phys.*, **17**, 1373 (1984).
68. Bienstock, S., Heil, T. G., Bottcher, C., and Dalgarno, A., *Phys. Rev. A*, **25**, 2850 (1982).
69. Cooper, D. L., and Wilson, S., *J. Chem. Phys.*, **78**, 2456 (1983).
70. Roothaan, C. C. J., *Rev. Mod. Phys.*, **23**, 69 (1951).
71. Weinstein, H., Pauncz, R., and Cohen, M., *Adv. At. Mol. Phys.*, **7**, 97 (1971).
72. See, e.g., Ramsden, E. N., *A-Level Chemistry*, Stanley Thornes, Cheltenham, 1985.
73. Hückel, E., *Z. Phys.*, **70**, 204 (1931); *Int. Conf. Phys., Phys. Soc.*, **2**, 9 (1935).
See also Streitwieser, A., *Molecular Orbital Theory for Organic Chemists*, Wiley, New York, 1961.
74. Pariser, R., and Parr, R. G., *J. Chem. Phys.*, **21**, 466 (1953).
Pople, J. A., *Trans. Faraday Soc.*, **49**, 1375 (1953).
75. Pauncz, R., *The Alternant Molecular Orbital Method*, W. B. Saunders, Philadelphia, 1967.
76. Norbeck, J. M., and Gallup, G. A., *J. Am. Chem. Soc.*, **96**, 3386 (1974).
77. Tantardini, G., Raimondi, M., and Simonetta, M., *J. Am. Chem. Soc.*, **99**, 2913 (1977).
78. Coulson, C. A., *Valence*, 2nd Edn, Clarendon Press, Oxford, 1961.
79. Allan, N. L., Cooper, D. L., Gerratt, J., and Raimondi, M., *J. Electron Spectrosc.*, submitted.
80. Allan, N. L., and Cooper, D. L., *J. Chem. Phys.*, **86**, 5594 (1986).
81. Cooper, D. L., and Allan, N. L., *J. Chem. Soc., Faraday Trans. II*, submitted.
82. Williams, B. G. (Ed.), *Compton Scattering*, McGraw-Hill, New York, 1977.
Cooper, M. J., *Rep. Prog. Phys.*, **48**, 415 (1985).
83. Moshinsky, M., and Seligman, T. Y., *Ann. Phys.*, **66**, 311 (1971).
84. Paldus, J., in *Theoretical Chemistry; Advances and Perspectives*, (Eds H. Eyring and J. Henderson), Vol. 2, Academic Press, New York, 1976.
85. Cooper, D. L., and Gerratt, J., unpublished work.

86. Gallup, G. A., Vance, R. L., Collins, J. R., and Norbeck, J. M., *Adv. Quantum Chem.*, **16**, 229 (1982).
87. Prosser, F., and Hagstrom, S., *J. Chem. Phys.*, **48**, 4807 (1968).
88. King, H. F., and Stanton, R. E., *J. Chem. Phys.*, **48**, 4808 (1968).
89. King, H. F., Stanton, R. E., Kim, H., Wyatt, R. E., and Parr, R. G., *J. Chem. Phys.*, **47**, 1936 (1967).
90. Prosser, F., and Hagstrom, S., *Int. J. Quantum Chem.*, **2**, 89 (1968).
91. Raimondi, M., Campion, W., and Karplus, M., *Mol. Phys.*, **34**, 1483 (1977).
92. Raimondi, M., Tantardini, G. F., and Simonetta, M., *Mol. Phys.*, **30**, 703 (1975). Raimondi, M., *Mol. Phys.*, **53**, 161 (1984).
93. Raimondi, M., Tantardini, G. F., and Simonetta, M., *Mol. Phys.*, **30**, 797 (1975).
94. Hayes, I. C., and Stone, A. J., *Mol. Phys.*, **53**, 69 (1984).
95. Aitken, A. C., *Determinants and Matrices*, Oliver and Boyd, Edinburgh, 1967.
96. Leasure, S. C., and Balint-Kurti, G. G., *Phys. Rev.*, **31**, 2107 (1985).
97. Gianinetti, E., and Raimondi, M., *Int. J. Quantum Chem.*, submitted.
98. Figari, G., and Magnasco, V., *Mol. Phys.*, **55**, 319 (1985).
99. Gerrity, D. P., and Valentini, J. J., *J. Chem. Phys.*, **82**, 1323 (1985). Marinero, E. E., Rettner, C. T., and Zare, R. N., *J. Chem. Phys.*, **80**, 4142 (1984).
100. Neumark, D. M., Wodke, A. M., Robinson, G. N., Hayden, C. C., and Lee, Y. T., *J. Chem. Phys.*, **82**, 3045 (1985).
101. See, e.g., Duggan, J. J., and Grice, R., *J. Chem. Phys.*, **78**, 3842 (1983); and Duggan, J. J., and Grice, R., *J. Chem. Soc., Faraday Trans. II*, **80**, 729, 739, 795, 809 (1984).
102. Moffit, W., *Proc. R. Soc. A*, **210**, 245 (1951).
103. Ellison, F. O., *J. Am. Chem. Soc.*, **85**, 3540 (1963).
104. Tully, J. C., *Adv. Chem. Phys.*, **42**, 63 (1980).
105. Kuntz, P. J., *Ber. Bunsenges. Phys. Chem.*, **86**, 367 (1982).
106. Faist, M. B., and Muckerman, J. T., *J. Chem. Phys.*, **71**, 225, 233 (1979).
107. Radlein, D. St. A. G., *J. Chem. Phys.*, **78**, 3084 (1983).
108. Meier, P. F., Hayes, E. F., and Ellison, F. O., *J. Chem. Phys.*, **71**, 1948 (1979).
109. Isaacson, A. D., and Muckerman, J. T., *J. Chem. Phys.*, **73**, 1729 (1980).
110. Kuntz, P. J., and Roach, A. C., *J. Chem. Phys.*, **74**, 3420 (1981). Roach, A. C., and Kuntz, P. J., *J. Chem. Phys.*, **74**, 3435 (1981).
111. Olson, J. A., and Garrison, B. J., *J. Chem. Phys.*, **81**, 1355 (1984) (Erratum: **81**, 5221 (1984)).
112. Schreiber, J. L., and Kuntz, P. J., *J. Chem. Phys.*, **76**, 1872 (1982). Kuntz, P. J., and Schreiber, J. L., *J. Chem. Phys.*, **76**, 4120 (1982).
113. Last, I., and Baer, M., *J. Chem. Phys.*, **80**, 3246 (1984).
114. Baer, M., and Last, I., *Chem. Phys. Lett.*, **119**, 393 (1985).
115. Gallup, G. A., and Gerratt, J., to be published.
116. Wang, H.-T., Felps, W. S., and McGlynn, S. P., *J. Chem. Phys.*, **67**, 2614 (1977).
117. Docker, M. P., Hodgson, A., and Simons, J. P., *Mol. Phys.*, **57**, 129 (1986) and references therein.
118. Runau, R., Peyerimhoff, S. D., and Buenker, R. D., *J. Mol. Spectrosc.*, **68**, 253 (1977).
119. Zeiri, Y., and Shapiro, M., *J. Chem. Phys.*, **75**, 1170 (1981) and references therein.
120. Goldfield, E. M., Gislason, E. A., and Sabelli, N. H., *J. Chem. Phys.*, **82**, 3179 (1985).
121. Goldfield, E. M., Kosmas, A. M., and Gislason, E. A., *J. Chem. Phys.*, **82**, 3191 (1985).
122. Zeiri, Y., Shapiro, M., and Tenne, R., *Chem. Phys. Lett.*, **99**, 11 (1983).
123. Zeiri, Y., and Balint-Kurti, G. G., *J. Mol. Spectrosc.*, **99**, 1 (1983).

124. Bienstock, S., Cooper, D. L., and Dalgarno, A., *J. Chem. Phys.*, submitted.
125. Bottcher, C., and Dalgarno, A., *Proc. R. Soc. A*, **340**, 187 (1974).
126. Partridge, H., Bauschlicher, C. W., Jr, Walch, S. P., and Liu, B., *J. Chem. Phys.*, **79**, 1866 (1983).
127. Müller, W., Flesch, J., and Meyer, W., *J. Chem. Phys.*, **80**, 3297 (1984).
Müller, W., and Meyer, W., *J. Chem. Phys.*, **80**, 3311 (1984).
128. Jeung, G. H., Malrieu, J. P., and Daudey, J. P., *J. Chem. Phys.*, **77**, 371 (1932).
129. See, for example, a comparison with experiment of Jeung's calculations on the double minimum in the $(2)^1\Sigma_u^-$ state of Na_2 in: Cooper, D. L., Barrow, R. F., Verges, J., Effantin, C. C., and d'Incan, J., *Can. J. Phys.*, **62**, 1543 (1984).
130. Cremaschi, P., Morosi, G., Raimondi, M., and Simonetta, M., *Mol. Phys.*, **38**, 1555 (1979).
131. Cremaschi, P., Morosi, G., Raimondi, M., and Simonetta, M., *Chem. Phys. Lett.*, **109**, 442 (1984).
132. Wormer, P. E. S., van Berkel, T., and van der Avoird, Ad., *Mol. Phys.*, **29**, 1181 (1975).
133. Raimondi, M., *Mol. Phys.*, **53**, 161 (1984).
134. Gallup, G. A., and Gerratt, J., *J. Chem. Phys.*, **83**, 2316 (1985).
135. Stone, A. J., and Hayes, I. C., *Faraday Disc., Chem. Soc.*, **73**, 19 (1984).
136. Gallup, G. A., and Collins, J. R., *Mol. Phys.*, **49**, 871 (1983).
137. Geurts, P. J. M., Wormer, P. E. S., and van der Avoird, Ad., *Chem. Phys. Lett.*, **35**, 444 (1975).
138. Wormer, P. E. S., and van der Avoird, Ad., *J. Chem. Phys.*, **62**, 3326 (1975).
139. Berns, R. M., and van der Avoird, Ad., *J. Chem. Phys.*, **72**, 6107 (1980).
140. Wormer, P. E. S., Bernards, J. P. C., and Gribnau, M. C. M., *Chem. Phys.*, **81**, 1 (1983).
141. Wormer, P. E. S., and van der Avoird, Ad., *J. Chem. Phys.*, **72**, 6107 (1980).
142. Reeves, C. M., *Commun. ACM*, **9**, 276 (1966).
143. Löwdin, P. O., *Adv. Quantum Chem.*, **5**, 185 (1970).
144. Daudey, J. P., Claverie, P., and Malrieu, J. P., *Int. J. Quantum Chem.*, **8**, 1 (1974).
145. Lischka, H., *Chem. Phys. Lett.*, **20**, 448 (1973).
146. Rodwell, W. R., Jiu Fai Lam, L. T., and Watts, R. O., *Mol. Phys.*, **44**, 225 (1983).
147. Murrell, J. N., Randić, M., and Williams, D. R., *Proc. R. Soc. A*, **284**, 566 (1965).
148. Das, G., and Wahl, A. C., *Phys. Rev. A*, **4**, 825 (1971).
149. Sironi, M., Tesi di Laurea in Chimica, University of Milan, 1987.
150. Musso, G. F., and Magnasco, V., *Mol. Phys.*, **53**, 615 (1984) and references therein.
151. Hayes, I. C., and Stone, A. J., *Mol. Phys.*, **53**, 83 (1984).
152. Hayes, I. C., Hurst, G. J. B., and Stone, A. J., *Mol. Phys.*, **53**, 107 (1984).
153. Magnasco, V., Musso, G. F., Costa, C., and Figari, G., *Mol. Phys.*, **56**, 1249 (1985).
154. Gerratt, J., and Papadopoulos, M., *Mol. Phys.*, **41**, 1071 (1980).
155. Gallup, G. A., and Gerratt, J., *Chem. Phys. Lett.*, **117**, 589 (1985).
156. Gallup, G. A., and Gerratt, J., *J. Chem. Phys.*, **83**, 2323 (1985).
157. Murrell, J. N., and Shaw, G., *J. Chem. Phys.*, **49**, 4731 (1968).
158. Briggs, M. P., Murrell, J. N., and Stamper, J. G., *Mol. Phys.*, **17**, 381 (1969).
159. Gallup, G. A., and Gerratt, J., *Chem. Phys. Lett.*, **112**, 228 (1984).
160. See, e.g., Bethe, H. A., and Salpeter, E. E., *Quantum Mechanics of One- and Two-Electron Atoms*, Plenum/Rosetta, New York, 1977.
161. Burke, P. G., and Robb, W. D., *Adv. At. Mol. Phys.*, **11**, 144 (1975).
162. Lipscomb, W. N., *Adv. Magn. Reson.*, **2**, 137 (1966).
163. Alexander, M. H., *J. Chem. Phys.*, **52**, 3354 (1970).
164. Dalgarno, A., *Adv. Chem. Phys.*, **12**, 143 (1967).

165. Doran, M. B., *J. Phys. B: At. Mol. Phys.*, **7**, 558 (1974).
166. Muenter, J. S., *J. Chem. Phys.*, **56**, 5409 (1972).
167. Fowler, P. W., and Madden, P. A., *Mol. Phys.*, **49**, 913 (1983).
168. Fowler, P. W., and Buckingham, A. D., *Mol. Phys.*, **50**, 1349 (1983).
169. Wells, B. H., and Wilson, S., *Mol. Phys.*, **50**, 1295 (1983).
170. Boys, S. F., and Bernardi, F., *Mol. Phys.*, **19**, 553 (1970).
171. van Lenthe, J. H., van Duijneveldt, T., and Kroon-Batenburg, L. M. J., *Faraday Symp., Chem. Soc.*, **19**, 125 (1984).
172. Raimondi, M., *Faraday Symp., Chem. Soc.*, **19**, (1984), section on general discussion.
173. Schwenke, D., and Truhlar, D. G., *J. Chem. Phys.*, **82**, 2418 (1985).

Ab Initio Methods in Quantum Chemistry—II
Edited by K. P. Lawley

THE COMPLETE ACTIVE SPACE SELF-CONSISTENT FIELD METHOD AND ITS APPLICATIONS IN ELECTRONIC STRUCTURE CALCULATIONS

BJÖRN O. ROOS

Department of Theoretical Chemistry, Chemical Centre, PO Box 124, S-221 00 Lund, Sweden

CONTENTS

I. Introduction . . . 399
II. Optimization of a Complete Active Space Self-consistent Field Wavefunction . . . 405
A. The Orbital Space and the Complete Active Space Self-consistent Field Wavefunction . . . 405
B. Optimization of the Wavefunction . . . 409
C. The Configuration-interaction Secular Problem . . . 415
D. Simplified Optimization Procedures . . . 416
III. Applications . . . 420
A. The C_2 Molecule. . . . 422
B. The BH Molecule . . . 429
C. The N–N Bond in N_2O_4 . . . 432
D. Transition-metal Chemistry . . . 434
1. The NiH Molecule . . . 435
2. The Nickel–Ethene Complex . . . 437
IV. Summary and Conclusions . . . 440
References . . . 442

I. INTRODUCTION

Modern applications of the electronic structure theory for molecular systems based are almost entirely upon the molecular-orbital (MO) concept, introduced in molecular spectroscopy by Mulliken[1]. This is not surprising. The success of the Hartree–Fock method[2,3], in describing the electronic structure of most closed-shell molecules has made it natural to analyse the wavefunction in terms of the molecular orbitals. The concept is simple and has a close relation to experiment through Koopmans' theorem[4]. The two fundamental building blocks of Hartree–Fock (HF) theory are the molecular orbital and its occupation number. In closed-shell systems each occupied molecular orbital carries two electrons, each with opposite spin. The orbitals themselves are only defined as an occupied one-electron subspace of the space spanned by the eigenfunctions of the Fock operator. Transformations between them leave the total HF wavefunction invariant. Normally the orbitals are obtained in a delocalized form as solutions to the Hartree–Fock equations. This formulation is the most relevant one in studies of spectroscopic properties of the molecule, that is, excitation and ionization. The invariance property, however, makes a transformation to localized orbitals possible. Such localized orbitals can be of value for an analysis of the chemical bonds in the system.

The Hartree–Fock method can be extended to open-shell systems in two ways. In the restricted Hartree–Fock (RHF) method[5], the open-shell orbitals are added to the closed-shell orbitals and the resulting wavefunction is then projected to have the correct spin and space symmetry. In some cases this leads to a wavefunction comprising more than one Slater determinant. An alternative formulation is the unrestricted Hartree–Fock (UHF) method[6], where the spin orbitals for a closed-shell electron pair are no longer assumed to be equal. As a result the method is capable of describing the spin polarization of paired electrons in the presence of unpaired spins. The method has for this reason been used extensively in studies of the spin polarization of radicals. In contrast to the closed-shell Hartree–Fock method the UHF method also gives a qualitatively correct description of bond dissociation[7]. The serious drawback of the UHF method is, however, that it cannot easily be projected to give a wavefunction which is an eigenfunction of the total spin. If such a projection is attempted *a priori* to a variational treatment, very complex equations for the molecular orbitals are obtained. The simplicity of the Hartree–Fock method is lost and the gain in accuracy is not large enough to make such an approach useful in practical applications.

All variants of the Hartree–Fock method lead to a wavefunction in which all the information about the electron structure is contained in the occupied molecular orbitals (or spin orbitals) and their occupation numbers, the latter being equal to 1 or 2.

The concept of the molecular orbital is, however, not restricted to the Hartree–Fock model. A set of orbitals can also be constructed for more complex wavefunctions, which include correlation effects. They can be used to obtain insight into the detailed features of the electronic structure. These are the natural orbitals, which are obtained by diagonalizing the spinless first-order reduced density matrix[8]. The occupation numbers (η) of the natural orbitals are not restricted to 2, 1 or 0. Instead they fulfil the condition

$$0 < \eta < 2 \tag{1}$$

If the Hartree–Fock determinant dominates the wavefunction, some of the occupation numbers will be close to 2. The corresponding MOs are closely related to the canonical Hartree–Fock orbitals. The remaining natural orbitals have small occupation numbers. They can be analysed in terms of different types of correlation effects in the molecule[9,10]. A relation between the first-order density matrix and correlation effects is not immediately justified, however. Correlation effects are determined from the properties of the second-order reduced density matrix. The most important terms in the second-order matrix can, however, be approximately defined from the occupation numbers of the natural orbitals. Electron correlation can be qualitatively understood using an independent electron-pair model[11]. In such a model the correlation effects are treated for one pair of electrons at a time, and the problem is reduced to a set of two-electron systems. As has been shown by Löwdin and Shull[9,10] the two-electron wavefunction is determined from the occupation numbers of the natural orbitals. Also the second-order density matrix can then be specified by means of the natural orbitals and their occupation numbers. Consider as an example the following simple two-configurational wavefunction for a two-electron system:

$$\Phi = C_1(\varphi_1)^2 - C_2(\varphi_2)^2 \qquad (C_2 > 0) \tag{2}$$

where $(\varphi_1)^2$ is the HF wavefunction. The MOs φ_1 and φ_2 are the natural orbitals and have occupation numbers $\eta_1 = 2C_1^2$ and $\eta_2 = 2C_2^2$. The diagonal of the second-order density matrix, $\rho_2(1,2)$, is with real orbitals obtained as

$$\rho_2(1,2) = \tfrac{1}{2}\eta_1\varphi_1(1)^2\varphi_1(2)^2 + \tfrac{1}{2}\eta_2\varphi_2(1)^2\varphi_2(2)^2 - (\eta_1\eta_2)^{1/2}\varphi_1(1)\varphi_2(1)\varphi_1(2)\varphi_2(2) \tag{3}$$

The last term is the one which describes the correlation of the electron pair by the orbital φ_2. If φ_1 is a bonding MO and φ_2 the corresponding antibonding orbital, the last term in (3) ensures that there is one electron on each atom when the bond dissociates. If, as another example, $\varphi_1 = 2s$ and $\varphi_2 = 2p$ in the Be atom, the last term in (3) decreases the probability of finding both electrons on the same side of the nucleus and increases the probability of finding the electrons on opposite sides (angular correlation). This effect is very large in beryllium and accounts for 95% of the correlation energy in the valence shell.

For systems with more than two electrons, the simple picture illustrated above obviously breaks down. The approximate validity of the independent electron-pair model, however, also makes it possible to estimate different correlation effects in many-electron systems from an inspection of the natural-orbital occupation numbers.

The correlation error is normally defined as the difference between the exact eigenvalue of the non-relativistic Hamiltonian for the molecule and the HF energy. While this definition works well for closed-shell systems, it becomes less meaningful when degeneracies, or near-degeneracies, occur between different electronic configurations. For example, the HF energy for the H_2 molecule at large internuclear distances is in error by more than 7 eV. Obviously, the electrons of two non-interacting hydrogen atoms are not correlated. The reason behind this is, of course, that the Hartree–Fock model breaks down in cases where several configurations become degenerate or near-degenerate. This is the case in most bond dissociation processes, and also along the reaction path for a symmetry-forbidden chemical reaction, to mention just two examples. Thus while the Hartree–Fock model is a valid approximation for most (but not all) molecules around their equilibrium geometry, it cannot in general be used as a qualitatively correct model for energy surfaces.

The complete active space self-consistent field (CASSCF) model[12–14] is an attempt to generalize the HF model to such situations, while trying to keep as much of the conceptual simplicity of the HF approach as possible. Technically, the CASSCF is by necessity a more complex model, since it is based on a multiconfigurational wavefunction. The building blocks are, as in the Hartree–Fock method, the occupied molecular orbitals (the inactive and active orbitals). The number of orbitals is, however, in general larger than $(N + 1)/2$ (N being the number of electrons). The number of electron configurations generated by the orbital space is therefore larger than unity. The total electronic wavefunction is formed as a linear combination of all the configurations, in the N-electron space, that fulfil the given spin and space symmetry requirements. It is 'complete' in the configurational space spanned by the active orbitals. The inactive orbitals are kept doubly occupied in all these configurations; they represent an 'SCF sea' in which the active electrons move. These orbitals then have occupation numbers exactly equal to 2, while the active orbitals have occupation numbers varying between 0 and 2. The choice of the inactive and active orbital subspaces will be discussed in detail later, but it is obvious that the inactive orbitals should be chosen as orbitals which are not expected to contribute to near-degeneracy correlation effects (e.g. core orbitals).

The conceptual simplicity of the CASSCF model lies in the fact that once the inactive and active orbitals are chosen, the wavefunction is completely specified. Such a model also leads to certain simplifications in the computational procedures used to obtain optimal orbitals and configuration-interaction (CI) coefficients, but that is of less importance in the present context. The major technical difficulty inherent to the CASSCF model is the size of the complete CI expansion, N_{CAS}. It is given by the so-called Weyl formula, that is, the dimension of the irreducible representation of the unitary group U(n) associated with n active orbitals, N active electrons and a total spin S[15]:

$$N_{CAS} = \frac{2S+1}{n+1}\binom{n+1}{N/2-S}\binom{n+1}{N/2+S+1} \tag{4}$$

Obviously N_{CAS} increases strongly as a function of the size n of the active orbital space. In practice this means that there is a rather strict limit on the size of this space. Experience shows that this limit is normally reached for n around 10–12 orbitals, except for case with few active electrons or holes. As shown by the large number of CASSCF calculations performed during the last six years, this limitation does not create any serious problem in most applications. It should be remembered that the CASSCF is an extension of the Hartree–Fock model. As such it is supposed to produce a good zeroth-order approximation to the wavefunction, when near-degeneracies are present. This goal can in most cases be achieved with a few active orbitals. The CASSCF model has not been developed for treating dynamical correlation effects, but to provide a good starting point for such studies. There are, however, cases where more active orbitals are needed than it is possible to handle with the computational methods available. Cr_2 is such a case, where a full valence (3d and 4s active) calculation fails to describe the nature of the bond[16]. Only a weak 4s–4s bond is formed with a bond distance around 2.5 Å. The true bond length is 1.68 Å[17], indicating considerable 3d–3d bonding. In order to account for this bonding, dynamical correlation in the 3d shell must be included in the wavefunction. The 4s,4p near-degeneracy also has to be included. This results in an active orbital space including 3d, 4s, 4p and 4d on both atoms: 12 electrons distributed in 28 orbitals. Such a calculation is impossible. There does not seem to exist today an *ab initio* quantum-chemical method which is capable of a consistent treatment of the Cr–Cr bond, and this molecule therefore represents a major challenge in the development of new computational methods.

It should be pointed out, however, that the size of the CI expansion used in a CASSCF calculation is almost always much larger than those normally used in earlier applications of the multiconfigurational SCF (MCSCF) method[18]. It was only when the graphical unitary group approach (GUGA) for full CI calculations was invented in the years 1975–78[15,19] that an efficient computational procedure for CASSCF calculations could be developed[12–14].

The idea behind the CASSCF model is, of course, not at all new. It was realized very early that full CI calculations in a valence orbital basis would be a valid model for studies of potential curves, e.g. for the diatomic molecules. Such calculations were already being used 20 years ago for qualitative studies of excited-state potential curves[20,21].

The generalized valence bond (GVB) method developed by Gaddard and coworkers[22] does not employ a full CI in the valence shell, but includes only the configurations needed to describe proper dissociation of a chemical bond. As will be demonstrated later, such a wavefunction represents a restricted form of the CASSCF wavefunction, where the active subspace is partitioned into subsets with a fixed number of electrons occupying the orbitals in each subset (acutally the GVB function is further restricted by allowing only specific spin couplings within each subset).

The fully optimized reaction space (FORS) model suggested by Ruedenberg and coworkers[23,24] is equivalent to the CASSCF model in its basis concepts, while differing in the technical implementation. The final wavefunction is expanded in the full list of configurations generated by a set of 'reaction orbitals' (in CASSCF language: active orbitals). Normally the MCSCF orbital optimization is performed in a smaller configuration space, presumably due to limitations inherent to the computational procedure used. It is pointed out that in the cases studied only 5% of the full set of configurations were necessary in determining molecular orbitals which, when used in a full CI calculation, yields an energy less than a millihartree from the fully optimized value[24]. On the other hand it is emphasized that the final full CI calculation is necessary in order to obtain reliable energies for an energy surface for chemical reactions: 'Experience has shown that there exists no justifiable selection of a small number of configurations, such as 5% of the FRS, to represent a system without energy bias' (Ref. 24, p. 74).

It is not clear, however, how valid the notion of optimizing the MOs in a smaller CI basis is in the general case. It involves a selection of the most important configurations in a first step and a subsequent enlargement of the basis in a stepwise procedure, until all important configurations have been included. Even though this might be a straightforward procedure in simple cases (like those illustrated in Ref. 24), it may lead to difficulties in more complex and unknown situations where the primary selection of configurations is not so easy. Examples of such cases will be given later. The possibility for general reductions of the configuration space has not been implemented into the CASSCF programs, since they destroy the simple structure of the model. They are also not necessary. The limitation of the CASSCF method is set by one's ability to do large complete CI calculations. The orbital optimization can always be performed without any serious extra effort (apart from the fact that MCSCF is an interative procedure; thus the CI secular problem has to be solved several times). Recently CASSCF calculations including as many as 178 916 configurations have been performed with full optimization of the orbitals[25].

The CASSCF model and its implementation into efficient computer programs will be discussed in more detail in the following sections. The main emphasis will, however, not be put on the technical problems concerning the search for efficient MCSCF optimization schemes. A comprehensive discussion of these problems has recently been published in this series[26]. The present contribution will instead concentrate around problems concerning the application of the CASSCF method to different chemical problems, ranging from accurate calculations of properties of small molecules to studies of the chemical bond in transition-metal compounds. Nevertheless a discussion on the computational strategies developed to obtain a CASSCF wavefunction is necessary and will be given below. However, the main emphasis will be on the problems encountered when the CASSCF method is used to solve chemical problems. One section includes a discussion of the choice of active orbitals. In most applications this is not a trivial problem. A number of examples from recent applications will be included in a later section.

II. OPTIMIZATION OF A COMPLETE ACTIVE SPACE SELF-CONSISTENT FIELD WAVEFUNCTION

A CASSCF study starts with the selection of the inactive and active orbital subspaces. This must be done with great care, as will be discussed in detail later. The major problem is to construct a wavefunction which gives a balanced description of near-degeneracy and dynamical correlation effects for the section of the energy surface studied.

Once the orbital space has been given, the total wavefunction is completely specified. What remains is to find the optimal molecular orbitals and CI coefficients, by application of the variation principle. In the last five years we have seen an outstanding development of efficient methods for optimizing the variables of an MCSCF wavefunction. Before 1980 almost all MCSCF optimization schemes in use were of the first-order type based on extensions of the Hartree–Fock iterative method[18,27], or on the so-called 'super-CI' method developed originally by Grein and coworkers[28] and also used by Ruedenberg *et al.* in applications of the FORS method[23,29]. A second-order optimization procedure was actually proposed and used (in an approximated form) by Levy[30] before 1970, but its large-scale implementation into general MCSCF programs had to wait until the late 1970s. An efficient second-order optimization scheme was reported by Yaffe and Goddard[31] in 1976. The development of similar and also higher-order procedures in the general MCSCF case includes contributions from a number of different research groups and the reader is referred to the review by Olsen *et al.*[26] for a comprehensive list of references. Two other recent contributions should be added to this list, since they give important improvements of the second-order algorithms in the non-local regions[32,33]. The work of Werner and Knowles deserves special attention. The CASSCF program developed by them has outstanding convergence properties; in fact no calculation reported has needed more than three iterations in order to converge to less than 10^{-7} a.u. in the energy. The CI scheme used can handle expansions comprising more than 10^5 terms.

A. The Orbital Space and the Complete Active Space Self-consistent Field Wavefunction

In order to proceed, a few preliminaries are needed. They will be given in this section. The basic quantity of interest is the molecular orbital. Therefore, the calculation normally starts by defining an orthonormal orbital space

$$\{\varphi_i(\mathbf{r}); i = 1, \ldots, m\} \tag{5}$$

Normally these molecular orbitals are obtained as expansions in a set of atom-centred basis functions (the linear combination of atomic orbitals (LCAO) method), m being the number of such functions. Recently, two-dimensional numerical integration methods have been developed to solve the MCSCF equations for linear molecules[34]. The dimension m is then, in principle, infinite (practice, it is determined by the size of the grid used in the numerical integration). The molecular-orbital space is further divided into three subspaces: the *inactive*, the *active* and the *external* orbitals. The inactive and active subspaces constitute the *internal* (occupied) orbital subspace, while the external orbitals are unoccupied. The CASSCF wavefunction is formed as a linear combination of configuration state functions (CSFs) generated from these orbitals in the following way.

1. The inactive orbitals are doubly occupied in all CSFs.
2. The remaining (active) electrons occupy the active orbitals. Using these electrons and orbitals, a *full* list of CSFs which have the required spin and space symmetry is constructed. The CASSCF wavefunction is written as a linear combination of all these CSFs, comprising a complete expansion in the active orbital subspace.

The optimization step then consists of finding those expansion coefficients and molecular orbitals that make the energy stationary with respect to all parameters. It has been suggested[33] that the deep-lying inactive orbitals (the 'core' orbitals) should be determined in a preceding SCF calculation and kept frozen during the optimization step. In second-order optimization procedures, this will reduce the computational efforts. If a first-order procedure is used, this problem does not arise, since the work done in the orbital optimization is practically independent of the number of inactive orbitals. Sometimes, however, there are other reasons to keep the core orbitals frozen in a calculation. For instance, large parts of the 'basis-set superposition error (BSSE)' can be avoided in this way. The core orbitals are then free-atom orbitals, orthogonalized against each other and kept frozen. This technique is, for example, used to avoid the BSSE in CASSCF calculations on transition-metal compounds[35]. It is especially important when moderately small atomic basis sets are used. CASSCF calculations have also been performed with the core electrons replaced by an effective core potential (ECP)[35,36]. Using this technique the CASSCF model can be extended to systems containing several heavy atoms, a current example being transition-metal[35] clusters.

As pointed out in the introduction, the CASSCF configuration space quickly becomes unmanageably large when the number of active orbitals is increased. While this does not create any serious problems in most applications, there are cases when a larger number of active orbitals is needed. In such cases a restricted form of the CASSCF wavefunction may be used. The active orbitals are divided into subgroups and the number of electrons is kept fixed within each subgroup. An example[16] of a calculation where such a partitioning of the active space was found necessary is Cr_2. A full valence calculation of the chromium dimer corresponds to distributing 12 electrons among 12 active orbitals (formed from the atomic 3d and 4s orbitals). With no symmetry restrictions, 226 512 singlet ($S = 0$) CSFs can be formed in this orbital space. For the $^1\Sigma_g^+$ ground state of Cr_2 the number of CSFs is reduced to 28 784. Even if possible with the programs available today, such a calculation would be time-consuming. A simple way to reduce the size of the CSF space is to divide the valence orbitals into the following three blocks:

$$\{3d\sigma_g, 3d\sigma_u, 4s\sigma_g, 4s\sigma_u\} \qquad \{3d\pi_u, 3d\pi_g\} \qquad \{3d\delta_g, 3d\delta_u\} \tag{6}$$

The $^1\Sigma_g^+$ ground state of Cr_2 has the following orbital occupation (excluding all core orbitals):

$$(4s\sigma_g)^2(3d\sigma_g)^2(3d\pi_u)^4(3d\delta_g)^4 \tag{7}$$

Thus four electrons occupy σ orbitals, four occupy π orbitals and four occupy δ orbitals. The above configuration represents the wavefunction in an SCF picture. It is worth noting that this configuration has a weight of less than 50% of the total CASSCF configuration for Cr_2 at the experimental bond distance[16]. Correct dissociation can be obtained by allowing only excitations which keep the number of electrons in the orbital groups (6) fixed at the values given by (7). With four electrons in each of the groups (6), the number of CSFs is reduced from 28 784 to 3088, a number which is much easier to handle in the calculation. It is actually possible to divide the first group into two (one for $3d\sigma$ and one for 4s), which would reduce the number further. A wavefunction of this restricted type resembles a GVB treatment of the system. The GVB method[22] is, however, further restricted by allowing only specific spin couplings of the active orbitals.

It can be assumed that the configurations left out in a constrained wavefunction of the type given above will be less important for the dissociation process. This may, however, not be the case close to equilibrium. Obviously a calculation on Li_2, which does not include excitations from the bonding σ orbitals to the π orbitals, would not be very meaningful[37]. A test was performed on the less obvious case of the nitrogen molecule at the experimental geometry. An extended contracted Gaussian AO basis was used (13s, 8p, 3d and 2f contracted to 8s, 5p, 2d and 1f). First a CASSCF calculation was performed with the active orbitals (see Section III for a motivation of this choice)

$$(2\sigma_g, 2\sigma_u, 3\sigma_g, 3\sigma_u, 4\sigma_g, 4\sigma_u, 1\pi_g, 1\pi_u) \tag{8}$$

TABLE I
Some properties of N_2 computed with CASSCF wavefunctions (for details see text).

Method	r_e (Å)	D_e (eV)	Θ_e (a.u.)[a]
CASSCF ($2\sigma_g, 2\sigma_u, 3\sigma_g, 3\sigma_u, 4\sigma_g, 4\sigma_u, 1\pi_u, 1\pi_g$)	1.102	9.58	−1.288
Constrained CASSCF	1.105	9.30	−1.337
MR-CI[b]	1.101	9.46	–
MBPT(4)[c]	–	–	−1.149
SCF	1.069	–	−0.903[d]
Expt	1.098[e]	9.91[e]	-1.09 ± 0.06[f]

[a]Quadrupole moment.
[b]Ref. 38.
[c]Full fourth-order MBPT result from Ref. 39.
[d]Ref. 39.
[e]Ref. 40.
[f]Ref. 41.

Constraints were then applied, such that the number of electrons in σ orbitals was fixed at six and the number of electrons in π orbitals at four. The results of the two calculations are presented in Table I, where the effects on some of the properties of the nitrogen molecule are given. For comparison the corresponding SCF values are also presented. As can be seen from these results, the effects of the constraints on the CASSCF wavefunction are not negligible. They are, however, considerably smaller than the difference between the CASSCF and the SCF values. Better agreement with experiment can only be obtained by including dynamical correlation effects, for example, by means of a large multireference CI calculation[38] or a many-body perturbation theory (MBPT) calculation[39].

Thus, in many cases it is possible to reduce the computational effort by adding constraints to the CASSCF wavefunction in the ways discussed above. The simple structure of the model is then lost to some extent. The selection of an active orbital space is extended to include several active subspaces with a fixed number of electrons attributed to each of them. It is not difficult to proceed one step further and allow limited excitations between the different subspaces. The number of electrons is then not fixed, but is allowed to vary between given limits.

The wavefunctions generated using the model discussed above will in the following be called constrained complete active space (CCAS) wavefunctions. In the examples given above the different active subspaces were of different symmetry. Rotations between active orbitals are redundant variables in the CASSCF orbital optimization process. This is no longer the case when the active CSF space is reduced. In principle, it is necessary to introduce rotations between the active orbitals. With the blocking illustrated above, these extra rotations are zero by symmetry, and in this respect there is no difference between a CAS and a CCAS orbital optimization process. As illustrated for Cr_2 it might, however, also be of interest to block active orbitals within a given symmetry species. Obviously this will always be the case when the molecule does not possess any symmetry. In such cases, orbital rotations between active subspaces has to be introduced into the optimization process (see below for details).

B. Optimization of the Wavefunction

The CAS (or CCAS) wavefunction, $|0\rangle$, is obtained as a superposition of the CSFs, $|\mu\rangle$, generated by the active orbital subspace(s)

$$|0\rangle = \sum_{\mu=1}^{M} C_{\mu}^{(0)} |\mu\rangle \tag{9}$$

The configuration state functions are in the CAS case most easily generated using the graphical unitary group approach (GUGA)[19]. All necessary information about the CSFs and their relative ordering (the 'lexical ordering') is contained in a compressed table, the distinct row table (DRT). Reduction of the CSF space from CAS to CCAS can be performed relatively easily, by deleting certain vertices and paths in the DRT (see Ref. 19 for details on the DRT).

The Hamiltonian is assumed to be spin-independent. It can then be written, in second quantized form, in terms of the spin-averaged excitation operators (the generators of the unitary group[42])

$$\hat{E}_{pq} = \sum_{\sigma} \hat{a}_{p\sigma}^{\dagger} \hat{a}_{q\sigma} \tag{10}$$

where $\hat{a}_{p\sigma}$ ($\hat{a}_{p\sigma}^{\dagger}$) are the normal annihilation (creation) operators for an electron in the molecular orbital φ_p with spin σ. The generators (10) fulfil the following commutation relations:

$$[\hat{E}_{pq}, \hat{E}_{rs}] = \delta_{qr} \hat{E}_{ps} - \delta_{sp} \hat{E}_{rq} \tag{11}$$

The Hamiltonian is, in the algebraic approximation defined by the finite MO basis, given as

$$\hat{H} = \sum_{p,q} h_{pq} \hat{E}_{pq} + \tfrac{1}{2} \sum_{p,q,r,s} (pq|rs)(\hat{E}_{pq}\hat{E}_{rs} - \delta_{qr}\hat{E}_{ps}) \tag{12}$$

where h_{pq} and $(pq|rs)$ are the normal one- and two-electron integrals. The total energy of the system

$$E_0 = \langle 0|\hat{H}|0\rangle / \langle 0|0\rangle \tag{13}$$

is, with $|0\rangle$ normalized to unity, obtained as

$$E_0 = \sum_{p,q} h_{pq} D_{pq}^{(00)} + \sum_{p,q,r,s} (pq|rs) P_{pqrs}^{(00)} \tag{14}$$

where $D_{pq}^{(00)}$ and $P_{pqrs}^{(00)}$ are the first- and second-order reduced density matrices, which according to (12) are obtained as

$$D_{pq}^{(00)} = \langle 0|\hat{E}_{pq}|0\rangle \tag{15a}$$

$$P_{pqrs}^{(00)} = \tfrac{1}{2}\langle 0|\hat{E}_{pq}\hat{E}_{rs} - \delta_{qr}\hat{E}_{ps}|0\rangle \tag{15b}$$

In the energy expression (14), the orbitals occur in the integrals h_{pq} and $(pq|rs)$. The density matrix elements (15) contain the CI expansion coefficients. Using (9) we obtain

$$D_{pq}^{(00)} = \sum_{\mu,\nu} C_\mu^{(0)} C_\nu^{(0)} A_{pq}^{\mu\nu} \tag{16a}$$

$$P_{pqrs}^{(00)} = \sum_{\mu,\nu} C_\mu^{(0)} C_\nu^{(0)} A_{pqrs}^{\mu\nu} \tag{16b}$$

where $A_{pq}^{\mu\nu}$ and $A_{pqrs}^{\mu\nu}$ are the one- and two-electron coupling coefficients which are used in forming the Hamiltonian matrix in the CSF basis:

$$A_{pq}^{\mu\nu} = \langle \mu|\hat{E}_{pq}|\nu\rangle \tag{17a}$$

$$A_{pqrs}^{\mu\nu} = \tfrac{1}{2}\langle \mu|\hat{E}_{pq}\hat{E}_{rs} - \delta_{qr}\hat{E}_{ps}|\nu\rangle \tag{17b}$$

with the Hamiltonian matrix element given as

$$H_{\mu\nu} = \langle \mu|\hat{H}|\nu\rangle = \sum_{p,q} h_{pq} A_{pq}^{\mu\nu} + \sum_{p,q,r,s} (pq|rs) A_{pqrs}^{\mu\nu} \tag{18}$$

The formulation presented above is based on the assumption that molecular orbitals and CI expansion coefficients are real.

The energy expression (14) is a function of the molecular orbitals, appearing in the one- and two-electron integrals, and of the CI expansion coefficients through the first- and second-order reduced density matrices (15). In an MCSCF optimization procedure the CI coefficients and the parameters determining the MOs (normally the LCAO expansion coefficients) are varied until the energy reaches a stationary value. A number of procedures for performing the optimization have been described in the literature (see Ref. 26 for an extensive review). Here only the basic features of these procedures will be outlined, and the reader is referred to the literature for further details[26,43].

In order to describe the variations of the CI coefficients, the orthogonal complement to (9), $|K\rangle$, is also introduced:

$$|K\rangle = \sum_{\mu=1} C_\mu^{(K)}|\mu\rangle \qquad (K = 1, \ldots, M-1) \tag{19}$$

The variation of the MC function is described by a unitary operator, $\exp(\hat{S})$, where $\hat{S}$ is an anti-Hermitian operator

$$\hat{S} = \sum_{K \neq 0} S_{K0}(|K\rangle\langle 0| - |0\rangle\langle K|) \tag{20}$$

Similarly orbital rotations can be described by the unitary operator, $\exp(\hat{T})$, with $\hat{T}$ defined as

$$\hat{T} = \sum_{p>q} T_{pq}(\hat{E}_{pq} - \hat{E}_{qp}) \tag{21}$$

The anti-Hermiticity of $\hat{T}$ follows from the fact that $\hat{E}_{qp} = \hat{E}^{\dagger}_{pq}$ (see Eq. (10)). A general variation of the wavefunction can now be written as the combined operation of $\hat{S}$ and $\hat{T}$:

$$|0'\rangle = \exp(\hat{T})\exp(\hat{S})|0\rangle \tag{22}$$

Since the operators $\hat{T}$ and $\hat{S}$ do not commute, the order of the two operations is not irrelevant. The order used above is the one commonly used in the formulation of the problem. The reverse order can be shown to lead to much more complex algebra, needed to determine the transformed state $|0'\rangle$[44]. Owing to the definitions of $\hat{S}$ and $\hat{T}$, the varied wavefunction (27) will remain normalized and the new molecular orbitals defined as

$$\varphi' = \exp(\hat{T})\varphi \tag{23}$$

will remain orthonormal. The parameters in $\hat{S}$ and $\hat{T}$ thus constitute a set of variables that can be used to determine the stationary point for the energy. These parameters are, however, not in general linearly independent and care has to be taken to delete redundant variables from the parameter space. This is a trivial problem for CAS (and CCAS) wavefunctions, as will be illustrated below.

The transformed energy expression can, with the aid of (22), be written as

$$E' = \langle 0|\exp(-\hat{S})\exp(-\hat{T})\hat{H}\exp(\hat{T})\exp(\hat{S})|0\rangle \tag{24}$$

The conditions for a stationary value of the energy as a function of **S** and **T** are now easily obtained following a power expansion of the exponential factors in (24). The first-order terms should vanish at the stationary point, which immediately gives the relations

$$\langle 0|[\hat{H}, \hat{S}]|0\rangle = 0 \tag{25a}$$

$$\langle 0|[\hat{H}, \hat{T}]|0\rangle = 0 \tag{25b}$$

or with the use of Eqs (20) and (21),

$$\langle 0|\hat{H}|K\rangle = 0 \qquad K \neq 0 \tag{26a}$$

$$\langle 0|\hat{H}(\hat{E}_{pq} - \hat{E}_{qp})|0\rangle = 0 \tag{26b}$$

The first condition simply states that the reference state $|0\rangle$ should not interact with the orthogonal complement, that is, it is a solution to the secular problem

$$(\mathbf{H} - E\mathbf{1})\mathbf{C} = 0 \tag{27}$$

The second condition is the generalized Brillouin theorem, first derived by Levy and Berthier[45].

Condition (26b) is, for CASSCF wavefunctions, automatically fulfilled if both p and q belong to the same orbital space (inactive, active or external). From now on the following index labelling will be used:

i, j, k, l	inactive orbitals
t, u, v, x	active orbitals
a, b, c, d	external orbitals
p, q, r, s	all orbitals

For rotations in the inactive and external subspaces we trivially obtain:

$$\hat{E}_{ij}|0\rangle = 2\delta_{ij}|0\rangle \tag{28a}$$

$$\hat{E}_{ab}|0\rangle = 0 \tag{28b}$$

since the inactive orbitals are doubly occupied in all CSFs, while the external orbitals are empty. For the active subspace on the other hand,

$$\hat{E}_{tu}|0\rangle = \sum_K \langle K|\hat{E}_{tu}|0\rangle |K\rangle \tag{29}$$

since the CSF space is complete in the active orbital space. Also for real orbitals,

$$\hat{E}_{ut}|0\rangle = \sum_K \langle 0|\hat{E}_{tu}|K\rangle |K\rangle \tag{30}$$

Using (29) and (30) in (26b), it follows immediately by using also (26a) that

$$\langle 0|\hat{H}(\hat{E}_{tu} - \hat{E}_{ut})|0\rangle = 0 \tag{31}$$

Since the energy is invariant to rotations within a given orbital subspace, they can be excluded from the orbital rotations. The only rotations which have to be included are those which occur between the three orbital subspaces, described by the generators $\hat{E}_{ai}$, $\hat{E}_{at}$ and $\hat{E}_{ti}$. No redundant variables will remain in the calculation. Such variables are sometimes a problem in general MCSCF calculations, where the MC expansion is not complete[26,44]. They have to be deleted from the calculation, in order to avoid singularities. The problem becomes even worse if some rotation parameters are near-redundant. Such situations can lead to difficult convergence problems, which, however, sometimes can be overcome by introducing higher-order terms in the expansion of the energy (see for example Werner and Knowles' treatment of internal–internal rotations[33]). This problem is less severe in CASSCF calculations since only rotations between the orbital subspaces are applied. There are, however, two cases which can also lead to convergence problems here. If an active orbital has an occupation number that is very close to 2, the energy will depend only weakly upon rotations between this orbital and the inactive orbitals. The same situation occurs when an orbital has a very small occupation number. Rotations between this orbital and the external subspace then become near-redundant. The easiest way to avoid this problem is to move the corresponding orbital into the inactive (or external) subspace. This is, however, not always possible. The situation may, for example, arise on one part of an energy surface, but not on others. Optimization methods like those recommended by Werner and Knowles[33] may then have to be used. It should be emphasized, however, that situations of this kind are rather unusual. Calculations with active orbital occupation numbers as high as 1.995 and as low as 0.001 have been found to converge without problems without using any special techniques.

In constrained CASSCF calculations, rotations between the different sub-blocks of the active orbital space have to be included as variational parameters. This is, however, only necessary when orbitals in different sub-blocks have the same symmetry.

The most commonly adopted procedure for optimizing the variational parameters in (24) is the non-linear Newton–Raphson procedure[26,30,46]. The energy expression is expanded to second order in the parameters **S** and **T**. By assuming the first-order derivatives of this expression to be zero, a linear equation system is obtained in **S** and **T**[26,43]:

$$\begin{bmatrix} \mathbf{g}^{(c)} \\ \mathbf{g}^{(o)} \end{bmatrix} + \begin{bmatrix} \mathbf{H}^{(cc)} & \mathbf{H}^{(co)} \\ (\mathbf{H}^{(co)\dagger})^{+} & \mathbf{H}^{(oo)} \end{bmatrix} \begin{bmatrix} \mathbf{S} \\ \mathbf{T} \end{bmatrix} = 0 \qquad (32)$$

where $\mathbf{g}^{(c)}$ and $\mathbf{g}^{(o)}$ are the gradients (26) (c for the configuration and o for the orbital part). The Hessian matrix **H** consists of three parts: the congfiguration–configuration, the orbital–configuration and the orbital–orbital parts. Explicit expressions for these matrices will not be presented here. They are available in the literature[26,43]. Special formulae for the orbital–orbital part corresponding to the different types of rotations occurring in a CASSCF calculation have been given by Siegbahn *et al.*[47].

In a straightforward application of the Newton–Raphson approach, Eq. (32) is solved iteratively for **S** and **T** until the convergence criteria (26) are fulfilled to the desired accuracy. This process converges nicely if the initial choice of the orbitals and the CI coefficients are close to the final result. The energy is then in the 'local region' where the second-order approximation is valid. Obviously such situations will not be very common in actual applications. In practice the starting orbitals are often obtained from a preceding SCF calculation, or even estimated from atomic densities[48], while the CI coefficients are obtained by solving Eq. (27) using the starting orbitals. It then often happens that the Hessian matrix has many negative eigenvalues. In order to force convergence in such situations, level-shifting and mode-damping procedures must be used[26]. The most efficient of these procedures is probably the augmented Hessian method with step restriction, which was originally introduced by Lengsfield[49]. The linear equations (31) are in this method replaced by an eigenvalue equation, which automatically introduces a level shift into the Hessian. A damping factor can also easily be introduced which sets an upper limit to the norm of the rotation vector[50]. Even using such methods it is not certain that convergence can be obtained in an acceptable number of iterations, if the starting values of the parameters place the energy far from the local region. Werner and Meyer[51] recognized that the small radius of convergence inherent to the Newton–Raphson approach stems from the fact that the energy is a periodic, rather than quadratic, function of the orbital rotations $\mathbf{T}$. They instead proposed the use of an expansion of the energy to second order in $\mathbf{V} = \mathbf{U} - 1$, where $\mathbf{U}$ is the unitary transformation matrix for the orbitals, $\mathbf{U} = \exp(\mathbf{T})$. The orthonormality conditions are then exactly accounted for, and the energy is now a periodic function of the orbital parameters $\mathbf{T}$. This method has been successfully applied in very extensive CASSCF calculations and shows, at least for the cases treated so far, beautiful convergence properties[25,33]. Three to four iterations are normally sufficient for a convergence to seven decimal places in the total energy.

The solution of the system of equations (31) (or the corresponding augmented Hessian secular problem) is not trivial when the dimension of the CI expansion becomes large. If the coupling terms are neglected the problem splits into two parts, a secular problem for the CI coefficient (Eq. (27)) and a separate problem for the orbital rotations. The CI problem can be efficiently treated for large CI expansions using direct CI methods[13,25,52], as will be discussed in more detail below. Such direct iteration schemes can, however, also be used when the coupled problem (31) is solved. Direct MCSCF was first introduced in an MCSCF program written by Lengsfield[53]. It was later introduced into CASSCF programs[25,26,33,54]. A possible extension of such methods that avoids the two-electron transformation step has been discussed by Almlöf and Taylor[55]. In all these approaches the linear (or, in the case of the method of Werner and Meyer, the non-linear) equation system (or secular problem) is solved using an iterative process, where the essential step consists of performing the matrix multiplication $\mathbf{Hc}$ where $\mathbf{c}$ is the solution vector and $\mathbf{H}$ is the Hessian matrix (e.g. as in Eq. (31)). In the direct methods, the Hessian is never explicitly constructed. Instead the matrix multiplication above is obtained using the explicit expressions for the matrix elements of $\mathbf{H}$ in terms of the basic quantities involved, that is, the one- and two-electron integrals and the first- and second-order reduced density matrices (16).

C. The Configuration-interaction Secular Problem

The CAS-CI expansion can be very long. It is therefore essential to have an efficient procedure for solving the corresponding secular problem. Such procedures must necessarily be based on the direct CI concept[52], since the storage and handling of a Hamiltonian matrix would severely limit the possibility of using long CI expansions. The special features of the CAS-CI allow certain simplifications, which are not possible in normal MR-CI calculations. Thus, the number of two-electron integrals appearing in the Hamiltonian matrix elements (18) is limited by the small size of the active orbital space. It is therefore no problem to store these integrals in the primary memory of the computer. As a result they can be randomly accessed. The problematic parts are the one- and two-electron coupling coefficients $A^{\mu\nu}_{pq}$ and $A^{\mu\nu}_{pqrs}$. An automatic procedure for computing these coefficients efficiently is necessary. The number of coupling coefficients can become large and is actually the limiting factor for the size of the CI expansion. The first of these problems was solved by the graphical unitary group approach (GUGA) by Paldus and Shavitt[15,19]. The GUGA is especially efficient in the CAS case, since it was originally based on the graphical representation of the complete configurational space. The one-electron coupling coefficients are obtained as a product of attached segments of a loop generated by the two interacting configurations:

$$A_{pq} = \prod_{k=p}^{q} \omega(T_k, b_k) \tag{33}$$

with one factor for each orbital in the range $\{p, q\}$. T_k and b_k specify the shape and orientation of the segment in the graph. For more details see the papers by Shavitt[19]. The two-electron coupling coefficients can be obtained in a similar way, as shown by Paldus and Boyle[56]:

$$A_{pqrs} = \tfrac{1}{2}\prod_{k\in s_1} \omega(T_k, b_k) \sum_{x} \prod_{l\in s_2} \omega_x(T_l, b_l) \tag{34}$$

where s_1 and s_2 are sets of orbital levels which are within one or both of the ranges $\{p, q\}$ and $\{r, s\}$. The summation x is over intermediate spin coupling.

The above approach represents an efficient way of computing the coupling coefficients. It does not, however, solve the problem of storage and handling of the large number of two-electron coupling coefficients generated for large CAS-CI expansions. A direct implementation of the above procedure sets the practical limit to around 10^4 CSFs[13].

The original idea of Paldus[57] was to generate the two-electron coefficients as a matrix product of the one-electron coefficients

$$A^{\mu\nu}_{pqrs} = \tfrac{1}{2}\sum_{\kappa} A^{\mu\kappa}_{pq} A^{\kappa\nu}_{rs} - \tfrac{1}{2}\delta_{qr} A^{\mu\nu}_{ps} \tag{35}$$

where the sum runs over the complete set of CSFs. Eq. (35) follows directly from the definiton (17b) by using the resolution of identity. This method was used by Robb and Hegarty[58] to construct a formula tape for the CI calculation.

In direct CI calculation the update vector for the CI coefficients, $\boldsymbol{\sigma}$, is given by

$$\sigma_\mu = \sum_\nu H_{\mu\nu} C_\nu \tag{36}$$

Siegbahn[59] realized that (35) could be used efficiently in (36) by ordering the coupling coefficients A_{pq} after the intermediate index. In his notation the first term in (35) gives rise to the following contribution to $\boldsymbol{\sigma}$:

$$\Delta\sigma_\mu = \tfrac{1}{2}\mathrm{Tr}\,(\mathbf{A}^\mu \mathbf{I}\mathbf{D}) \tag{37}$$

where $\mathbf{A}^\mu$ is a matrix of coupling coefficients ($A^\mu_{\kappa,pq} = A^{\mu\kappa}_{pq}$), $\mathbf{I}$ is a matrix of two-electron integrals ($I_{pq,rs} = (pq|rs)$) and $\mathbf{D}$ is the product of $\mathbf{A}$ with the coefficient vector $\mathbf{C}$ ($D_{rs,\kappa} = \Sigma_\nu A^{\kappa\nu}_{rs} C_\nu$). Eq. (37) can efficiently be adapted to a vector processor, thereby making complete CI calculations very effective on computers like the CRAY. Prior to this work the CI problem was the main bottleneck in vectorizing CASSCF program systems.

A radically different use of the factorization (35) was made by Knowles and Handy[60]. They realized that, if the CI expansion was based on determinants instead of spin-adapted CSFs, it was possible to perform the CI calculations without a precomputed list of coupling coefficients. The reason for this is simply that for determinants the one-electron coupling coefficients take the trivial values ± 1 or 0. Using a 'canonical' addressing scheme for the CI vectors, it was possible to construct an effectively vectorized code for CAS-CI calculations. The power of the program was demonstrated in a recent CASSCF calculation on the $^5\Delta$ ground state of FeO[25]. The number of Slater determinants used in the largest calculation was 230 045 corresponding to 178 910 CSFs. The calculation of the residual vector (36) took 36 seconds of CPU time on a CRAY-1S.

D. Simplified Optimization Procedures

The discussion above has centred around full second-order optimization methods where no further approximations have been made. Computationally such procedures involve two major steps which consume more than 90% of the computer time: the transformation of two-electron integrals and the update of the CI vector. The latter problem was discussed, to some extent, in the previous section. In order to make the former problem apparent, let us write down the explicit formula for one of the elements of the orbital–orbital parts of the Hessian matrix (31), corresponding to the interaction between two inactive–external rotations T_{ia} and T_{jb}:

$$\tfrac{1}{2}H^{(00)}_{ia,jb} = 2[4(ai|bj) - (ab|ij) - (aj|bi)] + 2\delta_{ij}F'_{ab} - 2\delta_{ab}F'_{ij} \tag{38}$$

where F'_{ab} and F'_{ij} are elements of a Fock matrix

$$F'_{pq} = h_{pq} + \sum_{r,s} D_{rs}[(pq|rs) - \tfrac{1}{2}(pr|qs)] \tag{39}$$

with r and s being restricted to the internal (inactive plus active) orbital space. D_{rs} are the elements of the first-order density matrix. In order to compute the matrix element (38) the Fock matrix (39) must be known and also the two-electron integrals $(ai|bj)$, etc. The Fock matrix is easily obtained using the well established procedures of the closed-shell Hartree–Fock method. The molecular two-electron integrals are obtained by a four-index transformation from the integrals over the AO basis functions. The transformation is called 'second order' since molecular integrals with two indices in the usually large external space have to be formed. In a direct procedure the Hessian matrix is not formed explicitly but is directly multiplied with the vector **T**, leading to expressions of the form

$$\sum_{j,b} [4(ai|bj) - (ab|ij) - (aj|bi)]T_{jb} \tag{40}$$

This expression can be expressed as a sum over AO integrals, multiplied by a 'density matrix', thus avoiding two-electron transformations[55]. Similar transformations can be made for other parts of the Hessian. The price to be paid is that the AO integral list has to be read once for every update (4), which is a very I/O intensive process. It is therefore doubtful whether such a procedure is advantageous in a conventional scheme, where the AO integrals are precomputed. The situation would be different if the integrals were computed when needed, as suggested in Ref. 55.

For large AO basis sets, (most CASSCF calculations are today performed with basis sets in the range of 80–150 Gaussian-type orbitals (GTOs) and for a moderately small CI expansion (less than 1000 CSFs)) the transformation step will dominate the calculation in a second-order procedure. It might therefore be worth while to look for methods in which this transformation is avoided as much as possible. Such procedures were actually first used in MCSCF calculations. They are based on the generalized Brillouin theorem (26b). One, the so-called first-order method, which has proven to have near-quadratic convergence in practical applications, is the super-CI method, first development by Grein and coworkers[28] and later used by Ruedenberg *et al.*[29]. The orbital rotations and CI coefficients are here found by an iterative solution of a secular problem in the variational space spanned by the CI states $|0\rangle$, $|K\rangle$ (or alternatively the CSF space), and the configurations generated in the generalized Brillouin theorem (the Brillouin states)

$$(\hat{E}_{pq} - \hat{E}_{qp})|0\rangle \tag{41}$$

The coefficients of the Brillouin states are used as the rotation parameters for the orbitals[28]. Alternatively the first-order density matrix is generated and diagonalized. The corresponding natural orbitals (selected via an overlap criterion) are used as input for the next iteration[12,29].

The super-CI method can be regarded as an approximation in the augmented Hessian variant of the Newton–Raphson (NR) procedure (see for example Ref. 43). However, in its original formulation, it does not constitute any simplification when compared with the NR method. The matrix elements between the Brillouin states (41) are actually more difficult to compute than the corresponding Hessian matrix elements, since they involve third-order density matrix elements[47].

The orbital optimization was, in the first implementation of the CASSCF method, performed using an approximate version of the super-CI method, which avoided the calculation of the third-order density matrix[12]. This was later[14] developed as a procedure entirely based on the average MCSCF Fock operator[61]:

$$F_{mn} = \sum_{p} D_{mp}^{(00)} h_{pn} + \sum_{p,q,r} P_{mpqr}^{(00)} (np|qr) \tag{42}$$

Contributions from the inactive orbitals are easily obtained in (42) from atomic two-electron integrals (or super-matrix elements) as was the case for the matrix $\mathbf{F}'$ defined in Eq. (39). The second-order density matrix can be expressed using the first-order density matrix, if any of the indices m, p, q, or r refer to inactive orbitals. The remaining part has all four indices in the active subspace. Thus the only molecular two-electron integrals explicitly occurring in (42) have three indices in the active orbital space. The fourth index, n, runs over all orbitals. To transform these integrals from AO to MO basis requires around $n_a n_b^4/2$ operations, where n_a is the number of active orbitals and n_b is the number of AO basis functions. The second-order transformation needed to form the Hessian matrix elements requires on the other hand around $n_o n_b^4$ operations, where n_o is the number of occupied orbitals (inactive and active)[29]; n_o can be much larger than n_a in calculations on systems with many electrons. A first-order transformation can further be made entirely in core memory (except for the read of a sorted and symmetry-blocked list of AO integrals). This is generally not possible for a second-order transformation, where the half-transformed integrals have to be written to mass storage. Thus a first-order transformation runs much more effectively on vector processors like CRAY-1 and even more so on a single-user computer like the FPS-164.

In Ref. 14 an approximate super-CI approach was developed, where the Hamiltonian was replaced by the one-electron Hamiltonian

$$\hat{H}_0 = \sum_{p,q} \hat{E}_{pq} F_{pq} \tag{43}$$

This Hamiltonian was used to compute the matrix elements between the

Brillouin states

$$\hat{E}_{ai}|0\rangle \tag{44a}$$

$$\hat{E}_{at}|0\rangle \tag{44b}$$

$$\hat{E}_{ti}|0\rangle \tag{44c}$$

which corresponds to the only non-redundant rotations appearing in a CASSCF calculation. The matrix elements could be expressed entirely in first- and second-order density matrices together with the Fock matrix (42). The interaction between the Brillouin states and the reference state $|0\rangle$ is given exactly in terms of the Fock matrix (42)

$$\langle 0|\hat{H}(\hat{E}_{pq} - \hat{E}_{qp})|0\rangle = F_{pq} - F_{qp} \tag{45}$$

The solution of the super-CI secular problem takes very little time compared to the remaining parts of the calculation. It can be implemented with or without CI coupling.

As already pointed out, the super-CI procedure can be regarded as an approximate version of the augmented Hessian approach. This is of course also true for the method described above, the only difference being that the approximations made are more severe.

This is, however, of little importance as long as the MCSCF process converges within a satisfactory number of iterations. It can be estimated that the first-order transformation is 3–5 times as fast as the second-order transformation. Thus a first-order procedure could compete with second-order procedures if the number of iterations needed are not more than three times as many, provided of course that the calculation is not dominated by the CI step. The most efficient second-order procedure seems to be that of Werner and Knowles[25,33]. They use three macro-iterations and about 30–50 micro-iterations to converge to less than 10^{-7} a.u. in the energy. A first-order transformation is performed in each micro-iteration and a second-order one in each macro-iteration. It seems that first-order procedures can compete with this performance in many cases. With reasonable starting vectors, convergence to 10^{-6} a.u. is often reached in less than 10 iterations.

First-order methods often show good convergence in calculations on ground-state energy surfaces (that is, the lowest state in each symmetry), and compete well with the Newton–Raphson procedure in the non-local regions. It is often a useful compromise, therefore, to start a calculation with a first-order scheme, and switch to second order when the local region has been reached. The situation is different for excited states where CI coupling often becomes necessary for a calculation to converge at all. Here second-order schemes with full CI coupling are liable to be the more efficient procedures. CI coupling can also be introduced into super-CI methods. Such calculations have, however, not been performed so far, so experience about their convergence behaviour is still lacking.

Experience in a variety of applications of the CASSCF method has shown it to be a valuable tool for obtaining good zeroth-order approximations to the wavefunctions. Attempts have been made to extend the treatment to include also the most important dynamical correlation effects. While this can be quite successful in some specific cases (see below for some examples), it is in general an impossible route. Dynamical correlation effects should preferably be included via multireference CI calculations. It is then rarely necessary to perform very large CASSCF calculations. Degeneracy effects are most often described by a rather small set of active orbitals. On the other hand experience has also shown that it is important to use large basis sets including polarization functions in order to obtain reliable results. The CASSCF calculations will in such studies be dominated by the transformation step rather than by the CI calculation. A mixture of first- and second-order procedures, as advocated above, is then probably the most economic alternative.

III. APPLICATIONS

In this section, some illustrative examples of the CASSCF method will be given. The emphasis will not be on the technical problem of solving the MCSCF equations but rather on the chemical problems encountered when using the CASSCF method as a tool for electronic structure calculations. Most of the examples have been selected from recently published results.

CASSCF, like most other MCSCF procedures, is a method which makes it possible to study systems where degeneracies or near-degeneracies occur between different configurations. In such cases the Hartree–Fock model is not a valid zeroth-order approximation. The CASSCF model studies these effects with as little bias as possible, since the only decision made by the user is the selection of the inactive and active orbitals, and of course the AO basis set. The method is not primarily aiming at studies of dynamical correlation effects.

However, a sharp limit does not exist between near-degeneracy and dynamical correlation effects. A simple example is H_2 where the $1\sigma_u$ orbital describes near-degeneracy effects for large bond distances, while it is conventionally considered as contributing to the dynamical correlation for bond distances around equilibrium. A simple solution to this dilemma is formally to define near-degeneracy correlation as the effect due to all configurations generated within the valence orbital space. A formal definition of the dynamical correlation energy E_{DC} can then be obtained as

$$E_{DC} = E_{exact} - E_{VCAS} \tag{46}$$

where E_{exact} is the exact eigenvalue of the Hamiltonian and E_{VCAS} is the CASSCF energy of a calculation with all valence orbitals active. The choice of the valence orbitals would for H and He be 1s, for Li–Ne 2s and 2p, for Na–Ar, 3s and 3p, and for K–Zn 3d, 4s and 4p, etc. It is obviously not possible to compute E_{VCAS} for most molecules. The definition above is therefore rather formal. However, there is another reason for writing it down. Dynamical correlation effects are intuitively considered as creating a Coulomb hole around each electron, making the average distance between a pair of electrons slightly larger. The magnitude of these correlations are probably measured well by the diagonal of the second-order density function $\rho_2(\mathbf{r},\mathbf{r})$. Near-degeneracy effects are, on the other hand, often of a different nature. They separate the electrons in a pair much more effectively. An obvious example is again H_2 at large internuclear separation. The effect of the $1\sigma_u$ orbital here leads to a total separation of the two electrons, such that one electron is localized at each of the two atoms. The beryllium atom constitutes a similar example. An SCF calculation gives the total energy − 14.572 a.u.[62]. A corresponding valence CASSCF calculation (including 2s and 2p in the active space) gives an energy lowering of − 0.0455 a.u., accounting for almost all of the correlation energy for the $(2s)^2$ electron pair in Be. The remaining dynamical valence correlation energy was estimated to be only − 0.0022 a.u., which is only 5% of the total value. This estimate was also done by a CASSCF calculation with the active orbitals 2s, 2p, 3s, 3p, 3d, 4s and 4p. The reason for the small dynamical correlation energy is the effective separation of the two electrons effected by the 2p orbital, leading to a sharp decrease in $\rho_2(\mathbf{r},\mathbf{r})$. The separation was illustrated by Eq. (3) in the introduction. In beryllium, the coefficient in front of the last term in (3) is as large as 0.3. (Total separation is obtained with the value 1.0.)

The important conclusion to be drawn from the discussion above is that there exists a strong interplay between near-degeneracy effects and dynamical correlation. Strong near-degeneracy effects lead to effective separation of the electron pair, and thus to a reduction in the dynamical correlation effects. A method which only takes into account near-degeneracy effects will in general not give a balanced description of the total correlation effects over an energy surface, for example along a dissociation channel. This is, of course, well known and is the reason why full valence MCSCF calculations in general give binding energies which are too low. It is sometimes possible to counter-balance this effect by adding a few more active orbitals. The principle behind this is that the active space should contain one correlating orbital for each strongly occupied orbital. The N_2 molecule is an example. The SCF configuration in N_2 is

$$(2\sigma_g)^2(2\sigma_u)^2(3\sigma_g)^2(1\pi_u)^4 \tag{47}$$

where the 1s orbitals have not been given. The $3\sigma_g$ and $1\pi_u$ orbitals are correlated to the corresponding antibonding orbitals $3\sigma_u$ and $1\pi_g$ in a full valence calculation. No correlation is, however, introduced into $2\sigma_g$ and $2\sigma_u$. A FORS calculation by Ruedenberg *et al.* gave for this type of wavefunction a binding energy of 9.06 eV[24]. The most important dynamical structure-dependent correlation effects can be accounted for by adding two more orbitals to the active subspace, $4\sigma_g$ and $4\sigma_u$. At small internuclear distances these orbitals introduce left–right correlation into the pairs $(2\sigma_g)^2$ and $(2\sigma_u)^2$, while at large distances they introduce the corresponding angular correlation in $(2s)^2$ on each of the two atoms. The calculated binding energy with this active space is 9.57 eV[63], a value much closer to the experimental value, 9.90 eV[40]. Obviously, there is no chance to obtain an exact balance between atomic and molecular correlation energies in this way. Unfortunately it does not seem to be much easier to reach this goal with the multireference CI method, as shown for example by Siegbahn's work[38] on N_2. Convergence in dissociation energies can only be achieved simultaneously with convergence in the total correlation energy.

A similar situation as for N_2 occurs for the water molecule. Here the SCF configuration is in C_{2v} symmetry

$$(2a_1)^2(3a_1)^2(1b_2)^2(1b_1)^2 \tag{48}$$

The orbitals $3a_1$ and $1b_2$ are correlated by the corresponding antibonding orbitals $4a_1$ and $2b_2$ in a full valence description of the molecule. This is, however, an unbalanced situation, since only one of the OH bonds is correlated (together with the oxygen lone pair $3a_1$). A bond angle of only 103° is for example obtained[14]. Balance is approximately restored by adding two more orbitals, $5a_1$ and $2b_1$, which includes the corresponding correlation effects into $2a_1$ and $1b_1$. The computed angle is now 104.8° (expt 104.5°), and also other properties of the molecule are improved (see Ref. 14 for more details). It should be noted that these 'extra valence' orbitals have rather large occupation numbers, 0.013 for $5a_1$ and 0.021 for $2b_1$. A more intricate example of the same problem of balance in the dynamical and near-degeneracy correlations will be given below for the C_2 molecule.

In calculations on larger systems it will, of course, not be possible to include all valence orbitals into the active subspace. It is also not necessary. Studies of an energy surface, i.e. for an chemical reaction, will normally be concentrated in regions where only one or two of the chemical bonds are broken. The calculation can then be performed by choosing as active orbitals only those taking part in the bond-breaking process. However, only semi-quantitative results can be expected from such calculations. Additional dynamical correlation effects have to be include via configuration-interaction calculations.

A. The C_2 Molecule

The C_2 molecule offers an example where near-degeneracy effects have large amplitudes even near the equilibrium internuclear separation. The ground state is $^1\Sigma_g^+$ with the electronic configuration (excluding the 1s orbitals)

$$(2\sigma_g)^2(2\sigma_u)^2(1\pi_u)^4 \tag{49}$$

The electrons in $2\sigma_u$ are essentially non-bonding, giving a biradical character to the molecule. Strong interaction with the configuration

$$(2\sigma_g)^2(3\sigma_g)^2(1\pi_u)^4 \tag{50}$$

can be expected, leading to a separation of the two electrons on different atoms. This effect is well known[64,65]. The relative weight of the two configurations $(C_I/C_{II})^2$ is in the present calculation 0.19 at R(C–C) = 2.35 a.u. (the experimental value is 2.348 a.u.[66]). This is a large value, even if it is still far from the value 1.0 which corresponds to two completely separated electrons. One would therefore assume that an SCF treatment of the C_2 molecule would not give very accurate results for spectroscopic constants. Such a calculation was performed by Dupuis and Liu[67]. Surprisingly enough they obtained much better results with the single-configuration wavefunction (49) than with an MCSCF wavefunction including both configurations (49) and (50) (MC-2). The R_e and ω_e values were in the former case 2.341 a.u. and 1905 cm^{-1} (expt 2.348 and 1855)[67], while the two-configurational treatment gave the results 2.261 a.u. and 2188 cm^{-1}, respectively. The surprising results is not so much the poor values obtained with MC-2 but the good values obtained in the SCF treatment. Obviously, configuration (50) is balanced by other configurations with an opposite effect on the potential curve, finally making the SCF wavefunction a reasonable approximation around $R = R_e$. That this is the case will be illustrated below.

The $^1\Sigma_g^+$ is nearly degenerate with the $^3\Pi_u$ state with the electronic configuration

$$(2\sigma_g)^2(2\sigma_u)^2(3\sigma_g)(1\pi_u)^4 \tag{51}$$

The experimental T_e value is only 0.08 eV[67]. The wavefunction for this state can be expected to be more dominated by the SCF configuration (51) than was the case for the $^1\Sigma_g^+$ state, since the excitation $(2\sigma_u)^2 \rightarrow (3\sigma_g)^2$ is not possible here. In an attempt to compute the T_e value by the CASSCF method one is then faced with an imbalance in the near-degeneracy correlation effects which must be accounted for either by extending the active subspace outside the valence shell or by including dynamical correlation effects via a CI calculation. It is clear from the preceding discussion that a more balanced description, on the CAS level, can be achieved by adding a $4\sigma_g$ orbital to the active space, thus accounting for the larger dynamical correlation in the $2\sigma_u$ electron pair in the $^3\Pi_u$ state. For a correct description of dissociation it is then also necessary to add a $4\sigma_u$ orbital.

Using the criteria for selecting the active subspace which were discussed above, it should not be necessary to include this orbital, since $2\sigma_g$ is correlated

TABLE II
CASSCF results for the $X^1\Sigma_g^+$ and a $^3\Pi_u$ states of C_2.

State		DZ basis[a] $n_a = 8$[d]	$n_a = 10$[e]	Large basis[b] $n_a = 10$	Expt[c]
$^1\Sigma_g^+$	R_e(Å)	1.284	1.284	1.253	1.243
	D_e(eV)	5.56	5.57	6.06	6.32
	ω_e(cm^{-1})				
$^3\Pi_u$	R_e(Å)	1.357	1.362	1.325	1.312
	D_e(eV)	5.01	5.28	5.89	6.24
	ω_e(cm^{-1})				
	T_e(eV)	0.55	0.29	0.17	0.08

[a] 9s, 5p contracted to 4s, 2p (Ref. 68).
[b] 13s, 8p contracted to 7s, 5p (Ref. 68) adding one diffuse s-type (exponent 0.04) and one p-type (exponent 0.025) function. Polarization functions: 4 d-type functions (exponents 2.179, 0.865, 0.362, 0.155) contracted (2, 1, 1) (Ref. 69).
[c] Ref. 67.
[d] Active orbital space: $2\sigma_g$, $2\sigma_u$, $3\sigma_g$, $3\sigma_u$, $1\pi_u$ and $1\pi_g$.
[e] Active space as in *d* with $4\sigma_g$ and $4\sigma_u$ added.

by $3\sigma_u$. However, at dissociation, $4\sigma_g$ and $4\sigma_u$ become degenerate. Both have to be included in order to make the calculation size-consistent, that is, the energies at $R = \infty$ should be equal to twice the energy obtained in a corresponding CASSCF calculation on the free atom. This criterion has not always been met in earlier calculations, probably resulting in dissociation energies[14,70] somewhat too large.

Two sets of preliminary CASSCF calculations were preformed in order to test the above assumption. A double-zeta (DZ) AO basis was used. One calculation was made with a full valence active space, while the two orbitals $4\sigma_g$ and $4\sigma_u$ were added in the second calculation. Some of the results of these calculations are reported in Table II. The addition of the two extra valence orbitals had a very small effect on the potential curve for the $^1\Sigma_g^+$ state, but the T_e value of the $^3\Pi_u$ state dropped from 0.55 to 0.29 eV with the addition of these active orbitals. The rather poor overall agreement with experiment obtained in these calculations is, of course, due to the limited basis set used, but they illustrate the importance of the $4\sigma_g$ orbital for a balanced description of the two potential curves.

The larger active space was then used in a new set of calculations which used a considerably larger AO basis. The 13s, 8p basis of van Duijneveldt[68] was contracted (7, 1, 1, 1, 1, 1, 1/4, 1, 1, 1, 1). Additional diffuse functions were added, one of s-type (exponent 0.04) and of one p-type (exponent 0.025). Three 3d polarization functions were also included, contracted from four primitive GTOs with exponents 2.179, 0.865, 0.362 and 0.155. The exponents and the contraction coefficients were chosen according to the polarized basis-set

TABLE III
Occupation numbers for the active orbitals in C_2 around equilibrium and at large internuclear separation (large basis, $n_a = 10$).

State	R (a.u.)[a]	$2\sigma_g$	$3\sigma_g$	$4\sigma_g$	$2\sigma_u$	$3\sigma_u$	$4\sigma_u$	$1\pi_u$	$1\pi_g$
$^1\Sigma_g^+$	2.30	1.980	0.406	0.005	1.589	0.013	0.002	3.792	0.212
	2.35	1.979	0.392	0.005	1.603	0.014	0.002	3.779	0.225
	2.40	1.979	0.398	0.005	1.616	0.015	0.002	3.765	0.238
	∞	1.950	1.000	0.007	1.950	1.000	0.007	1.044	1.044
$^3\Pi_u$	2.45	1.981	1.038	0.010	1.905	0.019	0.007	2.887	0.152
	2.50	1.980	1.039	0.010	1.904	0.020	0.007	2.882	0.158
	2.55	1.979	1.039	0.010	1.903	0.021	0.007	2.875	0.165
	∞	1.950	0.522	0.007	1.950	0.522	0.007	1.521	1.521

[a]In atomic units (1 Å = 1.889 76 a.u.).

concept[69]. No f-type functions were used in these calculations. While f-type functions certainly are important in calculations of dynamical correlation effects (i.e. by MR-CI methods), they seem to have only small effects on results obtained in calculations which mainly include-degeneracy correlations[70]. The final basis set consists of 130 primitive GTOs contracted to 88 basis functions. The first-order approximate super-CI method was used to optimize the orbitals. Between five and 10 iterations were necessary to obtain an energy which converged to 10^{-7} a.u. when using starting vectors from a nearby point on the potential curve. The first point was obtained with starting vectors from an SCF calculation on the closed-shell configuration (49). Here 16 iterations were needed to meet the same convergence threshold.

The results of these calculations are presented in the last column of Table II and in Tables III and IV. The computed values for the dissociation energies are consistent with the results obtained for N_2 in a similar calculation[63]. The C_2 molecule has a triple bond ($\cdot C{\equiv}C\cdot$) like N_2. The dissociation energy for the ground state is in error by 0.26 eV, while the corresponding error for N_2 is 0.33 eV, in both cases amounting to about 0.1 eV per bond. In view of the well known difficulty in calculating dissociation energies with higher accuracy, even using large-scale multireference CI methods (see for example Ref. 38), these results must be considered as very satisfactory. Table III gives the occupation numbers for the natural orbitals of the active subspace. The two extra valence orbitals, $4\sigma_g$ and $4\sigma_u$, have as expected only small occupation numbers in the $^1\Sigma_g^+$ state. They were also found to have very little effect on the computed potential curve, as the smaller (DZ basis) calculation showed. The situation is, as expected, different in the $^3\Pi_u$ state, where the occupation numbers for these orbitals are more than doubled.

The general structure of the correlation effects can be easily deduced from Table III. Consider first the $^1\Sigma_g^+$ state. Adding together the occupation numbers for $2\sigma_g$ and $3\sigma_u$ gives a number very close to 2. Similarly $2\sigma_u$ and $3\sigma_g$ contain two electrons, while $1\pi_u$ and $1\pi_g$ have a total occupation number of 4. An overall description of the wavefunction can be formalized as

$$(2\sigma_g, 3\sigma_u)^2(2\sigma_u, 3\sigma_g)^2(1\pi_u, 1\pi_g)^4 \tag{52}$$

The above occupations correspond to a constrained CASSCF wavefunction with the active subspace divided accordingly. This description is, however, not valid for the entire potential curve. At internuclear distances larger than 3 a.u. (around 1.6 Å), the bonding changes completely and the dominant configuration is now

$$(2\sigma_g)^2(2\sigma_u)^2(3\sigma_g)^2(1\pi_u)^2\ {}^1\Sigma_g^+ \tag{53}$$

Configuration (53) has a weight of 74.4% at the internuclear separation 3.2 a.u., where the weight of configuration (49) has dropped to only 9.3%. This behaviour is not difficult to understand. The (weak) bonding at the larger distances is more effective with σ orbitals. The orbitals are, at these distances, only weakly hybridized. Thus $2\sigma_g$ and $2\sigma_u$ are well represented as carbon 2s orbitals, while $3\sigma_g$ and $1\pi_u$ are composed of 2p atomic orbitals. The change of the dominant configuration takes place around 3 eV above the minimum, which is halfway towards dissociation. It is thus clear that a constrained CASSCF (or a generalized valence bond) wavefunction according to (52) cannot be used in a complete description of the potential curve. At large internuclear distances the occupation of the carbon 2p shell is 2.044 on each atom. The active space here describes the well known 2s–2p near-degeneracy effect in the carbon atom, corresponding to the double excitation

$$(2s)^2(2p)^2\ {}^3P \rightarrow (2p)^4\ {}^3P \tag{54}$$

Turning now to the ${}^3\Pi_u$ state, a more complex structure is found. Here the sum of all σ occupation numbers is 4.96 instead of 5.00 as configuration (51) would indicate. Thus there exists in the wavefunction for this state non-negligible contributions from configurations with more than four π electrons. The most important of these configurations corresponds to the double excitation $(2\sigma_u)^2 \rightarrow (1\pi_g)^2$, which gives angular correlation to the electron pair in the $2\sigma_u$ orbital. Why are these configurations not equally important in the wavefunction for the ${}^1\Sigma_g^+$ state? The answer to this question has already been given. The $2\sigma_u$ electrons are, in the ground state, effectively correlated by the $3\sigma_g$ orbital. Further dynamical correlation effects then become less important. Thus we see here another example of the intricate balance between different types of correlation effects. As a result it is not possible to use the same simple spin-pairing picture for the ${}^3\Pi_u$ wavefunction around equilibrium as could be done for the ${}^1\Sigma_g^+$ wavefunction.

Table IV presents the weights of the most important configurations for the two spectroscopic states of C_2. The success of the SCF calculation done by

TABLE IV
Weights (%) of the most important configurations for the $X^1\Sigma_g^+$ and a $^3\Pi_u$ states of C_2 (large basis, $n_a = 10^a$).

		R (a.u.)		
		2.30	2.35	2.40
$^1\Sigma_g^+$	$(2\sigma_g)^2(2\sigma_u)^2(1\pi_u)^4$	70.7	70.9	71.1
	$(2\sigma_g)^2(3\sigma_g)^2(1\pi_u)^4$	14.4	13.5	12.6
	$(2\sigma_g)^2(2\sigma_u)^2(1\pi_u)^2(1\pi_g)^2$	2.6	2.8	3.0
	$(2\sigma_g)^2(3\sigma_g)^2(1\pi_u)^2(1\pi_g)^2$	1.4	1.4	1.5
	$(2\sigma_g)^2(2\sigma_u)(3\sigma_g)^2(1\pi_u)^3(1\pi_g)$	6.5	6.8	7.0
	Total	95.6	95.4	95.2
		R (a.u.)		
		2.45	2.50	2.55
$^3\Pi_u$	$(2\sigma_g)^2(2\sigma_u)^2(3\sigma_g)(1\pi_u)^3$	88.0	87.6	87.2
	$(2\sigma_g)^2(2\sigma_u)(3\sigma_g)^2(1\pi_u)^2(1\pi_g)$	3.5	4.0	4.1
	$(2\sigma_g)^2(2\sigma_u)^2(3\sigma_g)(1\pi_u)(1\pi_g)^2$	2.7	2.8	3.0
	Total	94.2	94.4	94.3

[a]The weight is given as $\sum C_i^2$ where the sum is over all CSFs corresponding to the same orbital occupation.

Dupuis and Liu[67] and the corresponding failure of the two-configurational MCSCF treatment for the $^1\Sigma_g^+$ state is clearly explained by the occupation numbers presented in the upper part of the table. The SCF configuration (49) has a weight which varies rather slowly with the distance R. The configuration (50) added by Dupuis and Liu in their MC treatment has a weight that decreases with R. Three other configurations appear in the table with a total weight close to that of configuration (50). They all describe bond breaking, and their summed weight will increase with R. The effect of the strongly bonding configuration (50) is thus almost completely balanced, and the SCF configuration will therefore give reasonable values for the spectroscopic constants, as obtained by Dupuis and Liu. Of course, such a behaviour is completely fortuitous, and does not mean that the SCF wavefunction gives a valid description of the $^1\Sigma_g^+$ state of the C_2 molecule.

Finally, the results obtained above on the CASSCF level of approximation were checked by a set of multireference CI calculations using the externally contracted CI (CCI) method of Siegbahn[38]. Two sets of calculations were performed, which differed in the choice of the reference configurations. In the first set the reference space was chosen as the CASSCF CSFs with coefficients larger than 0.1 around equilibrium. For the $^1\Sigma_g^+$ state this gave seven reference configurations and for $^3\Pi$ eight. The calculation at large internuclear distance

TABLE V
MR-CCI results for the $X^1\Sigma_g^+$ and $a^3\Pi_u$ states of C_2 (large basis)[a].

State		$n_{ref} = 7^b$	$n_{ref} = 10^c$	Expt
$^1\Sigma_g^+$	R_e (Å)	1.244 (1.255)	1.249 (1.260)	1.243
	D_e (eV)	5.79 (5.76)	5.94 (5.83)	6.32
	ω_e (cm^{-1})			
		$n_{ref} = 8^b$		Expt
$^3\Pi_u$	R_e (Å)	1.319 (1.319)		1.312
	D_e (eV)	5.85 (5.86)		6.24
	ω_e (cm^{-1})			
	T_e (eV)	−0.06 (−0.10)		+0.08

[a]Values within parentheses have been obtained by adding Davidson's correction[71] to the CCI results.
[b]The reference states are selected from the CASSCF configurations with a weight greater than 1%.
[c]As *a* but with the threshold 0.25%.

used the C(^{3}P) HF configuration as the reference state. The results of these calculations are presented in the second column of Table V. Comparing with the CASSCF result (Table II, large basis) we note an improvement in the equilibrium properties but not in the D_e and T_e values (the CCI calculation places $^3\Pi_u$ slightly below $^1\Sigma_g^+$). The calculations for the $^1\Sigma_g^+$ state were repeated with the threshold for the reference configurations lowered to 0.05. This resulted in an increase in D_e, but it still does not represent an improvement over the CASSCF results. Adding a Davidson correction[71] to the CCI results deteriorates the results even further for the $^1\Sigma_g^+$ state, while the $^3\Pi_u$ state is not affected. Obviously the Davidson correction, which was originally derived for a closed-shell reference CI, does not work for the highly degenerate $^1\Sigma_g^+$ state. The results obtained from these CI calculations seem to indicate that it is equally difficult to obtain a balanced description of the dynamical correlation effects as it is to achieve this goal for the near-degeneracy correlation in the CASSCF method. It should, however, be pointed out that the CI calculations were rather small (193 944 single and double excitations were included in the $^3\Pi_u$ wavefunction). Further studies of the convergence of the results with a lowering of the threshold could therefore have been carried out, probably leading to better agreement with experiment than was obtained in the present study. However, the purpose of this study of the C_2 molecule has been to illustrate the usefulness of the CASSCF method, rather than pressing the calculations towards the basis-set limit. A comprehensive account of the calculations on the C_2 and also the C_2^+ systems will be presented elsewhere[72].

B. The BH Molecule

The BH molecule represents a challenge for the CASSCF model quite different from that of C_2. In an eight-electron system like C_2 (with the four 1s electrons inactive), the number of active orbitals cannot exceed 10–12. The problem is then to find an active orbital space that represents the correlation effects, both near-degenerate and dynamical, in a balanced way. This situation is rather general since systems with 6–10 active electrons are common in CASSCF calculations. The BH molecule on the other hand can, to a very good approximation, be treated with four active electrons, leaving the two 1s electrons in an inactive orbital. A much larger active orbital space can then be used, and it becomes possible to include a large fraction of the dynamical correlation effects in to the CASSCF wavefunction. One might argue that for a four-electron system a full CI approach could just as well be used, the only remaining problem being the choice of an appropriate basis set. Such calculations have also been performed on BH, but with a smaller AO basis than was employed in the study to be discussed below[73]. A full CI with the present basis would comprise around 200 000 CSFs, which is not a very large number. Full CI calculations with 944 348 CSFs have been performed by Harrison and Handy for the FH molecule[73]. Obviously, a CASSCF calculation can in this case only be an approximation to the full CI result, albeit a very accurate one. The CASSCF method offers, however, several advantages compared to the brute-force CI method. It presents a wavefunction which is much more compact (645 CSFs was used for the $^1\Sigma^+$ states of BH). CASSCF calculations do not depend on the basis set in the same way as a full CI calculation does. Maybe the difference in approach is best summarized by noting that the CASSCF method will concentrate the information from a full CI treatment into a smaller, but optimized, set of molecular orbitals. This is of course only true if the active subspace can be sufficiently extended so that the results will converge towards the full CI limit. This does not seem to be a problem in a four-electron case. CASSCF calculations yielding results close to the full CI limit will of course very quickly become impossible when the number of electrons increases. The calculations on FH by Harrison and Hand, mentioned above, were done with eight electrons distributed in 18 orbitals[73]. With this CI calculation inserted into an MCSCF program, one probably has a good estimate of the upper limit of the CASSCF method in any of its present implementations.

A detailed account of the CASSCF results for the $X^1\Sigma^+$, $B^1\Sigma^+$ and $A^1\Pi$ potential curves of the BH molecule was given in 1981[74]. The main features of these results will be repeated here, together with some additional data concerning the structure of the CASSCF wavefunction. The calculations were performed using an AO basis set of 56 contracted Gaussian functions: B, 8s, 6p, 3d; H, 6s, 3p. The first calculations were made with 10 active orbitals:

6σ and 2π. Remaining correlation effects were estimated using second-order perturbation theory[75]. As a result of this study the final active subspace included 14 orbitals ($6\sigma, 3\pi$ and 1δ), resulting in a wavefunction comprising 645 CSFs of $^1\Sigma^+$ symmetry.

The total correlation energy of the BH molecule in the $X^1\Sigma^+$ ground state at equilibrium internuclear distance has been estimated by Meyer and Rosmus to be 0.152 a.u.[76]. Of this 0.048 a.u. correspond to 1s–1s and 1s–valence correlation. The valence correlation energy in BH is thus 0.104 a.u. The CASSCF calculation, utilizing the basis set and active space given above, recovers 89% (0.093 a.u.) of the total valence correlation energy, the remaining error being 0.30 eV. These values can be compared to the pair natural-orbital CI (PNO-CI) valence correlation energy (0.091 a.u.) obtained by Meyer and Rosmus, using a similar basis set[76]. The coupled electron-pair approximation (CEPA) corrected value was 0.100 a.u., but this is of course not an upper bound. The natural-orbital occupation numbers are presented in Table VI. The values reported show that the active space used includes in the wavefunction natural orbitals with occupation numbers down to around 10^{-3}. The most important correlation effect in the $^1\Sigma^+$ states is the angular correlation of the σ electrons affected by the 1π natural orbital. To a smaller extent, the 2π orbital plays the same role in the $^1\Pi$ state. Horizontal correlation is concentrated into one antibonding σ orbital, 4σ in the $X^1\Sigma^+$ and $A^1\Pi$ states and 5σ in the $B^1\Sigma^+$ state. The 4σ orbital is, in the B state, a Rydberg orbital with mainly B(3s) character. The remaining σ and π orbitals have small occupation numbers. They describe higher-order radial and angular correlation effects. The δ orbitals have very low occupation numbers in the $^1\Sigma^+$ states (the value is

TABLE VI
Natural-orbital occupation numbers of the CAS wavefunction for BH (at $R = 2.40$ a.u.).

	State		
Orbital	$X^1\Sigma^+$	$B^1\Sigma^+$	$A^1\Pi$
2σ	1.956	1.929	1.926
3σ	1.864	0.996	0.984
4σ	0.025	0.955	0.037
5σ	0.009	0.021	0.008
6σ	0.007	0.004	0.007
7σ	0.002	0.001	0.002
$1\pi^a$	0.123	0.081	0.016
$2\pi^a$	0.014	0.008	0.014
$3\pi^a$	0.002	0.005	0.003
$1\delta^a$	0.000	0.000	0.005

[a]Summed over the two components.

TABLE VII
Some properties of the BH molecule computed with the CASSCF method. Experimental values within parentheses[a].

Property	$X^1\Sigma^+$	$B^1\Sigma^+$	$A^1\Pi$
R_e (Å)	1.233	1.216	1.230
	(1.232)	(1.216)	(1.219)
D_e (eV)	3.52	1.88	0.50
	(3.57)	(2.05)	(0.70)[b]
T_e (cm^{-1})	–	52512	24017
		(52336)	(23135)
Dipole moment (D)	1.310	–4.677	0.591
	(1.270)	(–)	(0.58)
$\Delta G^c_{v+\frac{1}{2}}$			
$v=0$	2265.9	2237.2	1993.2
	(2269.3)	(2248.2)	(2086.2)
1	2171.0	2052.0	1642.2
	(2173.8)	(2073.7)	(1830.4)
2	2077.3	1855.0	–
	(2080.5)	(1875.8)	(1479.6)

[a]Refs 66 and 77.
[b]Experimental value estimated in Ref. 78.
[c]Intervals between the first vibrational levels.

0.0002), but are, as expected, of considerably greater importance in the $A^1\Pi$ state, where one π orbital is occupied.

Some selected properties of the three potential curves for BH are presented in Table VII. A more complete account can be found in Ref. 74. The agreement with experiment is excellent for the $X^1\Sigma^+$ ground state, and it is difficult to see how it could be improved by a more extensive calculation. The $B^1\Sigma^+$ state is characterized by a double minimum potential resulting from the crossing between an ionic, B^+H^-, state with the Rydberg state dissociating into boron $(2s)^2(3s)$ and hydrogen. The inner minimum, which is dominantly a Rydberg state, is the only one observed experimentally and the results presented in Table VII correspond to properties of this part of the potential curve only. The agreement with experiment is satisfactory. The remaining errors are most probably due to a less accurate description of the ionic configuration. The diffuse character of the hydrogen negative ion is pathologically difficult to describe with Gaussian basis sets.

Also the $A^1\Pi$ state involves two dominant configurations. The interaction between the boron $(2s)^2$ (2p) ground state and hydrogen is repulsive in this symmetry. A attractive potential between the excited boron (2s) $(2p)^2$ configuration and hydrogen crosses over the repulsive curve, resulting in an adiabatic potential with a maximum and a shallow minimum. The repulsive part of the potential is probably better described by the present basis set, which was optimized for the boron $(2s)^2$ (2p) ground state. The result is too large a T_e value and too small a dissociation energy. Even if the errors are small they have a large effect on the calculated vibrational levels. Four bands have been found experimentally (cf. Table VII) but for the theoretical curve only three levels were found below the maximum. These results illustrate some of the difficulties encountered in calculations on potential curves involving electronic configurations of very different character. For BH this is mainly a basis-set problem, but for larger systems the problem of a balanced treatment of the different correlation effects also becomes crucial. The C_2 molecule was an illustration of such a situation.

C. The N–N Bond in N_2O_4

It has already been stated several times that the primary purpose of the CASSCF method lies in producing a qualitatively correct zeroth-order wavefunction for a molecular system, rather than yielding accurate numerical results. The C_2 and BH molecules represent cases where it was possible to go beyond this more limited goal, for C_2 by a careful balance of atomic and molecular correlation effects, and for BH by including a large fraction of the dynamical correlation effects. The present example, the N–N bond in dinitrogen tetroxide, is of a different nature. The aim here is not to obtain results of high accuracy, but to explain why the N–N bond in this molecule is so long, a problem which has remained unsolved in spite of numerous experimental and theoretical investigations over the past 30 years.

The experimental value for the N–N bond length in N_2O_4 varies between 1.76 and 1.78 Å[79,80], which is much longer than the bond length found, for example, in hydrazine, N_2H_4, 1.47 Å[81]. Several theoretical investigations have been performed in order to explain the long N–N bond in N_2O_4. The SCF optimized values have been obtained with AO sp basis sets and also with basis sets including d-type polarization functions. The former basis set gives the value 1.67 Å, while the inclusion of polarization functions shortens the bond length to 1.59 Å, almost 0.2 Å smaller than experiment[82]. Obviously, this molecule cannot be described on the SCF level of approximation. The bond is weak and it might be expected that inclusion of the N–N antibonding σ^* orbital, to correlate the bonding σ electron pair, could have a large effect on the bond distance. Such a calculation has also been performed using a one-pair GVB wavefunction and a split valence 4-31G basis set[80]. The resulting N–N distance was 1.80 Å, in apparent agreement with experiment. The problem is, however, that inclusion of polarization functions in the basis set decreases this value to 1.67 Å[82], which is again 0.1 Å shorter than the experimental value. The GVB treatment is equivalent to a CASSCF calculation with two active electrons in two orbitals, σ and σ^*. Such a calculation thus recovers half the discrepancy between the SCF value and experiment.

The planar structure of the molecule[80] is most probably due to an attractive interaction between the π systems on the two NO_2 fragments. The π system in NO_2 is known to be nearly degenerate, involving, apart from the SCF configuration $(1b_1)^2\,(1a_2)^2$, a large contribution from the double excitation of the non-bonding electrons in $1a_2$ into the antibonding orbital $2b_1$. It might be suspected that these effects will contribute to the interaction of the two π systems in N_2O_4. The first natural extension of the two-configurational CASSCF treatment of N_2O_4, therefore, includes the π system into the active subspace, resulting in a wavefunction based on 10 electrons distributed among eight orbitals. This not very large CASSCF wavefunction contains 176 ^1A_g configurations (D_{2h} symmetry). A calculation using a double-zeta plus polarization (DZP) basis set (9s5p1d→4s2p1d for N and O) gives an N–N bond distance of 1.58 Å. The near-degeneracy effects in the π system thus increase the interaction and shorten the bond. The discrepancy compared to experiment is now close to 0.2 Å.

It is not immediately clear how to proceed from this point. The number of valence electrons in N_2O_4 is 34. Even if some of them can safely be left inactive (oxygen 2s, for example), it is still virtually impossible to perform a CASSCF calculation using as active orbitals a large fraction of the valence space. The problem is not resolved by performing a singles and doubles CI with SCF as the reference configuration. Such a calculation yields a bond length of 1.62 Å; including Davidson's correction to account for higher-order excitations increases the value to 1.68 Å, which is still around 0.1 Å too small.

However, it is possible to perform a CASSCF calculation on NO_2 which includes all valence orbitals, except oxygen 2s, into the active subspace. An analysis of the resulting wavefunction immediately gives an explanation for the failure to predict a correct N–N bond length using the wavefunctions described above. An inherent assumption behind all these attempts is the σ radical character of NO_2 with the odd electron mainly localized on the nitrogen atom. This is, however, only one possible resonance structure. Other important structures are obtained by double excitations of the type $(\pi, \mathrm{On}) \rightarrow (\pi^*, \mathrm{N}\sigma)$, where On is an oxygen lone-pair orbital and Nσ is the nitrogen radical orbital. These excitations were found to be even more important than the $\pi^2 \rightarrow \pi^{*2}$ correlating excitations. The corresponding valence structures have the odd electrons localized in an oxygen lone-pair orbital, with the Nσ orbital doubly occupied. As a result they cannot be used to form an N–N σ bond in N_2O_4. A proper analysis of the N–N bond has to include these repulsive structures into the CASSCF wavefunction for the dimer, which is achieved by adding the four oxygen lone-pair orbitals and the corresponding eight electrons to the active subspace. As a result the active subspace will contain 18 electrons distributed among 12 orbitals. It is possible, without losing any of the described characteristics, to constrain the subspace into a σ and π block, with six σ orbitals (and 10 electrons) and six π orbitals (and eight electrons). The CI expansion then contains 976 symmetry-adapted CSFs. CASSCF calculations using such an active subspace and a DZP basis set yield an N–N bond distance of 1.80 Å, and a dissociation energy into two NO_2 fragments of 7 kcal mol^{-1}. These results are very satisfactory. The bond distance is slightly too long, which is expected from the level of approximation used. The experimental dissociation energy has been estimated[83] to be around 13 kcal mol^{-1}. Better agreement with this value can only be expected with a substantially larger AO basis set, and a more extensive level of correlation.

An analysis of the natural orbitals shows that the repulsive resonance structures will indeed contribute to the wavefunction also in N_2O_4. The occupation fo the σ, σ^* orbital pair was, in the 10-electron calculation, almost exactly 2. In the 18-electron calculation the occupation is 2.06, showing the presence of configurations of the type $(\text{On}, \pi) \rightarrow (\sigma^*, \pi^*)$. The weight of these configurations is smaller in N_2O_4 than in two non-interacting NO_2 fragments. Since they stabilize the NO_2 radicals and are repulsive towards dimer formation, their effect will be to weaken and lengthen the N–N bond.

The NO_2 dimer is a beautiful example of the power of the CASSCF method in elucidating complex electronic structures. A detailed account of the calculations has been presented elsewhere[84].

D. Transition-metal Chemistry

Chemical bonds involving transition-metal atoms are often of a complex nature. Several factors contribute to this complexity. Many atoms contain several unpaired electrons, resulting in many close-lying spectroscopic states. The spin and space symmetry of the lowest molecular state is then difficult to postulate in advance. Most transition-metal atoms have a $(3d)^n (4s)^2$ ground-state electronic configuration. In molecules the bonding state is often $(3d)^{n+1} (4s)$. The energy difference between these two configurations is then a crucial parameter. It is often very difficult to compute and large errors are sometimes encountered on the SCF or MCSCF level of approximation. Taking Ni as an example, one finds at the HF level a splitting of 1.25 eV[85] between the $(3d)^8 (4s)^2\,{}^3F$ ground state and $(3d)^9 (4s)\,{}^3D$, while the experimental value is -0.03 eV. Including the dynamical correlation effects of the 3d shell, using standard basis sets, reduces the splitting to 0.4 eV[86]. Also, the difference in shape between the atomic orbitals in the two configurations causes problems.

The most severe problem is most probably the large dynamic correlation effects inherent in a 3d shell with many electrons. Omission of these effects sometimes even leads to an incorrect qualitative description of the chemical bond. A drastic example is the Cr_2 molecule, mentioned in the introduction, where a full valence CCAS calculation predicts a bond length which is 0.8 Å too long[16], and a bond energy which is only a small fraction of the true value. Large structure-dependent correlation effects in the 3d shell are responsible for the failure of the simpler model to give a proper description of the multiple bond in Cr_2.

As a consequence it is mandatory to use a higher level of theory in studies of 3d systems, for example, by combining CASSCF and multireference CI calculations. A number of studies of transition-metal systems published during the last three years have followed this strategy. They range from calculations on diatomic molecules to large transition-metal complexes like ferrocene[87] and (Mo_2Cl^{4-})[88]. Some examples of such calculations can be found in Refs 86–96. It would be outside the scope of this review to include a comprehensive discussion of all these studies. Therefore only two illustrative examples have been selected: the NiH molecule and the complex between a nickel atom and an ethene molecule.

1. The NiH Molecule

This molecule was the first system containing a transition-metal atom to be studied with the CASSCF method[86]. It is the simplest molecule containing a nickel atom, and is therefore a suitable prelude to studies of more complex nickel-containing systems. Some molecular parameters are also known experimentally, and can be used to judge the quality of the calculations.

The lowest states of NiH are formed from Ni$(d^9s\,^3D)$ interacting with a hydrogen atom. Three doublet states, $^2\Delta$, $^2\Sigma^+$ and $^2\Pi$, can be formed, of which $^2\Delta$ is known to be the ground state. This state is characterized as having an NiH σ bond and a 3dδ open shell, the remaining eight electrons being in closed shells. The simplest CAS wavefunction giving a proper description of the dissociation process would consider only two electrons in two orbitals, the NiH bonding and antibonding pair. For technical reasons the open-shell orbital $3d\delta_{xy}$ also has to be include into the active subspace, which then includes three electrons in three orbitals. It is, however, unclear as to what extent the 3dσ orbital is involved in the bonding, and the preliminary calculations, therefore, also included this orbital, together with two electrons. In the σ symmetry there are then four electrons distributed among three orbitals. Only one of them consequently is weakly occupied. Such an active subspace is ambiguous, since the weakly occupied orbital can be used to correlate only one of the strongly occupied orbitals. At large internuclear separations the choice will obviously be the NiH bonding orbital, but at smaller bond distances this choice is not immediately clear. The correlation energy gained by correlating the 3dσ orbital is now of the same order. Thus there exist two close-lying local minima on the energy surface. This is a typical situation where convergence problems can be expected in a CASSCF calculation. The remedy is, of course, to introduce a second weakly occupied orbital into the active subspace. The ambiguity is then removed, and a consistent treatment of the bond-breaking mechanism is obtained. This is not a unique case. Convergence problems in CASSCF calculations are often due to an unbalanced choice of the active subspace leading to ambiguities in the character of the weakly occupied orbitals.

The final active subspace chosen for NiH ($^2\Delta$) thus consists of five electrons distributed among five orbitals. The corresponding CASSCF calculations give the molecular parameters: $r_e = 1.525$ Å, $D_e = 1.45$ eV and $\omega_e = 1642\ \text{cm}^{-1}$. These results are not very accurate. The experimental values are[66]: $r_e = 1.475$ Å and $\omega_e = 1927\ \text{cm}^{-1}$. The dissociation energy is not known, except for an upper limit of 3.07 eV[66]. A subsequent MR-CI calculation, including correlation of all the valence electrons, improves the results to: $r_e = 1.470$ Å, $D_e = 2.28$ eV and $\omega_e = 1911\ \text{cm}^{-1}$, in very good agreement with experiment.

The study of NiH also included calculations of a number of excited states of both d^9s and d^8s^2 origin. It was not possible to base the CI calculations of all these states on independent CASSCF determinations of the orbitals, owing to convergence problems in the Newton–Raphson procedure. These problems occurred for some states which were not the lowest of its symmetry, and were most probably due to an incomplete inclusion of CI–orbital coupling. The CI calculations were therefore performed with orbitals obtained from CASSCF calculations using average density matrices for the states of interest. It was not possible, however, to use this procedure for full potential curve calculations, since some of the states change character adiabatically due to avoided crossings. Also, averaging the densities for d^8s^2 and d^9s in the nickel atom does not give a good orbital basis for the different spectroscopic states of the atom. The spectrum of NiH around the equilibrium internuclear separation is on the other hand quite insensitive to the choice of the molecular-orbital basis. Using average orbitals in CI calculations of molecular spectra is therefore a procedure which should be used with much care.

Is it possible to improve the results for NiH on the CASSCF level of accuracy by extending the active subspace? The answer to this question is most probably 'no'. The next important feature to include would be the radial correlation effects in the 3d shell of the nickel atom. The active subspace then has to include two sets of 3d orbitals together with the NiH σ and σ^* orbitals: 11 electrons distributed among 12 orbitals. Such a calculation is well within the limits of the present capabilities, but it is not at all certain that it would give a balanced description of the correlation effects of the entire potential curve:

A simpler molecule which exhibits the same features as NiH, of large intra-atomic correlation effects, is FH. The SCF configuration for this molecule is

$$(2\sigma)^2(3\sigma)^2(1\pi)^4 \tag{55}$$

where 2σ is the fluorine 2s orbital, 3σ the FH bond and 1π the fluorine $2p\pi$ orbitals. It is well known that inter-pair correlations are important for the description of the bonding in this molecule. A CASSCF description of the most important dynamical correlation effects in the molecule is obtained with an active space where the orbitals $4\sigma, 5\sigma$ and 2π are added to the SCF orbitals (55). At large separations 4σ will become the hydrogen 1s orbital, 5σ the fluorine 3s orbital and 2π the 3p orbital pair. A CASSCF calculation using such an active subspace gives a D_e value of 6.00 eV and $r_e = 0.921$ Å[97], to be compared with the experimental values 6.12 eV and 0.917 Å[66]. Although these values are quite satisfactory, they are clearly the result of an unbalanced treatment of the 2p–2p correlations in fluorine. The $3p_z$ orbital is not included in the active subspace at large separations. The double excitation

$$(3\sigma)(1\pi) \rightarrow (4\sigma)(2\pi) \tag{56}$$

describes the corresponding inter-pair correlation effect at small internuclear separations, but vanishes at dissociation where 4σ becomes the singly occupied hydrogen 1s orbital[98]. This type of imbalance is very difficult to correct for in an MCSCF calculation with a small number of active orbitals. It is tempting to attempt to remedy the imbalance by adding one more σ orbital to the active space. Such a calculation gives a dissociation energy of 5.56 eV and $r_e = 0.922$. The inter-pair correlations at dissociation are now much better described, but the effect of the new orbital is much smaller at equilibrium. The occupation number varies from 0.0055 at large separations to only 0.0017 at equilibrium. Actually, the smaller calculation does give a more balanced treatment of the inter-pair correlation effects, in the negative sense of neglecting about the same amount in the molecule and the separate atoms.

It is clear that an enlargement of the active subspace in NiH, which tries to account for the 3d pair correlation effects, will run into balance problems similar to those experienced in FH. The conclusion seems rather clear: structure-dependent dynamical correlation effects in systems with high electron density cannot be accounted for in an MCSCF treatment in a balanced way. Large CI or MBPT treatments then become necessary and the calculations have to include a large fraction of the total correlation energy, in order to give reliable results for relative energies.

2. *The Nickel–Ethene Complex*

The weak chemical bond formed between a nickel atom (in the d^9s electronic configuration) and an ethene (C_2H_4) molecule will be considered as a final example. The results to be discussed below are not unique to this system. The main features have been found in a number of transition-metal compounds of the type TM–X where X is a ligand molecule[93–95,99–103].

Ni $d^9s\,^3D$ forms only a weak bond of the van der Waals type with C_2H_4. The reason is the repulsion between the 4s electron of Ni and the closed shells of the ligand, which cannot be counterbalanced by a charge transfer (CT) from the 3d orbitals, which are much more contracted than the diffuse 4s orbital. The situation is quite different for the singlet state, $^1A_1$ with C_{2v} symmetry, which dissociates into Ni $d^9s\,^1D$ plus ethene. Here it becomes possible to unshield the Ni 3d orbitals by a hybridization of the 4s and one 3d orbital, thus reducing the repulsion and making bonding via back-donation possible. The elucidation of this bonding mechanism in the low-spin transition-metal ligand bond is another illustration of the strength of the CASSCF method in predicting the basic features of complex electronic structures.

The bonding between a transition metal and an olefin is usually discussed in terms of the Dewar–Chatt–Duncanson (DCD) model[104]. In this model, bonding occurs via a simultaneous donation of electrons from the olefin π orbital to the metal and a back-donation from the metal $3d_{xz}$ orbital to the empty π^* orbital. The complex is assumed to have C_{2v} symmetry with Ni on the positive z axis, the C–C bond on the x axis and the plane of the ethene molecule perpendicular to the Ni–C–C plane. Such a model involves four orbitals: the π and π^* orbitals of C_2H_4 and the 4s and $3d_{xz}$ orbitals of nickel. This would also be the natural starting point in a GVB treatment. The corresponding SCF electron configuration would be $(a_1)^2(b_2)^2$, where a_1 is a linear combination of 4s, and π, and b_2 a linear combination of $3d_{xz}$ and π^*. The inactive subspace includes the remaining four 3d orbitals. The formal valence state for nickel is, in this model, d^{10} with the 4s acceptor orbital empty and the $3d_{xz}$ donating orbital doubly occupied.

CASSCF calculations were performed on $Ni(C_2H_4)(^1A_1)$, but without the (as we shall see) biased selection of active orbitals, which the above discussion would suggest[94]. All 3d orbitals were included in the active subspace, together with the 4s orbital and the π and π^* orbitals of C_2H_2 (actually the C–C σ and σ^* orbitals were also included, but this is of less importance for the following discussion). The results of the calculations revealed that the simple CT picture of the bonding is incomplete. It turns out that the atomic character of the $Ni(^1D)$ atom is to some extent preserved also in the complex. The electronic configuration for the ^1D state of nickel can be written as (excluding the core electrons 1s–3p):

$$(3d\pi)^4(3d\delta)^4(3d\sigma, 4s)_S \tag{57}$$

where the $M = 0$ component has been chosen. The other four components can be obtained by switching the 3d hole to one of the other four orbitals. The singlet coupled pair $(3d, 4s)_S$ can alternatively be obtained as a linear combination of two closed shells

$$(3d, 4s)_S = (sd_+)^2 - (sd_-)^2 \tag{58}$$

where

$$(sd_+) = (1/\sqrt{2})(3d + 4s) \tag{59a}$$

$$(sd_-) = (1/\sqrt{2})(3d - 4s) \tag{59b}$$

The results of the CASSCF calculation shows that (58) is a natural starting point for a discussion of the bonding mechanism. When the molecule is formed, a continuous shift of the occupations of (sd_+) and (sd_-) takes place and (58) changes to

$$C_1(sd_+)^2 - C_2(sd_-)^2 \tag{60}$$

where C_1 is now larger than C_2. The s character of sd_+ also decreases somewhat. The 3d orbital that combines with 4s to form these hybrids is $3d_{y^2-z^2}$. Thus the (sd_+) orbital concentrates the charge along a line perpendicular to the Ni–C–C plane, with a diminished density in the bonding region. On the other hand (sd_-) is pointing along the line through Ni and bisecting the C–C bond. This orbital acquires some 4p character, which moves the charge to the back side of the Ni atom. The occupation numbers for (sd_+) and (sd_-) in $Ni(C_2H_4)$, at equilibrium, are 1.81 and 0.20 respectively. Occupation numbers for corresponding hybrid orbitals in some other NiX compounds are given in Table VIII.

The bonding mechanism is now clear. It involves five orbitals: (sd_+), (sd_-), $3d_{xz}$, π and π^*. When Ni (^{1}D) approaches the ligand the electrons move from (sd_-) to (sd_+), which reduces the repulsion between nickel and the ligand π orbital. A much more effective interaction between the π system of C_2H_4 and the nickel 3d orbitals becomes possible, leading to the formation of a weak bond between $3d_{xz}$ and π^* and some delocalization of the π orbital onto the nickel atom. The total donation of charge to nickel is 0.42 electrons while 0.61 electrons are transferred in the opposite direction.

The sd hybridization of d^9s nickel is not possible in the triplet states, which explains the weak bonding. The singlet state is, on the other hand, bound with almost 20 kcal mol^{-1} with respect to Ni (^{1}D) plus ethene[94]. A similar bond

TABLE VIII

Occupation numbers for the sd_+ and sd_- hybrids in some NiX complexes (singlet states).

X	sd_+	sd_-	Ref.
C_2H_4	1.81	0.20	94
C_2H_2	1.80	0.21	95
CO	1.94	0.06	101
H_2O	1.27	0.73	101
PH_3	1.72	0.29	101
N_2 (end on)	1.8	0.2	100

energy is found for $Ni(C_2H_2)$[95]. The same bonding mechanism is also present in NiCO, which for a long time was believed to have a triplet ground state ($^3\Delta$). CASSCF and CI studies have shown that the ground state in NiCO is $^1\Sigma^+$ with a binding energy[101] of 29 kcal mol^{-1}. The constrained space orbital variation (CSOV) method[105] has been used to give a detailed analysis of the bonding in XCO for X = Fe, Ni and Cu[103]. Also, low-spin complexes of other transition metals experience the same type of hybridization. Examples which have been studied with the CASSCF methods are FeCO[103] and FeN_2[100].

IV. SUMMARY AND CONCLUSIONS

The complete active space (CAS) SCF method has been reviewed. Current methods for optimization of an MCSCF wavefunction have been discussed with special reference to the CASSCF method. The strength of the method in solving complex electronic structure problems has been illustrated with examples from the current literature. The strength of the method lies in its simplicity. It is a pure orbital method in the sense that the user only has to worry about selecting an appropriate inactive and active orbital space in order to define the wavefunction. That this selection is far from trivial has been illustrated in some of the examples. FH, N_2O_4 and $Ni(C_2H_4)$ give different aspects to this problem.

The present review has presented some illustrations of the CASSCF method. The method has been applied to a number of problems not considered here. Core ionization and shake-up spectra have been successfully analysed in terms of near-degeneracy effects in the ionized states. The shake-up spectra of *p*-nitroaniline[106] and *p*-aminobenzonitrile[107] highlight this type of application. Studies of valence ionization spectra have been done for N_2[108], ozone[109] and acetylene[110].

One obvious use of the CASSCF method is in studies of energy surfaces for chemical reactions. A number of such calculations have been reported in the literature. Some of the studies in transition-metal chemistry have already been mentioned. In this context, a study of the elimination and addition reactions of methane and ethane with nickel is also worth mentioning[110].

A number of reactions including only first- and second-row atoms have also been studied. Some illustrative examples are: the intramolecular transformation of methylene peroxide to dioxirane[111]; the photolytic decomposition of oxathiirane[112]; the reaction of singlet molecular oxygen with ethene[113]; and the dissociation of diimide[114].

The CASSCF method has also been applied in studies of molecular properties. Thus a careful analysis of the electrical properties of LiH has been undertaken, with respect to both correlation and basis-set effects. For larger systems the CASSCF can recover a considerable portion of the correlation contribution to molecular properties, but cannot provide very high accuracy because of the limitation in the number of active orbitals. A comparison of CASSCF, CI(SD) and MBPT methods for the calculation of some molecular properties can be found in Ref. 115.

The CASSCF method is able to give a correct zeroth-order description of a wavefunction in situations where the simple independent-particle model breaks down. It is, however, not possible to use the method for accurate quantitative determinations of molecular parameters, except in special cases. The BH and C_2 molecules here serve as good illustrations. In general, however, a specific treatment of the dynamical correlation effects is necessary, in order to obtain results of chemical accuracy. Such calculations are today usually based on some form of the configuration-interaction technique. The most commonly used method is the externally contracted CI scheme[38]. The computational effort for such calculations is strongly dependent on the number of reference configurations used as a basis for the singles and doubles CI expansion. In most cases it is not possible to include all CASSCF CSFs into the reference space, but a rather restricted selection has to be made, based on the relative weights. In most cases such a selection is straightforward, but cases exist where it becomes difficult or even impossible to select a small reference space for the CI step. Cr_2 is a drastic example of such a case.

An alternative method, named internally contracted CI, was suggested by Meyer[116] and was applied by Werner and Reinsch[117] in the MCSCF self-consistent electron-pair (SCEP) approach. Here only one reference state is used, the entire MCSCF wavefunction. The CI expansion is then in principle independent of the number of configurations used to build the MCSCF wavefunction. In practice, however, the complexity of the calculation also strongly depends on the size of the MCSCF expansion. A general configuration-interaction scheme which uses, for example, a CASSCF reference state, therefore still awaits development. Such a CI wavefunction could preferably be used on the first-order interacting space, which for a CASSCF wavefunction can be obtained from single and double substitutions of the form[118]:

$$|pq\rangle = \hat{E}_{pq}|0\rangle \tag{61a}$$

$$|pqrs\rangle = \hat{E}_{pq}\hat{E}_{rs}|0\rangle \tag{61b}$$

where $|0\rangle$ is the CASSCF wavefunction. The single substitutions (61a) do not interact directly with $|0\rangle$, but they are important for the correlation corrections to the first-order density matrix. The calculation of the matrix elements between the CSFs (61) leads to expressions in terms of density matrices up to order 6, which is still an unsolved complication. The CI expansion based on the configuration space (61) is, however, independent of the size of the CASSCF wavefunction. It also constitutes a consistent basis for a cluster theory based on CASSCF.

In conclusion CASSCF has during its six years of existence proved to be a very general and valuable tool for electronic structure calculations. Like all other methods, it has its limitations and drawbacks. There are, however, only a few cases where CASSCF could not be used as an appropriate starting approximation. This does not mean that its present state is without problems. There is still room for improvement in the CI step, which for large expansions is the most time-consuming part of a CASSCF calculation. A method to constrain the CAS expansion in consistent ways will also be of great value, making it possible to increase the active space further. However, the most important problem for future developments is, in my opinion, the search for unbiased methods to treat dynamical correlation effects either using CI methods, as suggested above, or preferably a cluster expansion based on the CASSCF as the reference state.

References

1. Mulliken, R. S., *Phys. Rev.*, **41**, 49 (1932).
2. Hartree, D. R., *Proc. Camb. Phil. Soc.*, **24**, 89, 111, 426 (1928).
3. Fock, V., *Z. Phys.*, **61**, 126 (1930).
4. Koopmans, T. A., *Physica*, **1**, 104 (1933).
5. Roothaan, C. C. J., *Rev. Mod. Phys.*, **32**, 179 (1960).
6. Pople, J. A., and Nesbet, R. K., *J. Chem. Phys.*, **22**, 571 (1954).
7. Coulson, C. A., and Fischer, I., *Phil. Mag.*, **40**, 386 (1949).
8. Löwdin, P.-O., *Phys. Rev.*, **97**, 1474 (1955).
9. Löwdin, P.-O., and Shull, H., *Phys. Rev.*, **101**, 1730 (1956).
10. Shull, H., *J. Chem. Phys.*, **30**, 1405 (1959).
11. For a discussion of this and other pair correlation theories, see, for example: Hurley, A. C., *Electron Correlation in Small Molecules*, Academic Press, London, 1976.
12. Roos, B. O., Taylor, P. R., and Siegbahn, P. E. M., *Chem. Phys.*, **48**, 157 (1980).
13. Siegbahn, P. E. M., Almlöf, J., Heiberg, A., and Roos, B. O., *J. Chem. Phys.*, **74**, 2384 (1981).
14. Roos, B. O., *Int. J. Quantum Chem. Sym.*, **14**, 175 (1980).
15. Paldus, J., in *Theoretical Chemistry: Advances and Perspectives* (Eds H. Eyring and D. G. Henderson), Vol. 2, Academic Press, New York, 1976.
16. Walch, S. P., Bauschlicher, C. W., Jr, Roos, B. O. and Nelin, C. J., *Chem. Phys. Lett.*, **103**, 175 (1983).
17. Bondybey, V. E., and English, J. H., *Chem. Phys. Lett.*, **94**, 443 (1983).
18. For a survey of such calculations, see: Detrich, J., and Wahl, A. C., in *Recent Developments and Applications of Multiconfiguration Hartree–Fock Methods* (Ed. M. Dupuis), NRCC Proc. No. 10, Report LBL-12151, Lawrence Berkeley, University of California, Berkeley, CA, 1981.
19. Shavitt, I., *Int. J. Quantum Chem. Symp.*, **11**, 133 (1977); *ibid.*, **12**, 5 (1978).
20. Harris, F. E., and Michels, H. H., *Int. J. Quantum Chem. Symp.*, **1**, 329 (1967).
21. Schaefer, H. F., and Harris, F. E., *J. Chem. Phys.*, **48**, 4946 (1968).
22. Goddard, W. A., III, *Phys. Rev.*, **157**, 81 (1967); Ladner, R. C., and Goddard, W. A., III, *J. Chem. Phys.*, **51**, 1073 (1969); Hunt, W. J., Hay, P. J., and Goddard, W. A., III, *J. Chem. Phys.*, **57**, 738 (1972).

23. Ruedenberg, K., and Sundberg, K. R., in *Quantum Science* (Eds J.-L. Calais, O. Goscinski, J. Linderberg and Y. Öhrn), Plenum, New York, 1976.
24. Ruedenberg, K., Schmidt, M. W., Gilbert, M. M., and Elbert, S. T., *Chem. Phys.*, **71**, 41, 51, 65 (1982).
25. Knowles, P. J., and Werner, H.-J., *Chem. Phys. Lett.*, **115**, 259 (1985).
26. Olsen, J., Yeager, D. L., and Jørgensen, P., *Adv. Chem. Phys.*, **54**, 1 (1983).
27. For a review of these methods, see, for example: Wahl, A. C., and Das, G., in *Modern Theoretical Chemistry*, Vol. 3, *Methods of Electronic Structure Theory* (Ed. H. F. Schaefer, III), Plenum, New York, 1977.
28. Grein, F., and Chang, T. C., *Chem. Phys. Lett.*, **12**, 44 (1971); Grein, F., and Banerjee, A., *Int. J. Quantum Chem. Sym.*, **9**, 147 (1975); *J. Chem. Phys.*, **66**, 1054 (1977).
29. Ruedenberg, K., Cheung, L. M., and Elbert, S. T., *Int. J. Quantum Chem.*, **16**, 1069 (1979).
30. Levy, B., Thesis, CNRS No. A05271, Paris, 1971; *Int. J. Quantum Chem.*, **4**, 297 (1970); *Chem. Phys. Lett.*, **4**, 17 (1969).
31. Yaffe, L. G., and Goddard, W. A., III, *Phys. Rev. A*, **13**, 1682 (1976).
32. Igawa, A., *Int. J. Quantum Chem.*, **28**, 203 (1985).
33. Werner, H.-J., and Knowles, P. J., *J. Chem. Phys.*, **82**, 5053 (1985).
34. Laaksonen, L., Pyykkö, P., and Sundholm, D., *Int. J. Quantum Chem.*, **23**, 309, 319 (1983); *Chem. Phys. Lett.*, **96**, 1 (1983).
35. See, for example: Siegbahn, P. E. M., Blomberg, M. R. A., and Bauschlicher, C. W., Jr, *J. Chem. Phys.*, **81**, 2103 (1984).
36. Broch-Mathisen, K., Wahlgren, U., and Pettersson, L. G. M., *Chem. Phys. Lett.*, **104**, 336 (1984).
37. Jönsson, B., Roos, B. O., Taylor, P. R., and Siegbahn, P. E. M., *J. Chem. Phys.*, **74**, 4566 (1981).
38. Siegbahn, P. E. M., *Int. J. Quantum Chem.*, **23**, 1869 (1983).
39. Cernusak, I., Diercksen, G. H. F., and Sadlej, A. J., *Chem. Phys.*, **108**, 45 (1986).
40. Lofthus, A., and Krupenie, P. A., *J. Phys. Chem. Ref. Data*, **6**, 113 (1977).
41. Buckingham, A. D., Graham, C., and Williams, J. H., *Mol. Phys.*, **49**, 703 (1983).
42. Moshinsky, M., *Group Theory and the Many-Body Problem*, Gordon and Breach, New York, 1968.
43. Roos, B. O., in *Methods of Computational Molecular Physics* (Eds G. H. F. Diercksen and S. Wilson), NATO ASI Series, Vol. 113, Reidel, Dordrecht, 1983.
44. Jørgensen, P., and Simons, J., *Second Quantization-Based Methods in Quantum-Chemistry*, Academic Press, New York, 1981.
45. Levy, B., and Berthier, G., *Int. J. Quantum Chem.*, **2**, 307 (1968).
46. Dalgaard, E., and Jørgensen, P., *J. Chem. Phys.*, **69**, 3833 (1978); Yeager, D. L., and Jørgensen, P., *J. Chem. Phys.*, **71**, 755 (1979); Dalgaard, E., *Chem. Phys. Lett.*, **65**, 559 (1979).
47. Siegbahn, P. E. M., Heiberg, A., Roos, B. O., and Levy, B., *Phys. Scr.*, **21**, 323 (1980).
48. Wahlgren, U., private communication.
49. Lengsfield, B. H., III, *J. Chem. Phys.*, **73**, 382 (1980).
50. Jensen, H. J. Aa., and Jørgensen, P., *J. Chem. Phys.*, **80**, 1204 1984).
51. Werner, H.-J., and Meyer, W., *J. Chem. Phys.*, **73**, 2342 (1980); *ibid.*, **74**, 5794 (1981).
52. Roos, B., *Chem. Phys. Lett.*, **15**, 153 (1972).
53. Lengsfield, B. H., III, *J. Chem. Phys.*, **77**, 4073 (1982); Lengsfield, B. H., III, and Liu, B., *J. Phys.*, **75**, 478 (1981).

54. Jensen, H. J. Aa., and Ågren, H., *Chem. Phys. Lett.*, **110**, 140 (1984).
55. Almlöf, J., and Taylor, P. R., in *Advanced Theories and Computational Approaches to the Electronic Structure of Molecules* (Ed. C. E. Dykstra), NATO ASI Series C, Vol. 133, Reidel, Dordrecht, 1984.
56. Paldus, J., and Boyle, M. J., *Phys. Scr.*, **21**, 295 (1980).
57. Paldus, J., *J. Chem. Phys.*, **61**, 5321 (1974).
58. Robb, M. A., and Hegarty, D., in *Correlated Wave Functions* (Ed. V. R. Saunders), Science Research Council, 1978.
59. Siegbahn, P. E. M., *Chem. Phys. Lett.*, **109**, 417 (1984).
60. Knowles, P. J., and Handy, N. C., *Chem. Phys. Lett.*, **111**, 315 (1984).
61. Hinze, J., *J. Chem. Phys.*, **59**, 6424 (1973).
62. Roos, B. O., unpublished results: the AO basis used in these calculations consisted of 12s, 8p and 4d Gaussian functions contracted to 7s, 8p and 4d.
63. Roos, B. O., and Sadlej, A. J., unpublished results: the AO basis set used was 13s, 8p, 3d, 2f, GTOs contracted to 8s, 5p, 2d, 1f CGTOs.
64. Verhaegen, G., Richards, W. G., and Moser, C. M., *J. Chem. Phys.*, **46**, 160 (1967).
65. Hay, P. J., Hunt, W. J., and Goddard, W. A., III, *J. Am. Chem. Soc.*, **92**, 8293 (1972).
66. Huber, K. P., and Herzberg, G., *Constants of Diatomic Molecules*, Van Nostrand, Princeton, NJ, 1979.
67. Dupuis, M., and Liu, B., *J. Chem. Phys.*, **73**, 337 (1980).
68. van Duijneveldt, F. B., IBM Research Report RJ945 (1971).
69. Roos, B. O., and Sadlej, A. J., *Chem. Phys.*, **94**, 43 (1985).
70. Blomberg, M. R. A., and Siegbahn, P. E. M., *Chem. Phys. Lett.*, **81**, 4 (1981).
71. Davidson, E. R., in *The World of Quantum Chemistry* (Eds R. Daudel and B., Pullman), Reidel, Dordrecht, 1974.
72. Kraemer, W. P., and Roos, B. O., to be published.
73. Harrison, R. J., and Handy, N. C., *Chem. Phys. Lett.*, **95**, 386 (1983).
74. Jaszunski, M., Roos, B. O., and Widmark, P.-O., *J. Chem. Phys.*, **75**, 306 (1981).
75. Roos, B. O., Linse, P., Siegbahn, P. E. M., and Blomberg, M. R. A., *Chem. Phys.*, **66**, 197 (1982).
76. Meyer, W., and Rosmus, P., *J. Chem. Phys.*, **63**, 2356 (1975).
77. Johns, J. W. C., Grimm, F. A., and Porter, R. F., *J. Mol. Spectrosc.*, **22**, 435 (1967).
78. Blint, R. J., and Goddard, W. A., III, *Chem. Phys.*, **3**, 297 (1974).
79. McClelland, B. W., Gundersen, G., and Hedberg, K., *J. Chem. Phys.*, **56**, 4541 (1972).
80. Kvick, Å., McMullan, R. K., and Newton, M. D., *J. Chem. Phys.*, **76**, 3754 (1982).
81. Morino, Y., Iijima, T., and Murata, Y., *Bull. Chem. Soc. Japan.*, **33**, 46 (1959).
82. Ahlrichs, R., and Keil, F., *J. Am. Chem. Soc.*, **96**, 7615 (1974).
83. Snyder, R. G., and Hisatsune, I. C., *J. Mol. Spectrosc.*, **1**, 139 (1957); Hisatsune, I. C., *J. Phys. Chem.*, **65**, 2249 (1961).
84. Bauschlicher, C. W., Jr, Komornicki, A., and Roos, B. O., *J. Am. Chem. Soc.*, **105**, 745 (1983).
85. Froese Fischer, C., *Comput. Phys. Commun.*, **4**, 107 (1972).
86. Blomberg M. R. A., Siegbahn, P. E. M., and Roos, B. O., *Mol. Phys.*, **47**, 127 (1982).
87. Luthi, H. P., Siegbahn, P. E. M., Almlöf, J., Faegri, K., and Heiberg, A., *Chem. Phys. Lett.*, **111**, 1 (1984).
88. Broch Mathisen, K., Wahlgren, U., and Pettersson, L. G. M., *Chem. Phys. Lett.*, **104**, 336 (1984).
89. Bauschlicher, C. W., Jr, Nelin, C. J., and Bagus, P. S., *J. Chem. Phys.*, **82**, 3265 (1985).

90. Hotokka, M., Kindstedt, T., Pyykkö, P., and Roos, B. O., *Mol. Phys.*, **84**, 23 (1984).
91. Blomberg, M. R. A., Brandemark, U., and Siegbahn, P. E. M., *J. Am. Chem. Soc.*, **105**, 5557 (1983).
92. Blomberg, M. R. A., and Siegbahn, P. E. M., *J. Chem. Phys.*, **78**, 986 (1983).
93. Siegbahn, P. E. M., Blomberg, M. R. A., and Bauschlicher, C. W., Jr, *J. Chem. Phys.*, **81**, 1373 (1984).
94. Widmark, P.-O., Roos, B. O., and Siegbahn, P. E. M., *J. Phys. Chem.*, **89**, 2180 (1985).
95. Widmark, P.-O., Sexton, G. J., and Roos, B. O., *J. Mol. Struct.*, **135**, 235 (1986).
96. Bauschlicher, C. W., Jr, Walch, S. P., and Siegbahn, P. E. M., *J. Chem. Phys.*, **78**, 3347 (1983).
97. Roos, B. O., unpublished results: the CGTO basis used was (F: 14s, 9p, 4d/H: 8s, 4p) contracted to (F: 8s, 6p, 3d/H: 4s, 3p).
98. Dunning, T. H., in *Advanced Theories and Computational Approaches to the Electronic Structure of Molecules* (Ed. C. E. Dykstra), NATO ASI Series C, Vol. 133, Reidel, Dordrecht, 1984.
99. Blomberg, M. R. A., and Siegbahn, P. E. M., *J. Chem. Phys.*, **78**, 5682 (1983).
100. Siegbahn, P. E. M., and Blomberg, M. R. A., *Chem. Phys.*, **87**, 189 (1984).
101. Blomberg, M. R. A., Brandemark, U. B., Siegbahn, P. E. M., Broch Mathisen, K., and Karlström G., *J. Phys. Chem.*, **89**, 2171 (1985).
102. Blomberg, M. R. A., Siegbahn, P. E. M., and Strich, A., *Chem. Phys.*, **97**, 287 (1985).
103. Bauschlicher, C. W., Jr, Bagus, P. S., Nelin, C. J., and Roos, B. O., *J. Chem. Phys.*, **85**, 354 (1986).
104. Chatt, J., and Duncanson, L. A., *J. Chem. Soc.*, 2939 (1953); Dewar, M. J. S., *Bull. Soc. Chim. Fr.*, 79 (1951).
105. Bagus, P. S., Hermann, K., and Bauschlicher, C. W., Jr, *J. Chem. Phys.*, **80**, 4378 (1984).
106. Ågren, H., Roos, B. O., Bagus, P. S., Gelius, U., Malmqvist, P.-Å., Svensson, S., Maripuu, R., and Siegbahn, K., *J. Chem. Phys.*, **77**, 3893 (1982).
107. Malmqvist, P.-Å., Svensson, S., and Ågren, H., *Chem. Phys.*, **76**, 429 (1983).
108. Bagus, P. S., and Roos, B. O., *Chem. Phys. Lett.*, **82**, 158 (1981).
109. Malmqvist, P.-Å., Ågren, H., and Roos, B. O., *Chem. Phys. Lett.*, **98**, 444 (1983).
110. Blomberg, M. R. A., Brandemark, U., and Siegbahn, P. E. M., *J. Am. Chem. Soc.*, **105**, 5557 (1983).
111. Karlström, G., and Roos, B. O., *Chem. Phys. Lett.*, **79**, 416 (1981).
112. Karlström G., Roos, B. O., and Carlsen, L., *J. Am. Chem. Soc.*, **106**, 1557 (1984).
113. Hotokka, M., Roos, B. O., and Siegbahn, P. E. M., *J. Am. Chem. Soc.*, **105**, 5263 (1983).
114. Brandemark, U., and Siegbahn, P. E. M., *Theor. Chim. Acta*, **66**, 217 (1984).
115. Diercksen, G. H. F., Roos, B. O., and Sadlej, A. J., *Int. J. Quantum Chem. Symp.*, **17**, 265 (1983).
116. Meyer, W., in *Modern Theoretical Chemistry*, Vol. 3, *Methods of Electronic Structure Theory* (Ed. H. F. Schaefer, III), Plenum, New York, 1977.
117. Werner, H. J., and Reinsch, E. A., *J. Chem. Phys.*, **76**, 3144 (1982).
118. Roos, B. O., Linse, P., Siegbahn, P. E. M., and Blomberg, M. R. A., *Chem. Phys.*, **66**, 197 (1982).
119. Karlström, G., Roos, B. O., and Sadlej, A. J., *Chem. Phys. Lett.*, **86**, 374 (1982).
120. Roos, B. O., and Sadlej, A. J., *J. Chem. Phys.*, **76**, 5444 (1982).
121. Roos, B. O., and Sadlej, A. J., *Chem. Phys.*, **94**, 43 (1985).

Ab Initio Methods in Quantum Chemistry—II
Edited by K. P. Lawley

TRANSITION-METAL ATOMS AND DIMERS

DENNIS R. SALAHUB

*Département de Chimie, Université de Montréal,
CP 6128, Succ. A, Montréal, Québec H3C 3J7, Canada*

CONTENTS

I. Introduction . . . 448
II. Electronic Correlation-mixing Determinants or Density Functional Theory? . . . 449
A. The Local Spin Density Method . . . 451
B. Beyond the Local Approximation . . . 456
1. Self-interaction Corrections . . . 457
2. Hartree–Fock Exchange Plus Local (or SIC) Correlation . . . 461
3. Gradient Corrections . . . 463
C. Solving the Local Spin Density Equations . . . 464
1. An Overview . . . 464
2. The LCGTO-LSD-MP Method . . . 466
III. Atoms . . . 469
A. Hartree–Fock, Configuration-interaction and Multiconfiguration Self-consistent Field Calculations . . . 469
B. Density Functional Calculations . . . 474
IV. Dimers . . . 479
A. 3d Dimers . . . 484
1. Sc_2 . . . 484
2. Ti_2 . . . 485
3. V_2 . . . 486
4. Cr_2 . . . 487
5. Mn_2 . . . 492
6. Fe_2 . . . 493
7. Co_2 . . . 495
8. Ni_2 . . . 495
9. Cu_2 . . . 499
B. 4d Dimers . . . 501
1. Y_2 . . . 502
2. Nb_2 . . . 502
3. Mo_2 . . . 503
4. Ru_2 . . . 505
5. Rh_2 . . . 505
6. Pd_2 . . . 505

7. Ag_2 . . . 506
C. 5d Dimers . . . 507
1. W_2 . . . 507
2. Pt_2 . . . 507
3. Au_2 . . . 508
V. Concluding Remarks . . . 508
References . . . 510

Plus quam uno modo felis pellem diripiendi potest

I. INTRODUCTION

A large and ever-growing number of quantum chemists are attempting to calculate the properties of systems containing transition-metal atoms. As is usually the case with a burgeoning field, the increasing interest and activity can be attributed to the presence of challenges and to these challenges being met, at least in part. The list of specific barriers to progress is long, and I will bring a number of them to the forefront in this review. However, if one single word had to be chosen to describe the primary difficulty of transition-metal quantum chemistry, the choice (in common with many other subfields of the discipline) would be clear: *correlation.* The transition metals are characterized by a relatively compact d valence shell containing up to 10 electrons. Simply stated, one has to deal with a lot of electrons whose space and spin distribution is intricate. When transition metals are involved, the cornerstone of quantum chemistry, the Hartree–Fock wavefunction, must be viewed, at best, as a not always convenient starting point for more elaborate correlated calculations.

While the correlation problem is all-pervasive and its importance can hardly be overstated, this does not mean that the 'minor' challenges of the transition metals are of a trivial nature. For an overall view of transition-metal systems one has to confront a number of problems besides correlation, or sometimes as part of it: open-shell systems involving a large number of electrons, magnetism, ligand lability, close-lying states, metal–metal bonds, core or 'semi-core' electrons that occupy the same part of space as the valence shell, fast electrons requiring relativistic corrections, etc. In addition, many of the transition-metal systems of practical interest (catalysts, photographic clusters, industrial alloys, metallo-enzymes, etc.) are highly complex and often incompletely characterized, so that modeling must be added to the list of difficulties.

Since many of the complexities are already manifest in the smallest systems, the atoms, and intensified in the smallest metal–metal bonded systems, the dimers, and since a considerable body of work has been dedicated to these small systems, I will limit the present review to these simplest cases—though, as we shall see, they are not so simple—leaving the more complex molecules, clusters and models for another time. I will attempt to show, in this context, by what means and to what extent the challenges of the transition metals are being met by state-of-the-art quantum chemistry. A large fraction of this chapter is concerned with calculations of the density functional (DF) family. This reflects my own interests and my belief that, at the present time, these methods are the most propitious for a correct overall view of transition-metal chemistry, including the more complex systems. Since the DF methods are less familiar to many chemists than are the more conventional *ab initio* approaches, the salient features of density functional theory (DFT) and the local (spin) density approach to the correlation problem are presented in the next section along with a discussion of some of the methodology involved in practical calculations. The atoms and dimers reviewed in the following sections provide the best available opportunity for in-depth analyses of both DFT and conventional *ab initio* approaches and for comparisons among them and with experiment. It is this type of analysis and comparison which, I am sure, will ultimately lead to improvements in both types of approach.

The review on atoms and dimers is extensive. I have attempted to include in the reference list, if not all, at least all the currently discussed papers. However, the bulk of the discussion focuses on a few key issues and, for the dimers, the few key molecules that have been extensively studied by several techniques. I have attempted to give enough details about those studies chosen for detailed discussion so as to provide a clear view of what can and what cannot be done with current methods and to highlight the insights that quantum chemistry has provided. Inevitably, this choice has meant that some equally interesting work has been given short shrift for want of space. No slight is intended by this.

II. ELECTRONIC CORRELATION-MIXING DETERMINANTS OR DENSITY FUNCTIONAL THEORY?

The correlation problem can be solved in principle by configuration interaction (CI) or one of its variants. The exact wavefunction is expanded as a linear combination of Slater determinants:

$$\Psi_i(1, 2, \ldots, N) = \sum_p C_{pi}\Phi_p(1, 2, \ldots, N) \tag{1}$$

Once the coefficients have been determined via the variation principle, the energy and other properties may be calculated in a straightforward manner as expectation values over the appropriate operators:

$$\langle \hat{\mathrm{O}} \rangle = \langle \Psi_i | \hat{\mathrm{O}} | \Psi_i \rangle \tag{2}$$

Of course, this has been well known for decades, as has the fact that the practical realization of an accurate CI calculation can be anything but straightforward. (CI is used here in a general sense and includes all the variants—multiconfiguration self-consistent field (MCSCF), generalized valence bond (GVB), many-body perturbation theory (MBPT) methods, etc.) The success or failure of a CI calculation depends not on its solid formal foundations but rather on the technical details, primarily on the choice of determinantal functions to be included in expansion (1). This choice is inevitably constrained by computing resources. It is a tribute to the technical expertise and ingenuity of the *ab initio* quantum-chemistry community that CI calculations can now be performed for molecules containing up to half a dozen or so atoms of the second (Li–Ne) or third (Na–Ar) rows in a manner which, while not always routine, is rational in the sense that the performance of various one-electron basis sets has been established and the most important configurations in the correlation energy calculation are chosen according to some reasonable and controllable criteria. There is enough accumulated experience in the area of CI calculations for 'small, light' molecules for a practitioner to estimate reasonably, at the outset, the effort required to attain a given accuracy for a given property of a given molecule in a given state.

The technology and applications of CI to 'small, light' molecules have been reviewed many times[1–8], including several chapters of these volumes, so I need not dwell on them here. The text by Szabo and Ostlund[5] provides an excellent introduction.

If 'small-molecule' CI is in a rather advanced and healthy state, the same cannot be said for applications to transition-metal systems. The field has not yet seen the accumulation of experience required for the formulation of sufficient rules of thumb, guidelines and the associated technology which would allow *a priori* estimates of the effort associated with a desired accuracy. Sure progress is being made toward this goal and I will review some of it in the following section; however, at the present time it is clear that alternate approaches are needed if quantum chemistry is to make its proper contribution in interpreting the exciting new data that are being obtained in many areas of transition-metal chemistry.

In an area such as this, where *ab initio* techniques are not yet sufficiently developed to confront experimental information directly, the traditional recourse of quantum chemistry has been to the semi-empirical methods, and indeed several groups have proposed variants of the extended Hückel or complete neglect of differential overlap–intermediate neglect of differential overlap (CNDO-INDO) type methods adapted to treat transition metals. While much helpful information can emerge from such calculations, they fall outside the scope of these volumes so are not discussed here.

At the present time, by far the most useful non-empirical alternatives to CI are the methods based on density functional theory (DFT)[9]. The development of DFT can be traced from its pre-quantum-mechanical roots in Drude's treatment of the 'electron gas' in metals[10], and Sommerfeld's quantum-statistical version of this, through the Thomas–Fermi–Dirac model of the atom[11], Slater's $X\alpha$ method[12], the laying of the formal foundations by

Hohenberg and Kohn[9] and Kohn and Sham[13] in the early 1960s, to a host of applications primarily involving the local (spin) density[13,14] (LSD) (or $X\alpha$) approximation (for some representative reviews, see Refs 12 and 15–22).

A. The Local Spin Density Method*

Following Hohenberg and Kohn[9], consider an N-electron system moving in an external potential, $v(\mathbf{r})$, provided by M nuclei. The Hamiltonian may be written:

$$\hat{H} = \hat{T} + \hat{V} + \hat{U} \tag{3}$$

where

$$\hat{T} = \sum_{i=1}^{N} -\tfrac{1}{2}\nabla_i^2 \tag{4}$$

$$\hat{V} = \sum_{i=1}^{N} v(\mathbf{r}_i) = \sum_{i=1}^{N} \sum_{k=1}^{M} \frac{Z_k}{|\mathbf{r}_i - \mathbf{R}_k|} \tag{5}$$

$$\hat{U} = \sum_{i>j}^{N} \frac{1}{r_{ij}} \tag{6}$$

They proved that the full many-particle ground state $\Psi(1, 2, \ldots, N)$ is a unique functional of the electron density, $\rho(\mathbf{r})$:

$$\rho(\mathbf{r}) = N \int |\Psi|^2 \, d\mathbf{r}_2 \cdots d\mathbf{r}_N \, ds_1 \cdots ds_N \tag{7}$$

Through the definition of the universal functional

$$F[\rho(\mathbf{r})] \equiv \langle \Psi | \hat{T} + \hat{U} | \Psi \rangle \tag{8}$$

and the energy functional

$$E_v[\rho(\mathbf{r})] \equiv \int v(\mathbf{r})\rho(\mathbf{r}) \, d\mathbf{r} + F[\rho(\mathbf{r})] \tag{9}$$

a variational principle was proved, that is, E_v has a minimum for the true ground-state density. (See Refs 23 and 24 for a discussion of the so-called v-representability problem and the removal of the restriction to non-degenerate ground states. See also Ref. 25 for interesting reviews.) Although $F[\rho(\mathbf{r})]$ exists and is formally defined in Eq. (8) in terms of the wavefunction which is determined by $\rho(\mathbf{r})$, the explicit functional dependence of F and E on ρ is not known. Approximations must be sought.

Kohn and Sham[13] provided one route to a set of working equations. They

*Parts of the following overview of the local spin density method and the computational methods associated with it have been taken over from Ref. 22.

first separated the 'classical' Coulomb energy in $F[\rho(\mathbf{r})]$:

$$F[\rho(\mathbf{r})] = \tfrac{1}{2}\int \frac{\rho(\mathbf{r})\rho(\mathbf{r}')}{|\mathbf{r}-\mathbf{r}'|}\,\mathrm{d}\mathbf{r}'\,\mathrm{d}\mathbf{r} + G[\rho(\mathbf{r})] \tag{10}$$

where the new universal functional $G[\rho(\mathbf{r})]$ contains the kinetic energy and all terms due to exchange and correlation. $G[\rho(\mathbf{r})]$ was then separated into two terms, namely

$$G[\rho(\mathbf{r})] \equiv T_s[\rho(\mathbf{r})] + E_{xc}[\rho(\mathbf{r})] \tag{11}$$

where T_s is the kinetic energy of a system of *non-interacting* electrons of density $\rho(\mathbf{r})$, and $E_{xc}[\rho(\mathbf{r})]$ contains the exchange and correlation energies of the interacting system (the 'usual' exchange and correlation energies plus the difference in kinetic energy between interacting and non-interacting systems of density $\rho(\mathbf{r})$). If the energy is now varied subject to the normalization constraint

$$\int \delta\rho(\mathbf{r})\,\mathrm{d}\mathbf{r} = 0 \tag{12}$$

the following Euler equation results:

$$\frac{\delta T_s[\rho(\mathbf{r})]}{\delta\rho(\mathbf{r})} + v(\mathbf{r}) + \int \frac{\rho(\mathbf{r}')\,\mathrm{d}\mathbf{r}'}{|\mathbf{r}-\mathbf{r}'|} + \frac{\delta E_{xc}[\rho(\mathbf{r})]}{\delta\rho(\mathbf{r})} = 0 \tag{13}$$

Eq. (13) is exact; however, it still contains an unknown functional, $E_{xc}[\rho(\mathbf{r})]$. Now, if DFT is applied to a system of *non-interacting* particles moving in an 'external' potential defined by the last three terms of Eq. (13), then clearly the same Euler equation will result, since a non-interacting system has no exchange or correlation and T_s was defined as the non-interacting kinetic energy. But for non-interacting particles Schrödinger's equation is separable so that DFT can be expressed in the following Hartree-like equations:

$$\left(-\tfrac{1}{2}\nabla^2 + v(\mathbf{r}) + \int \frac{\rho(\mathbf{r}')}{|\mathbf{r}-\mathbf{r}'|}\,\mathrm{d}\mathbf{r}' + v_{xc}(\rho(\mathbf{r}))\right)\psi_i = \varepsilon_i\psi_i(\mathbf{r}) \tag{14}$$

where

$$v_{xc}(\rho(\mathbf{r})) = \frac{\delta E_{xc}[\rho(\mathbf{r})]}{\delta\rho(\mathbf{r})} \tag{15}$$

and

$$\rho(\mathbf{r}) = \sum_{i=1}^{N} |\psi_i(\mathbf{r})|^2 \tag{16}$$

The Kohn–Sham equations (14)–(16) are still exact. To proceed further, approximations have to be introduced. If $\rho(\mathbf{r})$ varies 'sufficiently' slowly, the local density approximation (LDA) for $E_{xc}[\rho(\mathbf{r})]$ may be introduced, i.e.

$$E_{xc}[\rho(\mathbf{r})] = \int \rho(\mathbf{r})\varepsilon_{xc}(\rho(\mathbf{r}))\,d\mathbf{r} \tag{17}$$

where $\varepsilon_{xc}(\rho(\mathbf{r}))$ is the exchange and correlation energy (including the residual kinetic energy term) per particle of an (interacting) homogeneous electron gas of density $\rho(\mathbf{r})$. The exchange–correlation potential entering the LDA Kohn–Sham equations is

$$v_{xc}(\rho(\mathbf{r})) = \frac{d[\rho(\mathbf{r})\varepsilon_{xc}(\rho(\mathbf{r}))]}{d\rho(\mathbf{r})} \tag{18}$$

The LDA or its spin-polarized generalization, the local spin density (LSD) approximation[14], provides a means of folding exchange and correlation effects, calculated on the basis of the local behavior of a uniform electron gas, into a set of self-consistent Hartree-like equations which contain only local operators for the potential. This procedure is represented schematically in Fig. 1.

The specific form of the LSD equations depends on the treatment of exchange and correlation used in the electron-gas calculation. If only exchange is considered[13,26] then

$$v_x^{\uparrow}(\rho^{\uparrow}(\mathbf{r})) = -\frac{2}{3}\left(\frac{81}{4\pi}\right)^{1/3}(\rho^{\uparrow}(\mathbf{r}))^{1/3} \tag{19}$$

with a similar expression for $v_x^{\downarrow}(\rho^{\downarrow}(\mathbf{r}))$. This exchange potential differs from the

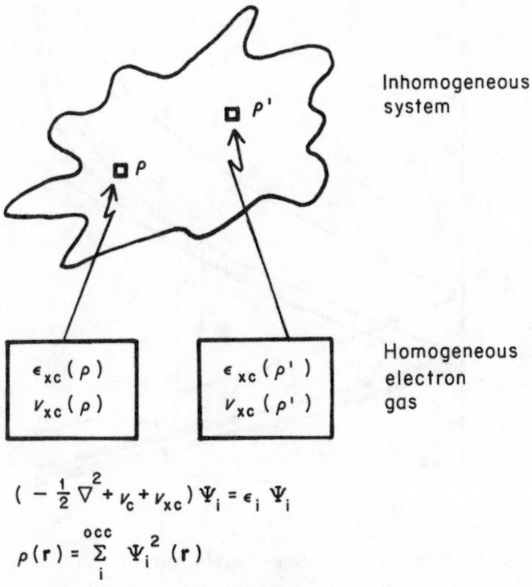

Fig. 1. Schematic representation of the local density approximation.

well known Slater[12,27] exchange potential by a factor of $\frac{2}{3}$ and from the $X\alpha$ potential[12] by a factor of $2/3\alpha$. Slater exchange was derived from the one-electron Hartree–Fock equations by averaging the Fermi hole and introducing the LSD approximation. Gaspar[26] made these approximations in the Hartree–Fock total energy expression and then applied the variation principle, yielding the potential of Eq. (19). Unfortunately, this work went unnoticed for a number of years. The 'rediscovery' of Eq. (19) by Kohn and Sham in 1965 led to a number of comparisons of Slater and Gaspar–Kohn–Sham exchange, and ultimately to the introduction of a scaling parameter α to yield the $X\alpha$ method.

Of course, the Kohn–Sham paper was much more than a mere rediscovery of an approximate Hartree–Fock method. It is soundly based on density functional theory and has paved the way for the approximate treatment of correlation effects, for example, through the use of the LSD approximation in conjunction with correlated electron-gas calculations.

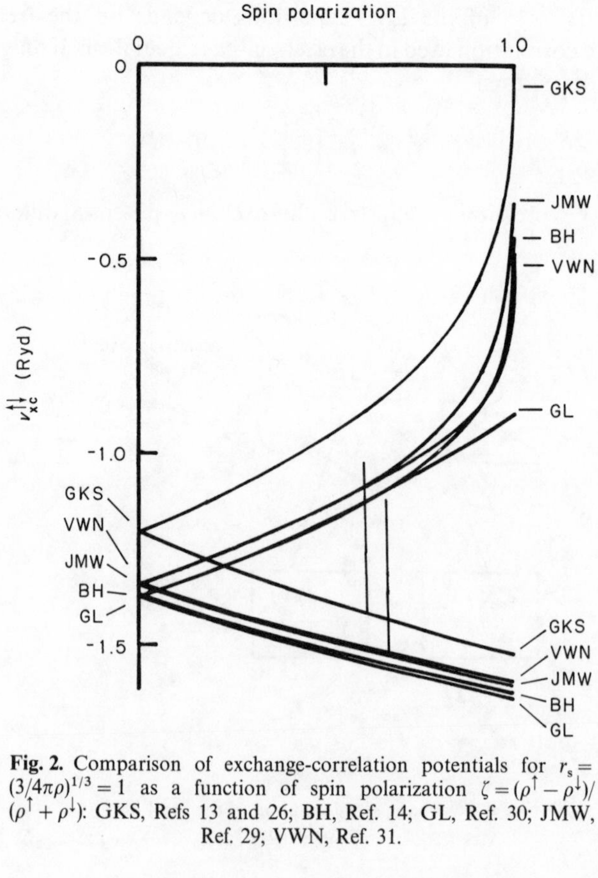

Fig. 2. Comparison of exchange-correlation potentials for $r_s = (3/4\pi\rho)^{1/3} = 1$ as a function of spin polarization $\zeta = (\rho^\uparrow - \rho^\downarrow)/(\rho^\uparrow + \rho^\downarrow)$: GKS, Refs 13 and 26; BH, Ref. 14; GL, Ref. 30; JMW, Ref. 29; VWN, Ref. 31.

A number of exchange–correlation potentials have been proposed over the years[12,14,28–39] including some[33–37] based on relativistic treatments. Those reported in Refs 31 and 32 are parametrizations of accurate Monte Carlo calculations for the electron gas[40] and are believed to represent closely the limit of the LSD approximation.

Several of these potentials are compared in Fig. 2, as a function of spin polarization, for a value of the density appropriate to the valence region of a transition-metal atom. The $X\alpha$ potential may be obtained from the curve marked GKS (exchange only) by multiplying by $3\alpha/2$.

Several features of Fig. 2 are noteworthy. First, the correlated potentials BH[14], GL[30], JMW[29] and VWN[31] are reasonably similar, with the exception of the erroneous behavior of GL in the high polarization limit. Apart from the overall stabilization due to the inclusion of correlation, the most important difference between the correlated and the GKS or $X\alpha$ potentials is a markedly reduced difference between $v^{\uparrow}$ and $v^{\downarrow}$ for the former. The two vertical lines in Fig. 2 are of equal length and are meant to guide the eye in comparing this 'exchange splitting'. In a sense, the GKS and $X\alpha$ potentials are too magnetic. Inclusion of correlation reduces this tendency. The balance between exchange and correlation is a crucial aspect of the binding in systems containing spin-polarized transition-metal atoms, a topic to which we will return. Fig. 2 also provides a rationalization for an old observation[42] that in $X\alpha$-band calculations for Cr the usual α value yields magnetic moments which are much too high. A severely reduced value of α, around 0.5, which is even smaller than the GKS value, is required to obtain the correct moments. Use of a correlated LSD potential[43,44], on the other hand, yields good results.

Relatively few direct comparisons have been made between results of $X\alpha$ calculations and those based on the more elaborate potentials. For many properties of 'simple' systems (atomic ionization potentials[45], equilibrium spacings, vibrational frequencies and binding energies of first-row diatomics[46], clusters of a simple metal, K[47], the adsorption of O_2 on Ag clusters[48]), the differences among the results provided by the various potentials are far from overwhelming. This is not surprising in view of the vast body of reasonable results which have been obtained with the $X\alpha$ method.

However, because of the pathological behaviour of $X\alpha$ for some ('magnetic') transition-metal systems (see below) and because the other potentials include correlation in a more explicit manner and are parameter-free (except for parameters needed to fit electron-gas results), there seems to be little reason to retain the $X\alpha$ potential, except perhaps for the sake of compatibility with previous calculations. Since the VWN[31] or the PZ[32] potential represents the local limit, it should be used as the standard of reference. If an LSD-VWN problem has been accurately solved, any remaining errors may be attributed to non-local effects (Section II.B).

The accuracy and utility of LSD calculations are best judged, in a pragmatic sense, through hard comparisons with experiment, many of which are discussed in later sections. Certainly for the more complex systems involving several 'heavier' atoms, LSD is the method of choice at the present time. However, analysis of simpler systems (primarily light atoms and small molecules) has led to considerable insight concerning the accuracy and the limitations of the LSD approximation. Exchange-only ($X\alpha$ with $\alpha = \frac{2}{3}$) LSD calculations underestimate the exchange energy, typically by 10% or more[49]. The LSD correlation energy is overestimated, usually by a factor of 2 or more. The absolute error in the exchange energy is greater, by about a factor of 3, than that in the correlation energy. Of course, the largest contributions to these errors arise from the core electrons, whose density varies rapidly, so that the energy differences of interest in chemistry benefit from cancellation of errors.

Jones and Gunnarsson[50,51] have recently presented an interesting analysis of the local approximation to exchange. They find that the most serious errors accompany changes in the angular nodal structure of the wavefunctions for the two states involved in an energy difference. The point of reference for exchange is, of course, the Hartree–Fock treatment and, again from a pragmatic point of view, it is not necessarily desirable to reproduce this behaviour. Indeed, it has often been argued[13,52] that the local approximation for the exchange terms in exchange-only LSD or $X\alpha$ calculations is at the heart of their practical success; the 'leveling' of the exchange interactions among the various orbitals provided by the local exchange potential perhaps mimics the effects of configuration interaction (see Sabin and Trickey[53] for an interesting, if not necessarily final, discussion).

B. Beyond the Local Approximation

The LSD approximation occupies a position in density functional theory roughly equivalent to that of the Hartree–Fock approximation in wavefunction theory in that it provides a useful level of approximation for a variety of systems and properties. It has distinct advantages over HF theory in that it is computationally less demanding and some account of correlation is included through the electron-gas model. The principal disadvantage of local DFT at the present time is that there is no rigorous and practical procedure for correcting the inherent errors of the local approximation. While considerable effort is being expended in the search for more accurate functionals, most of the formal work is not yet at the stage where practicable computational schemes may be envisaged. The existing techniques which are specifically aimed at computations may be classified into three categories. Self-interaction corrected (SIC) methods concentrate on removing one error of the LSD approximation, the inexact cancellation of self-Coulomb terms by self-exchange and/or self-correlation terms. These are not, strictly speaking, density functional methods since individual orbital corrections are involved. The second category uses exact (HF) exchange and a local (or local plus SIC) approximation for the correlation energy. Finally, the third category involves a gradient expansion of the exchange and correlation energies to account for inhomogeneities in the electron distribution.

1. *Self-interaction Corrections*

In Hartree–Fock theory, the simplest wavefunction theory involving an antisymmetric wavefunction, the electron repulsion energy of an N-electron system is given by

$$\langle V_{\mathrm{ee}}^{\mathrm{HF}} \rangle = \sum_{i<j} (J_{ij} - K_{ij}) \tag{20}$$

with Coulomb, J_{ij}, and exchange, K_{ij}, integrals defined over spin orbitals as

$$J_{ij} = \iint \psi_i^*(1)\psi_j^*(2) \frac{1}{|\mathbf{r}_1 - \mathbf{r}_2|} \psi_i(1)\psi_j(2)\, \mathrm{d}\tau_1\, \mathrm{d}\tau_2 \tag{21}$$

$$K_{ij} = \iint \psi_i^*(1)\psi_j^*(2) \frac{1}{|\mathbf{r}_1 - \mathbf{r}_2|} \psi_j(1)\psi_i(2)\, \mathrm{d}\tau_1\, \mathrm{d}\tau_2 \tag{22}$$

Since $K_{ii} = J_{ii}$, one can rewrite (20) as an unrestricted sum:

$$\langle V_{\mathrm{ee}}^{\mathrm{HF}} \rangle = \tfrac{1}{2} \sum_{i,j}^{N} (J_{ij} - K_{ij}) \tag{23}$$

The 'self-exchange' integrals K_{ii} serve only to cancel the 'self-Coulomb' integrals J_{ii}.

Since the electron density is given by

$$\rho(\mathbf{r}_1) = \sum_i^N \psi_i^*(\mathbf{r}_1)\psi_i(\mathbf{r}_1) \tag{24}$$

$\langle V_{\mathrm{ee}}^{\mathrm{HF}} \rangle$ may be written as

$$\langle V_{\mathrm{ee}}^{\mathrm{HF}} \rangle = \tfrac{1}{2} \iint \rho(\mathbf{r}_1) \frac{1}{|\mathbf{r}_1 - \mathbf{r}_2|} \rho(\mathbf{r}_2)\, \mathrm{d}\mathbf{r}_1\, \mathrm{d}\mathbf{r}_2 - \tfrac{1}{2} \sum_{i,j} K_{ij} \tag{25}$$

In density functional theory it is natural to define the (direct) Coulomb energy E_C by the first term on the right of (25). The exact exchange–correlation energy functional $E_{\mathrm{xc}}[\rho(\mathbf{r})]$ cancels the spurious self-Coulomb repulsions as well as taking care of the effects of 'true' exchange and of correlation. The approximate functional used in the LSD method does not do so exactly. As a result, in an LSD treatment of a one-electron system, the Coulomb energy is not exactly canceled by the exchange–correlation energy.* For example, for the hydrogen atom the self-interaction is 8.5 eV and 93% of this spurious energy is canceled within the LSD approximation[32,54].

Perdew and Zunger (PZ)[32] have listed a number of inadequacies of the LSD approximation which they attributed primarily to the spurious self-interaction terms. They then proposed an orbital self-interaction correction scheme. The list of LSD 'failures' given by PZ is:

(i) While the LSD total energy of a metal surface is *too low* when compared with the exact value[55,56], the LSD energy for atoms is *too high*[57]. Furthermore, the lowest-order correction to the LSD exchange–correlation energy predicted by many-body theory, i.e. the density gradient correction[58], is positive and so its inclusion can only worsen the agreement between the calculated energies of atoms and experiment.

(ii) For atoms[57] the magnitude of the exchange is consistently underestimated by 10–15% in LSD theory. The magnitude of the correlation energy is overestimated by as much as 100–200%.

(iii) The experimentally stable negative ions (e.g. H^-, O^-, F^-) are predicted to be unstable in LSD theory[59,60].

(iv) Self-consistent LSD band structure calculations systematically underestimate the one-electron energy gaps of insulators by as much as 40% (e.g. Refs 61–65). More generally, the LSD one-electron energy eigenvalues are not close to physical removal energies from bound states. There are often large deviations from experiment when LSD energy eigenvalues are identified with the positions of surface states, deep defect levels in solids, core bands in solids, etc.

(v) The long-range behavior of the LSD one-electron potential for ions of charge Q is $-Q/r$, rather than the electrostatically correct limit of $-(Q+1)/r$. Among other problems, this leads to an erroneous description of charged point-defect states in solids.

(vi) The LSD calculated spin splitting of the energy bands in ferromagnetic metals (e.g. Ni[66]) is often much larger than the observed value.

(vii) LSD total energies unduly favor the $d^{n-1}s^1$ configuration over the $d^{n-2}s^2$ configuration in 3d transition atoms[67]. In addition, the LSD ordering of s and d levels in the $d^{n-1}s^1$ configuration of the 3d elements Sc to Mn is reversed relative to HF.

To keep these 'failures' in perspective relative to the numerous successes of the theory, a few comments on points (i) to (vii) may be in order.

(i) Much of the error comes from core electrons, for which both the SIC and any other non-local (e.g. gradient) corrections are the largest. For energy

*There is no rigorous separation of exchange and correlation in DFT since one does not work directly with a wavefunction. In the LSD approximation the two can be separated in a natural way by reference to an HF calculation for the homogeneous electron gas.

differences involving changes in the valence shells, these errors tend to cancel.

(ii) From a pragmatic point of view, the separation into exchange and correlation contributions is unimportant. Indeed, having exact Hartree–Fock exchange is often a very bad starting point (exchange-only LSD or $X\alpha$ theory is often more accurate than HF).

(iii) While this is a genuine failure, it can be side-stepped in at least two reasonable ways. Electron affinities can be estimated by Slater's transition state method[12] in which only half an electron is added[12,68,69] to the neutral system. As has been well known for many years[12,68,69], this method removes the self-interaction terms involved in orbital energy differences and allows the calculation of ionization potentials, electron affinities and transition energies including an account of physical relaxation.

The error for negative ions is due to the wrong asymptotic form of the LSD potential. An LSD electron far from a neutral atom still 'sees' itself and the extra repulsion leads to instability. Reasonable estimates of electron affinities can be obtained by preventing the electron from entering this poorly described asymptotic region, either by imposing a potential barrier[60] or through the use of a finite basis-set expansion with no very diffuse functions[70].

(iv) There is no *a priori* reason to expect LSD eigenvalues to correspond to physical removal energies (or their differences to correspond to excitation energies or band gaps). This has been known since the early days of $X\alpha$ theory; the realization that $X\alpha$ eigenvalues did not correspond to ionization potentials led to the introduction of Slater's transition state method. See also (vi).

(v) See (iii).

(vi) The observed value comes from photoemission spectroscopy and, hence, involves positive holes. An appropriate comparison should allow the possibility of localization of this hole[71], i.e. breaking Bloch symmetry.

(vii) The LSD calculations typically involve a spherical average approximation. In reality, the transition-metal atoms typically have a complex term structure, involving non-spherical states. In an *ab initio* approach a multi-configurational calculation would be necessary. As will be seen in Section IV such 'multiconfigurational' situations (e.g. Cr_2) can sometimes be reasonably handled by lowering the symmetry constraints imposed on the LSD equations, in a spin-unrestricted formulation. Such considerations may affect the $d^{n-1}s^1$–$d^{n-2}s^2$ separation. (A discussion of some relevent aspects is offered in Section III.)

Notwithstanding these comments in 'defence' of the LSD approximation, the residual self-interaction certainly represents an error which should be removed as one develops more and more accurate functionals.

The procedure proposed by PZ (following earlier work by Lindgren[72]) involves a straight subtraction of the self-interaction error, orbital by orbital, to yield a self-interaction corrected exchange–correlation energy.

$$E_{xc}^{SIC} = E_{xc}^{LSD}[\rho^{\uparrow}, \rho^{\downarrow}] - \sum_{i,\sigma} \delta_{i\sigma} \tag{26}$$

where $\delta_{i\sigma}$ is the SIC for orbital i of spin σ:

$$\delta_{i\sigma} = E_C[\rho_{i\sigma}] + E_{xc}^{LSD}[\rho_{i\sigma},0] \tag{27}$$

$\rho_{i\sigma}$ being the density associated with $\psi_{i\sigma}$ ($\rho_{i\sigma} = \psi_{i\sigma}^*\psi_{i\sigma}$). Minimizing E with E_{xc}^{SIC} leads to the addition, to the orbital Schrödinger-like equation, of an SIC term:

$$v^{SIC}(i) = -\int \frac{\rho_{i\sigma}(2)}{|\mathbf{r}_1 - \mathbf{r}_2|} d\mathbf{r}_2 + v_{xc}^{\uparrow,LSD}([\rho_{i\sigma}, 0]) \tag{28}$$

One has a separate operator for each spin orbital so the equation has to be solved several times and (controllable) problems of orthogonality have to be dealt with. Unlike the LSD energy, the LSD-SIC energy is not invariant to a unitary transformation among the occupied orbitals. For example, in a solid the SIC is zero for Bloch functions but not for Wannier functions. This clearly leads to arbitrariness in the application of LSD-SIC in situations involving wavefunctions which are delocalized by symmetry—a topic discussed further below.

The SIC leads to some clear improvements over LSD for the shape of the exchange–correlation hole (Fig. 1 of Ref. 32). (The energy depends on the spherical average of the exchange–correlation hole, and the success of the LSD approximation has been attributed[73] to the fact that this spherical average is reasonably given since the LSD exchange–correlation hole contains one electron, as it should, and properly cancels the spin density at a given point[12,27].)

Some of the results of PZ for transition-metal atoms, for which the SIC was calculated using spherically averaged orbital densities, will be discussed in Section III.B. To summarize their conclusions on points (i) to (vii)—the SIC improves exchange, correlation and total energies, accurate binding energies are found for negative ions since the long-range behavior of the potential is correct, and orbital eigenvalues are close to physical removal energies. On the other hand the s–d interconfigurational energies and the spin splittings are unimproved over the LSD results in the spherical approximation considered. Similar conclusions were reached by Gunnarsson and Jones[74].

Further work on the SIC has been performed by Harrison, Lin and coworkers[75–81]. Harrison *et al.*[76] proposed a unified Hamiltonian approach to take care of the off-diagonal Lagrange multipliers associated with the orthogonality constraints. Harrison[78] pointed out that PZ had used sphericalized expressions for the total energy and showed that, while inconsequential in HF theory, in SIC-LSD calculations it may represent an appreciable approximation, rooted in the fact that the SIC is not invariant to a unitary transformation of the orbitals (even for a closed shell). He suggested calculating the SIC for central field (i.e. non-spherical) orbitals and then incorporating these into the usual spherical approximation for the rest of the energy functional. While this led to improved results over those of PZ for first-row and transition-metal atoms (see also Section III.B), it remains unsettling to have results dependent on an arbitrary localization transformation (this aspect remains for molecular[80,81] or solid-state[77] applications). It would be interesting to evaluate an approach in which the degree of localization is determined variationally rather than being imposed in an *ad hoc* manner. Of course, in cases where symmetry is involved this will imply broken-symmetry solutions.

2. *Hartree–Fock Exchange Plus Local (or SIC) Correlation*

Several authors[49,82–98] have attempted to use density functional type approaches for only the correlation energy. If the Hartree–Fock expression for exchange is kept, this of course ensures that the self-Coulomb integrals are properly cancelled by self-exchange; one goes back (for better or worse) to the HF level as the point of reference. The computational demands are of the same order as those of the HF calculation itself. Results of early attempts of this nature have been summarized by Stoll *et al.*[95] If the local density approximation to correlation,

$$\mathrm{E_c} = \int \rho\varepsilon_\mathrm{c}[\rho]\,\mathrm{d}\tau \tag{29}$$

is used, where ε_c is the correlation energy density of the homogeneous electron gas, then (a) the correlation energies are overestimated by roughly a factor of 2, (b) correlation energy differences between closed-shell systems are described quite well and (c) unsatisfactory results are obtained for correlation energy differences between closed- and open-shell systems.

The main source of the error was attributed to a residual self-correlation energy (the electron is treated as a continuous fluid and the correlation energy is calculated between different parts of this density[88]). Various 'fixes' have been attempted[86,87,90–94]. That of Stoll *et al.*[94] involves using the LSD approximation:

$$E_\mathrm{c} = \int (\rho_\uparrow + \rho_\downarrow)\varepsilon_\mathrm{c}(\rho_\uparrow + \rho_\downarrow)\,\mathrm{d}\tau \tag{30}$$

but subtracting out the contributions coming from electrons of the same spin, hence excluding the self-correlation terms (as well as any 'real' correlation between like-spin electrons). Their final expression is

$$E_\mathrm{c} = \int (\rho_\uparrow + \rho_\downarrow)\varepsilon_\mathrm{c}(\rho_\uparrow, \rho_\downarrow)\,\mathrm{d}\tau - \int \rho_\uparrow\varepsilon_\mathrm{c}(\rho_\uparrow, 0)\,\mathrm{d}\tau - \int \rho_\downarrow\varepsilon_\mathrm{c}(0, \rho_\downarrow)\,\mathrm{d}\tau \tag{31}$$

This approach typically reduced the correlation energy errors for atoms to the 10% range using the Gunnarsson–Lundqvist correlation potential which, unfortunately, is not the most accurate[49].

Vosko and Wilk[49] investigated a number of self-interaction corrected correlation energy functionals constructed to yield zero correlation energy for a one-electron system and to reduce to LSD in the limit of slowly varying electron density. Their preferred expression is

$$E_c^A[\rho_\uparrow, \rho_\downarrow] = E_c^{LSD}[\rho_\uparrow, \rho_\downarrow] - \sum_\sigma N_\sigma E_c^{LSD}[\rho_\sigma/N_\sigma, 0]$$

$$= \sum_\sigma \int \rho_\sigma[\varepsilon_c(\rho_\uparrow, \rho_\downarrow) - \varepsilon_c(\rho_\sigma/N_\sigma, 0)] \tag{32}$$

where N_σ is the total electron number for spin σ. This may be thought of as removing the self-correlation of N_σ electron distributions each of density ρ_σ/N_σ, similar in spirit to the Fermi–Amaldi[99] correction to Thomas–Fermi theory. This approach was applied to a number of spherical atoms. Vosko and Wilk conclude that the HF plus local correlation (LC) approach is superior to LSD for calculating total energies and $\rho_\sigma(\mathbf{r})$ but that non-local effects are still substantial for $\rho(\mathbf{r})$ even when SIC terms are included. For physical properties which depend on differences, the HF + LC approach appears to be a viable and simpler alternative to CI.

Baroni and Tuncel[96] proposed an exact exchange treatment in which either the LSD or a SIC version was used for correlation. The correlation potential was treated in the paramagnetic approximation, spin-polarization effects being added perturbatively. Details of the correlation SIC were not given. Their LSDX-SIC results for some atomic total energies are in better agreement with experiment than either LSD or LSD-SIC results. Baroni's[97] results for (non-spherical) transition-metal atoms will be discussed in Section III.B.

Very recently Kemister and Nordholm[98] have modified a standard Gaussian quantum-chemical program package to include local approximations for exchange and correlation or for correlation only (the approximation of Stoll and coworkers[94]) and applied it to H_2, HF, N_2, F_2 and OH^-. They find that the bond lengths calculated using the local approximation for correlation only are contracted relative to the HF values whereas, as is well known, the experimental bond lengths are longer. They attribute this flaw to the inconsistent treatments of exchange and correlation.

Overall, the work on exact exchange plus density functional correlation has led to considerable insight into the nature of the LSD approximation, providing some rationalization for why it works as well as it does, and pointing out some of its limitations. It has also served to underline the crucial role played by various assumptions and constraints on the symmetry employed in the various phases of solving atomic problems, a point discussed in several later sections. However, the computational overhead associated with these methods is considerably greater than that of straight LSD (or SIC-LSD), because they involve a HF calculation plus something extra.

3. *Gradient Corrections*

The exact (Kohn–Sham) exchange–correlation potential, while unknown, is certainly non-local; that is, it depends on the entire electron distribution rather than the value of ρ at a given point. To go beyond the local approximation it seems natural to develop a theory involving the spatial derivatives of ρ along with ρ itself. Such gradient expansions go back at least to the von Weizsäcker[100] kinetic energy correction for the electron gas. Herman *et al.*[101,102] developed the $X\alpha\beta$ method in which the energy is given by

$$E_{X\alpha\beta} = E_{X\alpha} - \beta \int \left(\frac{(\nabla\rho_\uparrow)^2}{\rho_\uparrow^{4/3}} + \frac{(\nabla\rho_\downarrow)^2}{\rho_\downarrow^{4/3}} \right) d\mathbf{r} \tag{33}$$

β being a new parameter with a value around 0.003 (Ref. 103). This has most recently been applied by Becke[103] along with a self-interaction-free local curvature approximation for correlation, leading to improved (over $X\alpha$) dissociation energies for light diatomics, but to no improvement for bond lengths or vibrational frequencies. He has also suggested[291] a 'semi-empirical' gradient correction for exchange,

$$\varepsilon_x^\sigma(1) = f_x^\sigma(1) - \frac{\beta[\nabla\rho^\sigma(1)]^2}{[\rho^\sigma(1)]^{7/3}} \left(1 + \frac{\gamma[\nabla\rho^\sigma(1)]^2}{[\rho^\sigma(1)]^{8/3}} \right)^{-1} \tag{34}$$

with $\beta = 0.0036$ and $\gamma = 0.004$, f_x being the local, exchange-only approximation. Tschinke and Ziegler[104] have applied this, along with the VWN treatment of correlation and the correlation SIC of Stoll *et al.*[94], to calculate bond energies for a number of diatomic and polyatomic molecules including a series of transition-metal carbonyls at their experimental equilibrium geometries. Improvement over $X\alpha$ or LSD results was found for the 3d complexes but not for the 4d or 5d molecules.

Langreth and Mehl (LM)[105] carefully studied the convergence properties of the gradient expansion and suggested a practical scheme which, for spherical atoms, yielded improved densities and correlation energies. The LM correction is

$$a \int d\mathbf{r} [\nabla\rho(\mathbf{r})]^2 [\rho(\mathbf{r})]^{-4/3} (e^{-F} + 9f^2) \tag{35}$$

where $F = b|\nabla\rho(\mathbf{r})|[\rho(\mathbf{r})]^{-7/6}$, $a = \pi/8(3\pi^2)^{4/3}$, $b = (9\pi)^{1/6} f$ and f is an adjustable parameter with a suggested value of about 0.15. Savin *et al.*[106] evaluated this correction numerically using HF densities for a number of small diatomic and polyatomic molecules. They find that the quality of the LM correlation energies is similar to that of the SIC of Stoll *et al.* (SPP)[94] (LM being better for diatomics and worse for polyatomics). LM or SPP differences in correlation energies are not necessarily better than LSD.

Since gradient corrections have not yet been systematically applied to transition-metal systems, let us be content with this brief overview (but see Cr_2, Mo_2 and W_2 in Section III).

While the SIC, exact exchange plus local correlation, and gradient expansion techniques have offered some improvements and several directions for new research, the LSD approximation remains the workhorse of the field and we turn, in the next section, to a discussion of some of the methods for solving the LSD equations.

C. Solving the Local Spin Density Equations

1. An Overview

The SCF-LSD equations

$$\left(-\tfrac{1}{2}\nabla^2 + v(\mathbf{r}) + \int \frac{\rho(\mathbf{r}')}{|\mathbf{r}-\mathbf{r}'|}\,d\mathbf{r}' + v_{xc}(\rho(\mathbf{r})) \right)\psi_i(\mathbf{r}) = \varepsilon_i \psi_i(\mathbf{r}) \tag{36}$$

may be approximately solved in a number of ways. The salient features of the main methods currently in use are summarized in Table I along with key references to the methodology and some typical applications.

The scattered-wave (SW) method involves a shape approximation for the potential, the so-called muffin-tin approximation (a spherical potential inside atomic spheres and outside an outer sphere, combined with a constant potential between the spheres). Numerical solution of the radial Schrödinger equation in the spheres, and a rapidly convergent partial-wave expansion between them, lead to a set of secular equations which can be solved rapidly, allowing complex systems to be treated. The SW method has been widely used in the area of transition-metal clusters and chemisorption models. Several reviews exist[12,15–17,20–22]. While the method is extremely rapid and has provided a vast array of useful results, it suffers from one serious drawback of great importance for many interesting problems: accurate values of the total energy are not readily accessible. This prevents direct geometry optimizations and the calculation of vibrational properties, dissociation energies and other energy differences which are needed to characterize the potential energy surface of a complex system.

The linear combination of atomic orbitals (LCAO) methods (LCGTO, DVM, LMTO) overcome this drawback at the expense of more onerous computations. All three of these approaches should yield identical results (the LSD limit), provided all the technical details such as basis sets, sampling points, fitting procedures, numerical integrations, etc., are properly controlled. It is healthy, during this developmental stage, to have these competing technologies; each has its own pros and cons and its own group of proponents.

TABLE I
Characteristics of various local spin density (LSD) methods.

$X\alpha$(LSD)-SW[107]	Scattered wave Multiple scattering Muffin-tin potential Numerical solution in spheres Partial waves Most rapid method Accurate orbitals and spectroscopic quantities Does not yield accurate total energy surfaces E.g. Cu_{19} [108], $Ni_{13} + CO$[109]
DVM-LSD[110,111]	Discrete variational method LCAO (Slater orbitals, or numerical atomic orbitals) Fit ρ and $\rho^{1/3}$ with auxiliary functions numerical sampling for matrix elements Quite rapid—work goes up about as m^2 Very large number of sampling points needed for total energy E.g. Cu_{13} [112], Cr_2 [113], $Co_4(CO)_{12}$ [114]
LMTO[115–117]	Basis of muffin-tin orbitals $\phi_l^i(\varepsilon, r) Y^L(\hat{r}_i)$ and $\delta(\phi_l^i(\varepsilon, r))/\delta\varepsilon$ Hankel functions outside spheres Muffin-tin components of matrix elements evaluated 'semi-analytically' Non-muffin-tin parts done by partial-wave expansion of potential and Gaussian quadrature outside spheres Coulomb fit E.g. Cr_2 [118], NH_3 [119], $Al_9 + H_2O$ [120]
LCGTO-$X\alpha$(LSD)[121,122]	LCAO (Gaussian type orbitals) Fit ρ, v_{xc} and ε_{xc} with auxiliary functions Analytical integrals Accurate total energies E.g. Ni_4 [123], $Pd_{10} + H$[124], Cr_2 [125,126]
Numerical[127,128]	2D numerical integration of LSD equation Limited to linear molecules Very expensive for transition metals E.g. first-row dimers[127], Cr_2 [129]

Whether one of these methodologies, or a different approach, will ultimately prevail remains to be seen.

In my group we have chosen to work with the LCGTO approach for a number of reasons, both historical and rational. First, the early work of Dunlap *et al.*[122] showed that the method could provide remarkably accurate spectroscopic constants for the first-row diatomics. Comparisons with completely numerical calculations[127] showed that the technical details could be handled. Of the three methods, DVM, LMTO and LCGTO, the latter is the most similar to the standard Gaussian techniques of *ab initio* quantum chemistry. This has provided some advantage for such matters as the optimization of basis sets[130,131] and the introduction of model potentials[132], and should facilitate future extensions of the programs to calculate various properties, analytical energy gradients, etc.

2. *The LCGTO-LSD-MP Method*

Since a number of the results in the following sections were calculated with this approach, a brief description of the LCGTO method is presented here. The molecular orbitals are expanded by the usual LCAO expression:

$$\psi_i = \sum_{p=1}^{m} C_{pi}\chi_p \tag{37}$$

The computer codes of Sambe and Felton (SF)[121] and Dunlap *et al.* (DCS)[122] are based on the choice of a Hermite–Gaussian[133] expansion set*. Applying the variational theorem with the trial function of Eq. (37) and the LSD Hamiltonian of Eq. (36) leads to the usual matrix pseudo-eigenvalue problem:

$$\mathbf{HC} = \mathbf{SC}\boldsymbol{\varepsilon} \tag{38}$$

The overlap and Hamiltonian matrix elements over Hermite Gaussians must then be evaluated. The one-electron operators present no special problems and need not be discussed further. The Coulomb integrals are identical to those in Hartree–Fock theory, i.e.

$$\begin{aligned}\langle \chi_p | v_C | \chi_q \rangle &= \int \chi_p(\mathbf{r}) v_C(\mathbf{r}) \chi_q(\mathbf{r})\, d\mathbf{r} \\ &= \int \chi_p(\mathbf{r}) \int \frac{\rho(\mathbf{r}')}{|\mathbf{r}-\mathbf{r}'|}\, d\mathbf{r}'\, \chi_q(\mathbf{r})\, d\mathbf{r}\end{aligned} \tag{39}$$

If $\rho(\mathbf{r}')$ is expanded, then the usual four-index two-electron integrals result:

$$\langle \chi_p | v_C | \chi_q \rangle = \sum_{i}^{\text{occ}} \sum_{r} \sum_{s} C_{ri} C_{si} \int \chi_p(\mathbf{r}) \int \frac{\chi_r(\mathbf{r}')\chi_s(\mathbf{r}')}{|\mathbf{r}-\mathbf{r}'|}\, d\mathbf{r}'\, \chi_q(\mathbf{r})\, d\mathbf{r} \tag{40}$$

One index can be saved if the charge density in the Coulomb operator is

* For s and p functions the Hermite Gaussians are identical to the familiar Cartesian Gaussians. For higher *l*, differences occur. For example some of the Hermite 'd' functions contain spherically symmetric components. These differences must be remembered when comparing results with, say, *ab initio* calculations based on Cartesian Gaussians. While calculations can be set up to use Cartesian Gaussians effectively by taking appropriate linear combinations, this is somewhat inconvenient. We hope eventually to rewrite the programs to use Cartesian Gaussians in order to be somewhat more in line with mainstream quantum chemistry.

fitted to an auxiliary set of functions. (In a Hartree–Fock calculation this would confer no advantage since all m^4 integrals are needed in any event for the exchange terms.) We write

$$\tilde{\rho}(\mathbf{r}') = \sum_s a_s f_s(\mathbf{r}') \tag{41}$$

where f is a second set of Hermite Gaussians, the charge density basis (CDB). The fitting coefficients a_s are determined by a least-squares procedure[122] which minimizes

$$D' = \int \frac{[\rho(\mathbf{r}) - \tilde{\rho}(\mathbf{r})][\rho(\mathbf{r}') - \tilde{\rho}(\mathbf{r}')]}{|\mathbf{r} - \mathbf{r}'|} \, \mathrm{d}\mathbf{r} \, \mathrm{d}\mathbf{r}' \tag{42}$$

In the end one has to evaluate integrals of the type

$$\left\langle \chi_p(\mathbf{r}) \left| \frac{f_s(\mathbf{r}')}{|\mathbf{r} - \mathbf{r}'|} \right| \chi_q(\mathbf{r}) \right\rangle \tag{43}$$

There are roughly m^3 integrals to evaluate, assuming the same number of functions in the CDB as in the orbital basis. At most three centers are involved in the integrals.

Integrals involving the exchange–correlation potential v_{xc} or the exchange–correlation energy density $\varepsilon_{\mathrm{xc}}$ cannot be evaluated analytically so that further sets of auxiliary functions are introduced. (In practice v_{xc} and $\varepsilon_{\mathrm{xc}}$ behave similarly so that a common set is used to fit both functions.) The exchange–correlation basis (XCB) also consists of Hermite Gaussians

$$\tilde{v}_{\mathrm{xc}}(\mathbf{r}') = \sum_s b_s g_s(\mathbf{r}') \tag{44}$$

The least-squares procedure used to evaluate the coefficients b_s involves sampling $v_{\mathrm{xc}}(\mathbf{r}')$ on a grid of points, typically a radial distribution based on that of Herman and Skillman[134] (every tenth point) coupled with an angular mesh consisting of the 12 vertices of a regular icosahedron or 14 points defined by the corners and face centers of a cube. Both of these grids are accurate through fifth-order spherical harmonics.

For the heavier elements, considerable computational advantage can be gained by introducing a model potential to represent the field of the core electrons. Besides the computational advantage, the use of a model potential can also reduce basis-set superposition errors which are often exacerbated in all-electron calculations by the presence of poorly described core orbitals (which can be 'patched up' by the tails of functions on neighboring centers). The model potential we have chosen to implement in our program[132] is an adaptation of that introduced into *ab initio* methods by Huzinaga and coworkers[135–137]. It is a very flexible model potential based on a rigorous separation of core and valence orbitals, which is possible for both restricted

TABLE II
Potential sources of error in the LCGTO-LSD method.

Errors inherent in the LSD approximation
Orbital basis-set incompleteness
Auxiliary basis-set incompleteness
Near-linear dependencies in auxiliary basis sets
Orbital basis-set superposition error
Auxiliary basis-set superposition error
Exchange–correlation sampling grid
Model potential for core electrons (if employed)

and unrestricted LSD calculations. The full nodal structure of the valence orbitals may be retained if necessary. This model potential is in general more expensive than the various 'nodeless' pseudo-potentials. However, we have found that it is very reliable and, importantly, it can be systematically improved, when necessary, by expanding the valence basis set and/or explicitly including some of the higher-lying core levels in the valence shell.

There are several potential sources of error in the method just sketched. These are summarized in Table II. While there is a need for systematic study of each of these potential error sources, the results presented in Section IV and elsewhere (e.g. Refs 22, 124 and 125) indicate that the errors can be controlled.

Several other variants of the Gaussian-LSD approach have been proposed. Kitaura *et al.*[138] used a Gaussian basis, evaluated the Coulomb integrals exactly and evaluated the local exchange integrals by scaling those calculated numerically on a DVM grid, using analytical values of integrals for a reference operator (obtained by fitting a superposition of atomic values of $\rho^{1/3}$ with spherical Gaussians). They studied first-row diatomics, H_2O, NH_3 and C_2H_4. Katsuki and coworkers[139–143] have implemented a model-potential $X\alpha$ method, using only s functions for the fits, and applied it to diatomics[139] and to Ni–CO and Pd–CO systems[140–142]. Painter and Averill[144] studied first-row diatomics with a method in which all the Coulomb integrals were calculated analytically and the exchange–correlation contributions by numerical quadrature. A similar scheme, based on GAUSSIAN70, was implemented by Grimley and Bhasu[145] and applied at the STO-3G level to atoms and some light diatomics. Bernholc and Holzwarth[146,147] developed a Gaussian LSD method incorporating a ('nodeless') pseudo-potential and have applied it to systems such as Cr_2 and Mo_2[146] (see Section IV) and $Mo_2O_2S_2(S_2)_2^{2-}$ (Ref. 147). Callaway and coworkers have examined several Gaussian LSD schemes. For diatomic molecules[148,149] all of the Coulomb integrals were evaluated analytically and the exchange–correlation terms by numerical integration. For larger, symmetrical, systems (e.g. Fe_{15}[150], Fe_{13}[151], $FeCu_{12}Cu_6$[152], SF_6[153]) the density was fitted to s-type Gaussians and the symmetry was used to reduce the numerical exchange–correlation integrations to a fraction of space (1/48 for octahedral symmetry). The work of Lin, Harrison and coworkers[75–78] with a Gaussian approach was already mentioned in connection with the SIC. The groups of both Callaway and Lin have been performing Gaussian band calculations on solids for a number of years.

Clearly, there exists a healthy level of interest in implementing and refining Gaussian LSD methods. The computer programs currently in use are not as highly refined as typical *ab initio* packages. However, the practical results for complex systems which are now beginning to appear certainly indicate that further effort will be worth while. I would like to turn now to a survey of results on the transition-metal atoms and dimers in order to put the LCGTO and other LSD methods in perspective with respect to the more conventional *ab initio* methods and with respect to experiment.

III. ATOMS

In addition to their intrinsic interest, calculations of the electronic structure of atoms are important, for several reasons, to those studying molecules. The wealth of available atomic spectral data[154] provides a testing ground for computational methods. Since most atomic states, including the ground states, are open shells involving more or less complex term and multiplet structures, which are often poorly described by the simpler approaches, the tests provided are usually severe, indeed, more severe than a 'typical' closed-shell molecule near its equilibrium geometry. Since an atom in a molecule can be viewed as receiving contributions from ground, excited and ionic states of the isolated atom, the performance of a method for these atomic states provides some indication of how it will perform for molecules. Atomic calculations are also often used to define at least partially the input for molecular calculations (optimized basis sets, model potential parameters, etc.). Finally, if one wishes to calculate dissociation energies then the separated atoms become an integral part of the problem and must be treated at a level of theory consistent with that used for the molecular calculations. This balance is not always easy to achieve.

The transition-metal atoms present definite computational and conceptual challenges. I will now review some recent progress using conventional *ab initio* (HF, CI or MCSCF) or density functional techniques.

A. Hartree–Fock, Configuration–interaction and Multiconfiguration Self-consistent Field Calculations

The (HF) radial 3d function of a first transition series atom is compact and not accurately representable by a single exponential ($r^2e^{-\zeta r}$). This is strikingly illustrated in Table III, which compares 3d orbital energies from single-zeta

TABLE III
STO-SZ and HF 3d orbital energies (in a.u.) for third-row atoms. (*From Ref. 155. Reproduced by permission of the authors and the American Institute of Physics.*)

	STO-SZ[a]	HF[a]
Sc(^{2}D)	−0.18988	−0.34357
Ti(^{3}F)	−0.24338	−0.44071
V(^{4}F)	−0.25642	−0.50966
Cr(^{5}D)	−0.25083	−0.56920
Mn(^{6}S)	−0.25159	−0.63884
Fe(^{5}D)	−0.15690	−0.64689
Co(^{4}F)	−0.08085	−0.67548
Ni(^{3}F)	−0.00054	−0.70693
Cu(^{2}S)	0.50637[b]	−0.49074

[a]Ref. 156.
[b]STO-DZ result is −0.66185.

(SZ) STO calculations with their numerical HF counterparts. The errors are huge, much more so than for first-row elements where, say, STO-3G at least makes some sense. The most striking examples are provided by Cu whose STO-SZ 3d level is unbound by 14 eV and by Ni where this level is barely bound. A double-zeta (DZ) calculation improves things but the errors are still appreciable (1.2 eV for Ni). In an oft-cited paper, Hay[157] has suggested using the Wachters[158] (14s9p5d) primitive set augmented with an additional diffuse d function. In molecular applications, extra p functions are often added and the resulting primitive set (e.g. (14s11p6d)) may be contracted to the triple-zeta level for the d functions (e.g. (5s/4p/3d)). Several other basis sets of varying quality have been proposed[159–167]. A general conclusion is that, for many properties, extensive, well optimized basis sets are required to obtain results with acceptable errors relative to the Hartree–Fock limit.

The relationship of the Hartree–Fock limit to experimental reality is yet another matter. The near-degeneracy of the 3d and 4s orbitals results in a number of low-lying states. Several studies[168–176] have focused on the energy differences between $d^{n-2}s^2$, $d^{n-1}s$ and d^n configurations and on the effect of correlation on these differences. Because of its relevance to catalytic problems, the nickel atom received some of the first detailed attention[157,169]. For Ni, states arising from the d^8s^2 (^{3}F) and d^9s (^{3}D) manifolds overlap, with the center of gravity of the latter lying lower by 0.03 eV. The d^{10} (^{1}S) state is 1.7 eV higher. Hartree–Fock theory yields a d^8s^2 ground state with d^9s at 1.28 eV and d^{10} at 5.47 eV. Clearly, correlation is crucial for an accurate description of these atomic states and, by inference, for molecular states formed as superpositions of them. Martin[169] performed CI calculations involving all single and double orbital replacements (CI(SD)). Correlating the 3d and 4s electrons reduced the excitation energies to 0.46 eV for d^9s and to 2.87 eV for d^{10}. Adding f functions to the basis, correlating the 'semi-core' 3s and 3p electrons and estimating the effect of 'unlinked cluster' quadruple excitations via the Davidson[177] correction led to further small changes, the most extensive calculations yielding 0.2 eV for d^9s^1 and 2.5 eV for d^{10}. (In passing, it is worth while to note that while the (3s, 3p) contribution to the correlation energy is, to a good approximation (~ 0.1 eV), transferable among the three configurations of Ni, the direct contribution of these semi-core electrons at the SCF level is a sensitive function of the valence configuration, increasing by about 5 eV per d electron. This, of course, has implications for the use of pseudo-potentials; for high accuracy these electrons should either be included explicitly in the valence shell[132,178] or, at a minimum, core-polarization terms should be included in the model potential.)

The experimental $d^{n-1}s$–$d^{n-2}s^2$ energy differences for the 3d series are summarized in Fig. 3a. Fig. 3b shows that the HF errors are substantial; for Sc–Cr the $d^{n-1}s$ state is unduly favored by about 0.3 eV whereas for Mn–Cu the $d^{n-2}s^2$ state is relatively too stable by about 1.0 eV. MCSCF/CI(SD) calculations[172,173,175] (Table IV) reduce the discrepancies with experiment to

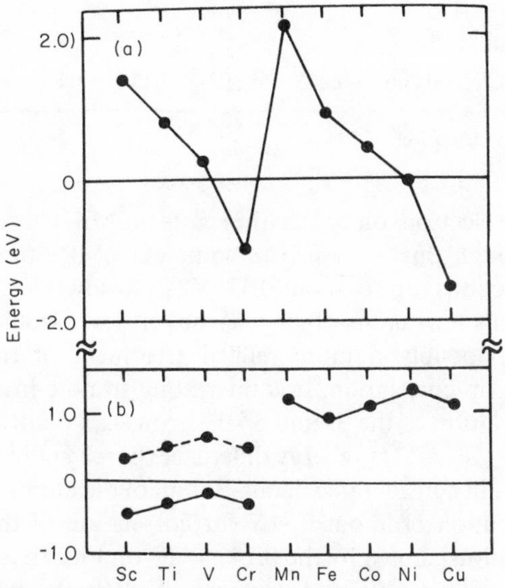

Fig. 3. $4s3d^{n+1}$–$4s^23d^n$ excitation energies of scandium to copper $[E(sd^{n+1}) - E(s^2d^n)]$. All units are in eV. (a) Experimental values (Ref. 154). (b) Error in the numerical HF excitation energy $(\Delta E_{HF} - \Delta E_{exp})$ (full line). The broken line is the HF error corrected for the $(4s^2, 4p^2)$ near-degeneracy effects $\Delta E_{HF} + 0.78$ eV (ave.). (*Reproduced from Ref. 173 by permission of the authors and the American Institute of Physics.*)

TABLE IV
Summary of d^n4s^2–$d^{n+1}4s^1$ separations. All values are in eV and all CI results include Davidson's correction. (*From Ref. 172. Reproduced by permission of the authors and the American Institute of Physics.*)

	Sc ^2D–^4F	Ti ^3F–^5F	V ^4F–^6D	Cr ^7S–^5D	Mn ^6S–^6D	Fe ^5D–^5F	Ni ^3F–^5D	Cu ^2S–^2D
SCF	1.03	0.56	0.15	1.24	3.34	1.80	1.27	0.39
f exponent	1.4	0.55	1.8	0.55	2.2	1.3	2.3	2.5
CI(SD)	1.79	0.98	0.50	0.95	2.86	1.26	0.30	1.33
CI(SD) (3s, 3p)	1.60	0.89	0.35	1.02	2.64	1.15	0.23	1.33
Opt. *f*		1.55		2.00		2.5	3.1	
CI(SD) (3s, 3p)		0.93		0.96		1.26	0.26	
2*f*								
CI(SD)	1.77	1.02	0.46	0.95		1.25		
CI(SD) (3s, 3p)	1.51	0.81	0.27	1.04		1.13		
3*f*								
CI(SD)	1.75							
CI(SD) (3s, 3p)	1.48							
Exp[a]	1.437	0.806	0.245	1.003	2.144	0.875	−0.32	1.49

[a]Average of M; values from Ref. 154.

a few tenths of an electronvolt or less (the errors for Mn–Cu being greater than for Sc–Cr) although this is spoiled to some extent if estimated differential relativistic corrections (up to about 0.37 eV (see below)) are included[174,176]. To go beyond this few-tenths-of-an-electronvolt level would require higher excitations and possibly a more refined treatment of relativity and its interdependence on correlation. It is interesting to note from Table IV that differential correlation of the 3s and 3p electrons can contribute as much as 0.27 eV to the $d^{n-1}s - d^{n-2}s^2$ energy difference. Botch *et al.*[173] have analyzed their calculations in considerable detail. For all of the atoms they find a near-constant contribution of about 0.8 eV for correlation of the s^2 pair in the $d^{n-2}s^2$ configuration (shown by the broken line on Fig. 3b for Sc–Cr) arising from $4s^2 \rightarrow 4p^2$ excitations (near-degeneracy effect). In the orbital language of generalized valence bond (GVB) theory[179,180] this would correspond to a mixing of 4s and 4p orbitals: the orbitals are of broken symmetry even though the total wavefunction is not. Further, Botch *et al.* find that for the d electrons radial correlation ($3d^2 \rightarrow 4d^2$ excitations) makes the most important contribution to the energy. For Sc to Cr the radial correlation of (same-spin) d electrons ranges from 0 to 0.52 eV for $d^{n-2}s^2$ and from 0.21 to 1.21 eV for

$d^{n-1}s$ (Table IV of Ref. 173), the differential stabilization of $d^{n-1}s$ being less than the stabilization of $d^{n-2}s^2$ arising from the angular correlation of the s electrons (cf. Fig. 3b). For Mn to Cu the 3d correlation also receives contributions from opposite-spin electrons and ranges from 0.79 to 2.99 eV for $d^{n-2}s^2$ and from 1.85 to 4.59 eV for $d^{n-1}s$. The differential stabilization of $d^{n-1}s$ is always larger than the approximate 0.8 eV stabilization of $d^{n-2}s^2$ due to s-electron correlation (cf. Fig. 3b). Botch *et al.* also propose a GVB-style orbital interpretation in which one d electron of d^{n-1} is placed in an orbital which is inequivalent to (more diffuse than) that of the other $(n-2)$ electrons. While the completeness[175] and some of the details[176] of this interpretation have been contested, it does provide an appropriate and appealing explanation of at least some of the important correlation contributions, in orbital language. The fact that the GVB-like orbitals are of reduced symmetry and allow 'hybridization' in the atomic ground or low-lying excited states may ultimately prove to be of some relevance for the development and application of orbital theories such as the Kohn–Sham formulation of density functional theory (at or beyond the LSD approximation). I will return to this point in the next section.

In addition to correlation, for the transition metals one has to start worrying about relativity. The common wisdom is that relativistic corrections will have little effect on the properties of 3d systems, a small but non-negligible influence for the 4d series and dramatic, even qualitative, consequences for 5d. For the atoms, Martin and Hay[174] have evaluated the scalar (mass-velocity and

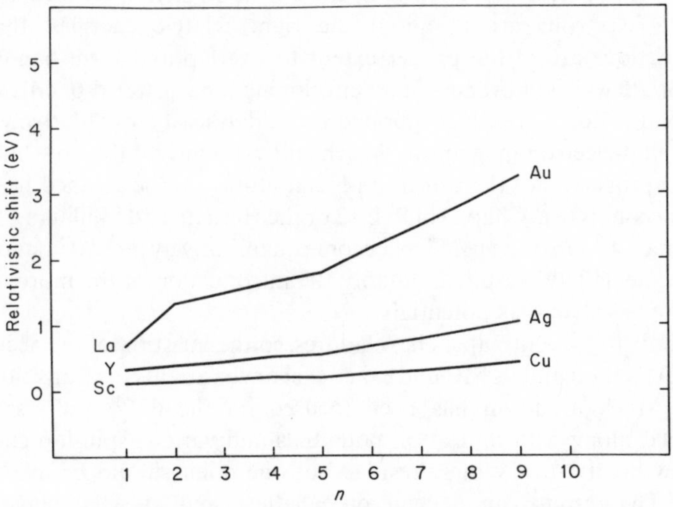

Fig. 4. Relativistic shifts for s^2d^n–s^1d^{n+1} excitation energies of transition-metal atoms. Differences between excitation energies from Hartree–Fock and relativistic Hartree–Fock calculations are plotted. (*Reproduced from Ref. 174 by permission of the authors and the American Institute of Physics.*)

Darwin) corrections at the Hartree–Fock level. Their results for the $d^{n-1}s$–$d^{n-2}s^2$ separation are shown on Fig. 4. While the common wisdom is roughly verified, the relativistic shifts for the later members of the 3d series are probably larger than many would have expected. They certainly cannot be neglected if very high accuracy is required. (A similar statement applies to molecules (see e.g. Cu_2 below).) Of course Fig. 4 assumes that the relativistic correction can be separated from correlation effects. Martin and Hay reasoned that, since the relativistic correction scales roughly as the valence s-orbital population, correlation effects which change this population would have the most important influence. By using the CI coefficient for the most important ($s^2 \rightarrow p^2$) contribution they estimated that, for Ni, correlation could reduce the relativistic correction of 0.35 eV by about 0.07 eV (probably an upper limit).

B. Density Functional Calculations

Density functional type calculations on atoms date back as far as the Thomas–Fermi[11,181] model. The use of an electron-gas model for exchange was taken up by Slater in his famous 1951 paper[27] and later on developed into the $X\alpha$ method[12]. Much of the early atomic work carried out by Slater and his school was aimed at providing reasonably accurate atomic potentials for use in band structure calculations. Indeed, modern band theory owes a large part of its development and present success to the fact that atomic calculations using Slater or $X\alpha$ exchange provide a good first-order account of atomic energy levels and wavefunctions. For the transition metals this means that the d and s electrons are at about the right relative energies, that their wavefunctions are of the proper extent to overlap and form bands of the appropriate width (a broad s band enveloping a narrower d band), and that spin polarization can be incorporated to yield at least a qualitatively correct picture of d-electron magnetism. Much still valid insight (by now 'intuition') into the properties of the transition-metal atoms can be gleaned from these earlier accounts (e.g. Chap. 3 of Ref. 12 or the Herman and Skillman[134] book). With time, of course, the field has become quantitatively more demanding and improvements have resulted, notably the introduction of the more accurate correlated electron-gas potentials.

A number of recent papers have addressed the question of the accuracy of the LSD method and its SIC and exact-exchange variants for transition-metal atoms. Much attention has been focused on the $d^{n-1}s^1$–$d^{n-2}s^2$ energy difference, along with ionization potentials and various spin-flip energies. I will now briefly review these results, but one point should be made at the outset. The results can depend on whether, and at what stage of the calculations, spherical averaging is invoked (see Section II.B).

Harris and Jones[118,182] performed central-field LSD calculations for the 3d and 4d atoms using the Gunnarsson–Lundqvist[30] potential. Their results for

$\Delta E_{ic}(d^{n-1}s\text{–}d^{n-2}s^2)$ of the 3d series are compared with experiment in Fig. 5. The experimental results correspond to the lowest-lying term for each configuration, averaged over multiplet states with different J. On the other hand, the central-field calculations often correspond to an average over states of different L arising from the open-shell configurations, so that the comparison is not entirely clean. Ideally, of course, one would like to compare state-by-state but it is not clear how to do this within a theory based strictly on orbitals (it is not obvious how to work with eigenfunctions of $\widehat{S^2}$ and $\widehat{L^2}$ if one is not working directly with any N-electron wavefunction at all). This 'conceptual difficulty' underlines the fact that more work is needed on the density functional treatment of open-shell systems[183–185] and excited states as well as on the relationship of DFT to wavefunction theory for these cases. My own feeling is that one can go a long way by removing symmetry constraints and allowing the (spin) orbitals to localize (or not) in order to optimize the exchange and correlation interactions 'seen' on some reasonably local scale by a given electron. If this could be done, it is at least conceivable that the consequences for the energy and other properties would be more important than the global necessity of having space and spin eigenfunctions.

Having presented the above caveat, we return to Fig. 5. The trends in the interconfigurational energy across the series are interpreted very well by the central-field LSD calculations including the break between Cr and Mn.

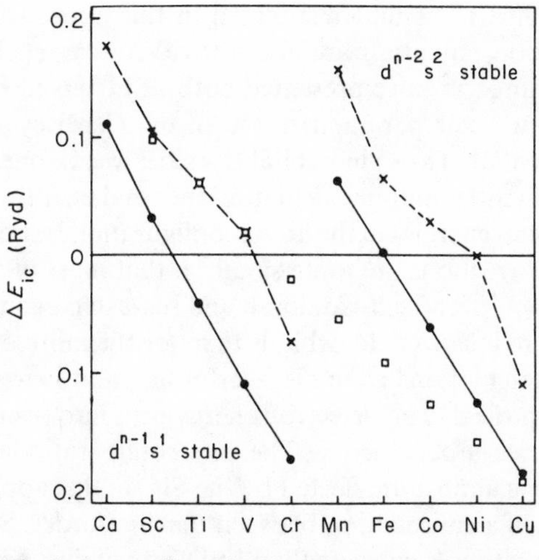

Fig. 5. Interconfigurational energy difference $\Delta E_{ic} \equiv E[3d^{n-1}4s^1] - E[3d^{n-2}4s^2]$, for atoms of 3d series. Squares: non-spin-polarized results; full circles: spin-polarized results; crosses: experimental values (Ref. 154). (*Reproduced from Ref. 118 by permission of the authors and the American Institute of Physics.*)

However, the LSD results unduly favor the $d^{n-1}s$ configuration by roughly 1 eV. This also holds for the 4d series. Interestingly, this is quite similar in magnitude to the contribution of 0.8 eV to the correlation energy attributed to angular correlation ($4s^2 \rightarrow 4p^2$) of the 4s electrons in the MCSCF/CI calculations discussed above. It is tempting to attribute the LSD error to the same source. This would imply, following the analysis of Botch *et al.*[173], that the radial correlation of the d electrons is reasonably handled by LSD but the angular correlations of the s electrons are not, since within a central-field orbital model one cannot mix functions of different angular character. It might be possible to mimic the angular correlation effects in a broken-symmetry calculation which would allow s–p mixing, effectively localizing one '4s' electron with, say, spin up on one side of the atom and the other with spin down on the other side. It may be of some importance to note that in a molecular environment the symmetry is 'automatically' broken and such hybridization is allowed. Hence, imposing severe symmetry constraints on atomic calculations within an orbital model may introduce a bias into the calculation of dissociation energies (some examples are discussed in Section IV).

Harris and Jones also presented results for spin-flip energies and for ionization potentials. Again, trends are reasonably well interpreted. For the spherical GL potential used, the energy to flip an s spin is overestimated for all of the atoms except Ni and Cu, the largest discrepancies, of about 0.5 eV, occurring near the middle of the series. On the other hand, the spin-flip energy for a d electron in Cr^+ is underestimated in this approach by about 1 eV. Errors in ionization potentials are about 0.4 eV.

Perdew and Zunger[32] have presented both LSD and LSD-SIC results for the 3d series using their parametrization of the Ceperley–Alder correlated electron-gas potential. The spin-orbital densities were sphericalized both in evaluating the potential and in calculating the total energy. A breakdown of the 4s and 3d orbital energies in the $d^{n-1}s$ configuration is shown in Fig. 6. The self-correlation corrections are quite small, so that most of the SIC is due to inexact cancellation of the self-Coulomb and self-exchange terms. The SIC is larger for the more localized 3d orbitals than for the more diffuse 4s orbitals. Exchange-only calculations were also performed and it was found that the SIC-LSD brought the d–s one-electron energy gaps into good agreement with those from Hartree–Fock theory. The interconfigurational, $d^{n-1}s$–$d^{n-2}s^2$, energy difference is almost unaffected by the SIC in this approximation. This arises as a result of a cancellation between the 'first-order' SIC, which favors the $d^{n-1}s$ configuration even more than LSD, and higher-order effects, which lower the energy of the more polarizable $d^{n-2}s^2$ configuration. A similar conclusion was reached by Gunnarsson and Jones[74]. Finally, Perdew and Zunger find that the spin splittings for atomic Fe, Co and Ni are increased by the SIC, which is the opposite direction to the correction needed for LSD

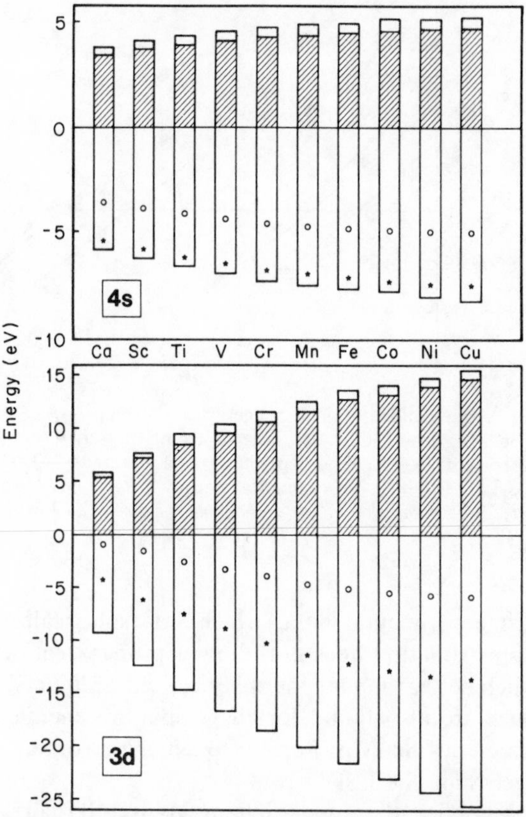

Fig. 6. Breakdown of the orbital energies in the SIC-LSD formalism for spin-up 4s and 3d orbitals of transition atoms in the $d^{n-1}s^{\uparrow}$ configuration. White area: self-Coulomb; dashed area: self-exchange; dotted area: self-correlation. Open circles: LSD eigenvalues; asterisks: SIC eigenvalues. (*Reproduced from Ref. 32 by permission of the authors and the American Physical Society.*)

calculations on the solids if one wishes to compare eigenvalue differences directly with the results of photoemission experiments.

Harrison[78] has performed SIC-LSD calculations for the 3d series but without sphericalizing the spin-orbital densities. The SIC is calculated for Cartesian orbitals and then the central-field approximation is reintroduced by spherically averaging the contributions to the energy and to the potential arising from a given shell. As Harrison pointed out, the errors introduced by sphericalizing the orbital densities are present even for spherical atoms because of the non-linear orbital density dependence of the SIC (see also Ref. 186). His results for the $d^{n-1}s$–$d^{n-2}s^2$ separation are shown in Fig. 7. The

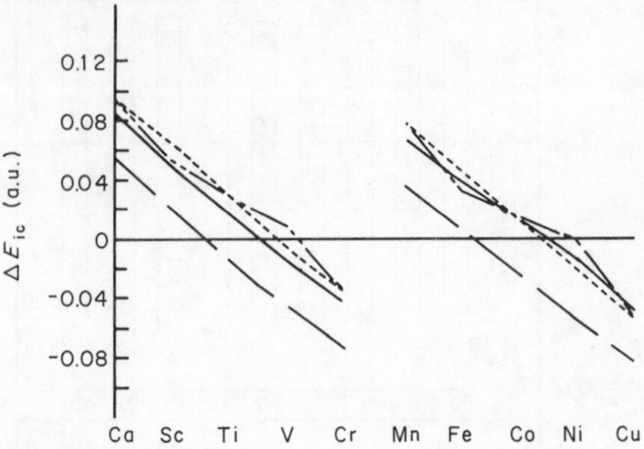

Fig. 7. Interconfigurational energy difference, $\Delta E_{ic} \equiv E[3d^{n-1}4s^1] - E[3d^{n-1}4s^2]$, for atoms of 3d series; -----, experiment; ————, experiment (sphericalized); —--—, SIC-LSD; — — —, SIC (spherical density), (*Reproduced from Ref. 79 by permission of the author and the American Institute of Physics.*)

agreement with experimental data, which was sphericalized because the central-field approximation does not allow a proper treatment of the term structure, is much better than for the sphericalized LSD or SIC-LSD results. Improved values are also found for the s spin-flip energies for the early members of the series Sc–Mn. For Fe to Ni the errors are similar to, or somewhat larger than, the LSD errors.

Baroni[97] has applied the exact exchange plus LSD correlation method (LSDX) to the 3d atoms. He solved the LSDX equations numerically, treating the correlation effects as a perturbation. No spherical averaging of the total or orbital electron densities was performed. Atomic states were chosen which can be represented by a single determinant. The LSD correlation corrections improve the HF $d^{n-1}s^1$–$d^{n-2}s^2$ separation, the correction being slightly overestimated for Sc–Cr and definitely underestimated for Mn–Cu, the final discrepancies for the latter being in the range 0.5–0.9 eV. The s spin-flip energies are underestimated by 0.05 to 0.35 eV. Ionization potentials are significantly improved over HF. Overall, the quality of these results is similar to that found in the SIC-LSD calculations of Harrison.

Finally, electron affinities of all three transition series have been calculated[187] using the sphericalized SIC-LSD method. The SIC is important for negative ions since in its absence many occupied eigenvalues are unbound. Agreement with experiment, while clearly much better than LSD, is only moderate. The calculated affinities for 3d electrons are too large by as much as 1 eV. The effects of removing the spherical constraints are currently under investigation[188].

In summary, the available calculations on transition-metal atoms confirm their status as complex many-electron systems, a correct description of which requires detailed consideration of electron correlation. The most extensive *ab initio* calculations so far performed still involve errors of several tenths of an electronvolt for important energy differences such as that between $d^{n-1}s$ and $d^{n-2}s^2$ configurations. The LSD method generally presents a reasonable semi-quantitative picture and interprets trends correctly; however, it yields quantitative errors in relative energies as large as 1 eV or so. Some of these errors can be reduced by considering (non-sphericalized) self-interaction corrections or by using HF exchange, both at the cost of more elaborate computations. Having molecular applications and dissociation energies in mind, I have argued that it might be worth while to examine the effects of the symmetry constraints on the LSD or SIC-LSD orbitals by making broken-symmetry calculations. In the final analysis, however, a large part of the utility of a molecular method depends on the information it provides for molecules. Given the state of atomic calculations, it is clear that molecular dissociation energies cannot yet be calculated with very high accuracy. We turn now to an evaluation of the performance of the various methods for the homonuclear transition-metal dimers.

IV. DIMERS

Calculations and experimental studies of the transition-metal dimers have previously been reviewed by Weltner and Van Zee[189] and by Shim[190]. The present section may be considered as a commented update of those reviews. Two very recent reviews[385,386] should also be mentioned. While I have tried to make the reference list complete, I will limit my comments for the most part to the most accurate available 'conventional' *ab initio* or density functional calculations.

For convenience, the known experimental data for the homonuclear dimers are gathered in Table V. The spectroscopic constants have been measured either in low-temperature matrices or in the gas phase, most recently in elegant molecular-beam experiments. Precise gas-phase constants are available for only a few of the dimers, V_2, Cr_2, Ni_2, Cu_2, Mo_2, Ag_2 and Au_2 (and vibrational frequencies for Fe_2, Co_2 and Re_2, although the moderate photoelectron resolution introduces quite large (15–20 cm^{-1}) uncertainties). If one discounts the noble metals this means that only four 'true' transition-metal dimers have been characterized with high precision in the absence of possible perturbing effects of a matrix. Of these, there are overlapping data from matrix and gas-phase studies for the vibrational frequencies of V_2, Cr_2 and Mo_2. V_2 and Mo_2 show only small shifts due to the matrix whereas the frequency of Cr_2 is displaced by about 40 cm^{-1} ($\sim 10\%$). Data, gas-phase or matrix, are very sparse for the 4d and 5d series. Given the intrinsic interest of these molecules and their importance as a demanding testing ground for computational methods, there is a clear need to complete the table. As we will see below, the

TABLE V

Experimental spectroscopic constants of transition-metal dimers. Adapted from Refs 189 and 190.

	G/M[a]	G.s.[b]	r_e (Å)	r^c_{bulk} (Å	ω_e (cm^{-1})	$\omega_e x_e$ (cm^{-1})	D^d_{spec} (eV)	D^e_{MS} (eV)	D^f_{bulk} (eV)	Refs[g]
Sc_2	M	$^5\Sigma$		3.25	238.91	0.93		1.65	3.93	193; 194; 195
Ti_2	M			2.91	407.9	1.08		1.41	4.86	–; 196, 197; 198
V_2	M			2.62	537.5	4.2		2.47	5.30	–; 196, 197; 198
	G	$^3\Sigma_g^-$	1.76		535		< 1.85			199; 199; –
Cr_2	M			2.49	427.5^h	15.75^h		2.0^q	4.10	–; 200; 201, 202
	M				438.0^i	14.5^i				–; 203; –
	M				78.6^j	0.4^j				–; 204; –
	G		1.68							–; 205–207; –
	G	$^1\Sigma_g^+$	1.68							–; 208; –
	G	$^1\Sigma_g^+$	1.68		$\approx 470^k$	$\approx 9^k$				–; 209; –
	G				$\approx 470^k$	$\approx 10^k$				–; 210; –
					$\approx 490^l$	$\approx 20^l$				
Mn_2	M	$^1\Sigma$		2.26^m	124.7	0.24		0.43	2.98	211–213; 194; 214
Fe_2	M		1.87^n	2.48				0.78	4.29	–; –; 215–217
	M		2.02^o							
	M				300.26^p	1.4				–; 194, 218;–
	G				300^s					–; 219; –
Co_2	M			2.50	290			0.95	4.39	–; 220; 221, 222
	G				280^s					–; 219; –
Ni_2	M			2.49	380.9	1.08		2.00	4.44	–; 223; 224, 225
	G	$^1\Gamma_g$ or $^3\Gamma_u$	2.20				2.07			226; 226; –
Cu_2	G	$^1\Sigma_g^+$	2.22	2.55	264.55	1.025		2.04	3.50	227–229; 227–229; 214
	G		2.22		266.1					–; 230, 231;–
Y_2				3.56				1.62		–; –; 232
Zr_2				3.16						
Nb_2				2.85	421	1.4		5.21	7.47	–; 233; 234
Mo_2	M			2.72	475.7			4.18	6.81	–; 235; 236
	G	$^1\Sigma_g^+$	1.93		477.1	1.51	4.12			207; 207; –
	G	$^1\Sigma^+$	1.94							237; 237; –

Tc_2				2.70						
Ru_2				2.65					6.62	
Rh_2				2.69				2.92		–; –; 238
Pd_2				2.75				1.03	3.94	–; –; 239
Ag_2	M			2.88	194			1.65	2.96	–; 240; 214
	G	$^1\Sigma_g^+$			192.4	0.643				241–243; 241–243; –
	G		2.48[r]							–; 244; –
Re_2	G				340[s]					–; 245; –
Pt_2	M			2.77	216[p]			3.71	5.85	–; 246; 247
Au_2	G	$^1\Sigma_g^+$	2.47	2.88	191	0.42		2.3		248, 249; 248, 249; 250, 251

[a]G = gas phase, M = matrix isolated.
[b]Ground state.
[c]Bulk nearest-neighbor separation—see Ref. 191.
[d]Spectroscopic value of D_e.
[e]D_0° determined from high-temperature mass spectrometric studies. Typical uncertainties cited are in the ± 0.2 eV range although some caution is necessary since the spectroscopic information on low-lying electronic excited states necessary for an accurate evaluation of the partition function is often lacking. See Ref. 192. The values shown have been corrected using results of *ab initio* calculations (see Ref. 190).
[f]Bulk cohesive energy per atom, taken from Table 1 of Ref. 189.
[g]Refs for G. s; $r_e, \omega_e, \omega_e x_e, D_{spec}; D_{MS}$.
[h]Argon matrix.
[i]Xenon matrix; results for a Kr matrix closely resemble those for argon.
[j]Progression observed when pumping $X^1\Sigma_g^+ \rightarrow A^1\Sigma_u$ transition; suggested as corresponding to either a member of the ground-state manifold of states with high spin multiplicity or more likely a long-bond form of Cr_2 corresponding to the molecule trapped in the outer minimum of a double minimum ground-state potential.
[k]Estimates from $v'' = 0$ and $v'' = 1$ using Pekeris relation and Morse curve; $\Delta G1/2 = 452.34\,cm^{-1}$.
[l]Estimated by ignoring quartic term in potential and obtaining $\omega_e x_e$ from α_e, which determines the cubic term.
[m]Mn has a complicated (simple cubic) crystal structure with four inequivalent atomic positions and 58 atoms in the unit cell. See Table 25–3 of Ref. 191.
[n]Extended x-ray absorption fine structure (EXAFS) in argon.
[o]EXAFS in neon.
[p]$\Delta G1/2''$.
[q]Kant and Strauss[201] reported 1.56 ± 0.3 eV, based on old assumptions for r_e, ω_e and spin. Goodgame and Goddard[202] have corrected this using $r_e = 1.68$ Å, $\omega_e = 470\,cm^{-1}$ and an electronic degeneracy of 1.
[r]Value assumed in the spectroscopic analysis rather than a directed determination. The good fits obtained corroborate the value.
[s]From negative-ion photoelectron spectroscopy.
[t]Bondybey (private communication, 1986) has compelling evidence that the spectrum reported in Ref. 223 is due to Se_2.

electronic structure of the transition-metal dimers can be very subtle indeed, depending on fine balances among various components of the total energy. I am sure that the considerable effort and ingenuity which will be required to complete the picture using beam spectroscopy or other techniques would be well rewarded by a number of surprises and by the new insight such measurements would force theory to provide. I hope, notwithstanding the enormous current interest and excitement about the chemistry of more complex cluster beams[252–254], that some of the energy of the field will be devoted to the isolated dimers.

Before discussing calculations on the individual molecules, it may be helpful to have in mind several general features and trends. The first is that an adequate treatment of correlation is absolutely crucial to obtaining even a proper qualitative description of the binding and other properties of these molecules. The Hartree–Fock wavefunction is, in general, a poor starting point. For example, Cr_2 is unbound by about 20 eV[255]. On the density functional side, exchange-only or $X\alpha$ potentials are inadequate. $X\alpha$ calculations on Cr_2 yield a bond which is much too weak and much too long[256–258]. The question of how best to include the essential correlation effects for the whole series is, despite the considerable progress outlined below, still very much an open one.

As we saw in the last section, the transition-metal atoms are characterized by low-lying $d^{n-2}s^2$ and $d^{n-1}s$ configurations. Many of the intriguing properties of molecules or solids involving bonds between these atoms arise from the different range of the d and s electrons. The s orbitals are diffuse and the longest-range interactions result from their overlap. In the $d^{n-2}s^2$ configuration this leads to a 'closed-shell:closed-shell' repulsion so that molecular or solid-state configurations are typically closer to $d^{n-1}s$. The d electrons are more compact and, very importantly, their range decreases along with the screening efficiency as one proceeds from the left to the right or from the bottom to the top of the Periodic Table. Furthermore, d–d bonding can lead to bond lengths which are too short for optimum s–s overlap so that, in MO language, the simple fact that the bonding $\sigma_g(s)$ orbital is occupied does not necessarily indicate a stabilizing contribution of the s electrons to the bond energy. Through hybridization or configurational mixing this may even lead to a significant involvement of the higher-lying d^n atomic configuration. Indeed, it has been proposed that the d^{10} configuration of Ni contributes significantly in the NiN_2 molecule[259].

The presence of magnetism for the solid transition metals provides vital information on the nature of the metal–metal bonding. Ferromagnetism is observed in bulk Fe, Co and Ni. Bulk Cr and Mn have complex magnetic structures with dominant antiferromagnetic interactions. The remaining transition metals are non-magnetic ('Pauli paramagnetic' in the language of the solid-state physicists), although palladium has a very high magnetic susceptibility and is essentially on the verge of being magnetic. $X\alpha$-SW and

LSD-SW calculations on small clusters[20,21,260–263] have yielded similar, usually slightly larger, magnetic moments and the same type of ordering as for the bulk materials, indicating that the fundamental interactions are determined on a molecular scale.

If one views the transition-metal block globally, magnetism is present to the right and to the top, that is, where the d electrons are most contracted. I believe it is useful to view this as a result of a competition between, on the one hand, the magnetic ('exchange') stabilization associated with high-spin atoms and, on the other, the stabilization associated with chemical bonds, which pair spins and depend crucially on correlation. The ground-state and other properties of a given dimer depend on a delicate balance.

The transition-metal atoms typically have several unpaired electrons in a high-spin arrangement, as high as ^7S for Cr. If efficient d overlap can be achieved (in a correlated fashion) then the magnetism can be quenched in favor of chemical bonding. Otherwise, the molecule can retain some of its ('atomic') high-spin character and help to minimize its energy through these terms.

To the left and the bottom of the d block, the d orbitals are relatively extended and, all other things being equal*, one would expect the 'magnetism' to be quenched. For these cases molecular-orbital language (though not necessarily HF calculations) provides a reasonable basis for discussion. To the right and the top of the d block where the d orbitals are compact, 'magnetic' aspects must be dealt with. This can mean either high-spin molecular states (ferromagnetism) or some 'antiferromagnetic' character for singlet states. In the latter case a symmetry-adapted MO calculation (either HF or local density) does not provide a useful zeroth-order approximation. In terms of calculations, this implies the necessity of some sort of orbital localization. In CI or GVB type calculations the simplest such case is the left–right correlation obtained by including double excitations from bonding to antibonding orbitals. The GVB orbitals localize to an optimum extent on one center or the other. The 'antiferromagnetic' spin arrangement ($\alpha\beta$) is properly symmetrized to form a singlet pair $(1/\sqrt{2})(\alpha\beta - \beta\alpha)$.

Within the framework of the local density method, which is strictly an orbital theory, the antiferromagnetic state can be attained by reducing the symmetry constraints imposed on spin-polarized calculations, hence allowing the spin orbitals to localize and local magnetic moments to persist, if it is variationally favorable to do so. While I do not know of any formal justification for this type of symmetry breaking (one cannot just mix determinants within DFT to obtain proper spin and space eigenfunctions), the results discussed below for Cr_2, Mo_2, and Mn_2 certainly indicate that it is a reasonable approach. A rough rationalization can be obtained if one reasons

*The first rule of thumb of transition-metal quantum chemistry is that, more often than not, 'all other things' are not equal.

that in an LSD calculation a given electron 'sees' reasonable exchange and correlation potentials on some local scale but is indifferent to the question of the global spin and space eigenstates, except for those aspects which are communicated through the occupations and forms of the Kohn–Sham orbitals.

The final general trend which should be mentioned concerns the effects of special relativity on the geometry and electronic structure of the dimers. Non-relativistic calculations are acceptable for most purposes for the 3d series. For the 4d dimers, small but non-negligible relativistic contractions of bond lengths have been calculated[132,264,265], typical values lying in the 0.05 to 0.10 Å range. (For some recent reviews or key papers on relativistic calculations see Refs 266–271.) These contractions can be faithfully incorporated into a model core potential by adjusting parameters to fit the results of relativistic atomic calculations. For the 4d series, neglect of the spin–orbit coupling, that is inclusion of only the scalar ('mass–velocity' and Darwin) corrections, represents a good approximation. For the 5d series, the scalar relativistic corrections, the atomic spin–orbit coupling and chemical binding energies are all of the same order of magnitude so *a priori* none of them can be ignored. There has so far been very little work on the 5d transition-metal dimers. In particular, I do not know of any attempt to include spin–orbit effects in an *ab initio* fashion. Given, among other things, the importance of the 5d metals in catalysis, I believe that efforts in this direction would be well rewarded.

With these general comments in mind we may now turn to a review of the calculations so far performed for the dimers.

A. 3d Dimers

1. Sc_2

(See Refs 118 and 272–280.) Electron spin resonance (ESR) measurements[193] of Sc_2 at 4 K in neon and argon matrices have revealed a $^5\Sigma$ state. This agrees with the density functional calculations of Harris and Jones (HJ)[118] who included spin-polarization effects perturbatively. They found $^5\Sigma_u^-$ $1\sigma_g^2 1\sigma_u^1 1\pi_u^2 2\sigma_g^1$ as the lowest state, bound by 1.8 eV. Given the tendency of LSD to favor d-electron-rich configurations for the atoms (see Section III), they argued that the ground state might be $^3\Sigma_g^-$ $1\sigma_g^2 1\sigma_u^2 1\pi_u^2$. They state further that, for technical reasons, the singlet $1\sigma_g^2 1\pi_u^4$ state could not be traced to its minimum but is certainly strongly bound. The vibrational frequencies, 200 cm^{-1} for $^5\Sigma_u^-$ and 235 cm^{-1} for $^3\Sigma_g^-$, are both reasonably close to the experimental value (239 cm^{-1}).

The early CI calculations of Wood *et al.*[272] indeed predicted a $^5\Sigma_u^-$ state as being the most stable of those states considered. However, others[190,276] have pointed out that this result was not convincing because of errors in the atomic limits. The $^5\Sigma$ state was ignored[274,275] until the appearance of the experimental work. In recent complete active space SCF (CASSCF) calculations, Walch and Bauschlicher[276] have found that a $^5\Sigma_u^-$ state of dominant $1\sigma_g^2 1\sigma_u^1 2\sigma_g^1 1\pi_u^2$ character is bound by about 0.44 eV. Their calculated value for r_e (2.79 Å) agrees roughly with the LSD value (2.70 Å) of HJ. The calculated vibrational frequency (184 cm^{-1}) is only in moderate agreement with the experimental (matrix-isolated) value.

In a recent DVM-$X\alpha$ study, Fursova *et al.*[277] argue that the ground state of Sc_2 should be $^1\Sigma_g^+$ $1\sigma_g^2 1\pi_u^4$ (the configuration which HJ could not follow to its minimum), with a bond length of 2.21 Å. They point out that the disagreement with the ESR $^5\Sigma$ ground state could be due either to matrix perturbations or to the fact that a $^1\Sigma$ state would be invisible to ESR and that the observed spectrum might be due to a minority of molecules in a $^5\Sigma$ excited state. Some aspects of these calculations can be criticized. It is not clear whether proper spin-polarized calculations were performed and the use of the $X\alpha$ potential is questionable. I believe, however, given the approximations made in the density functional calculations so far performed, along with the usual tendency of *ab initio* approaches to favor high-spin states if the correlation treatment is not sufficiently extensive and balanced, that the question of the ground state of isolated Sc_2 is not yet closed.

Very recently Jeung[278] has reported on multi-reference doubles-CI (MRDCI) calculations for the $^5\Sigma_u^-$ and $^1\Sigma_g^+$ states. In these calculations the $^1\Sigma_g^+$ state has a short bond (2.28 Å without and 2.36 Å with the Davidson correction) and lies about 1 eV above the $^5\Sigma_u^-$ state. Trinquier and Barthelat[279] have also performed calculations on these states, although details have not yet been published.

2. Ti_2

(See Refs 118, 273, 277 and 280.) There have been only a few non-empirical calculations for Ti_2. HJ[118] found several low-lying states of high spin, the lowest, a $^7\Sigma_u^+$ state, corresponding to the configuration $1\sigma_g^2 2\sigma_g^1 1\pi_u^2 1\delta_g^2 1\sigma_u^1$. Given the deficiency of this approach for the next member of the series, V_2 (see below), where the stability of high-spin states is greatly overestimated, the HJ results for Ti_2 should probably be disregarded. More recently the DVM-$X\alpha$ calculations of Fursova *et al.*[277] yielded a $^1\Sigma_g^+$ state ($1\pi_u^4 1\sigma_g^2 2\sigma_g^2$) with an equilibrium distance of 1.96 Å. RHF calculations[273] yield $r_e = 1.87$ Å and $\omega_e = 580$ cm^{-1}. The difference between the latter value and the observed (matrix-isolated) frequency of 408 cm^{-1} provides some indication of the need to include correlation. The only correlated calculations I am aware of are in a brief report by Walch and Bauschlicher[280]. They considered two states: $^7\Sigma_u^+$ $1\sigma_g^2 2\sigma_g^1 1\sigma_u^1 1\pi_u^2 1\delta_g^2$ and $^1\Sigma_g^+$ $1\sigma_g^2 2\sigma_g^2 1\pi_u^4$. At the CI level (more than 180 000 configuration state functions) the former lies lower by 0.4 eV and has $r_e = 2.63$ Å, $\omega_e = 205$ cm^{-1} and $D_e = 1.53$ eV with respect to the $d^3s + d^2s^2$

asymptote (about 0.7 eV with respect to two ground-state, d^2s^2, atoms). The $^1\Sigma_g^+$ state has a triple 3d bond, $r_e = 1.97$ Å, $\omega_e = 438\ cm^{-1}$ and $D_e = 1.94$ eV with respect to two d^3s atoms (about 0.3 eV with respect to ground-state atoms). Since the vibrational frequency calculated for the $^1\Sigma_g^+$ state agrees rather well with the observed value, and since CI calculations usually describe high-spin states better than low-spin states, the authors tentatively assign the ground state as $^1\Sigma_g^+$. Clearly, much more work, both experimental and theoretical, is needed for this molecule.

3. V_2

(See Refs 118, 273 and 281–285.) The vanadium dimer is the first of the series for which theory must confront the hard facts of beam spectroscopy[199]. The ground state is $^3\Sigma_g^-$ with an equilibrium spacing of 1.76 Å, a vibrational frequency of 535 cm^{-1} ($\Delta G''_{1/2} = 529.5\ cm^{-1}$ for $^3\Sigma_{0g}^-$ and 540.5 cm^{-1} for $^3\Sigma_{1g}^-$) and a dissociation energy less than, but probably close to, 1.85 eV. There is evidence for a low-lying $^1\Sigma_g^+$ state.

The early LSD calculations of HJ do very poorly, predicting a $^9\Sigma_u^-$ ground state with a bond length of 2.65 Å and a vibrational frequency of 230 cm^{-1}. With the benefit of hindsight this can probably be attributed to the same technical problem mentioned above for Sc_2 (essentially the inability to handle situations in which the semi-core 3s or 3p electrons extend beyond the muffin-tin radius) which prevented examination of the true ground-state configuration in the region of the minimum.

Walch *et al.*[281] performed CASSCF calculations, constraining the orbital occupancies to four σ electrons, two π_y, two π_x and one each in δ_{xy} and $\delta_{x^2-y^2}$. The dominant configuration is $1\sigma_g^2 2\sigma_g^2 1\pi_u^4 1\delta_g^2$, which correlates with two d^4s^1 vanadium atoms, the ground state being d^3s^2 (^4F) with d^4s (^6D) lying 0.25 eV higher. In accordance with Hund's rule the open δ shell leads to a triplet ground state, $^3\Sigma_g^-$, in agreement with experiment. The calculated values, $r_e = 1.77$ Å and $\omega_e = 593.6\ cm^{-1}$, are in very acceptable agreement with their experimental counterparts. At $r = 3.00a_0$ (1.59 Å) the contribution of the dominant configuration is 76%, which can be taken as some measure of the large effect of correlation for this molecule. The calculated binding energy is only 0.33 eV, indicating that by no means all of the important correlation effects are included at this level of theory although, as discussed in Section III, some of the problem may lie with the atoms. It was also found that 4f polarization functions have a considerable effect on the results, a point to which we will return in the discussion of Cr_2 (see also Ref. 285 and references therein).

Das and Jaffe[282] proposed corrections to the CASSCF results in a rather complex scheme involving partially localized orbitals (PLOs). They took the CASSCF results as a starting point and, at each nuclear separation, developed a 'PLO analog' of the CASSCF wavefunctions. The predominant configurations were selected, leading to a zeroth-order wavefunction, ψ_0. Then the various higher-order corrections (ostensibly those not included in CASSCF) were computed using ψ_0 and added to the CASSCF energies of Walch *et al.*[281]. The wavefunction ψ_0 contained 'antiferromagnetic' and 'first- and second-order charge-transfer' terms. The final estimates of $r_e = 1.79$ Å, $D_e = 1.45$ eV and $\omega_e = 556.7\ \mathrm{cm}^{-1}$ represent a slight worsening for r_e, and considerable improvements for D_e and ω_e. It remains to be seen whether such configurations could be incorporated into a viable complete computational scheme.

Fully self-consistent spin-polarized LSD calculations have been performed[283] using the LCGTO formalism. The correct triplet ($\alpha\alpha$) ($1\pi_u^4 1\sigma_g^2 1\delta_g^2 2\sigma_g^2$) ground state was found with $r_e = 1.75$ Å and an estimated vibrational frequency of $594\ \mathrm{cm}^{-1}$, both in good agreement with experiment. The corresponding $\alpha\beta$ configuration (half singlet, half triplet if one thinks in determinantal terms) is found about 0.4 eV higher. The latter calculations were performed with reduced, $C_{\infty v}$, symmetry constraints to allow the orbital localization necessary for proper dissociation. The symmetry broke at around 2 Å, that is, well outside the minimum so that within the LSD framework the ground state is well described in 'ordinary' molecular-orbital terms—of course with correlation incorporated through the LSD approximation. The calculated dissociation energy is 3.85 eV with respect to spherical atoms in the calculated LSD ground-state configuration, d^4s within the spherical approximation. This configuration lies about 1.45 eV below the spherical approximation for d^3s^2. That much of the overestimate of D_e may be attributed to problems with the atoms is indicated by the fact that calculations on a spherical, d^5, state of vanadium, coupled with the experimental d^5–d^3s^2 spacing, reduce the estimate of D_e to 2.15 eV. The above results are for the JMW potential which, for Cr_2, yielded the value of D_e closest to experiment for the various correlated potentials examined. Results for V_2 using other potentials will be published in due course.

Finally, it should be mentioned that Klyagina and Levin[284] have performed DVM-$X\alpha$ calculations for V_2, apparently with good results.

4. *Cr*$_2$

(See Refs 113, 118, 129, 146, 202, 255–258, 272, 273, 281, 282 and 284–291.) The Cr_2 molecule has become somewhat of a *cause célèbre* (or a *bêre noire*, depending on how one's favorite computational method performs). The balance between exchange and correlation, or between magnetism and chemical bonding, is particularly delicate in this region of the Periodic Table and it is highly dependent on the internuclear spacing. For the following discussion it will be helpful to have in mind the two extreme cases. The Cr atom has a high-spin, d^5s (^7S), ground state (a magnetic moment of $6\mu_B$). At long internuclear distance one can usefully described Cr_2 as involving an antiferromagnetic arrangement of these atomic moments. At the opposite extreme, if one assumes that chemical bonding (spin pairing) will dominate the magnetic aspects then, for the region around the minimum, molecular-orbital language (though not necessarily the HF method) may be used. This leads to the description of Cr_2 as a hexuple bond in which all of the bonding MOs are fully occupied, $1\sigma_g^2 2\sigma_g^2 1\pi_u^4 1\delta_g^4$. In reality, as we shall see, Cr_2 involves both of these aspects and accurate computations must allow for the optimal mixture of them.

Finding this optimal mixture of extremely different binding situations has proven to be of enormous difficulty for 'conventional' *ab initio* techniques. In the restricted Hartree–Fock (RHF) approximation, Cr_2 is unbound by about 20 eV at the experimental internuclear distance[255,286]. The unrestricted Hartree–Fock (UHF) approach, which allows for the antiferromagnetic state, leads to a long very weakly bound molecule[286]. Perfect-pairing GVB results are also unbound, by about 10 eV for all distances studied[286]. In their early GVB work, Goodgame and Goddard[286] (GG) found that it was necessary to add van der Waals (vdW) configurations. In this manner they obtained a weak ($D_e = 0.3_5$ eV), long ($r_e = 3.0_6$ Å) bond with a vibrational frequency of 110 cm^{-1}. (These results agree quite closely with those of an MCSCF study by Atha and Hillier[287].) GG argued that the discrepancy with the mass spectrometric value[201] of D_e (1.56 eV) was due to errors in the partition function and that the disagreement with the value of Efremov *et al.*[205–207] for r_e (1.68 Å) could be attributed to an improper assignment of the observed spectrum to Cr_2. They[202] offered their results on Cr_2 and Mo_2 as a challenge for the local density approximation based on comparisons with the results of HJ[118] (see below).

The beam spectroscopic results of Michalopoulos *et al.*[208] removed any possibility that Efremov *et al.*[205–207] has observed a species other than Cr_2. The goal for computations was now clear, $r_e = 1.68$ Å, $\omega_e = 470$ cm^{-1} and $D_e = 2.0$ eV (see footnote *q* of Table V). The very high anharmonicity (10–20 cm^{-1}) is also an important aspect of the binding in Cr_2, which must be accounted for in an accurate description.

Several *ab initio* studies of Cr_2 have subsequently been performed. Kok and Hall[288] applied the generalized molecular-orbital (GMO) method. They found quite nice results for r_e (1.73 Å) and ω_e (396 cm^{-1}) although this method does not allow for proper dissociation. Given the results from MCSCF calculations at a similar level (see below) it is likely that Cr_2 is unbound in the GMO approximation. Shim[190] has also raised the possibility of a large basis-set superposition error.

McLean and Liu[255] examined the role of f and g polarization functions at the Hartree–Fock level. They find that these polarization functions make a contribution of 0.93 eV, which they compare with the experimental dissociation energy of Kant and Strauss[201] (1.56 eV). Such a large effect, if it were also present in the real Cr_2 molecule, would be surprising, to say the least, since it implies a binding energy contribution of f and g functions which surpasses that of the d and s orbitals. It is likely that this large polarization contribution is an artefact of the model chosen. Indeed, the RHF energy is unbound by about 20 eV, putting this model in an energy region appropriate to highly excited orbitals. Since the treatment of correlation is so inadequate, it is likely that the RHF method attempts to compensate, to keep electrons apart as best it can, by overemphasizing the weight of high l functions with their more complicated angular factors. Indeed, Walch *et al.*[281] find that the effect of f functions is slightly smaller at the CASSCF level than at the SCF level. Correlated LSD calculations[126] (see below) show only minor changes when f functions are included in the basis.

Walch *et al.*[281] performed CASSCF calculations involving 3088 configurations. At this level of theory, Cr_2 is unbound and does not show a minimum. The authors estimate that, to achieve a reasonable description within this type of approach, a multireference singles and doubles CI based on the CASSCF function would involve 57 million configurations. The PLO corrections of Das and Jaffe[282] are also insufficient to yield a bound molecule.

Very recently Walch[280] has applied a similar, though more extensive, 'additive' scheme and does generate a 'reasonable' curve ($r_e = 1.78$ Å, $\omega_e = 383\ \mathrm{cm}^{-1}$ and $D_e = 0.71$ eV). He states that this is due at least in part to cancellation of errors, i.e. the overestimation of 3d′ (in–out) correlation and the omission of other molecular correlation effects.

The latest installment from Goodgame and Goddard (GG)[290] involves a semi-empirical correction of the atomic self-Coulomb energy. GG reason that the correlation error in GVB calculations arises primarily from the negative ionic terms. Decreasing the one-center s–s and d–d self-Coulomb integrals by 0.67 eV and 2.55 eV respectively and calculating all other integrals exactly, they obtain an equilibrium distance of 1.61 Å and a dissociation energy of 1.86 eV. While these results are certainly better than the previous GVB-vdW results, there is still much room for improvement on r_e. Moreover, while GG do not report values for ω_e or $\omega_e x_e$, it is clear from their figure that the calculated frequency is far too high and that the curve is much too harmonic. GG also find a second minimum, at long distance, which corresponds to the antiferromagnetic situation from the previous GVB-vdW calculations. (Interestingly the previously predicted double minimum for Mo_2 disappears in the corrected version of the GVB calculations.) A double-minimum curve has also been proposed by Moskovits *et al.*[204] as a possible explanation for an observed band of Cr_2 in a matrix (see footnote *j* of Table V). However, given that matrix perturbations are particularly large for Cr_2 and that explanations for the spectrum other than a double minimum have also been proposed, it would be premature to invoke the matrix spectroscopy results as support for or against the existence of a double minimum for gas-phase Cr_2.

To summarize, at the present time, no 'conventional' *ab initio* calculation has been able to yield accurate spectroscopic constants for Cr_2. The source of the problem is clear in general terms—more correlation is needed and the types of correlation terms needed have been at least partly defined[281,282,290]. However, the non-empirical implementation of calculations including these terms appears to be beyond the limits of present-day technology. (See also Ref. 388.)

We now turn to the alternative offered by density functional theory.

The early calculation of HJ[118] (which were challenged by GG) did not allow the possibility of an antiferromagnetic configuration, although such a possibility was envisaged in footnote 28 of their paper. The challenge of Goodgame and Goddard was quickly taken up by a number of groups working with local density theory.

Dunlap[256] performed $X\alpha$ calculations using reduced symmetry constraints to allow for the antiferromagnetic configuration. He found results reasonably consistent with the GVB-vdW picture; that is, a long weak bond. In retrospect, this shows the importance of symmetry breaking, as well as the inadequacy of the $X\alpha$ potential.

Several calculations[113,125,126,146] have now been performed using correlated LSD potentials. Delley *et al.*[113] were the first to publish LSD results that confronted not only Goodgame and Goddard's challenge but also the experimental data of Efremov *et al.* The correlated LSD calculations have been performed by four different groups using three different computational techniques (LCGTO[125,126], DVM[113] and a Gaussian pseudo-potential approach[146]). The overall good agreement amongst the various techniques provides reassurance that the technical aspects (basis sets, fitting functions, sampling points, etc.) are under control. Since there is agreement, I will use our own results as illustrations.

Calculated binding energy curves are shown in Fig. 8 for a variety of exchange–correlation potentials. If we focus on the LSD limit (the VWN potential) then the calculated values of $r_e = 1.68$ Å and $\omega_e = 441\ \text{cm}^{-1}$ are in very satisfying agreement with their experimental counterparts. Values for other correlated potentials are similar; those for $X\alpha$ are very different. The curves are clearly anharmonic, a rough estimate giving $\omega_e x_e = 15\ \text{cm}^{-1}$, although the precise value is sensitive to the curve-fitting procedure. The binding energy of 2.6 eV at the LSD limit is overestimated by about 0.6 eV (25%). Since the Cr atom is spherical (^7S), this cannot be attributed to the central-field approximation for the atoms and provides an indication of the room for improvement left for non-local corrections.

Beyond these quantitative results, the picture of the binding in Cr_2 which has emerged from these calculations, while subtle and delicate, is appealing in its simplicity. If one starts with infinitely separated atoms, then the antifer-

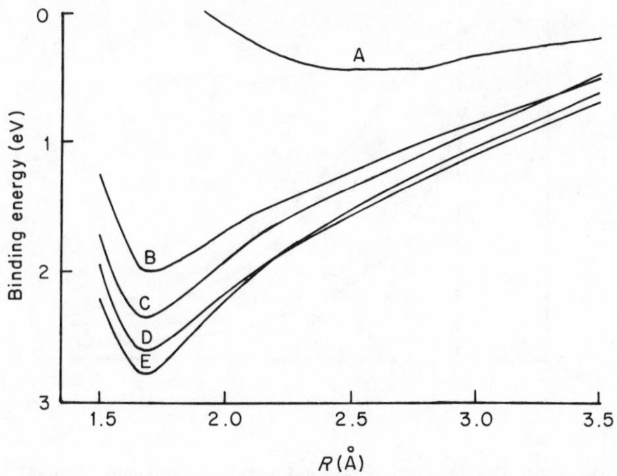

Fig. 8. Calculated binding energy versus internuclear distance for Cr_2 for various exchange–correlation potentials: (A) $X\alpha$, Ref. 12; (B) JMW, Ref. 29; (C) GL, Ref. 30; (D) VWN, Ref. 31; and (E) BH, Ref. 14. (*Reproduced from Ref. 125 by permission of Taylor & Francis Ltd.*)

romagnetic description applies. As the atoms approach each other, the spin orbitals leak over onto the other center, where the opposite spin is predominant. This delocalization and spin pairing corresponds to bond formation, which is in competition with the 'magnetic' interactions favoring high-spin atoms. The degree of delocalization depends crucially on the treatment of correlation employed. For example, $X\alpha$ has an exchange term which is far too strong relative to any correlation that is 'snuck in' by adjusting α. As a result, the orbitals remain too localized and the local magnetic moments remain too high. Bonding, basically, has no chance to occur before the repulsive terms which dominate at short internuclear separations take over. The degree of localization also varies from orbital to orbital, as is shown in Fig. 9 for calculations using the correlated JMW potential. At r_e, the σ and π orbitals are almost completely delocalized whereas the δ orbital is still symmetry-broken. Moving away from r_e leads to a rather rapid localization of the orbitals, especially δ, and we have attributed the observed large anharmonicity to this rapid change in character of the wavefunctions near r_e. Essentially, the 'availability' of the 'magnetic' stabilization keeps the potential curve from rising, as r increases, as quickly as it would in the absence of such strong magnetic effects.

We also performed symmetry-adapted calculations for the hexuply bonded configuration. While the calculated values of r_e and D_e are reasonably close to the broken-symmetry results, the vibrational frequency is far too high and the anharmonicity far too low. In the final analysis, Cr_2 should be regarded neither as a hexuply bonded nor as an antiferromagnetic dimer. Its correct description

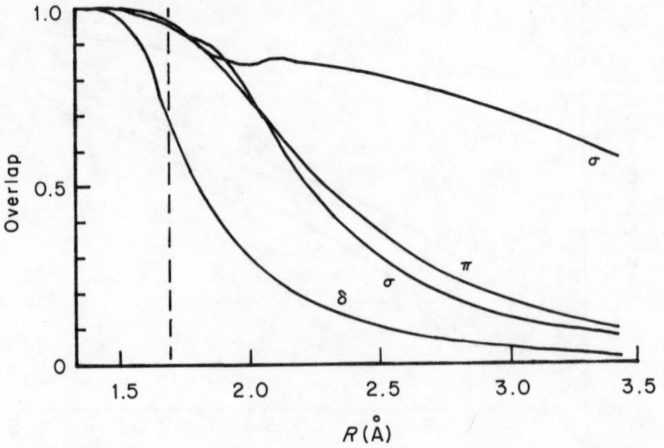

Fig. 9. Overlap between corresponding spin-up and spin-down molecular orbitals versus internuclear distance for Cr_2 using the JMW (Ref. 29) exchange–correlation potential. The broken line marks the calculated equilibrium distance. (*Reproduced from Ref. 125 by permission of Taylor & Francis Ltd.*)

contains both magnetic and multiple bonding aspects.

Finally, mention should be made of some very recent calculations which go beyond the local level. Becke[291] has applied his fully numerical technique to Cr_2 at its experimental internuclear distance. He calculates a dissociation energy of 3.0 eV at the LSD level, 0.9 eV with the gradient correction of Langreth and Mehl[105] and 1.7 eV with his semi-empirical gradient correction. Ziegler *et al.*[292], using the latter potential in DVM calculations, found $r_e = 1.65$ Å and $D_e = 1.75$ eV. Further work along these lines would be most welcome for this and the other dimers.

5. Mn_2

(See Refs 118, 273, 283, 293 and 294.) The earliest *ab initio* calculations for Mn_2, or for that matter for any transition-metal dimer, were performed by Nesbet[293] in 1964. He calculated a $^1\Sigma_g^+$ $1\sigma_g^2 1\pi_u^4 1\delta_g^4 2\sigma_g^2 1\sigma_u^2$ state at the RHF level and then used these orbitals to calculate the energies of various high-spin states derived by spin-flip excitations to the empty antibonding levels. The lowest state obtained was $^9\Sigma_g^+$ $1\pi_g^2 1\delta_u^2 1\pi_u^2 1\delta_g^2$ for the three distances considered ($4.5a_0$, $5.0a_0$ and $5.5a_0$). This corresponds to a situation in which the σ orbitals remain delocalized but the π and δ orbitals localize on either center. A Heisenberg-like treatment of the resulting $S = 2$ atoms yielded an (antiferromagnetic (AF)) $^1\Sigma_g^+$ ground state with an equilibrium spacing of 2.88 Å and an estimated binding energy of 0.79 eV. The Heisenberg exchange parameter J was estimated to be $-4\,\text{cm}^{-1}$. This state has since been observed by matrix spectroscopy; by electron spin resonance spectroscopy (ESR), yielding $J = -9 \pm 3\,\text{cm}^{-1}$ (Refs 211 and 212); by magnetic circular dichroism (MCD), providing $J = -10.3 \pm 0.6\,\text{cm}^{-1}$ (Ref. 213); and by resonance Raman spectroscopy[194], furnishing a vibrational frequency of $125\,\text{cm}^{-1}$.

The only other 'conventional' *ab initio* results so far reported are in the RHF study of Wolf and Schmidtke[273] who found $r_e = 1.52$ Å and $\omega_e = 680\,\text{cm}^{-1}$ for a $^1\Sigma_g^+$ state. There is no reason to expect any resemblance between these results and reality.

On the LSD side, HJ[118] found two states $^{11}\Pi_u$ and $^{11}\Sigma_u^+$ with nearly the same binding energies (1.25 eV), similar values of r_e (2.66 and 2.70 Å respectively) and similar vibrational frequencies (220 and $210\,\text{cm}^{-1}$). The $^9\Sigma_g^+$ state, which was previously studied by Nesbet, had an r_e of 2.62 Å, in fair agreement with Nesbet's value (2.88 Å), and a vibrational frequency of $235\,\text{cm}^{-1}$. HJ found this state to be unbound by about 0.05 eV.

Fully self-consistent, broken-symmetry LSD calculations[283] on the AF $(1\sigma^2 1\delta^4 1\pi^4 2\sigma^2 3\sigma^2)$ configuration yielded $r_e = 2.52$ Å, $\omega_e = 144\,\text{cm}^{-1}$ and $D_e = 0.86$ eV. The r_e value is in the same range as those of Nesbet and of HJ and the frequency is in very reasonable agreement with the matrix Raman value. At the equilibrium distance, all of the orbitals are of broken symmetry (see Fig. 2 of Ref. 283), the dδ functions being almost completely localized and dσ and dπ highly so.

Several other configurations have also been examined[283,294] in which the last two electrons are put in δ or π orbitals rather than in 3σ. While the results have not yet been analyzed in detail, the preliminary indications are extremely intriguing. A triplet ($\alpha\alpha$) state is found for the $\cdots 2\delta^2$ configuration with $r_e = 1.67$ Å, $D_e = 0.98$ eV, $\omega_e = 729\,\text{cm}^{-1}$ and symmetry-adapted orbitals, whereas the AF ($\alpha\beta$) state of $\cdots 2\pi^2$ has $r_e = 2.15$ Å, $D_e = 0.72$ eV and $\omega_e = 233\,\text{cm}^{-1}$ with broken-symmetry orbitals. Hence, the LSD model predicts at least three low-lying states, all within 0.3 eV of each other, but with vastly different characteristics (these are not just the various spin states of a Heisenberg manifold). The last mentioned states correlate with the d^6s asymptote. Given the overall success of the LSD approach and the known idiosyncracies of Mn–Mn bonds, as evidenced by the unusual structural and magnetic properties of the solid, I believe that there is every reason to expect the unexpected and that an experimental search for these states would be well worth while.

6. *Fe*$_2$

(See Refs 118, 188, 190, 273 and 295–298.) Apart from some SCF calculations[273,297], the only conventional *ab initio* work on Fe_2 has been performed by Shim and Gingerich[190,296]. In their 1982 paper[296] they used Wachters' basis set[158], without a diffuse d function, optimized the MOs for a

$^7\Sigma_u^+$ state ($1\sigma_g^1 1\pi_u^4 1\delta_g^2 1\delta_u^2 1\pi_g^4 1\sigma_u^1 2\sigma_g^2$) and performed a CI which allowed reorganization of the d electrons while the $2\sigma_g$(s) orbital was kept doubly occupied. A plethora of low-lying states resulted; about 50 lying within a tenth of an electronvolt of the ground state ($^7\Delta_u$) for an internuclear distance corresponding to that of bulk iron, 2.48 Å. At this level of theory, an equilibrium distance of 2.40 Å and a vibrational frequency of 204 cm^{-1} were calculated. The molecule is 0.69 eV more stable than the d^7s^1 (^5F) limit from which it is derived, but unbound with respect to ground-state, d^6s^2 (^5D), iron atoms. Inclusion of a diffuse d function in the basis[190] changed the bond length to 2.64 Å (even longer than bulk iron) and the frequency to 134 cm^{-1}. The molecule is now bound by only 0.04 eV with respect to the d^7s^1 limit and (even more?) unbound with respect to d^6s^2. Both the bond length and the frequency differ considerably from their experimental (matrix) counterparts. The two values of r_e derived from extended x-ray absorption fine structure (EXAFS) studies on argon and neon matrices differ by 0.15 Å and this may be taken as an indication of significant matrix perturbations. Presumably the value for the less polarizable neon matrix would more closely resemble the bond length of the isolated molecule. In fact LSD calculations (see below) on a $^7\Delta$ state yield a bond length of 2.01 Å, in close agreement with the EXAFS result for the neon matrix. In view of these results and of the relatively modest level of CI included in the *ab initio* calculation, as evidenced by the lack of binding, the presence of all these very low-lying states must be questioned. Indeed, it is hard to reconcile the crowded spectrum of Fe_2 predicted by these CI calculations with the very clean photoelectron spectrum of Fe_2^- recently reported by Leopold and Lineberger[219], which shows only two bands in the region of interest. In all likelihood, a more adequate treatment of correlation would allow the d orbitals to overlap more, leading to a decrease of the bond length and a relative stabilization of some of the states. Indeed, this is precisely what happens in Shim's calculations[190] for Ru_2, the 4d congener of Fe_2, for which the d electrons are relatively more diffuse. The ground state, $^7\Delta_u$, splits off the bottom of the band by about 0.15 eV at this level of theory. I cannot agree with Shim's statement[190] for Fe_2 that 'we expect that larger CI calculations would make the molecule bound without changing significantly the description of the chemical bond'. A bond length change of over 0.5 Å is clearly significant, as is the participation of the d electrons in the bonding which the above arguments imply. Clearly, more work is needed before an adequate *ab initio* description of this 'microferromagnet' can be claimed.

The picture is also far from complete from the density functional point of view. The early calculations of HJ[118] predict a $^7\Delta_u$ $1\sigma_g^2 2\sigma_g^2 1\pi_u^4 1\delta_g^3 1\delta_u^2 1\pi_g^2 1\sigma_u^1$ ground state with $r_e = 2.10$ Å, $\omega_e = 390$ cm^{-1} and a binding energy of 3.45 eV (relative to spherical atoms). This state is separated by about 0.5 eV from a dense 'band' of nonets and quintets (cf. above). The values of r_e and ω_e are in moderate agreement with their experimental (matrix) counterparts. We have recently performed[188] fully self-consistent LCGTO-LSD calculations for a $^7\Delta_u$ state and found $r_e = 2.01$ Å and $\omega_e = 402\,cm^{-1}$ with $D_e = 4.0$ eV (again relative to spherical atoms). A more complete study would be in order.

Finally, mention should be made of two attempts to derive information on the ground state of Fe_2 by comparing DVM-$X\alpha$[295] or $X\alpha$-SW[298] calculations with experimental Mössbauer parameters[299–301].

7. Co_2

(See Refs 118, 190, 273 and 302.) Here again, very few studies have been carried out. Shim and Gingerich[302] performed CI calculations at a level similar to that just described for Fe_2. Again they find a huge number of low-lying states, nine within 0.1 eV and 84 within about 0.4 eV of the ground state, $^5\Sigma_g^+$, at the bulk internuclear separation. The molecule is unbound with respect to ground-state d^7s^2 (^4F) Co atoms and is calculated to have a bond length of 2.56 Å (again larger than the bulk metal) and a vibrational frequency of $162\,cm^{-1}$. The latter is much lower than the experimental value of $280\,cm^{-1}$. Based on the bond length errors found in similar calculations for Ni_2 and Cu_2 (see below), the authors estimated that the true bond length of Co_2 should be shorter by 0.2–0.25 Å. The recently observed[219] photoelectron spectrum of Co_2^- is nearly as clean as that of Fe_2^- so again it seems that, at the true equilibrium distance, the vast majority of these excited states lie at much higher energy than predicted.

The LSD calculations of HJ[118] predict a $^5\Delta_g$ $1\sigma_g^2 2\sigma_g^2 1\pi_u^4 1\delta_g^4 1\delta_u^3 1\pi_g^2 1\sigma_u^1$ ground state with a second state derived by transferring an electron from δ_u to π_g about 0.2 eV higher. The calculated bond length, 2.07 Å, is a reasonable interpolant of the experimental values for Ni_2 and Fe_2 while the calculated frequency, $360\,cm^{-1}$, is only in moderate accord with the experimental value. No fully self-consistent spin-polarized LSD calculations have yet been performed for Co_2.

8. Ni_2

(See Refs 118, 190, 272, 273 and 303–312.) The nickel dimer has received considerable attention, at least partially because of the importance of nickel clusters in catalysis. Several *ab initio* calculations have been performed[190,272,273,303–308] and a semiquantitative consensus on the nature of the ground and low-lying excited states appears to have been reached. The work of Noell *et al.*[308] represents the most accurate computational realization of this consensus. The ground state of Ni_2 correlates with two d^9s^1 atoms. The two s electrons pair to form a σ bond. The 18 d electrons occupy all but two of the 20 bonding and antibonding spin orbitals. Assigning the two d holes in various manners to the antibonding σ, π and δ orbitals leads to a dense manifold of

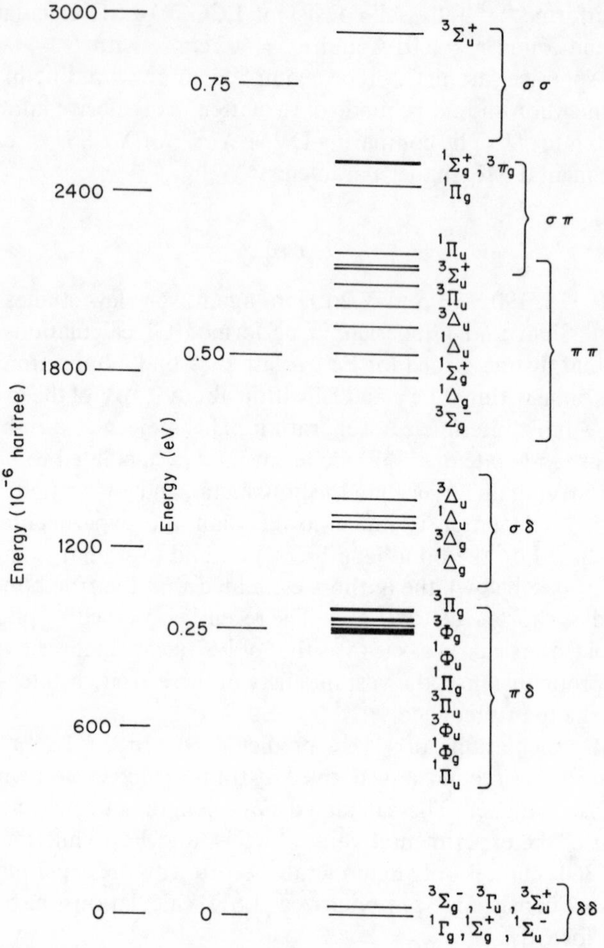

Fig. 10. Spectrum of all states of Ni_2 which asymptotically dissociate to two ^3D atoms. Ordering based on the GVBCI results at $r = 4.6$ a.u. (*Reproduced from Ref. 308 by permission of the authors and the American Institute of Physics.*)

low-lying states, 30 of them within about 0.75 eV (Fig. 10). The six states with two δ holes lie lowest. While this array of states may be taken to indicate essentially localized and weakly interacting d electrons, the results shown in Fig. 11 demonstrate that this is an oversimplification. At the RHF level the molecule is bound by only 0.48 eV (2 eV expt). Allowing left–right correlation (GVB) of the s electrons increases this to 0.77 eV. The polarization singles and doubles CI(POLSDCI) calculation, 'largely as a result of adding angular correlation to the s bond' yields 1.30 eV. The added (primarily d electron) correlation present in the most extensive CI(SD) calculations leads to a further

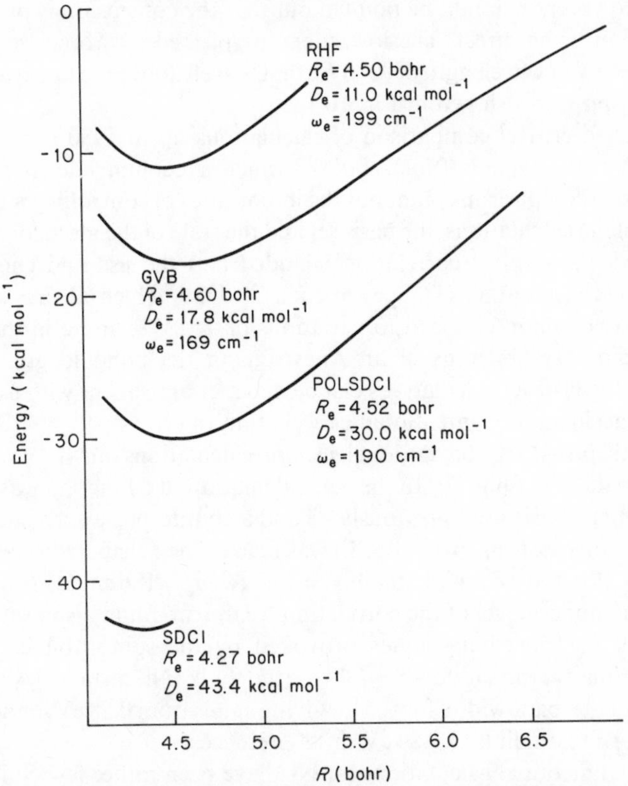

Fig. 11. Potential curves for ground state of Ni_2 at different levels of correlation. Since the GVBCI curve is very similar to the GVB curve, it was omitted from the figure. (*Reproduced from Ref. 308 by permission of the authors and the American Institute of Physics.*)

increase of D_e to 1.88 eV along with a decrease of the bond length by over 0.13 Å to 2.26 Å. These are in close agreement with their experimental counterparts of 2.07 eV and 2.20 Å, although particularly the latter value still leaves some room for improvement. With the points calculated at the CI(SD) level, Noell *et al.* were unable to obtain an accurate estimate of the vibrational frequency. The frequencies for the lower-level calculations are all below 200 cm^{-1}, much less than the experimental (matrix) value of 380.9 cm^{-1}. (see footnote *t* of table V.)

The Ni_2 spectrum[226] is consistent only with a lower state of $^1\Gamma_g$, $^3\Gamma_u$ or $^3\Phi$ symmetry. The first two of these arise only from the $\delta\delta$ hole configuration whereas $^3\Phi$ arises from a configuration with one π and one δ hole. While the devil's advocate might argue for $^3\Phi$ as the ground stare, there is no reason to expect an imbalance in the calculations that would lead to a wrong ordering. The crowded nature of the observed Ni_2 spectrum has been invoked [190,226] as experimental support for the existence of the array of closely spaced low-lying states. However, it should be pointed out that the conclusion is by inference, rather than being direct. The lowest energy probed by Morse *et al.*[226] was about 1.38 eV (a wavelength of 9000 Å), that is, well above the uppermost of the 30 low-lying states shown in Fig. 10.

A detailed critical comparison of calculations up to 1980 is contained in Section V of the paper by Noell *et al.*[308], which is recommended to the reader. The various comparisons point out the importance of controlling a number of aspects of the calculations: the basis set and the state of the atom for which it is optimized, the level of correlation included and the use and choice of an effective core potential. Changes in each of these, which at first sight may appear to be minor, can lead to variations of 0.5 eV or more in the binding energy and several tenths of an Ångström in the bond length. As often happens, lower-level calculations can be in better agreement with experiment through accidental or intentional cancellation of errors.

The only post-1980 conventional *ab initio* calculations on Ni_2 of which I am aware are those of Shim[190]. In the new calculations a d function was added to the Wachters' basis used previously[306] and a counterpoise correction for the basis-set superposition error (BSSE) was made. The calculated spectroscopic constants (for the $^1\Sigma_g^+$ ($\delta\delta$) state), $r_e = 2.47$ Å, $\omega_e = 186\,\mathrm{cm}^{-1}$, $D_e = 0.95$ eV, reflect the modest level of the correlation treatment. Shim also examined the mixing of the close-lying states provoked by the spin–orbit coupling. A perturbation treatment for $r = 2.49$ Å (the bulk Ni distance) showed an increase of the 'bandwidth' for the low-lying states from 0.58 eV in the absence of spin–orbit coupling to 0.84 eV in its presence.

Density functional calculations for Ni_2 have been rather few. Snijders and Baerends[310] used the non-spin-polarized molecule as a testing ground for the development of a pseudo-potential. Harris and Jones (HJ)[118] examined a number of low-lying states, though apparently not the Γ ($\delta\delta$ hole) state found to lie lowest in the *ab initio* studies. The three lowest states ($^3\Sigma_g^-$, $^3\Delta_g$ and $^3\Phi_u$) are calculated to have r_e values of 2.18, 2.21 and 2.22 Å, vibrational frequencies of 320, 310 and 320 cm^{-1} and dissociation energies of 2.70, 2.65 and 2.55 eV respectively. While it is not clear exactly to what extent the perturbational treatment of spin polarization will affect the results, I believe that the close agreement with experiment for r_e is probably not fortuitous and that these spectroscopic constants would be relatively little modified in a fully self-consistent calculation. Such calculations are in progress[311]. There are significant differences between the HJ results and those calculated[312] with the LCGTO technique using the $X\alpha$ potential, which predict a $^3\Pi_u$ ground state, $r_e = 2.03$ Å, $\omega_e = 356\,\mathrm{cm}^{-1}$ and $D_e = 2.91$ eV. At least some of the difference may be attributed to the use of the $X\alpha$ potential which undoubtedly exaggerates the extent of spin polarization for this 'ferromagnet'. Viewed globally, the available LSD results are not radically at odds either with the most extensive CI calculations or with experiment.

9. Cu_2

(See Refs 118, 190, 273 and 313–348.) Having made it from Sc_2 to Ni_2 (the real transition metals), at Cu_2 (a noble metal) one should be able to breathe a sigh of relief. The last d holes have been filled and while the d electrons by no means disappear entirely from the scene, for Cu_2 they have definitely turned in their cards for the bad actors' union. In fact it is only a slight overstatement to maintain the impossibility of making a *really* bad calculation for Cu_2, certainly not on the scale of the disasters seen for Cr_2. The long list of references bears witness to this—things work for Cu_2. This, of course, does not mean that a quantitative calculation of the spectroscopic constants is a trivial matter (any more than it is for H_2) but state-of-the-art CI techniques do closely approach the experimental values.

In the simplest approximation, Cu_2 can be viewed as a two valence-electron system, a σ bond being formed with the 4s electrons. The d electrons are, however, not inert. In order to obtain reasonable spectroscopic constants within a one-electron model-potential approach, core–valence polarization effects must be included in the model potential. Within such schemes, Jeung and Barthelat[329] found $r_e = 2.14$ Å, $\omega_e = 269\ \mathrm{cm}^{-1}$ and $D_e = 1.54$ eV and Stoll *et al.*[330] found $r_e = 2.26$ Å, $\omega_e = 262\ \mathrm{cm}^{-1}$ and $D_e = 1.95$ eV (versus 2.22 Å, 266.1 cm^{-1} and 2.04 eV experimentally).

Direct attacks, either all-electron or 22 valence-electron model-potential approaches, have clearly pointed out the need for extended basis sets, an extensive treatment of d-electron correlation and the inclusion of relativistic corrections if one aspires to high accuracy. Since there are so many papers on Cu_2 and since many of them, of course, arrive at similar conclusions, I will be content with a brief review of five of the *ab initio* papers which I believe present the essential features.

Bauschlicher *et al.*[328] carried out a variety of SCF and CI calculations. They used the (14s9p5d) basis of Wachters[158] augmented with a diffuse d and two diffuse p functions, contracted to [8s6p4d]. For some of the calculations an f polarization function was added. They reported values of the bond length. At the SCF level 2.43 Å and 2.40 Å were obtained without and with the f function, respectively. Hence, correlation must account for a shortening of about 0.2 Å. As expected, correlation of the 4s σ_g bond by mixing in the antibonding σ_u^2 configuration (in a two-electron MCSCF treatment) lengthened the bond to 2.49 Å. Including the π_u^2 configuration shrunk the bond to 2.44 Å, just slightly longer than the SCF prediction. Hence the 0.2 Å bond length discrepancy should be attributed to some combination of d-electron correlation, relativistic effects and, perhaps, correlation of the semi-core, 3s and 3p, electrons. Correlating the dσ electrons in a CI calculation yields a shortening to 2.40 Å, adding the π electrons gives 2.34 Å and a 22-electron CI yields 2.32 Å. Releasing the semi-core electrons has no effect on the bond length. At the 22-

electron CI level, the f function causes a decrease of 0.02 Å. Hence, at this level of theory the Cu_2 bond is too long by about 0.10 Å (0.08 Å if the Davidson correction[177] for the effect of quadruple excitations is applied). Bauschlicher *et al.* observed that this discrepancy corresponds very closely to twice the relativistic contraction of the 4s orbitals of atomic copper calculated by Desclaux[350].

Bauschlicher[333] compared CASSCF, CI(SD) and (POLCI) approaches using the polarized basis set from Ref. 328. The spectroscopic constants r_e, ω_e and D_e were calculated to be 2.44 Å, 184 cm^{-1} and 1.25 eV at the CASSCF level; 2.34 Å, 220 cm^{-1} and 1.51 eV for CI(SD); 2.32 Å, 219 cm^{-1}, and 1.61 eV with the Davidson correction; and 2.35 Å, 227 cm^{-1} and 1.99 eV for POLCI. Bauschlicher argues that the POLCI result for D_e is better than the CI(SD) value primarily because of a large size-consistency error in the latter, due to the inclusion of d–d correlation. The bond is still over a tenth of an ångström too long by either method.

Martin[334] was the first to estimate the effects of relativity on the spectroscopic constants of Cu_2. The scalar relativistic (mass–velocity and Darwin) terms were evaluated perturbatively using Hartree–Fock or GVB (Two configuration SCF ($\sigma_g^2 \rightarrow \sigma_u^2$)) wavefunctions. At these levels the relativistic corrections for r_e, ω_e and D_e were found to be -0.05 Å, $+15$ cm^{-1} and $+0.06$ eV for SCF, and -0.05 Å, $+14$ cm^{-1} and $+0.07$ eV for GVB. The shrinking of the bond length is less than half of the estimate based on the contraction of the 4s atomic orbital.

Very recently the howitzers have been brought out. Scharf *et al.*[340] used a relativistically corrected size-consistent modification of CI(SD), the coupled pair functional (CPF) method[351]. Basis sets including up to g functions were included. When the dust had settled, they found $r_e = 2.24$ Å, $D_e = 1.84$ eV, in satisfying agreement with experiment. No value of ω_e was reported, presumably since an insufficient number of points were calculated. Werner and Martin[341] examined the effects of unlinked clusters (coupled electron-pair approximation (CEPA) calculations)[352] along with the relativistic contributions. Their final value for r_e, 2.23 Å, is in excellent agreement with experiment. The contraction of the bond length due to correlation is 0.17 Å (and 0.04 Å relative to CI(SD)), whereas the relativistic contribution is -0.04 Å (cf. -0.05 Å at the SCF level). The calculated vibrational frequency 263 cm^{-1}, and dissociation energy, 1.80 eV, are also in satisfying agreement with their experimental counterparts, significantly better than the CI(SD) values of 242 cm^{-1} and 0.23 eV (the Davidson correction yields 253 cm^{-1} and 1.24 eV). Analysis of the relativistic contribution to the bond contraction showed that it could be attributed about two-thirds to the 4s–4s bonding orbital, the remainder being concentrated in the $n = 3$ shell, primarily in 3s and 3p.

There have also been a number of density functional calculations for Cu_2. The bonding picture is analogous to that derived from the CI treatments, a single σ(4s–4s) bond, and a 'closed' d shell which now is correlated by the exchange-correlation functional. This works quite well. HJ[118] calculated $r_e = 2.28$ Å and $\omega_e = 280\,cm^{-1}$. The calculations showed the by now familiar overbinding of the LDA, $D_e = 2.30$ eV versus the experimental value of 2.04 eV. Ziegler *et al.*[344] included a perturbational treatment of the scalar relativistic corrections in DVM-$X\alpha$ ($\alpha = 0.7$) calculations. Their non-relativistic values for r_e, ω_e and D_e were 2.26 Å, 268 cm^{-1} and 2.20 eV; relativity changes these to 2.24 Å, 274 cm^{-1} and 2.29 eV. Post and Baerends[345] pointed out the necessity of allowing for hole localization in calculations of the d-electron ionization potentials (see also Refs 309, 346 and 353). Delley *et al.*[347], in a DVM study of a number of Cu clusters, presented results for Cu_2 using both analytical and numerical bases. They examined the Gunnarsson–Lundqvist[30] potential as well as $X\alpha$ with $\alpha = 0.7$. For the (presumably more accurate) numerical basis they calculate $r_e = 2.22$ Å and $D_e = 2.10$ eV for $X\alpha$ and $D_e = 2.30$ eV for a fixed distance of 2.25 Å using the GL potential. Our own[348] LCGTO-VWN calculations yield $r_e = 2.21$ Å, $\omega_e = 248\,cm^{-1}$ and $D_e = 2.4$ eV.

Very recently, Wang[349] has calculated Cu_n and Cu_n^+ ($n = 1$–3) clusters using a one-electron (relativistic) pseudo-potential LSD (GL) approach, including an examination of the self-interaction correction. A core–valence polarization term was included in the model potential. She calculates $r_e = 2.26$ Å and $D_e = 1.845$ eV at the LSD level. It is not clear whether the discrepancy with our LCGTO results is due to the model potential or to the use of the GL potential, or both. The SIC decreases r_e to 2.22 Å and increases D_e to 2.049 eV, both in excellent agreement with experiment, although perhaps to some extent fortuitously so. The calculated (vertical) ionization potential of Cu_2 at the LSD level (7.987 eV) agrees slightly better with the experimental value (7.894 eV) than does the LSD-SIC value (8.237 eV). (See also Ref. 388).

B. 4d Dimers

Compared with the situation for the 3d series, the 4d orbitals are relatively more diffuse and hence can overlap more efficiently. One result of this diffuseness is the absence of permanent magnets amongst the 4d solids; the 4d wavefunctions cannot support local moments. This can be rationalized in terms of the Stoner theory[354,355] of itinerant magnetism which also has a cluster or molecular analog[261]. Briefly, a magnetic instability may occur if

$$IN(\varepsilon_F) \geqslant 1$$

where $N(\varepsilon_F)$ is the density of states at the Fermi level of the non-spin-polarized system (one needs low-lying empty majority spin states in order to create a magnetic moment) and I is an exchange integral (a measure of the stabilization associated with a spin flip). Diffuse functions lead to low values of both I and

$N(\varepsilon_F)$, the first directly and the second because large overlaps lead to broad 'bands' and, hence, to a lowering of the density of states. So, in this respect the 4d dimers should be somewhat less demanding than their 3d congeners. We will see, for the cases so far studied, that this is so. On the other hand, all-electron calculations are complicated by the extra shell of core electrons, although these can readily be incorporated into a model potential for most of the properties of current interest. Also on the 'more difficult' side of the ledger are the increased relativistic corrections which, while still reasonably small for the 4d atoms, should be incorporated for high accuracy.

Consistent with the experimental situation shown in Table V, with the exception of Mo_2 and Ag_2, the 4d dimers have received relatively little attention from non-empirical quantum chemistry. Given the increased possibilities of measuring the properties of these molecules provided by the new beam spectroscopy techniques, I think it would be very interesting to have more before-the-facts predictions from the various state-of-the-art quantum-chemical techniques.

Since most of the theoretical studies on the 4d dimers run parallel to those on their 3d congeners, a few brief comments should suffice.

1. Y_2

(See Ref. 280.) The only theoretical study on Y_2 of which I am aware is described briefly in a very recent report by Walch and Bauschlicher[280]. As was the case for Sc_2, their CASSCF/CI calculations predict a $^5\Sigma_u^-$ ground state. However, owing to the increased d–d bonding in the 4d series, the $^1\Sigma_g^+\ 1\sigma_g^2 1\pi_u^4$ state is now relatively low-lying, 0.87 eV above $^5\Sigma_u^-$. Since, contrary to observations for Sc_2, Knight *et al.*[356] did not observe an ESR spectrum for Y_2, and since high-spin states may be favored in this type of calculation, the authors raise the possibility that $^1\Sigma_g^+$ is the true ground state. Their predicted spectroscopic constants for the $^5\Sigma_u^-$ state are $r_e = 3.03$ Å, $\omega_e = 171\ \mathrm{cm}^{-1}$ and $D_e = 2.44$ eV with respect to the $d^1s^2 + d^2s^1$ asymptote (about 1.1 eV with respect to ground-state atoms); those for $^1\Sigma_g^+$ are $r_e = 2.74$ Å, $\omega_e = 206\ \mathrm{cm}^{-1}$ and $D_e = 2.93$ eV with respect to two d^2s^1 atoms (about 0.2 eV with respect to ground-state atoms). The need for corresponding experimental values is apparent.

2. Nb_2

(See Refs 357 and 358.) Cotton and Shim[357] have performed low-level CI calculations. As is typical of this type of calculation (see above), the calculated bond length is very long (about 3 Å) and, as a result, there is a jungle of low-lying states. One is really approaching asymptotic conditions at these bond lengths and it is doubtful whether any reliable analysis can be made.

We have recently[358] performed LCGTO-LSD calculations for Nb_2. While the analysis is not yet complete the results to date are very different from those of Cotton and Shim. Using the VWN potential (results for other potentials will be reported in due course[358]), for the $1\sigma_g^2 1\pi_u^4 1\delta_g^2 2\sigma_g^2$ configuration, we have examined both the triplet ($\alpha\alpha$) arrangement of the δ spins and the 'half-singlet, half-triplet' ($\alpha\beta$) arrangement. These states have nearly the same binding energy, 5.96 and 6.05 eV, relative to sphericalized d^4s^1 Nb atoms. Adjusting these through calculations on a real spherical, d^5, state reduces them to 4.99 and 5.08 eV, respectively, in good agreement with the mass spectrometric value, 5.21 eV. This is a very strong bond. It is also short. We calculate 2.09 Å for the ($\alpha\beta$) state and 2.05 Å for ($\alpha\alpha$). Relativistic corrections would likely reduce this by around 0.05 Å. So Nb_2 is only slightly longer than the 'hexuply' bonded Mo_2 (see below), which is consistent with the removal of two weakly bonding δ electrons from the hexuple bond. Rough estimates of the vibrational frequencies are 500 cm^{-1} for ($\alpha\alpha$) and 600 cm^{-1} for ($\alpha\beta$). The bond is rigid. These predictions await experimental comparisons as well as further theoretical analysis.

After the preceding paragraph was written I became aware of the CASSCF calculations for Nb_2 by Walch and Bauschlicher[280]. Similar to V_2, (ref. 281) they calculate a $^3\Sigma_g^-$ $1\sigma_g^2 2\sigma_g^2 1\pi_u^4 1\delta_g^2$ ground state with a very low-lying $^1\Gamma_g$ state, 0.09 eV higher. Two further low-lying states were located at 0.12 eV ($^3\Delta_g$ $1\sigma_g^2 2\sigma_g^1 1\pi_u^4 1\delta_g^3$) and at 0.96 eV ($^3\Phi_g$ $1\sigma_g^2 2\sigma_g^2 1\pi_u^3 1\delta_g^3$). The increased stability of $^3\Delta_g$ as compared to V_2 was attributed to stronger dδ bonding. The calculated spectroscopic constants r_e, ω_e and D_e were 2.10 Å, 448 cm^{-1} and 2.24 eV for $^3\Sigma_g^+$; 2.11 Å, 427 cm^{-1} and 2.15 eV for $^1\Gamma_g$; 2.01 Å, 501 cm^{-1} and 2.12 eV for $^3\Delta_g$; and 2.19 Å, 340 cm^{-1} and 1.28 eV for $^3\Phi_g$.

3. Mo_2

(See Refs 113, 146, 202, 255, 257, 258, 272, 280, 287, 289, 290, 292, 359–364.) The considerations for the bonding in Mo_2 are the same as those for Cr_2. The main difference, rooted in the greater extent of the 4d electrons, is that the dominance of the (antiferro-)magnetic aspects does not set in until somewhat longer distances, which makes the region around the minimum a little easier to describe by *ab initio* techniques. At the Hartree–Fock level[255] the hexuply bonded configuration has too short a bond length, 1.80 Å (1.94 Å expt), too high a vibrational frequency, 717 cm^{-1} (477.1 cm^{-1} expt) and is unbound by about 18 eV. At the GVB-vdW level[202], however, things are much better, though still not perfect. Goodgame and Goddard[202] found $r_e = 1.97$ Å and $\omega_e = 455$ cm^{-1}. The binding energy is only 1.41 eV. Results at a similar (slightly lower) level had previously been obtained by Bursten *et al.*[359] and Atha *et al.*[360]. The CASSCF calculations of Walch and Bauschlicher[280] yield $r_e = 1.99$ Å, $\omega_e = 399$ cm^{-1} and $D_e = 0.77$ eV. GG also predicted a second minimum at long r, 3.09 Å, with $\omega_e = 80\,\text{cm}^{-1}$ and $D_e = 0.49$ eV, dominated by the 5s–5s overlap, the non-interacting d electrons being coupled antiferromagnetically. In the recent semi-empirical correction[290] of the GVB results, GG obtain $r_e = 1.92$ Å and $D_e = 3.94$ eV, the latter indicating the importance of the ionic terms within this picture (and by inference the difficulty of their direct calculation). In this modified GVB approach the outer minimum disappears although an inflection remains. No value was reported for the vibrational frequency or the anharmonicity.

Mo_2 has also been examined by several local density techniques[113,146,257,258,289,292,361–363]. The earlier work on the hexuply bonded configuration[361–363] used the $X\alpha$-SW method so the binding curve could not be elucidated. More recently, broken-symmetry DVM[113], Gaussian pseudopotential[146] and LCGTO[257,258] calculations have been performed. The results are in satisfying agreement both with each other and with experiment. For example, using the VWN potential we[257] calculated $r_e = 1.97$ Å, $\omega_e = 423\,\text{cm}^{-1}$ and $D_e = 5.0$ eV. Relativistic corrections would likely decrease the bond length by a few hundredths of an ångström. The main difference between

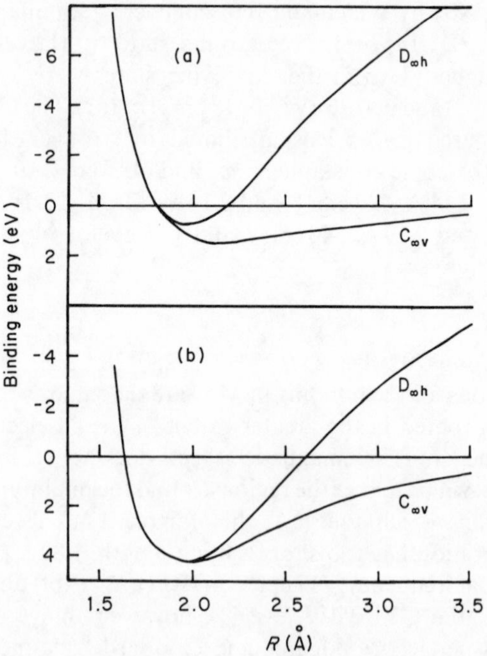

Fig. 12. Calculated binding energy versus internuclear distance for Mo_2 using $D_{\infty h}$ and $C_{\infty v}$ symmetry constraints and the (a) $X\alpha$ (Ref. 12) and (b) JMW (Ref. 29) exchange–correlation potentials. (*Reproduced from Ref. 125 by permission of Taylor & Francis Ltd.*)

the binding picture of Mo_2 and that of Cr_2, as they emerge from the LSD calculations, is that, for the more diffuse 4d system, the symmetry breaks at a slightly longer distance (relative to r_e). This is illustrated in Fig. 12, which compares results for the hexuply bonded configuration ($D_{\infty h}$) with the broken-symmetry ($C_{\infty v}$) curve. At r_e, all of the MOs are almost fully delocalized, whereas for Cr_2 the δ orbitals were highly localized. As was the case for Cr_2, Mo_2 cannot be described either as a 'hexuple' bond or as an antiferromagnet. A correct description contains elements of both.

Ziegler *et al.*[292] have very recently reported results of a DVM study using Becke's[291] semi-empirical gradient potential. They calculate $r_e = 1.95$ Å and $D_e = 4.03$ eV. (See also Ref. 388.)

4. Ru_2

(See Refs 132, 186 and 365.) The only conventional *ab initio* results so far reported for Ru_2 are those of Cotton and Shim[190,365], who used a CI approach similar to Shim's previous work on other dimers. They calculated a $^7\Delta_u$ ground state (the same as for Fe_2) with $r_e = 2.72$ Å, $\omega_e = 116\ cm^{-1}$ and $D_e = 0.64$ eV, which are undoubtedly far too long (in fact, longer than the bulk interatomic distance), far too floppy and far too weak. There is one aspect of their results, mentioned above, which deserves note and which may help to clarify the situation for Fe_2, relative to the 'clean' photoelectron spectrum observed for Fe_2^-. For Ru_2, owing to the longer range of the d orbitals, the $^7\Delta_u$ state is separated from the dense manifold of higher-lying states by about 0.15 eV. A more adequate correlation treatment would be expected to decrease r_e and increase the separation between states. Similar considerations for Fe_2 would 'clean up' its theoretical spectrum.

All electron LCGTO-LSD(VWN) calculations[132] using a compact basis set yield, for a $^7\Sigma$ state, $r_e = 2.41$ Å, $\omega_e = 380\ cm^{-1}$ and $D_e = 2.7$ eV (relative to sphericalized atoms). Relativity would likely decrease the distance by 0.05 Å or so. A model-potential calculation, including the semi-core 4s and 4p electrons explicitly in the valence shell, changed the above values to 2.42 Å, 330 cm^{-1} and 3.0 eV. While these results are preliminary, they do represent the most trustworthy theoretical estimates available.

5. Rh_2

(See Ref. 190) Shim's[190] CI calculations yield $r_e = 2.86$ Å (vs 2.69 Å for bulk Rh), $\omega_e = 118\ cm^{-1}$ and $D_e = 0.85$ eV (vs 2.92 expt) for the $^5\Sigma_g^+$ ground state. Comments similar to those for Ru_2 apply.

6. Pd_2

(See Refs 190, 239, 318 and 366–368.) Pd_2 has received very little *ab initio* attention and none of state-of-the-art quality. The early MCSCF calculations of Basch *et al.*[368] used a relativistic effective core potential. For the $^3\Gamma_u$ $(\delta\delta)$ hole state they calculate $r_e = 2.81$ Å, $\omega_e = 216\ cm^{-1}$ and $D_e = 0.76$ eV. Shim[190,239] finds the $^1\Sigma^{+g}$ $(\delta\delta)$ state lowest, with $r_e = 2.91$ Å, $\omega_e = 104\ cm^{-1}$, unbound by 0.97 eV.

LCGTO-LSD(VWN)-MP calculations[366] using a relativistic model potential lead to $r_e = 2.30$ Å, $\omega_e = 320\ cm^{-1}$ and $D_e = 1.1$ eV for the $(\sigma\sigma)$ hole configuration (not necessarily the LSD ground state since not all possibilities have been considered.)

7. Ag_2

(See Refs 132, 318, 330, 331, 336, 337, 343, 344 and 369–375.) The *ab initio* history of Ag_2 is quite similar to that of Cu_2 except that the last step (the howitzers) has not yet been taken. As was the case for Cu_2, the first-order picture of a pseudo-alkali dimer is a serious oversimplification—the d electrons are far from inert.

The most extensive all-electron calculations to date have been performed by McLean[370], using a 'triple-zeta valence' Slater basis. At the SCF level he finds $r_e = 2.79$ Å, $\omega_e = 129\ cm^{-1}$ and $D_e = 0.382$ eV. MCSCF calculations at the lowest (two-configurational) level needed for proper dissociation (left-right correlation of the s bond) change these to 2.86 Å, 107 cm^{-1} and 0.669 eV. Freeing the 4d as well as the semi-core 4s and 4p electrons in MCSCF calculations provides spectroscopic constants of 2.86 Å, 108 cm^{-1} and 0.674 eV, very little changed from the previous values. A 22-electron CI(SD), including all single excitations and only those double excitations where either one or two bond (4s) electrons are excited, gave $r_e = 2.72$ Å and $D_e = 1.132$ eV. These differ from their experimental counterparts by 0.24 Å and 0.52 eV respectively, indicating the importance of some combination of higher order correlation, polarization functions and relativistic corrections.

The most extensive (relativistic) effective core potential calculations have recently been reported by Hay and Martin[373]. Their best (CI(SD)) results are $r_e = 2.62$ Å, $\omega_e = 172\ cm^{-1}$ and $D_e = 1.32$ eV. At the Hartree–Fock level the relativistic corrections are about -0.06 Å for r_e, and $+15\ cm^{-1}$ for ω_e. Clearly, attaining good quantitative *ab initio* spectroscopic constants for Ag_2 will require very extensive calculations. (See also Refs 387 and 388.)

Ag_2 has also been calculated several times with local density methods[132,343,344,374,375]. The relativistic DVM-$X\alpha$ results of Ziegler *et al.*[344], $r_e = 2.52$ Å, $\omega_e = 203\ cm^{-1}$ and $D_e = 2.03$ eV, are in satisfactory agreement with experiments. The relativistic corrections are -0.15 Å, $+19\ cm^{-1}$ and $+0.3$ eV. The first of these is significantly larger than both the HF estimate and our own[132] estimate from LCGTO-LSD(VWN)-MP calculations. These latter calculations yield $r_e = 2.48$ Å, $\omega_e = 186\ cm^{-1}$ and $D_e = 2.1$ eV including relativistic corrections of -0.07 Å, $+14\ cm^{-1}$ and $+0.1$ eV. Martins and

Andreoni[374,375] have applied a pseudo-potential LSD method and find $r_e = 2.51$ Å, $\omega_e = 212\ \text{cm}^{-1}$ and $D_e = 2.5$ eV.

C. 5d Dimers

This section must, alas, be extremely short, for the simple reason that very little is known either experimentally or from theory about the diatomics formed from the platinum-row elements. This scarcity of data is certainly not because of a lack of practical applications for the corresponding metals; Re, Os, Ir and Pt particles are all of great importance in catalysis. The main roadblock as far as theory is concerned has undoubtedly been the lack, until recently, of appropriate methods for incorporating relativistic corrections. For 5d electrons these corrections are no longer minor. The energies associated with the mass–velocity, Darwin and spin–orbit operators approach the chemical bond range (roughly 1 eV), so that relativity can have profound effects on chemical binding and on molecular properties.

Since relativistic methods are reviewed elsewhere[266–269] including a review in these volumes[376], I will be content with a brief mention of the few available calculations for 5d dimers.

1. W_2

(See Ref. 292.) Ziegler *et al.*[292] have very recently performed relativistic DVM calculations for W_2 using the semi-empirical gradient potential of Becke[291]. They predict $r_e = 2.03$ Å, $\omega_e \approx 300\ \text{cm}^{-1}$ and $D_e = 4.41$ eV, which may be compared with their non-relativistic values of 2.07 Å, $\approx 300\ \text{cm}^{-1}$ and 3.54 eV.

2. Pt_2

(See Refs 368, 377 and 378.) Basch *et al.*[368] used a relativistic effective core potential in low-level MCSCF calculations. Their predicted spectroscopic constants are $r_e = 2.58$ Å, $\omega_e = 267\ \text{cm}^{-1}$ and $D_e = 0.93$ eV for a $(\delta\delta)$ hole state, which may be compared with the experimental values, $\omega_e = 216\ \text{cm}^{-1}$ and $D_e = 3.71$ eV.

Relativistic $X\alpha$ scattered-wave calculations have been reported by Yang[377], who used a Dirac, four-component, approach, and by Morin[378], using a two-component (Wood-Boring) SCF treatment followed by diagonalization of the spin-orbit matrix. The results of both approaches are nearly identical. The scalar relativistic terms lead to a stabilization of the valence levels by about 2 eV. Spin–orbit coupling completely changes the arrangement of the levels, leading to two groups which can be (very approximately) labelled $d_{3/2}$ and $d_{5/2}$ to show their parentage in the spin–orbit split atomic d level.

3. Au_2

(See Refs 266, 318, 344, 379 and 380.) Au_2 has been chosen as a target by Pitzer and coworkers in their efforts to develop accurate relativistic effective core potentials. The results in Refs 379 and 380 were made with a version of the method 'now known to yield somewhat too short bond distances'[266]. Nevertheless the agreement with experiment was good: $r_e = 2.37$ Å (2.47 expt), $\omega_e = 165\,\text{cm}^{-1}$ (191 expt) and $D_e = 2.27$ eV (2.31 expt) and the analysis was illuminating. Relativity accounts for 1 eV of the dissociation energy, a decrease of about 0.3 Å in the bond length and a 25% increases in the vibrational frequency.

On the local density side, Ziegler *et al.*[344] calculated $r_e = 2.44$ Å, $\omega_e = 201\,\text{cm}^{-1}$ and $D_e = 2.51$ eV, all in very satisfactory agreement with experiment. The relativistic contributions are -0.46 Å, $108\,\text{cm}^{-1}$ and 1.34 eV.

V. CONCLUDING REMARKS

Despite the considerable activity just outlined and the ongoing work in several laboratories, the portrait of our knowledge of these smallest transition-metal systems is far from complete.

The ground and low-lying excited states of the atoms are in reasonably good shape. Experimental spectroscopic data are extensive, CI calculations have provided an analysis of the important correlation effects and relativistic corrections have been estimated. This does not mean that the transition-metal atoms represent a closed book—higher accuracy, more work on the fine and hyperfine structure, interactions with external magnetic and electric fields, more accurate 'many-electron' relativistic theories and other advances are sure to come. However, at its present level, *ab initio* atomic theory has provided much useful information on the atoms and, from the point of view of molecular physics, on at least some of the ingredients necessary for successful calculations on molecules. Since molecular states can be usefully viewed as receiving contributions primarily from the ground and low-lying excited states of the atoms, a successful computational approach for molecules should yield acceptable relative energies and wavefunctions for these states when applied to the atoms (so as not to bias the mix in the molecule). Therefore, much attention has been focused on the positions of states arising from the $d^{n-2}s^2$ and $d^{n-1}s$ configurations of the transition-metal atoms (and, to a somewhat lesser extent, on d^n). To get these states about right requires extended basis sets, a far-reaching treatment of correlation, for the s and even more so for the d electrons, and, for the heavier atoms, inclusion of relativistic corrections.

Much of the success of local density functional theory for transition-metal system can be attributed to its good description of the d and s orbitals and their relative energies in a molecule or solid. However, applications of the LSD method to the atoms, while showing the correct trends across the Periodic Table, showed a bias toward d-electron-rich configurations. While we have seen that some of this bias might be due to the imposition of spherical symmetry on the calculations and that there could be merit in further examining broken-symmetry solutions, including self-interaction and gradient corrections, it remains a fact that the term and multiplet structures of open-shell systems are not properly treated within what is operationally a Hartree–like, orbital, approach. Hence, in its present form, *local* density functional theory must be viewed as incomplete, albeit the most useful and reliable technique currently available for transition-metal molecules. The transition-metal atoms will continue to serve as a severe testing ground as new DFT methods develop.

For the dimers, only some of the main lines of the picture have so far been sketched. The hard-earned experimental data, particularly those from beam spectroscopy, are still sparse. I hope that this will change over the coming months and years. In particular, the long-awaited infrared lasers would provide the exciting possibility of observing (or not) some of the low-lying states predicted by theory. Development of further structural, vibrational, magnetic and other probes is also highly needed.

New developments in theory are also much needed. So far, fully *ab initio* approaches have been, with a few notable exceptions, arduous and disappointing. The level of correlation required to describe these molecules properly is typically beyond the reach of the 'standard' technologies. New techniques have to be developed. As a DFT advocate, I hope these will soon arrive, for I believe that in order to improve on the current DFT schemes one must understand the correlation in these systems better and (for the time being?) we need wavefunctions to do this.

So far, the density functional methods have provided the most accurate theoretical values of the equilibrium properties, r_e and ω_e, of the dimers. The predictions of these properties for those cases where no experimental values exist are likely reliable to about 0.03 Å or so for r_e and about 10–20% for ω_e. In order to achieve this accuracy, in many cases, symmetry breaking had to be allowed in order to let the spin density functional have its full effect at a local level. Dissociation energies are typically overestimated. This overestimation can be seriously exaggerated if sphericalized atoms are usd for the asymptotic limit. In those cases involving real spherical atoms, S states, or if an excited S state can be used to adjust D_e, a typical error is in the 0.5 eV range and this may be taken as the differential energy which would have to be recovered through non-local corrections. Both the self-interaction correction and the 'gradient' corrections are expected to be larger for the more localized and hence more rapidly varying electron densities of the separated atoms. While preliminary results on such corrections (for Cr_2 and W_2 using a semi-empirical gradient correction) are encouraging, more work will be required to evaluate their effect properly. New and practicable non-local potentials will also have to be developed if applied DFT is to continue to prosper. The 'term and multiplet structure' problem is also present for many of the dimers (those with partially filled π or δ shells as well as the broken-symmetry cases). This has not yet received systematic attention (partly because, for the case of 'weakly coupled' d electrons, the various terms and multiplets arising from a given configuration are thought to have similar spectroscopic constants). The problem deserves more attention.

One distinct advantage of the LSD methods is that they may be applied to more complex systems. Indeed, my initial intention when asked to write this review was to include two more sections, one on metal–metal bonded complexes and one on transition-metal clusters and their interactions with adsorbates. Space, time and humanitarian considerations prevent me from so doing. Suffice it to say (with a few leading references) that the performance of the LSD method for these larger systems is highly encouraging. For example, Ziegler[381] has successfully applied the DVM-$X\alpha$ method to multiply metal–metal bonded complexes[382] of the type $Cr_2(CO_2H)_4$. Comparisons with conventional *ab initio* approaches[383] are enlightening. The largest transition-metal cluster so far treated with the LCGTO-LSD-MP method is a $Pd_{14}+CO$ aggregate which models chemisorption on Pd(00)[384]. The agreement with experimental results from surface science is at the same high level as we have seen for the dimers. The first inroads into dynamic aspects, diffusion barriers for hydrogen in palladium[122] using a $Pd_{10}H$ cluster, are also highly encouraging.

These are exciting times for transition-metal quantum chemistry. Some of the many challenges are being met, at least partially; each month brings new experimental and theoretical advances. If, as the epigraph maintains, there is more than one way to skin a cat, there are many, many cats waiting to be skinned.

References

1. Schaefer, H. F., III (Ed.), *Methods of Electronic Structure Theory*, Plenum, New York, 1977.
2. Peyerimhoff, S. D., and Buenker, R. J., in *Excited States in Chemistry* (Eds C. A. Nicolaides and D. R. Beck), Reidel, Dordrecht, 1978.
3. Bartlett, R. J., *Annu. Rev. Phys. Chem.*, **32**, 359 (1981).
4. Hurley, A. C., *Electron Correlation in Small Molecules*, Academic Press, New York, 1976.
5. Szabo, A., and Ostlund, N. S., *Modern Quantum Chemistry: Introduction to Advanced Electronic Structure Theory*, MacMillan, New York, 1982.
6. Diercksen, G. H., and Wilson, S. (Eds), *Methods in Computational Molecular Physics*, NATO ASI, Ser. A, Vol. 113, Reidel, Dordrecht, 1983.
7. Dykstra, C. E. (Ed.), *Advanced Theories and Computational Approaches to the Electronic Structure of Molecules*, NATO ASI, Ser. A, Vol. 133, Reidel, Dordrecht, 1984.

8. Bartlett, R. J. (Ed.), *Comparison of Ab Initio Quantum Chemistry with Experiment for Small Molecules*, Reidel, Dordrecht, 1985.
9. Hohenberg, P., and Kohn, W., *Phys. Rev.*, **136**, B864 (1964).
10. Ashcroft, N. W., and Mermin, N. D., *Solid State Physics*, Holt, Rinehart and Winston, Philadelphia, 1976.
11. March, N. H., *Self-Consistent Fields in Atoms—Hartree and Thomas–Fermi Atoms*, Pergamon, New York, 1975.
12. Slater, J. C., *Adv. Quantum Chem.*, **6**, 1 (1972); *The Self-Consistent Field for Molecules and Solids*, Vol. 4, McGraw-Hill, New York, 1974.
13. Kohn, W., and Sham, L. J., *Phys. Rev.*, **140**, A1133 (1965).
14. von Barth, U., and Hedin, L., *J. Phys. C: Solid State Phys.*, **5**, 1629 (1972).
15. Johnson, K. H., *Crit. Rev. Solid State Mater. Sci.*, **7**, 101 (1978).
16. Messmer, R. P., in *Nature of the Surface Chemical Bond* (Eds T. N. Rhodin and G. Ertl), North-Holland, Amsterdam, 1978.
17. Case, D. A., *Annu. Rev. Phys. Chem.*, **33**, 151 (1982).
18. Lundqvist, S., and March, N. H. (Eds), *Theory of the Inhomogeneous Electron Gas*, Plenum, New York, 1983.
19. Dahl, J. P., and Avery, J. (Eds), *Local Density Approximations in Quantum Chemistry and Solid State Physics*, Plenum, New York, 1984.
20. Salahub, D. R., in *Entre l'Atome et le Cristal: les Agrégates* (Ed. F. Cyrot-Lackmann), p. 59, Les Éditions de Physique, les Ulis, 1981.
21. Salahub, D. R., in *Contributions of Clusters Physics to Materials Science and Technology* (Eds J. Davenas and P. Rabette), Nijhoff, Amsterdam, 1986.
22. Salahub, D. R., in *Applied Quantum Chemistry Symp.* (Eds V. H. Smith, H. F. Schaefer, III and K. Morokuma, Jr), Reidel, Dordrecht, 1986.
23. Levy, M., *Proc. Natl. Acad. Sci. USA*, **76**, 6062 (1979).
24. Levy, M., in *Density Functional Methods in Physics* (Eds R. M. Dreizler and J. da Providencia), Plenum, New York, 1984.
25. Parr, R. G., *Annu. Rev. Phys. Chem.*, **34**, 631 (1983); Callaway, J., March, N. H., *Solid State Phys.*, **38**, 135 (1984).
26. Gaspar, R., *Acta Phys. Acad. Sci. Hung.*, **3**, 263 (1954).
27. Slater, J. C., *Phys. Rev.*, **81**, 385 (1951).
28. Hedin, L., and Lundqvist, B. I., *J. Phys. C: Solid State Phys.*, **4**, 2064 (1971).
29. Janak, J. F., Moruzzi, V. L., and Williams, A. R., *Phys. Rev. B*, **12**, 1257 (1975).
30. Gunnarsson, O., and Lundqvist, B. I., *Phys. Rev. B*, **13**, 4274 (1976).
31. Vosko, S. H., L., Wilk, L., and Nusair, M., *Can. J. Phys.*, **58**, 1200 (1980).
32. Perdew, J. P., and Zunger, A., *Phys. Rev. B*, **23**, 5048 (1981).
33. Rosen, A., and Ellis, D. E., *J. Phys. B: At Mol. Phys.*, **10**, 1 (1977).
34. MacDonald, A. H., and Vosko, S. H., *J. Phys. C: Solid State Phys.*, **12**, 2977 (1979).
35. MacDonald, A. H., *J. Phys. C: Solid State Phys.*, **16**, 3869 (1983).
36. Ramana, M. V., and Rajagopal, A. K., *Adv. Chem. Phys.*, **54**, 231 (1983).
37. Rajagopal, A. K., in Ref. 24, p. 159, 1984.
38. Strange, P., Staunton, J., and Gyorffy, B. L., *J. Phys. C: Solid State Phys.*, **17**, 3355 (1984).
39. Cortona, P., Doniach, S., and Sommers, C., *Phys. Rev. A*, **31**, 2842 (1985).
40. Ceperley, D. M., and Alder, B. J., *Phys. Rev. Lett.*, **45**, 566 (1980).
41. McMaster, B. N., unpublished.
42. Wakoh, S., and Yamashita, J., *J. Phys. Soc. Japan*, **21**, 1712 (1966).
43. Kübler, J., *J. Magn. Magn. Mater.*, **20**, 279 (1980).
44. Skriver, H. L., *J. Phys. F: Metal Phys.*, **11**, 97 (1981).
45. Schwarz, K., *J. Phys. B: At. Mol. Phys.*, **11**, 1339 (1978).
46. Painter, G. S., and Averill, F. W., *Phys. Rev. B*, **26**, 1781 (1982).
47. Pellegatti, A., McMaster, B. N., and Salahub, D. R., *Chem. Phys.*, **75**, 83 (1983).

48. Selmani, A., Sichel, J. M., and Salahub, D. R., *Surface Sci.*, **157**, 208 (1985).
49. Vosko, S. H., and Wilk, L., *J. Phys. B: At. Mol. Phys.*, **7**, 3687 (1983).
50. Jones, R. O., and Gunnarsson, O., *Phys. Rev. Lett.*, **55**, 107 (1985).
51. Gunnarsson, O., and Jones, R. O., *Phys. Rev. B*, **31**, 7588 (1985).
52. Salahub, D. R., Lamson, S. H., and Messmer, R. P., *Chem. Phys. Lett.*, **85**, 430 (1982).
53. Sabin, J. R., and Trickey, S. B., in Ref. 19, p. 333, 1984.
54. Perdew, J. P., *Chem. Phys. Lett.*, **64**, 127 (1979).
55. Langreth, D. C., and Perdew, J. P., *Solid State Commun.*, **17**, 1425 (1975); *Phys. Rev. B*, **15**, 2884 (1977).
56. Langreth, D. C., and Perdew, J. P., *Solid State Commun.*, **31**, 567 (1979); *Phys. Rev. B*, **21**, 5469 (1980).
57. Tong, B. Y., and Sham, L. J., *Phys. Rev.*, **144**, 1 (1966).
58. Geldart, D. J. W., and Rasolt, M., *Phys. Rev. B*, **13**, 1477 (1976).
59. Schwarz, K., *Chem. Phys. Lett.*, **57**, 605 (1978).
60. Shore, H. B., Rose, J. H., and Zaremba, E., *Phys. Rev. B*, **15**, 2858 (1977).
61. Rössler, U., in *Rare Gas Solids* (Eds M. L. Klein and J. A. Venables), p. 505, Academic Press, New York, 1976.
62. Hamrin, K., Johansson, G., Gelius, U., Nordling, C., and Siegbahn, K., *Phys. Scr.*, **1**, 277 (1970).
63. Trickey, S. B., and Worth, J. P., *Int. J. Quantum Chem.*, **S11**, 529 (1977).
64. Boring, M., *Int. J. Quantum Chem.*, **S8**, 451 (1974).
65. Zunger, A., and Freeman, A. J., *Phys. Rev. B*, **16**, 2901 (1977) and references therein; *Phys. Lett.*, **57A**, 453 (1976).
66. Eastman, D. E., Himpsel, F. J., and Knapp, J. A., *Phys. Rev. Lett.*, **44**, 95 (1980).
67. Harris, J., and Jones, R. O., *J. Chem. Phys.*, **68**, 3316 (1978).
68. E.g. Gopinathan, M. S., *J. Phys. B: At. Mol. Phys.*, **12**, 521 (1979).
69. E.g. Post, D., and Baerends, E. J., *Chem. Phys. Lett.*, **86**, 176 (1982).
70. Andzelm, J., and Salahub, D. R., unpublished.
71. E.g. Anisimov, V. I., Katsnelson, M. I., Kurmaev, E. Z., Liechtenstein, A. I., and Gubanov, V. A., *Solid State Commun.*, **40**, 927 (1981).
72. Lindgren, I., *Int. J. Quantum Chem.*, **5**, 411 (1971).
73. Gunnarsson, O., Jonson, M., and Lundqvist, B. I., *Phys. Rev. B*, **20**, 3136 (1979).
74. Gunnarsson, O., and Jones, R. O., *Solid State Commun.*, **37**, 249 (1981).
75. Heaton, R. A., Harrison, J. G., and Lin, C. C., *Solid State Commun.*, **41**, 827 (1982).
76. Harrison, J. G., Heaton, R. A., and Lin, C. C., *J. Phys. B: At. Mol. Phys.*, **16**, 2079 (1983).
77. Heaton, R. A., Harrison, J. G., and Lin, C. C., *Phys. Rev. B*, **28**, 5992 (1983).
78. Harrison, J. G., *J. Chem. Phys.*, **78**, 4562 (1983).
79. Harrison, J. G., *J. Chem. Phys.*, **79**, 2265 (1983).
80. Pederson, M. R., Heaton, R. A., and Lin, C. C., *J. Chem. Phys.*, **80**, 1972 (1984).
81. Pederson, M. R., Heaton, R. A., and Lin, C. C., *J. Chem. Phys.*, **82**, 2688 (1985).
82. Sinanoglu, O., *J. Chem. Phys.*, **36**, 706 (1962).
83. Kelley, H. P., *Phys. Rev.*, **131**, 684 (1963).
84. Lundqvist, S., and Ufford, C. W., *Phys. Rev.*, **139A**, 1 (1965).
85. Tong, B. Y., and Sham, L. J., *Phys. Rev.*, **144**, 1 (1966).
86. Tong, B. Y., *Phys. Rev. A*, **3**, 1027 (1974).
87. Tong, B. Y., *Phys. Rev. A*, **4**, 1375 (1975).
88. Kim, Y. S., and Gordon, R. G., *J. Chem. Phys.*, **60**, 1842 (1974).
89. Stoll, H., Wagenblast, G., and Preuss, H., *Theor. Chim. Acta*, **49**, 67 (1978).
90. Schneider, W. J., *J. Phys. B: At. Mol. Phys.*, **11**, 2589 (1978).

91. Lie, G. C., and Clementi, E., *J. Chem. Phys.*, **60**, 1275 (1974).
92. Lie, G. C., and Clementi, E., *J. Chem. Phys.*, **60**, 1288 (1974).
93. McKelvey, J. M., and Streitwieser, A., *J. Am. Chem. Soc.*, **99**, 7121 (1977).
94. Stoll, H., Pavlidou, C. M. E., and Preuss, H., *Theor. Chim. Acta*, **49**, 143 (1978).
95. Stoll, H., Golka, E., and Preuss, H., *Theor. Chim. Acta*, **55**, 29 (1980).
96. Baroni, S., and Tuncel, E., *J. Chem. Phys.*, **79**, 6140 (1983).
97. Baroni, S., *J. Chem. Phys.*, **80**, 5703 (1984).
98. Kemister, G., and Nordholm, S., *J. Chem. Phys.*, **83**, 5163 (1985).
99. Fermi, E., and Amaldi, E., *Accad. Ital. Rome*, **6**, 119 (1934).
100. von Weizsäcker, C. F., *Z. Phys.*, **96**, 431 (1935).
101. Herman, F., Van Dyke, J. P., and Ortenburger, I. B., *Phys. Rev. Lett.*, **22**, 807 (1969).
102. Herman, F., Ortenburger, I. B., and Van Dyke, J. P., *Int. J. Quantum Chem.*, **S3**, 827 (1970).
103. Becke, A., *Int. J. Quantum Chem.*, **27**, 585 (1985); *Int. J. Quantum. Chem.*, **23**, 1915 (1983).
104. Tschinke, V., and Ziegler, T., *Proc. Coleman Symp. on Density Matrices and Density Functionals*, Kingston, Ontario, August 1985, Reidel, Dordrecht, in press.
105. Langreth, D. C., and Mehl, M. J., *Phys. Rev. B*, **28**, 1809 (1983); erratum **29**, 2310 (1984).
106. Savin, A., Wedig, U., Preuss, H., and Stoll, H., *Phys. Rev. Lett.*, **53**, 2087 (1984).
107. Johnson, K. H., *Adv. Quantum Chem.*, **7**, 143 (1973).
108. Messmer, R. P., Knudson, S. K., Johnson, K. H., Diamond, J. B., and Yang, C. Y., *Phys. Rev. B*, **13**, 1396 (1976).
109. Raatz, F., and Salahub, D. R., *Surf. Sci.*, **146**, L609 (1984); *Surf. Sci.*, **176**, 219 (1986).
110. Painter G. S., and Ellis, D. E., *Phys. Rev. B*, **1**, 4747 (1970).
111. Ellis, D. E., and Painter, G. S., *Phys. Rev. B*, **2**, 1887 (1970).
112. Delley, B., Ellis, D. E., Freeman, A. J., Baerends, E. J., and Post, D., *Phys. Rev. B*, **27**, 2132 (1983).
113. Delley, B., Freeman, A. J., and Ellis, D. E., *Phys. Rev. Lett.*, **50**, 1451 (1983).
114. Holland, G. F., Ellis, D. E., and Trogler, W. C., *J. Chem. Phys.*, **83**, 3507 (1985).
115. Andersen, O. K., and Woolley, R. G., *Mol. Phys.*, **26**, 905 (1973).
116. Andersen, O. K., *Phys. Rev. B*, **12**, 3060 (1975).
117. Gunnarsson, O., Harris, J., and Jones, R. O., *Phys. Rev. B*, **15**, 3027 (1977).
118. Harris, J., and Jones, R. O., *J. Chem. Phys.*, **70**, 830 (1979).
119. Müller, J. E., Jones, R. O., and Harris, J., *J. Chem. Phys.*, **79**, 1874 (1983).
120. Müller, J. E., and Harris, J., preprint.
121. Sambe, H., and Felton, R. H., *J. Chem. Phys.*, **62**, 1122 (1975).
122. Dunlap, B. I., Connolly, J. W. D., and Sabin, J. R., *J. Chem. Phys.*, **71**, 3386, 4993 (1979).
123. Messmer, R. P., and Lamson, S. H., *Chem. Phys. Lett.*, **90**, 31 (1982).
124. Baykara, N. A., Andzelm, J., Salahub, D. R., and Baykara, S. Z., *Int. J. Quantum Chem.*, **29**, 1025 (1986).
125. Baykara, N. A., McMaster, B. N., and Salahub, D. R., *Mol. Phys.*, **52**, 891 (1984).
126. Messmer, R. P., *J. Vac. Sci. Technol. A*, **2**, 899 (1984).
127. Becke, A. D., *J. Chem. Phys.*, **76**, 6037 (1982); *J. Chem. Phys.*, **78**, 4787 (1983); *Phys. Rev. A*, **33**, 2786 (1986).
128. Laaksonen, L., Pyykkö, P., and Sundholm, D., *Int. J. Quantum Chem.*, **23**, 309, 319 (1983); *Int. J. Quantum Chem.*, **27**, 601 (1985).
129. Sundholm, D., Pyykkö, P., and Laaksonen, L., *Finn. Chem. Lett.*, 51 (1985).

130. Andzelm, J., Radzio, E., and Salahub, D. R., *J. Comput. Chem.*, **6**, 520 (1985).
131. Radzio, E., Andzelm, J., and Salahub, D. R., *J. Comput. Chem.*, **6**, 533 (1985).
132. Andzelm, J., Radzio, E., and Salahub, D. R., *J. Chem. Phys.*, **83**, 4573 (1985).
133. Zivkovic, T., and Maksic, Z. B., *J. Chem. Phys.*, **49**, 3083 (1968).
134. Herman, F., and Skillman, S., *Atomic Structure Calculations*, Prentice-Hall, Englewood Cliffs, NJ, 1963.
135. Bonifacic, V., and Huzinaga, S., *J. Chem. Phys.*, **60**, 2779 (1974).
136. Huzinaga S., Klobukowski, M., and Sakai, Y., *J. Phys. Chem.*, **88**, 4880 (1984).
137. Sakai, Y., and Huzinaga, S., *J. Chem. Phys.*, **76**, 2537 (1982).
138. Kitaura, K., Saito, C., and Morokuma, K., *Chem. Phys. Lett.*, **65**, 206 (1979).
139. Katsuki, S., and Taketa, H., *Int. J. Quantum Chem.*, **18**, 25 (1980).
140. Katsuki, S., and Taketa, H., *J. Phys. Soc. Japan*, **50**, 855 (1981).
141. Katsuki, S., and Taketa, H., *Solid State Commun.*, **39**, 711 (1981).
142. Katsuki, S., Taketa, H., and Inokuchi, M., *J. Phys. Soc. Japan*, **52**, 2156 (1983).
143. Katsuki, S., and Inokuchi, M., *J. Phys. Soc. Japan*, **51**, 3652 (1982).
144. Painter, G. S., and Averill, F. W., *Phys. Rev. B*, **26**, 1781 (1982).
145. Grimley, T. B., and Bhasu, V. C. J., *J. Phys. B*, **16**, 1125 (1983).
146. Bernholc, J., and Holzwarth, N. A. W., *Phys. Rev. Lett.*, **50**, 1451 (1983).
147. Bernholc, J., and Holzwarth, N. A. W., *J. Chem. Phys.*, **81**, 3987 (1984).
148. Dhar, S., and Callaway, J., preprint.
149. Dhar, S., Ziegler, A., Kanhere, D. G., and Callaway, J., *J. Chem. Phys.*, **82**, 868 (1985).
150. Lee, K., and Callaway, J., preprint.
151. Lee, K., Callaway, J., Kwang, K., Tang, R., and Ziegler, A., *Phys. Rev. B*, **31**, 1796 (1985).
152. Blaha, P., and Callaway, J., *Phys. Rev. B*, **33**, 1706 (1986).
153. Tang, R., and Callaway, J., preprint.
154. Moore, C. E., *Atomic Energy Levels*, NBS Circular 467, Vols 1 and 2, National Bureau of Standards, Washington DC, 1949.
155. Tavouktsoglu, A. N., and Huzinaga, S., *J. Chem. Phys.*, **72**, 1385 (1980).
156. Clementi, E., and Roetti, C., *At. Data Nucl. Data Tables*, **14**, 177 (1974).
157. Hay, P. J., *J. Chem. Phys.*, **66**, 4377 (1977).
158. Wachters, A. J. H., *J. Chem. Phys.*, **52**, 1033 (1970).
159. Roos, B., Veillard, A., and Vinot, C., *Theor. Chim. Acta*, **20**, 1 (1971).
160. Hyla-Kryspin, I., Demuynck, J., Strich, A., and Bénard, M., *J. Chem. Phys.*, **75**, 3954 (1981).
161. Veillard, A., and Dedieu, A., *Theor. Chim. Acta*, **65**, 215 (1984).
162. Friedlander, M. E., Howell, J. M., and Snyder L. G., *J. Chem. Phys.*, **77**, 1921 (1982).
163. Pietro, W. J., and Hehre, W. J., *J. Comput. Chem.*, **4**, 241 (1983).
164. Huzinaga, S., *J. Chem. Phys.*, **66**, 4245 (1977).
165. Tatewaki, H., Sakai, Y., and Huzinaga, S., *J. Comput. Chem.*, **2**, 278 (1981).
166. Huzinaga, S., Andzelm, J., Klobukowski, M., Radzio-Andzelm, E., Sakai, Y., and Tatewaki, H., *Gaussian Basis Sets for Molecular Calculation*, Elsevier, Amsterdam, 1984.
167. Seijo, L., Barandiaran, Klobukowski, M., and Huzinaga, S., *Chem. Phys. Lett.*, **117**, 151 (1985).
168. Guse, M., Ostlund, N. S., and Blyholder, G. D., *Chem. Phys. Lett.*, **61**, 526 (1979).
169. Martin, R. L., *Chem. Phys. Lett.*, **75**, 290 (1980).
170. Bauschlicher, C. W., *J. Chem. Phys.*, **73**, 2510 (1980).
171. Dunning, T. H., Jr, Botch, B. H., and Harrison, J. F., *J. Chem. Phys.*, **72**, 3419 (1980).

172. Bauschlicher, C. W., and Walch, S. P., *J. Chem. Phys.*, **74**, 5922 (1981).
173. Botch, B. H., Dunning, T. H., Jr, and Harrison, J. F., *J. Chem. Phys.*, **75**, 3466 (1981).
174. Martin, R. L., and Hay, P. J., *J. Chem. Phys.*, **75**, 4539 (1981).
175. Fischer, C. F., *J. Chem. Phys.*, **76**, 1934 (1982).
176. Bauschlicher, C. W., Walch, S. P., and Partridge, H., *J. Chem. Phys.*, **76**, 1033 (1982).
177. Davidson, E. R., and Silver, D. W., *Chem. Phys. Lett.*, **52**, 403 (1977).
178. Hay, P. J., and Wadt, W. R., *J. Chem. Phys.*, **82**, 270, 284, 299 (1985).
179. Goddard, W. A., III, *Phys. Rev.*, **157**, 81 (1967).
180. Hunt, W. J., Hay, P. J., and Goddard, W. A., III, *J. Chem. Phys.*, **57**, 738 (1972).
181. Thomas, L. H., *Proc. Camb. Phil. Soc.*, **23**, 542 (1927); Fermi, E., *Rend. Acad. Naz. Lincei*, **6**, 602 (1927).
182. Harris, J., and Jones, R. O., *J. Chem. Phys.*, **68**, 3316 (1978).
183. Ziegler, T., Rauk, A., and Baerends, E. J., *Theor. Chim. Acta*, **43**, 261 (1977).
184. von Barth, U., *Phys. Rev. A*, **20**, 1693 (1979).
185. Lannoo, M., Baraff, G. A., and Schluter, M., *Phys. Rev. B*, **24**, 943 (1981).
186. Andzelm, J., unpublished.
187. Cole, L. A., and Perdew, J. P., *Phys. Rev. A*, **25**, 1265 (1982).
188. Andzelm, J., and Salahub, D. R., unpublished.
189. Weltner, W., Jr, and Van Zee, R. J., *Annu. Rev. Phys. Chem.*, **35**, 291 (1984) and in Ref. 8, p. 1.
190. Shim, I., *Mat.-Fys. Meddr. Danske Vidensk. Selsk.* (*16 Res. Rep. of Niels Bohr Fellows*), **41**, 147 (1985).
191. Donohue, J. *The Structures of the Elements*, Wiley-Interscience, New York, 1974.
192. Gingerich, K. A., *Curr. Top. Mater. Sci.*, **6**, 345 (1980).
193. Knight, L. B., Jr, Van Zee, R. J., and Weltner, W., Jr, *Chem. Phys. Lett.*, **94**, 296 (1983).
194. Moskovits, M., DiLella, D. P., and Limm, W., *J. Chem. Phys.*, **80**, 626 (1984).
195. Drowart, J., in *Phase Stability in Metals and Alloys* (Eds P. S. Rutman, J. Stringer and R. I., Jaffee), pp. 305–17, McGraw-Hill, New York, 1967.
196. Cossé, C., Fouassier, M., Mejean, T., Tranquille, M., DiLella, D. P., and Moskovits, M., *J. Chem. Phys.*, **73**, 6076 (1980).
197. Moskovits, M., and DiLella, D. P., in *Metal Bonding and Interactions in High Temperature Systems* (Eds J. L. Gole and W. C. Stwalley), ACS Symp. Ser. 179, p. 153, American Chemical Society, Washington DC, 1982.
198. Kant, A., and Lin, S.-S., *J. Chem. Phys.*, **51**, 1644 (1969).
199. Langridge-Smith, P. R. R., Morse, M. D., Hansen, G. P., Smalley, R. E., and Merer, A. J., *J. Chem. Phys.*, **80**, 593 (1984).
200. DiLella, D. P., Limm, W., Lipson, R. H., Moskovits, M., and Taylor, K. V., *J. Chem. Phys.*, **77**, 5263 (1982).
201. Kant, A., and Strauss, B., *J. Chem. Phys.*, **45**, 3161 (1966).
202. Goodgame, M. M., and Goddard, W. A., III, *Phys. Rev. Lett.*, **48**, 135 (1982).
203. Moskovits, M., Limm, W., and Mejean, T., *J. Phys. Chem.*, **89**, 3886 (1985).
204. Moskovits, M., Limm, W., and Mejean, T., *J. Chem. Phys.*, **82**, 4875 (1985).
205. Efremov, Yu. M., Samoilova, A. N., and Gurvich, L. V., *Opt. Spectrosc.*, **36**, 381 (1974).
206. Efremov, Yu. M., Samoilova, A. N., and Gurvich, L. V., *Chem. Phys. Lett.*, **44**, 108 (1976).
207. Efremov, Yu. M., Samoilova, A. N., Kozhukhovsky, V. B., and Gurvich, L. V., *J. Mol. Spectrosc.*, **73**, 430 (1978).

208. Michalopoulos, D. L., Geusic, M. E., Hansen, S. G., Powers, D. E., and Smalley, R. E., *J. Phys. Chem.*, **86**, 3914 (1982).
209. Bondybey, V. E., and English, J. H., *Chem. Phys. Lett.*, **94**, 443 (1983).
210. Riley, S. J., Parks, E. K., Pobo, L. G., and Wexler, S., *J. Chem. Phys.*, **79**, 2577 (1983).
211. Baumann, C. A., Van Zee, R. J., Bhat, S. V., and Weltner, W., Jr, *J. Chem. Phys.*, **74**, 6977 (1981).
212. Baumann, C. A., Van Zee, R. J., Bhat, S. V., and Weltner, W., Jr, *J. Chem. Phys.*, **78**, 190 (1983).
213. Rivoal, J.-C., Shakhs Emampour, J., Zeringue, K. J., and Vala, M., *Chem. Phys. Lett.*, **92**, 313 (1982).
214. Kant, A., Lin, S.-S., and Strauss, B., *J. Chem. Phys.*, **49**, 1983 (1968).
215. Montano, P. A., and Shenoy, G. K., *Solid State Commun.*, **35**, 53 (1980).
216. Purdum, H., Montano, P. A., Shenoy, G. K., and Morrison, T., *Phys. Rev. B*, **25**, 4412 (1982).
217. Shim, I., and Gingerich, K. A., *J. Chem. Phys.*, **77**, 2490 (1982).
218. Moskovits, M., and DiLella, D. P., *J. Chem. Phys.*, **73**, 4917 (1980).
219. Leopold, D. G., and Lineberger, W. C., *J. Chem. Phys.*, **85**, 51 (1986).
220. Ford, T. A., Huber, H., Klotzbücher, W., Kundig, E. P., Moskovits, M., and Ozin, G. A., *J. Chim. Phys.*, **66**, 524 (1977).
221. Kant, A., and Strauss, B., *J. Chem. Phys.*, **41**, 3806 (1964).
222. Shim, I., and Gingerich, K. A., *J. Chem. Phys.*, **78**, 5693 (1983).
223. Ahmed, F., and Nixon, E. R., *J. Chem. Phys.*, **71**, 3547 (1979).
224. Kant, A., *J. Chem. Phys.*, **41**, 1872 (1964).
225. Noell, J. O., Newton, M. D., Hay, P. J., Martin, R. L., and Bobrowicz, *J. Chem. Phys.*, **73**, 2360 (1980).
226. Morse, M. D., Hansen, G. P., Langridge-Smith, P. R. R., Zheng, L.-S., Geusic, M. E., Michalopoulos, D. L., and Smalley, R. E., *J. Chem. Phys.*, **80**, 5400 (1984).
227. Åslund, N., Barrow, R. F., Richards, W. G., and Travis, D. N., *Ark. Fys.*, **30**, 171 (1965).
228. Lochet, J., *J. Phys. B: At. Mol. Phys.*, **11**, L55 (1978).
229. Kleman, B., and Lundqvist, S., *Ark. Fys.*, **8**, 333 (1954).
230. Preuss, D. R., Pace, S. A., and Gole, J. L., *J. Chem. Phys.*, **71**, 3553 (1979).
231. Gole, J. L., English, J. H., and Bondybey, V. E., *J. Phys. Chem.*, **86**, 2560 (1982).
232. Verhaegen, G., Smoes, S., and Drowart, J., *J. Chem. Phys.*, **40**, 239 (1964).
233. Moskovits, M., and Limm, W., *Ultramicroscopy*, in press (1986).
234. Gupta, S. K., and Gingerich, K. A., *J. Chem. Phys.*, **70**, 5350 (1979).
235. Pellin, M. J., Foosnaes, T., and Gruen, D. M., *J. Chem. Phys.*, **74**, 5547 (1981).
236. Gupta, S. K., Atkins, R. M., and Gingerich, K. A., *Inorg. Chem.*, **17**, 3211 (1978).
237. Hopkins, J. B., Langridge-Smith, P. R. R., Morse, M. D., and Smalley, R. E., *J. Chem. Phys.*, **78**, 1627 (1983).
238. Gingerich, K. A., and Cocke, D. L., *J. Chem. Soc., Chem. Commun.*, 536 (1972).
239. Shim, I., and Gingerich, K. A., *J. Chem. Phys.*, **80**, 5107 (1984).
240. Schulze, W., Becker, H. U., Minkwitz, R., and Manzel, K., *Chem. Phys. Lett.*, **55**, 59 (1978).
241. Ruamps, J., *C. R. Acad. Sci. Paris*, **238**, 1489 (1954).
242. Ruamps, J., *Ann. Phys. Paris*, **4**, 1111 (1959).
243. Kleman, B., and Lundqvist, S., *Ark. Fys.*, **9**, 385 (1955).
244. Srdanov, V. I., and Pesic, D. S., *J. Mol. Spectrosc.*, **90**, 27 (1981).
245. Leopold, D. G., Miller, T. M., and Lineberger, W. C., *J. Am. Chem. Soc.*, **108**, 178 (1986).

246. Jansson, K., and Scullman, R., *J. Mol. Spectrosc.*, **61**, 299 (1976).
247. Gupta, S. K., Nappi, B. M., and Gingerich, K. A., *Inorg. Chem.*, **20**, 966 (1981).
248. Ames, L. L., and Barrow, R. F., *Trans. Faraday Soc.*, **63**, 39 (1967).
249. Kleman, B., Lundqvist, S., and Selin, L. E., *Ark. Fys.*, **8**, 505 (1954).
250. Kordis, J., Gingerich, K. A., and Seyse, R. J., *J. Chem. Phys.*, **61**, 5114 (1974).
251. Gingerich, K. A., in Ref. 197, p. 109, 1982.
252. Morse, M. D., Geusic, M. E., Heath, J. R., and Smalley, R. E., *J. Chem. Phys.*, **83**, 2293 (1985).
253. Liu, K., Parks, E. K., Richtsmeier, S. C., Pobo, L. G., and Riley, S. J., *J. Chem. Phys.*, **83**, 2882 (1985).
254. Whetton, R. L., Cox, D. M., Trevor, D. J., and Kaldor, A., *Phys. Rev. Lett.*, **54**, 1494 (1985).
255. McLean, A. D., and Liu, B., *Chem. Phys. Lett.*, **101**, 144 (1983).
256. Dunlap, B. I., *Phys. Rev. A*, **27**, 2217 (1983).
257. Baykara, N. A., McMaster, B. N., and Salahub, D. R., *Mol. Phys.*, **52**, 891 (1984).
258. Messmer, R. P., *J. Vac. Sci. Technol. A*, **2**, 899 (1984).
259. Kao, C. M., and Messmer, R. P., *Phys. Rev. B*, **31**, 4835 (1985).
260. Messmer, R. P., Knudson, S. K., Johnson, K. H., Diamond, J. B., and Yang, C. Y., *Phys. Rev. B*, **13**, 1396 (1976).
261. Yang, C. Y., Johnson, K. H., Salahub, D. R., Kaspar, J., and Messmer, R. P., *Phys. Rev. B*, **24**, 5673 (1981).
262. Salahub, D. R., and Messmer, R. P., *Surf. Sci.*, **106**, 415 (1981).
263. Kaspar, J., and Salahub, D. R., *J. Phys. F: Metal Phys.*, **13**, 311 (1983).
264. E.g. Ziegler, T., Snijders, J. G., and Baerends, E. J., *J. Chem. Phys.*, **74**, 1271 (1981).
265. E.g. Igel, G., Wedig, U., Dolg, M., Fuentealba, P., Preuss, H., and Stoll, H., *J. Chem. Phys.*, **81**, 2737 (1984).
266. Pitzer, K. S., *Int. J. Quantum Chem.*, **25**, 131 (1984).
267. Pyykkö, P., *Adv. Quantum Chem.*, **2**, 353 (1978).
268. Malli, G. (Ed.), *Relativistic Effects in Atoms, Molecules and Solids*, NATO ASI, Ser. B, Plenum, New York, 1983.
269. Cowan, R. D., and Griffin, D. C., *J. Opt. Soc. Am.*, **66**, 1010 (1976).
270. Martin, R. L., and Hay, P. J., *J. Chem. Phys.*, **75**, 4539 (1981).
271. Kahn, L. R., *Int. J. Quantum Chem.*, **25**, 149 (1984).
272. Wood, C., Doran, M., Hillier, I. H., and Guest, M. F., *Faraday Symp. Chem. Soc.*, **14**, 159 (1980).
273. Wolf, A., and Schmidtke, H.-H., *Int. J. Quantum Chem.*, **18**, 1187 (1980).
274. Das, G., *Chem. Phys. Lett.*, **86**, 482 (1982).
275. Walch, S. P., and Bauschlicher, C. W., *Chem. Phys. Lett.*, **94**, 290 (1983).
276. Walch, S. P., and Bauschlicher, C. W., *J. Chem. Phys.*, **79**, 3590 (1983).
277. Fursova, V. D., Klyagina, A. P., Levin, A. A., and Gutsev, G., *Chem. Phys. Lett.*, **116**, 317 (1985).
278. Jeung, G. H., *Chem. Phys. Lett.*, **125**, 407 (1986).
279. Trinquier, G., and Barthelat, J. C., unpublished, cited in Ref. 278, 1986.
280. Walch, S. P., and Bauschlicher, C. W., in Ref. 8, p. 17, 1985.
281. Walch, S. P., Bauschlicher, C. W., Roos, B. O., and Nelin, C. J., *Chem. Phys. Lett.*, **103**, 175 (1983).
282. Das, G. P., and Jaffe, R. L., *Chem. Phys. Lett.*, **109**, 206 (1984).
283. Salahub, D. R., and Baykara, N. A., *Surf. Sci.*, **156**, 605 (1985).
284. Klyagina, A. P., and Levin, A. A., *Koord. Khim.*, **10**, 579 (1984).
285. Langhoff, S. R., and Bauschlicher, C. W., *J. Chem. Phys.*, **84**, 4485 (1986).
286. Goodgame, M. M., and Goddard, W. A., III, *J. Phys. Chem.*, **85**, 215 (1981).

287. Atha, P. M., and Hillier, I. H., *Mol. Phys.*, **45**, 285 (1982).
288. Kok, R. A., and Hall, M. B., *J. Phys. Chem.*, **87**, 715 (1983).
289. Klyagina, A. P., Gutsev, G. L., Fursova, V. S., and Levin, A. A., *Zh. Neorg. Khim.*, **29**, 2765 (1984).
290. Goodgame, M. M., and Goddard, W. A., III, *Phys. Rev. Lett.*, **54**, 661 (1985).
291. Becke, A. D., *J. Chem. Phys.*, **84**, 4524 (1986); *Proc. Coleman Symp. on Density Matrices and Density Functionals*, Kingston, Ontario, August 1985.
292. Ziegler, T., Tschinke, V., and Becke, A., *Polyhedron*, in press (1986) and personal communication.
293. Nesbet, R. K., *Phys. Rev.*, **135**, A460 (1964).
294. Baykara, N. A., and Salahub, D. R., unpublished.
295. Guenzburger, D., and Baggio Saitovich, E. M., *Phys. Rev. B*, **24**, 2368 (1981).
296. Shim, I., and Gingerich, K. A., *J. Chem. Phys.*, **77**, 2490 (1982).
297. Goldstein, E., Flores, C., and Hsia, Y. P., *J. Mol. Struct. Theochem.*, **124**, 191 (1985).
298. Nagarathna, H. M., Montano, P. A., and Naik, V. M., *J. Am. Chem. Soc.*, **105**, 2938 (1983).
299. McNab, T. K., Micklitz, H., and Barrett, P. H., *Phys. Rev. B*, **4**, 3787 (1971).
300. Montano, P. A., Barrett, P. H., and Shanfield, Z., *J. Chem. Phys.*, **64**, 2896 (1976).
301. Montano, P. A., *Faraday Symp. Chem. Soc.*, **14**, 79 (1980).
302. Shim, I., and Gingerich, K. A., *J. Chem. Phys.*, **78**, 5693 (1983).
303. Ratner, M., *Surf. Sci.*, **59**, 279 (1976).
304. Herman, K., and Bagus, P. S., *Phys. Rev. B*, **16**, 4195 (1977).
305. Upton, T. H., and Goddard, W. A., III, *J. Am. Chem. Soc.*, **100**, 5659 (1978).
306. Shim, I., Dahl, J. P., and Johansen, H., *Int. J. Quantum Chem.*, **73**, 311 (1979).
307. Basch, H., Newton, M. D., and Moskowitz, J. W., *J. Chem. Phys.*, **73**, 4492 (1980).
308. Noell, J. O., Newton, M. D., Hay, P. J., Martin, R. L., and Bobrowicz, F. W., *J. Chem. Phys.*, **73**, 2360 (1980).
309. Newton, M. D., *Chem. Phys. Lett.*, **90**, 291 (1982).
310. Snijders, J. G., and Baerends, E. J., *Mol. Phys.*, **33**, 1651 (1977).
311. Fournier, R., and Salahub, D. R., unpublished.
312. Dunlap, B. I., and Yu, H. L., *Chem. Phys. Lett.*, **73**, 525 (1980).
313. Joyes, P., and Leleyter, M., *J. Phys. B: At. Mol. Phys.*, **6**, 150 (1973).
314. Bachmann, C., Demuynck, J., and Veillard, A., *Gazz. Chim. Ital.*, **108**, 389 (1978).
315. Dixon, R. N., and Robertson, I. L., *Mol. Phys.*, **36**, 1099 (1978).
316. Bachmann, C., Demuynck, J., and Veillard, A., *Faraday Symp. Chem. Soc.*, **14**, 170 (1980).
317. Bachmann, C., Demuynck, J., and Veillard, A., in *Growth and Properties of Metal Clusters* (Ed. J. Bourdon), Elsevier, Amsterdam, 1980.
318. Basch, H., *Faraday Symp. Chem. Soc.*, **14**, 149 (1980).
319. Tatewaki, H., and Huzinaga, S., *J. Chem. Phys.*, **72**, 399 (1980).
320. Pelissier, M., *J. Chem. Phys.*, **75**, 775 (1981).
321. Tatewaki, H., Sakai, Y., and Huzinaga, S., *J. Comput. Chem.*, **2**, 96 (1981).
322. Tatewaki, H., Sakai, Y., and Huzinaga, S., *J. Comput. Chem.*, **2**, 278 (1981).
323. Bauschlicher, C. W., Walch, S. P., and Siegbahn, P. E. M., *J. Chem. Phys.*, **76**, 6015 (1982).
324. Jeung, G. H., Barthelat, J. C., and Pelissier, M., *Chem. Phys. Lett.*, **91**, 81 (1982).
325. Tatewaki, H., Miyoshi, E., and Nakamura, T., *J. Chem. Phys.*, **76**, 5073 (1982).
326. Witko, M., and Beckmann, H.-O., *Mol. Phys.*, **47**, 945 (1982).
327. del Conde, G., Bagus, P. S., and Novaro, O., *Phys. Rev. A*, **26**, 3653 (1982).
328. Bauschlicher, C. W., Walch, S. P., and Siegbahn, P. E. M., *J. Chem. Phys.*, **78**, 3347 (1983).

329. Jeung, G. H., and Barthelat, J. C., *J. Chem. Phys.*, **75**, 2097 (1983).
330. Stoll, H., Fuentealba, P., Dolg, M., Szentpaly, L. V., and Preuss, H., *J. Chem. Phys.*, **79**, 5532 (1983).
331. Stoll, H., Fuentealba, P., Schwerdtfeger, P., Flad, J., Szentpaly, L. V., and Preuss, H., *J. Chem. Phys.*, **81**, 2732 (1984).
332. Pauling, L., *J. Chem. Phys.*, **78**, 3346 (1983).
333. Bauschlicher, C. W., Jr, *Chem. Phys. Lett.*, **97**, 204 (1983).
334. Martin, R. L., *J. Chem. Phys.*, **78**, 5840 (1983).
335. Pelissier, M., *J. Chem. Phys.*, **79**, 2099 (1983).
336. Shim, I., and Gingerich, K. A., *J. Chem. Phys.*, **79**, 2903 (1983).
337. Cingi, M. B., Clemente, D. A., and Foglia, C., *Mol. Phys.*, **53**, 301 (1984).
338. Flad, J., Igel-Mann, G., Preuss, H., and Stoll, H., *Chem. Phys.*, **90**, 257 (1984).
339. Sunil, K. K., Jordan, K. D., and Raghavachari, K., *J. Phys. Chem.*, **89**, 457 (1985).
340. Scharf, P., Brode, S., and Ahlrichs, R., *Chem. Phys. Lett.*, **113**, 447 (1985).
341. Werner, H.-J., and Martin, R. L., *Chem. Phys. Lett.*, **113**, 451 (1985).
342. Walch, S. P., and Laskowski, B. C., *J. Chem. Phys.*, **84**, 2734 (1986).
343. Ozin, G. A., Huber, H., McIntosh, D. F., Mitchell, S. A., Norman, J. G., Jr, and Noodleman, L., *J. Am. Chem. Soc.*, **101**, 3504 (1979).
344. Ziegler, T., Snijders, J. G., and Baerends, E. J., *J. Chem. Phys.*, **74**, 1271 (1981).
345. Post, D., and Baerends, E. J., *Chem. Phys. Lett.*, **86**, 176 (1982).
346. Guenzburger, D., *Chem. Phys. Lett.*, **86**, 316 (1982).
347. Delley, B., Ellis, D. E., Freeman, A. J., Baerends, E. J., and Post, D., *Phys. Rev. B*, **27**, 2132 (1983).
348. Radzio, E., Andzelm, J., and Salahub, D. R., unpublished.
349. Wang, S.-W., *J. Chem. Phys.*, **82**, 4633 (1985).
350. Desclaux, J. P., *At Data Nucl. Data Tables*, **12**, 311 (1973).
351. Paldus, J., Cizek, J., and Shavitt, I., *Phys. Rev. A*, **5**, 50 (1972).
352. Meyer, W., *J. Chem. Phys.*, **64**, 2901 (1976).
353. Messmer, R. P., Caves, T. C., and Kao, C. M., *Chem. Phys. Lett.*, **90**, 296 (1982).
354. Stoner, E. C., *Proc. R. Soc. A*, **154**, 656 (1936).
355. Stoner, E. C., *Proc. R. Soc. A*, **169**, 339 (1939).
356. Knight, L. B., Woodward, R. W., Van Zee, R. J., and Weltner, W., Jr, *J. Chem. Phys.*, **79**, 5820 (1983).
357. Cotton, F. A., and Shim, I., *J. Phys. Chem.*, **89**, 952 (1985).
358. Baykara, N. A., and Salahub, D. R., unpublished.
359. Bursten, B. E., Cotton, F. A., and Hall, M. B., *J. Am. Chem. Soc.*, **102**, 6348 (1980).
360. Atha, P. M., Hillier, I. H., and Guest, M. F., *Chem. Phys. Lett.*, **75**, 84 (1980).
361. Klotzbücher, W., Ozin, G. A., Norman, J. G., Jr, and Kolari, J. H., *Inorg. Chem.*, **16**, 2871 (1977).
362. Bursten, B. E., and Cotton, F. A., *Faraday Symp. Chem. Soc.*, **14**, 180 (1980).
363. Norman, J. G., Jr, Kolari, H. J., Gray, H. B., and Trogler, W. C., *Inorg. Chem.*, **16**, 987 (1977).
364. Castro, M., Keller, J., and Mareca, P., *Int. J. Quantum Chem.*, **15S**, 429 (1981).
365. Cotton, F. A., and Shim, I., *J. Am. Chem. Soc.*, **104**, 7025 (1982).
366. Andzelm, J., and Salahub, D. R., unpublished.
367. Garcia-Prieto, J., and Novaro, O., *Int. J. Quantum Chem.*, **18**, 595 (1980).
368. Basch, H., Cohen, D., and Topiol, S., *Isr. J. Chem.*, **19**, 233 (1980).
369. Basch, H., *J. Am. Chem. Soc.*, **103**, 4657 (1981).
370. McLean, A. D., *J. Chem. Phys.*, **79**, 3392 (1983).
371. Klobukowski, M., *J. Comput. Chem.*, **4**, 350 (1983).
372. Flad, J., Igel-Mann, G., Preuss, H., and Stoll, H., *Surf. Sci.*, **156**, 379 (1985).
373. Hay, P. J., and Martin, R. L., *J. Chem. Phys.*, **83**, 5174 (1985).

374. Andreoni, W., and Martins, J. L., *Surf. Sci.*, **156**, 635 (1985).
375. Martins, J. L., and Andreoni, W., *Phys. Rev. A*, **28**, 3637 (1983).
376. Pitzer, K. S., in *Ab Initio Methods in Quantum Chemistry* (Ed. K. P. Lawley), Wiley, Chichester, 1987.
377. Yang, C. Y., in Ref. 268, p. 335, 1983.
378. Morin, M., M.Sc. Thesis, Université de Montréal, 1986.
379. Lee, Y. S., Ermler, W. C., Pitzer, K. S., and McLean, A. D., *J. Chem. Phys.*, **70**, 288 (1979).
380. Ermler, W. C., Lee, Y. S., and Pitzer, K. S., *J. Chem. Phys.*, **70**, 293 (1979).
381. Ziegler, T., *J. Am. Chem. Soc.*, **107**, 4453 (1985).
382. Cotton, F. A., and Walton, R. A., *Multiple Bonds Between Metal Atoms*, Wiley, New York, 1982.
383. Wiest R., and Bénard, M., *Chem. Phys. Lett.*, **98**, 102 (1983).
384. Andzelm, J., and Salahub, D. R., *Int. J. Quantum Chem.*, **29**, 1091 (1986) and unpublished.
385. Kovtecky, J., and Fantucci, P., *Chem. Rev.*, **86**, 539 (1986).
386. Morse, M. D., *Chem. Rev.*, in press.
387. Walch, S. P., Bauschlicher, C. W., and Langhoff, S. R., *J. Chem. Phys.*, **85**, 5900 (1986).
388. von Niessen, W., *J. Chem. Phys.*, **85**, 337 (1986).
389. Rochefort, A., and Salahub, D. R., unpublished.

Ab Initio Methods in Quantum Chemistry—II
Edited by K. P. Lawley

WEAKLY BONDED SYSTEMS

J. H. VAN LENTHE, J. G. C. M. VAN DUIJNEVELDT-VAN DE RIJDT
and F. B. VAN DUIJNEVELDT

Rijkuniversiteit Utrecht, Vakgroep Theoretische Chemie, Padualaan 8, De Uithof, Utrecht, The Netherlands

CONTENTS

I. Introduction . . . 522
II. Components of the Interaction Energy . . . 524
A. Long-range Interactions . . . 524
B. Short-range Effects: Penetration . . . 526
C. Short-range Effects: Exchange . . . 527
D. Charge-transfer Energy . . . 527
E. Intramolecular Correlation Effects . . . 528
F. Energy Components in the Supermolecular Approach . . . 528
1. The SCF Case . . . 529
2. The Better-than-SCF Case . . . 530
III. Supermolecular Methods for Calculating Interaction Energies. . . . 531
A. Self-consistent Field plus Dispersion Method . . . 531
B. Second-order Møller–Plesset Theory . . . 532
C. Coupled Electron-pair Approximation and Related Methods . . . 533
D. Limitations of the Coupled Electron-pair Approach . . . 534
IV. Basis Sets . . . 535
A. Long-range Terms . . . 536
B. Short-range Terms . . . 540
V. The Basis-set Superposition Error: Historical Background . . . 544
A. The Counterpoise Method . . . 544
B. Different Interpretations of the Basis-set Superposition Error . . . 545
C. Alternative Counterpoise Scheme . . . 546
D. Experiences With the Full and Virtuals-only Counterpoise Schemes 547
E. Ways of Avoiding Basis-set Superposition Errors . . . 551
VI. Justification of the Full Counterpoise Procedure . . . 552
A. The Heitler–London Interaction Energy ΔE^{HL} May Contain Basis-set Superposition Errors . . . 553
B. Monomer A in Dimer AB Can Recover the Full Counterpoise Correction δ^{CP} . . . 555
C. Practical Considerations . . . 557

VII. Conclusions 558
Acknowledgements 559
References 560

I. INTRODUCTION

The foundations for our present understanding of intermolecular forces were laid in the first decades of this century. First, Keesom[1], Debye[2] and Falckenhagen[3] elucidated the role played by permanent electric moments and polarizabilities. After the advent of quantum mechanics, Heitler and London[4] identified the exchange forces which keep molecules apart, and London[5,6] discovered the dispersion forces which explained such puzzling phenomena as the condensation of noble gases.

The quantitative evaluation of the corresponding interaction energies had to await the development of computers and *ab initio* systems in the 1960s. By the early 1970s it was apparent that self-consistent field (SCF) theory provides a reasonably accurate description of hydrogen-bonded complexes like $(H_2O)_2$ [7–9], while theories that explicitly account for electron correlation[10–13] must be used for systems which are predominantly bound by dispersion forces, such as $He–H_2$ and He_2. Rapid developments in both hardware and software have since taken place and *ab initio* calculations on weakly bound systems are now routinely being carried out. Useful information is gathered in this way and the potential surfaces obtained find application in simulation studies of liquids, solids and various solvation problems[14,15].

The precision of present-day *ab initio* results is not very high, however. This is evident from the discrepancies between results obtained in different calculations on the same system. Worse still, calculations which introduce methodological improvements do not always lead to better agreement with experiment. The water dimer provides a case in point. Hartree–Fock (HF) calculations in a 6-31 + G(d) basis, which by many would be regarded as of good quality, yield $\Delta E = 5.4\,\mathrm{kcal\,mol^{-1}}$ at $R = 2.96$ Å[16], both in good agreement ($5.4 \pm 0.7\,\mathrm{kcal\,mol^{-1}}$, 2.98 ± 0.01 Å[17]). Improvement of the basis set (at the HF level) spoils the agreement, yielding $3.8\,\mathrm{kcal\,mol^{-1}}$ at 3.05 Å[18]. Keeping the same basis, but introducing correlation at the second-order Møller–Plesset (MP2) level spoils the agreement the other way, yielding about $7\,\mathrm{kcal\,mol^{-1}}$ at 2.90 Å[16]. (Clearly a combination of these two improvements yields more perspective[18].)

The conclusion to be drawn from these and many similar observations is that good agreement frequently comes about through a cancellation of errors, and in the absence of foreknowledge about the behaviour of these errors in any particular example, this severely detracts from the predictive power of *ab initio* theory.

There are a number of applications for which a theory with increased predictive power would be most welcome. The wealth of experimental data on weakly bonded systems that has been gathered, primarily in various molecular-beam experiments, can only be adequately interpreted if precise potential energy surfaces are available. It is a laborious (and somewhat hazardous) task to assemble such potentials in an empirical manner, and a truly predictive theory would speed up the interpretation process considerably. The possibility of calculating potential energy surfaces as a function of both intra- and intermolecular degrees of freedom will be especially helpful in studying various vibration and relaxation processes, leading to a better understanding of the kinetics and thermodynamics of the system. Many weakly bonded systems have potential surfaces with several local minima and only a theory in which all errors are under control may be expected to yield a correct ordering of the associated binding energies, etc. Finally, the theory should be able to provide accurate information on three-body effects in view of their relevance to bulk simulations.

Our aim in this review is to highlight the main problems in current *ab initio* methods, and to discuss ways of surmounting these problems, in the hope that this will contribute to the formulation of a theory with larger predictive power. The problems to be discussed are the following. First, which of the current methods of allowing for electron correlation, e.g. Møller–Plesset (MP) theory, many-body perturbation theory (MBPT), coupled electron-pair approximations (CEPA), are adequate? Secondly, which errors are caused by the use of limited basis sets? Finally, can basis-set superposition errors be avoided or corrected for? One might sympathize with the authors of the following quotation[19]: 'In general except for the case of He_2, there is no immediate prospect of calculating an accurate potential.' Paradoxically, it turns out that He_2 is one of the more difficult systems to treat accurately, and on the whole, our outlook, expressed in the final section, is definitely more optimistic.

Since we focus on methodological problems, little space will be devoted to the properties and peculiarities of specific systems. Many of these are discussed in the book by Hobza and Zahradnik[20]. *Ab initio* calculations on hydrogen-bonding systems have recently been reviewed by Beyer *et al*[21]. Interactions between non-polar molecules were treated by van der Avoird *et al.*[13], while noble-gas interactions were covered in the monograph by Maitland *et al.*[19].

The ability to obtain accurate potential energy surfaces is only a first step towards a full description of weakly bonded systems. A next step must be the determination of the intermolecular vibrational modes, in order to interpret the wealth of spectroscopic data available for these systems[22,23], and in order to estimate the zero-point vibrational energy correction that must be applied before calculated binding energies can be compared to experimental enthalpies of formation[24]. The appropriate force constants and frequencies are routinely produced at the harmonic level by the more recent *ab initio* program packages that employ gradient techniques, such as GAUSSIAN 80[25] and

GAMESS[26]. The vibrational data obtained in this way should, however, be used with care, since many weakly bonded systems exhibit large-amplitude motions (especially when simple hydrides are involved) which cannot be accurately modelled at the harmonic level. Indeed, the very concept of the 'equilibrium structure' of such complexes, while formally valid, loses much of its significance, and a much larger section of the surface must be sampled than is customary for ordinary molecules. Appropriate techniques for evaluating vibrational wavefunctions beyond the harmonic level have been reviewed by Le Roy *et al.*[27] and very recently by Briels *et al.*[15] and the interested reader is referred to these.

A final aspect not considered here is the search for analytical fits to potential energy surfaces given in numerical form on a grid of points, or, more generally, the development of simple models that can be applied to larger systems where the direct evaluation of ΔE by *ab initio* methods becomes impracticable. Recent activities in this field include the use of the distributed multipole model[28,29] for predicting the equilibrium structures of van der Waals complexes[30,31]. The legitimacy of applying electrostatics has been questioned[32], but in view of the well documented performance of the electrostatic model in rationalizing geometries of complexes[9,33–36] these criticisms seem to be unjustified. Mention should also be made of the so-called test-particle model[37,38] in which a nitrogen atom is used as a probe to determine the parameters of the repulsive $\exp(-R)$ terms in a site–site potential model that should be applicable to large molecules. Work along these lines can contribute to a better understanding of molecular liquids[39] and molecular solids[15].

II. COMPONENTS OF THE INTERACTION ENERGY

Knowledge of the separate terms that make up the interaction energy of a given complex can be useful for a variety of reasons. It helps to understand why the complex is formed, in relation to properties of the separate molecules. But it also offers a way of checking the accuracy of a calculation by requiring that not only the total energy but also its components must become stable with respect to further improvements in the description of the system. Finally, a knowledge of the separate terms facilitates the search for accurate fitted analytical potential energy functions, which is a necessary step if the results are to find further use in simulations of larger systems. Detailed reviews on these energy terms are available[12,13,40–42].

A. Long-range Interactions

The energy components are most naturally divided into long-range and short-range terms. The long-range terms may be defined in terms of London's[5,6] perturbation theory leading to electrostatic, induction and dispersion energies. For molecules with Hamiltonians H^{A} and H^{B} and unperturbed ground-state wavefunctions ψ_0^A and ψ_0^B the first-order electrostatic Coulomb energy is given by

$$E_{\text{Coul}} = \langle \psi_0^A \psi_0^B | V | \psi_0^A \psi_0^B \rangle \tag{1}$$

where $V = H^{AB} - H^{A} - H^{B}$ collects all intermolecular electrostatic interaction operators. At long range E_{Coul} may be written as a sum of the dipole–dipole, dipole–quadrupole, ... interactions between the molecular permanent multipole moments, by invoking the multipole expansion of V with respect to one origin in A and one in B. Several definitions for these moments are currently being used; see Refs 13, 43 and 44 for recent summaries.

Alternatively, distributed multipole expansions or distributed point-charge descriptions may be employed in which charges or multipoles are placed at a variety of centres in the molecule[28,29,45–49]. These have the advantage that a nearly converged multipole energy can be obtained even at a relatively short intermolecular separation, and without requiring multipoles of very high order. For example, models with a charge on each hydrogen atom, a charge and a dipole on each bond centre, and a charge plus dipole plus quadrupole on each first-row atom can give excellent results[28,29,46,49].

The induction and dispersion energies arising from mutual polarization are in second-order perturbation theory given by

$$E_{\text{ind}} = \sum_s{}' \frac{|\langle \psi_0^A \psi_0^B | V | \psi_0^A \psi_s^B \rangle|^2}{E_0^B - E_s^B} \qquad \text{(B by A)}$$

$$+ \sum_r{}' \frac{|\langle \psi_0^A \psi_0^B | V | \psi_r^A \psi_0^B \rangle|^2}{E_0^A - E_r^A} \qquad \text{(A by B)} \tag{2}$$

which describes the energy lowering due to polarization of B in the field of A and vice versa, and

$$E_{\text{disp}} = \sum_r{}' \sum_s{}' \frac{|\langle \psi_0^A \psi_0^B | V | \psi_r^A \psi_s^B \rangle|^2}{E_0^A - E_r^A + E_0^B - E_s^B} \tag{3}$$

which describes the energy lowering associated with polarization by instantaneous fluctuations in the charge distributions of A and B. Here, ψ_r^A and ψ_s^B are unperturbed singly excited wavefunctions of the monomers. At long range, the multipole expansion of V leads to expressions for E_{ind} (B by A) in terms of the permanent moments of A and the static polarizabilities of B[13,50]. (When the perturbation theory is carried beyond second order, the hyperpolarizabilities come into play as well[40].) Similarly, E_{disp} may be reduced, by applying the Casimir–Polder relation, to a closed expression involving the dynamic (i.e. frequency-dependent) polarizabilities $\alpha(i\omega)$ of the separate molecules[40]. This approach offers the advantage that once the $\alpha(i\omega)$ have been obtained for a series of molecules, the corresponding C_n coefficients in the R^{-1} expansion of the dispersion energy for any given pair of these molecules can be evaluated by a simple integration. In the past, experimental data were used in constructing the $\alpha(i\omega)$ functions[51], but more recently efficient *ab initio* techniques have been developed to evaluate $\alpha(i\omega)$ at the (TDCHF) and at the correlated level[52–55].

The applicability of these one-centre multipole-expanded methods to larger molecules is limited, since the convergence will be poor. Descriptions involving bond–bond or atom–atom interactions have therefore been proposed in the past[11,12]. A rigorous method to distribute a molecule's polarizability over more than one centre has been described very recently[56]. Similar in spirit is a method in which the V-expanded numerators in Eq. (3) are replaced by interactions between transition multipoles at well chosen centres[57]. Finally, we mention a recent method to evaluate (3) at the TDCHF level, but avoiding the expansion of V[58,59]. This should in principle be applicable to large molecules, but the computational effort involved is large unless further approximations are made[58,59].

B. Short-range Effects: Penetration

In the study of weakly bound systems the range of distances around the van der Waals minimum is usually the one of primary interest. Here the performance of the long-range terms in their multipole-expanded forms is somewhat problematic. First, the multipole series may not converge sufficiently fast or not at all and, secondly, penetration effects become non-negligible.

While the convergence problem may be somewhat alleviated by using distributed multipole descriptions, the penetration effects can only be recovered by switching to expressions invoking V in its unexpanded form. The origin of the penetration effects is the different R dependence of electron–electron, electron–nuclear and nuclear–nuclear electrostatic interactions as the charge clouds penetrate one another. For $R \rightarrow 0$ the latter term tends to infinity, whereas the other interaction terms remain finite, and hence the sum of these terms deviates from the multipole result, in which all charges are treated on an equal footing. Although the precise form taken by the penetration energy would seem to depend on the description chosen for the multipole series[60] there is agreement that penetration effects are far from negligible. For example, E_{Coul} for the cyclic geometry of HCOOH dimer at R_e, which equals 36.2 kcal mol^{-1}, contains about 5.5 kcal mol^{-1} penetration attraction[61]. Also, from the data in Table 2 of Ref. 62 one may deduce that for linear $(HF)_2$ with $R = 5.0$ a.u., penetration accounts for about 0.0015 hartree out of a total E_{Coul} of 0.0101 hartree. However, penetration effects fall off exponentially with distance and at $R = 6.0$ a.u., just outside the van der Waals minimum, they account for less than 3% of E_{Coul} in $(HF)_2$. For extremely diffuse systems, such as LiH, but also for systems with lone pairs pointing in each other's direction like linear N_2–N_2, the effects are larger[62].

Penetration effects also occur in the second-order long-range terms[63,64] and in E_{disp} they lead to reduced attraction compared to the multipole-expanded result. The effects may be accounted for by using Eqs (2) and (3) in non-expanded form or (in the case of E_{disp}) by applying suitable damping functions to the long-range C_n/R^n terms[65–67].

C. Short-range Effects: Exchange

London's perturbtation theory was formulated on the basis of the unperturbed product states $\psi_r^A \psi_s^B$ that are eigenfunctions of $H^A + H^B$. These states do not satisfy the Pauli principle (the electrons of A being distinguishable from those of B) and hence at short R one must adopt a different approach. The remedy is to use antisymmetrized states $\mathscr{A}\psi_r^A \psi_s^B$ which, however, are not eigenfunctions of $H^A + H^B$. The form of perturbation theory that can deal with these problems is now known as symmetry-adapted perturbation theory (SAPT)[41]. It may be formulated in such a way that ΔE comprises, in addition to each long-range term, a corresponding exchange term. The most important of these is the first-order exchange energy E_{exch} which in the majority of complexes is the main repulsive term and takes the form

$$E_{exch} = \frac{\langle \mathscr{P}\psi_0^A \psi_0^B | V | \psi_0^A \psi_0^B \rangle - \langle \psi_0^A \psi_0^B | V | \psi_0^A \psi_0^B \rangle \langle \mathscr{P}\psi_0^A \psi_0^B | \psi_0^A \psi_0^B \rangle}{\langle \mathscr{A}\psi_0^A \psi_0^B | \psi_0^A \psi_0^B \rangle} \tag{4}$$

where $\mathscr{P} = (\mathscr{A} - 1)$ is a sum of permutation operators exchanging one, two or more pairs of electrons between A and B. Each additional exchange introduces two additional overlap densities in the integrals and so the $O(S^2)$ single-exchange terms dominate in (4). Owing to the presence of these overlap densities E_{exch} decreases exponentially with increasing distance, and it is this rapidly varying repulsion which keeps molecules apart. For later reference it should be mentioned that the total first-order interaction energy may be written in the form

$$E^{(1)} = E_{Coul} + E_{exch} = \frac{\langle \mathscr{A}\psi_0^A \psi_0^B | V | \psi_0^A \psi_0^B \rangle}{\langle \mathscr{A}\psi_0^A \psi_0^B | \psi_0^A \psi_0^B \rangle} \tag{5}$$

In second order one has the $E_{exch\text{-}ind}$ and $E_{exch\text{-}disp}$ energies which were found to be repulsive in the few systems for which they have been evaluated[68]. For example, in Ne_2 they quench some 5% of the non-expanded E_{disp} (as given by (3)) in the region of the van der Waals minimum.

D. Charge-transfer Energy

The terms defined in (1)–(5) provide a complete description (to second order) of ΔE in any complex. Now there are many complexes in which the ground state of AB acquires some ionic character A^+B^-. Formally, this transfer of a single electron may be viewed as a special case of polarization of A by B, and hence the resulting energy lowering, the charge-transfer energy E_{CT}, forms part of E_{ind}, Eq. (2). However, E_{CT} depends quadratically on overlap densities between A and B[42] and hence its R dependence is more like that of $E_{exch-ind}$ than that of E_{ind} itself. For these reasons E_{CT} is often discussed separately from E_{ind}, in combination with the $E_{exch-ind}$ terms.

E. Intramolecular Correlation Effects

Most of the previous discussion has tacitly assumed the availability of the exact eigenfunctions of A and B. These are usually not available, and (1)–(5) may thus in principle be calculated by using wavefunctions ψ^{A}, etc., determined in the self-consistent field (SCF) or configuration-interaction (CI) approximations. Indeed, E_{Coul} evaluated from SCF monomer wavefunctions is one of the terms arising in the current energy partitioning schemes for SCF interaction energies[36,69,70]. Estimates of E_{disp}, to be added to SCF interaction energies, are as a rule obtained by using $\psi_0^{A,SCF}\psi_0^{B,SCF}$ as the unperturbed product state[71]. A formal justification of these procedures has been given in the double-perturbation variant of SAPT[41] in which the sum of the monomer Fock operators is used as H^0, rather than $H^A + H^B$. The difference ($H^A + H^B - H^0$(Fock)) is treated as a second perturbation, and can be used to estimate correlation corrections to (1)–(5) as given at the SCF level. Alternatively, these correlation corrections may be determined by inserting CI wavefunctions directly into (1)–(5). In this way it has been shown, for example, that intramolecular correlation in He leads to an increase of about 10% in the He–He first-order exchange energy[72]. Similar calculations have been performed for Be–Be[73] and for H_2–H_2[74] and depending on the case both increase and decrease of E_{exch} were observed.

F. Energy Components in the Supermolecular Approach

In the supermolecular approach the interaction energy ΔE of a complex is obtained by evaluating the total energy E_{AB}^{X} of the AB supermolecule using method X (X = SCF, CI, CEPA, MPn, CPF, ...) and subtracting the energy of the monomers:

$$\Delta E^{X} = E_{AB}^{X} - E_{A}^{X} - E_{B}^{X} \tag{6}$$

The connection with the perturbation method may be made by assuming the exact eigenfunctions ψ_0^A and ψ_0^B to be available. Using $\mathscr{A}\psi_0^A\psi_0^B$ as a zeroth-order dimer wavefunction the dimer energy is obtained as

$$E_{AB}^{HL} = \frac{\langle \mathscr{A}\psi_0^A\psi_0^B | H | \mathscr{A}\psi_0^A\psi_0^B \rangle}{\langle \mathscr{A}\psi_0^A\psi_0^B | \mathscr{A}\psi_0^A\psi_0^B \rangle} \tag{7}$$

This quantity will be referred to as the dimer Heitler–London energy. Using $[H, \mathscr{A}] = 0$ and $H = H^0 + V$ it may be reduced to

$$E_{AB}^{HL} = \frac{\langle \mathscr{A}\psi_0^A\psi_0^B | H^0 + V | \psi_0^A\psi_0^B \rangle}{\langle \mathscr{A}\psi_0^A\psi_0^B | \psi_0^A\psi_0^B \rangle} \tag{8}$$

from which the Heitler–London (first-order) interaction energy is found to be, using $H^0\psi_0^A\psi_0^B = (E_A + E_B)\psi_0^A\psi_0^B$,

$$\Delta E^{HL} = E_{AB}^{HL} - E_A - E_B = \frac{\langle \mathscr{A}\psi_0^A\psi_0^B | V | \psi_0^A\psi_0^B \rangle}{\langle \mathscr{A}\psi_0^A\psi_0^B | \psi_0^A\psi_0^B \rangle} \tag{9}$$

which is seen to be identical with the perturbation result $E^{(1)}$ of Eq. (5).

1. *The SCF Case*

Going one step further, one may analyse the case of an SCF (i.e. single-configuration) dimer calculation. The zeroth-order wavefunction $\mathscr{A}\psi_0^A\psi_0^B$ is equivalent to a single configuration and hence is a valid first step to consider before going on to the converged dimer SCF wavefunction. Substituting it in (7) one now finds (with $E_A + E_B = E^0$)

$$\Delta E^{HL} = \frac{\langle \mathscr{A}\psi_0^A\psi_0^B | V | \psi_0^A\psi_0^B \rangle}{\langle \mathscr{A}\psi_0^A\psi_0^B | \psi_0^A\psi_0^B \rangle} + \frac{\langle \mathscr{A}\psi_0^A\psi_0^B | H^0 - E^0 | \psi_0^A\psi_0^B \rangle}{\langle \mathscr{A}\psi_0^A\psi_0^B | \psi_0^A\psi_0^B \rangle} = E^{(1)} + \Delta \tag{10}$$

where $E^{(1)}$ takes the same form as (5), but the second term (Δ) is not zero because SCF functions are not eigenfunctions of H. The presence of Δ means that ΔE^{HL}, unlike $E^{(1)}$, cannot be regarded as a pure interaction energy, and so the use of ΔE^{HL} seems problematical. However, this view is too pessimistic. By expanding $\mathscr{A}$ it can be shown that for exact Hartree–Fock solutions the so-called Landshoff terms in Δ[75] are zero, and only some very small terms of $O(S^4)$ remain[76]. Hence at the Hartree–Fock limit ΔE^{HL} and $E^{(1)}$ will be almost indistinguishable[77]. For the more common case that ψ_0^A is determined in a finite set at A and ψ_0^B in a finite set at B, Δ will be of $O(S^2)$ and so will not necessarily be small[78]. However, it has been shown[79] that Δ may be regarded as a correction for the incompleteness of the monomer basis sets, ΔE^{HL} being a better approximation to the exact ΔE^{HL} than is $E^{(1)}$ evaluated in the same finite bases. (Extensions of this result are possible which involve a reinterpretation of E_{AB}^{HL}(SCF) leading to $\Delta = O(S^4)$, but this point will be taken up in Section VI.)

From the previous discussion it follows that at the SCF level ΔE^{HL} is a reasonable tool for partitioning SCF interaction energies obtained from Eq. (6) into terms which correspond to those obtained in the perturbation approach. In the usual partitioning schemes[36,69], ΔE^{HL} itself is partitioned according to

$$\Delta E^{HL} = E_{Coul} + E_{exch} \tag{11}$$

where E_{Coul} is evaluated from (1) by inserting monomer SCF functions. Next, the energy gained in the SCF process, starting from $E_{\text{AB}}^{\text{HL}}$ (Eq. (7)) is called the 'second-order energy' or 'delocalization energy'

$$\Delta E^{(2)} = E_{\text{AB}}^{\text{SCF}} - E_{\text{AB}}^{\text{HL}} \tag{12}$$

which may be further partitioned by separate calculation of E_{ind} and, in some schemes, E_{CT} [36]. In the present review we use

$$\Delta E^{(2)} = E_{\text{ind}} + E_{\text{CTX}} \tag{13}$$

where E_{ind} is evaluated perturbationally (Eq. (2)) and E_{CTX} collects E_{CT} and $E_{\text{exch-ind}}$, all of which are short-range terms. Note that since we are dealing with a variational method, $\Delta E^{(2)}$ will contain not only second-order but higher-order effects as well.

The total SCF interaction energy is given by

$$\Delta E^{\text{SCF}} = \Delta E^{\text{HL}} + \Delta E^{(2)} \tag{14}$$

In the form derived here it will be contaminated with the basis-set superposition error (BSSE) (cf. Section V). Moreover, since dispersion energy is a correlation effect, ΔE^{SCF} will not contain E_{disp}.

2. The Better-than-SCF Case

Here we discuss methods X that allow for electron correlation. In most of these, an SCF calculation is performed first, and some degree of CI is performed using ψ_{SCF} as the reference configuration. Supermolecule calculations at this level have the advantage over SCF that E_{disp} can be accounted for, and it becomes possible to obtain correlation corrections to the various terms in Eqs (1)–(4). The rigorous analysis of ΔE^{HL} at this level of theory is a complicated problem which has so far received little attention. In fact ΔE^{HL} may not even be the appropriate first-order energy to consider in the context of methods with a non-variational energy expression (e.g. MP2, MP4, CEPA, ...). Another problem is that the zeroth-order wavefunction $\mathscr{A}\psi_0^{\text{A,X}}\psi_0^{\text{B,X}}$ (e.g. an antisymmetrized product of two singles and doubles CI (CI(SD)) wavefunctions) will usually contain higher excitations than are present in the X description of the dimer, and so an analysis of ΔE^{HL} along the lines discussed above is not a valid first step in interpreting the final ΔE^{X} answer. This problem can only be rigorously solved by employing a full CI description throughout, but this is clearly impracticable. It may also be approximately solved by employing a size-consistent description (CEPA, (CPF), (CCD)), for here the missing excitations would implicitly be allowed for in calculating the dimer energy. We take up this point in Section III.

Because of these complexities it is customary to partition CI dimerization energies in a different way, namely as

$$\Delta E^{\mathrm{CI}} = \Delta E^{\mathrm{SCF}} + \Delta E^{\mathrm{corr}} \tag{15}$$

where ΔE^{SCF} can be analysed as indicated above, and ΔE^{corr} lumps together the correlation corrections that are achieved in the chosen CI approach. They can be further analysed by partitioning ΔE^{corr} into intra- and intermolecular contributions to the interaction energy.

III. SUPERMOLECULAR METHODS FOR CALCULATING INTERACTION ENERGIES

A. Self-consistent Field plus Dispersion Method

The simplest *ab initio* method used to study van der Waals molecules is the Hartree–Fock method. To the Hartree–Fock interaction energy one should add the dispersion energy, which ranges from some 20% of the dissociation energy for the water dimer to twice the van der Waals well depth for He_2, and is thus not negligible. The combined method may be termed SCF + Disp.

Some methods to calculate the dispersion energy were discussed in Section II. The integrals from the supermolecular Hartree–Fock calculation may be used in the evaluation of formula (3) using the non-expanded potential[6,71,80,81], yielding the polarization dispersion energy[82]. Either orbital energies (Møller–Plesset partitioning[83]) or state energies (Epstein–Nesbet partitioning[84,85]) may be used in the denominators. This method has recently been extended to open-shell molecules[86,87].

The inclusion of exchange corrections to the dispersion energy, not accounted for in the polarization approximation, is achieved in more expensive valence-bond (VB)[88–91] and CI (dispersion CI)[92,93] techniques, in which antisymmetrized excited configurations are employed. In these approaches only those configurations are included in the calculation that result from the simultaneous excitation of one electron out of the properly localized occupied spaces of each monomer. The (repulsive) exchange corrections are rather small (see Section II) and the complexity of these approaches approximates that of more complete CI or VB calculations. Both the polarization approach and the dispersion VB and dispersion CI methods[92] yield dispersion energies at the uncoupled Hartree–Fock (UCHF) level, the only differences being the choice of expansion functions and zeroth-order Hamiltonian. Though the corresponding infinite order, provided it converges, yields the same result in all these methods, the second-order perturbation dispersion energies may be quite different in different approaches; c.f. Ref. 94. More accurate results are obtained in the variational coupled Hartree–Fock (CHF) method, where the wavefunction adapts in a self-consistent way to the, in this case time-dependent, perturbation. The difference between CHF and

TABLE I
C_6 dispersion coefficients for He_2 and Ar_2.

	He–He	Ar–Ar
'Experiment'	1.46[256]	65[96]
MP2	1.12[257]	74[260]
EN	1.45[257]	77[97]
Disp CI/Tamm–Dancoff	1.55[98]	85[258]
TDCHF	1.37[98]	61[259]
CEPA-1	1.37[258]	58[258]

UCHF is called higher-order self-consistency or apparent correlation[68,95]. To illustrate the differences in the C_6 van der Waals coefficients in the various approaches we present some values for He_2 and Ar_2 in Table I. It is seen that the Epstein–Nesbet variant gives results close to experimental values for He_2 [99] and Ne_2 [68], but that it overestimates C_6 for heavier systems[100] where the Møller–Plesset variant is closer to experiment. Although an extension of the CHF method avoiding the multipole expansion of V has been proposed[58,59], most existing applications of CHF theory are for long-range dispersion coefficients[53–55], so damping functions have to be applied.

The SCF + Disp method is frequently applied since it is relatively cheap and thus allows the calculation of many points on the potential energy surface. A useful potential may be obtained by empirical adjustments to the potential. For example, a damping function may be adjusted using experimental data[101,102] or various components of a fitted dispersion energy may be scaled to more reasonable values[103]. The Hartree–Fock interaction energy may also be improved, by replacing its multipole and induction components by energies derived from multipoles obtained from better calculations (e.g. CI) or experiment; cf. Refs 78 and 104–107. While these procedures may yield reasonable potentials, we feel that the predictive power of the SCF + Disp approach is limited.

B. Second-order Møller–Plesset Theory

The next level of sophistication is to include inter- as well as intra-correlation effects using, as advocated by Pople[108], Møller–Plesset perturbation theory[83], a true supermolecular method subject to BSSE (cf. Section V). We restrict our discussion here to the frequently applied second-order variant (MP2).

The dispersion energy (inter-monomer correlation) in the MP2 appraoch is obtained at the uncoupled Hartree–Fock level[109] as discussed above. At longe range the result is identical to that of the polarization approach using the Møller–Plesset partitioning. No coupling between correlation effects is included in the MP2 approach, so dispersion interaction between uncorrelated (Hartree–Fock) monomers is described.

The intra-monomer correlation affects the (intermolecular) interaction energy through matrix elements of the form[110]

$$\langle \mathscr{A}\Psi_A^{(0)}\Psi_B^{(1)}|V|\mathscr{A}\Psi_A^{(0)}\Psi_B^{(1)}\rangle \tag{16}$$

where $\Psi^{(0)}$ is the Hartree–Fock wavefunction and $\Psi^{(1)}$ is the wavefunction at the MP2 level. This indicates that the MP2 method yields the interaction between Hartree–Fock monomer A and MP2-correlated monomer B and vice versa, which should account for the bulk of the correlation correction to ΔE. From the results of Diercksen *et al.*[111] one may infer that MP2 wavefunctions (i.e. the first-order wavefunction) give in response to a perturbation an interaction energy corresponding to reasonable multipole moments. More detailed research, using large basis sets, is required to establish firmly the reliability of the MP2 approach.

C. Coupled Electron-pair Approximation and Related Methods

The methods discussed so far require a computational time proportional to n^4, where n is the number of basis functions used. To go up one level to more sophisticated methods, CI, CEPA or higher-order MBPT, increases the computation time to an n^5–n^6 dependence. It is therefore not surprising that few studies at this level on larger van der Waals molecules have appeared, e.g. $(H_2O)_2$ [16,93,112–115], N_2–N_2 [103], Ne–Na_2 [116], N_2–HF, CO–HF, HCN–HF [117], Ar–HCl [100], $(AH_n)_2$ complexes (A = N, O, F, P, S, Cl)[16] and He–O_2 [118]. We may expect many more calculations at this level in the near future, as computers become rapidly more powerful.

A major requirement that any method should satisfy, and which is satisfied by all non CI- or VB-like methods discussed previously, is that of size extensivity[119,120]. This implies that the energy of n non-interacting pairs of electrons should be equal to the sum of the n energies of the electron pairs, e.g. a calculation on the supersystem at infinite distance between the monomers should yield the sum of the monomer energies (size consistency). An intrinsically size-extensive method is a coupled cluster (CC) or CEPA related method[94,121,122]. The higher-order MBPT approaches may be viewed as approximations to coupled pair methods[94]. For example, a MBPT calculation employing only double excitations is, if summed to infinite order, equivalent to L-CPMET (CEPAO)[123]. It is not obvious that a few higher orders of Møller–Plesset perturbation theory yield significantly more accurate results, since the convergence of Møller–Plesset perturbation theory is slow[124], the perturbation ('electron correlation') being not small. Furthermore, Frisch *et al.*[16] show that the third- and fourth-order Møller–Plesset contributions cancel each other approximately. The familiar method of

TABLE II
Effect of size-consistency corrections on the interaction energy.

		CI	'Size-consistent'	
$(H_2O)_2$ [113]		SCF BSSE corr.	Davidson corr.	
	D_e (hartree)	−0.008 618	−0.008 962	
	(kcal mol^{-1})	−5.41	−5.62	
$(H_2O)_2$ [115]— geometry of Ref. 128		Full BSSE corr.	Pople corr.	CEPA
DZP′ basis	D_e (hartree)	−0.006 245	−0.006 329	−0.006 337
	(kcal mol^{-1})	−3.92	−3.97	−3.98
DZPP basis	D_e (hartree)	−0.006 406	−0.006 446	−0.006 453
	(kcal mol^{-1})	−4.02	−4.04	−4.05
EZPP basis	D_e (hartree)	−0.006 494	−0.006 523	−0.006 520
	(kcal mol^{-1})	−4.07	−4.09	−4.09
Ar–HCl [128]		Full BSSE corr.	Pople corr.	
Basis 1	D_e (μhartree)	−0.000 122	−0.000 383	

configuration interaction (cf. Ref. 125) involving all single and double excitations out of a Hartree–Fock configuration does not satisfy the requirement of size extensitivity. In principle this also applies to more restricted forms of CI like the aforementioned dispersion CI.

The Davidson correction[126] is the most widely used size-consistency correction. However, we would like to draw attention to a correction formula derived by Pople *et al.*[127], which is only slightly more complicated to implement and more accurate[100]. Neither correction yields exactly size-consistent energies so the supermolecular calculation at large distance is to be the reference, i.e.

$$\Delta E = E(R = R_e) - E(R = \infty) \tag{17}$$

The effect of size-consistency corrections on the van der Waals well depth is quite pronounced if no dimer SCF precedes the CI-type calculation, but not negligible otherwise even in $(H_2O)_2$ (Table II). Considering that $(H_2O)_2$ is among the more strongly bound van der Waals complexes, the relative influence on weaker complexes will be much more serious.

D. Limitations of the Coupled Electron-pair Approach

Since no bonds are broken in the dissociation of a van der Waals molecule, the Hartree–Fock determinant is a correct reference function for, e.g., a CEPA calculation. As Werner and Meyer[129] show, a CEPA calculation is able to reproduce experimental dipole moments and polarizabilities very well, provided a sufficiently large basis is employed. The electrostatic and induction components are thus well taken care of. The only remaining problem is the dispersion energy. The CEPA method, as a pair method, should be quite good at describing a dispersion electron pair. As is seen from Table I, the C_6 value from CEPA is close to the CHF value. From this we may infer that the dispersion energy in the CEPA approach is still close to that of uncorrelated Hartree–Fock atoms (for He_2).

The influence of intra-correlation on the dispersion energy, the real intra–inter coupling, is only introduced through triple excitations[130–132]. According to Riemenschneider and Kestner[130] three-electron contributions contribute about 9% to the C_6 coefficient of He_2. Meyer *et al.*[131] show for He–H_2 that CEPA does include approximately some effect of triples, but not the whole intra–inter coupling by far. The performance of the CEPA method for heavier systems as well as the analysis of the formal performance of the method merits further research.

Since the actual inclusion of all triples is prohibited by their great number, a good way to include the most important ones is to perform a multireference CI or CEPA (cf. Ref. 132). An example of this approach, where the CI is limited to inter-correlation contributions, is the interacting correlating fragments (ICF) method. Liu and McLean obtained with this method an interaction energy for He_2 of 10.76 K at $R = 5.6$ bohr[133], to be compared with a CEPA-1 limit estimate of 9.62 K[134] and a recent CEPA-2 result of 9.54 K[135] and an experimental well depth of 10.57 K[136]. These results clearly show that the CEPA methods underestimate the dispersion energy by some 5% at the van der Waals minimum, yielding a 10% error in the dissociation energy. The ICF result of Liu and McLean is unfortunately insufficiently documented to allow a proper judgement. We suspect that their result is too low due to insufficient account of the intra-atomic correlation and possibly some BSSE.

The intra-correlation contribution to the interaction energy is not without pitfalls either. Its contribution to E_{exch} in He_2 due to s (radial) correlation is $+2$ K whereas p and d (angular) correlation provides -1 K, a total for a properly correlated He of $+1$ K[137,261]. If the correlating orbitals are not correctly chosen any intra-correlation correction between -1 and $+2$ K may be found. Collins and Gallup[138] for instance find only 0.16 K extra repulsion for this reason.

The He dimer proves to be particularly difficult to describe since the dispersion energy is about twice the van der Waals minimum. For systems like the H_2O dimer where the dispersion energy is only a quarter of the interaction energy, an error of 5% in the dispersion is quite acceptable. A CEPA method may thus be expected to be quite adequate for van der Waals complexes where the dispersion energy is not the major source of attraction.

IV. BASIS SETS

There is at present a considerable body of experience on how the choice of atomic orbital (AO) basis set affects the final accuracy of calculated interaction energies. If the resulting potential energy surface is to be of predictive value (i.e. relative energy of isomers, location of saddle points, vibrational properties), the final calculated interaction energies for geometries of interest should probably be accurate to within about 5%. This figure of 5% amounts to 0.25 kcal mol^{-1} for dimers like $(H_2O)_2$, and to 0.5 K for He_2. Although present-day calculations rarely attain this, it is not unrealistic to expect such accuracy within the next few years.

The requirements that must be met by the basis follow more or less directly from the partitioning of the interaction energy. Briefly, long-range terms (E_{Coul}, E_{ind}, E_{disp}) lead to the requirement that electric multipole moments and electric multipole polarizabilities must be accurately represented. The short-range terms, being a direct consequence of overlap effects, require an adequate description of the valence-electron density both close to the nuclei and far away from them. Finally, though the bulk of BSSE can be removed by applying the counterpoise (CP) method (see Section V), there may be lingering artefacts due to the basis-set extension effect[139] and in order to reduce these one may wish to choose the basis such that these effects are as small as possible.

Before discussing these requirements in more detail, it should be noted that a very precise description of the core electrons is usually not required, and this has led to various methods in which these are simulated by pseudo-potential methods[140–142] and to the freezing of core orbitals in CI calculations. The total energy, being very sensitive to errors in the core region, must certainly not be used as a criterion for choosing basis sets for interaction energy calculations[143].

A. Long-range Terms

Let us consider the long-range terms first. An accurate description of these requires the monomer moments and polarizabilities to be described accurately. However, the precise role played by the relative importance of the dipole, quadrupole, ..., etc., moment depends on the system studied. In polar–polar interactions the dipole–dipole term in E_{Coul} has a much larger weight in the final ΔE_{total} than E_{disp}. Hence for a 5% accuracy in ΔE_{total}, the dipole moment μ may have to be accurate to 1–2%, while the error in the polarizability α may be as large as 10%. As the systems get less polar, the relative weight of E_{Coul} diminishes while that of E_{disp} increases. Hence for a relative accuracy of 5% in ΔE_{total}, α may now have to be accurate to 2% or even better. A summary of these requirements is given in Table III.

How should one choose the basis to meet these requirements? The use of carefully balanced minimal basis sets has been advocated as a means for producing reasonable dipole moments[144,145]. Apart from the fact that such sets cannot yield a meaningful estimate of E_{disp}, the dipole moments obtained in several minimal sets were rather poor. Decontracting the basis, such as to

TABLE III
Accuracy requirements for molecular electric multipole properties.

Type of complex	Typical ΔE_{total} (μhartree) at equilibrium geometry	Example	Relative importance of long-range terms	Acceptable relative errors (%) in monomer properties[a]				
				μ	Θ	Ω	α_d	α_q
Polar–polar	10000	$HOH \cdots OH_2$	Coul $\gg$ ind $\approx$ disp	2	5	15	10	50
Polar–non-polar	1000	$Cl{-}H \cdots Ar$	Coul $\ll$ ind $<$ disp	5	15	50	3	15
Non-polar–non-polar	100	$He \cdots He$	Coul $\approx$ ind $\ll$ disp	–	–	–	2	10

[a]These errors correspond to an accuracy of 5% in ΔE_{total} at equilibrium geometry.

give split valence or double-zeta (DZ) type basis sets, provides for more flexibility and hence a better description of induction and charge redistribution, but molecular electric properties are inaccurate. For example, SCF DZ dipole moments differ typically by about 0.15 a.u.[146] from the Hartree–Fock limit, whereas for accurate work on polar systems the errors should be less than this (Table III). Increased accuracy can be obtained by adding polarization functions to the basis. For atoms with occupied p shells, dipole properties require d functions, quadrupole properties require f functions, etc. Similarly, H, He, Li and Be need p, d, ... functions. The simplest and most common procedure is to add one set of d functions to each 'heavy' atom (C, N, O, F) and a set of p functions on each H, and to optimize the orbital exponents by minimization of the total energy. Both at the SCF level[147,148] and at the CI level[120,149] this yields exponent values of roughly 1.0 for α_P^H, roughly 1.0 for $\alpha_d^{C,N,O,F}$ and roughly 0.5 for $\alpha_d^{P,S,Cl}$, yielding basis sets known as Dunning's DZP basis, 6–31G**, etc. Cations require higher and anions require lower d exponents[150]. Optimizing the correlation energy itself also yields higher exponents[129].

It is not always recognized that the improvement of the electric properties by adding these energy-optimized polarization functions is not spectacular. This is illustrated in Table IV, which shows SCF dipole properties for the HF molecule in a number of basis sets. The double-zeta plus polarization (DZP) dipole moment is still 0.05 a.u. too high, and the polarizability perpendicular to the bond direction is less than half the correct value. Far better results can be obtained by adding a second diffuse set of polarization functions[150,151], especially when its exponents are determined by maximizing the dipole polarizability or the dispersion energy[146,152–154]. This is illustrated by the (DZPP) basis in Table IV, which yields μ and α_d values good enough to allow accurate calculations on dimers. These results are in line with the observation of Kochanski[155] and others[156] that a single set of polarizability-optimized

TABLE IV
SCF dipole properties[a] for the HF molecule ($R = 1.7328a_0$) in different basis sets[b].

Basis set	Polarization function exponents	μ	$\alpha_{\parallel}$	$\alpha_{\perp}$
DZ		0.939	–	–
DZP	$\alpha_d = 1.0, \alpha_p = 1.0$[c]	0.811	4.35	1.71
	$\alpha_d = 0.40, \alpha_p = 0.30$	0.759	5.10	3.46
(DZP′)	$\alpha_d = 0.25, \alpha_p = 0.15$	0.783	5.44	3.83
	$\alpha_d = 0.15, \alpha_p = 0.08$	0.842	5.42	3.66
DZPP	$\alpha_d = 1.0/0.15, \alpha_p = 1.0/0.08$	0.764	5.59	4.23
LPP		0.765	5.62	4.34
HF limit[129]		0.765	5.70	4.47

[a]All values are given in atomic units (a.u.).
[b]Taken from Ref. 146.
[c]Energy-optimized.
DZ = (9s5p/4s) → [4s2p/2s].
L = (13s8p/8s) → [8s5p/5s].

polarization functions of a given symmetry can account for about 90% of the maximum contribution to E_{disp} that can result from that symmetry. In fact, in these papers not α_d but E_{disp} itself was used to optimize the exponents. This leads to exponents slightly higher thah those resulting from α_d optimization. The use of two polarizability optimized sets was found to yield about 99%[88,157], and this may be the appropriate choice when studying very weak interactions. The triply polarized sets used by Werner and Meyer[129] were derived with a similar high accuracy in mind. Other doubly and triply polarized sets have been proposed recently[158] but the exponents of these are not low enough to give accurate polarizabilities.

To conclude this section on SCF dipole properties it is instructive to inspect the variation of the DZP results in Table IV with orbital exponent. As mentioned above, energy-optimized exponents ($\alpha_p = \alpha_d = 1.0$) yield poor results, and very low exponents ($\alpha_d = 0.15, \alpha_p = 0.08$) yield reasonable polarizabilities, but poor moments. There exists, however, an intermediate choice ($\alpha_d = 0.25, \alpha_p = 0.15$) for which μ and α_d are *simultaneously* quite acceptable. The same property was demonstrated to hold for a number of other small molecules and this so-called 'moment-optimized' DZP′[146] basis was therefore advocated as the smallest set yielding reasonable E_{Coul} and E_{disp} energies for polar dimers.

To illustrate how these basis sets behave in actual calculations, Table V shows the results of SCF + Disp (non-expanded second-order perturbation theory) calculations on the water dimer at a near-equilibrium geometry[107].

TABLE V
Interaction energies and their components (kcal mol^{-1}) of the water dimer[a] for different basis sets[b].

	4-31G	DZP	DZP′	DZPP
E_{Coul}	−10.32	−9.10	−8.24	−8.26
E_{exch}	5.36	5.99	6.16	6.37
E_{ind}	−0.61	−0.67	−1.28	−1.35
E_{CTX}[c]	−1.18	−0.90	−0.61	−0.53
E_{disp}	−0.43	−1.01	−1.82	−1.93
ΔE_{total}[c]	−7.17	−5.69	−5.80	−5.69
BSSE	1.27	0.51	1.17	0.32

[a] $R = 5.50a_0$; $\theta = 45°$. Monomer wavefunctions in the monomer's own basis used to define E_{Coul} and E_{exch}.
[b] Taken from Ref. 107.
[c] Corrected for BSSE, using δ^{CP}.

The results confirm the inadequacy of the unpolarized 4–31G basis and the energy-optimized singly polarized DZP basis set, while DZP′ is seen to give results close to the ones obtained by the DZPP basis. Table V shows a large BSSE for the DZP′ basis because the polarization function exponents are not energy-optimized. One therefore has to correct for this BSSE in order to get meaningful interaction energies.

Very accurate dipole polarizabilities require not only the use of suitable polarization functions, but also an enlargement of the so-called 'isotropic' part of the basis. This may be achieved by replacing the 6–31G or DZ part of the basis by larger energy-optimized sets such as 6–311G[159], (11s7p)[160], (13s8p)[161] or one of the larger even-tempered basis sets[162]. A common alternative is to add diffuse s and p functions to a smaller basis[92,152]. The effect of this basis-set enlargement on the SCF dipole properties of HF is illustrated by the (LPP) entry in Table IV. The changes are rather small, and so it remains to be seen whether these functions are really necessary in dimer calculations, where the orbitals of the partner molecule may (partially) compensate for their absence[139].

A very similar discussion applies to quadrupole properties as well. The f functions needed for these properties should have their exponents optimized with respect to the contribution of quadrupole transitions to the dispersion energy[156] or with respect to the quadrupole polarizability α_q, which relates the induced quadrupole moment to an inducing field gradient. Optimal single f exponents have been reported as $\alpha_f^{\text{O}} = 0.18$, $\alpha_f^{\text{F}} = 0.275$[163,164] and $\alpha_f^{\text{Ne}} = 0.28$[92]. These values are 6–8 times lower than the exponents which maximize the contribution of f functions to the total correlated monomer energy[149,158], and so, as for d exponents, energy-optimized values are unsuitable. Static quadrupole moments are changed by added f functions[163,164] but, since the effects are small (e.g. for HF the effect appears to be a reduction of Θ by about 0.03 a.u.), this puts no further constraints on the exponents that should be chosen.

Our discussion has so far neglected correlation effects on molecular properties. Correlation is known to change dipole moments of small molecules by amounts up to 0.20 a.u. (H_2CO [146]), while dipole polarizability components may change by up to $2.3a_0^3$(NH_3 [129]), and so allowing for correlation effects in these properties is necessary in accurate work. Highly correlated wavefunctions require the presence of polarization functions with fairly high exponents in the basis set[129] and so one might expect that such functions are necessary also in dealing with the effects of correlation on electric properties. This turns out not to be the case, however, and reasonable correlation corrections have been obtained in basis sets of the DZPP type[151,152,154] and even in the DZP′ basis[107]. We should add that correlation corrections to properties are sensitive not only to the basis but also to the method chosen, MP2 and CI(SD) corrections typically deviating by about 20% from the effect found using more accurate methods like MP4 or CEPA[129,165].

B. Short-range Terms

These include the penetration parts of and the exchange corrections to each of the long-range terms (see Section II). By far the most important of these is the first-order exchange energy, E_{exch}, which is mainly responsible for the forces that keep molecules apart. Detailed expressions[166] show that E_{exch} depends on overlap densities between the occupied orbitals of the separate molecules, as well as on the value of the monomer electrostatic potentials in the overlap region. These potentials will be accurate enough if the basis-set requirements for the long-range terms are met, and so the additional requirement raised by E_{exch} is that overlap densities be described accurately. Now an overlap density (like $1s_A(1)1s_B(1)$ in He_2) has a high value all the way from nucleus A to nucleus B (since for each point P on the internuclear axis $\exp(-r_{AP})\exp(-r_{BP}) = \exp(-R_{AB})$) and so a correct description of the overlap density requires the monomer orbitals to be numerically accurate over the whole intermolecular region (and not only over the region where the molecules 'touch' each other). But in practice the main problems come from the values of the orbitals far from the nucleus. These tend to be underestimated (particularly if Gaussian atomic orbitals are used), since one usually employs energy-optimized basis sets and these are biased towards a good description close to the nuclei.

Several strategies are available to enhance the orbital density in regions far from the nuclei (we only discuss the case of Gaussian type orbitals (GTOs)).

1. An energy-optimized basis of given size (e.g. 9s5p or 4–31G) may be replaced by a large energy-optimized set (e.g. 11s7p or 6–311G) (note that

6–311G is roughly equivalent to 9s5p). This strategy may be carried to great lengths by switching to energy-optimized even-tempered sets, which are available up to very large size (e.g. 20s10p for Ne[162]). The main problem with this approach is that most of the added functions are used up to improve the description of the core electrons, and so the 'tail' region of the orbital improves only slowly. By the time that the tail is adequate, a large number of functions (closely spaced in exponent space) are present not only in the core region but also in the valence region. The usual technique of segmented contraction[167] will not be effective in reducing the basis to a manageable size since it must not be used for the valence region[168].

2. Instead of using energy-optimized sets one might use sets obtained by fitting to accurate Hartree–Fock orbitals[169–172]. These sets have both the exponents and the expansion coefficients determined by the criterion of a least-squares fit. Relatively small fits can give orbitals with good tails, as was demonstrated in calculations of the first-order interaction energies of He_2 [170], Be_2 [171] and Ne_2 [172]. However, in calculating higher-order energies involving charge redistributions, as well as in CI calculations, the expansion coefficients must be given variational freedom and under these circumstances the advantage of these sets over energy-optimized sets of the same size seems to be lost (see below).
3. Instead of using ever larger energy-optimized sets one could use a fairly small energy-optimized set and add diffuse s, p,... sets whose exponents are chosen as in even-tempered sets, i.e. by extrapolating the geometric progression of exponent values in the original set. As an addition to Huzinaga's 9s5p sets this has proved useful in dealing with anions and with Rydberg orbitals[173]. The size of the basis after contraction is [4s3p] or [5s3p] for first-row atoms. More recently, similar extensions with s and p sets on first-row atoms and s on H have been proposed for the 6–31G and 6–311G basis sets[158], yielding sets with a size [4s3p] and [5s4p] for first-row atoms. These extensions gave significant changes in calculations on weakly bound complexes involving H bonding or cation–molecule interactions[16]. For example, the extensions resulted in reduced interaction energies and in changed dimer geometries. The role of the added functions may partly have been to reduce the BSSE and so their effect on BSSE-corrected interaction energies remains to be assessed.
4. An alternative basis combining moderate size with the accuracy of the atomic outer region provided by large energy-optimized basis sets was recently proposed for the O atom in He–O_2 [87]. This so-called extended-zeta (EZ) basis is formed from an energy-optimized 9s5p set[161] by replacing the outer s function by two even-tempered more closely spaced s functions, the most diffuse of which is slightly more diffuse than the most diffuse s in the energy-optimized 13s8p set. Similarly the outer two p functions were replaced by three p functions. The resulting set (unlike 13s8p) can be

contracted without significant loss of accuracy to [5s3p], and yields electric moments and polarizabilities of similar accuracy[174] as the (13s8p) set in Ref. 146.

5. The final strategy for improving the tail of monomer wavefunctions is of a different kind. It is based on the consideration that the accurate calculation of E_{exch} will usually not be the final goal. Rather, one aims at the accurate evaluation of E_{total}, having an accurate E_{exch} as one of its components. From the counterpoise principle (see Section V) it follows that the basis set that decides the accuracy of E_{exch} in the final result is *not* the monomer basis set, as described in points 1–4 above, but the full basis of the dimer. In fact, the presence of the other molecule's orbitals will improve the tail precisely where it matters, viz. in the region where the exchange repulsion comes about. Since basis sets lacking good tails underestimate E_{exch}, it is to be expected that E_{exch} will increase when the dimer set is used in constructing the monomer wavefunctions. This is one of the factors causing changes in the first-order interaction energies for a number of systems when switching from monomer (MBS) to dimer (DBS) basis sets[104,175–177]. Note, however, that these energies contain a changed E_{Coul} as well, which may or may not be an improvement[139].

The relative merits of strategies 1–5 are illustrated in Table VI, which shows E_{Coul} and E_{exch} values for He_2. These results were calculated using Eqs (1) and (4), employing SCF monomer wavefunctions obtained in 10 different monomer basis sets. Entries 1–5 illustrate the use of ever larger energy-optimized sets (strategy 1). E_{Coul} and E_{exch} are seen to converge to values of -4.94 and 35.63 μhartree, respectively. Using MBS gives very slow convergence, but using DBS gives good results (converged to within 5% already at the 8s level). Entry 6 (strategy 2) employs an 8s basis fitted to an accurate Hartree–Fock orbital for He[170]. This is seen to give excellent results already for MBS. However, after redetermining the 8s expansion coefficients variationally, as one is forced to do in a full dimer calculation, the good results are spoilt (entry 7). Both E_{Coul} and E_{exch} are now overestimated, indicating that the bias in this basis towards low emponents exaggerates the tail density. Presumably, this type of defect cannot be remedied using DBS, but this was not checked. Strategy 3, the addition of diffuse functions, is illustrated in entries 8 and 9. At the MBS level, there is a clear improvement with respect to the original sets, viz. energy-optimized 5s and 10s. At the DBS level, however, the '10 + 1' basis behaves irregularly[76] and the original 10s DBS data seem more accurate. Although one cannot simply extrapolate these results to more strongly bound systems, it seems fair to say that addition of diffuse functions does not necessarily give a better description.

Finally, entry 10 illustrates the extended-zeta (EZ) basis of strategy 4, a 6s basis in this example. It is seen to give better results than the equally large

TABLE VI
E_{Coul} and E_{exch} (μhartree) for He_2 ($R = 5.6a_0$) evaluated from SCF monomer wavefunctions[a].

Calc.	GTO basis	Monomer energy (hartree)	Most diffuse exponent	MBS		DBS		Ref.	δ(μhartree) from Eq. (22)
				E_{Coul}	E_{exch}	E_{Coul}	E_{exch}		
1	4s[b]	2.855 160	0.2980	−0.79	−0.27	−4.35	26.85	178	173.1
2	5s[b]	2.859 895	0.2446	−1.91	3.15	−4.92	32.76	178	44.30
3	8s[b]	2.861 625	0.1642	−4.38	25.27	−5.03	35.38	178	0.59
4	10s[b]	2.861 672	0.1387	−4.88	33.10	−4.94	35.60	178	0.11
5	20s[c]	2.861 680	0.0989	−4.92	35.38	−4.94	35.63	178	0.00
6	8s[d]	2.858 875	0.0605	−4.94	35.73	–	–	This work	–
7	8s[e]	2.859 462	0.0605	−5.04	37.88	–	–	This work	–
8	5s[b] + 1s	2.859 967	0.0990	−2.89	11.94	−4.39	36.81	This work	4.67
9	10s[b] + 1s	2.861 673	0.0562	−4.98	34.37	−5.05	35.47	This work, 76	0.03
10	6s[f]	2.860 088	0.1500	−4.76	31.07	−5.05	35.56	This work	0.21

[a]Except in calculation 6, all sets were left uncontracted. In the DBS calculations of 1–5, the dimer set contained p sets on both atoms; see Ref. 178.
[b]Energy-optimized 161.
[c]Regularized even-tempered 162.
[d]Fitted to HF AO (quoted in Ref. 170).
[e]As entry 6, but coefficients determined variationally.
[f]EZ basis (see text) with exponents 98.124 267, 14.768 906, 3.318 825, 1.176, 0.420 and 0.150.

'5+1' set obtained by simply adding a diffuse function. It is also much better than energy-optimized 6s and, in fact, its quality resembles that of energy-optimized 9s, to which it was modelled. We conclude that this type of basis set shows some promise for high-accuracy work. However, for many purposes, the improvement offered by simply employing the dimer basis set allows one to use more standard basis sets with relatively little risk of inaccuracies in E_{exch}.

V. THE BASIS-SET SUPERPOSITION ERROR: HISTORICAL BACKGROUND

In a standard dimer calculation the energy of the supersystem (AB) is calculated with the union of the subsystem one-electron basis sets

$$\chi_{AB} = \chi_A \oplus \chi_B \tag{18}$$

In practice, the monomer basis sets are never complete, an obvious exception being the $^3\Sigma_u$ H_2 van der Waals molecule[179]. The extension of the basis set in the dimer then results in a relatively better description of the dimer with respect to the monomers, yielding a correspondingly better energy. If the interaction energy is obtained as the difference of the separately calculated dimer and monomer energies, each with their own basis,

$$\Delta E_{total}(R_{AB}) = E_{AB}(\chi_A \oplus \chi_B) - E_A(\chi_A) - E_B(\chi_B) \tag{19}$$

this energy contains, in addition to the effect of the physical interactions, the result of the basis-set extension for each monomer. This non-physical energy contribution is commonly known as the basis-set superposition error (BSSE)[180]. It was first explicitly considered by Kestner[181] as an explanation for the spurious minimum in the potential energy curve for He_2 calculated within the Hartree–Fock approximation by Ransil[182]. Earlier the problem was noted by Clementi[183] and cleverly avoided by Jansen and Ros[184], who used what came to be known as the Boys–Bernardi function counterpoise recipe (CP method).

A. The Counterpoise Method

Boys and Bernardi[185] argue that the effect of a perturbation is most accurately calculated if all other parameters in the calculation are kept the same ('counterpoise'), so that a maximal cancellation of errors is obtained. In the case of a van der Waals molecule, the perturbation is the effect of the nuclei and electrons of one monomer on the other and vice versa. Following Boys and Bernardi's concept that 'the full set of expansion functions used in the dimer calculation must also be used in the monomer calculations', the usual implementation of these ideas is as follows. One calculates the van der Waals interaction energy by computing both the 'perturbed monomers' (the dimer calculation) and the unperturbed monomers (the isolated monomers) with the same (dimer) basis set:

$$\Delta E_{\text{total}} = E_{\text{AB}}(\chi_{\text{A}} \oplus \chi_{\text{B}}) - E_{\text{A}}(\chi_{\text{A}} \oplus \chi_{\text{B}}) - E_{\text{B}}(\chi_{\text{A}} \oplus \chi_{\text{B}}) \tag{20}$$

The calculation of the monomer A energy $E_{\text{A}}(\chi_{\text{A}} \oplus \chi_{\text{B}})$ is now identical to the dimer calculation except for the number of electrons, which is that of A only, and the charges of the nuclei of monomer B (the ghost), which are set to zero. The monomer energies in the basis $(\chi_{\text{A}} \oplus \chi_{\text{B}})$ depend on the geometry of the complex, so they must be computed for every point on the van der Waals potential energy surface. The counterpoise procedure requires an increased computational time, which is less than three times the time needed for the original dimer calculation, since the (AO) two-electron integrals are identical in both dimer and monomer calculations and only have to be evaluated once. 'But a less than three times effort is well worth while if it gives the possibility that much more accurate values of intermolecular interactions may be obtained'[185].

The procedures described above are only valid when a size-consistent method is used to calculate the various energies. Special formulae have therefore been developed[186,187] to deal with BSSE in CI(SD) calculations, since these lack size consistency. However, it seems preferable to use Eq. (20) as it stands, and to apply size-consistency corrections (cf. Section III.C) to all energies.

B. Different Interpretations of the Basis-set Superposition Error

Although it is possible to use Eq. (20) directly, without any explicit consideration of a quantity called BSSE, it is customary to start from Eq. (19) and to consider the correction δ that would remove the BSSE in (19), viz.

$$\begin{aligned}\Delta E(\text{corrected}) &= \Delta E(\text{Eq. (19)}) + \delta \\ &= E_{\text{AB}}(\chi_{\text{A}} \oplus \chi_{\text{B}}) - E_{\text{A}}(\chi_{\text{A}}) - E_{\text{B}}(\chi_{\text{B}}) + \delta\end{aligned} \tag{21}$$

If ΔE(corrected) is taken to be ΔE(Eq. (20)) then the BSSE correction δ is seen to be

$$\delta^{\text{CP}} = \delta^{\text{A}} + \delta^{\text{B}} = E_{\text{A}}(\chi_{\text{A}}) - E_{\text{A}}(\chi_{\text{A}} \oplus \chi_{\text{B}}) + E_{\text{B}}(\chi_{\text{B}}) - E_{\text{B}}(\chi_{\text{A}} \oplus \chi_{\text{B}}) \tag{22}$$

Now the right-hand side of Eq. (21) can be read in two different ways. First, δ may be interpreted as a correction to the monomer energies, i.e. to the reference energy used in estimating the interaction energy. Adopting Eq. (20) as the corrected interaction energy then merely asserts, in true Boys–Bernardi spirit, that $E_{\text{A}}(\chi_{\text{A}} \oplus \chi_{\text{B}})$ is the proper reference for estimating the effect of the perturbation. The second interpretation of Eq. (21) is that δ is a correction which removes the unphysical lowering of the monomer energies *present in the dimer energy* $E_{AB}(\chi_A \oplus \chi_B)$. This second interpretation has led to widespread hesitation in accepting Eq. (20), since 'intuitively it seems clear that the Pauli principle will prevent one component to fully utilize the basis set of the other component in the dimer calculation'[188]. This interpretation has led to a variety of proposals (see below) in which the occupied orbitals of B are excluded from the calculation of E_A and E_B. The correction δ will then usually be smaller than δ^{CP}.

Johansson *et al.*[189] were the first to conclude that the counterpoise procedure overestimates the basis-set superposition error. They proposed to scale the correction by a factor reflecting the reduced number of ghost virtual orbitals available to a monomer in the dimer situation and obtained good agreement with larger basis sets for their STO-3G SCF calculations on hydrogen-bonded systems. Their idea has had few followers and little success[190,191]. Nevertheless, their conclusion that the BSSE is overestimated has received widespread acclaim and many authors believe the counterpoise procedure to yield a (high) upper bound[18,103,112,157,188,190–207].

C. Alternative Counterpoise Scheme

A rigorous implementation of the idea that the ghost's occupied orbitals should not be available in the monomer calculation was proposed by Daudey *et al.*[192]. In their scheme, which one may call virtuals-only CP, the BSSE for monomer A is obtained as

$$\begin{aligned} \delta_A^V &= E_A(\chi_A) - E_A(\chi_A \oplus \chi_B^V) \\ \chi_B &= \chi_B^{occ} \oplus \chi_B^V \end{aligned} \tag{23}$$

Fig. 1 is an attempt to illustrate the differences between the original CP scheme (often denoted as full CP) and the virtuals-only CP scheme. For the sake of argument the dimer wavefunction is assumed to be of the CI(SD) type but the following arguments apply equally to the SCF case, providing the dimer SCF wavefunction is considered to arise by mixing in excited configurations in a reference configuration built from free-monomer MOs.

In the full CP method (Fig. 1a) the dimer and monomer CI calculations are performed in the same manner. The occupied and virtual orbitals for each system are determined in an SCF calculation employing the complete basis set. These orbitals are subsequently employed in the CI calculation.

In the virtuals-only scheme (Fig. 1b) the monomer calculation is performed in the basis of the monomer augmented by the virtual orbitals of the ghost, which are orthogonal to the (absent) occupied orbitals of the ghost. Note that the monomer's own virtuals are not orthogonalized to these (absent) ghost occupied orbitals. The dimer calculation may be either identical to the one of the full CP scheme above or, more in the Boys–Bernardi spirit of using identical function spaces in both dimer and monomer calculations, it may

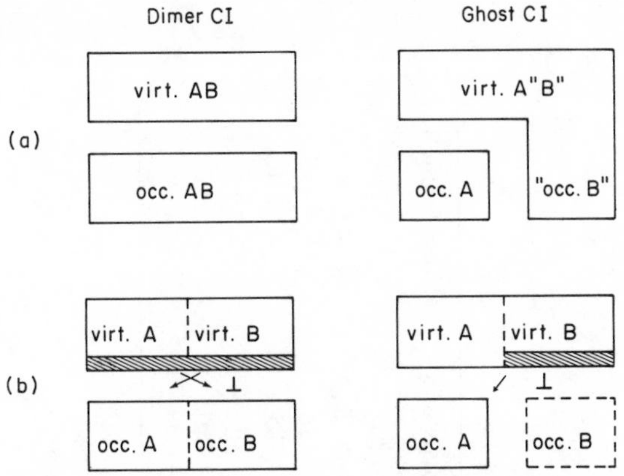

Fig. 1. Illustration of the orbital spaces employed in dimer and monomer plus ghost CI calculations: (a) full counterpoise; (b) virtuals-only counterpoise. The hatched areas symbolize changes (with respect to free-monomer SCF MOs) by orthogonalization ($\perp$).

employ suitably orthogonalized monomer orbitals. In practice, there appears to be little difference provided size-consistency corrections are applied[128]. Note that the full CP and virtuals-only schemes both satisfy the requirement that, in the limit of complete monomer basis sets, δ goes to zero.

The idea of using identical function spaces and thus identical excitation possibilities in a CI calculation in dimer and monomer is even better fulfilled if exactly identical virtual orbitals, therefore orthogonal to all occupied orbitals in the dimer, are used for both. This type of virtuals-only CP method was first considered by Spiegelmann and Malrieu[199]. As occupied orbitals for the dimer they chose the SCF localized dimer MOs and the orbitals used for the monomer calculation were matched to these by projection of the monomer SCF orbitals into the dimer Fock space. This raises the monomer energy. For the Ar dimer their SCF potential energy curve was consequently less repulsive than that calculated with a large basis[199]. Alternatively one may construct the reference configuration by using the monomer occupied orbitals for the monomer and orthogonalized combined monomer orbitals for the dimer. This procedure was shown to lead to negative BSSE corrections[128] and an exaggerated well depth. In fact, for an infinite basis, where the BSSE is zero, both schemes will yield a negative BSSE correction and they should therefore be discarded.

D. Experiences With the Full and Virtuals-only Counterpoise Schemes

A large numer of studies are now available in which one or both CP schemes were used to correct for the BSSE. We here present some of the conclusions

TABLE VII

Full CP, virtuals-only CP and uncorrected interaction energies.

		Basis	Method	Full	Virtuals-only	No correction
$(H_2O)_2$ [36]	ΔE	STO-3G	SCF	− 1.73	− 4.15	− 5.16
	(kcal mol^{-1})[a]	6-31G**		− 4.50	− 5.18	− 5.58
Ar–HCl[128]	ΔE	Basis I	Size-consistency	− 383	− 646	− 1527
	(μhartree)[b]	Basis II	corrected CI	–	− 785	− 1804
$(H_2O)_2$ [115]	ΔE	DZP′	CEPA-1	− 3.98	− 5.51	− 7.22
	(kcal mol^{-1})[c]	DZPP		− 4.05	− 4.61	− 5.38
		EZPP		− 4.09	− 4.39	− 4.91

[a] $R = 5.63$ a.u., $\theta = 60°$.
[b] $R(\text{Ar} \ldots. \text{Cl}) = 7.6$ a.u., $\theta = 0°$.
[c] $R = 5.60$ a.u., $\theta = 40°$.

and opinions that have been put forward in these studies. Some selected results are shown in Table VII. Without discussing these in detail, it is apparent that BSSE corrections can easily be of the same magnitude as the interaction energy itself. Also, even the differences between δ^{CP} and δ^{V} can be of that magnitude.

A feature not apparent from Table VII is that for larger basis sets the CI BSSE may be larger than the SCF BSSE, and it is difficult to remove it by enlarging the basis[18,115,116,128,134,202]. Its value may easily reach 50% of the correlation interaction energy. An example of this was provided by Clementi and Habitz[113] who, correcting only for the SCF BSSE in their Davidson-corrected CI calculations of the water dimer potential energy surface, found a well depth of 5.5 $kcal\,mol^{-1}$. The second virial coefficient calculated for their surface was a factor of 2 too high, indicating too much attraction. By decreasing the correlation contribution to the interaction energy by an arbitrary factor 0.5 the calculated virial coefficient was brought into good agreement with experiment and the well depth was reduced to 4.7 kcal mol^{-1}. Correcting is thus absolutely necessary both in SCF and CI calculations.

We stress that the purpose of a good BSSE correction scheme has to be to yield the true interaction energy corresponding to the quality of the basis set and the method. One should not expect *all* basis dependencies to be removed[143] nor the experimental result to be reproduced. Comparison with, e.g., the experimental dissociation energy may be misleading. Since the main defect of most basis sets employed in practical calculations is a lack of sufficient polarization functions (cf. Section IV) to reproduce the dispersion energy adequately, the true interaction for those basis sets will be above the experimental result and any scheme that underestimates the BSSE will be favoured in such a comparison[128].

These considerations are frequently overlooked in studies that claim, on numerical grounds, that δ^{CP} is an overcorrection, or simply unhelpful. For example, in an SCF study on $FH \ldots OH_2$ Kocjan *et al.*[157] observed that δ^{CP} correction of the STO-2G result at the calculated R_e removed all attraction. Their conclusion that δ^{CP} overcorrects is debatable, however, since R_e was probably so short that a repulsive corrected ΔE should have been anticipated.

In a recent CI study of Ar–HC1[128] the present authors compared δ^{CP} and δ^{V} corrected energies with the experimental ΔE and concluded that δ^{CP} was too large. However, as argued above, the comparison should have been with the (unknown) true ΔE in the given method and basis. Finally, a somewhat different approach was taken in a recent SCF study on $(HF)_2$[143]. It was argued that δ^{CP} correction was unhelpful, since it failed to reduce the range of interaction energies. However, since basis sets ranging from minimal to large as well as polarized were considered, the electric properties were very different from case to case and so a wide range of 'true' interaction energies should have been anticipated.

There have been a number of SCF studies on polar systems[78,107,208,209] in which 'the important contribution of the monomer dipole moment to the attractive interaction'[209] was recognized. By avoiding polarity differences[107,208] or by correcting for them[78,209] these studies arrived at the conclusion that δ^{CP} correction brings small-basis results in close agreement with large-basis results. The same conclusion was reached in SCF work on non-polar systems[175,209,210]. However, in the absence of similar work (SCF and CI), using the δ^{V} correction, no objective conclusion can be drawn (however, see Section VI).

As noted before, the majority of authors regard δ^{CP} as an overcorrection, at least in principle, and so it is not surprising that the δ^{V} correction—even though it is somewhat harder to implement—has found increasing use in recent years[36,90,128,143,192,195,197,199,203,206,211–214]. An approximate version is sometimes employed, intended to reduce the dimension of the ghost calculation already at the AO level. It consists of using only the polarization functions of the ghost[143,197,203] plus possibly the diffuse AOs[215]. This variant has mainly been applied to cluster calculations[197,203,215] where basis-set size is at a premium. Although in a few cases[128,197] δ^{V} was claimed to give better agreement with experiment than δ^{CP}, it seems fair to say that the results of all δ^{V} studies leave the question of the correctness of δ^{CP} or δ^{V} fully open.

This opinion is shared by a number of authors who regard the full counterpoise corrected and the virtuals-only corrected energies as results bracketing the real interaction energy for the basis[199,203] or the experimental interaction energy[214,216].

Fundamental support for the δ^{V} correction was expressed by Morokuma *et al.*[36], who observed that in energy-partitioned SCF calculations the E_{CT} term became excessive for basis sets subject to large BSSE. They argued that a virtuals-only scheme is the proper one to correct for this. It was noted that the virtuals of the ghost to be included should really be those of the counterpoise-corrected instead of the isolated ghost. This recipe is difficult to implement. However, in their example the difference from the simple recipe was extremely small. Similarly, Hayes and Stone[90,213] state that, while only the virtuals of the partner are available in their non-orthogonal second-order perturbation theory, the occupied orbitals may become available in higher orders.

An implicit assumption made in the above work[36,90,213] is that the first-order interaction energy ΔE^{HL}, when evaluated using monomer wavefunctions in the monomer's own basis, will be free of BSSE. This assumption was formulated explicitly in a recent letter by Collins and Gallup[217]. In their opinion δ should therefore be smaller than $\Delta E^{(2)}$. However, δ^{CP} was found to be larger than $\Delta E^{(2)}$ in several cases and so they concluded that δ^{CP} is an overcorrection. We return to this point in Section VI.

E. Ways of Avoiding Basis-set Superposition Errors

To this time no general agreement has been reached in the literature and the problem of how to determine the basis-set superposition error is considered unsolvable by many. Thus methods have been devised to avoid the BSSE either completely or to a large extent.

Non-orthogonal valence-bond (VB) techniques[89,91,138,179,213,218–220] are most suited to this purpose since the orbital spaces of the two monomers may be separated exactly. The configurations (structures) included in the calculation may be chosen such that no BSSE arises. However, as Wormer *et al.*[221] pointed out, configurations classified as charge-transfer type in VB calculations also contribute to the BSSE. To omit these configurations is too crude a solution since, as Gallup showed for the BSSE-free $^3\Sigma_u^+$ H_2 van der Waals system[179], the charge-transfer energy can be an important component of the interaction energy. In their He_2 calculations Collins and Gallup[138] avoid the BSSE when a small basis is used, by using for the dimer a reference configuration built from monomer occupied orbitals calculated in the dimer basis set[78,175], a recipe in complete agreement with the full counterpoise scheme at SCF level, and obtain a charge-transfer energy in good agreement with large-basis-set results. The BSSE at CI level, caused by double excitations with at least one of the virtuals of the partner involved, is avoidable at little penalty in the VB scheme[138]. Obviously one may also avoid CI BSSE simply by taking only inter-monomer correlation (dispersion) into account[88,89,91,92,218,219,221], but then the important contributions of the intra-monomer correlation to the interaction energy, both in the exchange repulsion and in the multipole interaction, are neglected.

Using orthogonal orbitals, the interacting correlated fragments (ICF) method[133,180,200,222] attempts to avoid the BSSE by combining correlated monomers using a few carefully optimized important configurations for each monomer, and allowing only inter-monomer correlation on top of that. The definition of inter-correlation is a bit troublesome, however, since the monomer orbitals are mixed in the dimer calculation, owing to the orthogonality constraint. The BSSE of the correlated fragments cannot be excluded completely but seems to be negligible where checked[200,222].

A less rigorous approach is to hope for cancellation of the BSSE and the effect of omitted (polarization) functions on the dispersion energy[103,223] or for cancellation of the BSSE and the total dispersion energy. The latter is implicitly assumed if the geometry of a van der Waals complex is optimized within the Hartree–Fock framework[16,108,224–230] or when uncorrected small-basis SCF results are preferred based on a comparison with experimental data[157,231]. The former is utilized if no BSSE correction is applied to CI or Møller–Plesset perturbation theory results for van der Waals molecules[16,93,112,232–243].

Such a cancellation of errors may occur for certain special basis sets, but it can never be complete[106,244], since the geometrical dependence of the BSSE may not be expected to be similar to that of the missing interaction energy, e.g. the R^{-6}, R^{-8}, etc., terms of the dispersion energy[106]. The BSSE does not even always decrease as the inter-monomer distance increases[245]. To quote Bolis *et al.*[244]: 'Fortuitously the basis-set superposition error can yield interaction energies in agreement with experimental data, however, for internuclear distances that generally disagree with the experimental data'. A right distance combined with the wrong dissociation energy is also found[225].

Finally, in view of the current uncertainty with regard to the proper scheme to correct for BSSE, some authors regard the full counterpoise-corrected and the uncorrected interaction energies as results bracketing the real interaction energy for the basis[188] or the experimental interaction energy[18,204]. The counterpoise method is sometimes only used as an indication for the reliability of the calculation[114,206,246,247]. Note, however, that a small BSSE may be due to either a good basis set of the monomer or to a basis on the ghost which is not fitted to improve the energy of the monomer. Wells and Wilson[202] find that, as the basis set is systematically increased, the BSSE may not monotonically decrease but may instead increase and pass through a maximum before decreasing to its limiting value of zero.

Summarizing, since the basis-set superposition error may be large (that is, comparable to the interaction energy) and since it depends in an erratic way upon basis set and geometry, and affects both CI and SCF calculations and is extremely difficult to avoid without reducing the quality of the calculation, it is of vital importance to arrive at a reliable accepted scheme to *correct* for it.

VI. JUSTIFICATION OF THE FULL COUNTERPOISE PROCEDURE

In the previous section it became clear that in accurate calculations the BSSE must be corrected for. Although this may be done using the full CP procedure, there are a large number of authors who maintain that δ^{CP} is an overcorrection. The following arguments for this seem worthy of further consideration:

1. ΔE^{HL}, involving only occupied MOs cannot contain BSSE, and since δ^{CP} is sometimes larger than $\Delta E^{(2)}$, it must be an overcorrection.
2. δ corrects for the unphysical energy lowering present in E^{AB} (the total dimer energy) and since occupied MOs are no longer available in the dimer, δ^{CP} must be an overcorrection. This may be formulated as: in calculating ΔE the function spaces must be the same, and since in the dimer there are no excited configurations where electrons of A enter the occupied orbitals of B, these configurations must also be excluded from the monomer calculation.

Now it is hard to prove formally that these arguments are wrong. But a number of recent calculations[134,178], both at the SCF and the CEPA level, have provided compelling numerical evidence that δ^{CP} is the proper correction to apply, and that δ^{V} undercorrects. This was achieved by studying ΔE for He_2 in a sequence of basis sets subject to widely different BSSE, while the true interaction energy for a given basis was monitored in independent calculations using perturbation theory.

On the basis of these numerical findings, we offer in this section some observations which in our opinion refute the above objections against the δ^{CP} correction. A central role is played by the ghost's occupied orbitals, since it is the inclusion of these in δ^{CP} that has aroused so much criticism.

A. The Heitler–London Interaction Energy ΔE^{HL} May Contain Basis-set Superposition Errors

In Section II.F it was shown that the total dimer Heitler–London energy E_{AB}^{HL} for a dimer wavefunction $\mathscr{A}\psi^{A}\psi^{B}$, where ψ^{A} and ψ^{B} are SCF wavefunctions for A and B, may be written as (cf. Eq. (10)):

$$E_{AB}^{HL} = E_A + E_B + E^{(1)} + \Delta \tag{24}$$

The corresponding interaction energy ΔE^{HL} (Eq. (10)) is obtained by subtracting the reference energy $E_A + E_B$. Here $E^{(1)}$ is the first-order energy (5), and $\Delta = O(S^4)$, hence small, if ψ^{A} and ψ^{B} are at the Hartree–Fock limit, but $\Delta = O(S^2)$ when these SCF wavefunctions are obtained in finite basis sets, one at A and one at B. E_A and ψ^{A} depend only on the occupied orbitals of A; hence with an obvious extension of the notation (24) becomes

$$E_{AB}^{HL} = E_A(\chi_A^{occ}) + E_B(\chi_B^{occ}) + E^{(1)}[\psi^{A}(\chi_B^{occ}), \psi^{B}(\chi_B^{occ})] + O(S^2) \tag{25}$$

We now use a theorem, proved elsewhere[76,78], that $\Delta = O(S^4)$, hence small, not only at the Hartree–Fock limit but also when ψ^{A} and ψ^{B} are SCF wavefunctions obtained in a finite *common (dimer) basis set*. Suppose we employ for this purpose the basis $(\chi_A^{occ} \oplus \chi_B^{occ})$. This leaves $\mathscr{A}\psi^{A}\psi^{B}$ and hence E_{AB}^{HL} unchanged, since it involves mixing of orbitals within a single-determinant wavefunction. Thus instead of (25) one has

$$E_{AB}^{HL} = E_A(\chi_A^{occ} \oplus \chi_B^{occ}) + E_B(\chi_A^{occ} \oplus \chi_B^{occ}) + E^{(1)}[\psi^{A}(\chi_A^{occ} \oplus \chi_B^{occ}), \psi^{B}(\chi_A^{occ} \oplus \chi_B^{occ})] + O(S^4) \tag{26}$$

This shows that one and the same dimer energy, viz. E_{AB}^{HL}, can be interpreted in two different ways. The Heitler–London interaction energy ΔE^{HL} can correspondingly be obtained by subtracting from E_{AB}^{HL} the monomer energies in (25) (this is the usual procedure in Morokuma's energy partitioning[36]) or those in (26). The latter procedure is clearly preferable, since the dubious Δ term—not a pure interaction energy—is smallest in (26).

Comparing the Heitler–London interaction energies corresponding to (25) and (26) one readily finds[178,245]

$$\Delta E^{\mathrm{HL}}(\text{Eq. (26)}) = \Delta E^{\mathrm{HL}}(\text{Eq. (25)}) + \delta^{\mathrm{occ}} \tag{27}$$

where

$$\delta^{\mathrm{occ}} = E_{\mathrm{A}}(\chi_{\mathrm{A}}^{\mathrm{occ}}) + E_{\mathrm{B}}(\chi_{\mathrm{B}}^{\mathrm{occ}}) - E_{\mathrm{A}}(\chi_{\mathrm{A}}^{\mathrm{occ}} \oplus \chi_{\mathrm{B}}^{\mathrm{occ}}) - E_{\mathrm{B}}(\chi_{\mathrm{A}}^{\mathrm{occ}} \oplus \chi_{\mathrm{B}}^{\mathrm{occ}}) \tag{28}$$

is a first-order BSSE in ΔE^{HL}(Eq. (25)) due to its use of the inappropriate reference energy $E_{\mathrm{A}}(\chi_{\mathrm{A}}^{\mathrm{occ}}) + E_{\mathrm{B}}(\chi_{\mathrm{B}}^{\mathrm{occ}})$ (cf. also Ref. 104). This δ^{occ} is identical with Δ^{EX} in Ref. 245.

We note in passing that $E_{\mathrm{AB}}^{\mathrm{HL}}$ allows even a third interpretation, namely in terms of full CI wavefunctions for A and B, each obtained in the common basis $(\chi_{\mathrm{A}}^{\mathrm{occ}} \oplus \chi_{\mathrm{B}}^{\mathrm{occ}})$ defined above. When combined in the form $\mathscr{A}\psi^{\mathrm{A}}\psi^{\mathrm{B}}$ this will yield the same wavefunction as before. Now Δ vanishes completely[248]. (Also, if the monomer CI functions are complete up to *n*-fold excitations then the contributions to Δ can be related only to higher than *n*-fold excitations[248].) In this case $E_{\mathrm{AB}}^{\mathrm{HL}}$ reduces to monomer energies plus a *pure* interaction energy. These monomer CI energies $E_{\mathrm{A}}(\chi_{\mathrm{A}}^{\mathrm{occ}} \oplus \chi_{\mathrm{B}}^{\mathrm{occ}})$ will be lower than those in (26) and so the corresponding ΔE^{HL} energy will be more repulsive. This is a true interaction phenomenon, reflecting a decrease in correlation effects when bringing A and B together. In this particular case, the entire correlation energy $E_{\mathrm{A}}^{\mathrm{CI}}(\chi_{\mathrm{A}}^{\mathrm{occ}} \oplus \chi_{\mathrm{B}}^{\mathrm{occ}}) - E_{\mathrm{A}}^{\mathrm{SCF}}(\chi_{\mathrm{A}}^{\mathrm{occ}} \oplus \chi_{\mathrm{B}}^{\mathrm{occ}})$ is lost upon dimerization as a result of the Pauli principle, giving rise to increased exchange repulsion in ΔE^{HL}. Exactly the same result would be obtained from a non-orthogonal valence-bond study, where each individual monomer is described by a CI function in the basis $(\chi_{\mathrm{A}}^{\mathrm{occ}} \oplus \chi_{\mathrm{B}}^{\mathrm{occ}})$. This example illustrates that a reduction of excited configurations in the *dimer* situation is not a valid reason to omit configurations from *monomer* calculations.

Returning to the SCF case, we note that the Δ term in $E_{\mathrm{AB}}^{\mathrm{HL}}$ can also be made small (i.e. $O(S^4)$) by using the full dimer basis $(\chi_{\mathrm{A}} \oplus \chi_{\mathrm{B}})$ in calculating the monomer SCF wavefunctions, i.e. full CP[175]. This leads to a different $E_{\mathrm{AB}}^{\mathrm{HL}}$, and a different ΔE^{HL}:

$$\begin{aligned} E_{\mathrm{AB}}^{\mathrm{HL}}(\chi_{\mathrm{A}} \oplus \chi_{\mathrm{B}}) = E^{\mathrm{A}}(\chi_{\mathrm{A}} \oplus \chi_{\mathrm{B}}) + E^{\mathrm{B}}(\chi_{\mathrm{A}} \oplus \chi_{\mathrm{B}}) + E^{(1)}[\psi^{\mathrm{A}}(\chi_{\mathrm{A}} \oplus \chi_{\mathrm{B}}), \\ \psi^{\mathrm{B}}(\chi_{\mathrm{A}} \oplus \chi_{\mathrm{B}})] + O(S^4) \end{aligned} \tag{29}$$

This recipe may be the only practical option, for example, when bond functions are used on the A–B axis, or when the numerical Hartree–Fock method is employed.

Values of ΔE^{HL} corresponding to (29) have been obtained by several authors[78,144,175–178,249]. They are sometimes higher, sometimes lower, than those of (25). The difference has been termed BSSE$^{(1)}$ (Ref. 144), but this is confusing since most of the difference may reside in the different $E^{(1)}$ energies in

(25) and (29). For example, in Section IV it was noted that wavefunctions obtained in the dimer basis set yield significantly better E_{exch} energies, and this would enter $E^{(1)}$. This explains also why Hayes *et al.*[206] find such good agreement with large-basis results when using only a split-valence basis, something they deem fortuitous. The change in $E^{(1)}$ may in fact be much larger even than the full δ^{CP} (cf. Table VI).

As long as the full δ^{CP} correction is applied to the final ΔE^{SCF}, it is a matter of taste whether one prefers ΔE^{HL} (Eq. (26)) or ΔE^{HL} (Eq. (29)), although of course the latter uses the most complete basis to describe the occupied orbitals. In both schemes the wavefunctions of the participating monomers are distance- and orientation-dependent, and this goes for their electrical moments too[139]. Using δ^{occ} to correct only the exchange component of ΔE^{HL} (Eq. (26))[247] is not consistent therefore.

To conclude this section we note that the observation that $\delta^{\text{CP}} > \Delta E^{(2)}$[61,217,223] that led to the rejection of δ^{CP} in Ref. 217 can now be understood as resulting from the choice of an inappropriate reference energy in interpreting ΔE^{HL}. Even in Cammi's ($\chi^{\text{occ}} \oplus \chi^{\text{occ}}$) scheme[245] such anomalies can arise in principle. They will only be absent when the full dimer basis is used to obtain ψ^{A} and ψ^{B}, i.e. when one employs ΔE^{HL}(Eq. (29)) to partition the final ΔE^{SCF}.

B. Monomer A in Dimer AB Can Recover the Full Counterpoise Correction δ^{CP}

The counterpoise recipe can be applied by using (20) directly, without explicit consideration of a quantity called BSSE (cf. Section V.B). If, on the other hand, one sets up the dimer calculation by starting from monomer wavefunctions obtained in the monomer's own basis set, then BSSE will enter the calculation at some stage, and a correction δ must be applied to remove it. In the previous section a small portion of δ^{CP} (viz. δ^{occ}) was seen to creep in already in ΔE^{HL}. We now examine how the remainder of δ^{CP} can enter the dimer calculation, even though the occupied orbitals seem 'unavailable'.

By performing the monomer plus ghost calculation in steps, first adding only the ghost occupied MOs, one obtains δ^{occ} (Eq. (28)); likewise, adding only the virtuals, one obtains δ^{virt} (Eq. (23)). Adding the full space of the ghost one finds δ^{CP}(Eq. (22)), which proves to be larger than the sum of δ^{occ} and δ^{virt}

$$\delta^{\text{total}} = \delta^{\text{CP}} = \delta^{\text{occ}} + \delta^{\text{virt}} + \delta^{\text{occ-virt}}$$

Typical results for these δ contributions are shown in Table VIII. The direct contribution of the ghost occupieds (δ^{occ}) is seen to be quite small, but the indirect contributions, viz. the mixing term $\delta^{\text{occ-virt}}$ can be as large as 60% of δ^{CP}[78,134,178,245]. This may be understood as follows.

In a monomer plus ghost calculation, the ghost occupieds and virtuals,

TABLE VIII
Separate contributions to the BSSE[a].

			δ^{occ}	δ^{virt}	$\delta^{occ-virt}$	δ^{total}	δ^{CT}
SCF	$H_2O....HOH$[245]				(kcal mol^{-1})		
	($R = 2.98$ Å)	STO-3G	0.77	1.0	1.64	3.41	
		4-31G	0.22	0.39	0.84	1.45	
	($R = 3.58$ Å)	STO-3G	0.11	0.16	0.21	0.48	
		4-31G	0.05	0.58	0.32	0.95	
SCF	He....He[178]				(K)		
	($R = 5.6a_0$)	4s2p	1.26	48.3	5.1	54.7	
		5s2p	0.83	10.3	2.9	14.0	
		10s2p	0.002	0.03	0.006	0.04	
CI	He....He[134]				(K)		
	($R = 5.6a_0$)	CR0*DS3	0.2	20.2	2.14	22.54	22.79
		CR0*DS5	0.24	33.29	9.37	42.90	43.08
		CR2*DS3	0.007	1.22	0.183	1.41	1.85

[a]In Refs 178 and 134, δ^{occ} for molecule A was calculated in the basis $(\chi_A \oplus \chi_B^{occ})$.

taken separately, do not have the right shape to act as an improvement to the monomer's wavefunction, since they have cusps at the ghost nuclear positions, nodal planes in the wrong places, etc. The monomer requires functions that behave smoothly at these positions, so the combination of ghost occupied and virtual orbitals is much better suited to improve the monomer wavefunction and $\delta^{occ-virt}$ is large. The monomer in the *dimer* situation is a different matter. Here we find a perturbing potential at the position of the other monomer (caused by its nuclei and electrons), so now functions with cusps are perfect for the physical situation the monomer is in, allowing it to recover the full BSSE. Formally, one may say that the monomer can still use the entire function space of the ghost, and only a physical perturbation (the Pauli principle) forces it away from the ghost occupieds, into a function space orthogonal to these. So in both dimer and monomer calculations the extra orbitals are adapted to the physical requirements of the system. Using only the virtual orbitals of the ghost in the monomer counterpoise calculation introduces restrictions on the orbital space which are *only* justified in the dimer calculation.

To check this point of view we set up a calculation to estimate the amount of BSSE actually recovered by the monomer in a dimer situation at the CI level. The BSSE δ^{CT} is now obtained via[250]

$$\delta_A^{CT} = E(\Psi_B^{frozen}, \Psi_A(\chi_A)) - E(\Psi_B^{frozen}, \Psi_A(\chi_A \oplus \chi_B^{virt})) \qquad (30)$$

$$\delta^{CT} = \delta_A^{CT} + \delta_B^{CT}$$

The energy of monomer A in the presence of the frozen monomer B is calculated once with its own basis set (this energy contains the small energy lowering $-\delta^{occ}$), and once with the total dimer basis set, where the occupied orbitals of B are already used. The difference is the extra energy gained by monomer A if it is offered the orbitals of B in the dimer situation. Since it is impossible to avoid charge transfer in this extra energy, the BSSE so defined will contain some charge-transfer energy and is therefore named δ_A^{CT}. Even allowing for some charge transfer, which for He_2 is estimated as 0.16 K[138], it seems from Table VIII that the total $\delta^{virt} + \delta^{occ-virt}$ is reproduced by the δ^{CT} estimate, rather than δ^{virt} alone.

C. Practical Considerations

A few practical points remain to be discussed. Acceptance of the counterpoise principle implies that all energy terms contained in the final ΔE must be interpreted in terms of monomer wavefunctions and properties calculated in the full dimer basis set. As mentioned before, the final representation of E_{exch} may in fact be better than would be expected from an estimate of E_{exch} in terms of the $\psi_A(\chi_A)$ monomer wavefunctions. On the other hand, the final representation of E_{Coul} and E_{ind} may contain undesirable artefacts[139,198], such as an unphysical dipole–dipole contribution to E_{Coul} of He_2. These effects have been termed higher-order BSSE[139], and they will not be removed by applying the δ^{CP} correction. In certain applications it may be desirable to remove these artefacts[178,198] for example by adding a correction $E_{Coul}(\psi_A(\chi_A), \psi_B(\chi_B)) - E_{Coul}(\psi_A(\chi_A \oplus \chi_B), \psi_B(\chi_A \oplus \chi_B))$ to the final ΔE. Likewise, if in polar complexes the multipole energy is corrected by inserting better values of the multipole moments[78,104,105,107], then the dimer-basis multipole energy is the proper reference. The same applies to the calculation of differential properties of van der Waals molecules. Counterpoise corrections have been applied to electron densities[198,251], multipole moments[139,198,201,205] and polarizabilities[139,252].

If, instead of dimer interactions, many-body effects are calculated, the Boys–Bernardi recipe is in a straightforward way extended to read 'each subsystem is to be calculated in the complete basis of the supersystem', a recipe that has been applied to trimer interactions[262,263] and, albeit approximately, to cluster calculations[197,203,204]. Wells and Wilson[196] call this the site–site counterpoise function method and formulate it nicely for two-body, three-body, etc., interactions. Computing only a counterpoise correction for pairs of monomers and assuming additivity proved to overestimate the BSSE even for small systems[196].

Extra complications occur if a full geometry optimization is performed, distorting the internal geometrical parameters of the monomers. Two monomer calculations are now needed to calculate an interaction energy with respect to the undistorted monomers. A counterpoise calculation with both the monomer and the ghost at their distorted geometry R' yields the reference energy for the interaction between the distorted monomers

$$\Delta E' = E_{AB}(\chi_A \oplus \chi_B; R'_A R'_B) - E_A(\chi_A \oplus \chi_B; R'_A R'_B) - E_B(\chi_A \oplus \chi_B; R'_A R'_B) \quad (31)$$

Then the deformation energy, needed to distort each monomer from its original geometry R to its dimer geometry R', must be calculated, e.g. according to

$$E^A_{\text{def}} = E_A(\chi_A, R'_A) - E_A(\chi_A, R_A) \quad (32)$$

Since the basis χ_A will usually be designed for describing interactions rather than monomer deformations, there may be a need to use a different basis in this step. One might even use a different calculational method than in the dimer calculation, or one could use experimental force constant and geometry information for the monomer. The total interaction energy is now given by

$$\Delta E = \Delta E' + E^A_{\text{def}} + E^B_{\text{def}} \quad (33)$$

and the optimized geometry is that which minimizes ΔE. This recipe has been applied rarely yet and only as far as we are aware using the monomer basis to calculate the deformation energy[190,253,254]. Since analytical gradient techniques have become widely available[25,26,255] the simultaneous optimization of many geometrical parameters has become feasible, so formulae (31)–(33) should find widespread use.

VII. CONCLUSIONS

Although the calculation of a potential energy surface for a van der Waals molecule is fraught with difficulties, a reliable *ab initio* surface should not be out of reach.

The basis set employed should be geared towards a proper description of all components of the interaction energy, suggesting a basis of triple-zeta quality augmented with polarization functions allowing a proper description of electrical moments, intra-correlation and inter-correlation (dispersion) effects. If the dispersion energy is not the major component of the interaction energy, a single moment-optimized ($1d^X1p^H$) set or an energy/polarizability-optimized ($2d^X2p^H$) set of polarization functions, possibly augmented by ($1f^X1d^H$), may suffice. SCF and MP2 may then be used as a cost-effective method to obtain a reasonable potential energy surface. For better results one could turn to a size-consistent CI method, like CEPA. If the dispersion energy is the major factor in the van der Waals interaction, like in He_2, the basis set should be quite extended and an approach better than single-reference CEPA is called for. A simple recipe is not available as yet in these cases.

The counterpoise procedure must be applied in order to avoid the basis-set superposition error.

We recommend that the components of the interaction energy be monitored routinely in order to avoid local fortuitous cancellation of errors, which could lead to distortion of the potential energy surface.

Note Added in Proof

The debate on the BSSE problem (section VI) has continued since we completed our manuscript.

Szcześniak and Scheiner have applied the ideas developed in Refs. 134 and 178 to an MP2 study on $(HF)_2$ [264]. The full δ^{CP} correction scheme was shown to produce interaction energies that behave as one should expect for the given basis, thus confirming the validity of the function counterpoise method at the correlated level of theory.

Perhaps the most direct theoretical argument in favour of including all (virtual and occupied) ghost orbitals in the monomer + ghost calculation and hence in the correction δ^{CP} is the following[178]: Suppose the dimer calculation is performed in a complete set of (one-electron) basis functions, composed of two incomplete sets, one centered at monomer A and one at monomer B. The exact interaction energy (for the given method) will then only be recovered if the same complete basis is used in calculating the monomer energies. Nobody would argue that the 'to be occupied' orbitals of the ghost at B should be removed when calculating the energy of A, since this would render the A-basis incomplete.

Alternative counterpoise correction schemes were proposed by Olivares *et al.*[265] and by Dykstra *et al.*[266]. Neither of these schemes produces the exact correction in the hypothetical case of a complete dimer basis, as sketched above, and so they must be rejected.

A rather novel objection[143,266,267] against the function counterpoise method is that it does not increase reliability, since δ^{CP} does not remove the remaining errors in ΔE. Thus 'the extra expense of performing counterpoise calculations is not warranted and it is better to increase the basis-set to the maximum affordable'[143]. This argument tacitly assumes that increasing the basis will simultaneously reduce both the BSSE and the remaining errors in ΔE. While this may be true in some special cases (e.g. see Refs. 266 and 178), there are now several well-documented examples where increases in the basis set lead to increased BSSE[134,202,264,268].

Acknowledgements

We thank G. Chałasiski and M. Gutowski for many stimulating debates and many insights into the various errors in calculations on van der Waals molecules. We thank V. R. Saunders and P. E. S. Wormer for helpful discussions. We thank J. Verbeek and R. J. Vos for providing some of the data presented here, and J. A. Vliegenthart and D. E A. Pozzi for assistance during the preparation of the manuscript.

References

1. Keesom, W. H., *Phys. Z.*, **22**, 129, 643 (1921); *Phys. Z.*, **23**, 225 (1922).
2. Debye, P., *Phys. Z.*, **21**, 178 (1920); *Phys. Z.*, **22**, 302 (1921).
3. Falckenhagen, H., *Phys. Z.*, **23**, 87 (1922).
4. Heitler, W., and London, F., *Z. Phys.*, **44**, 455 (1927).
5. London, F., *Z. Phys. Chem. (B)*, **11**, 222 (1930).
6. London, F., *Trans. Faraday Soc.*, **33**, 8 (1937).
7. Kollman, P. A., and Allen, L. C., *Chem. Rev.*, **72**, 283 (1972).
8. Schuster, P., Zundel, G., and Sandorfy, C. (Eds), *The Hydrogen Bond*, North-Holland, New York, 1976.
9. Morokuma, K., *Acc. Chem. Res.*, **10**, 294 (1977).
10. Certain, P. R., and Bruch, L. W., in *Theoretical Chemistry*, MTP Int. Rev. Sci., Phys. Chem. Ser. 1, Vol. 1, Butterworths, London, 1972.
11. Amos, A. T., and Crispin, R. J., in *Theoretical Chemistry, Advances and Perspectives* (Eds Vol. 2, H. Eyring and D. Henderson) p. 1, Academic Press, New York, 1976.
12. Claverie, P., in *Intermolecular Interactions: From Diatomics to Biopolymers* (Ed. B. Pullman), p. 69, New York, Wiley, 1978.
13. van der Avoird, A., Wormer, P. E. S., Mulder, F., and Berns, R. M., *Top. Curr. Chem.*, **93**, 1 (1980).
14. Clementi, E., *Lecture Notes in Chemistry*, Vol. 19, Springer, Berlin, 1980.
15. Briels, W. J., Jansen, A. P., and van der Avoird, A., *Adv. Quantum Chem.*, **18**, 131 (1986).
16. Frisch, M. J., Pople, J. A., and Del Bene, J. E., *J. Phys. Chem.*, **89**, 3664 (1985).
17. Curtiss, L. A., Frurip, D. J., and Blander, M., *J. Chem. Phys.*, **71**, 2703 (1973); Dyke, T. R., Mack, K. M., and Muenter, J. S., *J. Chem. Phys.*, **66**, 498 (1977).
18. Newton, M. D., and Kestner, N. R., *Chem. Phys. Lett.*, **94**, 198 (1983).
19. Maitland, G. C., Rigby, M., Smith, E., and Wakeham, W. A., *Intermolecular Forces*, Oxford University Press, Oxford, 1981.
20. Hobza, P., and Zahradnik, R., *Weak Intermolecular Interactions in Chemistry and Biology*, Elsevier, Amsterdam, 1980.
21. Beyer, A., Karpfen, A., and Schuster, P., *Top. Curr. Chem.*, **120**, 1 (1984).
22. Sandorfy, C., *Top. Curr. Chem.*, **120**, 41 (1984).
23. Dyke, T. R., *Top. Curr. Chem.*, **120**, 85 (1984).
24. Hobza, P., and Zahradnik, R., *Int. J. Quantum Chem.*, **23**, 325 (1983).
25. Binkley, J. S., Whiteside, R. A., Krishnan, R., Seeger, R., DeFrees, D. J., Schlegel, H. B., Topiol, S., Kahn, L. R., and Pople, J. A., GAUSSIAN 80, *Quantum Chemistry Program Exchange*, **13**, 406 (1981).
26. Dupuis, M., Spangler, D., and Wendoloski, J. J., NRCC Program QGDI, 1980; Guest, M. F., and Kendrick, J., GAMESS User Manual, part I, Daresbury Laboratory, Warrington, 1985.
27. Le Roy, R. J., and Carley, J. S., *Adv. Chem. Phys.*, **42**, 353 (1980).
28. Stone, A. J., *Chem. Phys. Lett.*, **83**, 233 (1981).
29. Stone, A. J., and Alderton, M., *Mol. Phys.*, **56**, 1047 (1985).
30. Buckingham, A. D., and Fowler, P. W., *J. Chem. Phys.*, **79**, 6426 (1983).

31. Buckingham, A. D., and Fowler, P. W., *Can. J. Chem.*, **63**, 2018 (1985).
32. Baiocchi, F. A., Reiher, W., and Klemperer, W., *J. Chem. Phys.*, **79**, 6428 (1983).
33. Bonaccorsi, R., Petrongolo, C., Scrocco, E., and Tomasi, J., *Theor. Chim. Acta (Berl.)*, **20**, 331 (1971).
34. Smit, P. H., Derissen, J. L., and van Duijneveldt, F. B., *J. Chem. Phys.*, **67**, 274 (1977).
35. Tomasi, J., in *Molecular Interactions* (Eds H. Ratajczak and W. J. Orville-Thomas), Vol. 3, p. 119, Wiley, Chichester, 1982.
36. Morokuma, K., and Kitaura, K., in *Chemical Applications of Atomic and Molecular Electrostatic Potentials* (Eds P. Politzer and D. G. Truhlar), p. 215, Plenum, New York, 1981.
37. Böhm, H.-J., and Ahlrichs, R., *J. Chem. Phys.*, **77**, 2028 (1982).
38. Böhm, H.-J., Ahlrichs, R., Scharf, P., and Schiffer, H., *J. Chem. Phys.*, **81**, 1389 (1984).
39. Böhm, H.-J., and Ahlrichs, R., *Mol. Phys.*, **54**, 1261 (1985).
40. Arrighini, P., *Lecture Notes in Chemistry*, Vol. 25, Springer, Berlin, 1981.
41. Jeziorski, B., and Kolos, W., in *Molecular Interactions* (Eds H. Ratajczak and W. J. Orville-Thomas), Vol. 3, p. 1, Wiley, Chichester, 1982.
42. Murrell, J. N., in *Orbital Theories of Molecules and Solids* (Ed. N. H. March), Clarendon, Oxford, p. 311, 1974.
43. Gray, C. G., and Lo, B. W. N., *Chem. Phys.*, **14**, 73 (1976).
44. Price, S. L., Stone, A. J., and Alderton, M., *Mol. Phys.*, **52**, 987 (1984).
45. Hall, G. G., and Martin, D., *Theor. Chim. Acta (Berl.)*, **59**, 281 (1981).
46. Dovesi, R., Pisani, C., Ricca, F., and Roetti, C., *J. Chem. Soc., Faraday Trans. II*, **70**, 1381 (1974).
47. Bonaccorsi, R., Cimiraglia, R., Scrocco, E., and Tomasi, J., *Theor. Chem. Acta (Berl.)*, **33**, 97 (1974).
48. Hirshfeld, F. L., *Theor. Chim. Acta (Berl.)*, **44**, 129 (1977).
49. Brobjer, J. T., and Murrell, J. N., *J. Chem. Soc., Faraday Trans. II*, **78**, 1853 (1982).
50. Stone, A. J., and Tough, R. J. A., *Chem. Phys. Lett.*, **110**, 123 (1984).
51. Dalgarno, A., and Davison, W. D., *Adv. At. Mol. Phys.*, **2**, 1 (1966).
52. Maeder, F., and Kutzelnigg, W., *Chem. Phys.*, **42**, 95 (1979).
53. Koide, A., Meath, W. J., and Allnatt, A. R., *J. Phys. Chem.*, **86**, 1222 (1982).
54. Visser, F., Wormer, P. E. S., and Stam, P., *J. Chem. Phys.*, **79**, 4973 (1983).
55. Amos, R. D., Handy, N. C., Knowles, P. J., Rice, J. E., and Stone, A. J., *J. Phys. Chem.*, **89**, 2186 (1985).
56. Stone, A. J., *Mol. Phys.*, **56**, 1065 (1985).
57. Karlström, G., *Theor. Chim. Acta (Berl.).*, **55**, 233 (1980).
58. McWeeny, R., *Croat. Chim. Acta*, **57**, 865 (1984).
59. Jaszunski, M., and McWeeny, R., *Mol. Phys.*, **55**, 1275 (1985).
60. Hoinkis, J., Ahlrichs, R., and Böhm, H.-J., *Int. J. Quantum Chem.*, **23**, 821 (1983).
61. Smit, P. H., Derissen, J. L., and van Duijneveldt, F. B., *Mol. Phys.*, **37**, 501 (1979).
62. Ng, K. C., Meath, W. J., and Allnatt, A. R., *Mol. Phys.*, **33**, 699 (1977).
63. Murrell, J. N., and Shaw, G., *J. Chem. Phys.*, **49**, 4731 (1968).
64. Kreek, H., and Meath, W. J., *J. Chem. Phys.*, **50**, 2289 (1969).
65. Ahlrichs, R., Penco, P., and Scoles, G., *Chem. Phys.*, **19**, 119 (1976).
66. Koide, A., Meath, W. J., and Allnatt, A. R., *Mol. Phys.*, **39**, 895 (1980).
67. Neumann, D. B., and Krauss, M., *J. Chem. Phys.*, **75**, 315 (1981).
68. Chałasiński, G., *Mol. Phys.*, **49**, 1353 (1983).
69. Dreyfus, M., and Pullman, A., *Theor. Chim. Acta (Berl.)*, **19**, 20 (1970).
70. Kollman, P. A., and Allen, L. C., *Theor. Chim. Acta (Berl.)*, **18**, 399 (1970).

71. Kochanski, E., *J. Chem. Phys.*, **58**, 5823 (1973); *Chem. Phys. Lett.*, **10**, 543 (1973).
72. Conway, A., and Murrell, J. N., *Mol. Phys.*, **23**, 1143 (1972).
73. Chałasiński, G., *Chem. Phys.*, **82**, 207 (1983).
74. Chałasiński, G., *Mol. Phys.*, **57**, 427 (1986).
75. Landshoff, R., *Z. Phys.*, **102**, 201 (1936).
76. Gutowski, M.,Chałasiński, G., and van Duijneveldt-van de Rijdt, J. G. C. M., *Int. J. Quantum Chem.*, **26**, 971 (1984).
77. van Duijneveldt-van de Rijdt, J. G. C. M., and van Duijneveldt, F. B., *Chem. Phys. Lett.*, **17**, 425 (1972).
78. Groen, Th. P., and van Duijneveldt, F. B., A comparative study of basis set effects in *ab initio* SCF Calculations on the HF dimer, quoted in Ref. 208.
79. Amos, A. T., and van den Berghe, C. S., *Mol. Phys.*, **47**, 897 (1982).
80. Jeziorski, B., and van Hemert, M., *Mol. Phys.*, **31**, 713 (1976).
81. van Duijneveldt, F. B., *Intermolecular Forces and the Hydrogen Bond*, Ph.D. Thesis, Utrecht, 1969.
82. Hirschfelder, J. O., *Chem. Phys. Lett.*, **1**, 325 (1967).
83. Møller, C., and Plesset, M. S., *Phys. Rev.*, **56**, 618 (1934).
84. Epstein, P. S., *Phys. Rev.*, **28**, 695 (1926).
85. Nesbet, R. K., *Proc. R. Soc. A*, **230**, 312 (1955).
86. Seger, G., and Kochanski, E., *Chem. Phys. Lett.*, **76**, 568 (1980).
87. van Lenthe, J. H., and van Duijneveldt, F. B., *J. Chem. Phys.*, **81**, 3168 (1984).
88. Mulder, F., Geurts, P., and van der Avoird, A., *Chem. Phys. Lett.*, **33**, 215 (1975).
89. Cremaschi, P., Morosi, G., Raimondi, M., and Simonetta, M., *Mol. Phys.*, **38**, 1555 (1979).
90. Hayes, I. C., and Stone, A. J., *Mol. Phys.*, **53**, 83 (1984).
91. Raimondi, M., *Mol. Phys.*, **53**, 161 (1984).
92. Wormer, P. E. S., Bernards, J. P. C., and Gribnau, M. C. M., *Chem. Phys.*, **81**, 1 (1983).
93. Matsuoka, O., Clementi, E., and Yoshimine, M., *J. Chem. Phys.*, **64**, 1351 (1976).
94. Kutzelnigg, W., in *Modern Theoretical Chemistry* (Ed. H. F. Schaefer, III), Vol. 3, p. 129, Plenum, New York, 1977.
95. Sadlej, A. J., *J. Chem. Phys.*, **75**, 320 (1981).
96. Dalgarno, A., *Adv. Chem. Phys.*, **12**, 143 (1967).
97. Kochanski, E., *Chem. Phys. Lett.*, **31**, 301 (1975).
98. Arrighini, G. P., Biondi, F., and Guidotti, G., *Chem. Phys.*, **2**, 85 (1973).
99. Mulder, F., *Ab Initio Calculations of Molecular Multipoles, Polarisabilities and van der Waals Interactions*, Ph.D. Thesis, Nijmegen, 1978.
100. van Dam, T., *Ab Initio Calculations on the Ar–HCl van der Waals Molecule*, Ph.D. Thesis, Utrecht, 1984.
101. Buck, U., Kohlhase, A., Secrest, D., Phillips, T., Scoles, G., and Grein, F., *Mol. Phys.*, **55**, 1233 (1985).
102. Buck, U., Kohl, K. H., Kohlhase, A., Faubel, M., and Staemmler, V., *Mol. Phys.*, **55**, 1255 (1985).
103. Böhm, H. J., and Ahlrichs, R., *Mol. Phys.*, **55**, 1159 (1985).
104. Karlström, G., On the evaluation of intermolecular potentials, in *Proc. 5th Semin. on Computational Methods in Quantum Chemistry*, Groningen, 1981, p. 333.
105. Poulsen, L. L., *Chem. Phys.*, **68**, 29 (1982).
106. Karlström, G., Linse, P., Wallqvist, A., and Jönsson, B., *J. Am. Chem. Soc.*, **105**, 3777 (1983).
107. Kroon-Batenburg, L. M. J., and van Duijneveldt, F. B., *J. Mol. Struct., Theochem*, **121**, 185 (1985).
108. Pople, J. A., *Faraday Disc. Chem. Soc.*, **73**, 7 (1982).

109. Szabo, A., and Ostlund, N. S., *J. Chem. Phys.*, **67**, 4351 (1977).
110. Chałasiński, G., private communication.
111. Diercksen, G. H. F., Roos, B. O., and Sadlej, A. J., *Int. J. Quantum Chem.*, **S17**, 265 (1983).
112. Diercksen, G. H. F., Kraemer, W. P., and Roos, B. O., *Theor. Chim. Acta (Berl.)*, **36**, 249 (1975).
113. Clementi, E., and Habitz, P., *J. Phys. Chem.*, **87**, 2815 (1983).
114. Baum, J. O., and Finney, J. L., *Mol. Phys.*, **55**, 1097 (1985).
115. Kroon-Batenburg, L. M. J., *Hydrogen Bonding and Molecular Conformation*, Ph.D. Thesis, Utrecht, 1985.
116. Schinke, R., Müller, W., and Meyer, W., *J. Chem. Phys.*, **76**, 895 (1982).
117. Benzel, M. A., and Dykstra, C. E., *J. Chem. Phys.*, **78**, 4052 (1983).
118. Jaquet, R., and Staemmler, V., *Chem. Phys.*, **101**, 243 (1986).
119. March, N. H., Young, W. H., and Sampanther, S., *The Many-Body Problem in Quantum Mechanics*, Cambridge University Press, Cambridge, 1967.
120. Pople, J. A., Binkley, J. S., and Seeger, R., *Int. J. Quantum Chem.*, **S10**, 1 (1976).
121. Ahlrichs, R., Scharf, P., and Ehrhardt, C., *J. Chem. Phys.*, **82**, 890 (1985).
122. Ahlrichs, R., *Adv. Chem. Phys.*
123. Purvis, G. D., III, and Bartlett, R. J., *J. Chem. Phys.*, **68**, 2114 (1978); *J. Chem. Phys.*, **71**, 548 (1979).
124. Knowles, P. J., Somasundram, K., Handy, N. C., and Hirao, K., *Chem. Phys. Lett.*, **112**, 8 (1985).
125. Shavitt, I., in *Modern Theoretical Chemistry* (Ed. H. F. Schaefer, III), Vol. 3, p. 189, Plenum, New York, 1977.
126. Langhoff, S. R., and Davidson, E. R., *Int. J. Quantum Chem.*, **S9**, 183 (1975).
127. Pople, J. A., Seeger, R., and Krishnan, R., *Int. J. Quantum Chem.*, **S11**, 149 (1977).
128. van Lenthe, J. H., van Dam, T., van Duijneveldt, F. B., and Kroon-Batenburg, L. M. J., *Faraday Symp. Chem. Soc.*, **19**, 125 (1984).
129. Werner, H. J., and Meyer, W., *Mol. Phys.*, **31**, 855 (1976).
130. Riemenschneider, B. R., and Kestner, N. R., *Chem. Phys.*, **3**, 193 (1974).
131. Meyer, W., Hariharan, P. C., and Kutzelnigg, W., *J. Chem. Phys.*, **73**, 1880 (1980).
132. Kutzelnigg, W., *Faraday Disc. Chem. Soc.*, **62**, 185 (1977).
133. Liu, B., and McLean, A. D., *J. Chem. Phys.*, **72**, 3418 (1980).
134. Gutowski, M., van Lenthe, J. H., Verbeek, J., van Duijneveldt, F. B., and Chałasiński, G., *Chem. Phys. Lett.*, **124**, 370 (1986).
135. Senff, U. E., and Burton, P. G., *Mol. Phys.*, **58**, 637 (1986).
136. Brugmans, A. L. J., Farrar, J. M., and Lee, Y. T., *J. Chem. Phys.*, **64**, 1345 (1976).
137. Chałasiński, G., van Smaalen, S., and van Duijneveldt, F. B., *Mol. Phys.*, **45**, 1113 (1982).
138. Collins, J. R., and Gallup, G. A., *Mol. Phys.*, **49**, 871 (1983).
139. Karlström, G., and Sadlej, A. J., *Theor. Chim. Acta (Berl.)*, **61**, 1 (1982).
140. Votava, C., Ahlrichs, R., and Geiger, A., *J. Chem. Phys.*, **78**, 6841 (1983).
141. Andzelm, J., and Huzinaga, S., *Chem. Phys.*, **100**, 1 (1985).
142. Popkie, H. E., and Kaufman, J. J., *Int. J. Quantum Chem.*, **10**, 47 (1976).
143. Schwenke, D. W., and Truhlar, D. G., *J. Chem. Phys.*, **82**, 2418 (1985).
144. Sokalski, W. A., Hariharan, P. C., and Kaufman, J. J., *J. Comput. Chem.*, **4**, 506 (1983).
145. Kolos, W., *Theor. Chim. Acta (Berl.)*, **54**, 187 (1980).
146. van Duijneveldt-van de Rijdt, J. G. C. M., and van Duijneveldt, F. B., *J. Mol. Struct.*, **89**, 185 (1982).
147. Dunning, T. H., Jr, *J. Chem. Phys.*, **55**, 3958 (1971).
148. Urban, M., Kellö, V., and Carsky, P., *Theor. Chim. Acta (Berl.)*, **45**, 205 (1977).

149. Ahlrichs, R., Driessler, F., Lischka, H., and Staemmler, V., *J. Chem. Phys.*, **62**, 1235 (1975).
150. Clementi, E., and Popkie, H., *J. Chem. Phys.*, **57**, 1078 (1972).
151. Rosenberg, B. J., and Shavitt, I., *J. Chem. Phys.*, **63**, 2162 (1975).
152. Zeiss, G. D., Scott, W. R., Suzuki, N., and Langhoff, S. R., *Mol. Phys.*, **37**, 1543 (1979).
153. Mulder, F., van Dijk, G., and van der Avoird, A., *Mol. Phys.*, **39**, 407 (1980).
154. Gready, J. E., Bacskay, G. B., and Hush, N. S., *Chem. Phys.*, **31**, 467 (1978).
155. Prissette, J., and Kochanski, E., *Chem. Phys. Lett.*, **47**, 391 (1977).
156. Chałasiński, G., van Lenthe, J. H., and Groen, Th. P., *Chem. Phys. Lett.*, **110**, 369 (1984).
157. Kocjan, D., Koller, J., and Azman, A., *J. Mol. Struct.*, **34**, 145 (1976).
158. Frisch, M. J., Pople, J. A., and Binkley, J. S., *J. Chem. Phys.*, **80**, 3265 (1984).
159. Krishnan, R., Binkley, J. S., Seeger, R., and Pople, J. A., *J. Chem. Phys.*, **72**, 650 (1980).
160. Huzinaga, S., and Sakai, Y., *J. Chem. Phys.*, **50**, 1371 (1969).
161. van Duijneveldt, F. B., *IBM Techn. Res. Rep.* RJ 945 (1971).
162. Schmidt, M. W., and Ruedenberg, K., *J. Chem. Phys.*, **71**, 3951 (1977).
163. Diercksen, G. H. F., and Sadlej, A. J., *Theor. Chim. Acta (Berl.)*, **63**, 69 (1983).
164. Diercksen, G. H. F., Kellö, V., and Sadlej, A. J., *Mol. Phys.*, **49**, 711 (1983).
165. Diercksen, G. H. F., Roos, B. O., and Sadlej, A. J., *Int. J. Quantum Chem.*, **S17**, 265 (1983).
166. Williams, D. R., Schaad, L.J., and Murrell, J. N., *J. Chem. Phys.*, **47**, 4916 (1967).
167. Raffenetti, R. C., *J. Chem. Phys.*, **58**, 4452 (1973).
168. Dunning, T. H., Jr, *J. Chem. Phys.*, **53**, 2823 (1970).
169. Stewart, R. F., *J. Chem. Phys.*, **52**, 431 (1970).
170. Lés, A., quoted by Chałasiński, G., and Jeziorski, B., *Mol. Phys.*, **32**, 81 (1976).
171. Lés, A., quoted by Bulski, M., *Mol. Phys.*, **4**, 1171 (1975).
172. Lés, A., quoted by Bulski, M.,Chałasiński, G., and Jeziorski, B., *Theor. Chim. Acta (Berl.)*, **52**, 93 (1979).
173. Dunning, T. H., Jr, and Hay, P. J., in *Modern Theoretical Chemistry* (Ed. H. F. Schaefer, III), Vol. 3, Chap. 1, Plenum, New York, 1977.
174. van Schaik, M. M. M., unpublished results.
175. Urban, M., and Hobza, P., *Theor. Chim. Acta (Berl.)*, **36**, 215 (1975).
176. Kurdi, L., Kochanski, E., and Diercksen, G. H. F., *Chem. Phys.*, **92**, 287 (1985).
177. Sokalski, W. A., Hariharan, P. C., and Kaufman, J. J., *J. Phys. Chem.*, **87**, 2803 (1983).
178. Gutowski, M., van Duijneveldt, F. B., Chałasiński, G., and Piela, L., *Mol. Phys.* (in press).
179. Gallup, G. A., *Mol. Phys.*, **49**, 865 (1983).
180. Liu, B., and McLean, A. D., *J. Chem. Phys.*, **59**, 4557 (1973).
181. Kestner, N. R., *J. Chem. Phys.*, **48**, 252 (1968).
182. Ransil, B. J., *J. Chem. Phys.*, **34**, 2109 (1961).
183. Clementi, E., *J. Chem. Phys.*, **46**, 3851 (1967).
184. Jansen, H. B., and Ros, P., *Chem. Phys. Lett.*, **3**, 140 (1969).
185. Boys, S. F., and Bernardi, F., *Mol. Phys.*, **19**, 553 (1970).
186. Dacre, P. D., *Mol. Phys.*, **37**, 1529 (1979).
187. Price, S. L., and Stone, A. J., *Chem. Phys. Lett.*, **65**, 127 (1979).
188. Jönsson, B., and Nelander, B., *Chem. Phys.*, **25**, 263 (1977).
189. Johansson, A., Kollman, P., and Rothenberg, S., *Theor. Chim. Acta (Berl.)*, **29**, 167 (1973).
190. Maggiora, G. M., and Williams, I. H., *J. Mol. Struct.*, **88**, 23 (1982).

191. Olivares Del Valle, F. J., Tolosa, S., Lopez Pineiro, A., and Requena, A., *J. Comput. Chem.*, **6**, 39 (1985).
192. Daudey, J. P., Claverie, P., and Malrieu, J. P., *Int. J. Quantum Chem.*, **8**, 1 (1974).
193. Kochanski, E., in *Intermolecular Forces* (Ed. B. Pullman), p. 15, Reidel, Dordrecht, 1981.
194. Bentley, J., *J. Am. Chem. Soc.*, **104**, 2754 (1982).
195. Pettersson, L., and Wahlgren, U., *Chem. Phys.*, **69**, 185 (1982).
196. Wells, B. H., and Wilson, S., *Chem. Phys. Lett.*, **101**, 429 (1983).
197. Fowler, P. W., and Madden, P. A., *Mol. Phys.*, **49**, 913 (1983).
198. Fowler, P. W., and Buckingham, A. D., *Mol. Phys.*, **50**, 1349 (1983).
199. Spiegelmann, F., and Malrieu, J. P., *Mol. Phys.*, **40**, 1273 (1980).
200. Lengsfield, B. H., III, McLean, A. D., Yoshimine, M., and Liu, B., *J. Chem. Phys.*, **79**, 1891 (1983).
201. Allavena, M., Silvi, B., and Cipriani, J., *J. Chem. Phys.*, **76**, 4573 (1982).
202. Wells, B. H., and Wilson, S., *Mol. Phys.*, **50**, 1295 (1983).
203. Miyoshi, E., Tatewaki, H., and Nakamura, T., *J. Chem. Phys.*, **78**, 815 (1983).
204. Kestner, N. R., Newton, M. D., and Mathers, T. L., *Int. J. Quantum Chem.*, **S17**, 431 (1983).
205. Andzelm, J., Kłobukowski, M., and Radzio-Andzelm, E., *J. Comput. Chem.*, **5**, 146 (1984).
206. Hayes, I. C., Hurst, G. J. B., and Stone, A. J., *Mol. Phys.*, **53**, 107 (1984).
207. Senff, U. E., and Burton, P. G., *J. Phys. Chem.*, **89**, 797 (1985).
208. Kołos, W., *Theor. Chim. Acta (Berl.)*, *51*, 219 (1979).
209. Ostlund, N. S., and Merrifield, D. L., *Chem. Phys. Lett.*, **39**, 612 (1976).
210. Bulski, M., and Chałasiński, G., *Theor. Chim. Acta (Berl.)*, **44**, 399 (1977).
211. Daudey, J. P., Malrieu, J. P., and Rojas, O., *Int. J. Quantum Chem.*, **8**, 17 (1974).
212. Daudey, J. P., *Int. J. Quantum Chem.*, **8**, 29 (1974).
213. Stone, A. J., and Hayes, I. C., *Faraday Disc. Chem. Soc.*, **73**, 19 (1982).
214. Pullman, A., Sklenar, H., and Ranganathan, S., *Chem. Phys. Lett.*, **110**, 346 (1984).
215. Tomonari, M., Tatewaki, H., and Nakamura, T., *J. Chem. Phys.*, **80**, 344 (1984).
216. Meunier, A., Levy, B., and Berthier, G., *Theor. Chim. Acta (Berl.)*, **29**, 49 (1973).
217. Collins, J. R., and Gallup, G. A., *Chem. Phys. Lett.*, **123**, 56 (1986).
218. Gerratt, J., and Papadopoulos, M., *Mol. Phys.*, **41**, 1071 (1980).
219. Surjan, P. R., Mayer, I., and Lukovits, I., *Chem. Phys. Lett.*, **119**, 538 (1985).
220. Balint-Kurti, G. G., and Karplus, M., in *Orbital Theories of Molecules and Solids* (Ed. N. H. March), p. 250, Clarendon, Oxford, 1974.
221. Wormer, P. E. S., van Berkel, T., and van der Avoird, A., *Mol. Phys.*, **29**, 1181 (1975).
222. Pouilly, B., Lengsfield, B. H., and Yarkony, D. R., *J. Chem. Phys.*, **80**, 5089 (1984).
223. Kochanski, E., and Flower, D. R., *Chem. Phys.*, **57**, 217 (1981).
224. Del Bene, J. E., *J. Comput. Chem.*, **2**, 188 (1981).
225. Popowicz, A., and Ishida, T., *Chem. Phys. Lett.*, **83**, 520 (1981).
226. Hinchliffe, A., *Chem. Phys. Lett.*, **85**, 531 (1982).
227. Huber, H., Hobza, P., and Zahradnik, R., *J. Mol. Struct.*, **103**, 245 (1983).
228. Sapse, A. M., and Jain, D. C., *J. Phys. Chem.*, **88**, 4970 (1984).
229. Toyonaga, B., Peterson, M. R., Schmid, G. H., and Csizmadia, I. G., *J. Mol. Struct.*, **94**, 363 (1983).
230. Latajka, Z., Sakai, S., Morokuma, K., and Ratajczak, H., *Chem. Phys. Lett.*, **110**, 464 (1984).
231. Jorgensen, W. L., *J. Am. Chem. Soc.*, **101**, 2016 (1979).
232. Jorgensen, W. L., *J. Am. Chem. Soc.*, **102**, 543 (1980).

233. Bouteiller, Y., Allavena, M., and Leclercq, J. M., *Chem. Phys. Lett.*, **84**, 361 (1981).
234. Sapse, A. M., *J. Chem. Phys.*, **78**, 5733 (1983).
235. Sapse, A. M., and Howell, J. M., *J. Chem. Phys.*, **78**, 5738 (1983).
236. Chandra Singh, U., and Kollman, P. A., *J. Chem. Phys.*, **80**, 353 (1984).
237. Hinchliffe, A., *J. Mol. Struct.*, **108**, 307 (1984).
238. Tatewaki, H., Tanaka, K., Ohno, Y., and Nakamura, T., *Mol. Phys.*, **53**, 233 (1984).
239. Szcześniak, M. M., and Scheiner, S., *J. Chem. Phys.*, **80**, 1535 (1984).
240. Latajka, Z., and Scheiner, S., *J. Chem. Phys.*, **81**, 2713 (1984).
241. Szcześniak, M. M., Scheiner, S., and Bouteiller, Y., *J. Chem. Phys.*, **81**, 5024 (1984).
242. Szczesniak, M. M., and Scheiner, S., *J. Chem. Phys.*, **83**, 1778 (1985).
243. Benzel, M. A., and Dykstra, C. E., *Chem. Phys.*, **80**, 273 (1983).
244. Bolis, G., Clementi, E., Wertz, D. H., Scheraga, H. A., and Tosi, C., *J. Am. Chem. Soc.*, **105**, 355 (1983).
245. Cammi, R., Bonaccorsi, R., and Tomasi, J., *Theor. Chim. Acta (Berl.)*, **68**, 271 (1985).
246. Leclercq, J. M., Allavena, M., and Bouteiller, Y., *J. Chem. Phys.*, **78**, 4606 (1983).
247. Bauschlicher, C. W., Jr, *Chem. Phys. Lett.*, **74**, 277 (1980).
248. Chałasiński, G., and Gutowski, M., *Mol. Phys.*, **54**, 1173 (1985).
249. Sokalski, W. A., Roszak, S., Hariharan, P. C., and Kaufman, J. J., *Int. J. Quantum Chem.*, **23**, 847 (1983).
250. Saunders, V. R., private communication.
251. Osman, R., Topiol, S., and Weinstein, H., *J. Comput. Chem.*, **2**, 73 (1981).
252. Dacre, P. D., *Mol. Phys.*, **45**, 1 (1982).
253. Emsley, J., Hoyte, O. P. A., and Overill, R. E., *J. Am. Chem. Soc.*, **100**, 3303 (1978).
254. Smit, P. H., Derissen, J. L., and van Duijneveldt, F. B., *J. Chem. Phys.*, **69**, 4241 (1978).
255. Pulay, P., in *Modern Theoretical Chemistry* (Ed. H. F. Schaefer, III), Vol. 4, p. 153, Plenum, New York, 1977.
256. Thakkar, A. J., *J. Chem. Phys.*, **75**, 4496 (1981).
257. Gutowski, M., Verbeek, J., van Lenthe, J. H., and Chałasiński, G., *Chem. Phys.* (1986) in press; five p-polarization functions were employed.
258. Vos, R. J., unpublished results using a [5s, 3p, 2d] gaussian basis for He and a [9s, 7p, 2d ($\alpha_d = .75, .14$)] basis for Ar.
259. Knowles, P. J., and Meath, W. J., *Chem. Phys. Lett.*, **124**, 164 (1986).
260. Andzelm, J., Huzinaga, S., Klobukowski, M., and Radzio, E., *Mol. Phys.*, **52**, 1495 (1984).
261. Cremaschi, P., Morosi, G., Raimondi, M., and Simonetta, M., *Chem. Phys. Lett.*, **109**, 442 (1984).
262. Clementi, E., Kołos, W., Lie, G. C., and Ranghino, G., *Int. J. Quantum Chem.*, **17**, 377 (1980).
263. Bulski, M., *Chem. Phys. Lett.*, **78**, 361 (1981).
264. Szcześniak, M. M., and Scheiner, S., *J. Chem. Phys.*, **84**, 6328 (1986).
265. Olivares del Valle, F. J., Tolosa, F., Esperilla, J. J., Ojalvo, E. A., and Requena, A., *J. Chem. Phys.*, **84**, 5077 (1986).
266. Loushin, S. K., Liu, S.-Y., and Dykstra, C. E., *J. Chem. Phys.*, **84**, 2720 (1986).
267. Frisch, M. J., del Bene, J. E., Binkley, J. S., and Schaefer III, H. F., *J. Chem. Phys.*, **84**, 2279 (1986).
268. Hobza, P., Schneider, B., Čársky, P., and Zahradnik, R., *J. Mol. Struct.*, **138**, 377 (1986).

AUTHOR INDEX

Abrikosov, A. A., 216, 236
Adamowicz, L., 252, 278
Adams, G. F., 257, 262, 267, 284
Adams, N., 189, 200
Agren, H., 16, 61, 115, 197, 414, 439, 445
Ahlrichs, R., 3, 4, 60, 154, 199, 252, 278, 281, 432, 444, 499, 519, 524, 526, 527, 532, 533, 536, 560
Ahmed, F., 480, 516
Aitken, A. C., 366, 395
Aizman, A., 307, 310, 317
Albertsen, P., 224, 225, 231, 236
Alder, B. J., 455, 511
Alderton, M., 524, 525, 560
Alexander, M. H., 390, 396
Allan, N. L., 358, 394
Allavena, M., 256, 280, 281, 546, 551, 565
Allen, L. C., 522, 524, 528, 559
Allnatt, A. R., 526, 560
Almlof, J., 4, 61, 169, 173, 198, 257, 262, 273, 281, 282, 402. 414, 415, 435, 442
Amaldi, E., 297, 316, 462, 513
Amemiya, A., 325, 340, 393
Ames, L. L., 480, 516
Amos, A. T., 522, 529, 559
Amos, R. D., 242, 266, 277, 281, 282, 526, 532, 560
Andersen, A., 133, 198
Andersen, O. K., 465, 513
Anderson, P. W., 316
Andreoni, W., 506, 519
Andzelm, J., 459, 465, 470, 477, 499, 505, 510, 512, 520
Anisimov, V. I., 459, 512
Arrighini, G. P., 222, 236, 532, 561
Ashcroft, N. W., 450, 510
Atha, P. M., 487, 503, 519
Averill, F. W., 455, 468, 511, 514
Avery, J., 451, 511
Azman, A., 538, 549, 563

Bachmann, C., 499, 518
Backsay, G. B., 172, 199, 281, 537, 563
Baer, M., 374, 395
Baerends, E. J., 294, 302, 303, 310, 311, 313, 316, 317, 318, 459, 465, 475, 484, 490, 495, 499, 504, 512
Baggio Saitovich, E. M., 493, 495, 518
Bagus, P. S., 345, 346, 394, 435, 437, 444, 495, 518
Baiocchi, F. A., 524, 560
Bair, R. A., 169, 199
Balduz, J. L., 289, 300, 306, 317
Balint-Kurti, G. G., 79, 196, 367, 376, 395, 551, 564
Ball, M. A., 202, 238
Banerjee, A., 79, 120, 128, 189, 195, 196, 272, 281, 405, 417, 443
Baraff, G. A., 475, 515
Barandiaran, K., 470, 514
Baroni, S., 462, 478, 513
Barrett, P. H., 495, 518
Barrow, R. F., 379, 396, 480, 516
Barth, A., 231, 236
Barthelat, J. C., 450, 484, 517
Bartlett, R. J., 2, 60, 133, 187, 198, 231, 236, 252, 275, 278, 281, 533, 562
Bas, G., 172, 199
Basch, H., 495, 505, 518
Bauche, J., 106, 197
Baum, J. O., 533, 552, 562
Baumann, C. A., 480, 516
Bauschlicher, C. W., 120, 163, 169, 198, 378, 396, 403, 406, 407, 417, 434, 435, 437, 439, 443, 470, 484, 499, 514, 552, 565
Bayfield, J. E., 352, 294
Baykara, N. A., 307, 310, 316, 465, 482, 502, 503, 519
Baykara, S. Z., 465, 513
Becke, A. D., 298, 300, 316, 465, 487, 492, 503, 505, 507, 518
Becker, H. U., 480, 516
Beebe, N. H. F., 133, 198
Benard, M., 470, 510, 514, 520
Bender, C. F., 195, 200

Bentley, J., 546, 564
Benzel, M. A., 533, 551, 562
Berkowitz, M., 306, 316
Bernardi, F., 391, 397, 544, 564
Bernards, J. P. C., 380, 396, 531, 539, 561
Bernholc, J., 468, 504
Berns, R. M., 380, 396, 522, 559
Bernstein, R. B., 322, 392
Berthier, G., 106, 197, 412, 443, 550, 564
Bethe, H. A., 390, 396
Beyer, A., 523, 559
Bhasu, V. C. J., 468, 514
Bhat, K., 480, 516
Biagi, A., 222, 236
Bienstock, S., 352, 377, 396
Bingel, W. A., 154, 199
Binkley, J. S., 169, 199, 252, 264, 271, 276, 284, 523, 533, 537, 559
Biondi, F., 222, 236, 553, 561
Bishop, D. M., 246, 281
Blaha, P., 468, 514
Blair, R. A., 169, 199
Blander, M., 522, 559
Blint, R. J., 345, 363, 393, 431, 444
Bloembergen, N., 206, 209, 236
Blomberg, M. R. A., 406, 424, 434, 435, 437, 444
Blyholder, G. D., 470, 514
Bobrowicz, F. W., 139, 198, 309, 316
Bobrowicz, J., 480, 516
Bobrowicz, W., 261, 281
Bohan, S., 222, 236
Bohm, H.-J., 169, 196, 524, 532, 533, 560
Bolis, G., 552, 565
Bonaccorsi, R., 524, 525, 552, 565
Bonybey, V., 403, 442, 480, 515
Bonifacic, V., 467, 513
Boring, M., 458, 512
Born, G., 216, 238
Born, M., 324, 393
Botch, B. H., 470, 514
Bottcher, C., 352, 378, 394
Bouman, T. D., 202, 222, 236
Bouteiller, Y., 551, 565
Boys, S. F., 391, 397, 544, 564
Brandas, E. J., 266, 281
Brandemark, U., 435, 437, 439, 445
Bratoz, S., 256, 265, 266, 280, 281
Brattsev, V. F., 222, 236
Brazuk, A., 352, 394
Briels, W. J., 522, 559
Briggs, M. P., 389, 396
Brobjer, J. T., 525, 560
Broch Mathison, K., 406, 435, 444
Brode, S., 499, 519
Brooks, B. R., 4, 60, 101, 197, 261, 277, 281
Brown, F. B., 4, 60, 99, 115, 123, 129, 133, 197
Bruch, L. W., 522, 559
Brugmans, A. L. J., 535, 562
Buck, U., 532, 561
Buckingham, A. D., 391, 397, 408, 443, 524, 546, 564, 557, 560
Buenker, R. J., 2, 60, 128, 198, 375, 395, 450, 510
Bulski, M., 550, 564
Bunch, J. R., 170, 199
Bunge, A., 38, 62
Burke, P. G., 390, 396
Burns, G., 295, 316
Bursten, B. E., 302, 316, 503, 519
Burton, P. G., 535, 562
Butcher, W., 2, 60
Byers Brown, W., 248, 282

Calais, J.-L., 303, 306, 318
Calazans, J. M., 304, 317
Callaway, J., 468, 514
Cammi, R., 552, 565
Camp, R. N., 135, 157, 198, 257, 267, 281
Carley, J. S., 524, 560
Carlsen, L., 439, 445
Carsky, P., 439, 445, 537, 563
Cartwright, D. C., 139, 198
Case, D. A., 296, 302, 307, 310, 316, 451, 464, 511
Castro, M., 503, 519
Caşula, F., 302, 316
Caves, T. C., 501, 519
Cederbaum, L. S., 202, 215, 216, 217, 218, 234, 236
Ceperley, D. M., 455, 511
Cernusak, I., 408, 443
Certain, P. R., 522, 559
Chalasinsky, G., 528, 532, 535, 537, 550, 554, 555, 561
Chandra Singh, U., 551, 565
Chang, T. C., 32, 62, 120, 197, 405, 417, 443

Chatt, J., 438, 445
Cheung, L. M., 4, 32, 61, 120, 133, 198, 405, 417, 443
Chiles, R. A., 3, 60
Chipman, D. M., 139, 145, 198
Cimiraglia, R., 525, 560
Cingi, M. B., 499, 519
Cipriani, J., 546, 564
Ciric, D., 352, 394
Cizek, J., 234, 236, 252, 275, 281, 500, 519
Claverie, P., 381, 396, 522, 546, 559, 564
Clementi, E., 461, 470, 513, 522, 531, 533, 537, 552, 559
Cocke, D. L., 480, 516
Cohen, D., 505, 519
Cohen, M., 356, 394
Cole, L. A., 478, 515
Coleman, A. J., 227, 236
Collins, J. R., 323, 362, 380, 384, 392, 535, 550, 555, 562, 564
Condon, E. U., 82, 296
Connolly, J. W. D., 291, 316, 465, 510, 513
Conrad, M., 270, 283
Conway, A., 528, 561
Cook, D. B., 270, 281
Cook, M., 316
Cooper, D. L., 330, 345, 353, 358, 363, 377, 379, 393
Cooper, I. L., 62
Cooper, M. J., 358, 394
Corson, R. M., 328, 393
Cortona, P., 455, 511
Cosse, C., 480, 515
Costa, C., 386, 396
Cotton, F. A., 502, 510, 519
Coulson, C. A., 306, 316, 358, 394, 400, 442
Cowan, R. D., 484, 517
Cox, D. M., 482, 517
Crandall, D. H., 352, 394
Cremaschi, P., 379, 385, 391, 396, 531, 551, 561
Crispin, R. J., 522, 559
Crist, B. V., 222, 237
Crooks, J. B., 322, 393
Csavinsky, P., 289, 316
Csizmadia, I. G., 551, 565
Curtiss, L. A., 522, 559

Dacre, P. D., 272, 281, 544, 557, 565
Dahl, F., 231, 236
Dahl, J. P., 451, 495, 511
Dalgaard, E., 4, 61, 64, 74, 88, 91, 103, 118, 196, 211, 213, 236, 413, 443
Dalgarno, A., 352, 377, 390, 394, 526, 532, 560
Das, G., 32, 62, 65, 172, 196, 382, 396, 405, 443, 484, 486, 517
Daudey, J. P., 378, 380, 396, 546, 550, 564
Davidson, E. R., 17, 30, 62, 71, 150, 195, 196, 266, 271, 282, 294, 316, 471, 515, 534, 562
Davis, M. J., 169, 199
Davison, W. D., 526, 560
De Heer, P. J., 352, 394
DeFrees, D. J., 523, 559
Deb, B. M., 254, 282
Debye, P., 522, 559
Dedieu, A., 470, 514
Del Bene, J. E., 522, 524, 555, 559
Del Conde, G., 499, 518
Delley, B., 311, 316, 465, 490, 499, 504, 513
Demuynck, J., 470, 499, 514
Derissen, J. L., 524, 526, 528, 560
Desclaux, J. P., 500, 519
Detrich, J. H., 64, 103, 196, 403, 442
Dhar, S., 468, 514
DiLella, D. P., 480, 515
Diamond, J. B., 296, 316, 465, 483, 517
Diercksen, G. H. F., 230, 234, 236, 408, 440, 445, 450, 510, 533, 539, 542, 554, 562
Dijkkamp, D., 352, 394
D'Incan, J., 379, 396
Ditchfield, R., 258, 270, 280, 282
Dixon, R. N., 499, 518
Dlemente, D., 499, 519
Docken, K. K., 32, 62
Docker, M. P., 375, 395
Dolg, M., 484, 499, 517
Domcke, W., 202, 215, 216, 236
Dongarra, J. J., 170, 199
Doniach, S., 455, 511
Donnelly, R. A., 306, 317
Donohue, J., 480, 515
Donovan, R. J., 322, 392
Doran, M. B., 390, 397, 484, 495, 517
Douady, L. G., 172, 199
Dovesi, R., 525, 560
Dreyfus, M., 528, 561

Driessler, F., 537, 563
Drowart, J., 480, 515
Duch, W. 4, 60
Duggan, J., 370, 395
Duncanson, L. A., 438, 445
Dunlap, B. I., 294, 302, 306, 310, 313, 316, 465, 482, 495, 510, 513
Dunning, T. H., 133, 136, 139, 145, 198, 330, 393, 437, 445, 470, 514, 537, 541, 563
Dupuis, M., 257, 258, 270, 273, 282, 424, 444, 524, 560
Dyke, T. R., 523, 559
Dykstra, C. E., 3, 60, 169, 199, 245, 282, 450, 510, 533, 551, 562
Dzyaloskhiniskii, I. E., 216, 236

Eades, R. A., 169, 199
Eastman, D. E., 458, 512
Effantin, C. C., 379, 396
Efremov, Yu., 480, 515
Ehrenfest, P., 207, 236
Ehrhardt, C., 169, 199, 533, 562
Elander, N., 147, 198, 212, 228, 231, 236
Elbert, S. T., 4, 32, 61, 120, 133, 195, 198, 404, 405, 417, 443
Ellinger, Y., 111, 150, 172, 197
Ellis, D. E., 294, 302, 311, 316, 455, 465, 499, 511
Ellison, F. O., 370, 373, 395
Emsley, J., 558, 565
Englisch, H., 292, 316
Englisch, R., 292, 316
English, J. H., 403, 442, 480, 515
Epstein, P. S., 531, 561
Epstein, S. T., 247, 248, 282
Erhardt, C., 252, 278, 281
Ermler, W. C., 508, 520
Eyler, J. R., 222, 236

Faegri, K., 435, 444
Faist, M. B., 370, 395
Falkenhagen, H., 522, 519
Farrar, J. M., 535, 562
Faubel, M., 532, 561
Felps, W. S., 375, 395
Felton, R. H., 302, 318, 465, 513
Fenske, R. F., 302, 316
Fermi, E., 297, 316, 462, 474, 513
Feynman, R. P., 202, 236, 254, 282
Figari, G., 368, 386, 395
Finney, J. L., 533, 552
Fischer, C. F., 434, 444, 471, 515
Fischer, I., 306, 316, 400, 442
Fitzgerald, G., 8, 61, 162, 199, 257, 278, 279, 282
Flad, J., 499, 506, 518
Flament, J. P., 232, 236
Flesch, J., 3, 60, 378, 396
Fletcher, R., 9, 61, 124, 198
Flores, C., 493, 518
Flower, D. R., 555, 564
Fock, V., 399, 442
Fogarasi, G., 242, 281, 282
Foglia, C., 499, 519
Ford, M. J., 345, 393
Ford, T. A., 480, 516
Fouassier, M., 480, 515
Fournier, R., 495, 518
Fowler, P. W., 391, 397, 524, 546, 560, 564
Fox, D. J., 4, 8, 60, 162, 199, 262, 268, 279, 282
Freed, K. F., 202, 215, 237
Freeman, A. J., 311, 316, 458, 465, 490, 499, 504, 509, 513
Friedlander, M. E., 470, 514
Friedrich, B., 347, 394
Frisch, M. J., 282, 522, 539, 559
Fronzoni, G., 232, 236
Frurip, D. J., 522, 559
Fuentealba, P., 484, 499, 517
Fukui, K., 276, 285
Fukutome, H., 291, 303, 316
Fursova, V. D., 484, 517

Galasso, V., 231, 232, 236
Gallup, G. A., 323, 356, 362, 375, 379, 385, 388, 392, 535, 544, 550, 555, 562
Garcia-Proeto, J., 505, 519
Gardner, L. D., 352, 394
Garrison, B. J., 373, 395
Gaspar, R., 297, 316, 453, 511
Gaw, J. F., 8, 61, 162, 199, 242, 257, 262, 264, 268, 279, 282
Gawronski, J. K., 222, 237
Gawronski, K., 222, 237
Geertsen, J., 221, 222, 231, 232, 234
Geiger, A., 536, 562
Geldart, D. J. W., 458, 512
Gelius, U., 439, 445, 458, 512

Gerratt, J., 244, 256, 264, 280, 282, 320, 324, 328, 330, 341, 342, 348, 363, 375, 379, 385, 388, 391, 392, 396, 551, 564
Gerrity, D. P., 369, 395
Gervais, A. B., 232, 236
Geurts, P., 380, 396, 531, 551, 561
Geurts, P. J. M., 380, 396
Geusic, M. E., 480, 488, 515
Ghandour, F., 232, 236
Ghosh, S. K., 306, 316
Gianinetti, E., 367, 395
Gilbert, M. M., 404, 443
Gilbody, H. B., 352, 363, 394
Gingerich, K. A., 480, 493, 495, 515
Giscinski, O., 147, 199
Gislason, E. A., 376, 395
Goddard, J. D., 256, 282
Goddard, W. A., 4, 11, 61, 139, 145, 172, 195, 198, 261, 281, 307, 309, 316, 330, 345, 363, 393, 403, 431, 444, 474, 480, 487, 488, 503, 518
Golab, J. T., 4, 8, 10, 19, 61, 111, 124, 197, 218, 237
Goldfield, E. M., 376, 395
Goldstein, E., 493, 518
Gole, J. L., 480, 516
Golebiewski, A., 112, 197
Golka, E., 461, 513
Goodgame, M. M., 139, 195, 198, 307, 309, 480, 488, 515
Gopinathan, M. S., 459, 512
Gordon, R. G., 461, 512
Gorkov, L. P., 216, 236
Goscinski, O., 212, 215, 226, 227, 228, 236, 266, 281
Graham, C., 408, 443
Gray, C. G., 525, 560
Gray, H. B., 503, 519
Gready, J. E., 537, 563
Green, S., 345, 346, 394
Grein, F., 79, 120, 197, 532, 561
Grev, R. S., 8, 61, 162, 199, 262, 268, 282
Gribnau, M. C. M., 380, 396, 531, 539, 561
Grice, R., 370, 395
Grien, F., 405, 417, 443
Griffin, D. C., 484, 517
Grimley, T. B., 468, 514
Grimm, F. A., 431, 444
Groen, Th. P., 529, 550, 554, 561
Gruner, N. E., 222, 230, 234, 236
Gubanov, V. A., 459, 512
Guenzburger, D., 493, 495, 518
Guest, M. F., 484, 495, 503, 517, 524, 558, 560
Guidotti, G., 532, 561
Gundersen, G., 432, 444
Gunnarsson, O., 299, 304, 313, 316, 455, 456, 460, 465, 476, 501, 511
Gupta, S. K., 480, 516
Gurvich, L. V., 480, 515
Guse, M., 470, 514
Gutowski, M., 529, 535, 543, 554, 555, 542
Gutsev, G., 484, 517
Gyorffy, B. L., 455, 511

Habitz, P., 533, 562
Hada, M., 255, 283
Hagstrom, S., 362, 366, 395
Hall, G. G., 255, 282, 525, 560
Hall, M. B., 458, 487, 503, 519
Hamermesh, M., 326, 393
Hamming, R. W., 73, 196
Handy, N. C., 4, 28, 60, 99, 115, 131, 162, 197, 242, 251, 253, 264, 271, 277, 282, 323, 392, 416, 429, 444, 526, 532, 533, 560
Hansen, Aa. E., 202, 222, 236
Hansen, G. P., 480, 515
Hansen, S. G., 480, 488, 515
Harding, L. B., 139, 169, 198
Hariharan, P. C., 536, 542, 554, 563
Harris, F. E., 344, 345, 393, 403, 442
Harris, J., 307, 313, 316, 458, 465, 474, 512
Harris, R. A., 221, 237
Harrison, J. F., 471, 514
Harrison, J. G., 460, 469, 512
Harrison, R. J., 278, 282, 429, 444
Hartree, D. R., 244, 282, 399, 442
Hay, J., 470, 474, 484, 506, 514
Hay, P. J., 330, 393, 403, 423, 442, 444, 542, 563
Hayden, C. C., 369, 395
Hayes, E. F., 373, 395
Hayes, I. C., 366, 380, 385, 386, 395, 531, 546, 550, 561
Heaon, R. A., 460, 469
Heath, J. R., 482, 517
Heaton, R. A., 301, 317
Hedberg, K., 432, 444

Hedin, L., 202, 237, 303, 307, 313, 318, 451, 453, 455, 491, 511
Hegarty, D., 270, 282, 416, 444
Hehre, W. J., 258, 282, 470, 514
Heiberg, A., 4, 32, 61, 121, 173, 174, 198, 402, 414, 418, 435, 442
Heil, T. G., 352, 394
Heitler, W., 309, 316, 322, 392, 522, 519
Helgaker, T. U., 257, 262, 268, 273, 281, 282
Hellmann, J., 254, 282
Herman, F., 302, 316, 463, 467, 514
Herman, K., 439, 445, 495, 518
Herman, M. F., 202, 215, 224, 231, 238
Herman, Z., 347, 394
Herzberg, G., 228, 237, 321, 347, 392, 394, 423, 436, 444
Hestenes, M. R., 266, 282
Hibbs, A. R., 202, 236
Hillier, I. H., 484, 487, 495, 503, 517
Himpsel, F. J., 458, 512
Hinze, J., 32, 62, 112, 172, 197, 418, 444
Hirao, K., 220, 237, 533, 547
Hiraya, A., 322, 392
Hirschfelder, J. O., 248, 282, 531, 561
Hirshfeld, F. L., 525, 560
Hisatsune, I. C., 434, 444
Hobza, P., 523, 542, 550, 551, 554, 559, 565
Hodgson, A., 375, 395
Hoffmann, M. R., 8, 61, 162, 199, 262, 268, 279, 282
Hohenberg, P., 288, 317, 450, 510
Hoinkis, J., 526, 560
Holland, G. F., 465, 503
Holzwarth, N. A. W., 468, 514
Hopkins, J. B., 480, 516
Hopper, D. G., 103, 197
Hotokka, M., 435, 439, 445
Howell, J. M., 470, 514, 551, 565
Hoyte, O. P. A., 558, 565
Hsia, Y. P., 493, 518
Huang, K., 324, 393
Hubbard, J., 294, 317
Huber, H., 270, 282, 480, 516, 551, 565
Huber, K. P., 228, 237, 347, 394, 423, 436
Huckel, E., 356, 394
Hunt, W. J., 139, 145, 198, 330, 393, 403, 417, 423, 444, 515
Hurley, A. C., 137, 139, 146, 154, 198, 282, 323, 330, 393, 401, 442, 450, 510
Hurst, G. J. B., 386, 396, 546, 564
Hush, N. S., 537, 563
Huzinaga, S., 20, 33, 467, 470, 499, 514, 539, 563
Hyla-Kryspin, I., 470, 514
Hylleraas, E. A., 196, 248, 282

Igawa, A., 405, 443
Igel, G., 484, 506, 517
Igel-Mann, G., 499, 519
Iijima, T., 432, 444
Imamura, A., 261, 285
Inokuchi, M., 468, 514
Isaacson, A. D., 373, 395
Ishida, K., 270, 283
Ishida, T., 551, 565
Ishiguro, E., 325, 340, 393

Jackels, C. F., 128, 150, 198
Jacon, M., 232, 237
Jaffe, R. L., 486, 517
Jahn, H. A., 329, 393
Jain, D. C., 551, 565
Janak, J. F., 289, 317, 455, 491, 504, 511
Jansen, A. P., 522, 559
Jansen, H. B., 544, 564
Jansson, A., 480, 516
Jaquet, R., 533, 562
Jasien, P. G., 245, 282
Jaszunski, M., 222, 231, 257, 283, 348, 394, 429, 444, 526, 560
Jensen, H. J., 4, 10, 61, 226, 227, 232, 237
Jensen, H. J., 414, 443
Jensen, J. O., 257, 272, 281
Jeung, G. H., 378, 396, 484, 499, 517
Jeziorski, B., 524, 531, 527, 560
Jiu Fai Lam, L. T., 381, 396
Johansen, H., 495, 518
Johansson, G., 458, 512
Johns, J. W. C., 431, 444
Johnson, K. H., 296, 302, 316, 451, 464, 465, 483, 513
Jones, A. C., 347, 394

Jones, R. O., 299, 307, 313, 314, 316, 456, 458, 460, 465, 474, 476, 512
Jonson, M., 460, 476, 512
Jonsson, B., 407, 443, 532, 544, 552, 562
Jordan, K. D., 137, 198, 499, 519
Jorgensen, P., 4, 8, 10, 13, 19, 61, 64, 74, 82, 88, 91, 103, 111, 118, 196, 202, 210, 215, 220, 222, 224, 225, 226, 231, 234, 236, 257, 262, 278, 281, 282, 404, 410, 413, 414, 443
Jorgensen, W. L., 551, 565
Joyes, P., 499, 518

Kahn, L. R., 258, 272, 484, 517, 524, 559
Kalapisch, M., 106, 197
Kaldor, A., 482, 517
Kaldor, U., 344, 345, 393
Kanda, K., 255, 283
Kant, A., 480, 515
Kao, C. M., 482, 501, 519
Kaplan, I. G., 326, 393
Karlstrom, G., 437, 439, 445, 526, 532, 536, 542, 557, 560
Karpfen, A., 523, 559
Karplus, M., 307, 316, 551, 564
Karwowski, J., 4, 60
Kaspar, J., 483, 517
Kato, S., 256, 283
Katsnelson, M. I., 459, 512
Katsuki, S., 468, 514
Kaufman, J. J., 347, 394, 536, 542, 554, 563
Keesom, W. H., 522, 559
Keil, F., 432, 444
Keller, J., 503, 519
Kelley, H. P., 461, 512
Kello, V., 537, 539, 563
Kemister, G., 462, 513
Kendrick, J., 524, 558, 560
Kertesz, M., 261, 283
Kestner, N. R., 522, 535, 546, 564
Kim, H., 363, 395
Kim, Y. S., 461, 512
Kimura, T., 325, 340, 393
Kindstedt, T., 435, 444
King, H. F., 135, 157, 198, 248, 257, 258, 266, 267, 273, 280, 281, 362, 381
Kirby, K., 347, 394
Kitaura, K., 468, 514, 524, 529, 550, 560
Klein, R., 348, 394
Kleman, B., 480, 516
Klemperer, W., 524, 560
Klobukowski, M., 467, 470, 506, 514
Klotzbucher, W., 480, 503, 516
Klyagina, A. P., 484, 517
Kmetko, E. A., 302, 317
Knapp, J. A., 458, 512
Knight, L. B., 480, 502, 519
Knobeloch, M. A., 222, 236
Knowles, P. J., 4, 28, 30, 32, 35, 55, 61, 99, 115, 125, 162, 188, 197, 283, 404, 405, 413, 416, 419, 443, 465, 483, 517, 526, 532, 533, 560
Koch, P. M., 352, 394
Kochanski, E., 528, 531, 532, 537, 542, 554, 561
Kocjan, D., 538, 549
Koelling, D. D., 294, 317
Kohl, K. H., 532, 561
Kohlhase, A., 532, 561
Kohn, W., 288, 289, 291, 292, 297, 317, 450, 451, 491, 511
Koide, A., 526, 527, 532, 561
Kok, R. A., 487, 517
Kolari, H. J., 503, 519
Koller, J., 538, 549, 563
Kollman, P. A., 522, 524, 528, 559
Kolos, W., 524, 527, 536, 550, 560
Komornicki, A., 248, 266, 271, 283, 434, 444
Koopmans, T. A., 400, 442
Kordis, J., 480, 516
Koski, W. S., 347, 394
Kosmas, A. M., 376, 395
Kotani, M., 325, 340, 393
Kozhukhovsky, V. B., 380, 515
Kraemer, W. P., 428, 444, 533, 551, 562
Krauss, M., 527, 561
Kreek, H., 527, 561
Krishnan, R., 187, 200, 283, 523, 559
Kroon-Batenburg, L. M. J., 392, 397, 532, 534, 540, 547, 550, 562
Krupenie, P. A., 408, 422, 443
Kubler, J., 455, 511
Kubo, R., 202, 237
Kumanova, M. D., 277, 283
Kundig, E. P., 480, 516

Kuntz, P. J., 323, 370, 373, 392
Kurdi, L., 552, 554, 563
Kurmaev, E. Z., 459, 512
Kurtz, H. A., 147, 199, 227, 228, 231, 236
Kutzelnigg, W., 144, 154, 198, 526, 531, 535, 560
Kvick, A., 432, 444
Kwang, K., 468, 514

Laaksonen, L., 298, 300, 316, 317, 406, 443, 465, 513
Lachlan, A. D., 202, 236
Ladik, J., 275, 284
Ladner, R. C., 145, 198, 345, 363, 442, 403, 442
Laidig, W. D., 4, 60, 195, 200, 252, 275, 278, 281
Lam, B., 195, 200
Lamson, S. H., 317, 456, 465, 512
Langhoff, S. R., 322, 392, 486, 517, 534, 537, 562
Langreth, D. C., 291, 317, 458, 463, 492, 513
Langridge-Smith, P. R. R., 480, 515
Lannoo, M., 475, 515
Laskowski, B. C., 499, 519
Last, I., 374, 395
Latajka, Z., 551, 565
Latham, W. A., 258, 282
Lazzeretti, P., 222, 237
Leasure, S. C., 367, 395
Leclerq, J. M., 551, 565
Lee, K., 468, 514
Lee, Y. S., 195, 200, 598, 520
Lee, Y. T., 369, 395, 535, 562
Lengsfield, B. H., 4, 61, 103, 115, 197, 257, 262, 267, 268, 283, 414, 418, 443, 546, 564
Lennard-Jones, J., 139, 146, 198, 330, 393
Leopold, D. G., 480, 516
LeRoy, R. J., 524, 560
Les, A., 541, 563
Levin, A. A., 494, 517
Levy, B., 4, 61, 106, 121, 172, 197, 251, 283, 405, 412, 413, 418, 443, 550, 564
Levy, M., 289, 292, 300, 306, 317, 451, 511
Li, K. C., 322, 392
Lie, G. C., 461, 513
Lieb, E. H., 291, 292, 298, 317
Liechenstein, A. I., 459, 512
Liegener, C. M., 204, 237
Lightner, D. A., 222, 237
Limm, W., 480, 515
Lin, C. C., 301, 317, 460, 469, 512
Lin, S. S., 480, 515
Linderberg, J., 93, 197, 202, 203, 220, 226, 237
Lindgren, I., 459, 512
Lindhard, J., 202, 238
Lineberger, W. C., 480, 516
Linse, P., 430, 440, 445, 532, 552, 562
Lipscomb, W. N., 209, 239, 330, 390, 396
Lipson, R. H., 480, 515
Lischka, H., 4, 60, 99, 129, 169, 197, 381, 396, 537, 563
Liu, B., 4, 38, 60, 103, 115, 197, 323, 345, 346, 347, 378, 392, 414, 418, 424, 443, 482, 488, 491, 517, 535, 544, 551
Liu, K., 482, 517
Lo, B. W. N., 525, 560
Lochet, J., 480, 516
Lofthus, A., 408, 422, 443
London, F., 309, 316, 522, 519
Lopez Pineiro, A., 546, 564
Lowdin, P. O., 75, 196, 212, 214, 238, 304, 317, 330, 380, 396, 400, 442
Lozes, R. L., 147, 198
Lukman, B., 212, 227
Lukovits, I., 551, 564
Lundqvist, B. I., 289, 292, 299, 306, 316, 455, 460, 476, 480, 501, 511
Lundqvist, S., 202, 237, 451, 455, 461, 511
Luthi, H. P., 435, 444
Lynch, D., 224, 231, 238

MacDonald, A. H., 455, 511
MacDonald, J. K. L., 196
Macias, A., 246, 281
Madden, P. A., 391, 397, 546, 564
Maeder, F., 526, 560
Maggiora, G. M., 546, 564
Magnasco, V., 368, 385, 395
Maitland, G. C., 523, 559
Maksic, Z. B., 466, 514
Malli, G., 484, 517
Malmqvist, P. A., 439, 445

Malrieu, J. P., 378, 381, 396, 546, 550, 564
Manley, J. C., 340, 393
Mann, J. B., 306, 318
Manne, R., 213, 238
Manzel, K., 480, 516
March, N. H., 289, 317, 450, 451, 510, 533, 562
Mareca, P., 503, 519
Marinelli, F., 222, 236
Marinero, E. E., 369, 395
Maripuu, R., 439, 445
Martin, D., 525, 560
Martin, R. L., 150, 199, 470, 480, 484, 499, 506, 514
Martins, J. L., 506, 519
Mathers, T., 546, 564
Mathews, J., 70, 196
Matsuoko, O., 531, 533, 561
Mattuck, R. D., 82, 86, 90, 197, 217, 238
Mayer, L., 551, 564
McCarthy, M. I., 33, 62
McClelland, B. W., 431, 432, 444
McCullogh, R. W., 352, 363, 394
McCurdy, C. W., 91, 197, 202, 223, 230, 238
McGlynn, S. P., 375, 395
McIntosh, D. F., 499, 519
McIver, J. W., 152, 157, 199, 249, 256, 257, 267, 281
McKelvey, J. M., 461, 513
McKoy, V., 202, 223, 230, 234, 238
McLachlan, A. D., 202, 238
McLean, A. D., 38, 62, 111, 150, 197, 323, 345, 346, 392, 482, 488, 491, 506, 517, 535, 544, 551, 562
McMaster, B. N., 307, 310, 316, 455, 465, 482, 490, 511, 513
McMurchie, L. E., 271, 283
McNab, T. K., 495, 518
McWeeny, R., 62, 64, 196, 222, 231, 309, 317, 526, 560
Mead, C. A., 78, 196
Meath, W., 526, 560
Mehl, M. J., 291, 317, 463, 492, 513
Meier, P. F., 373, 395
Mejean, T., 480, 489, 515
Melius, C. F., 345, 363
Merer, A. J., 480, 515
Mermin, N. D., 450, 510
Merrifield, D. L., 550, 564
Messmer, R. P., 317, 451, 456, 464, 465, 482, 503, 490, 501, 511
Meunier, A., 550, 564
Meyer, F. W., 352, 394
Meyer, W., 2, 4, 36, 39, 52, 60, 144, 154, 198, 246, 275, 277, 283, 378, 496, 414, 430, 440, 443, 533, 534, 535, 540, 549, 562
Michalopoulos, D. L., 480, 488, 515
Michels, H. H., 403, 442
Micklitz, H., 495, 518
Miller, T. M., 480, 516
Mills, I. M., 244, 256, 264, 280, 282
Minkwitz, R., 480, 516
Mitchell, S. A., 499, 519
Miyoshi, E., 546, 564
Moccia, R., 246, 256, 257, 268, 283
Moffit, W., 370, 395
Moler, C. B., 170, 199
Moller, C., 274, 283
Montano, P. A., 480, 493, 518
Moore, C. E., 352, 394, 469, 472
Morin, M., 507, 519
Morokuma, K., 256, 258, 270, 283, 468, 514, 522, 524, 529, 550, 560
Morosi, G., 379, 385, 391, 396, 531, 551, 561
Morrison, T., 480, 516
Morse, M. D., 480, 515
Moruzzi, V. L., 289, 317, 455, 491, 504, 511
Moser, C. M., 423, 444
Moshinsky, M., 361, 394, 409, 443
Moskovits, M., 480, 515
Muckerman, J. T., 370, 373, 395
Muenter, J. S., 390, 397
Mulder, F., 522, 531, 532, 537, 538, 551, 559, 561
Mullaly, D., 257, 267, 281
Muller, J. E., 465, 513
Muller, W., 378, 396, 533, 549, 562
Mulliken, R. S., 394, 442
Murata, Y., 432, 444
Murphy, D. R., 289, 317
Murrell, J. N., 381, 389, 396, 524, 525, 527, 528, 541, 560
Musso, G. F., 384, 386, 396

Nagarathna, H. M., 493, 518
Naik, V. M., 493, 518
Nakamura, T., 499, 518, 546, 551, 565
Nakatsuji, H., 220, 237, 255, 283

Nappi, B. M., 480, 516
Nelander, B., 544, 564
Nelin, C. J., 403, 407, 435, 437, 442, 484, 502, 517
Nesbet, R. K., 400, 442, 492, 518, 531, 561
Neumann, D. B., 527, 561
Neumark, D. M., 369, 395
Newton, M. D., 258, 282, 480, 495, 516, 522, 549, 559
Ng, K. C., 526, 560
Nichols, J. A., 91, 197, 218, 232, 237
Nielsen, E. S., 229, 230, 238
Nixon, E. R., 480, 516
Noell, J. O., 480, 516
Noodleman, L., 307, 310, 311, 313, 317, 499, 519
Norbeck, J. M., 356, 362, 395
Nordheim-Poschl, G., 392
Nordholm, S., 462, 513
Nordling, C., 458, 512
Norman, J. G., 301, 302, 309, 311, 313, 317, 499, 503, 519
Novaro, O., 499, 505, 519
Nusair, M., 455, 491, 511

O'Grady, B. V., 322, 392
Oddershede, J., 205, 214, 215, 219, 221, 222, 226, 231, 232, 234, 236
Ohno, Y., 551, 565
Ohrn, Y., 93, 147, 198, 216, 226, 227, 231, 232, 237
Olovares Del Valle, F. J., 546, 564
Olsen, J., 4, 8, 62, 64, 111, 196, 206, 210, 238, 404, 410, 413, 443
Olson, J. A., 373, 395
Ortenburger, I. B., 463, 513
Orth, F. B., 322, 392
Ortiz, J. V., 147, 198, 204, 226, 232, 237
Osamura, Y., 8, 61, 162, 199, 257, 262, 268, 279, 282
Osborne, J. H., 307, 310, 317
Osman, R., 557, 565
Ostlund, N. S., 202, 239, 450, 470, 510, 533, 550, 562
Ottinger, Ch., 347, 351, 394
Otto, P., 275, 284
Overill, R. E., 558, 565
Oxford, S., 298, 317
Ozin, G. A., 480, 499, 519
Pacansky, J., 111, 150, 197
Pace, S. A., 480, 516
Page, M., 257, 262, 267, 284
Painter, G. S., 455, 465, 468, 513
Paldus, J., 4, 60, 93, 197, 234, 238, 252, 266, 275, 281, 361, 394, 402, 415, 444, 500, 519
Palke, W. E., 306, 317, 345, 363
Paneuf, R. A., 352, 394
Papadopoulos, M., 387, 396, 552, 564
Pariser, R., 356, 394
Parks, E. K., 480, 515
Parks, J. M., 330, 393
Parr, R. G., 289, 300, 306, 317, 330, 356, 363, 393, 451, 511
Partridge, H., 322, 378, 396, 417, 471, 515
Pauling, L., 499, 519
Pauncz, R., 93, 97, 197, 328, 356, 394
Pavlidou, C. M. E., 461, 513
Pederson, M. R., 301, 317, 460, 512
Pegg, D. J., 352, 394
Pelissier, M., 499, 518
Pellegatti, A., 455, 511
Penco, P., 527, 561
Penotti, F., 331, 393
Perdew, J. P., 289, 300, 306, 308, 317
Perdew, J. P., 455, 458, 476, 478, 511
Pesic, D. S., 480, 516
Peterson, M. R., 551, 565
Peterson, R. S., 352, 394
Petterson, L. G. M., 406, 435, 444
Pettersson, L., 546, 564
Peyerimhoff, S. D., 2, 60, 128, 198, 375, 395, 450, 510
Phillips, T., 532, 561
Pickup, B. T., 212, 215, 238
Piela, L., 543, 555, 556, 563
Pietro, W. J., 470, 514
Pilar, F. L., 73, 77, 81, 196
Pisani, C., 525, 560
Pitzer, K. S., 484, 507, 508, 518, 517
Pitzer, R. M., 209, 239, 295, 317
Plesset, M. S., 274, 283, 531, 561
Pobo, L. G., 482, 517
Popkie, H. E., 536, 537, 563
Pople, J. A., 139, 146, 187, 257, 258, 262, 264, 282, 315, 317, 330, 356, 393, 400, 442, 522, 524, 532, 537, 539, 559
Popowicz, A., 551, 565
Porter, R. F., 431, 444

Post, D., 459, 465, 490, 499, 504, 512
Pouilly, B., 551, 564
Poulsen, L. L., 532, 562
Powers, D. E., 480, 488, 515
Preuss, H., 270, 284, 461, 463, 480, 484, 499, 506, 512
Price, S. L., 525, 544, 560
Prosser, F., 362, 366, 395
Prossette, J., 537, 563
Pterongolo, C., 524, 560
Pulay, P., 3, 60, 242, 256, 268, 270, 272, 275, 277, 279, 280, 281, 282, 558, 565
Pullman, A., 528, 550, 564
Purdum, H., 480, 516
Purvis, G. D., 133, 187, 198, 215, 218, 238, 275, 283, 533, 562
Pyper, N. C., 331, 339, 341
Pyykko, P., 298, 300, 316, 317, 406, 435, 443, 465, 484, 513

Raatz, F., 465, 513
Radlein, D. A. G., 370, 395
Radzio, E., 466, 470, 471, 499, 505, 513
Raffenetti, R. C., 541, 563
Raghavachari, K., 275, 276, 284, 499, 519
Raimondi, M., 323, 330, 341, 342, 345, 348, 356, 363, 379, 385, 392, 531, 551, 561
Rajagopal, A. K., 289, 317, 455, 511
Ramana, M. V., 455, 511
Ramsden, E. N., 356, 394
Randic, M., 381, 396
Ranganathan, S., 550, 564
Ransil, B. J., 544, 564
Rasolt, M., 458, 512
Ratajczak, H., 551, 565
Ratner, J., 495, 518
Ratner, M., 226, 237
Rauk, A., 303, 311, 318, 475, 515
Redmon, L. T., 218, 238
Reeves, C. M., 380, 396
Reichmuth, J., 347, 394
Reif, F., 306, 317
Reiher, W., 524, 560
Reinsch, E.-A., 3, 36, 60, 195, 200, 440, 445
Requena, A., 546, 564
Rescigno, T., 202, 223, 230, 238
Rettner, C. T., 369, 395
Rettrup, S., 220, 222, 236
Ricca, F., 525, 560
Rice, J. E., 242, 277, 284, 526, 532, 560
Richards, W. G., 423, 444, 480, 516
Riemenschneider, B. R., 535, 562
Rigby, M., 523, 559
Riley, S. J., 480, 482, 517
Rivoal, J. C., 480, 516
Roach, A. C., 373, 395
Robb, M. A., 416, 444
Robb, W. D., 390, 396
Robertson, I. L., 499, 518
Robinson, G. N., 369, 395
Rodwell, W. R., 381, 396
Roetti, C., 470, 514, 525, 560
Roos, B. O., 2, 4, 16, 32, 61, 64, 65, 121, 173, 196, 266, 284, 348, 394, 402, 403, 407, 410, 414, 418, 422, 424, 425, 429, 434, 435, 437, 438, 439, 442, 444, 470, 484, 502, 533, 539, 562
Roothaan, C. C. J., 103, 172, 197, 260, 284, 355, 394, 400, 442
Ros, P., 294, 302, 316, 544, 564
Rose, J. B., 234, 238
Rose, J. H., 458, 512
Rosen, A., 455, 511
Rosenberg, B. J., 537, 563
Rosmus, P., 4, 33, 61, 94, 348, 430, 444
Rossi, E., 222, 237
Rossler, U., 458, 512
Roszak, S., 554, 565
Rothenberg, S., 546, 564
Ruamps, J., 480, 516
Ruedenberg, K., 4, 32, 61, 62, 120, 133, 195, 198, 404, 405, 417, 443, 539, 543, 563
Rumer, G., 328, 393
Runau, R., 375, 395
Rutherford, D. E., 326, 393
Ruttink, P. J. A., 79, 196
Rys, J., 270, 284

Sabelli, N. H., 376, 395
Sabin, J. R., 202, 215, 222, 230, 232, 236, 302, 310, 316, 456, 465, 510, 513
Sadlej, A. J., 247, 257, 280, 282, 408, 440, 445, 531, 533, 536, 557, 561
Saebo, S., 3, 60, 264, 270, 281
Saito, C., 468, 514

Sakai, S., 551, 565
Sakai, Y., 467, 470, 499, 514, 537, 563
Salahub, D. R., 307, 310, 316, 451, 455, 464, 465, 471, 478, 490, 503, 505, 510, 511, 520
Salmon, W. I., 62
Salpeter, E. E., 390, 396
Sambe, H., 302, 318, 465, 513
Samoilova, A. N., 480, 515
Sampanther, S., 533, 562
Sandorfy, C., 522, 523, 559
Sangfelt, E., 147, 199, 228, 231, 236
Sapse, A. M., 551, 565
Satoko, C., 261, 285
Sauers, J., 347, 394
Saunders, V. R., 3, 16, 60, 257, 270, 271, 284, 556, 565
Savin, A., 463, 513
Saxe, P., 4, 60, 257, 261, 262, 268, 271, 272, 277, 281
Saxon, R. P., 347, 394
Schaad, L. J., 541, 563
Schaefer, H. F., 3, 60, 101, 162, 197, 201, 238, 257, 258, 261, 268, 272, 277, 281, 403, 442, 450, 510
Scharf, P., 169, 199, 252, 278, 281, 499, 519, 524, 533, 560
Schawlow, A. L., 321, 392
Scheiner, S., 551, 565
Scheraga, H. A., 552, 565
Schiffer, H., 169, 199, 524, 560
Schindler, M., 169, 199
Schinke, R., 533, 549, 562
Schirmer, J., 202, 217, 218, 231, 234, 236
Schlegel, H. B., 187, 200, 242, 248, 271, 275, 283, 523, 558, 559
Schluter, M., 475, 515
Schmid, G. H., 551, 565
Schmidt, M. W., 195, 404, 443, 539, 543, 563
Schmidtke, H. H., 484, 495, 517
Schneider, W. J., 461, 512
Schreiber, J. L., 373, 395
Schulten, K., 4, 60
Schulze, W., 480, 516
Schuster, P., 522, 523, 524, 559
Schwarz, K., 298, 302, 318, 455, 458, 511
Schwarz, W. H. E., 32, 62, 196
Schwenke, D. W., 391, 392, 397, 536, 549, 563
Schwerdtfeger, P., 499, 518
Scoles, G., 527, 532, 561
Scott, W. R., 537, 563
Scrocco, E., 524, 525, 560
Scullman, R., 480, 516
Secrest, D., 270, 285, 532, 561
Seeger, R., 252, 264, 276, 284, 523, 533, 537, 559
Seger, G., 531, 561
Seijo, L., 470, 514
Seligman, T. Y., 361, 394
Selin, L. E., 480, 516
Sellers, H. L., 249, 285
Sellini, I. A., 352, 394
Selmani, A., 455, 512
Semenova, L. I., 222, 236
Senff, U. E., 535, 546, 562
Sexton, G. J., 99, 115, 197, 283, 437, 439
Seyse, R. J., 480, 516
Shadwick, W. F., 301, 318
Shaks Emampour, J., 480, 516
Sham, L. J., 289, 291, 297, 317, 451, 458, 491, 511
Shapiro, M., 376, 395
Shavitt, I., 2, 4, 60, 71, 77, 93, 196, 252, 272, 281, 403, 415, 500, 519, 534, 537, 562
Shaw, G., 389, 396, 527, 561
Shenoy, G. K., 480, 516
Shepard, R., 4, 61, 75, 91, 103, 111, 115, 123, 137, 166, 169, 194, 196
Shibuya, T., 234, 238
Shim, I., 480, 493, 495, 498
Shim, L., 502, 519
Shobotake, K., 322, 392
Shore, H. B., 458, 512
Shortley, G. H., 82, 196
Shull, H., 400, 442
Sichel, J. M., 455, 512
Siegbahn, K., 458, 499, 512
Siegbahn, P. E. M., 2, 16, 36, 55, 60, 64, 121, 173, 174, 195, 196, 266, 284, 402, 406, 407, 418, 427, 435, 437, 438, 439, 442, 445
Siiver, D. W., 471, 515
Silver, M. D., 252, 272, 281
Silvi, B., 546, 564
Simandiras, E. D., 242, 282
Simonetta, M., 323, 356, 379, 385, 391, 392, 531, 551, 561

Simons, J., 4, 8, 19, 61, 75, 82, 91, 103, 111, 166, 194, 195, 196, 202, 213, 215, 217, 220, 225, 237, 248, 257, 262, 272, 281, 375, 395
Sinanoglu, O., 461, 512
Sironi, M., 384, 396
Skillman, S., 467, 514
Sklenar, H., 550, 564
Skriver, H. L., 455, 511
Slater, J. C., 289, 292, 302, 318, 454, 460, 511
Smalley, R. E., 480, 482, 488, 515
Smit, P. H., 524, 526, 528, 560
Smith, E. B., 523, 559
Smith, W. D., 212, 215, 238
Smoes, S., 480, 516
Snijders, J. G., 484, 495, 506, 508, 517
Snyder, L. G., 470, 514
Snyder, R. G., 434, 444
Sokalski, W. A., 536, 542, 554, 563
Somasundram, K., 533, 562
Sommers, C., 455, 511
Sondergaard, N. A., 347, 394
Spangler, D., 524, 560
Spiegelmann, F., 546, 564
Srdanov, V. I., 480, 516
Staemmler, V., 532, 533, 537, 561
Stam, P., 526, 560
Stamper, J. G., 389, 396
Stanton, R. E., 362, 381, 395
Staunton, J., 455, 511
Stevens, R. M., 209, 239
Stewart, G. W., 170, 199
Stewart, R. F., 270, 282, 541, 563
Stoll, H., 461, 463, 484, 499, 506, 512
Stone, A. J., 366, 380, 385, 386, 395, 524, 525, 526, 531, 532, 544, 550, 560
Stoner, E. C., 501, 519
Strane, P., 455, 511
Strauss, B., 480, 515
Streitweiser, A., 356, 394, 461, 513
Strich, A., 437, 445, 470, 514
Stroud, A. H., 271, 285
Stroyer-Hansen, T., 222, 231, 239
Stwalley, W. C., 322, 392
Subra, R., 172, 199
Sundberg, K. R., 120, 198, 404, 443
Sundholm, D., 298, 300, 316, 317, 406, 443, 465, 513
Sunil, K. K., 137, 198, 499, 519
Surjan, P. R., 552, 564
Sutcliffe, B. T., 64, 196
Suzuki, N., 537, 563
Svendsen, E. N., 222, 231, 232, 237
Svensson, S., 439, 445
Swanstrom, P., 4, 8, 10, 13, 61, 124, 198, 257, 285
Sykja, B., 303, 318
Szabo, A., 202, 239, 450, 510, 533, 562
Szczesniak, M. M., 551, 565
Szentpaly, L. V., 499, 518

Tachibana, A., 276, 285
Taketa, H., 468, 514
Talman, J. D., 301, 318
Tanaka, K., 551, 565
Tang, R., 468, 514
Tantardini, G. F., 323, 356, 363, 392
Tatewaki, H., 470, 499, 514, 546, 564
Tavan, P., 4, 60
Tavouktsoglu, A. N., 470, 514
Taylor, K. V., 480, 515
Taylor, P. R., 4, 60, 121, 169, 198, 257, 281, 402, 407, 415, 418, 442
Tenne, R., 376, 395
Teramae, H., 261, 285
Thies, B. S., 218, 237
Thomas, L. H., 474, 515
Thomsen, K., 257, 285
Thouless, D. J., 8, 61, 82, 88, 196, 202, 239
Thulstrup, E. W., 133, 198
Tolosa, S., 546, 564
Tomasi, J., 524, 552, 560
Tomonari, M., 550, 564
Tong, B. Y., 458, 461, 512
Topiol, S., 505, 519, 524, 558, 559
Torok, F., 256, 284
Tosi, C., 552, 565
Tough, R. J. A., 526, 560
Townes, C. H., 321, 392
Toyonaga, B., 551, 565
Tranquille, M., 480, 515
Travis, D. N., 480, 516
Trevor, D. J., 482, 517
Trickey, S. B., 310, 317, 456, 458, 512
Trinquier, G., 484, 517
Trogler, W. C., 465, 503, 519
Truhlar, D. G., 78, 196, 391, 392, 397, 536, 549, 563
Tschinke, V., 463, 492, 503, 505, 507, 518

Tully, J. C., 370, 395
Tuncel, E., 462, 478, 513

Ufford, C. W., 461, 512
Undheim, B., 196
Upton, T. H., 495, 518
Urban, M., 537, 542, 550, 554, 563

Vala, M., 480, 516
Valentini, J. J., 369, 395
Van Berkel, T., 379, 385, 396, 551, 564
Van Dam, T., 532, 534, 561
Van der Avoird, A., 379, 385, 396, 522, 531, 527, 551, 559
Van der Berghe, C. S., 529, 561
Van der Velde, G., 270, 282
Van Dijk, G., 537, 563
Van Duijneveldt, F. B., 424, 444, 516, 524, 529, 531, 534, 535, 539, 543, 550, 554, 558, 560
Van Duijneveldt, T., 392, 397
Van Duijneveldt-van de Rijdt, J. G. C. M., 529, 537, 542, 561
Van Dyke, J. P., 463, 513
Van Hemert, M., 531, 561
Van Lenthe, J. H., 3, 16, 60, 79, 196, 392, 397, 531, 534, 541, 561
Van Schaik, M. M. M., 542, 563
Van Smaalen, S., 535, 562
Van Vleck, J. H., 327, 393
Van Zee, R. J., 480, 502, 515
Vance, R. L., 362, 395
Vashista, P., 289, 317
Veillard, A., 470, 499, 514
Verbeek, J., 535, 555, 562
Verges, J., 379, 396
Verhaegen, G., 423, 444, 480, 516
Vincent, M. A., 262, 272, 283
Vinot, C., 470, 514
Visser, F., 266, 285, 526, 560
Voigt, B., 222, 236
Von Barth, U., 303, 307, 313, 318, 451, 453, 471, 491, 511
Von Niessen, W., 202, 218, 239
Von Weizsacker, C. F., 463, 513
Vosko, S. H. L., 455, 462, 491, 511
Votava, C., 536, 562

Wachters, A. J. H., 470, 514
Wagenblast, G., 461, 512
Wagner, A. F., 169, 199
Wahl, A. C., 64, 65, 169, 196, 382, 396, 403, 405, 442
Wahlgren, U., 406, 435, 444, 546, 564
Wakeham, W. A., 523, 559
Wakoh, S., 455, 511
Walch, S. P., 378, 396, 403, 407, 435, 471, 484, 515
Walker, R. L., 70, 196
Wallqvist, A., 532, 552
Walter, O., 217, 218, 234, 238
Walters, S. G., 331, 338, 393
Walton, R. A., 510, 520
Wang, H.-T., 375, 395
Wang, S.-W., 501, 519
Watts, R. O., 381, 396
Way, K. R., 322, 392
Wedig, U., 463, 484, 517
Weiner, B., 147, 198, 212, 226, 227, 231, 232, 237
Weinstein, H., 356, 394, 557, 565
Wells, B. H., 391, 397, 546, 564
Weltner, W., 480, 502, 515
Wendoloski, J. J., 524, 560
Werner, H.-J., 3, 4, 5, 11, 33, 36, 60, 65, 125, 195, 196, 348, 394, 404, 405, 413, 414, 419, 440, 443, 499, 519, 534, 540, 562
Wertz, D. H., 552, 565
Wetmore, R. W., 62
Wexler, S., 480, 515
Wheland, G. W., 323, 392
Whetton, R. L., 482, 517
Whiteside, R. A., 523, 559
Widmark, P.-O., 348, 394, 429, 437, 435, 445
Wiest, R., 510, 520
Wigner, E., 248, 285
Wigner, E. P., 325, 326, 393
Wijekoon, W. M., 222, 237
Wilk, L., 455, 491, 511
Wilkie, F. G., 352, 363, 394
Wilkinson, J. H., 69, 196
Williams, A. R., 289, 302, 317, 455, 491, 504, 511
Williams, B. G., 358, 394
Williams, D. R., 381, 396, 541, 569
Williams, I. H., 546, 564
Williams, J. H., 408, 443
Wilson, S., 353, 391, 394, 450, 510, 546, 564
Wilson, T. M., 306, 318
Winter, H., 352, 394

Witko, M., 499, 518
Wodke, A. M., 369, 395
Wolf, A., 484, 495, 517
Wolfe, S., 285
Wood, C., 484, 495
Wood, J. H., 302, 306, 307, 318
Woodward, R. W., 502, 519
Woolley, R. G., 465, 513
Wormer, P. E. S., 266, 285, 379, 380, 385, 522, 526, 531, 539, 551, 559
Worth, J. P., 458, 512
Wyatt, R. E., 363, 395

Yabushita, S., 91, 197
Yaffe, L. G., 4, 11, 61, 172, 199, 405, 443
Yamabe, T., 276, 285
Yamaguchi, Y., 162, 199, 257, 261, 264, 268, 271, 277, 279, 281
Yamanouchi, T., 329, 393
Yamashita, K., 276, 285, 455, 511
Yang, C. Y., 296, 302, 483, 507, 513, 519
Yang, W., 317
Yarkony, D. R., 120, 163, 198, 262, 283
Yeager, D. L., 4, 8, 10, 13, 19, 32, 61, 64, 74, 91, 103, 111, 196, 202, 215, 218, 224, 225, 231, 232, 236, 404, 410, 413, 443
Yonezawa, T., 255, 283
Yoshimine, M., 4, 60, 323, 345, 346, 392, 531, 533, 561
Young, W. H., 533, 562
Yu, H. L., 495, 518
Yurtsever, E., 112, 197

Zahradnik, R., 270, 282, 523, 559
Zanasi, R., 222, 237
Zare, R. N., 369, 395
Zaremba, E., 458, 512
Zeiri, Y., 376, 395
Zeiss, G. D., 537, 563
Zemke, W. T., 322, 392
Zheng, L.-S., 480, 497, 516
Zeringue, K. J., 480, 516
Ziegler, T., 303, 311, 318, 406, 463, 468, 475, 484, 508, 510, 513, 518
Zivkovic, T., 466, 514
Zubarev, D. N., 202, 205, 239
Zundel, G., 522, 559
Zunger, A., 300, 317, 455, 458, 476, 511

INDEX OF COMPOUNDS

These compounds are the subject of *ab initio* calculations referred to in the text. Van der Waals complexes are written thus: Ar–HF. Potential energy surfaces are indicated by the reaction coordinate, e.g. H–NH_2 or H–O–H.

Ag_2, 506
Al, 301
Al_9–H_2O, 465
AlH_4^-, 222
Ar–HCl, 386, 533, 534, 548
Ar–HF, 386
Ar_2, 532
Au_2, 508

B–H_2^+, 347
BH, 429
BH^+, 351
Be, 421
Be–HF, 373
BeH, 348
Be_2, 231, 322, 541
Butene, 222

CH, 323, 331
CH^+, 227, 231, 345
CH_2, 31, 36, 232
CH_2^+, 338
CH_2SiH_2, 233
CH_3CSCH_3, 222
CH_4, 222, 232, 256
C_2, 314, 422, 441
C_2H_2, 222, 232
C_2H_4, 138, 222, 233
$C_2H_5CSC_2H_5$, 222
C_2H_6, 222, 233, 355
C_3^+–H, 351
$C_3H_3^+$, 222
CN, 231
CO, 23, 231, 323
CO–HF, 533
CO_2, 222, 232
CS_2, 23
C_{12}, 232
Ca–HC1, 373
Chromone, 3-hydroxy-, 223
Co_2, 495
$Co_4(CO)_{12}$, 465
Cr_2, 307, 403, 407, 434, 441, 465, 487
$Cr_2(CO_2H)_4$, 510
Cu_2, 499
Cu_{19}, 465
Cyclohexadiene, 222
Cyclooctene, 222
Cyclopentene, 222

Diimide, 440

F_2, 462
Fe O, 31, 35, 416
Fe_2, 493
Fe_{13}, 468
Fe_{15}, 468
Ferredoxin, 310

H–NH_2, 375
H–OH, 375
HCHO, 19
HCN, 232
HCN–HF, 533
$HCNH^+$, 232
HCOOH dimer, 526
HCl, 232
HCl–Cl, 374
HF, 222, 429, 440, 462
HF–HF, 222, 392, 526, 533, 549
HNC, 232
H_2, 222, 231, 310, 315, 462
H_2–H_2, 232
H_2CO, 269, 540
H_2O, 37, 222, 232, 256, 375
H_2O–HF, 549
H_2O–H_2O, 522, 533, 534, 539, 548, 556
H_2S–H_2S, 533
He, 535
He^-, 382
He–Be, 386

$He-H-He^+$, 373
He–HF, 369, 381
$He-H_2$, 232, 380, 522
He–He, 222, 231, 379, 383, 390, 522, 532, 541, 543, 556, 558
He–Li, 381
He–LiH, 384
He–Ne, 390
$He-O_2$, 533

Li, 344
LiH, 222, 231, 322
Li_2, 228, 407

$M-Br_2$, 376
Methylene peroxide, 440
Mn_2, 492
Mo_2, 503

NH_3, 33, 256, 375, 465, 540
NH_3-NH_3, 533
NO, 20
N_2, 23, 222, 228, 231, 323, 408, 421, 462
N_2-CO, 323
N_2-HF, 533
N_2-N_2, 380, 526, 533
N_2O_4, 432
Nb_2, 502
Ne–HF, 386, 390
$Ne-H_2$, 380, 386
$Ne-Na_2$, 533
Ne–Ne, 390, 527, 541
$Ni-C_2H_4$, 437
NiCO, 468
NiH, 435
NiN_2, 483
Ni_2, 495
Ni_4, 465
$Ni_{13}-CO$, 465
Norbornenone, 223

OH, 37
OH^-, 462
O_2, 232, 323, 380
O_3, 37
Oxathiirane, 440

Pd–CO, 468
Pd_2, 505
$Pd_{10}-H$, 465
Propene, 222
Pt_2, 507

Rh_2, 505
Ru_2, 505

SF_6, 468
Sc_2, 484
SiC_2, 232
SiHCH, 232
SiH_3NH_2, 233
SiH_4, 222, 256
Si_2, 306
Si_2C, 232
Si_2H_2, 232
Si_2H_6, 233
Si_3, 232

Ti_2, 484

V_2, 484
Vinoxy, 32

W_2, 507

Y_2, 501

SUBJECT INDEX

Algebraic diagrammatic construction, 217, 234
Analytic energy gradients, 59, 241
Annihilation operator, 84
Aromaticity, 358
Atomic polarizability, 390
Atoms, excited states, 470
Avoided crossings, 129

Basis functions,
  field dependent, 280
  floating, 255
  Gaussian lobe, 270
Basis set superposition error, 362, 406, 498
Basis sets, diffuse, 541, 542
  dimer, 542
  even tempered, 541
  EZ (extended-zeta), 541, 542
  for molecular interaction, 536
  polarization, 558
  superposition error, 551
  triple polarized, 538
  triple-zeta, 558
Binary electron spectroscopy, 358
Bloch functions, 460
Bond lengths, 433, 435, 462
Born–Oppenheimer approximation, 287, 294
Bracketing theorem, 75
Branching diagrams, 338
Brillouin theorem, 412, 417
Broken spatial symmetry, 306, 313
Broken spin symmetry, 309

CASSCF method, 55, 134, 348, 399 *et seq*., 486, 503
CEPA, 500, 533
CI externally contracted, 427
CI internally contracted, 441
CI method, 132 *et seq*., 367, 392
  derivatives from, 276, 279
  frozen core, 277
CI with polarization, 496
CRAY computer, 266
CSF method, 248
Casimir–Polder relation, 525
Charge transfer processes, 352
Charge transfer states, 527
Charge transfer structures, 382
Chemical reaction, PES for, 374, 440
Chemisorption, 510
Chiroptical properties, 223
Clebsch–Gordon coefficients, 97
Cluster beams, 482
Complete active space, 27
Complex spherical harmonics, 299
Compton scattering, 358
Computational effort, 243
Configuration interaction, 290
Conical intersection, 33
Constrained space orbital variation method, 440
Correlated fragments (ICF), 551
Correlation energy, 131, 378, 528
  intermolecular, 531
  intramolecular, 382
Counterpoise method, 392, 544, 552, 555
Coupled cluster methods, 3, 275
Coupled cluster propagator method, 234
Coupled cluster wave functions, 251
Coupled electron pair approximation, 252, 430, 523, 530
Coupled perturbed HF method, 258, 263, 276
Creation operator, 84
Cubic force constants, 244
Cusps in surfaces, 78

Davidson correction, 428, 534
Davidson procedure, 30
Delocalization, 356
Density functional theory, 449, 474, 287
Density matrix, 401
  first order, 307
Dewar–Chatt–Duncanson model, 438
Diatomics, $X\alpha$ codes, 300

Diatomics-in-molecules method, 369
Dipole moment, 35, 381, 536
derivatives, 244, 280
dimers, 540
Direct CI, 2, 27
Discrete variational method, 465
Dispersion coefficient (C_6), 233
Dispersion energy, 524, 534
Distinct row table, 99, 409
Double-zeta basis set, 424
Dynamical correlation energy, 274, 420
Dynamical electron correlation, 274
Dyson equation, 216

Electrical anharmonicity, 244
Electron affinity, 459
Electron correlation, 279, 448, 522
core-valence, 378
Electron gas, 455, 463, 474
Exchange correlation basis, 467
Exchange correlation potentials, 455
Extended Independent Particle model, 133
External contraction, in CI, 36
External fields, 292

Fermi–Amaldi approximation, 297, 462
Fermi energy, 294
Ferromagnetism, 358, 482
Fletcher optimization, 9
Fock matrix, 294
Fock operator, 400
Four index transformation, 16, 369
Fractional occupation numbers, 305

GAUSSIAN70 package, 258
GAUSSIAN80 package, 270
GTO in $X\alpha$ method, 302
GUGA, 36
GVB, strongly orthogonal, 145
GVB method, 3, 144, 330, 403, 407, 432, 473, 488
Gelfand tableau, 361
Geminal power method, 225
Geminal product wavefunctions, 146
Genealogical spin eigen functions, 55
General linear group, 361
Generalized valence bond (GVB), 309
Ghost calculation, 555
Gradient techniques, 558
Graphical unitary group, 415
Green's function method, 215

Harriman orbitals, 308
Hartree–Fock theory, 400
Hartree problem, 296
Hellmann–Feynman forces, 254, 273, 280
Hermite–Gaussian functions, 466
Higher derivatives of PES, 268
Hohenberg–Kohn theorem, 287
Hückel theory, 450
Hydrogen bonding, 523
Hyperfine structure, 508
Hyperpolarizability tensor, 206
Hypervirial theorem, 234

Inertia of a matrix, 71
Infrared optical rotary power, 249
Intermolecular forces, 378, 522
Intermolecular perturbation theory, 385
Internal contraction, in CI, 36, 57
Internal–internal rotations, 413
Ionic structures, 323
Ionization potentials, 476

Jellium approximation, 287, 297

Kekulé structure, 358
Kohn–Sham construction, 287
Kohn–Sham equations, 452, 463
Koopman's theorem, 305, 400
Kotani basis, 327, 358

Lagrangian function, 250
Level-shifting of eigenvalues, 4, 9, 189
Local density approximation, 453
Local density functional, 291
Local spin density and correlation, 478
Local spin density method, 451

MCSCF, 77, 103, 127, 169
convergence of, 65
excited state, 78
higher derivatives, 267
internal contraction, 33
method, 223, 235, 256, 346, 404, 414, 449, 469
methods, 1 *et seq*.
MILANO package, 363, 367
MO VB mixed methods, 369

MR–CI, 4
Magnetic circular dichroism, 210
Magnetic properties, 502
Many-body perturbation theory, 274, 408, 523
Matrix partitioning, 107
Maxwell relation, 309
Metal clusters, 465
Micro-iterations, 16
Micro-iterative method, 124
Model core potential, 468
Model potential, 505
Molecular clusters, 550
Moller–Plesset theory, 264, 274
  second order, 276, 522, 532, 551
  third order, 251, 278
Momentum-space properties, 358
Muffin tin approximation, 300
Muffin tin potential, 464
Multicharged ions, PES for, 351
Multiconfiguration random phase approximation, 224
Multipole expansion, 526
Multipole models, 524
Multireference CI, 34

N representability, 227
NICR method, 296
NMR chemical shifts, 280
Natural orbitals, 130, 400
Negative ions, 458
Newton–Raphson optimization, 7, 118
Non-linear reference functions, 195
Non-orthogonal configuration basis, 59

One-electron models, 289
Open shell atoms, 469
Optimized valence configuration method, 131
Orbital rotation, 11
Oscillator strengths, 225
Outer valence Green's function, 217

Parallel processing, 195, 344
Paramagnetic molecules, 306
Particle-hole symmetry, 305
Photoelectron spectra, 218
Pipeline computer, 170
Polarizability, 381, 525
Polarizability derivatives, 244, 280
Potential energy surfaces, 440
Propagator, coupled cluster, 234
  one electron, 215
  particle-hole, 204
  polarization, 218
  retarded, 206
Propagator methods, 201 *et seq*.
Property derivatives, 280
Pseudopotentials, 536
Pseudo-states, 389

Quadrupole moments, 539
Quadrupole polarizability, 539, 557
Quartic force constants, 539

RHF method, 488
Radial coupling, 352
Raman intensities, 244
Random phase approximation, 218, 222
Rational function approximation, 121
Rayleigh–Schrödinger expansion, 216, 230
Reaction potential surface, 268
Redundant variables, 151 *et seq*.
Relativistic effects, 473, 484, 507
Response equations, 249
Response functions, 204
Root flipping problem, 32
Rotational invariance, 271
Rumer spin functions, 328, 358
Rydberg orbitals, 24
Rydberg states, 194, 322, 375
Rys polynomial method, 257, 270

Saddle points, 536
Scaling of solution vector, 189
Scattered wave method, 464
Schmidt orthogonalization, 369
Second derivatives, 261
Second order polarization propagator, 229
Second quantization, 82
Self-consistency, in $X\alpha$, 292
Self-consistent electron-pair method, 279, 441
Self-consistent electron pairs, 3
Self-Coulomb energy, 298
Self-interaction corrected methods, 456
Self-interaction correlation, 301
Serber basis, 339
Shake-up spectra, 440
Shell structure, in $X\alpha$, 289

Single-centre expansions, 256
Size-consistency, 274, 534
Slater determinant, 81, 369
Spin coupled orbitals, 356
Spin coupled wavefunction, 329
Spin density, atomic, 344
Spin eigenfunctions, 92
Spin recoupling, 321
Spin–spin coupling constant, 220
State averaging, 32
Stoner theory, 501
Super-CI method, 120, 405, 417
Supermolecule method, 379, 381, 528
Super-operator formalism, 212
Superposition error, 523, 544
  basis set, 362, 379, 382
Symmetry adapted perturbation theory, 527

Tamm–Dancoff approximation, 218, 227
Test-particle model, 524
Thermodynamics, of electron gas, 308
Thomas–Reiche–Kuhn sum rule, 225, 234
Time-dependent coupled HF, 516
Transition density matrix, 180
Transition metal atoms, 434
Transition metal complexes, 437
Transition metal dimers, 479
Transition metals, $X\alpha$, 311
Transition moments, 210
Triple excitations, 535
Trust radius, 190
Tunnelling, 353
Two-electron integral transformation, 192
Two-photon adsorption cross-section, 210

UHF method, 3, 261, 400, 488
Unitary group approach, 92
Unrestricted Hartree–Fock method, 301, 302, 310

V representability, 292
VB MO mixed methods, 369
VB method, 319 *et seq*., 531, 551
  valence shell, 378
VB theory, classical, 320
  semi-empirical, 376
  spin-coupled, 324
van der Waals coefficients, 390
Vector processing, 344
Vectorization, 28, 51, 55, 59, 416, 418
  computer, 169
Virial theorem, 292
Virtual orbitals, 343

Wannier functions, 460
Wavefunction force, 255
Weinstein residual norm theorem, 73
Weyl formula, 402

$X\alpha$ method, 287, 454
$X\alpha$ relativistic, 507
$X\alpha$ scattered-wave method, 296
$X\alpha$ with GTO, 466
$X\alpha\beta$ method, 463

Young–Yamanouchi basis, 327